Handbook of Experimental Pharmacology

Volume 157

Editor-in-Chief

K. Starke, Freiburg i. Br.

Editorial Board

G.V.R. Born, London
M. Eichelbaum, Stuttgart
D. Ganten, Berlin
H. Herken†, Berlin
F. Hofmann, München
L. Limbird, Nashville, TN
W. Rosenthal, Berlin
G. Rubanyi, Richmond, CA

Springer

*Berlin
Heidelberg
New York
Hong Kong
London
Milan
Paris
Tokyo*

Antidepressants: Past, Present and Future

Contributors
R.D. Alarcon, G.B. Baker, P. Bech, T. Bettinger, J.F. Bober,
W.F. Boyer, O. Brawman-Mintzer, M.J. Burke, M.L. Catterson,
T.J. Connor, J.P. Feighner, V. Garlapati, S. Glover, A. Holt,
W.D. Horst, S.H. Kennedy, B. Kleiber, A.J. Kolodny, M. Lader,
B.E. Leonard, I. Nalepa, S.H. Preskorn, R. Ross, L.J. Siever,
C.Y. Stanga, F. Sulser, M. Trivedi, M.L. Wainberg, D. Wheatley,
K.A. Yonkers

Editors
Sheldon H. Preskorn, John P. Feighner,
Christina Y. Stanga and Ruth Ross

Springer

Dr. Sheldon H. Preskorn
Department of Psychiatry
and Behavioral Sciences
University of Kansas School of Medicine
and Psychiatric Research Insitute
1010 N. Kansas
Wichita, Kansas 67214, USA
e-mail: spreskor@kumc.edu

Dr. John P. Feighner
President and Director of Research
Feighner Research Institute
5230 Carrol Canyon Rd # 220
San Diego, CA 92121, USA
e-mail: jpfmd@mindspring.com

Dr. Christina Y. Stanga
W8057 Moonlite Road
Neillsville, WI 54456, USA
e-mail: jas64@tds.net

Ruth Ross
Ross Editorial
228 Black Rock Mtn Ln
Independence, VA 24348, USA
e-mail: ross@ls.net

With 31 Figures and 44 Tables

ISSN 0171-2004

ISBN 3-540-43054-7 Springer-Verlag Berlin Heidelberg New York

Library of Congress Cataloging-in-Publication Data
Antidepressants : past, present, and future / contributors, R.D. Alarcon ... [et al.] ; editors,
S.H. Preskorn ... [et al.]. p. ; cm. ··· (Handbook of experimental pharmacology ; v. 157)
Includes bibliographical references and index.
ISBN 3-540-43054-7 (hard : alk. paper)
1. Antidepressants. 2. Depression, Mental–Chemotherapy. I. Alarcón, Renato D. II.
Preskorn, Sheldon III. Series. [DNLM: 1. Depressive Disorder–drug therapy.
2. Antidepressive Agents–therapeutic use. 3. Drug Therapy–trends. WM 171 A6293 2004
OP905.H3 vol. 157 [RM332] 615'.1s–dc21 [616.85'27061] 2003054297

This work is subject to copyright. All rights are reserved, whether the whole or part of the material is concerned, specifically the rights of translation, reprinting, re-use of illustrations, recitation, broadcasting, reproduction on microfilm or in any other way, and storage in data banks. Duplication of this publication or parts thereof is permitted only under the provisions of the German Copyright Law of September 9, 1965, in its current version, and permission for use must always be obtained from Springer-Verlag. Violations are liable for Prosecution under the German Copyright Law.

Springer-Verlag is a part of Springer Science+Business Media
springeronline.com

© Springer-Verlag Berlin Heidelberg 2004
Printed in Germany

The use of general descriptive names, registered names, etc. in this publication does not imply, even in the absence of a specific statement, that such names are exempt from the relevant protective laws and regulations and free for general use.

Product liability: The publishers cannot guarantee the accuracy of any information about dosage and application contained in this book. In every individual case the user must check such information by consulting the relevant literature.

Cover design: design & production GmbH, Heidelberg
Typesetting: Stürtz AG, 97080 Würzburg

Printed on acid-free paper   27/3150 hs – 5 4 3 2 1 0

Preface

This book is being published as we mark the end of the first 50 years of the modern antidepressant era. This era began with the chance discovery that tricyclic antidepressants and monoamine oxidase inhibitors had antidepressant properties. That discovery had three consequences. First, it brought simple and effective treatment to patients suffering from major depressive illnesses. Second, these discoveries, together with the discovery of lithium and chlorpromazine, began the remedicalization of psychiatry by making it clear that the treatment of many of the major psychiatric illnesses can be approached in the same way as other medical conditions. However, the most far-reaching effect was to provide the first clue as to what mechanisms might underlie antidepressant efficacy, a development that has led to an explosion in the number of available antidepressants and to their widespread use.

The goal of this book is to provide a thorough review of the current status of antidepressants—how we arrived at this point in their evolution and where we are going in both the near and the long term. The book employs both a scientific and historical approach to accomplish these goals. The book is intended for practitioners who use antidepressants on a daily basis in their practice as well as for the student and researcher. Each will find that the book provides a comprehensive and logical approach to this important group of medications.

The importance of the book is a function of its topic. Antidepressants are among the most widely used of all medications. Major depressive disorder is a significant health problem, being second only to cardiovascular disease in prevalence and impact on psychosocial functioning and quality of life. If not properly treated, major depression can result in serious disability and needless mortality. For these reasons, this book begins with a focus on our understanding of the basic science of antidepressants and then moves on to discuss the clinical application of that knowledge to optimize treatment of patients with major depressive disorder. While great strides have been made over the last 50 years to increase the safety, tolerability, and effectiveness of antidepressants, more is needed and is being done. For that reason, the book ends with chapters that peer into the near and far future and discuss where we are going from here. Thus, the book is divided into parts that each build upon what went before, paralleling the way antidepressant pharmacotherapy has evolved over the last 50 years.

The first part of the book describes the basic pharmacological principles underlying treatment strategies for depression. In the first chapter, Dr. Horst describes the basic physiology of neurotransmission in the brain relevant to the clinical effectiveness of antidepressants. In the second chapter, Drs. Preskorn and Catterson outline basic pharmacokinetic principles that are important for prescribers to understand in order to maximize the safety, tolerability, and effectiveness of antidepressant pharmacotherapy. In the final chapter of Part 1, Drs. Burke and Preskorn review the basic principles underlying therapeutic drug monitoring (TDM) and then examine the current role and usefulness of TDM in the therapeutic use of the different classes of antidepressants.

The second part of the book deals with clinical methodological issues relevant to both the study and the treatment of patients with major depressive disorder. In the first of these two chapters, Drs. Connor and Leonard describe the search for biological markers that can help the prescriber, student, and researcher understand the underlying biological basis of major depressive disorder and its different subtypes, with the ultimate goal of providing more carefully targeted treatments. In the second, Dr. Bech describes rating scales that can be used to measure the specific symptoms of major depressive disorder as well as scales that assess more subjective quality-of-life issues, since both types of evaluation are important in assessing the effectiveness of a treatment for depression.

The third part of the book provides an overview of what is currently known about the pharmacotherapy of major depressive disorder. In the first chapter, Dr. Preskorn and Ms. Ross provide an overview of the currently marketed antidepressants and a logical framework for understanding their clinical pharmacology. Each of the subsequent chapters in this part focuses on different classes of antidepressant: tricyclic antidepressants in the chapter by Dr. Lader; monoamine oxidase inhibitors in the chapter by Drs. Kennedy, Holt, and Baker; the selective serotonin reuptake inhibitors in the chapter by Dr. Preskorn, Dr. Stanga, and Ms. Ross; and other newer antidepressants in the chapter by Dr. Preskorn and Ms. Ross. To conclude Part 3, the current role of herbal preparations in the treatment of depression is described by Dr. Wheatley.

The fourth part of the book deals with the treatment of depression in special populations. Drs. Bober and Preskorn discuss the special considerations that arise in treating depression in children and adolescents. Next, Drs. Yonkers and Brawman-Mintzer discuss how the action of antidepressant medications may differ in women. Drs. Glover and Boyer then discuss depression in the elderly, among whom the diagnosis of depression is frequently missed and who frequently suffer from concurrent medical illness that may contribute to and complicate the treatment of depression. Dr. Alarcon then discusses the special considerations that arise in the use of antidepressants and mood stabilizers to treat bipolar depression and avoid triggering a manic episode. Following that chapter, Dr. Trivedi describes strategies for managing treatment-refractory depression. And finally, Drs. Wainberg, Kolodny, and Siever describe what is known about the use of antidepressants to treat personality disorders.

The final part of this textbook focuses upon future directions in the development of new antidepressants and describes new paradigms and approaches to antidepressant drug development. Drs. Nalepa and Sulser discuss how new hypotheses concerning the mechanism of action of antidepressants are leading to the discovery of a new generation of therapeutic agents. In the next chapter, Drs. Garlapati, Boyer, and Feighner describe promising new directions in antidepressant development, with a focus on current research underway in the search for antidepressants that affect a variety of neuropeptides. As described by Dr. Preskorn in the final chapter, our field will ultimately be influenced by pharmacogenetics and pharmacogenomics, and it is important that everyone involved in either research on or treatment of major depressive disorder be aware of the important new advances being made in this area.

This textbook will be ideal for those studying the basic science of depression, for the clinical researcher in this area, and for academicians, residents, and practitioners who are actively involved in treating patients with major depressive disorder.

We thank all of the contributors to this textbook for their excellent, comprehensive, and state-of-the-art presentations.

S.H. Preskorn, J.P. Feighner, C.Y. Stanga and R. Ross

List of Contributors

(Their addresses can be found at the beginning of their respective chapters.)

Alarcon, R.D.   421

Baker, G.B.   209
Bech, P.   149
Bettinger, T.   447
Bober, J.F.   355
Boyer, W.F.   393, 565
Brawman-Mintzer, O.   379
Burke, M.J.   87

Catterson, M.L.   35
Connor, T.J.   117

Feighner, J.P.   565

Garlapati, V.   565
Glover, S.   393

Holt, A.   209
Horst, W.D.   3

Kennedy, S.H.   209
Kleiber, B.   447
Kolodny, A.J.   489

Lader, M.   185
Leonard, B.E.   117

Nalepa, I.   519

Preskorn, S.H.   35, 87, 171, 241, 263, 355, 583, 597

Ross, R.   171, 241, 263

Siever, L.J.   489
Stanga, C.Y.   241
Sulser, F.   519

Trivedi, M.   447

Wainberg, M.L.   489
Wheatley, D.   325

Yonkers, K.A.   379

List of Contents

Part 1. Basic Principles: Pharmacologic Science

Biochemical and Physiological Processes in Brain Function
and Drug Actions 3
   W. D. Horst

General Principles of Pharmacokinetics 35
   S. H. Preskorn, M. L. Catterson

Therapeutic Drug Monitoring of Antidepressants 87
   M. J. Burke, S. H. Preskorn

Part 2. Basic Principles: Clinical Science

Biological Markers of Depression 117
   T. J. Connor, B. E. Leonard

Quality of Life and Rating Scales of Depression 149
   P. Bech

Part 3. Current Pharmacotherapy of Major Depressive Disorder

Overview of Currently Available Antidepressants 171
   S. H. Preskorn, R. Ross

Tricyclic Antidepressants 185
   M. Lader

Monoamine Oxidase Inhibitors 209
   S. H. Kennedy, A. Holt, G. B. Baker

Selective Serotonin Reuptake Inhibitors 241
   S. H. Preskorn, R. Ross, C. Y. Stanga

Other Antidepressants 263
   S. H. Preskorn, R. Ross

Current Role of Herbal Preparations 325
   D. Wheatley

Part 4. Use of Antidepressants in Special Populations

Children and Adolescents 355
   J. F. Bober, S. H. Preskorn

Women 379
   K. A. Yonkers, O. Brawman-Mintzer

Older Adults 393
   S. Glover, W. F. Boyer

Bipolar Mood Disorders 421
   R. D. Alarcon

Treatment-Refractory Depression 447
   M. Trivedi, T. Bettinger, B. Kleiber

Personality Disorders 489
   M. L. Wainberg, A. J. Kolodny, L. J. Siever

Part 5. Future Directions in the Treatment of Major Depressive Disorder

New Hypotheses to Guide Future Antidepressant Drug Development 519
   I. Nalepa, F. Sulser

Promising New Directions in Antidepressant Development 565
   V. Garlapati, W. F Boyer, J. P. Feighner

Role of Pharmacogenetics/Pharmacogenomics in the Development
of New Antidepressants 583
   S. H. Preskorn

Appendix 597

Subject Index 609

Part 1
Basic Principles: Pharmacologic Science

Biochemical and Physiological Processes in Brain Function and Drug Actions

W. D. Horst[1]

[1] Psychiatric Research Institute, Suite 200, 1100 N. St. Francis, Wichita, KS 67214, USA
e-mail: Dale_horst@via-christi.org

1	The Needle in the Haystack.	4
2	Neurobiology.	4
2.1	Glia	5
2.2	Neurons.	8
2.3	The Electrical Nature of Neurons.	8
2.4	The Synapse	9
2.5	Chemical Transmission	10
2.5.1	Neurotransmitter Synthesis.	11
2.5.2	Neurotransmitter Storage and Release.	12
2.5.3	Neurotransmitter Inactivation	13
2.5.4	Receptors.	17
2.5.5	G proteins	21
2.5.6	Second Messengers	23
	References.	28

Abstract An understanding of the basic elements of neurotransmission in the brain is an important foundation for any consideration of the clinical use and future development of antidepressants. However, attempts to describe the influences of drugs on brain and neuronal function have become increasingly complex, and it is now clear that neuronal processes are complex molecular events involving multiple control factors. The brain consists of approximately 100 million neurons, which account for half of the brain's volume, with the other half being made up of glial cells. New insights into the function of glial cells, the influence of phosphorylation on neuronal functions, and the regulation of genetic functions in synthesizing neuronal proteins have enhanced our appreciation of the complexity of neural function in the brain. Glial cells play an important role in recycling and conserving the neurotransmitters glutamate and GABA and also have an important effect on neurons in the brain via glial cell line-derived neurotrophic factor (GDNF). The basic function of neurons is to convey electrical signals in a highly organized and integrated way. The dominant means of neuron-to-neuron communication or transmission occurs by means of specific chemicals (i.e., neurotransmitters). Neuronal communication is a complex process that involves neurotransmitter storage and release, neurotransmitter inactivation, receptors, G proteins, and second messengers. A number of neuronal mechanisms have been identified that are thought to play important roles in the etiology of major depressive disorder (e.g., the serotonin transporter mechanism); an increasing understanding of the neuronal mechanisms that underlie psychiatric disorders will help to guide future drug development.

Keywords Neurotransmitters · Glial cells · Neurons · Synapse · Glial cell line-derived neurotrophic factor (GDNF) · G proteins · Second messengers

1
The Needle in the Haystack

From relatively simple beginnings a half century ago, our attempts to describe drug influences on the brain and neuronal functions have become quite complex. Whereas we once spoke of drugs blocking neurotransmitter uptake, causing neurotransmitter release or depletion, or influencing receptors, today we understand that all of these basic neuronal processes are, in reality, complex molecular events that involve multiple control factors. Identifying a specific molecular mechanism in a drug's action on neuronal functions is, in fact, very much like a search for a needle in a haystack.

New insights concerning glial functions, the influence of phosphorylation on neuronal functions, and, most recently, the regulation of genetic functions in synthesizing neuronal proteins, have significantly enhanced our appreciation of the complexity of neural functions in the brain. As our understanding of basic neuronal and synaptic processes increases, so does the number of potential sites or mechanisms for the expression of depressive behavior and for drug actions. The needle is still only a needle, but the haystack continues to grow larger.

This chapter summarizes and describes our understanding of the basic elements of neurotransmission in the brain and provides a foundation for subsequent discussions of the clinical use and future development of antidepressants.

2
Neurobiology

The brain consists of approximately 100 million neurons, which account for one half of the brain's volume, the other half being made up of glial cells. Glial cells guide the synaptic formation of neurons during brain development (Bacci et al. 1999), influence the extracellular environment of the neurons (Zahs 1998), synthesize neurotransmitter precursors (Martin 1992), and respond to, or in some cases, may cause brain damage (McGeer and McGeer 1998; Aschner et al. 1999; Raivich et al. 1999). In turn, neurons are known to produce factors that influence the development and function of glia (Melcangi et al. 1999; Vardimon et al. 1999).

2.1
Glia

For many years, glia were thought to have a somewhat limited or passive role in brain function, but more recent findings indicate that glia have receptors, uptake mechanisms, and enzymes for several neurotransmitters, suggesting that the functions of these cells are closely integrated with neuronal functions (Martin 1992; Ransom and Sontheimer 1992; Attwell1994; Inagaki and Wada 1994; Otero and Merrill 1994) (Table 1).

An important example of this type of interaction is shown in Fig. 1. Glutamate and gamma aminobutyric acid (GABA) are ubiquitous neurotransmitters in brain, serving as primary excitatory and inhibitory neurotransmitters. Glial cells (astrocytes) play a prominent role in recycling and conserving both glutamate and GABA by recapturing them from the synapse, converting them to glutamine, and then returning glutamine to the presynaptic neurons for conversion back to the appropriate transmitter. In addition to the conservation of transmitter, glia protect neurons by limiting synaptic levels of glutamate (Bacci et al. 1999; Vardimon et al. 1999). Prominent among glutamate's several cellular influences is the opening of specific calcium channels that allow the influx of calcium ions into neurons. The overstimulation of neurons via this mechanism has been

Table 1 Membrane elements of mammalian glia (from Ransom and Sontheimer 1992; Sontheimer 1994)

Receptors[a]	Response mode
Noradrenergic (α and β)	G protein
Adenosine	G protein
Acetylcholine (muscarinic)	G protein
Neuropeptides (substance P)	G protein
Glutamate (quisqualate, kainate)	Ion gating
Gamma aminobutyric acid (GABA A)	Ion gating
Ion channels (voltage-sensitive)	
Potassium	
Inwardly rectifying	
Outwardly rectifying	
Transient A-type	
Sodium	
Tetrodotoxin-sensitive–"neuronal"	
Tetrodotoxin-resistant–"glial"	
Calcium (L and T types)	
Chloride	
Transporters (uptake sites)	
GABA	
Glutamate	
Glycine	

[a] Although nearly all neuronal-type neuroreceptors have been shown to occur on glial membranes, only those that have been shown to produce a response are included here.

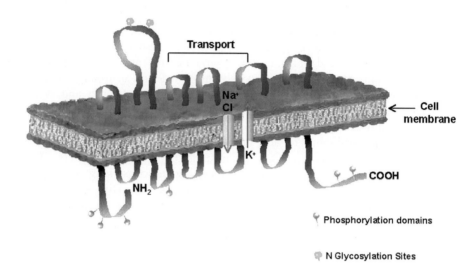

Fig. 1 The role of glia (astrocytes) in accumulating, metabolizing and conserving synaptic GABA and glutamate (Dale Horst 1995). Specific membrane transporters move GABA and glutamate into glia cells where GABA is carboxylated (CO_2) to form glutamate; glutamate in turn is aminated (NH_4) to create glutamine. Glutamine is then transported out of the glia and is available to GABAergic and glutamatergic presynaptic neurons for conversion to their respective transmitters. Thus, glia appear to play an important role in salvaging these two ubiquitous and important transmitters

associated with neurotoxicity and neuronal death. Thus, it is of considerable importance that extraneuronal concentrations of glutamate be controlled and that mechanisms exist to limit the synaptic activity of this transmitter. Glial cells have been shown to possess membrane transporters or uptake sites for glutamate; it has also been demonstrated that glutamate uptake into glial cells is the primary route for clearing synaptic glutamate. Once inside the glial cell, the glutamate is metabolized to glutamine, a neuronally inactive substance. The conversion of glutamate to glutamine occurs primarily in glial cells via the enzyme glutamine synthetase. Thus, glial cells play a critical roll in limiting the synaptic concentrations of glutamate and conserving the neurotransmitter for reuse. An important element in the above function is the maintenance of appropriate levels of glutamine synthetase activity in glial cells. Significant factors in the regulation of glutamine synthetase activity are glucocorticoid stimulation of gene expression and the absolute requirement for the astrocyte and neuron to be in juxtaposition. The critical neuronal factor required for gene expression of glutamine synthetase has not been identified (Vardimon et al. 1999).

Further evidence of the role astrocytes play in limiting synaptic levels of glutamate is seen in the influence that neuronal factors play in the regulation of glutamate transporter expression in glial membranes (Swanson et al. 1997; Schlag et al. 1998). A soluble, diffusible substance secreted by neurons has been shown to increase the expression of glutamate transporter molecules in astrocytes. Thus, it would appear that the level of neuronal activity plays a role in regulating the rate at which glutamate is transported into the glia for inactivation as described above. Although the substance has not been identified, the above effects can be mimicked by cyclic adenosine 5'-monophosphate (AMP) analogs (Swanson et al. 1997; Schlag et al. 1998).

Neuron dependent expression of two types of calcium channels has also been shown in astrocytes (Corvalan et al. 1990). The agent responsible for this intercellular communication has not been identified but may be cyclic AMP or a related substance (Corvalan et al. 1990).

Recent investigations have revealed a metabolic coupling between glia and neurons (Poitry-Yamate et al. 1995; Tsacopoulos and Magistretti 1996; Bacci et al. 1999). Considerable energy is consumed by the various processes of synaptic transmission. The preferred energy source for brain function is glucose. Astrocytes are capable of transporting glucose across the cell membrane via an active, carrier-assisted mechanism. Since astrocytes are well know to be in intimate contact with the brain's vascular system, it is assumed that glucose is transported directly into the glia from the circulation. Inside the glia, glycolysis transforms the glucose to lactose and, in the process, provides energy for the transport of transmitters and ions across the glial membrane. Lactose is then transferred out of the glia and accumulated by neurons where it is the preferred substrate for oxidative metabolism. This energy transfer process is stimulated by the uptake of neurotransmitters such as glutamate and gamma aminobutyric acid (GABA). Neurotransmitter uptake by glia is accompanied by the influx of sodium ions. The accumulation of sodium ions stimulates a Na^+/K^+ adenosine triphosphatase (ATPase) pump that consumes adenosine 5'-triphosphate (ATP) and exchanges intracellular sodium ions for extracellular potassium ions. The activity of the ATPase stimulates the metabolism of glucose resulting in increased lactose production. Thus the production of the metabolic precursor keeps pace with the overall synaptic activity.

In some cases astrocytes have been shown to possess functional ion channels controlled by neurotransmitter receptors. In this way, glia have been associated with long-distance signal transmission in brain via gap junctions across glial membranes (Cornell-Bell et al. 1990; Cornell-Bell and Finkbeiner 1991; Robinson et al. 1993). The observation that glial gap junctions are in part controlled by components of second messenger systems (Enkvist and McCarthy 1992) supports the active role of glia cells in brain function. It has been suggested that glial dysfunction plays a role in epilepsy and in the degenerative diseases Parkinson's and Huntington's (Ransom and Sontheimer 1992). Although several psychopharmacological agents interact with glial elements, as well as those same elements found on neurons, the contributions of these glial interactions to the agents' overall pharmacodynamics remain largely unknown and the subject of intensive investigation.

One exciting, potential therapeutic lead for the treatment of neurodegenerative disorders is found in research with glial cell line-derived neurotrophic factor (GDNF). GDNF was isolated from a culture of glial cells and found to stimulate the growth of embryonic dopamine neurons (Bohn 1999). Since this observation on dopaminergic neurons was made, GDNF has been found to elicit a trophic response on several other brain neuron types including motor neurons and noradrenergic, cholinergic, and serotonergic neurons (Bohn 1999). Other neurotrophic factors have been identified with activity similar to that of GDNF (Saarma and Sariola 1999); these include neurturin, persephin, and artemin. GDNF is a chain of 134 peptides synthesized from a larger propeptide. It exerts its biological activity through a series of complex receptor interactions, requiring cofactors and a tyrosine kinase receptor (Grondin and Gash 1998; Saarma and Sariola 1999). Although the exact role of GDNF in brain development and maintenance is not known, GDNF has been shown to be active in a variety of animal models for Parkinson's disease (Grondin and Gash 1998; Lapchak 1998; Bohn 1999) and has been proposed as a potential therapeutic agent for the treatment of Parkinson's disease, amyotrophic lateral sclerosis, and other neurodegenerative diseases. While the activity of GDNF is promising, in that it demonstrates efficacy in a variety of midbrain, dopamine-deficient models, many hurdles must be crossed before it can become a therapeutic reality. For example, what are the effects of long-term GDNF administration (dopamine enhancement may result in psychotic symptoms)? How will the substance be administered (it is a long chain peptide and will not cross the blood–brain barrier)? Are there subpopulations of patients who will not respond to GDNF (they may be deficient in receptors or essential components of the messenger system)? A small sampling of patients with Parkinson's disease did not reveal genetic abnormality in the GDNF gene (Wartiovaara et al. 1998).

2.2
Neurons

The basic function of neurons is to convey electrical signals in a highly organized and integrated way, each neuron receiving input from many other neurons and, in turn, providing input to many other neurons. This function is the product of complex chemical processes transmitting signals across neuronal synapses—a symphony of intra- and interneuronal events, layers of feedback, and control mechanisms assuring the correct or appropriate response. Although chemical transmission is a complex, multi-stepped system, it provides for maximum flexibility and unidirectional flow of neuronal signals.

2.3
The Electrical Nature of Neurons

Resting neurons maintain an electrical polarization between the inside and outside of the cellular membrane. This polarization is negative on the inside and positive on the outside of the neuron. Key elements in maintaining the polarized state include the presence of large (nondiffusible), intracellular, negatively charged, proteins and specific ion pumps, located in neuronal membranes that use cellular energy to pump ions against concentration gradients. Changes in the state of transmembranal polarization are affected by a system of specific ion channels activated by neurotransmitter substances or by the degree of transmembrane polarization (voltage sensitive). Ions move through the channels because of concentration (from high to low concentrations) or electrogenic (opposite charges attract) gradients. Activation of neurotransmitter controlled ion channels reduces the level of polarization to a critical level at which voltage-sensitive channels open and permit the rapid influx of cations (e.g., Na^+). This influx completely depolarizes the neuron and even reverses the polarization for a brief period. Membrane ion channels adjacent to the area of depolarization open, thus extending the depolarization along the neuron and causing the formation of an action potential. In this way, electrical signals are carried from one end of the neuron to the other. Repolarization occurs by the opening of voltage-sensitive K^+ channels. Since K^+ concentrations are high inside the neuron and low outside, K^+ carries positive charges to the outside of the membrane, making the inside more negative. The restoration of conditions in the resting state is completed by the exchange of intracellular Na^+ ions for extracellular K^+ ions. Ion exchange is accomplished by an energy-dependent pump (excellent reviews of the electrical nature of neurons may be found in Levitan and Kaczmarek 1997 and Shepherd 1994).

2.4
The Synapse

The synapse is defined as the juncture of two neurons: the neuron from which the signal is coming is known as the presynaptic neuron, while the receiving neuron is called the postsynaptic neuron. Signals are passed across the synapse by either of two mechanisms. The first is by direct connection of the pre- and postsynaptic neurons via gap junctions (similar to the connection of astrocytes described above). The physiological significance of this type of connection is that transmission is rapid, can occur in two directions, and can synchronize the activity of many neurons. Electrogenic coupling of neurons occurs in only a few populations of neurons located primarily in the brain stem (Baker and Llinas 1971; Korn et al. 1973; Llinas et al. 1974). Cortical precursor cells are also known to communicate via gap junctions (LoTurco and Kriegstein 1991), although this function is lost as the cells develop into mature functioning neurons. Other neurons in the suprachiasmatic nucleus are also known to transmit signals via gap junctions, which are at least partially influenced by the neurotransmitter GABA (Shinohara et al. 2000). These gap junctions may be unique in that they are electrogenic in nature but are influenced by a neurotransmitter substance; they may thus exhibit the advantages of both modes of interneuronal communication.

2.5
Chemical Transmission

By far the dominant means of neuron-to-neuron communication or transmission occurs by means of specific chemicals or neurotransmitters. Chemical transmission requires the presence of several elements to operate effectively (Table 2). These elements consist of specific proteins in the form of enzymes,

Table 2 Elements required for chemical transmission

Presynaptic
Enzymes for neurotransmitter synthesis
Mechanism for neurotransmitter storage
Mechanism for appropriate neurotransmitter release
Neurotransmitter receptors for feedback modification of neurotransmitter release
Synaptic
Mechanism for terminating neurotransmitter action
Postsynaptic
Neurotransmitter receptors to initiate response
Coupler proteins
Mechanism for response (ion channels/second messenger systems)

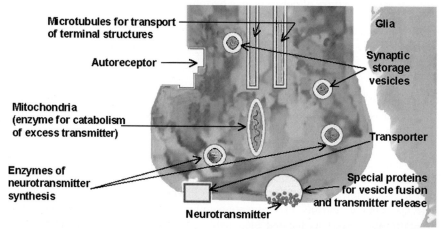

Fig. 2 Typical presynaptic neuron with key structures relevant to neurotransmission (Dale Horst 1995). Microtubules transport storage vesicles, enzymes, and a variety of proteins from the neuronal soma where they are synthesized to the nerve ending where they are required for carrying out their physiological functions. Storage vesicles maintain stores of neurotransmitter molecules for eventual release into the synaptic cleft. Mitochondria contain enzymes vital to providing energy to the neuron and, in many cases, such as the biogenic amines, they contain enzymes such as monoamine oxidase that help regulate neurotransmitter levels in the nerve ending. Calcium and a variety of special fusion proteins fuse the storage vesicle membranes with the neuronal membrane to release the transmitter substance into the synaptic cleft. Reuptake pumps are proteins incorporated into the neuronal membrane that transport the transmitter substance from the synapse into the neuron where it can be reincorporated back into the storage vesicle. Autoreceptors respond to neurotransmitter released from the nerve ending to provide feedback regulation of presynaptic depolarization

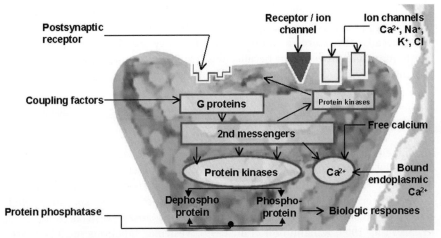

Fig. 3 Typical postsynaptic neuron with key structures relevant to neurotransmission (Dale Horst 1995). Neurotransmitter substances bind to postsynaptic receptors, which may be one of two major types, G protein or ion channel coupled. Other ion channels are regulated by intraneuronal ions such as calcium or potassium as well as transmembrane voltages. G proteins couple receptors with second messenger systems, which in turn regulate a variety of protein kinases that are responsible for initiating biological responses. Second messengers also directly regulate intraneuronal calcium levels. The various elements in the illustration are shown in their functional sequence, not in their anatomical domains. Postsynaptic receptors, G proteins, and second messengers are in fact neuronal membrane-associated elements

storage-binding proteins, uptake/membrane transport structures, receptors, and response systems (ion channels/second messenger systems). Each element represents an opportunity for malfunction (disease state) and/or a point for modulation of transmission through pharmacological intervention. In fact, the manipulation of these elements serves as the basis of modern psychopharmacology. The functional relationship of the elements of neurotransmission is illustrated in Figs. 2 and 3.

2.5.1
Neurotransmitter Synthesis

The final enzymatic steps in synthesizing a neurotransmitter generally occur in or near the storage site. This assures maximum efficiency in the neurotransmitter molecules getting to the storage sites. Neuropeptides are a notable exception to this role since their synthesis involves gene activation followed by DNA transcription and RNA translation to form large polypeptides. The polypeptides are then broken down into the component neuropeptides. All of this occurs within the neuronal soma so that the neuropeptides must be transported along the axon to the nerve terminal for storage. In several cases, neurotransmitter synthesizing enzymes are shared by more than one transmitter system. For example, the transmitters norepinephrine and dopamine share the enzymes tyrosine hydroxylase, and L-dopa decarboxylase, while several peptides share common peptidases. Although many drugs are known to inhibit specific neurotransmitter synthesizing enzymes, these drugs have not proven useful as therapeutic agents, either because the neurotransmitters they influence are ubiquitous and important to many life processes or because the enzymes influence multiple transmitters and thus produce broad, nonspecific effects.

2.5.2
Neurotransmitter Storage and Release

Many neurotransmitters are stored in organelles known as synaptic vesicles (Thiel 1995). These structures—constructed in the soma and transported along the axons, with their full complement of neurotransmitter—concentrate near the nerve terminal. The vesicular membranes contain many specific protein structures involved in the multiple functions of the storage vesicles (Kelly 1999; Krantz et al. 1999; Rahamimoff et al. 1999). Vesicular functions include neurotransmitter synthesis, neurotransmitter transport across the vesicular membrane, neurotransmitter binding inside the vesicle, docking proteins for attaching to the neuronal plasma membrane, calcium binding proteins for membrane fusion and neurotransmitter release, and special coating proteins for vesicular endocytosis and recycling processes.

Calcium ions are an essential element for the release of neurotransmitters. A major mechanism for calcium entry results from the activation of voltage-gated ion channels located in the plasma membrane (Rahamimoff et al. 1999; Zhang and Ramaswami 1999). Of course, the activation of the voltage-gated ion channels depends on many factors, such as the activity of numerous other membrane ion channels (Rahamimoff et al. 1999) and a variety of presynaptic ligand-gated ion channels (MacDermott et al. 1999) as well as autoreceptors (MacDermott et al. 1999). Given the absolute requirement for Ca^{++} in the neurotransmitter release process, it is not surprising that proteins that comprise the voltage-gated calcium channel are intimately bound to specific proteins associated with storage vesicle docking and fusion processes (Catterall 1999). This would ensure that transmitter release is occurring in a microenvironment containing an appropriate concentration of calcium.

Proteins involved in the storage and release mechanisms of neurotransmitters are well conserved in many different transmitter systems so that drugs that interfere with storage and release tend to have broad nonspecific effects. Reserpine is an example of such a drug. Reserpine destroys the ability of presynaptic storage vesicles to transport and store all biogenic amines such as norepinephrine, dopamine, serotonin, and histamine (Krantz et al. 1999). Since this effect is not reversible, recovery from the effects of reserpine requires the synthesis and transport of new storage vesicles to the nerve terminals, a process that requires several days. For these reasons, reserpine had a short-lived and limited use in psychiatry.

The pharmacology surrounding the control of neurotransmitter release via heterosynaptic, ligand-gated mechanisms presents some clinically significant examples. Transmitters for the presynaptic, ligand-gated channels include GABA, glutamate, acetylcholine, serotonin, and ATP (MacDermott et al. 1999).

Presynaptic GABA receptors are of the GABA-A type and thus relevant to the inhibitory actions of the benzodiazepines (Tallman et al. 1999). The presynaptic GABA-A receptors are channels for chloride ions, and their activation hyperpolarizes the presynaptic neuron, resulting in inhibition of presynaptic release of neurotransmitter. Presynaptic cholinergic receptors are of the nicotinic type and, when activated, tend to depolarize the neuron by admitting sodium and calcium ions. The activation of these presynaptic receptors by nicotine is the source of tobacco smoke's central pharmacological and addictive properties (Grady et al. 1992; Wonnacott 1997; Yeomans and Baptista 1997; MacDermott et al. 1999; Watkins et al. 1999). Presynaptic serotonin receptors are of the 5-hydroxytryptamine 3 (5-HT$_3$) type; when activated, they admit calcium into the neuron, thus promoting depolarization and transmitter release (Ronde and Nichols 1998). As with all other serotonergic receptors, the activity of the 5-HT$_3$ receptors is enhanced by drugs that block the reuptake of serotonin (MacDermott et al. 1999). In addition, ethanol has been demonstrated to influence several of these presynaptic receptors such as GABA-A, glutamate, nicotinic cholinergic, and 5-HT$_3$ serotonergic types (Lovinger 1997; Narahashi et al. 1999). The relevance of ethanol's influence on these structures to its pharmacological actions is the subject of continued investigations.

2.5.3
Neurotransmitter Inactivation

Once liberated into the synapse, neurotransmitters are available to transmit their signals until they are removed or inactivated in some manner. Thus, transmitter inactivation is an important element in controlling synaptic transmission. Rapid inactivation would attenuate transmission, while slow inactivation would accentuate signal transmission. Three major routes for neurotransmitter inactivation are known. First, transmitters may be removed by washout or turnover of the extraneuronal fluid. This occurs at a relatively slow rate in brain and would appear to be inadequate in most situations.

A second mode of transmitter inactivation is by enzymatic degradation of the neurotransmitter. A notable example of this mode is the function of acetylcholine, a major excitatory transmitter in brain, in which the time span in the synapse is vital for proper function. Acetylcholine is rapidly metabolized in the synapse by acetylcholinesterase, a process that splits the transmitter into two parts, acetate and choline. These two components are then transported back into the presynaptic neuron where another enzyme, choline acetyltransferase, rejoins them to form acetylcholine. Thus, these two enzymes and the transport of the precursors across the membrane form an effective and efficient means of rapidly terminating the synaptic activity of the transmitter, while protecting the availability of the precursors required for transmitter synthesis.

A third mechanism for transmitter inactivation is a variation of the second. In this case, rather than transporting the individual components of the neurotransmitter, the intact transmitter is transported across the presynaptic neuronal membrane and into the cytoplasm, where it is eventually accumulated by storage vesicles and becomes available for release once again. Transport or "reuptake" is accomplished by special transmembrane proteins, which serve as carriers. Since the concentrations of transmitter are generally greater in the presynaptic terminal than in the synaptic fluid, cellular energy in the form of ATP is expended in the reuptake process. Many major transmitters are primarily inactivated by this reuptake process, including the biogenic amines, norepinephrine, dopamine, and serotonin, and the amino acids, GABA, glycine, and glutamine. The transporter proteins for the majority of transmitters have similar properties (Kanner et al. 1994; Nelson 1998; Krantz et al. 1999), such as a requirement for sodium and chloride ions. The transporters characteristically have 12 transmembrane sections, with a large extracellular loop between the third and fourth sections. This third loop contains a site or sites for glycosylation (Blakely et al. 1997; Gegelashvili and Schousboe 1997; Nelson 1998; Krantz et al. 1999). Inhibition of glycosylation at these sites appears to diminish the function or efficiency of the transporter but does not influence substrate affinity (Melikian et al. 1996).

A second set of transporters, similar to, but genetically distinct from, those found in the neuronal membrane, is located within the membranes of the neurotransmitter storage vesicles. These transporters move the neurotransmitters from the neuronal cytoplasm to the interior of the vesicles, where they are ready for release once again (Krantz et al. 1999).

The serotonin transporter has been the subject of intensive investigation because it is the site of action for the selective serotonin reuptake inhibitors (SSRIs), the drugs of choice in the treatment of major depression. A transporter protein structure approximating the serotonin transporter is shown in Fig. 4. In common with other neurotransmitter transporters, the serotonergic transporter has 12 transmembrane sections with both the amino and carboxyl terminals within the neuron (Blakely et al. 1997; Nelson 1998). The serotonin transporter molecule includes intraneuronal sites for phosphorylation that are critical for the regulation of transporter activity (Blakely et al. 1997, 1998). These sites appear to be primarily phosphorylated by kinase C and dephosphorylated via the action of phosphatase 2A (Blakely et al. 1998). Phosphorylation inactivates the transporter molecules and therefore slows the clearing of serotonin from the synaptic cleft. This phosphorylation process likely serves as a kind of feedback control, since kinase C activity is regulated via neurotransmitter interaction with neuroreceptors (see the section "Second Messengers" later in this chapter and Fig. 7).

Recognition or binding sites for serotonin to the transporter are located at extraneuronal loops of the molecule. A precondition for substrate binding is the binding of one each of sodium and chloride ions to the transporter (Nelson 1998; Krantz et al. 1999). Presumably the binding of these ions places the tertia-

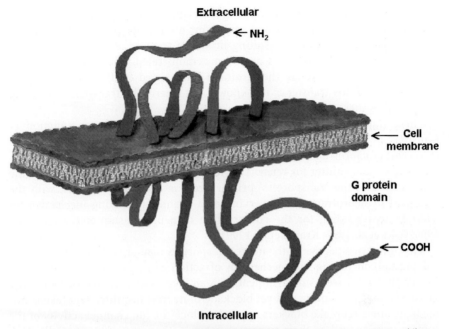

Fig. 4 A typical neurotransmitter protein (Dale Horst 2000). Neurotransmitter transporters, while exhibiting specificity for the neurotransmitter being transported, have many structural features in common. Each has 12 transmembrane sections, a large extraneuronal loop between the third and fourth transmembrane sections, intraneuronal sites for phosphorylation (P), and a requirement for ion binding. The figure approximates a serotonin transporter

ry configuration of the transporter in the best position for substrate binding. The exact location of the site of attachment of inhibitory drugs, such as many of the antidepressant compounds, is not known for all drugs, but it most likely occurs at a variety of external sites. The tricyclic antidepressants are known to bind to the central loops (Blakely et al. 1991; Nelson 1998) in the transport domain for serotonin.

The transporters for several types of neurotransmitters, including those for the amino acids, GABA and glutamate, the catecholamines, dopamine and norepinephrine, and the biogenic amine, serotonin, have long been known to require the presence of sodium and chloride ions in order to function. In general, binding of these monovalent ions appears to be required in order for the neurotransmitter to bind to the transporter. Recent information suggests that transporters for the neurotransmitters may actually serve a significant role as ion channels and thus play a part in neuronal function and regulation of neuronal activity (Lester et al. 1996; Galli et al. 1997; Nelson 1998; Krantz et al. 1999). Although the ion requirements may vary depending on which transporter is involved, the basic mechanisms appear to be similar for the entire family of transporters.

As an example, the serotonin transporter appears to function with neutral stoichiometry. Thus, Na^+, Cl^-, and $serotonin^+$ are transported into the neuron, while K^+ is transported out, resulting in no net transfer of charge (Galli et al. 1997; Krantz et al. 1999); yet it has been demonstrated that significant charge transfer does occur through the serotonin transported (Galli et al. 1997). The exact nature of this transfer is not known, but clearly the transporter is acting as an ion channel.

These findings raise interesting questions about the primary function of transporters. It has always been assumed that the primary role of transporters is to remove neurotransmitter from the synapse and to do so in a way that conserves neurotransmitter for reuse. This recently reported information raises the possibility that the transporters' primary role is as an ion channel, with the transport of neurotransmitters serving as a channel-regulating mechanism in addition to providing for the conservation of neurotransmitter (Lester et al. 1996; Galli et al. 1997; Krantz et al. 1999).

This novel perspective of the neurotransporters may provide a solution to an old puzzle concerning the mechanism of action of antidepressants. It is well known that a few weeks of treatment with antidepressants is needed before the onset of clinical response, and yet blockade of neurotransmitter reuptake is immediate. Since reuptake blockers appear to block the ion channel activity of the transporters as well as their transport function (Lester et al. 1996; Galli et al. 1997), it may be that the significant pharmacological action of the reuptake inhibitor antidepressants is related more to the presynaptic, intraneuronal, ionic milieu than to increased synaptic neurotransmitter concentrations.

Whether transporter or ion channel, these structures are important for proper brain function. It has been found that approximately 4% of the human population have a genetic defect that reduces the transcription process for the serotonin gene, resulting in individuals with reduced transporter function (Lesch et al. 1996). Such individuals have been found to demonstrate relatively high anxiety-related traits (Lesch et al. 1996). In addition, it has been suggested that this defect has a possible link to neurodevelopment and neurodegenerative disorders (Lesch and Mossner 1998).

In the case of the biogenic amines, while reuptake is the major means of limiting their synaptic activity, significant contributions to inactivation are made via the enzymes catechol-O-methyl transferase (Guldberg and Mardsen 1975; Mannisto et al. 1992) and monoamine oxidase (Singer and Ramsay 1995). Inactivation by these enzymes does not result in any known reusable or physiologically active products. The reuptake sites for the biogenic amines exhibit some specificity for each amine, but the sites for each amine appear to be identical on all neurons and in all brain regions containing that amine. Thus, it is possible to design a drug that will specifically block the reuptake of a single bioamine, and will influence reuptake to the same extent at all neurons containing that amine. As will be discussed later, this specificity is an important factor in therapeutic drug design.

Unlike the biogenic amines, the amino acid transmitters are taken up by both presynaptic neurons and by adjacent glial cells (Jursky et al. 1994; Kanai et al. 1994; Kanner et al. 1994) (Fig. 1). Within the glia, GABA and glutamate are metabolized to glutamine, which is then transported out of the glia and into GABA and glutamate presynaptic neurons where it is converted back to either GABA or glutamate (Shank et al. 1989). The physiological significance of this dual uptake system is not fully understood. Also, unlike the biogenic amines, amino acid transporters occur in multiple variations for each of the transmitters. Thus, three variations of the glutamate transporter (Kanai et al. 1994) have been identified, while four are known for GABA and three for glycine (Jursky et al. 1994). As more is learned about the physiological role of these various transporter subtypes, they may prove to be useful sites for specific modification by drugs.

The pharmacological manipulation of reuptake or transport sites has been an area of intensive activity in developing drugs for the treatment of major depression. The discovery that tricyclic antidepressants, such as imipramine and amitriptyline, had the ability to be potent reuptake blockers of norepinephrine and serotonin led to the early hypothesis that depression was the result of an insufficiency of these biogenic amines and that this insufficiency was corrected by reducing the rate at which these transmitters were removed from the synapse. This hypothesis was supported by other pharmacological and biological observations. Recent clinical success with a new class of antidepressants, the SSRIs, underscores the importance of amine reuptake as an appropriate mechanism of action for antidepressant activity, although it is now apparent that reuptake inhibition alone may not be directly responsible for the antidepressant activity. Rather, reuptake inhibition initiates changes or adjustments in receptors, ion fluxes, and intracellular messenger systems that alter neurotransmission in key pathways (Kilts 1994; Paul et al. 1994; Galli et al. 1997).

2.5.4
Receptors

Neuroreceptors are specific, membrane-bound proteins that bind neurotransmitter molecules and translate that molecular attachment into a physiological response. The amino acid sequence of each receptor type imparts a specificity for the particular neurotransmitter that will bind to it. Of particular importance in defining a receptor is that a physiological response results when the receptor is activated. Many proteins are capable of binding neurotransmitter substances but are not capable of eliciting a response. Such binding proteins are better referred to as acceptors.

Neuroreceptors may be placed into four broad categories, depending on the mode of action of their physiological response (Cooper et al. 1996). The most prevalent of these are those receptors that connect to a second messenger system through one of a family of proteins referred to as G proteins. This receptor type is characterized by having seven transmembrane sections of the protein

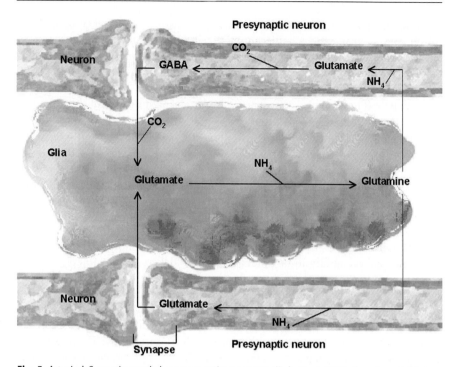

Fig. 5 A typical G protein-coupled neurotransmitter receptor (Dale Horst 1995). Receptors of this type have seven transmembrane sections with the carboxyl terminal on the inside of the neuron. The figure approximates a serotonin 1A receptor. Receptors of this type vary with regard to their amino acid composition and the lengths of the extra- and intracellular segments

with extracellular and intracellular loops that serve as neurotransmitter binding sites and sites for receptor regulation (Fig. 5).

A second class of receptors consists of those that form membrane ion channels or ionophores. Stimulation of receptors in this class opens ion channels specific for sodium, potassium, calcium, and chloride ions. The receptor/ion channel consists of five individual proteins, each with four transmembrane sections. The receptors are made up of a combination of protein subtypes; each subtype may exist in several variations. Thus, receptor/ion channels for a specific neurotransmitter may exist in several variations.

The third receptor class consists of receptors that attach to allosteric sites on other neuroreceptors that regulate receptor ligand affinities. Examples of receptors that act at allosteric sites include the benzodiazepine and associated receptors, which modify GABA receptor activity, and the glycine B receptor, which is associated with the N-methyl-D-aspartate (NMDA) subtype of glutamate receptor. The presence of glycine on the glycine B receptor is one of the requirements for glutamate activation of the NMDA receptor.

The fourth receptor class is made up of intraneuronal receptors best characterized by steroid transcription factor binding and the synthesis of various components of synaptic transmission, such as enzymes, receptors, and second messengers systems (Joels and de Kloet 1994). As might be expected, the effects mediated by these receptors have a slow onset and persist over an extended period of time.

Neuroreceptors generally exhibit specificity for neurotransmitters and are usually identified by the transmitter that binds to them and to which they elicit a response. The existence of subtypes of receptors for specific transmitters has been known for many decades. For example, acetylcholine receptors were traditionally thought of as being either nicotinic or muscarinic, based on their pharmacological response. It is now known that they not only exhibit different pharmacological characteristics but that they are entirely different in their functions and mode of responses. The nicotinic acetylcholine receptor is a five-protein, sodium ion channel that is opened in the presence of acetylcholine, while the muscarinic receptor is a single-protein unit coupled to a second messenger system (inositol phosphate) by a G protein.

Historically, receptor subtypes have been identified by pharmacological studies in which specific drugs are used to stimulate or block receptor activity. Such techniques are limited in that they do not differentiate among receptor subtypes as either two different proteins or one protein in different membrane configurations. Pharmacological techniques are also limited in the number of subtypes that can be identified, particularly as the pharmacological differences become increasingly subtle and if very specific pharmacological agents are not available.

In recent years, molecular biological techniques have been used to identify and characterize many new receptor subtypes. Specific genes have been identified that express neuroreceptors subtypes for many of the neurotransmitters. Through these techniques, populations of "pure" receptor types have been produced, which may then be used to identify specific ligands for the receptors. Specific ligands, in turn, are useful for determining the location and density of specific receptors. Methods also exist by which transcription paths are altered to prevent the expression of specific receptors (Lucas and Hen 1995), thus providing animal models that lack specific receptors. Such studies provide important clues to the physiological function of specific receptors (Thomas and Capecchi 1990; Wahlestedt et al. 1993; Furth et al. 1994; Lai et al. 1994; Saudou et al. 1994; Silvia et al. 1994; Standifer et al. 1994; Zhou et al. 1994; Tecott et al. 1995).

As shown in Table 3, receptors for various transmitters come in a variety of subtypes. For example, 14 subtypes have been identified for serotonin, with the possibility of more to be discovered (Lucas and Hen 1995). Nearly all receptors for the various neurotransmitters come in several subtypes. This fact is very important from a pharmacological perspective, since it means that there is the possibility of identifying compounds for a specific receptor subtype, thus limiting the pharmacological effects.

Neuroreceptors are important sites for pharmacological intervention in psychiatric disorders. For example, all antipsychotic medications are known to have antagonist activity at dopamine receptors. Antidepressant drugs are well known

Table 3 Some major types of neurotransmitter receptors

	Mode of response	Messenger system	Receptor subtypes	Multiple subunit variations
Amines				
Acetylcholine	Ionophore	$Na^+/K^+/Ca^{++}$ channels	Nicotinic	5
	G protein	cAMP IP_3	Muscarinic	6
Norepinephrine	Ionophore and G protein	Ca^{++} channels and cAMP	Alpha	3
	G protein	cAMP	Beta	5
Dopamine	G protein	cAMP		
Serotonin	G protein	cAMP	$5\text{-}HT_{1A\text{-}F}$	13
			$5\text{-}HT_{2A\text{-}C}$	13
			$5\text{-}HT_{4\text{-}7}$	13
			$5\text{-}HT_3$	1
	Ionophore	Cation channel		
Amino acids				
Glutamate (inotropic)	Ionophore	Na^+/K^+	Kainate/quisqualate	2
Glutamate	Ionophore	Ca^{++}	NMDA	1
Glutamate (metabotropic)	G protein	cAMP, IP_3	$MGLU_{1\text{-}7}$	7
Glycine	Ionophore	Cl^-		Multiple subunit variations
	Allosteric site on glutamate receptor	Required for glutamate (NMDA) receptor activation		
GABA	Ionophore	Cl^-	GABA A	1
	G protein	cAMP	GABA B	1
Benzodiazepine	Allosteric site	Occupation of this site increases efficacy of GABA receptor		
Purines				
Adenosine	G protein		$A_1\ A_{2\,a\text{-}b}\ A_3$	4
ATP/ADP/VDP	Ionophore			4
	G protein			
Peptides				
Opioid (Enkephalins/Endorphins)	G protein	cAMP	MU, SIGMH, Kappa	3
Angiotensin	G protein		AT_1 and AT_2	2
Cholecystokinin	G protein		CCK_a and CCK_B	2
Vaso Pressin/Oxytocin	G protein		V_{1A}, V_{1B}, V_2, OT	3/1
Somatostatin	G protein		$SST_{1\text{-}5}$	5
Neurotensin	G protein	cGMP		1
Steroids				
Corticosterone/cortisol	Gene transcription	Modification of neurotransmission elements	$Mr_5\ Gr_5$	2

to influence receptors either directly as antagonists or indirectly by up or down regulation of receptor populations (Kilts 1994). Many antidepressants, the tricyclic antidepressants in particular, are known to interact with receptors of several neurotransmitter systems. Anxiolytic agents, the benzodiazepines and buspirone, exert their pharmacological actions through interaction with benzodiazepine and serotonin (5-HT_{1A}) receptors, respectively.

Drugs may interact with receptors in one of several ways. They may bind to the receptor and cause a physiological response similar to that of a natural neurotransmitter; such a drug would be referred to as an agonist. Other drugs may also bind to the receptor but do not elicit a physiological response; rather they prevent agonists from binding to the receptor. Such drugs are described as antagonists. A third type of ligand–receptor interaction is referred to as inverse agonism. An inverse agonist is a drug that binds to a receptor but produces an effect opposite to that of agonist activity. This type of action has been described in studies of the benzodiazepine receptor (Stephens et al. 1986), where a single receptor mediates agonist, antagonist, and inverse agonist activities. Pharmaceutical agents are also known that appear to exhibit mixtures of these basic reactions (e.g., mixed antagonist/agonists that produce partial or limited agonist activities, but may behave as antagonists when in the presence of full agonists).

Another distinctive interaction between psychotropic drugs and receptors is well known to occur with chronic antidepressant treatment. Multiple, but not single, doses of many antidepressant compounds are known to down-regulate beta adrenergic receptors (Wolfe et al. 1978) and NMDA receptors in brain tissue (Paul et al. 1994) (i.e., they reduce the actual number of receptor sites). Since many of these drugs do not interact with these receptor populations directly, the induced changes may be the result of some activity in the second messenger system, although the precise mechanisms for these effects are not known. Since the slow onset of the receptor adaptations is similar to the timing of the onset of the clinical antidepressant effect (Oswald et al. 1972), it has been suggested that one or the other of these changes may be related to the antidepressant effect itself (Caldecott-Hazard et al. 1991; Paul et al. 1994).

2.5.5
G proteins

Serving as linking proteins between extracellular receptors and intracellular effector mechanisms (second messengers), the G proteins [regulatory guanosine 5′-triphosphate (GTP)-binding proteins] constitute a large family of related structures vital to the transmission of interneuronal signals. G proteins are actually heterotrimeric structures composed of one each of three protein subunits termed alpha, beta, and gamma (Rens-Domiao and Hamm 1995). To date, 18 specific alpha subunits, five beta subunits, and seven gamma subunits have been identified. The various subunits are not all interchangeable and some combinations of the subunits are not compatible.

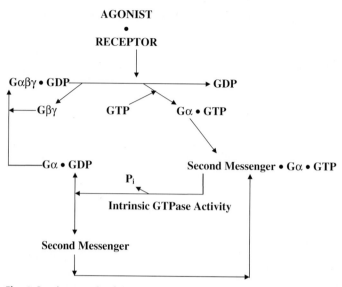

Fig. 6 Regulatory cycle of G protein signal coupling. G protein exists as a tri-protein with alpha, beta, and gamma subunits. Binding of an agonist to a receptor induces the release of the alpha subunit from the beta/gamma subunits and GDP from the alpha subunit. GTP binds to the alpha subunit; this complex then binds to a second messenger (adenylate cyclase or phospholipase C). The intrinsic GTPase converts GTP to GDP, which results in the uncoupling of the alpha subunit from the second messenger. The second messenger is then available for recoupling to an alpha/GTP complex. The alpha/GDP complex then binds to a beta/gamma subunit complex and the cycle is ready to begin again

The alpha subunits play a key role in the transduction process. The alpha unit binds the guanine nucleotide and exhibits intrinsic GTPase activity. It is also the alpha subunit that interacts with the neuroreceptor to initiate the transduction process. Frequently it is the alpha subunit that interacts with the effector proteins, although this function has been attributed to the beta/gamma subunit in some instances (Haga and Haga 1992; Pitcher et al. 1992; Clapham and Neer 1993). Although the alpha subunit disengages from the other two subunits at certain stages of transduction, the beta and gamma subunits remain bound to each other at all times.

The sequence of events for the G protein transduction cycle is illustrated in Fig. 6. The binding of an agonist to a neuroreceptor causes the release of guanosine 5′-diphosphate (GDP) from the alpha subunit with the subsequent binding of GTP. The binding of GTP releases the alpha subunit from the beta/gamma subunit complex and at the same time binds the alpha subunit to the effector protein. Following the interaction with the effector, the alpha subunit converts GTP to GDP (intrinsically) and recombines with the beta/gamma subunits to begin the cycle over again.

All of the neuroreceptors known at this time to stimulate G protein regulatory units are of the seven-transmembrane, helical type. G proteins are known to interact with a variety of effectors, including adenyl cyclase, phosphodiesterase (phosphatidylinositol turnover), calcium and potassium channels, and receptor-coupled kinases. Through these effectors, G proteins are involved in both excitatory and inhibitory roles. Through the stimulation of receptor-coupled kinases and the phosphorylation of specific intracellular domains of the receptor proteins, G proteins provide feedback control of receptor sensitivity (Hausdorff et al. 1990).

Relatively little is known about drug influences on G protein functions. Lithium is known to inhibit G protein function in the adrenergic stimulation of adenylate cyclase (Belmaker et al. 1990); however, the role of this effect in the therapeutics of lithium is not known. There is insufficient evidence at this time to suggest that G proteins would provide useful sites for drug interventions. Although the existence of multiple specific subtypes of subunits suggests that drugs with selected and limited activity could be identified, there is no evidence to suggest that specific G proteins are associated with specific neuroreceptors or transmitters. Thus, drugs that alter G protein function may produce broader effects than desired for therapeutic use. Likewise, no psychiatric disorders have been identified that are the result of defects in G protein regulatory systems. As more is learned about this vital link in neuronal transmission, opportunities for pharmacological manipulation may become more evident.

2.5.6
Second Messengers

As stated earlier, many types of neuroreceptors are connected via a family of G proteins to one of two classes of second messenger systems, the cyclic adenosine monophosphate (cAMP)/protein phosphorylation system or the inositol triphosphate/diacylglycerol system. Each of these two systems is regulated by the action of G proteins and the effectors for each system include protein kinases that catalyze the transfer of the terminal phosphate group of ATP to a wide variety of substrate proteins (Table 4). In addition to activating protein kinases, the inositol triphosphate pathway is directly involved in the regulation of intraneuronal calcium concentrations. Unlike the localized effects of changes in ions, second messenger actions are known to spread over long distances in neurons, thus influencing many types of neuronal functions (Kasai and Petersen 1994).

Table 4 Classes of proteins that are targets for phosphorylation by protein kinases

G proteins
Microtuble-associated proteins or neurofilaments
Synaptic vesicle proteins
Neurotransmitter-synthesizing enzymes
Neurotransmitter receptors
Ion channel proteins
Neurotransmitter transporters

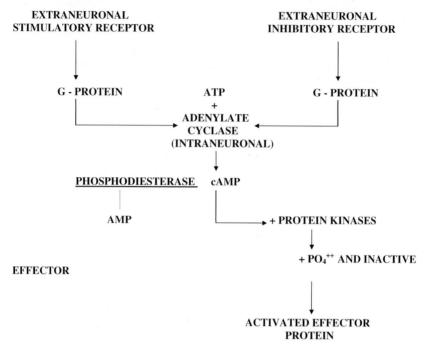

Fig. 7 Regulation and actions of the second messenger adenylate cyclase. Adenylate cyclase is either stimulated or inhibited in its production of cAMP (cyclic adenosine monophosphate) by specific receptors and G proteins. cAMP stimulates a variety of protein kinases, which in turn phosphorylate (PO_4^{++}) an effector which activates it and produces biological responses. cAMP is inactivated by the enzyme phosphodiesterase, which converts cAMP to AMP

Neuroreceptors, through specific G proteins, either stimulate or inhibit the enzyme adenylate cyclase that catalyzes the formation of cAMP (Gilman 1989). cAMP in turn binds to protein kinases that activate specific effector proteins through the process of phosphorylation (Fig. 7 and Table 4). A key element in this pathway is the intraneuronal concentration of cAMP. The rate of synthesis of cAMP is the ratio of stimulatory to inhibitory receptor input, while the rate of metabolic degradation of cAMP is determined by the activity of the enzyme, phosphodiesterase. Multiple genetic forms of adenylate cyclase (Cooper et al. 1995), and phosphodiesterase (McKnight 1991) have been identified. More than 300 specific forms of protein kinase are known (Walsh and Van Patten 1994), resulting in diverse activities.

In the case of the inositol/diacylglycerol system, extraneuronal signals are transmitted via a neurotransmitter receptor through a G protein to a phosphodiesterase, phospholipase C, which in turn hydrolyzes phosphatidylinositol-4,5-bisphosphate (PIP2), an intermembrane-bound phospholipid (Fig. 8). The products of this hydroxylation are inositol 1,4,5-trisphosphate (IP_3) and diacylglycerol, both of which serve second messenger roles (Hokin and Dixon 1993). IP_3 diffuses to the endoplasmic reticulum and stimulates a specific IP_3 receptor

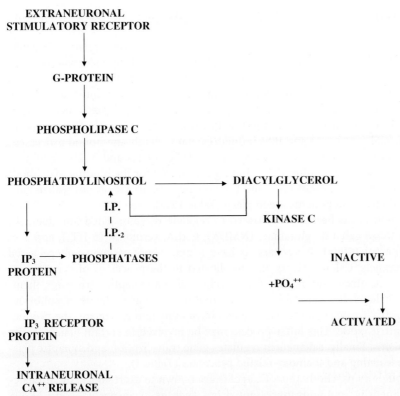

Fig. 8 Intraneuronal actions of phospholipase C activity. Phospholipase C is activated via a receptor stimulated G protein. Phospholipase C splits phosphatidylinositol into inositol triphosphate (IP$_3$) and diacylglycerol moieties. The diacylglycerol stimulates kinase C, which activates effector proteins through phosphorylation. The IP$_3$ binds to a receptor on the endoplasmic reticulum. Stimulation of this receptor releases bound calcium into the cytoplasm. IP$_3$ is inactivated by a series of phosphatases. Inositol is reincorporated into phosphatidylinositol. The anti-manic drug lithium is a potent inhibitor of phosphatase and blocks the reincorporation of inositol back into phosphatidylinositol

to release sequestered Ca^{++}. The IP$_3$ receptor is now known to exist in multiple subtypes (Marshall and Taylor 1993; Danoff and Ross 1994). The activity of IP$_3$ receptors are regulated by several allosteric sites for Ca^{++}, adenine nucleotides, and protean kinases. IP$_3$ is inactivated by the removal of phosphate through a series of phosphatase enzymes and the inositol moiety recycled back to phosphatidylinositol.

The diacylglycerol formed by the action of phospholipase C activates a widely distributed kinase, kinase C. Kinase C phosphorylates several proteins associated with a variety of neuronal membranes, such as those of synaptic vesicles, microtubules, receptor proteins (Nalepa 1994; Premont et al. 1995), and transporters (Blakely et al. 1998). The action of diacylglycerol is quite short, and it is rapidly recycled into phosphatidylinositol or metabolized to enter prostaglandin synthetic pathways.

The activity level of a protein that is activated by phosphorylation is determined by the relative rates of phosphorylation verses dephosphorylation. Calcineurin has been identified as a major factor in the dephosphorylation of a wide variety of proteins with key roles in synaptic transmission (Yakel 1997). These processes include ion channels (receptor and voltage gated), neuroreceptors, and neurotransmitter release. Calcineurin is approximately 50% bound to the neuronal membrane and thus influences many membrane-bound processes.

Calcineurin is composed of two subunits, designated A and B. The A subunit binds Ca^{++} and calmodulin; Ca^{++} binding is required for the binding of calmodulin. The B unit also binds Ca^{++} and full phosphatase activity is not realized unless all of these components are in place (Yakel 1997).

Calcineurin has been demonstrated to regulate receptor gated ion channels, such as those gated by glutamate (NMDA), GABA, serotonin (5-HT_3), and acetylcholine (nicotinic). It appears to have a major influence on voltage gated Ca^{++} channels. Calcineurin is also implicated in the processes of synaptic release of neurotransmitters and the recycling of the synaptic structures themselves. Nitric oxide synthetase is a substrate for calcineurin, the dephosphorylation of nitric oxide synthetase increasing its activity and enhancing the production of nitric oxide. This latter process may be involved in certain neurotoxicity mechanisms. Finally, calcineurin regulates gene transcription and synaptic plasticity in learning and memory-related processes (Table 5).

Although no psychotherapeutic agents are known to exert effects by influencing calcineurin, immunosuppressant drugs such as cyclosporin A and FK506 are known to be potent inhibitors of calcineurin (Yakel 1997). The clinical significance of these immunosuppressant drugs with respect to their inhibition of calcineurin is not known; however, these drugs are known to cause neurotoxici-

Table 5 Synaptic structures that are substrates for calcineurin regulation (from Yakel 1997)

Receptors
Serotonin (5-HT_3)
GABA-A
Glutamate (NMDA)
Acetylcholine
Ion channels (voltage-gated)
Calcium
Sodium
Potassium (M-current)
Other proteins
Nitric oxide synthase
Dynamin I (vesicle recycling)
Darpp-32 (regulator of protein phosphatase 1)
CREB (synaptic plasticity)
Inhibitor-1 (regulator of protein kinase A)

ty and sympathetic hypertension in in vivo animal studies (Hughes 1990; Lyson et al. 1993; Yakel 1997).

Traditionally, the components of second messenger systems have not been the primary targets of psychopharmacological agents. This is because second messengers are few in number and, therefore, are not specific or selective compared to neurotransmitter receptors or reuptake sites. For example, a drug that influences the intracellular levels of cAMP would be expected to have the same influence in many types of synapses simultaneously or, for that matter, in many kinds of tissues, since second messengers exist in many types of cells outside the nervous system (Nishizuka 1995). Of course, now that subtypes of such elements as phospholipase, phosphodiesterase, adenylate cyclase, and IP_3 receptors are known, the identification of more selective agents may be possible.

At this time, only one psychotherapeutic agent is known that is likely to have a mechanism of action that involves a second messenger system. Lithium, as an agent for treating mania, is well known to block inositol monophosphatase, a critical enzyme in the synthesis of phosphatidylinositol and the subsequent production of IP_3 (Hokin and Dixon 1993; Parthasarathy et al. 1994). It has been suggested that this is the mechanism of lithium's anti-manic action, since the influences on IP_3 occur at therapeutic doses (Baraban et al. 1989; Belmaker et al. 1990). Other pharmacological observations support this hypothesis (Kofman and Belmaker 1993). It is becoming more evident that there is a great deal of interaction among the various components of the second messenger systems and that some drug effects may be accounted for in this way. For example, the antidepressants are known to influence cAMP levels through their influences on adrenergic receptors; however, they also influence the inositol/diacylglycerol system by modifying the action of kinase C (Nalepa 1994). Thus, antidepressant effects may be produced through more than one neurotransmitter system.

The ultimate influence that an agonist exerts upon a neuronal systems depends on a complex series of interactions between the agonist and the receptor, the receptor and the G protein, and the G protein and the second messenger system (Kenakin 1995a, 1995b). Although it has not yet been demonstrated, different agonists may influence receptors in ways that alter the interaction of the receptors with a variety of G proteins. It is known that receptors may activate more than one kind of G protein, thus providing qualitatively differing biological responses. It has also been demonstrated that the relative concentrations of receptors to G proteins may be an important determinant in the qualitative response to agonist activity. High concentrations of receptors relative to G proteins cause an interaction with multiple G proteins with multiple effects. The roles that these factors play in disease states or in the mechanisms of action of drugs are not known at this time; however, psychotropic agents such as antidepressants and antipsychotics are certainly well known for their influence on receptor populations, and some of their pharmacological actions may result from changes in the interaction of these crucial elements in neurotransmission.

Acknowledgements. The author gratefully acknowledges permission from the American Psychiatric Press, Inc, Washington, D.C., to reprint the figures, tables, and much of the text of this chapter from *The American Psychiatric Press Textbook of Neuropsychiatry*, fourth edition (Yudosky and Hales 2002).

References

Aschner M, Allen JW, Kimelberg HK, LoPachin RM, Streit WJ (1999) Glial cells in neurotoxicity development. Annu Rev Pharmacol Toxicol 39:151–173
Attwell D (1994) Neurobiology: glia and neurons in dialogue. Nature 369:707–708
Bacci A, Verderio C, Pravettoni E, Matteoli M (1999) The role of glial cells in synaptic function. Philos Trans R Soc Lond B Biol Sci 354:403–409
Baker R, Llinas R (1971) Electronic coupling between neurons in the rat mesencephalic nucleus. J Physiol 212:45–63
Baraban JM, Worley PF, Snyder SH (1989) Second messenger systems and psychoactive drug action: focus on the phosphoinositide system and lithium. Am J Psychiatry 146:1251–1259
Belmaker RH, Livine A, Agam G, Moscovich DG, Grisaru N, Schreiber G, Avissar S, Danon A, Kofman O (1990) Role of inositol-1-phosphatase inhibition in the mechanism of action of lithium. Pharmacol Toxicol 66(suppl 3):76–83
Blakely RD, Berson HE, Fremeau RT Jr, Caron MG, Peek MM, Prince HK, Bradley CC (1991) Cloning and expression of a functional serotonin transporter from rat brain. Nature 354:66–70
Blakely RD, Ramamoorthy S, Qian Y, Schroeter S, Bradley CC (1997) Regulation of antidepressant-sensitive serotonin transporters. In: Reith MEA (ed) Neurotransmitter transporters: structure, function, and regulation. Humana Press, Totowa, NJ, pp 29–72
Blakely RD, Ramamoorthy S, Schroeter S, Qian Y, Apparsundaram S, Galli A, DeFelice LJ (1998) Regulated phosphorylation and trafficking of antidepressant-sensitive serotonin transporter proteins. Biol.Psychiatry 44:169–178
Bohn MC (1999) A commentary on glial cell line-derived neurotrophic factor (GDNF): from a glial secreted molecule to gene therapy. Biochem Pharmacol 57:135–142
Caldecott-Hazard S, Morgan DG, DeLeon-Jones F, Overstreet DH, Janowsky D (1991) Clinical and biochemical aspects of depressive disorders. II: transmitter/receptor theories. Synapse 9:251–301
Catterall WA (1999) Interactions of presynaptic Ca^{2+} channels and snare proteins in neurotransmitter release. In: Rudy B, Seeburg P (eds) Molecular and functional diversity of ion channels and receptors. The New York Academy of Sciences, New York, pp 144–159
Clapham DE, Neer EJ (1993) New roles for G-protein beta gamma-dimers in transmembrane signaling. Nature 365:403–406
Cooper DM, Mons N, Karpen JW (1995) Adenylyl cyclases and the interaction between calcium and cAMP signalling. Nature 374:421–424
Cooper JR, Bloom FE, Roth RH (1996) The biochemical basis of neuropharmacology, 7th edition. Oxford University Press, New York
Cornell-Bell AH, Finkbeiner SM (1991) Ca^{2+} waves in astrocytes. Cell Calcium 12:185–204
Cornell-Bell AH, Finkbeiner SM, Cooper MS, Smith SJ (1990) Glutamate induced calcium waves in cultured astrocytes: long-range glial signaling. Science 247:470–473
Corvalan V, Cole R, de Vellis J, Hagiwara S (1990) Neuronal modulation of calcium channel activity in cultured rat astrocytes. Proc Natl Acad Sci U S A 87:4345–4348

Danoff SK, Ross CA (1994) The inositol trisphosphate receptor gene family: implications for normal and abnormal brain function. Prog Neuropsychopharmacol Biol Psychiatry 18:1–16

Enkvist MOK, McCarthy KD (1992) Activation of protein kinase C blocks astroglial gap junction communication and inhibits the spread of calcium waves. J Neurochem 59:519–526

Furth PA, St Onge L, Boger H, Gruss P, Gossen M, Kistner A, Bujard H, Hennighausen L (1994) Temporal control of gene expression in transgenic mice by a tetracycline-responsive promoter. Proc Natl Acad Sci U S A 91:9302–9306

Galli A, Petersen CI, deBlaquiere M, Blakely RD, DeFelice LJ (1997) Drosophila serotonin transporters have voltage-dependent uptake coupled to a serotonin-gated ion channel. J Neurosci 17:3401–3411

Gegelashvili G, Schousboe A (1997) High affinity glutamate transporters: regulation of expression and activity. Mol Pharmacol 52:6–15

Gilman AG (1989) G proteins and regulation of adenylyl cyclase. JAMA 262:1819–1825

Grady S, Marks MJ, Wonnacott S, Collins AC (1992) Characterization of nicotinic receptor-mediated [^{3}H] dopamine release from synaptosomes prepared from mouse striatum. J Neurochem 59:848–856

Grondin R, Gash DM (1998) Glial cell line-derived neurotrophic factor (GDNF): a drug candidate for the treatment of Parkinson's disease. J Neurol 245:35–42

Guldberg HC, Mardsen CA (1975) Catechol-O-methyl transferase: pharmacological aspects and physiological role. Pharmacol Rev 27:135–206

Haga K, Haga T (1992) Activation by G protein $\beta\gamma$ subunits of agonist- or light-dependent phosphorylation of muscarinic acetylcholine receptors and rhodopsin. J Biol Chem 267:2222–2227

Hausdorff WP, Caron MG, Lefkowitz RJ (1990) Turning off the signal: desensitization of β-adrenergic receptor function. FASEB J 4:2881–2889

Hokin LE, Dixon JF (1993) The phosphoinositide signalling system. I. historical background. II. Effects of lithium on the accumulation of second messenger inositol 1,4,5-trisphosphate in brain cortex slices. Prog Brain Res 98:309–315

Hughes RL (1990) Cyclosporine-related central nervous system toxicity in cardiac transplantation [letter]. N Engl J Med 323:420–421

Inagaki N, Wada H (1994) Histamine and prostanoid receptors on glial cells. Glia 11:102–109

Joels M, de Kloet ER (1994) Mineralcorticoid and glucocorticoid receptors in the brain: implications for ion permeability and transmitter systems. Prog Neurobiol 43:1–36

Jursky F, Tamura S, Tamura A, Mandiyan S, Nelson H, Nelson N (1994) Structure, function and brain localization of neurotransmitter transporters. J Exp Biol 196:283–295

Kanai Y, Smith CP, Hediger MA (1994) A new family of neurotransmitter transporters: the high-affinity glutamate transporters. FASEB J 8:1450–1459

Kanner BI, Bendahan A, Pantanowitz S, Su H (1994) The number of amino acid residues in hydrophilic loops connecting transmembrane domains of the GABA transporter GAT-1 is critical for its function. FEBS Lett 356:191–194

Kasai H, Petersen OH (1994) Spatial dynamics of second messengers: IP3 and cAMP as long-range and associative messengers. Trends Neurosci 17:95–101

Kelly RB (1999) An introduction to the nerve terminal. In: Bellen HJ (ed) Neruotransmitter release. Oxford University Press, Oxford, pp 1–33

Kenakin T (1995a) Agonist-receptor efficacy I: mechanisms of efficacy and receptor promiscuity. Trends Pharmacol Sci 16:188–192

Kenakin T (1995b) Agonist-receptor efficacy II: agonist trafficking of receptor signals. Trends Pharmacol Sci 16:232–238

Kilts CD (1994) Recent pharmacologic advances in antidepressant therapy. Am J Med 97(suppl 6A):3S–12S

Kofman O, Belmaker RH (1993) Ziskind-Somerfeld Research Award 1993. Biochemical, behavioral, and clinical studies of the role of inositol in lithium treatment and depression. Biol Psychiatry 34:839–852

Korn H, Sotelo C, Crepel F (1973) Electronic coupling between neurons in rat lateral vestibular nucleus. Exp Brain Res 16:255–275

Krantz DE, Chaudhry FA, Edwards RH (1999) Neurotransmitter transporters. In: Bellen HJ (ed) Neurotransmitter release. Oxford University Press, Oxford, pp 145–207

Lai J, Bilsky EJ, Rothman RB, Porreca F (1994) Treatment with antisense oligodeoxynucleotide to the opioid delta receptor selectively inhibits delta 2-agonist antinociception. Neuroreport 5:1049–1052

Lapchak PA (1998) A preclinical development strategy designed to optimize the use of glial cell line-derived neurotrophic factor in the treatment of Parkinson's disease. Mov Disord 13(suppl 1):49–54

Lesch KP, Bengel D, Heils A, Sabol SZ, Greenberg BD, Petri S, Benjamin J, Muller CR, Hamer DH, Murphy DL (1996) Association of anxiety-related traits with a polymorphism in the serotonin transporter gene regulatory region [see comments]. Science 274:1527–1531

Lesch KP, Mossner R (1998) Genetically driven variation in serotonin uptake: is there a link to affective spectrum, neurodevelopmental, and neurodegenerative disorders? Biol Psychiatry 44:179–192

Lester HA, Cao Y, Mager S (1996) Listening to neurotransmitter transporters. Neuron 17:807–810

Levitan IB, Kacmarek LK (1997) The neuron: cell and molecular biology, 2nd edition. New York, Oxford University Press

Llinas R, Baker R, Sotelo C (1974) Electronic coupling between neurons in the cat inferior olive. J Neurophysiol 37:560–571

LoTurco JJ, Kriegstein AR (1991) Clusters of coupled neuroblasts in embryonic neocortex. Science 252:563–566

Lovinger DM (1997) Alcohols and neurotransmitter gated ion channels: past, present and future. Naunyn Schmiedebergs Arch Pharmacol 356:267–282

Lucas JJ, Hen R (1995) New players in the 5-HT receptor field: genes and knockouts. Trends Pharmacol Sci 16:246–252

Lyson T, Ermel LD, Belshaw PJ, Alberg DG, Schreiber SL, Victor RG (1993) Cyclosporine- and FK506-induced sympathetic activation correlates with calcineurin-mediated inhibition of T-cell signaling. Circ Res 73:596–602

MacDermott AB, Role LW, Siegelbaum SA (1999) Presynaptic ionotropic receptors and the control of transmitter release. Annu Rev Neurosci. 22:443–485

Mannisto PT, Ulmanen I, Lundstrom K, Taskinen J, Tenhunen J, Tilgmann C, Kaakkola S (1992) Characteristics of catechol O-methyl-transferase (COMT) and properties of selective COMT inhibitors. Prog Drug Res 39:291–350

Marshall IC, Taylor CW (1993) Regulation of inositol 1,4,5-trisphosphate receptors. J Exp Biol 184:161–182

Martin DL (1992) Synthesis and release of neuroactive substances by glial cells. Glia 5:81–94

McGeer PL, McGeer EG (1998) Glial cell reactions in neurodegenerative diseases: pathophysiology and therapeutic interventions. Alzheimer Dis Assoc Disord 12(suppl 2):S1–S6

McKnight GS (1991) Cyclic AMP second messenger systems. Curr Opin Cell Biol 3:213–217

Melcangi RC, Magnaghi V, Martini L (1999) Steroid metabolism and effects in central and peripheral glial cells. J Neurobiol 40:471–483

Melikian HE, Ramamoorthy S, Tate CG, Blakely RD (1996) Inability to N-glycosylate the human norepinephrine transporter reduces protein stability, surface trafficking, and transport activity but not ligand recognition. Mol Pharmacol 50:266–276

Nalepa I (1994) The effect of psychotropic drugs on the interaction of protein kinase C with second messenger systems in the rat cerebral cortex. Pol J Pharmacol 46:1–14

Narahashi T, Aistrup GL, Marszalec W, Nagata K (1999) Neuronal nicotinic acetylcholine receptors: a new target site of ethanol. Neurochem Int 35:131–141

Nelson N (1998) The family of Na+/Cl- neurotransmitter transporters. J Neurochem 71:1785–1803

Nishizuka Y (1995) Protein kinase C and lipid signaling for sustained cellular responses. FASEB J 9:484–496

Oswald I, Brezinova V, Dunleavy DLF (1972) On the slowness of action of tricyclic antidepressant drugs. Br J Psychiatry 120:673–677

Otero GC, Merrill JE (1994) Cytokine receptors on glial cells. Glia 11:117–128

Parthasarathy L, Vadnal RE, Parthasarathy R, Devi CS (1994) Biochemical and molecular properties of lithium-sensitive myo-inositol monophosphatase. Life Sci 54:1127–1142

Paul IA, Nowak G, Layer RT, Popik P, Skolnick P (1994) Adaptation of the N-methyl-D-aspartate receptor complex following chronic antidepressant treatments. J Pharmacol Exp Ther 269:95–102

Pitcher JA, Inglese J, Higgins JB, Arriza JL, Casey PJ, Kim C, Benovic JL, Kwatra MM, Caron MG, Lefkowitz RJ (1992) Role of beta gamma subunits of G proteins in targeting the beta-adrenergic receptor kinase to membrane-bound receptors. Science 257:1264–1267

Poitry-Yamate CL, Poitry S, Tsacopoulos M (1995) Lactate released by Muller glial cells is metabolized by photoreceptors from mammalian retina. J Neurosci 15:5179–5191

Premont RT, Inglese J, Lefkowitz RJ (1995) Protein kinases that phosphorylate activated G protein-coupled receptors. FASEB J 9:175–182

Rahamimoff R, Butkevich A, Duridanova D, Ahdut R, Harari E, Kachalsky SG (1999) Multitude of ion channels in the regulation of transmitter release. Philos Trans R Soc Lond B Biol Sci 354:281–288

Raivich G, Jones LL, Werner A, Bluthmann H, Doetschmann T, Kreutzberg GW (1999) Molecular signals for glial activation: pro- and anti-inflammatory cytokines in the injured brain. Acta Neurochir Suppl (Wien) 73:21–30

Ransom BR, Sontheimer H (1992) The neurophysiology of glial cells. J Clin Neurophysiol 9:224–251

Rens-Domiao S, Hamm HE (1995) Structural and functional relationships of heterotrimeric G-proteins. FASEB J 9:1059–1066

Robinson SR, Hampson E, Munro MN, Vaney DI (1993) Unidirectional coupling of gap junctions between neuroglia. Science 262:1072–1074

Ronde P, Nichols RA (1998) High calcium permeability of serotonin 5-HT3 receptors on presynaptic nerve terminals from rat striatum. J Neurochem 70:1094–1103

Saarma M, Sariola H (1999) Other neurotrophic factors: glial cell line-derived neurotrophic factor (GDNF). Microsc Res Tech 45:292–302

Saudou F, Amara DA, Dierich A, LeMeur M, Ramboz S, Segu L, Buhot MC, Hen R (1994) Enhanced aggressive behavior in mice lacking 5-HT1B receptor. Science 265:1875–1878

Schlag BD, Vondrasek JR, Munir M, Kalandadze A, Zelenaia OA, Rothstein JD, Robinson MB (1998) Regulation of the glial Na+-dependent glutamate transporters by cyclic AMP analogs and neurons. Mol Pharmacol 53:355–369

Shank RP, William JB, Charles WA (1989) Glutamine and 2-oxoglutarate as metabolic precursors of the transmitter pools of glutamate and GABA: correlation of regional uptake by rat brain synaptosomes. Neurochem Res 16:29–34

Shepard GM (1994) Neurobiology, 2nd edition. New York, Oxford University Press

Shinohara K, Hiruma H, Funabashi T, Kimura F (2000) GABAergic modulation of gap junction communication in slice cultures of the rat suprachiasmatic nucleus. Neuroscience 96:591–596

Silvia CP, King GR, Lee TH, Xue ZY, Caron MG, Ellinwood EH (1994) Intranigral administration of D2 dopamine receptor antisense oligodeoxynucleotides establishes a role for nigrostriatal D2 autoreceptors in the motor actions of cocaine. Mol Pharmacol 46:51–57

Singer TP, Ramsay RR (1995) Monoamine oxidases: old friends hold many surprises. FASEB J 9:605–610

Sontheimer H (1994) Voltage dependent ion channels in glial cells. Glia 11:156–172

Standifer KM, Chien CC, Wahlestedt C, Brown GP, Pasternak GW (1994) Selective loss of delta opioid analgesia and binding by antisense oligodeoxynucleotides to a delta opioid receptor. Neuron 12:805–810

Stephens DN, Kehr W, Duka T (1986) Anxiolytic and anxiogenic β-carbolines: tools for the study of anxiety mechanisms. In: Biggio G, Costa E (eds) GABAergic transmission and anxiety. Raven, New York, pp 91–106

Swanson RA, Liu J, Miller JW, Rothstein JD, Farrell K, Stein BA, Longuemare MC (1997) Neuronal regulation of glutamate transporter subtype expression in astrocytes. J Neurosci 17:932–940

Tallman JF, Cassela JV, White G, Gallager DW (1999) $GABA_A$ receptors: diversity and its implications for CNS disease. The Neuroscientist 5:351–361

Tecott LH, Sun LM, Akana SF, Strack AM, Lowenstein DH, Dallman MF, Julius D (1995) Eating disorder and epilepsy in mice lacking 5-HT2c serotonin receptors. Nature 374:542–546

Thiel G. (1995) Recent breakthroughs in neurotransmitter release: paradigm for regulated exocytosis? News in Physiological Science 10:42–46

Thomas KR, Capecchi MR (1990) Targeted disruption of the murine int-1 proto-oncogene resulting in severe abnormalities in midbrain and cerebellar development. Nature 346:847–850

Tsacopoulos M, Magistretti PJ (1996) Metabolic coupling between glia and neurons. J Neurosci 16:877–885

Vardimon L, Ben-Dror I, Avisar N, Oren A, Shiftan L (1999) Glucocorticoid control of glial gene expression. J Neurobiol 40:513–527

Wahlestedt C, Pich EM, Koob GF, Yee F, Heilig M (1993) Modulation of anxiety and neuropeptide Y-Y1 receptors by antisense oligodeoxynucleotides. Science 259:528–531

Walsh DA, Van Patten SM (1994) Multiple pathway signal transduction by the cAMP-dependent protein kinase. FASEB J 8:1227–1236

Wartiovaara K, Hytonen M, Vuori M, Paulin L, Rinne J, Sariola H (1998) Mutation analysis of the glial cell line-derived neurotrophic factor gene in Parkinson's disease. Exp Neurol 152:307–309

Watkins SS, Epping-Jordan MP, Koob GF, Markou A. (1999) Blockade of nicotine self-administration with nicotinic antagonists in rats. Pharmacol Biochem Behav 62:743–751

Wolfe BB, Harden TK, Sporn JR, Molinoff PB (1978) Presynaptic modulation of beta adrenergic receptors in rat cerebral cortex after treatment with antidepressants. J Pharmacol Exp Ther 207:446–457

Wonnacott S (1997) Presynaptic nicotinic ACh receptors. Trends in Neuroscience 20:92–98

Yakel JL (1997) Calcineurin regulation of synaptic function: from ion channels to transmitter release and gene transcription. Trends Pharmacol Sci. 18:124–134

Yeomans J, Baptista M (1997) Both nicotinic and muscarinic receptors in ventral tegmental area contribute to brain-stimulation reward. Pharmacol Biochem Behav 57:915–921

Yudosky SC, Hales RE (eds) (2002) The American Psychiatric Press Textbook of Neuropsychiatry, 4th edn. American Psychiatric Publishing. Washington, DC, pp 1123–1147

Zahs KR (1998) Heterotypic coupling between glial cells of the mammalian nervous system. Glia 24:85-96
Zhang B, Ramaswami M (1999) Synaptic vesicle endocytosis and recycling. In: Bellen HJ (ed) Neurotransmitter release. Oxford University Press, Oxford, pp 389-431
Zhou LW, Zhang SP, Qin ZH, Weiss B (1994) In vivo administration of an oligodeoxynucleotide antisense to the D2 dopamine receptor messenger RNA inhibits D2 dopamine receptor-mediated behavior and the expression of D2 dopamine receptors in mouse striatum. J Pharmacol Exp Ther 268:1015-1023

General Principles of Pharmacokinetics

S. H. Preskorn[1] · M. L. Catterson[2]

[1] Department of Psychiatry and Behavioral Sciences,
University of Kansas School of Medicine and Psychiatric Research Institute,
1010 N. Kansas, Wichita, KS 67214, USA
e-mail: spreskor@kumc.edu
[2] University of Kansas School of Medicine and Psychiatric Research Institute,
Wichita, KS, USA

1	Introduction	37
2	**Pharmacokinetic Principles**	38
2.1	Absorption and Distribution	38
2.2	Metabolism	41
2.3	Elimination	43
2.4	Half-Life and Steady-State Concentration	43
2.5	Linear Versus Nonlinear Pharmacokinetics	44
2.6	Factors That Can Affect Pharmacokinetics	45
2.6.1	Protein Binding	45
2.6.2	Cardiac Function and Hepatic Arterial Blood Flow	45
2.6.3	Hepatic Integrity	46
2.6.4	Renal Factors	46
2.6.5	Aging and Disease: Effects on Metabolism and Elimination	46
2.6.6	Gender: Effect on Metabolism and Elimination	47
2.7	Summary	47
3	**Tricyclic Antidepressants**	48
3.1	Absorption	49
3.2	Distribution	49
3.3	Metabolism and Elimination	50
3.4	Half-Life	51
3.5	Linear Versus Nonlinear Pharmacokinetics	51
3.6	Pharmacokinetically Mediated Drug–Drug Interactions	51
4	**Selective Serotonin Reuptake Inhibitors**	52
4.1	Absorption	52
4.2	Distribution	52
4.3	Metabolism and Elimination	53
4.4	Half-Life	58
4.5	Linear Versus Nonlinear Pharmacokinetics	59
4.6	Pharmacokinetically Mediated Drug–Drug Interactions	60
5	**Bupropion**	61
5.1	Absorption	62
5.2	Distribution	62
5.3	Metabolism and Elimination	62

5.4	Half-Life	63
5.5	Linear Versus Nonlinear Pharmacokinetics	63
5.6	Pharmacokinetically Mediated Drug–Drug Interactions	63
6	**Serotonin–Norepinephrine Specific Reuptake Inhibitors**	64
6.1	Absorption	65
6.2	Distribution	65
6.3	Metabolism and Elimination	66
6.4	Half-Life	66
6.5	Linear Versus Nonlinear Pharmacokinetics	66
6.6	Pharmacokinetically Mediated Drug–Drug Interactions	67
7	**Phenylpiperazine Agents**	67
7.1	Absorption	67
7.2	Distribution	68
7.3	Metabolism and Elimination	68
7.4	Half-Life	69
7.5	Linear Versus Nonlinear Pharmacokinetics	69
7.6	Pharmacokinetically Mediated Drug–Drug Interactions	70
8	**Monoamine Oxidase Inhibitors**	71
8.1	Absorption	71
8.2	Distribution	71
8.3	Metabolism and Elimination	71
8.4	Half-Life	72
8.5	Linear Versus Nonlinear Pharmacokinetics	72
8.6	Pharmacokinetically Mediated Drug–Drug Interactions	72
9	**Mirtazapine**	72
9.1	Absorption	73
9.2	Distribution	73
9.3	Metabolism and Elimination	73
9.4	Half-Life	73
9.5	Linear Versus Nonlinear Pharmacokinetics	74
9.6	Pharmacokinetically Mediated Drug–Drug Interactions	74
10	**Conclusion**	74
	Glossary	75
	References	78

Abstract A number of general pharmacokinetic principles and properties apply to all drugs; these include absorption, distribution, metabolism, elimination, half-life and steady-state concentration, and linear versus nonlinear pharmacokinetics. Factors that can affect the pharmacokinetics of a drug include protein binding, cardiac function and hepatic arterial blood flow, hepatic integrity, renal factors, aging, disease, and gender. An understanding of basic pharmacokinetic principles provides a background for understanding the pharmacokinetic properties of the major classes of antidepressants. The chapter reviews the absorption, distribution, metabolism and elimination, half-life, linear versus nonlinear pharmacokinetics, and pharmacokinetically mediated drug–drug interactions associated with each of the following classes of antidepressants: tricyclics, selective serotonin reuptake inhibitors, bupropion, serotonin–norepinephrine-specific reuptake inhibitors (venlafaxine and duloxetine), phenylpiperazine agents (trazodone and nefazodone), monoamine oxidase inhibitors, and mirtazapine.

Keywords Pharmacokinetics · Antidepressants · Absorption · Distribution · Metabolism · Elimination · Half-life · Steady-state concentration · Linear pharmacokinetics · Nonlinear pharmacokinetics · Drug–drug interactions

Portions of this chapter are adapted with permission from Chap. 3 on "Pharmacokinetics" in Janicak PG, Davis JM, Preskorn SH, Ayd FJ Jr. (2001) *Principles and Practice of Psychopharmacotherapy*, third edition. Lippincott Williams & Wilkins, Philadelphia, PA.

1
Introduction

The pharmacokinetics of a drug can be divided into four primary phases:

- Absorption
- Distribution
- Metabolism
- Elimination

Pharmacokinetics involves *what the body does to the drug*. In contrast, pharmacodynamics involves *what the drug does to the body*. The fundamental relationship between pharmacodynamics and pharmacokinetics is expressed in the following equation:

Effect = affinity for site of action × drug level
 × biological variance of the individual

Pharmacodynamics describes what the drug is theoretically capable of doing. For example, any drug that can block the central histamine-1 receptors is theoretically capable of causing sedation. However, to cause this effect, a sufficient amount of the drug must reach this target. The pharmacokinetics of a drug determines whether that amount is achieved under specific dosing conditions. Factors unique to each individual also affect the outcome. These include genetic differences, age-associated changes in physiology, disease states, and concomitant medications, which can affect related mechanisms of actions or the mechanisms that mediate the absorption, distribution, and elimination of the drug in question.

Generally speaking, there is an optimal receptor occupation (site of action) that will produce an optimal response. Below this critical level, the site of action will not be sufficiently engaged to achieve the desired response. Conversely, receptor occupation above an upper threshold is associated with an increased likelihood of adverse effects. This principle is fundamental to understanding the potential benefit of therapeutic drug monitoring (TDM). TDM is a means to determine drug level, the second variable in the equation presented above. Since TDM is the focus of the next chapter, only a few comments on TDM will be made here.

Drug level is determined by two variables: dosing rate and clearance. From this perspective, dose-finding clinical trials are essentially population pharmacokinetic studies in which the goal is to determine the dose of the drug needed to achieve the optimal drug level (i.e., concentration) in the average person enrolled in the study. Since individuals with medical conditions or diseases or who are taking multiple concomitant drugs are generally excluded from such studies, the effect of biological variance in the individual becomes an important variable that the astute clinician must factor into the equation.

This chapter first outlines basic pharmacokinetic principles to provide a background for the discussion of the properties of the different classes of antidepressants that follows. The rest of the chapter focuses on the pharmacokinetics of the following classes of antidepressants: tertiary and secondary amine tricyclic antidepressants (TCAs) and selective norepinephrine reuptake inhibitors (including both secondary amine TCAs such as desipramine and nonTCAs such as reboxetine), selective serotonin reuptake inhibitors (SSRIs), bupropion, combined serotonin–norepinephrine reuptake inhibitors (e.g., venlafaxine), phenylpiperazine agents (trazodone and nefazodone), monoamine oxidase inhibitors (MAOIs), and mirtazapine. Each section includes discussions of the following pharmacokinetic characteristics:

1. Absorption and distribution
2. Metabolism and elimination
3. Half-life
4. Linear versus nonlinear pharmacokinetics
5. Pharmacokinetically mediated drug–drug interactions

For ease of reference, a glossary of pharmacokinetic terms is provided at the end of the chapter.

2
Pharmacokinetic Principles

2.1
Absorption and Distribution

The principal route of administration for psychoactive drugs, including antidepressants, is oral, with absorption generally occurring in the small bowel. Drug then passes into the portal circulation and enters the liver. Cytochrome P450 (CYP) enzymes in the bowel wall and in the liver can metabolize drugs before they reach the systemic circulation (i.e., first-pass metabolism). Most psychiatric medications are highly lipophilic. This lipophilicity enables them to readily pass the blood–brain barrier and enter the central nervous system, (Hegarty and Dundee 1977; Greenblatt et al. 1989, 1996; Janicak and Davis 2000) and causes them to share other similarities, including:

1. Rapid absorption
2. Rapid and extensive distribution in tissue compartments
3. High first-pass effect
4. Large volume of distribution

The more polar (less lipophilic) a compound, the slower the absorption from the gastrointestinal tract and penetration into the brain from the systemic circulation.

First-pass effect can be altered by diseases (e.g., cirrhosis, portacaval shunting, persistent hepatitis, congestive heart failure) and by some drugs (e.g., alcohol, ketoconazole, fluoxetine), influencing the peak concentrations achieved and the ratio of the parent compound to its metabolites (Preskorn 1997; Klotz 1998; Thurmann and Hompesch 1998; DeVane and Pollock 1999).

Bioavailability refers to the portion of a drug absorbed from the site of administration. Fig. 1 illustrates drug concentration curves in plasma as a function of time for IM, IV, and oral administration of a drug. The area under the curve (AUC) is the total amount of drug in the systemic circulation available for distribution to the site(s) of action (i.e., its bioavailability). Because IV administration

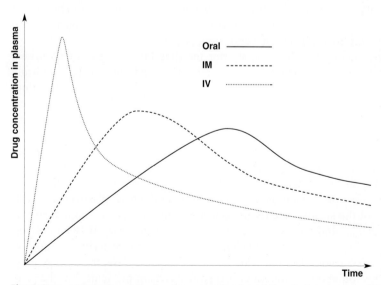

Fig. 1 Drug concentration curves in plasma as a function of time for intramuscular (*IM*), intravenous (*IV*), and oral administration of a drug (copyright Preskorn 2003)

of a drug produces 100% absorption, it is the reference site of administration. The same dose, if completely absorbed from one of the other routes, would produce an AUC identical to IV administration, although the shape of the curve would be different principally because of the delayed time to peak concentration (C_{max}).

Different routes of administration can affect the ratio of parent compound to its various metabolites as well as the rate of absorption. The concentration versus time curve is usually shifted to the right with IM compared to IV administration and generally shifted even more to the right with oral administration, due to the immediate entrance into the central compartment with IV administration as opposed to the more gradual (and often less complete) absorption with these other two routes of administration (Fig. 1). Hence, the time needed to reach the maximum plasma concentration following a dose (T_{max}) is shifted to the left (i.e., shortened), while the peak plasma drug concentration occurring after a dose (C_{max}) is higher, contracting the curve, even though the AUC may be unchanged.

Still other factors can affect the rate of absorption, independent of the site of administration. For example, some drugs are not stable when given in IM formulation and may crystallize at tissue pH, making them less bioavailable when given IM than orally.

A decrease in the AUC represents a decrease in bioavailability for that route of administration (Benet et al. 1996). Common factors that influence bioavailability include physicochemical properties of the drug, formulation of the product, disease states that influence gastrointestinal function or first-pass effect, and precipitation of the drug at the injection site.

Other clinically important parameters beyond the extent of absorption that have an impact on the single-dose plasma curve include peak concentration (C_{max}) and time to the peak concentration (T_{max}). Generally, C_{max} will be inversely correlated to T_{max} (i.e., the shorter the time for a drug to be absorbed, the higher the peak concentration). A higher C_{max} and shorter T_{max} typically mean a more rapid appearance of clinical activity following administration. Since T_{max} and C_{max} are generally a function of a compound's physicochemical properties, these parameters can also determine whether a particular drug is appropriate for a specific indication.

While fast absorption is desired for some psychotropic agents, that may not always be the case particularly since adverse effects may be a function of C_{max}. For example, cardiac toxicity due to the stabilization of excitable membranes, which is an effect of TCAs, is as much a function of the peak plasma concentration as it is of the steady-state tissue concentration. Thus, changing a formulation to delay T_{max} and reduce C_{max} may significantly increase safety and/or tolerability. Dividing the dose into smaller amounts and administering it more frequently can also accomplish the same result. In the latter case, the average plasma concentration and the amount absorbed will remain the same, but the peak concentration will be lower and the trough concentration higher.

Bioavailability, particularly the rate of absorption, can vary significantly among different formulations (i.e., products) of the same drug. The Food and Drug Administration considers a generic product to be comparable to a brand name if there is no more than a 20% difference (more or less) in bioavailability (i.e., T_{max} and C_{max}) (Schwartz 1985). Hence, there could theoretically be as much as a 40% difference between two generic preparations of the same drug. This might explain treatment failure or side effects in a patient who had previously tolerated and benefited from one preparation of a medication and is then switched to another preparation because of cost or availability.

Most psychotropic drugs are highly protein-bound (Vallner 1977; Routledge 1986; Campion et al. 1988; MacKichan 1989). Bound drug often accounts for more than 90% of the total plasma concentration. While the free-drug fraction is the smallest absolute amount, it is the most important, since its concentration determines the final equilibration with the site of action. While the difference between 95% and 90% bound drug may seem small, the corresponding change in the free fraction would be from 5% to 10%, thus doubling the effective drug concentration.

2.2
Metabolism

Most psychotropic agents, including antidepressants, undergo extensive oxidative biotransformation resulting in the formation of more polar metabolites, which are then excreted in the urine. The necessary biotransformation steps may involve one or several of the following:

1. Hydroxylation
2. Demethylation
3. Oxidation
4. Sulfoxide formation

Whereas most psychotropic drugs undergo extensive oxidative biotransformation (i.e., phase I metabolism) prior to elimination, some undergo simple conjugation with moieties such as glucuronic acid (i.e., phase II metabolism), and others are excreted unmetabolized (e.g., lithium) (Preskorn 1996). Since conjugation can occur in most organs, phase II metabolism is not dependent on liver function. Thus, the clearance of drugs that undergo only glucuronidation is generally not affected or affected to only a small degree by even significant liver impairment. However, most antidepressants undergo oxidative biotransformation mediated principally by CYP enzymes in the liver and hence can be affected to a clinically meaningful degree by significant liver impairment.

Oxidative biotransformation results in the formation of metabolites whose effects may be similar or dissimilar to the parent compound. For example, norfluoxetine has essentially the same activity as fluoxetine in terms of both serotonin uptake blockade and inhibition of several CYP enzymes, but is cleared more slowly (Preskorn et al. 1994; Preskorn 1997; Richelson 1998). As a result, norfluoxetine accumulates extensively in the body following chronic administration of fluoxetine, making it, rather than the parent compound, the principal determinant of clinical effect.

In contrast to fluoxetine, clomipramine's major metabolite, desmethylclomipramine, has a markedly different pharmacological profile from the parent drug (Preskorn 1996). While clomipramine is a potent inhibitor of serotonin uptake, desmethylclomipramine is a more potent inhibitor of norepinephrine uptake. If clomipramine's value in obsessive-compulsive disorder depends on its ability to block uptake of 5-hydroxytryptamine (5-HT) (and not block uptake of norepinephrine), then this effect should be a function of the relative ratio between clomipramine and desmethylclomipramine. Thus, this agent could lose its effectiveness for treatment of obsessive-compulsive disorder if a patient were an efficient demethylator of clomipramine such that desmethylclomipramine rather than the parent drug was the predominant molecule reaching the brain.

Another important clinical situation may arise when the parent drug is biotransformed into a less efficacious and possibly more toxic metabolite. For example, if the concentration of the hydroxylated metabolite of imipramine (2-hydroxyimipramine) were increased, this TCA could lose its effectiveness while simultaneously increasing in toxicity (Jandhyala et al. 1977).

CYP enzyme induction and inhibition also play a role a role in the metabolism of psychotropic agents. Alcohol, nicotine, and most anticonvulsants induce a number of CYP enzymes (Pippenger 1987; Spina et al. 1996). Barbiturates and carbamazepine induce their own metabolism (i.e., autoinduction) as well as that of other drugs.

Alcohol has a triphasic effect on the elimination rates of drugs such as TCAs that require extensive biotransformation (Weller and Preskorn 1984). Acute alcohol ingestion in combination with a TCA in a teetotaler who attempts suicide will significantly block the first pass metabolism of the TCA. This inhibition can triple the peak concentration of a TCA by increasing its bioavailability. These facts are why the consumption of alcohol in association with a TCA overdose increases lethality.

Drinking on a regular basis for several weeks to months can induce CYP enzymes, resulting in a lower TCA plasma concentration. Thus, subacute and subchronic alcohol consumption induces liver enzymes and causes lower plasma levels of drugs that undergo oxidative biotransformation as a necessary step in their elimination.

Chronic alcohol ingestion can cause cirrhosis, reducing hepatic CYP enzyme concentration and liver mass and causing portacaval shunting. These effects will result in increased plasma drug levels due to both greater bioavailability (due to reduced first-pass metabolism) and decreased clearance.

In contrast to anticonvulsants and alcohol, drugs such as bupropion, fluoxetine, quinidine, paroxetine, and some antipsychotics can inhibit specific CYP enzymes (Wright et al. 1982; Fonne-Pfister and Meyer 1988; Preskorn et al.

1994; Preskorn 1997; Richelson 1998; Baker 1999; Naranjo et al. 1999; Tanaka and Hisawa 1999). As a result, drugs such as fluoxetine can slow the metabolism of TCAs, certain benzodiazepines, bupropion, some steroids, and antipsychotics to a clinically meaningful degree. For example, 20 mg/day of fluoxetine produces, on average, a 500% increase in the levels of those co-prescribed drugs that are principally dependent on CYP 2D6 for their clearance. If the co-prescribed drug has a narrow therapeutic index (e.g., a TCA) and the dose is not adjusted for the change in clearance, serious or even life-threatening toxicity could result.

2.3
Elimination

The last step in a drug's clearance from the body is elimination. Elimination occurs via the kidneys for most psychotropic drugs. The drug is first converted by drug metabolizing enzymes into polar metabolites, which are more water- and less lipid-soluble than the parent compound. This process facilitates the clearance of these metabolites in urine. Following a single dose, this step is reflected on the plasma drug concentration versus time curve by the terminal elimination phase, which is typically a gradual and steady decline in the plasma drug level over time (Fig. 1). The slope of this clearance curve is a function of the rates of biotransformation and elimination. Thus, the slope (the kinetic constant of elimination [K_e]) represents a summation of the CYP enzymatic activity required for the biotransformation and the glomerular filtration rate that clears the polar metabolites from the blood.

2.4
Half-Life and Steady-State Concentration

Half-life ($t_{1/2}$) is the time needed to clear 50% of a drug from the plasma. It also predicts the length of time necessary to reach steady state. As a general rule, the time to steady-state concentration (C_{ss}) of a drug is five times the half-life of the drug. Parenthetically, that is not necessarily five times the dosing interval. For example, a drug may have a half-life of 6 h but be administered only once a day (i.e., dosing rate) as a hypnotic. It takes a comparable period of time to clear the drug from the body after steady state has been attained. During each half-life period, a patient either clears or accumulates 50% of the eventual C_{ss} produced by that dosing rate. Therefore, in one half-life, a patient will have reached 50% of the concentration that will eventually be achieved. In two half-lives, a patient achieves the initial 50% plus half of the remaining 50%, for a total of 75%. After three half-lives, a patient achieves the initial 50% and the next 25% plus half of the remaining 25%, for a total of 87.5%. At 97% (i.e., five half-lives), the patient is essentially at steady state, which is the rationale behind this general rule.

The half-life of a drug also has clinical relevance because the longer the half-life of a drug, the longer the interval between starting the drug and seeing its full effects, whether beneficial or adverse. The length of the half-life also is one parameter determining the time needed to assess for the full effect of the drug at that dose.

C_{ss} is that total concentration of a drug in plasma once steady state has been achieved meaning that this concentration will not change as long as the dosing rate (i.e., dose per interval of time) remains unchanged or other factors do not alter the rate of metabolism or elimination of the drug (e.g., the administration of a concomitant drug which induces or inhibits the metabolism of the drug of interest). When C_{ss} is achieved, the amount of drug excreted every 24 h will equal the amount taken every 24 h, as long as other medical illnesses, diet (e.g., grapefruit juice), personal habits (e.g., alcohol use, smoking), or the ingestion of concomitant medications (e.g., carbamazepine, fluoxetine) do not alter the elimination rate.

Under steady-state conditions, there is a proportional relationship between the tissue and the plasma compartments (Holford and Sheiner 1981; Glotzbach and Preskorn 1982; Benet et al. 1996). This is the underlying principle of TDM. Even though psychotropic agents do not exert their psychoactive effects in plasma, and tissue concentrations of the drugs (depending upon the organ) are 10–100 times greater than plasma concentrations, plasma concentrations provide an indirect or surrogate measurement of tissue concentrations (Glotzbach and Preskorn 1982; Seeman 1995). TDM thus uses drug concentration in plasma to assure adequacy of drug dose for efficacy while at the same time seeking to avoid toxicity.

2.5
Linear Versus Nonlinear Pharmacokinetics

When the amount of drug eliminated per unit of time is directly proportional to the amount ingested (i.e., there is a linear relationship between dose change and plasma level change), this condition is termed *first order* or *linear kinetics*. When only a fixed amount of drug is eliminated in a given interval of time, because mechanisms for biotransformation and elimination are saturated, the condition is terms *zero order kinetics*. With zero order kinetics, each dose increment produces a larger than proportional increase in plasma concentration. A classic example of zero order kinetics is alcohol, for which blood levels rise disproportionately as increased amounts are ingested.

While most drugs exhibit linear kinetics over their usual concentration ranges, a number of psychotropic drugs demonstrate nonlinear pharmacokinetics. With nonlinear pharmacokinetics, the drug concentration increases disproportionately at specific concentration of the drug as the mechanism responsible at lower concentrations becomes saturated and a less efficient but higher capacity mechanism takes over. CYP enzymes are usually the mechanisms in such nonlinear pharmacokinetics. These enzymes have different affinities and capacities to mediate the biotransformation of a drug at different concentrations. At a low concentration, the biotransformation is principally mediated by a CYP enzyme with a higher affinity for the drug but a lower capacity to metabolize it. If the drug saturates this capacity as its concentration increases (i.e., it autoinhibits its own metabolism), then the levels of the drug rise disproportionately with increasing dose until they reach a point at which a lower affinity but higher capacity CYP enzyme takes over the drug's biotransformation. For example, one third of patients taking TCAs demonstrate nonlinear kinetics when concentrations exceed approximately 200 ng/ml (Nelson and Jatlow 1987; Preskorn 1999). Other antidepressants with nonlinear pharmacokinetics include fluoxetine, nefazodone, and paroxetine (Preskorn 1997; Baker 1999; Preskorn 1999).

2.6
Factors That Can Affect Pharmacokinetics

2.6.1
Protein Binding

Most psychotropic medications, including antidepressants, are highly protein-bound. Any condition that shifts the ratio of bound to free drug can change the concentration of the drug at its site(s) of action and thus alter the magnitude of its effect. Conditions that lead to a functional decrease in the amount of circulating protein include:

- Malnutrition (e.g., severe anorexia)
- Wasting (e.g., the nephrotic syndrome)
- Aging (Cusack et al. 1985; Campion et al. 1988)
- Concomitant drugs that compete for protein-binding sites (Paxton 1980; Goulden et al. 1987)

Increasing the relative amount of the free fraction can increase toxicity. Most assays used in routine TDM do not distinguish between bound and free-drug fractions, and hence will not detect such changes unless specialized techniques are employed.

Inflammatory processes can increase the amount of circulating alpha 1-acid glycoprotein (which avidly binds a variety of psychotropic drugs), increasing the amount of bound drug while the free fraction remains unchanged. In such circumstances, the circulating total concentrations may seem excessive, but actually simply represent an increase in the bound (but biologically inert) circulating fraction (Abernethy and Kertzner 1984).

2.6.2
Cardiac Function and Hepatic Arterial Blood Flow

After a drug enters the systemic circulation, the delivery back to the liver depends on left ventricular function. Hence, drugs may also affect the clearance of other drugs indirectly, through an effect on hepatic arterial blood flow (Cohen

1995; Parker et al. 1995; DeVane and Pollock 1999). The rate of drug conversion is dependent on the rate of delivery to the liver, which is determined by arterial flow. Cimetidine and beta-blockers, such as propranolol, decrease arterial flow, slowing the clearance of various drugs that undergo extensive oxidative biotransformation.

While decreased left ventricular output can result in a decrease in hepatic arterial flow, right ventricular failure causes hepatic congestion, reducing the first-pass effect and delaying biotransformation.

2.6.3
Hepatic Integrity

Certain conditions can also directly or indirectly affect hepatic integrity or function. Diseases that directly affect hepatic integrity include cirrhosis, viral infections, and collagen vascular diseases. Diseases that indirectly affect function include metabolic disorders (e.g., azotemia secondary to renal insufficiency) and cardiac disease.

2.6.4
Renal Factors

Multiple renal factors can affect the elimination of psychotropic medications, including renal insufficiency, dehydration, plasma pH, and concurrent medications. Renal insufficiency can delay clearance and result in the accumulation of higher concentrations of polar metabolites. Depending on the pharmacological profile of these metabolites, patients may accumulate compounds that are less efficacious and/or more toxic than the parent compound. Likewise, dehydration diminishes glomerular filtration rate. Changing the plasma pH (e.g., giving cranberry juice to acidify it or sodium bicarbonate to make it more basic) can also hasten or retard the clearance of certain drugs (e.g., amphetamines). Concurrent medications may also affect the ability of the renal tubules to excrete a drug. For example, loop diuretics and nonsteroidal anti-inflammatory agents decrease renal clearance of lithium (Mehta and Robinson 1980; Ragheb et al. 1980).

2.6.5
Aging and Disease: Effects on Metabolism and Elimination

The absolute and the relative size of the body's fat compartment can change with normal aging or morbid obesity (Edelman and Leibman 1959; Cohen 1995; Parker et al. 1995; Klotz 1998; DeVane and Pollock 1999). The percentage of total body water and protein content decreases in the elderly, while the percentage of fat content increases, providing a relatively larger reservoir to store psychotropics. These changes partially explain why many psychotropic drugs have a more persistent effect in elderly than in younger patients. The morbidly obese patient also has an increased reservoir in which to store drug, with the effect being proportional to the size of the adipose reservoir.

2.6.6
Gender: Effect on Metabolism and Elimination

A number of differences in drug pharmacokinetics are found in females.

- *Absorption/bioavailability*
- Lower acidic environment increases absorption of weak bases (e.g., TCAs, benzodiazepines, some antipsychotics).
- Exogenous estrogen may also increase absorption through this same mechanism.
- Slower transit time in small intestine delays drug absorption and peak levels; lowers peak blood concentration.

- *Volume of distribution (VD)*
- Drug concentration and distribution are greater in young women than in men.
- Protein binding is lower in women than in men.
- Exogenous hormones and pregnancy can alter protein binding.

- *Metabolism and elimination*
- Estradiol and progesterone (used in oral contraceptives) reduce specific CYP enzyme activity (e.g., levels of antipsychotics such as haloperidol, clozapine, and risperidone may rise).

- *Effects of pregnancy*
- Increased endogenous hormones of the luteal phase cause decreased gastrointestinal motility and promote drug absorption.

2.7
Summary

As noted at the beginning of the chapter, pharmacokinetics is only one of three variables that determine a drug's effect. While a drug cannot exert its action without reaching a critical concentration at its site of action, achieving that concentration will not result in a clinically meaningful response if the site of action of the drug is not relevant to the disease the patient has. Also, there can be interindividual genetic variability at both the drug's site of action and in the mechanism mediating its pharmacokinetics (e.g., CYP enzymes, transporter proteins). Such mutations in receptors can alter the binding affinity of the drug so that a different concentration is needed to occupy the target to a clinically meaningful degree. Thus, for example, a number of different factors might explain why some patients appear to need a different concentration of a drug to achieve the same clinical benefit. These include both genetic and acquired factors:

- Inter-individual variation in CYP enzyme function
- Inter-individual variation in the site of action (e.g., a receptor protein)
- Inter-individual variation in ABC transport pumps that can affect the ratio of drug concentration in plasma to that in the brain
- Inter-individual variation in plasma protein binding
- Assay limitations
- Temporal dissociation between changes in drug concentration and changes in drug effect

Although the clinical relevance of pharmacokinetics has been emphasized in this chapter, it is important to keep these limitations in mind when applying pharmacokinetic principles.

A table summarizing the pharmacokinetic parameters of commonly used antidepressants is provided in the Appendix (Table 4).

3
Tricyclic Antidepressants

There are two major classes of TCAs: the older tertiary amine TCAs (e.g., amitriptyline, imipramine) and the subsequent secondary amine TCAs (nortriptyline, desipramine). The former have multiple mechanisms of action over their clinically relevant concentration range whereas the latter are relatively selective norepinephrine reuptake inhibitors. Tertiary amine TCAs are converted by N-demethylation to secondary amine TCAs in humans (e.g., amitriptyline into nortriptyline and imipramine into desipramine). Several CYP enzymes mediate this conversion including 1A2, 3A (usual major pathway), and 2C19. This fact will be important latter when considering pharmacokinetic drug–drug interactions (DDIs) involving TCAs. Unless otherwise specified, TCAs in this chapter will refer to tertiary amine TCAs, since they are the most popularly used members of the superfamily of TCAs.

The pharmacokinetics of the various TCAs are generally similar. Of critical importance to the clinical pharmacology of tertiary amine TCAs are their multiple mechanisms of action, some of which result in toxicity over a relatively narrow range in plasma concentration. For example, amitriptyline's most potent mechanism of action involves histamine-1 receptor blockade, which causes sedation (Richelson 1988). Direct membrane stabilization occurs at only a tenfold higher plasma concentration (Preskorn and Irwin 1982) and mediates central nervous system and cardiac toxicity including seizures, arrhythmias, and sudden death (Halper and Mann 1988; Preskorn 1989; Preskorn and Fast 1991). The fact that TCAs have so many different concentration-dependent mechanisms of action makes knowledge of their pharmacokinetics critical in avoiding their toxic effects and maximizing their therapeutic efficacy.

3.1
Absorption

TCAs are well absorbed from the gastrointestinal tract. Their absorption is not influenced to a clinically significant extent by disease states, age, or concomitant drug administration. However, only 50%–60% of an oral dose reaches the systemic circulation due to first-pass metabolism in the gut and the liver. Some of the metabolites produced by such metabolism do reach the systemic circulation in part as a result of enterohepatic recirculation.

Tertiary amine TCAs are absorbed relatively rapidly after an oral dose, with their maximum plasma concentration (C_{max}) occurring 0.5–4 h after an oral dose. Secondary amine TCAs are absorbed from the gastrointestinal tract more slowly, achieving C_{max} in 4–8 h after an oral dose. This difference may account, in part, for the generally better tolerability of secondary compared to tertiary amine TCAs.

Concentration-dependent side effects of TCAs such as sedation (histamine-1 blockade) and orthostatic hypotension (alpha-1 blockade) are most likely to occur around the time of T_{max}. Therefore, the timing of a single daily dose can be adjusted so that such effects occur when the patient is supine and asleep. However, elderly patients may be especially prone to falls leading to injuries when getting up to use the bathroom at night (Preskorn 1993b). Thus, a divided dose of a tricyclic would lead to a lower C_{max}, thereby decreasing the potential for an orthostatic drop in blood pressure mediated by alpha-1 receptor blockade. Although the average steady-state concentration (C_{ss}) will not change if the drug is given once daily or in divided doses, the difference between C_{max} and C_{min} will be exaggerated. Concentration-dependent adverse effects of TCAs, such as direct membrane stabilization, may also become clinically apparent at C_{max} resulting in intracardiac conduction delay and subsequent arrhythmias (Preskorn and Irwin 1982).

3.2
Distribution

TCAs have volumes of distribution (V_d) that exceed a normal adult's plasma volume (about 5 liters). This means that these drugs tend to accumulate extensively in tissues outside of the systemic circulation. The consequence of a large V_d is the relative ineffectiveness of dialysis in reducing the body's burden of drug (Pentel et al. 1982).

The TCAs are all highly bound to plasma proteins. This fact raises concern about possible pharmacokinetic interactions mediated by the displacement of other highly protein-bound drugs producing a higher free fraction of the displaced drug. DDIs due to displacement from binding proteins can manifest themselves in one of two ways: (1) drug A displaces the TCA leading to elevated TCA levels and toxicity, (2) the TCA may displace drug A leading to toxicity associated with drug A. In addition, acidosis due to TCA overdose results in decreased protein binding and increased free tricyclic availability (Levitt et al. 1986). Since TCAs are highly bound to plasma proteins, competition for binding sites can exist with other co-prescribed drugs such as aspirin, phenytoin, and phenothiazine that are also highly protein bound (Azzaro and Ward 1994). Conventional TDM of the TCA and co-prescribed drugs does not assess changes in the ratio of bound to free drug.

3.3
Metabolism and Elimination

The TCAs are extensively biotransformed prior to being eliminated from the body. The primary biotransformations involve hydroxylation of the ring structure or demethylation of the terminal nitrogen (Hammer and Sjoqvist 1967; Potter and Manji 1990). Many CYP enzymes are involved in the oxidative metabolism of TCAs including CYP 2D6, 2C19, 1A2, and 3A3-4 (Pollock 1994). The rate-limiting step in elimination of TCAs is biotransformation mediated by CYP 2D6 (Brosen et al. 1992). The fact that approximately 7% of Caucasians are genetically deficient in this enzyme complicates matters (Eichelbaum and Gross 1990). As a result of this deficiency, plasma concentrations can be fourfold higher than in nonaffected individuals receiving the same dose of the same TCA (Nelson and Jatlo 1987; Preskorn et al. 1993). Conventional doses of a TCA such as imipramine (150–250 mg/day) may lead to toxic side effects such as delirium and seizures in such individuals (Preskorn and Jerkovich 1990; Preskorn and Fast 1992).

While CYP 2D6 is the rate-limiting step in the clearance of all TCAs, tertiary amine TCAs are demethylated to secondary amine TCAs by multiple CYP enzymes: 1A2, 2C19, and 3A. In contrast to CYP 2D6, these CYP enzymes are inducible. That fact is relevant as discussed below in terms of pharmacokinetically mediated DDIs. This conversion is of clinical importance because tertiary amine TCAs such as imipramine and amitriptyline are more potent inhibitors of serotonin reuptake than their secondary amine TCA metabolites, desipramine and nortriptyline, respectively (Hyttel 1994). Desipramine and nortriptyline are 5–20 times more potent at norepinephrine reuptake than their parent compounds. Clomipramine, which is the TCA that is most potent at blocking serotonin uptake, is converted to desmethylclomipramine by hepatic enzymes. Desmethylclomipramine is 100 times more potent at blocking norepinephrine reuptake than serotonin (Hyttel 1994). Due to this variability in mechanism of action of metabolites and parent compounds, the clinician must consider the following:

- The physiology of the psychiatric illness being treated (i.e., will the depressive illness respond more favorably to a norepinephrine- or serotonin-mediated mechanism of action?)

- The tolerability of TCAs with troublesome side effects mediated by blockade of cholinergic, histaminic, and noradrenergic receptors (which are greater with tertiary amine than secondary amine TCAs)
- The narrow therapeutic window of the TCAs (i.e., inadvertent or intentional overdoses can translate into seizures or cardiac conduction problems)
- The individual's capacity to metabolize/eliminate TCAs (patients with impaired renal and/or hepatic function may accumulate toxic levels of TCAs, while, conversely, patients who are rapid metabolizers may not develop sufficiently therapeutic TCA levels)

3.4
Half-Life

The half-life of all secondary amine TCAs is approximately 24 h. Protriptyline is the longest lived of all TCAs, with a half-life of approximately 80 h. While the tertiary amine TCAs have half-lives less than 24 h, they are metabolized to secondary amine TCAs. The half-lives of all TCAs are sufficiently long to allow once-daily dosing. Plasma levels of amitriptyline and nortriptyline increase as a function of age consistent with a modest increase in half-life (Preskorn and Mac 1985). Prescribing the total daily dose of a TCA to be taken at bedtime is generally more convenient for patients and should improve compliance. Exceptions to once-daily dosing would be appropriate for patients in whom problems may arise as a result of effects occurring at the height of C_{max} as discussed earlier in the chapter.

3.5
Linear Versus Nonlinear Pharmacokinetics

The majority of patients taking TCAs over their clinically relevant dosing range demonstrate linear pharmacokinetics. For these patients, changing the dose of the TCA results in a proportionate change in plasma drug concentration (Nelson and Jatlo 1987; Preskorn 1989; Bupp and Preskorn 1991; Preskorn and Fast 1991). Nonlinear pharmacokinetics of TCAs occur when patients: (1) are genetically deficient in CYP 2D6; (2) take other drugs that inhibit this enzyme; or (3) have intercurrent disease such as hepatic or renal failure that significantly delays elimination. The consequence of nonlinear pharmacokinetics is that higher doses of TCAs will produce disproportionate increases in concentration-dependent adverse effects unless the TCA dose is adjusted.

3.6
Pharmacokinetically Mediated Drug–Drug Interactions

Drugs coadministered with TCAs may inhibit, induce, or have no impact on the metabolism of the TCA. Carbamazepine, a CYP 3A enzyme inducer, can cause decreases of 50%–100% in steady-state plasma concentrations of tertiary amine

TCAs by increasing the rate of their conversion to their respective secondary amine TCA metabolites (Preskorn et al. 1993). This induction can decrease the TCA plasma concentration to the point that antidepressant efficacy is compromised.

Drugs such as bupropion, fluoxetine, paroxetine, and thioridazine cause substantial inhibition of CYP 2D6 (Table 3; Shen and Lin 1991). Concomitant use of such drugs with TCAs can convert normal metabolizers into phenocopies of individuals who are genetically deficient in CYP 2D6. If such combinations of drugs are used, the TCA dose must be adjusted downward with use of TDM to avoid TCA toxicity.

4
Selective Serotonin Reuptake Inhibitors

The SSRIs include citalopram, escitalopram, fluoxetine, fluvoxamine, paroxetine, and sertraline. As a class, these drugs have become the first choice for many physicians in the treatment of conditions such as depression and obsessive-compulsive disorder. While members of this class are remarkably similar in their pharmacodynamic actions, they are quite different in their pharmacokinetics and their effects on the CYP enzymes that mediate most oxidative drug metabolism.

4.1
Absorption

All of the SSRIs are well absorbed after an oral dose, with 70%–100% reaching the systemic circulation. Coadministration of SSRIs with food may lead to slight and not clinically meaningful changes in their rate of absorption (DeVane 1994). The T_{max} comes later for most SSRIs (4–8 h) compared with TCAs. Acute adverse effects of SSRIs (e.g., restlessness) typically reach their peak or have their onset when the drug reaches its C_{max}. The fact that gastrointestinal side effects often occur within 1 h of an oral dose of an SSRI suggests that this effect may be mediated by direct serotonergic actions within the gut as opposed to an effect on the chemoreceptor trigger zone in the brain (Preskorn 1993a).

4.2
Distribution

Although SSRIs share a large V_d with TCAs, SSRIs are much safer than TCAs even when taken in overdose. Therefore the high V_d of SSRIs should not result in any clinically deleterious effects (Dechant and Clissold 1991; Henry 1991). Fluoxetine, paroxetine, and sertraline are highly bound to plasma proteins (>95%) (Preskorn 1993a), which always raises concern about the potential for DDIs as a result of another more toxic drug being displaced from its binding protein by a co-prescribed drug such as an SSRI. The protein binding of citalopram (50%), escitalopram (50%), and fluvoxamine (77%) is considerably less (Fredericson 1978; Palmer and Benfield 1994).

Despite these differences in protein binding, significant protein binding drug displacement interactions are generally not an issue with SSRIs (Apseloff et al. 1997). The reason is as follows: While these drugs are themselves highly protein bound, their binding is relatively weak, and even more importantly their concentrations are sufficiently low that protein binding is not saturated. For the latter reason, there is still adequately capacity to bind most other drugs unless their concentrations are unusually high as is the case with only a few drugs such as valproate. In addition, even the remote chance of an increase in the free fraction of drug would be mitigated by the parallel increase in its clearance. Unfortunately, the potential significance of protein binding displacement has not been tested in the extreme cases where it might be meaningful; such cases include the debilitated patient with protein wasting who is receiving a drug such as valproate that achieves a concentration (mcg/ml) under usual dosing conditions that saturates protein binding. Given all of these considerations, protein-binding displacement as a mechanism of DDIs with most psychotropic drugs including all of the SSRIs is unlikely to be meaningful. Nevertheless, caution is recommended when co-prescribing SSRIs with other highly protein-bound drugs such as warfarin or valproate in medically compromised, elderly patients.

4.3
Metabolism and Elimination

Table 1 lists some common drugs that are metabolized by different CYP enzymes; Table 2 summarizes the inhibitory effect of the SSRIs and some other newer antidepressants at their usually effective, minimum dose on specific CYP enzymes; and Table 3 presents a summary of formal in vivo studies of the effects of different SSRIs on CYP 2D6 model substrates.

All the SSRIs have metabolites, some of which are clinically relevant while others are not.

Both fluoxetine and citalopram are marketed as racemic mixtures of two enantiomers. Fluoxetine's metabolite, norfluoxetine, is equipotent to the parent compound in terms of serotonin reuptake inhibition and inhibition of several CYP enzymes, particularly CYP 2D6, CYP 2C9/10, and CYP 2C19 (Table 2). Fluoxetine and its metabolites are excreted chiefly in the urine, which raises suspicion that the drug may accumulate more extensively in patients with renal impairment. While single-dose studies of fluoxetine (Aronoff et al. 1984; Lemberger et al. 1985) do not reveal any effect of renal impairment on this drug's pharmacokinetics, such studies are inadequate due to the drug's extended half-life. One multiple-dose pharmacokinetic study of fluoxetine in this population (Bergstrom et al. 1993) concluded that the pharmacokinetics of fluoxetine are unaltered in renal failure. However, graphic data presented in this paper suggest fluoxetine levels may double in individuals with severe renal failure.

Table 1 Drugs metabolized by CYP enzymes (copyright Preskorn 2003)[a]

Metabolizing enzyme	Drug type	Drug
CYP 1A2	Antidepressants	Amitriptyline, clomipramine, imipramine
	Antipsychotics	Clozapine[b], olanzapine[b], thioridazine[b]
	β-Blockers	Propranolol
	Opiates	Methadone[b]
	Miscellaneous	Caffeine[b], paracetamol, tacrine[b], theophylline[b], R-warfarin[b]
CYP 2C9/10		Phenytoin[b], S-warfarin[b], tolbutamide[b]
CYP 2C19	Antidepressants	Citalopram[b], clomipramine, imipramine
	Antipsychotics	Thioridazine
	Barbiturates	Hexobarbital, mephobarbital, S-mephenytoin[b]
	β-Blockers	Propranolol
	Benzodiazepines	Diazepam
CYP 2D6	Antiarrhythmics	Encainide[b], flecainide[b], mexiletine, propafenone
	Antipsychotics	Haloperidol (minor), molindone, perphenazine[b], risperidone[b], thioridazine (minor)
	β-Blockers	Alprenolol, bufuralol, metoprolol[b], propranolol, timolol
	Opiates	Codeine[b], dextromethorphan[b], ethylmorphine
	SSRIs	Fluoxetine, N-desmethylcitalopram[b], paroxetine[b]
	TCAs	Amitriptyline, clomipramine[b], desipramine[b], imipramine[b], N-desmethylclomipramine[b]
	Other antidepressants	Venlafaxine[b], mCPP metabolite of nefazodone and trazodone[b]
	Miscellaneous	Debrisoquin[b], donepezil, 4-hydroxyamphetamine, perhexiline[b], phenformin[b], parteine[b]
CYP 3A3/4	Analgesics	Acetaminophen, alfentanil
	Antiarrhythmics	Amiodarone, disopyramide, lidocaine, propafenone, quinidine
	Anticonvulsants	Carbamazepine[b], ethosuximide
	Antidepressants	Amitriptyline, clomipramine, imipramine, nefazodone[b], sertraline[b], O-desmethylvenlafaxine[b]
	Antiestrogens	Docetaxel, paclitaxel, tamoxifen[b]
	Antihistamines	Astemizole[b], loratadine[b], terfenadine[b]
	Antipsychotics	Quetiapine[b], clozapine
	Benzodiazepines	Alprazolam[b], clonazepam, diazepam, midazolam[b], triazolam[b]
	Calcium channel blockers	Diltiazem[b], felodipine[b], nicardipine, nifedipine[b], niludipine, manidipine, isoldipine, nitrendipine, verapamil[b]
	Immunosuppressants	Cyclosporine[b], tacrolimus (FK506-macrolide)
	Local anesthetics	Cocaine, lidocaine
	Macrolide antibiotics	Clarithromycin, erythromycin, triacetyloleandomycin
	Steroids	Androstenedione, cortisol[b], dehydroepiandrosterone 3-sulfate, dexamethasone, estrogen[b], testosterone[b], estradiol[b], ethinylestradiol, progesterone
	Miscellaneous	Benzphetamine, cisapride[b], dapsone[b], donnepezil, lovastatin, omeprazole (sulfonation)

[a] For more detailed information readers are referred to the Cytochrome P450 Drug Interaction Table maintained by David A. Flockart, MD, PhD (http://medicine.iupui.edu/flockhart).

[b] Drug has a principal CYP enzyme pathway.

Such lists are not comprehensive, since the CYP enzyme(s) responsible for biotransformation is known for only approximately 20% of marketed drugs. Many drugs were developed before the necessary knowledge and technology existed. Some drugs are listed under more than one CYP enzyme. That does not necessarily mean that each of these enzymes contributes equally to the elimination of the drug. One enzyme may be principally responsible based on substrate affinity and capacity and abundance of the enzyme.

Table 2 The inhibitory effect of newer antidepressants at their usually effective minimum dose on specific CYP enzymes (adapted from Harvey and Preskorn 1996a,b; Preskorn 1996; Shad and Preskorn 2000; copyright Preskorn 2003)

	No or minimal effect (<20%)[a]	Mild (20%–50%)[a]	Moderate (50%–150%)[a]	Substantial (>150%)[a]
Bupropion[b]	?	?	?	2D6
Citalopram	1A2, 2C9/10, 2C19, 3A3/4	2D6	–	–
Fluoxetine	1A2	3A3/4	2C19	2D6, 2C9/10
Fluvoxamine	2D6	–	3A3/4	1A2, 2C19
Nefazodone	1A2, 2C9/10, 2C19, 2D6	–	–	3A3/4
Paroxetine	1A2, 2C9/10, 2C19, 3A3/4	–	–	2D6
Sertraline	1A2, 2C9/10, 2C19, 3A3/4	2D6	–	–
Venlafaxine	1A2, 2C9/10, 2C19, 3A3/4	2D6	–	–

Mirtazapine, based on in vitro modeling, is unlikely to produce clinically detectable inhibition of these five CYP enzymes. However, no in vivo studies have been done to confirm that prediction.

[a] Percentage increase in plasma levels of a coadministered drug dependent on this CYP enzyme for its clearance.

[b] The potential in vivo effects of bupropion on CYP enzymes other than 2D6 have not been studied and thus are unknown.

The racemic mixture of citalopram produces racemic desmethylcitalopram and didesmethylcitalopram. The active S-enantiomer of citalopram is generally only one-third of the total citalopram plasma level under steady-state conditions (Rochat et al. 1995). However, there is variability in this ratio among different patients that may be characteristic of patients who are genetically deficient in CYP 2C19, which is one of the enzymes principally responsible for the metabolism of citalopram (Rochat et al. 1995). Racemic citalopram is metabolized by both CYP 2C19 and CYP 2D6, based on a study comparing the steady-state levels of citalopram, desmethylcitalopram, and didesmethylcitalopram in individuals who were extensive metabolizers of sparteine (i.e., CYP 2D6) and mephenytoin (i.e., CYP 2C19), and individuals who were poor metabolizers of either sparteine or mephenytoin, respectively (Sindrup et al. 1993). Both the total clearance of citalopram and the clearance of desmethylcitalopram were significantly slower in poor versus extensive metabolizers of mephenytoin (i.e., deficient in CYP 2C19), while the demethylation clearance of desmethylcitalopram was significantly lower in poor versus extensive metabolizers of sparteine (i.e., deficient in CYP 2D6). Both of these CYP enzymes exhibit considerable genetic polymorphism in specific populations: approximately 5%–10% of white populations in Western Europe and North America are genetically deficient in CYP 2D6 (Evans et al. 1980; Steiner et al. 1988), whereas 20% of Orientals are genetically deficient in CYP 2C19 (Kupfer and Preisig 1984; Goldstein and de Morais 1994). Citalopram's dependence on these two enzymes would be expected to increase the interindividual variability in plasma levels of this drug.

Escitalopram, the S enantiomer of the citalopram, has now been approved in several European countries and marketed in the United States (Package insert

Table 3 Summary of formal in vivo studies of the effects of different SSRIs on CYP 2D6 model substrates (adapted from Preskorn 1998a; copyright Preskorn 2003)

SSRI	Author	n	SSRI treatment: dose (mg/day) × duration (days)	Substrate	Substrate dosing	(AUC2−AUC1)/AUC1 (% increase)	EMs to PMs
Citalopram	Gram et al. 1993	8	40 mg×10 days	Desipramine	Single dose	47%	
	Citalopram package insert (PI)[a]	NA[b]	40×22	Metoprolol	NA	100%	
Escitalopram	Escitalopram PI	NA	20×21	Metoprolol	NA	82%	
	Escitalopram PI	NA	20×21	Desipramine	NA	100%	
Fluoxetine	Alfaro et al. 1999	8	60×8[a]	Dextromethorphan	Single dose		63%
	Amchin et al. 2001	12	20×28[a]	Dextromethorphan	Single dose		25%
	Bergstrom et al. 1992	6	60×8[a]	Desipramine	Single dose	640%	
		6	60×8[a]	Imipramine	Single dose	Imipramine 235% Desipramine 430%	
	Preskorn et al. 1994	9	20×21	Desipramine	21 days	380%	
	Otton et al. 1993	19	37±17×21	Dextromethorphan	Single dose		95%
	Maes et al. 1997	11	20×28	mCPP	7 days	820% (270%)[b]	
	Alfaro et al. 2000	12	60×8	Dextromethorphan	Single dose		42%
Fluvoxamine	Alfaro et al. 1999	6	100×8	Dextromethorphan	Single dose		0%
	Spina et al. 1993	6	100×10	Desipramine	Single dose	14%	
Paroxetine	Alderman et al. 1997	17	20×9	Desipramine	9 days	421%	
	Brosen et al. 1993a	9	20×8	Desipramine	Single dose	364%	78%
	Albers et al. 1996	10	30×4	Imipramine	Single dose	Imipramine 74% Desipramine 327%	
	Alfaro et al. 1999	8	20×8	Dextromethorphan	Single dose		50%
	Ozdemir et al. 1997	8	20×10	Perphenazine	Single dose	595%	
	Alfaro et al. 2000	12	20×8	Dextromethorphan	Single dose		83%
	Hemeryck et al. 2000	8	20×6	Metoprolol (racemic)	Single dose	S-meteprolol 700% R-metoprolol 410%	

Table 3 (continued)

SSRI	Author	n	SSRI treatment: dose (mg/day) × duration (days)	Substrate	Substrate dosing	(AUC2−AUC1)/AUC1 (% increase)	EMs to PMs
Sertraline	Alderman et al. 1997	17	50×9	Desipramine	8 days	37%	
	Jann et al. 1995	4	50×7	Desipramine	7 days	0%	
	Preskorn et al. 1994	9	50×21	Desipramine	21 days	23%	
	Solai et al. 1997	13	50×>5	Nortriptyline	Chronic dosing	14%	
	Ozdemir et al. 1998	21	94±26×24±17	Dextromethorphan	Single dose	0%	0%
	Sproule et al. 1997	6	108±49×21	Dextromethorphan	Single dose	6%	0%
	Alfaro et al. 1999	7	100×8	Imipramine	Single dose	68%	
	Kurtz et al. 1997	6	150×8	Desipramine	Single dose	54%	
	Zussman et al. 1995	13	150×29	Desipramine	Single dose	70%	
	Alfaro et al. 2000	12	100×8	Dextromethorphan	Single dose		0%

NA, not available; *m*CPP, metachlorophenylpiperazine (a metabolite of nefazodone and trazodone); PI, package insert from Physician's Desk Reference (2003).
[a] 60 mg/day for 8 days is a loading dose strategy used to approximate the plasma levels of fluoxetine and norfluoxetine achieved under steady-state conditions on a dose of 20 mg/day.
[b] 820% is based on all the data. If the two highest increases are excluded, the average was 270%.

for escitalopram, Physicians' Desk Reference 2003, p 3532). In many European countries, approval is a two-step process involving first approval to market and then approval as of price. The marketing of escitalopram has been delayed in some European countries due to discussions about how much additional value the single enantiomer represents relative to the racemate, citalopram. Escitalopram at 10–20 mg/day is comparable to 20–40 mg/day of citalopram in terms of serotonin uptake inhibition, weak CYP 2D6 inhibition, antidepressant efficacy, and adverse effect profile. Its pharmacokinetics are comparable to citalopram in terms of half-life, linear pharmacokinetics, and modest effects on the metabolism of other drugs via the inhibition of CYP enzymes in contrast to fluoxetine, fluvoxamine, and paroxetine.

Fluvoxamine, paroxetine, and sertraline produce multiple metabolites without apparent clinically relevant activity in terms of serotonin uptake inhibition. Fluvoxamine and paroxetine are eliminated primarily by renal excretion, whereas sertraline elimination is divided equally between fecal and renal excretion (Kaye et al. 1989). Studies of fluvoxamine and sertraline have shown no effect of renal impairment on the pharmacokinetics of these SSRIs (Warrington 1991; Goodnick 1994), whereas renal impairment significantly delays elimination of paroxetine (Doyle et al. 1989). As a result, clinicians may wish to halve the usual daily dose of fluoxetine and paroxetine in patients with renal impairment. The nonlinear pharmacokinetics of fluoxetine and paroxetine suggest that significant elevations in plasma concentration may occur in renally impaired patients, especially with multiple doses. As a result, concentration-dependent adverse effects may emerge more quickly and be more severe in patients with renal impairment who are prescribed these drugs, unless the dose is adjusted.

4.4
Half-Life

The half-lives of the SSRIs vary substantially and in clinically important ways. The half-lives of fluvoxamine and paroxetine at low doses (or concentrations) are approximately 10–12 h. That sets the stage for both the need for divided daily dosing and for a higher incidence and greater severity of withdrawal symptoms. Paroxetine is usually dosed once a day, but that is because it has nonlinear pharmacokinetics: at low concentrations, paroxetine is preferentially cleared by CYP 2D6 which is a highly efficient but low-capacity enzyme such that the half-life of paroxetine is approximately 10 h at low concentrations, which are achieved early in the course of treatment with 20 mg/day or during washout. As paroxetine continues to accumulate, it saturates CYP 2D6 and its levels and its half-life increases as its clearance becomes dependent on another CYP enzyme, most likely CYP 3A, for its biotransformation prior to its elimination (Harvey and Preskorn 1996a,b). This phenomenon likely contributes to the increased risk of SSRI discontinuation symptoms when paroxetine is stopped, since its clearance accelerates as its level falls below the point at which it saturates CYP 2D6.

Fluoxetine is at the opposite extreme from fluvoxamine and paroxetine. In young healthy individuals, the half-life of fluoxetine is 2–4 days and the half-life of its active metabolite, norfluoxetine, is 7–15 days. In healthy elderly individuals, the half-life of fluoxetine and norfluoxetine is even longer — 3 weeks on average for its active metabolite, norfluoxetine (Harvey and Preskorn 2001). The extended half-life of norfluoxetine means that plasma drug concentrations will continue to rise for 1–2.5 months in young individuals at a fixed dose and for up to 4 months on average in healthy individuals over the age of 65. These long half-lives set the stage for the late-emergence of concentration-dependent adverse effects. Once fluoxetine is started, there is also a gradually increasing inhibition of multiple CYP enzymes; and a prolonged period is needed to recover the drug metabolizing capacity of CYP enzymes once fluoxetine is discontinued. This means that the patient is at risk for pharmacodynamic and pharmacokinetic DDIs mediated by the fluoxetine-induced inhibition of multiple CYP enzymes for a prolonged period of time. This phenomenon is the reason the FDA warns that MAOIs should not be started for several weeks after fluoxetine is discontinued to avoid the potentially lethal serotonin syndrome (Sternbach 1988).

One advantage of fluoxetine's extended half-life is its use as a depot antidepressant for patients with compliance problems. Once a steady-state concentration of fluoxetine is achieved, missing any given dose will have little, if any, clinical impact. The extended half-life of fluoxetine and norfluoxetine also decreases the risk of a withdrawal syndrome following fluoxetine discontinuation. These advantages do not override the disadvantages of the long half-life in most clinical situations. For this reason, fluoxetine is recommended primarily as a niche SSRI in situations where these advantages are important and outweigh the disadvantages for a specific patient.

Citalopram, escitalopram, and sertraline have half-lives of about 1–2 days. This half-life permits once-a-day dosing. Steady state with these SSRIs is achieved in approximately 5–10 days, so that the maximum effect occurs and clears at a manageable rate. For these reasons, the risk of SSRI discontinuation symptoms with these SSRIs is intermediate between the risk with fluvoxamine and paroxetine which have the shortest half-lives (i.e., increased risk) and the risk with fluoxetine which has the longest half-life (i.e., reduced risk).

4.5
Linear Versus Nonlinear Pharmacokinetics

Citalopram, escitalopram, and sertraline have linear pharmacokinetics over their clinically relevant dosing range, whereas fluoxetine, fluvoxamine, and paroxetine do not as a result of CYP enzyme inhibition. As a consequence, the disproportionate increases in plasma concentrations of these drugs that occur with dose increases may translate into disproportionate increases in concentration-dependent adverse effects. This difference is particularly important in elderly patients whose clearance of fluoxetine and paroxetine is already decreased

(Preskorn 1993b). The effect on the inhibition of CYP enzymes will also be magnified.

4.6
Pharmacokinetically Mediated Drug–Drug Interactions

There are many possible mechanisms of pharmacokinetically mediated DDIs. Of paramount importance are the differential effects SSRIs have on the inhibition of CYP enzymes. Table 2 lists inhibitory effects of SSRIs on CYP enzymes at their usually effective doses. Many drugs are dependent on CYP enzymes for their biotransformation into polar metabolites, which can be excreted by the kidneys. Such oxidative drug metabolism is most often dependent on the functional integrity of one or more CYP enzymes. A CYP enzyme can lose its functional capacity due to (1) genetic mutation, (2) concomitant ingestion of a drug that is an inhibitor, or (3) intercurrent disease affecting the liver. If this happens, drugs dependent on that particular CYP enzyme for their biotransformation will accumulate to a greater than usual extent, which can lead to the development of concentration-dependent adverse effects. This is particularly worrisome in the case of a drug with a narrow therapeutic index where toxicity may result from drug accumulation.

The capacity of the SSRIs to inhibit the functional integrity of specific CYP enzymes has been extensively studied (Tables 2 and 3; Jeppensen et al. 1996). Different SSRIs have different capacities to inhibit specific CYP enzymes. That action translates into clinically meaningful changes in the clearance rates of concomitantly prescribed drugs dependent on these CYP enzymes for their biotransformation into more polar metabolites. For example, fluoxetine and its metabolite, norfluoxetine, and paroxetine are potent inhibitors of CYP 2D6 (Crewe et al. 1992; Skjelbo and Brosen 1992; Otton et al. 1993; Otton et al. 1994; von Moltke et al. 1994; Alderman et al. 1997), whereas citalopram, escitalopram, fluvoxamine, and sertraline are not (Bergstrom et al. 1992; Preskorn 1997). Fluvoxamine and fluoxetine's metabolite, norfluoxetine, are modest inhibitors of CYP 3A4 (von Moltke et al. 1995) whereas the other SSRIs have little or no effect on CYP 3A4. Fluvoxamine is a much more potent inhibitor of CYP 1A2 in vitro and in vivo than other SSRIs (Thomson et al. 1992; Brosen et al. 1993b; Jeppenson et al. 1996).

As mentioned in the previous section, the clearance of TCAs is principally dependent on the oxidative drug-metabolizing enzyme CYP 2D6. As such, TCAs have been used as model substrates to test in vivo differences in the capacity of SSRIs to inhibit CYP 2D6. For example, sertraline at 50–100 mg/day has been shown to produce a mild inhibition of CYP 2D6 in vivo (Table 3) (Preskorn and Fast 1992; Preskorn et al. 1994; Jann et al. 1995; Sproule et al. 1997). Additional studies of sertraline at 150 mg/day (three times its usually effective antidepressant dose) have been done using either desipramine or imipramine as a model substrate. At these doses, sertraline produced a 54%–66% increase in AUC of these TCAs (Zussman et al. 1995; Kurtz et al. 1997). This is six to seven times less than the effect of fluoxetine or paroxetine at their usually effective antidepressant doses (20 mg/day). Paroxetine at 20 mg/day causes a doubling of the C_{max} and a fivefold decrease in clearance of desipramine (Brosen et al. 1993a). Fluoxetine at 20 mg/day leads to a fourfold increase in the C_{max} of desipramine when co-administered, an effect that is likely to result in TCA toxicity unless an appropriate dose adjustment is made (Preskorn et al. 1994). Fluvoxamine produces a modest inhibition of desipramine clearance that is of the same order of magnitude as sertraline at its usually effective daily dose (Spina et al. 1993). Citalopram at 40 mg/day and escitalopram at 20 mg/day are moderate inhibitors of CYP 2D6 comparable to sertraline at 150 mg/day but substantially less than the substantial inhibition seen with 20 mg/day of either fluoxetine or paroxetine. (Gram et al. 1993; Package inserts for citalopram and escitalopram, Physicians' Desk Reference 2003, pp 1344 and 3532; Jeppensen et al. 1996). Neither citalopram nor escitalopram have any appreciable effect on the other CYP enzymes.

Knowledge of the pharmacokinetics of SSRIs and of drugs prescribed with them will enable clinicians to predict potential drug–drug interactions based on the following paradigm:

– Drug A → affects → CYP enzyme X
– CYP enzyme X → metabolizes → Drugs B, C, D, and E
– then, Drug A → may affect → Drugs B, C, D, and E

If one knows which CYP enzyme an SSRI inhibits as well as the potency an SSRI has for such an effect, the clinical outcome can be predicted when other drugs known to be metabolized by specific CYP enzymes are prescribed with SSRIs. This is especially important when drugs with narrow therapeutic indexes are combined with SSRIs.

5
Bupropion

Bupropion is an antidepressant with a unique mechanism of action. Dopamine and norepinephrine uptake inhibition are its most potent pharmacodynamic actions, while its uptake inhibition of serotonin is relatively weak (Preskorn and Othmer 1984; Golden et al. 1988). This drug has a relatively narrow therapeutic index so that too low a dose is not efficacious and too high a dose may result in toxicity (Lineberry et al. 1990; Preskorn 1991). Knowledge of the pharmacokinetics of bupropion can help the clinician navigate this narrow therapeutic index. Due to its pharmacokinetics and its narrow therapeutic index, there are two formulations of bupropion: the original immediate release (IR) and a subsequent sustained release (SR).

There are a number of studies demonstrating a minimum threshold concentration for efficacy for the parent drug, bupropion, of approximately 25 ng/m: (Preskorn 1982, 1991; Goodnick 1992) and one study demonstrating poorer outcomes with higher levels of the three major metabolites, hydroxybupropion, erythrohydrobupropion, and threohydrobupropion (Golden et al. 1988).

5.1
Absorption

The bioavailability of bupropion is not known with certainty, but it is estimated to be approximately 80% based on urinary recovery following oral administration of radio-labeled bupropion (Smith et al. 1980). Bupropion and its metabolites have a T_{max} of 1–4 h, which correlates with the onset of side effects such as tremulousness (Goodnick 1994). Dose-dependent seizures are most likely to occur during this time interval (Davidson 1989). The dose-dependent nature of the risk of bupropion-induced seizures is the reason why the total daily IR and SR doses should not exceed 450 mg and 400 mg/day, respectively. No single IR and SR dose should exceed 150 mg and 200 mg/day, respectively. Doses of either formulation should not be given any closer than 4 h apart.

5.2
Distribution

The volume of distribution of bupropion is 19–21 l/Kg, which suggests significant accumulation in tissue. Bupropion is 85% bound to plasma proteins. As with the SSRIs and the TCAs, concern about DDIs due to plasma protein binding displacement is warranted.

5.3
Metabolism and Elimination

Bupropion is extensively metabolized in the liver to three metabolites: hydroxybupropion, threohydrobupropion, and erythrohydrobupropion. The hydroxy and threo metabolites are thought to produce clinical activity, whereas the erythro metabolite is thought to be clinically inactive (Goodnick 1994). Metabolite-to-bupropion ratios are 44:1 for hydroxybupropion and 18:1 for threohydrobupropion (Goodnick 1994). Accumulation of the hydroxy and threo metabolites occurs as a result of the longer half-lives of these metabolites. Knowledge of accumulation of bupropion and its metabolites is of value, since toxicity such as seizures may be mediated by the parent compound and/or its metabolites.

Unfortunately, little is known about the CYP enzymes involved in bupropion's metabolism, although the existence of the metabolites of bupropion suggests that CYP enzymes are involved. Case report data from DDIs involving bupropion and other drugs known to inhibit or induce CYP-mediated drug metabolism are also suggestive. This issue will be discussed in more detail in the section on DDIs below.

5.4
Half-Life

The parent drug bupropion has a relatively short half-life of 10–14 h. Because of its short half-life, as well as concern about the magnitude of C_{max} being proportional to the risk of bupropion-induced seizures, the dosing recommendation for the IR formulation is to give it in three divided doses of no more than 150 mg/dose, with at least 4 h between doses. To further minimize the risk of seizures, the dosage should not exceed 450 mg/day and cannot be collapsed into a single daily dose.

Due to its relatively short half-life and its dose-dependent (and hence pharmacokinetic-dependent) seizure risk, a SR formulation has been developed. The C_{max} for the SR preparation is reduced and the T_{max} modestly extended relative to the IR formulation. However, this preparation did not achieve the goal of once-a-day administration. Instead, the labeled dosing recommendation is for twice-a-day administration in equal doses. The rationale for twice-daily dosing comes from a study that compared bupropion IR 100 mg three times daily with bupropion SR 150 mg twice daily. At these dosages, the C_{max} and AUC for bupropion and all of its major metabolites were nearly equivalent in the two groups (GlaxoSmithKline data on file). Due to dosage strengths, the maximum dose with this formulation is 400 mg/day compared to 450 mg/day with the IR formulation. No single dose of the SR formulation should exceed 200 mg/day. Although there are pharmacokinetic reasons to suspect that the seizure risk is less with the SR versus the IR formulation, that has not been adequately studied to make a definitive claim.

5.5
Linear Versus Nonlinear Pharmacokinetics

Plasma levels of bupropion follow linear pharmacokinetics at doses between 100 and 250 mg (Preskorn 1993a). In animals, bupropion induces its own metabolism (Schroeder 1983). However, healthy volunteers who received bupropion at doses up to 450 mg/day for 14 consecutive days showed no evidence of induced metabolism (GlaxoSmithKline data on file). There are no data about the nature of the pharmacokinetics of bupropion's metabolites: linear, nonlinear, or inducible.

5.6
Pharmacokinetically Mediated Drug–Drug Interactions

There has been little systematic study of DDIs with bupropion either as the perpetuator or the victim. However, there are reasons to suspect that pharmacokinetically mediated DDIs may occur with bupropion. First, bupropion is extensively metabolized in the liver, which means that any one of several steps in its biotransformation could be inhibited or induced by other drugs. For example, a drug such as carbamazepine appears to induce the metabolism of bupropion, causing decreased levels of the parent compound and increased levels of the hydroxybupropion metabolite (Ketter et al. 1992). This conversion is mediated by a relatively obscure CYP enzyme, 2B6 (Hesse et al. 2000; Wang and Halpert 2002). While this enzyme may be inducible by carbamazepine, the ability of commonly used drugs to inhibit this CYP enzyme in vivo is largely unknown. There are in vitro data indicating that ritonavir, efavirenz, and nelfinavir can inhibit CYP 2B6 and thus have the potential for elevating levels of bupropion when co-administered with it (Hesse et al. 2001). The induction of the conversion of bupropion to its hydroxylated metabolites could lead to treatment failure if the efficacy is principally due to the parent compound and could increase the risk of adverse effects if they are principally mediated by the drug's metabolites as suggested by the report by Golden et al. (1988).

Other reasons to suspect pharmacokinetically mediated DDIs with bupropion are found in case report data, which suggest that coadministration of fluoxetine, a known inhibitor of CYP 2D6, can cause elevated levels of hydroxybupropion, resulting in toxicity. In one case, a 39-year-old woman developed a catatonic state within 9 days of discontinuing fluoxetine and initiating bupropion at 225 mg/day. She recovered when bupropion was discontinued. She was re-challenged with the same dose of bupropion after allowing fluoxetine to fully wash out. She tolerated this re-challenge well with a corresponding significant decrease in her hydroxybupropion levels (Preskorn 1991). In addition, Rosenblatt and Rosenblatt (1992) have analyzed post-marketing reports to the Food and Drug Administration regarding cases of grand mal seizures. In such cases, patients taking bupropion and fluoxetine in combination developed seizures at half the bupropion dose as patients taking bupropion alone. Until more is known about bupropion's metabolism, concomitant administration of drugs known to inhibit or induce hepatic enzymes should be done cautiously and TDM should be used to maximize efficacy and minimize toxicity.

There are also minimal data about the effects of bupropion on the pharmacokinetics of other drugs. Bupropion at a dose of 300 mg/day does cause substantial inhibition of CYP 2D6 as witnessed by a 500% increase in the levels of desipramine used as a model CYP 2D6 substrate (bupropion package insert, Physicians' Desk Reference 2003, p 1678). There are also case reports supporting the fact that bupropion can cause clinically significant interactions when co-administered with CYP 2D6 substrates that have a narrow therapeutic index such as TCAs (Shad and Preskorn 1997; Weintraub 2001).

6
Serotonin–Norepinephrine Specific Reuptake Inhibitors

While venlafaxine and duloxetine are structurally and pharmacologically distinct from the tertiary amine TCAs, they share with those older antidepressants the ability to inhibit the uptake pumps for both serotonin and norepinephrine. However, both venlafaxine and duloxetine avoid multiple neuroreceptors and fast sodium channels, which complicate the use of TCAs. For this reason, both venlafaxine and duloxetine are termed selective serotonin and norepinephrine reuptake inhibitors. As with other drugs, their pharmacodynamic effects are pharmacokinetically determined. Less has been published on the pharmacokinetics of duloxetine (Sharma et al. 2000) so it will be covered in more detail in the chapter "Other Antidepressants," and this section will focus on the pharmacokinetics of venlafaxine.

6.1
Absorption

Following an oral dose of venlafaxine, absorption is rapid and extensive. However, first-pass metabolism by the liver results in a relatively high proportion of the drug being converted and reaching the central compartment as the principal active metabolite, O-desmethylvenlafaxine (Schweizer et al. 1991). Food does not significantly affect venlafaxine's absorption. The T_{max} is about 2 h for venlafaxine and 2–3 h for O-desmethylvenlafaxine (Klamerus et al. 1992). Concentration-dependent adverse effects are most likely to occur at or near the drug's C_{max} and include such things as nausea, hypertension, tremor, and abnormal ejaculation.

Although originally marketed in an immediate release formulation with the recommendation for administration twice or three times a day, venlafaxine, like bupropion, is also marketed in an extended release preparation (Effexor XR). The XR formulation is designed to allow for slow diffusion of venlafaxine from the capsule into the gut and permits once-a-day administration. In a study comparing 150 mg of venlafaxine XR with 75 mg twice daily of venlafaxine immediate release (IR), initial doses of venlafaxine XR resulted in a generally lower C_{max} and longer T_{max}. With repeated dosing, the overall exposure of venlafaxine and O-desmethylvenlafaxine was similar in the two groups (Wyeth Pharmaceuticals data on file). In other words, venlafaxine XR results in a slower rate of absorption but an equivalent extent of absorption compared to venlafaxine IR. Accordingly, venlafaxine XR can be administered once daily with a mg-for-mg equivalence to venlafaxine IR.

6.2
Distribution

Venlafaxine has neither extensive tissue binding nor plasma protein binding. This means that, given venlafaxine's short half-life, the drug is cleared quickly after being discontinued. It also means that the risk of DDIs due to plasma protein binding displacement is negligible. However, the downside is that these same characteristics account for a significant risk of SRI discontinuation symptoms when venlafaxine is abruptly discontinued. This risk is not less and may arguably be greater with the XR compared to the IR formulation, since the XR formulation provides more stable plasma concentrations and hence more chance for tolerance to develop, but does not change the rate of elimination. For this reason, the FDA package insert advises that venlafaxine should be tapered rather than abruptly discontinued (Venlafaxine package insert, Physicians' Desk Reference 2003, pp 3392 and 3397).

6.3
Metabolism and Elimination

Venlafaxine is extensively metabolized by CYP 2D6 into two clinically insignificant metabolites (*N*-desmethyl and *N,O*-didesmethyl metabolites) and into *O*-desmethylvenlafaxine (ODV), the major metabolite of venlafaxine. ODV retains significant pharmacodynamic properties of the parent compound (Klamerus et al. 1992). ODV is further metabolized by CYP 3A into more polar metabolites. The clearance of these metabolites occurs primarily via renal excretion. As a result, patients with renal disease have significantly decreased clearance of venlafaxine and the dose of venlafaxine should accordingly be reduced by about half in such patients, especially those with creatinine clearance values below 30 ml/min (Troy et al. 1994).

6.4
Half-Life

The half-lives of venlafaxine and *O*-desmethylvenlafaxine are 5 and 11 h, respectively. For this reason, the immediate release formulation of venlafaxine is recommended to be given at least twice daily in equally divided doses. A three times per day dosing schedule may be even more desirable, especially for the compliant patient who may experience significant concentration-dependent adverse effects at higher doses. The prolonged absorption of the extended-release (XR) formulation permits once-a-day administration but it is important to note that, as discussed above, it does not alter the rate of elimination.

6.5
Linear Versus Nonlinear Pharmacokinetics

Venlafaxine follows linear pharmacokinetics up to doses of about 75 mg/day. Doses greater than 75 mg/day lead to nonlinear pharmacokinetics (Ellingrod and Perry 1994). The evidence for this effect is delayed clearance of venlafaxine and *O*-desmethylvenlafaxine at the higher dose range (Klamerus et al. 1992), which suggests that enzymes responsible for its metabolism become saturated. Increasing the dose above 75 mg/day may lead to disproportionate increases in plasma drug levels; such increases may result in a desirable outcome, such as antidepressant response, or an undesirable outcome, such as hypertension.

6.6
Pharmacokinetically Mediated Drug–Drug Interactions

Venlafaxine is 130 times less potent than fluoxetine as an inhibitor of cytochrome CYP 2D6 in vitro (Otton et al. 1994). In vivo studies support the in vitro findings that venlafaxine is a weak inhibitor of CYP 2D6 (Amchin et al. 2001). As such, venlafaxine is unlikely to create clinically meaningful inhibition of CYP 2D6 when combined with drugs dependent on CYP 2D6 metabolism. A number of other studies have examined the inhibitory effects that venlafaxine may have on other cytochromes, including but not limited to CYP 3A4, 1A2, 2E1, 2C9, and 2C19. Based on the results of these studies, venlafaxine produces weak to no inhibition of these human drug metabolizing enzymes and therefore is unlikely to contribute to clinically significant inhibition of co-prescribed drugs that are substrates for them (Ball et al. 1997; Amchin et al. 1998a,b, 1999a,b; Preskorn and Catterson 1998; Shad and Preskorn 2000).

7
Phenylpiperazine Agents

This class of agents is represented by trazodone and nefazodone. The antidepressant activity of these agents appears to be mediated through postsynaptic blockade of the 5-HT$_2$ receptor. Nefazodone has additional albeit weak activity as a serotonin uptake blocker (Fontaine 1993). The pharmacokinetics of this class of agents is complex. This section will describe the pharmacokinetic similarities and differences between trazodone and nefazodone so that their use can be optimized.

7.1
Absorption

Both nefazodone and trazodone are well absorbed after an oral dose and reach C_{max} in 1–3 h. In the case of nefazodone, only about 20% of a low, oral dose is bioavailable due to extensive first-pass metabolism in the bowel wall and liver mediated by CYP 3A. Since nefazodone saturates or inhibits CYP 3A, its bioavailability increases with increasing dose, and thus it demonstrates nonlinear pharmacokinetics.

The potent histamine-1 blockade by trazodone and its rapid absorption have led to frequent use of this drug as a soporific agent. Sedation, which is a concentration-dependent adverse effect, and its short half-life, limited the usefulness of trazodone as an antidepressant agent. In contrast, nefazodone is a much less potent blocker of histamine-1 receptors (Cusack et al. 1994). Nevertheless, its sedative effects coupled with the need to titrate the dose has seriously limited its use. Its use was further limited by the addition of a warning to the package insert about liver failure being a possible consequence of nefazodone use (Nefazodone package insert, Physicians' Desk Reference 2003, p 1106).

7.2
Distribution

The V_d of nefazodone and trazodone is relatively low compared to that of the SSRIs, although significant tissue penetration occurs with each drug. Plasma protein binding is extensive for both drugs, which raises the concern about possible DDIs resulting from more toxic drugs being displaced from their binding proteins. However, this mechanism as discussed above is rarely clinically relevant for numerous reasons. Consistent with that discussion, in vitro studies show no effect of nefazodone on the plasma protein binding of agents such as phenytoin, theophylline, and warfarin (Dockens et al. 1995; Salazaar et al. 1995; Marino et al. 1997; Duncan et al. 1998). Nefazodone has been shown to displace 5% of plasma protein-bound haloperidol, an effect that is unlikely to be clinically significant.

7.3
Metabolism and Elimination

Trazodone's main metabolite is *m*-chlorophenylpiperazine (*m*CPP). While the parent compound exerts its antidepressant action via 5-HT$_2$ blockade, *m*CPP is a postsynaptic 5-HT$_{2C}$ agonist and is anxiogenic when administered alone (Kahn et al. 1988). The clearance of trazodone is delayed in the elderly (Greenblatt et al. 1987). Thus, the histamine-1 blocking effects of trazodone and/or the anxiogenic effects of *m*CPP may be magnified in this population, depending on the metabolic status of the patient. The metabolite *m*CPP is of interest because studies have been done attempting to use it as a biological marker for such conditions as panic disorder, obsessive-compulsive disorder, and schizophrenia as well as depression. For example, depressed patients show a blunted growth hormone response to *m*CPP infusion compared to normal control subjects (Anand et al. 1994). Although *m*CPP has not been used outside the research arena, it is discussed here to highlight its importance as a metabolite of nefazodone and trazodone and the potential importance of metabolites in determining the overall clinical effect of a drug, especially in subsets of patients. *m*CPP is dependent on biotransformation by CYP 2D6 to be cleared from the body. In a study by Ferry et al. (1994), *m*CPP levels were found to be elevated four- to fivefold in poor metabolizers of dextromethorphan (a 2D6 substrate). This finding implies that individuals who are genetically deficient in CYP 2D6 or who are receiving substantial CYP 2D6 inhibitors such as fluoxetine can accumulate much higher levels of this metabolite and may potentially experience clinically meaningful anxiogenic effects. This may explain otherwise paradoxical reactions in such individuals to either nefazodone or trazodone.

The metabolism of nefazodone results in three clinically relevant metabolites: hydroxynefazodone, a triazole-dione metabolite, and *m*CPP. Hydroxynefazodone is equipotent to the parent compound in terms of 5-HT$_2$ antagonism and is probably a significant contributor to the therapeutic efficacy of nefazodone. The triazole-dione metabolite is one-seventh the potency of the parent compound, but achieves plasma concentrations that are ten times that of the parent compound; it may therefore be a significant contributor to nefazodone's antidepressant activity. The *m*CPP metabolite is usually less clinically important than these other metabolites because its plasma level is generally significantly lower than the level achieved in trazodone's metabolism (Mayol et al. 1994).

Renal insufficiency does not alter the clearance of nefazodone (Barbihaiya et al. 1995c). On the other hand, severe hepatic failure does cause clinically significant increases in half-life and AUC for nefazodone and hydroxynefazodone (Ferry et al. 1994; Barbihaiya et al. 1995b). As a result, caution is warranted if nefazodone is used in patients with severe hepatic impairment.

7.4
Half-Life

The half-life of trazodone, which is at 5–9 h, is fairly short for an antidepressant. Because of this short half-life, trazodone must be administered in multiple daily doses in order to reach steady-state concentrations that will result in antidepressant efficacy. However, the need for multiple daily doses may lead to noncompliance problems, resulting in reduced antidepressant efficacy.

Although the half-life of nefazodone, which is 2–4 h, is less than that of trazodone, its antidepressant efficacy has been established using a twice-daily dosing schedule (Rickels et al. 1994). The antidepressant efficacy of nefazodone is substantially bolstered by its active metabolites hydroxynefazodone (half-life 1.5–4 h) and triazole-dione (half-life 0.5–18 h), which may account for the need for less frequent dosing of nefazodone than trazodone. In fact, there is evidence that patients who have shown a favorable antidepressant response to nefazodone on a twice daily dosing schedule can be safely converted to a single daily dose without compromising antidepressant efficacy (Preskorn et al. 1997).

7.5
Linear Versus Nonlinear Pharmacokinetics

Trazodone may exhibit saturation of its first-pass metabolism, which may lead to nonlinear pharmacokinetics at increasing doses over its clinically relevant dosing range (Preskorn 1993a). As a result, the C_{max} may increase disproportionately with increasing doses. Correspondingly, concentration-dependent effects like sedation are amplified as the dose increases.

Nefazodone is metabolized by and inhibits CYP 3A. Therefore, nefazodone not surprisingly has nonlinear pharmacokinetics across its clinically relevant dosing range (Kaul et al. 1995; Barbihaiya et al. 1996a). At daily doses between 200 and 600 mg, plasma levels and AUC increase exponentially with dose increases. Thus, a small increase in total daily dose could cause a disproportionately greater effect whether desired or undesired. For example, a patient with a partial antidepressant response may develop concentration-dependent adverse effects with a small increase in total daily dose. Unfortunately, dosing studies have determined antidepressant efficacy at total daily doses between 400 and 600 mg/day (Fontaine et al. 1994). This has complicated the dose titration of nefazodone and has significantly limited its acceptance in clinical practice. Moreover, these effects are likely to be more pronounced in elderly patients, because these patients achieve higher plasma levels of nefazodone and hydroxynefazodone than younger individuals (Barbihaiya et al. 1996b).

7.6
Pharmacokinetically Mediated Drug-Drug Interactions

The wide therapeutic index of trazodone and its lack of systemic toxicity even in overdose make pharmacokinetically mediated DDIs in which it is the "victim" drug clinically inconsequential. However, since *m*CCP, a metabolite of trazodone, is a 2D6 substrate, drugs that are potent inhibitors of 2D6, such as fluoxetine and paroxetine, could cause worsening of anxiety symptoms if co-administered with nefazodone or trazodone. Due to the long half-life of fluoxetine, this potential may exist for weeks after it has been discontinued. This fact should be kept in mind when switching from fluoxetine to nefazodone as well as other drugs that are dependent on CYP 2D6 or CYP 2C for their clearance.

Nefazodone significantly inhibits CYP 3A4 and possibly P-glycoprotein (Bristol-Myers Squibb, data on file, 1992; Schmider et al. 1996; Stormer et al. 2001; Garton 2002). For this reason, the coadministration of nefazodone and drugs that depend on this enzyme for their metabolism must be done cautiously. For example, the coadministration of nefazodone and drugs such as astemizole, cisapride, and terfenadine was contraindicated because nefazodone via its inhibition of CYP 3A can elevate the levels of these drugs to the point that they could cause significantly delayed intracardiac conduction, setting the stage for the development of a potentially fatal *torsades de pointes* arrhythmia (Woosley et al. 1993; Abernethy et al. 2001). In fact, terfenadine, astemizole, and cisapride have been withdrawn from many markets because of this interaction not only with nefazodone but also with a number of other CYP 3A inhibitors (Preskorn 2002; also see information on Dr. Preskorn's website at www.preskorn.com). Given that CYP 3A is responsible for over 50% of all known human oxidative drug metabolism, there are numerous drugs whose pharmacokinetics can be affected by coadministration of nefazodone. In addition to astemizole, cisapride, and terfenadine, there are reports of clinically significant interactions between nefazodone and alprazolam, carbamazepine, cyclosporine, tacrolimus, and triazolam (Campo et al. 1998; Olyaei et al. 1998; Vella and Sayegh 1998; Wright et al. 1999; Garton 2002). For example, the central nervous system effects of drugs such as alprazolam and triazolam that depend on 3A4 metabolism can also be substantially potentiated by coadministration of nefazodone. Although not contraindicated for combined use, the dosage of alprazolam or triazolam should be decreased by 75% if co-prescribed with nefazodone (Barbhaiya 1995a; Green et al. 1995).

Parenthetically, nefazodone, like the SSRIs and venlafaxine, is capable of inhibiting serotonin uptake and therefore coadministration with MAOIs is contraindicated (i.e., a pharmacodynamically mediated drug–drug interaction).

8
Monoamine Oxidase Inhibitors

Limited data are available concerning the pharmacokinetics of this class of antidepressants, because of the era in which they were developed. They have not gained widespread acceptance as antidepressants due to concerns about hypertensive crises, the need to follow a tyramine-free diet, and the potential for orthostatic hypotension and weight gain.

MAOIs can be further subclassified based on pharmacodynamic activity. For example, inhibitors of MAO-A prevent deamination elimination predominately of serotonin and norepinephrine and include clorgyline, an irreversible inhibitor, and moclobemide, a reversible inhibitor. Inhibitors of MAO-B prevent deamination of predominantly dopamine and include selegiline and pargyline. Nonselective inhibitors of MAO-A and MAO-B include irreversible agents such as phenelzine, tranylcypromine, and isocarboxazid. Limited knowledge of the pharmacokinetic properties of the generally available MAOIs is summarized in the sections that follow.

8.1
Absorption

MAOIs such as phenelzine, tranylcypromine, and moclobemide reach peak plasma concentrations in 1–4 h, which correlates with the occurrence of adverse effects such as orthostatic hypotension and fatigue. Therefore, more frequent lower doses may be helpful for the patient who has had a favorable antidepressant response but is troubled by these concentration-dependent side effects.

8.2
Distribution

The volume of distribution of MAOIs is difficult to establish. Presumably these drugs penetrate CNS tissue well, where the MAO-inhibiting activity is desired.

8.3
Metabolism and Elimination

Many MAOIs are converted to compounds that are structurally similar to catecholamine and psychostimulants. To what extent these metabolites possess MAO-inhibiting capabilities is not known. Agents like tranylcypromine, selegiline, and phenelzine may produce stimulant effects that are mediated by their metabolites (Preskorn 1993a). Differences in efficacy of some MAOIs such as phenelzine may be mediated by differences in acetylation status (Davidson et al. 1978).

8.4
Half-Life

Most MAOIs have a short half-life of 2–4 h, meaning plasma levels of these drugs will become nondetectable within 1 day of discontinuing them. However, the effect of irreversible binding and deactivation of MAO has a half-life of about 2 weeks (Janicak et al. 2001). Thus, the pharmacodynamic effects persist until the body manufactures new MAO.

8.5
Linear Versus Nonlinear Pharmacokinetics

There is some evidence that phenelzine has nonlinear pharmacokinetics (Robinson et al. 1980, 1985). This has no clinical significance with respect to inhibition of monoamine oxidase, because no further MAO inhibition can occur once the enzyme is saturated. It may, however, become clinically relevant with respect to concentration-dependent side effects such as hypotension and fatigue.

8.6
Pharmacokinetically Mediated Drug–Drug Interactions

Reflecting the age of these drugs, little is known about their potential to be either the perpetrator or victim of pharmacokinetically mediated DDIs. Selegiline is in part metabolized by CYP 2D6 and adverse interactions have been reported with SSRIs that are also substantial CYP 2D6 inhibitors (e.g., fluoxetine) (Preskorn 1998b).

DDIs involving pharmacodynamic mechanisms are of such clinical relevance that they need to be mentioned here. As previously discussed, SSRIs should not be prescribed for at least 2 weeks after MAOI discontinuation. By the same token, SSRIs must be fully washed out prior to prescribing an MAOI. In the case of fluoxetine, 5 weeks must elapse after discontinuation prior to starting therapy with an MAOI. MAOIs may precipitate hypertensive crises if combined with tyramine or sympathomimetic amines. The hypoglycemic effects of insulin may be potentiated if combined with MAOIs. The combination of an MAOI and meperidine may be fatal (Goodnick 1994). These matters are more extensively covered in the chapter on MAOIs by Kennedy et al.

9
Mirtazapine

Mirtazapine is an antidepressant that offers a unique pharmacodynamic profile in terms of its antidepressant mechanism of action. It is a potent alpha-2 antagonist that produces increases in noradrenergic and serotonergic neurotransmission, which is the primary mechanism thought to underlie its antidepressant activity (deBoer et al. 1995). Mirtazapine also has a favorable pharmacokinetic profile.

9.1
Absorption

Mirtazapine is rapidly and completely absorbed following an oral dose, reaching its T_{max} in about 2 h. Food has a minimal impact on the rate and extent of absorption, and therefore does not require any dosage adjustment [Remeron (mirtazapine) package insert, Physicians' Desk Reference 2003, pp 2401–2404].

9.2
Distribution

The volume of distribution of mirtazapine is unknown. The plasma protein binding of mirtazapine is 85%. As discussed above, displacement interactions are rarely clinically significant.

9.3
Metabolism and Elimination

Multiple human CYP enzymes are involved in the biotransformation of mirtazapine. The 8-hydroxy metabolite is formed primarily by CYP 2D6 and CYP 1A2; the N-demethyl metabolite by CYP 1A2, CYP 3A4 and CYP 2D6; and the N-oxide metabolite by CYP 1A2, CYP 3A4, and CYP 2C9 (Montgomery 1995; Stormer et al. 2000; Timmers et al. 2000).

Demethylmirtazapine is the only metabolite that appears to be pharmacologically active and is three to four times less potent than the parent compound at the presumed sites of action mediating antidepressant response (Sitsen and Zivkov 1995). The metabolites of mirtazapine are excreted chiefly in the urine as sulfated or glucuronidated products (Dahl et al. 1997). Renal failure can lead to a 30%–50% decrease in the clearance of mirtazapine while hepatic impairment has been found to reduce clearance of mirtazapine by approximately 30% compared with clearance in normal subjects [Remeron (mirtazapine) package insert, Physicians' Desk Reference 2003, pp 2401–2404]. Accordingly, dosage adjustments may be needed for populations of depressed patients with liver or renal failure.

9.4
Half-Life

The half-life of mirtazapine is 20–40 h, which means that steady-state levels can be achieved with once daily dosing at bedtime, thus taking advantage of the sedative and sleep enhancing qualities of the drug (Voortman and Paanakker 1995).

9.5
Linear Versus Nonlinear Pharmacokinetics

Mirtazapine has linear kinetics across its clinically relevant dose range (15–80 mg/day) (Timmer et al. 1995).

9.6
Pharmacokinetically Mediated Drug–Drug Interactions

Given the multiple and approximately equal pathways for the biotransformation and elimination of mirtazapine, this drug is relatively protected against being the victim of a clinically meaningful CYP enzyme-mediated drug–drug interaction. Nevertheless, fluvoxamine has been reported to cause a three- to fourfold elevation in mirtazapine plasma levels (Antilla et al. 2001). Of interest, a case of serotonin syndrome has been reported following the coadministration of fluvoxamine and mirtazapine (Demers and Malone 2001).

There is limited in vivo data regarding mirtazapine's effects on various CYP enzymes. Based on in vitro studies, mirtazapine is ten times less potent in inhibiting CYP 2D6 than fluoxetine. Mirtazapine is also three orders of magnitude less potent in inhibiting CYP 1A2 and CYP 3A4 compared with fluvoxamine and ketoconazole, respectively (Dahl et al. 1997). The Dahl group also studied mirtazapine in poor metabolizers versus extensive metabolizers of debrisoquine, a substrate for CYP 2D6. In this study, no significant differences were found in the pharmacokinetic parameters of mirtazapine or demethylmirtazapine, suggesting that CYP 2D6 plays a relatively minor role in the total elimination of mirtazapine in humans in vivo. Based on these limited data, there is little reason to suspect that DDIs from mirtazapine would inhibit the metabolism of co-prescribed drugs that are substrates for CYP 1A2, CYP 2D6, or CYP 3A4. Consistent with this concept, mirtazapine has been reported to not have an effect on plasma levels of risperidone and 9-hydorxyrisperidone, which are CYP 2D6 and CYP 3A substrates, respectively (Loonen et al. 1999)

10
Conclusion

This chapter provides an overview of the pharmacokinetics of all the major classes of antidepressants to help clinicians select the most appropriate medications and doses for their patients and avoid pharmacokinetically mediated DDIs. Information on the use of therapeutic drug monitoring with antidepressants is discussed in the next chapter. The clinical profiles of all the major classes of antidepressants are reviewed in Part 3.

Glossary

Absorption	Process by which a drug passes from the site of administration to the systemic circulation.
Area under curve (AUC)	The total area under the plot-of-drug concentration versus time following either a single dose or multiple doses of a specific drug product (e.g., formulation) in a specific patient by a specific route of administration.
Autoinduction	The induction of enzymes responsible for the biotransformation of a drug by the drug itself, such that the half-life decreases with chronic exposure to the drug (e.g., carbamazepine).
Autoinhibition	The inhibition or saturation of the highest affinity enzymes responsible for the biotransformation of a drug by the drug itself, such that lower affinity enzymes become important in elimination and the half-life increases with chronic exposure to the drug (e.g., fluoxetine, paroxetine).
Bioavailability	Fraction of the dose administered that is absorbed and reaches the systemic circulation as active drug. This fraction will range between 0 and 1.0.
Biotransformation	The process by which a drug is converted to more polar substances (i.e., metabolites), which are then eliminated from the body either in the urine or in the stool (e.g., demethylated and hydroxylated metabolites of tricyclic antidepressants; the three metabolites of bupropion).
Clearance (Cl)	The volume of plasma completely cleared of drug per unit of time, generally per minute.
Concentration maximum (C_{max})	The highest or peak plasma drug concentration occurring after a dose.
Concentration minimum (C_{min})	The lowest or trough plasma drug concentration occurring before the next dose.
Distribution	Process by which a drug moves from the systemic circulation to the target site (e.g., brain) and to other tissue compartments (e.g., heart, adipose tissue).
Dosing rate	The amount of drug delivered to the body per interval of time by a specific route of administration [i.e., oral (po), intramuscular (IM), or intravenous (IV)].
Elimination half-life ($t_{1/2}$, plasma half-life)	The time necessary for the concentration of drug in a bodily compartment (usually plasma) to decrease by one half as a result of clearance. Half-life is generally measured in hours. For psychotropic medications, half-life can be as short as 2–4 h (e.g., triazolam) to as long as 20 days (e.g., norfluoxetine, the demethylated metabolite of fluoxetine).

Elimination rate constant (K_e)	Rate of decrease in drug concentration per unit of time as a result of elimination from the body.
Enterohepatic recirculation	The reabsorption of drug and/or metabolites from the small bowel into the portal circulation after first pass metabolism has occurred. Such recirculation contributes to the total amount of drug and/or metabolites, which eventually enters the systemic circulation (e.g., demethylated and hydroxylated metabolites of tricyclic antidepressants).
First-order kinetics	The amount of drug eliminated per unit of time is directly proportional to its concentration. In this state, the mechanisms for biotransformation and elimination are not saturated (e.g., benzodiazepines, tricyclic antidepressants, lithium carbonate).
First-pass metabolism (first-pass effect)	The passage of the drug from the portal circulation into hepatocytes and conversion there into metabolites. These metabolites may have a pharmacological profile different from that of the parent drug. They are typically then excreted by the hepatocytes into the biliary system and pass back into the small bowel where enterohepatic recirculation may occur (e.g., benzodiazepines, bupropion, nefazodone, neuroleptics, tricyclic antidepressants).
Percent protein binding	Percentage of the drug present in the plasma that is bound to plasma protein under physiological conditions. Many drugs, especially acidic agents such as phenytoin, are bound to albumin. Basic drugs, such as tricyclic antidepressants, may bind extensively to $\alpha 1$-acid glycoprotein. Levels of this protein can rise during acute inflammatory reactions (e.g., infections) and cause a temporary increase in plasma drug levels of no clinical consequence.
Pharmacodynamics	What the drug does to the body. The mechanism of action exerted by the drug such as cholinergic receptor blockage (e.g., benztropine).
Pharmacokinetics	What the body does to the drug. The process of drug absorption from the site of administration, distribution to the target organ and other bodily compartments, metabolism or biotransformation (if necessary), and eventual elimination.
Plasma drug concentration-time curve	A plot of the changes in plasma drug concentration as a function of time following single or multiple doses of a drug. This curve represents the series of events that follow the absorption of the drug into the systemic circulation: (1) the rate and the magnitude of the rise of plasma drug concentration; (2) the decline in plasma drug concentrations as a result of distribution to the target site

	(i.e., the brain for psychotropic drugs) and other tissue compartments where the drug may be active (e.g., the heart) or simply stored (e.g., the adipose tissue); and (3) the eventual decline due to biotransformation (if necessary) and elimination.
Steady-state concentration (C_{ss})	The condition under which drug concentration does not change as a result of continued drug administration at the same dosing rate. The amount administered per dosing interval equals the amount eliminated per dosing interval.
Therapeutic drug monitoring (TDM)	Based on the ability to quantitate the concentration of drug and its clinically relevant metabolites in a biological sample such as plasma. Using this information, the physician can rationally adjust the dose to achieve a plasma drug concentration within the range in which optimal response occurs for most patients, in terms of both efficacy and safety. Such monitoring is used primarily for drugs that have a narrow therapeutic index and wide interindividual variability in clearance rates (e.g., bupropion, clozapine, lithium, some anticonvulsants, tricyclic antidepressants).
Therapeutic index	The difference between the maximally effective dose or concentration and the toxic dose or concentration.
Therapeutic range	The concentration of drug in plasma that usually provides a therapeutically desirable response in the majority of individuals without substantial risk of serious toxicity. The target concentration for an individual patient is usually within the therapeutic range.
Time maximum (T_{max})	The time needed to reach the maximal plasma concentration following a dose. T_{max} can vary depending on the route of administration and formulation of the product (e.g., immediate versus sustained release forms).
Volume of distribution (V_d)	The apparent volume into which the drug must have been distributed to reach a specific concentration. Many psychotropic drugs have much large apparent volumes of distribution than would be expected based on physical size of the body, because the drugs dissolve disproportionately more in lipid and protein compartments (i.e., tissue) than in the body's water compartment.
Zero order kinetics	Situation in which only a fixed amount of drug eliminated per unit of time regardless of plasma concentration because all mechanisms mediating elimination are saturated.

References

Abernethy DR, Kertzner L (1984) Age effects on alpha-1-acid glycoprotein concentration and imipramine plasma protein binding. J Am Geriatr Soc 32:705–708

Abernethy DR, Barbey JT, Franc J, Brown KS, Feirrera I, Ford N, Salazar DE (2001) Loratadine and terfenadine interaction with nefazodone: Both antihistamines are associated with QTc prolongation. Clin Pharmacol Ther 69:96–103

Albers LJ, Reist C, Helmeste D, Vu R, Tang SW (1996) Paroxetine shifts imipramine metabolism. Psychiatry Res 59:189–196

Alderman J, Preskorn SH, Greenblatt DJ, Harrison W, Allison J, Chung M (1997) Desipramine pharmacokinetics when co-administered with paroxetine and sertraline. J Clin Psychopharmacol 17:284–291

Alfaro CL, Lam YWF, Simpson J, Ereshefsky L (1999) CYP2D6 Status of extensive metabolizers after multiple-dose fluoxetine, fluvoxamine, paroxetine or sertraline. J Clin Pharmacol 19:155–163

Alfaro CL, Lam YWF, Simpson J, Ereshefsky L (2000) CYP2D6 Inhibition by fluoxetine, paroxetine, sertraline, and venlafaxine in a crossover study: Intraindividual variability and plasma concentration correlations. J Clin Pharmacol 40:58–66

Amchin J, Zarycranski W, Taylor KP, Albano D, Klockowski PM (1998a) Effect of venlafaxine on the pharmacokinetics of alprazolam. Psychopharmacol Bull 34:211–9

Amchin J, Zarycranski W, Taylor KP, Albano D, Klockowski PM (1998b) Effect of venlafaxine on the pharmacokinetics of terfenadine. Psychopharmacol Bull 34:383–9

Amchin J, Zarycranski W, Taylor KP, Albano D, Klockowski PM (1999a) Effect of venlafaxine on CYP1A2-dependent pharmacokinetics and metabolism of caffeine. J Clin Pharmacol 39:252–9

Amchin J, Zarycranski W, Taylor KP, Albano D, Klockowski PM (1999b) Effect of venlafaxine on the pharmacokinetics of risperidone. J Clin Pharmacol 39:297–309

Amchin J, Ereshefsky L, Zarycranski W, Taylor K, Albano D, Klockowski PM (2001) Effect of venlafaxine versus fluoxetine on metabolism of dextromethorphan, a CYP2D6 probe. J Clin Pharmacol 41:443–51

Anand A, Charney DS, Delgado PL, McDangle CJ, Heninger GR, Price LH (1994) Neuroendocrine and behavioral responses to intravenous m chlorophenylpiperazine (mCPP) in depressed patients and healthy comparison subjects. Am J Psychiatry 151:1626–1630

Anttila S, Rasanen I, Leinonen E (2001) Fluvoxamine augmentation increases serum mirtazapine concentrations three to fourfold. Annals of Pharmacotherapy 35:1221–1223

Apseloff G, Wilner KD, Gerber N, Tremaine LM (1997) Effect of sertraline on the protein binding of warfarin. Clin Pharmacokinet 32(suppl 1):37–42

Aronoff GR, Bergstrom RF, Pottratz ST, Sloan RS, Wolen RL, Lemberger L (1984) Fluoxetine kinetics and protein binding in normal and impaired renal function. Clin Pharmacol Ther 36:138–144

Azzaro AJ, Ward HE (1994) Drugs used in mood disorders. In: Craig CR, Stitzel RE (eds) Modern pharmacology, 4th edition. Little, Brown and Company, Boston, New York, Toronto, London, pp 397

Baker GB (1999) Drug metabolism and psychiatry: introduction. Cell Mol Neurobiol 19:301–308

Ball SE, Ahern D, Scatina J, Kao J (1997) Venlafaxine: in vitro inhibition of CYP2D6 dependent imipramine and desipramine metabolism; comparative studies with selected SSRIs and effects on human hepatic CYP3A4, CYP2C9 and CYP1A2. Br J Clin Pharmacol 43:619–626

Barbhaiya RH, Shukla UA, Kroboth PD, Greene DS (1995a) Coadministration of nefazodone and benzodiazepines: II. A pharmacokinetic interaction study with triazolam. J Clin Psychopharmacol 15:320–326

Barbhaiya RH, Shukla UA, Natarajan CS, Behr DA, Greene DS, Sainati SM (1995b) Single and multiple dose pharmacokinetics of nefazodone in patients with hepatic cirrhosis. Clin Pharmacol Ther 58:390–398

Barbhaiya RH, Brady ME, Shukla UA, Greene DS (1995c) Steady-state pharmacokinetics of nefazodone in subjects with normal and impaired renal function. Eur J Clin Pharmacol 49:229–35

Barbhaiya RH, Shukla UA, Chaikin P, Greene DS, Marathe PH (1996a) Nefazodone pharmacokinetics: assessment of nonlinearity, intra-subject variability and time to attain steady-state plasma concentrations after dose escalation and de-escalation. Eur J Clin Pharmacol 50:101–7

Barbhaiya RH, Buch AB, Greene DS (1996b) A study of the effect of age and gender on the pharmacokinetics of nefazodone after single and multiple doses. J Clin Psychopharmacol 16:19–25

Benet LZ, Kroetz DL, Sheiner LB (1996) Pharmacokinetics: the dynamics of drug absorption, distribution, and elimination. In: Hardman JG, Limbird L, Molinoff P, et al. (eds) The pharmacological basis of therapeutics. McGraw-Hill, New York, pp 3–28

Bergstrom RF, Peyton AL, Lemberger L (1992) Quantification and mechanism of the fluoxetine and tricyclic antidepressant interaction. Clin Pharmacol Ther 51:239–248.

Bergstrom RF, Beasley CM Jr., Levy NB, Blumenfield M, Lemberger L (1993) The effect of renal and hepatic disease on the pharmacokinetics, renal tolerance, and risk-benefit profile of fluoxetine. Int Clin Psychopharmacol 8:261–266

Brosen K, Gram LF, Sindrup S, Nielsen KK (1992) Pharmacogenetics of tricyclics and novel antidepressants: recent developments. Clin Neuropharmacol 15(suppl 1):80A–81A

Brosen K, Hansen JG, Nielsen KK, Sindrup SH, Gram LF (1993a) Inhibition by paroxetine of desipramine metabolism in extensive but not in poor metabolizers of sparteine. Eur J Clin Pharmacol 44:349–355

Brosen K, Skjelbo E, Rasmussen BB, Poulsen HE, Loft S (1993b) Fluvoxamine is a potent inhibitor of cytochrome P450 1A2. Biochem Pharmacol 45:1211–1214

Bupp SJ, Preskorn SH (1991) The effects of age on plasma levels of nortriptyline, Ann Clin Psychiatry 3:61–65

Campion EW, deLabry LO, Glynn RJ (1988) The effect of age on serum albumin in healthy males: report from the normative aging study. J Gerontol 43:M18–M20

Campo JV, Smith C, Perel JM (1998) Tacrolimus toxic reaction associated with the use of nefazodone: Paroxetine as an alternative agent. Arch Gen Psychiatry 55:1050–1052

Charney DS, Miller H, Licinio J, Salomon R (1995) The pharmacological treatment of depression. In: Schatzberg AF, Nemeroff CB (eds) Textbook of Psychopharmacology, American Psychiatric Press, Washington, DC, pp 575–601.

Cohen LJ (1995) Principles to optimize drug treatment in the depressed elderly: practical pharmacokinetics and drug interactions. Geriatrics 50(suppl 1):S32–S40

Crewe HK, Lennard MS, Tucker GT, Woods FR, Haddock RE (1992) The effect of selective serotonin reuptake inhibitors on cytochrome P450 2D6 (CYP2D6) activity in human liver microsomes. Br J Clin Pharmacol 34:262–265

Cusack B, Nelson A, Richelson E (1994) Binding of antidepressants to human brain receptors: focus on newer generation compounds. Psychopharmacology 114:559–565

Cusack B, O'Malley K, Lavan J, Noel J, Kelly JG (1985) Protein binding and disposition of lidocaine in the elderly. Eur J Clin Pharmacol 29:232–239

Dahl ML, Voortman G, Alm C, Elwin CE, Delbressine L, Vos R, Bogaards JJP, Bertilsson L (1997) In vitro and in vivo studies on the disposition of mirtazapine in humans. Clin Drug Invest 13:37–46

Davidson J (1989) Seizures and bupropion: a review. J Clin Psychiatry 50:256–261

Davidson J, McLeod MN, Blum M (1978) Acetylation phenotype, platelet monoamine oxidase inhibition, and the effectiveness of phenelzine in depression. Am J Psychiatry 135:467–469

DeBoer T, Ruight GSF, Berendsen HHG (1995) The alpha-2 selective adrenoceptor antagonist Org 3770 (Mirtazapine, Remeron) enhances noradrenergic and serotonergic transmission. Human Psychopharmacology 10:S107-S118.

Dechant KL, Clissold SP (1991) Paroxetine: a review of its pharmacodynamic and pharmacokinetic properties and therapeutic potential in depressive illness. Drugs 41:225-253

Demers JC, Malone M (2001) Serotonin syndrome induced by fluvoxamine and mirtazapine. Ann Pharmacother 35:1217-20

DeVane CL (1994) Pharmacokinetics of newer antidepressants: clinical relevance. Am J Med 97(suppl 6A):13S-23S

DeVane CL, Pollock BG (1999) Pharmacokinetic considerations of antidepressant use in the elderly. J Clin Psychiatry 60(suppl 20):38-44

Dockens RC, Rapoport D, Roberts D, Greene DS, Barbhaiya RH (1995) Lack of an effect of nefazodone on the pharmacokinetics and pharmacodynamics of theophylline during concurrent administration in patients with chronic obstructive pulmonary disease. Br J Clin Pharmacol 40:598-601

Doyle GD, Laher M, Kelly JG, Byrne MM, Clarkson A, Zussman BD (1989) The pharmacokinetics of paroxetine in renal impairment. Acta Psychiatr Scand 80 (suppl 340):89-90

Duncan D, Sayal K, McConnell H, Taylor D (1998) Antidepressant interactions with warfarin. Int Clin Psychopharmacol 13:87-94

Edelman JS, Leibman J (1959) Anatomy of body water and electrolytes. Am J Med 27:256-277

Eichelbaum M, Gross AS (1990) The genetic polymorphism of debrisoquine/sparteine metabolism: clinical aspects. Pharmacol Ther 46:377-394

Ellingrod VL, Perry PJ (1994) Venlafaxine: a heterocyclic antidepressant, Am J Hosp Pharm 51:3033-3046

Evans DA, Mahgoub A, Sloan TP, Idle JR, Smith RL (1980) A family and population study of the genetic polymorphism of debrisoquine oxidation in a white British population. J Med Genet 17:102-105

Ferry N, Bernard N, Cuisinaud G, Rougier P, Trepo C, Sassard J (1994) Influence of hepatic impairment on the pharmacokinetics of nefazodone and two of its metabolites after single and multiple oral doses. Fundam Clin Pharmacol 8:463-473

Fonne-Pfister R, Meyer UA (1988) Xenobiotic and endobiotic inhibitors of cytochrome P-450dbl function, the target of the debrisoquine/sparteine type polymorphism. Biochem Pharmacol 37:3829-3835

Fontaine R (1993) Novel serotonergic mechanisms and clinical experience with nefazodone. Clin Neuropharmacol 16(suppl 2):45-50

Fontaine R, Ontiveros A, Elie R, Kensler TT, Roberts DL, Kaplita S, Ecker JA, Faludi G (1994) A double-blind comparison of nefazodone, imipramine, and placebo in major depression. J Clin Psychiatry 55:234-241

Fredericson OK (1978) Preliminary studies of the kinetics of citalopram in man. Eur J Clin Pharm 14:69-73

Garton T (2002) Nefazodone and CYP450 3A4 interactions with cyclosporine and tacrolimus. Transplantation 74:745

Glotzbach RK, Preskorn SH (1982) Brain concentrations of tricyclic antidepressants: single-dose kinetics and relationship to plasma concentrations in chronically dosed rats. Psychopharmacology 78:25-27

Golden RN, DeVane L, Laizure SC, Rudorfer MV, Sherer MA, Potter WZ (1988) Bupropion in depression. Arch Gen Psychiatry 45:145-149

Goldstein JA, de Morais SM (1994) Biochemistry and molecular biology of the human CYP 2C subfamily. Pharmacogenetics 4:285-299

Goodnick PJ (1992) Blood levels and acute response to bupropion. Am J Psychiatry 149:399-400

Goodnick PJ (1994) Pharmacokinetic optimization of therapy with newer antidepressants. Clin Pharmacokinet 27:307–330
Goulden KJ, Dooley JM, Camfield PR, Fraser AD (1987) Clinical valproate toxicity induced by acetylsalicylic acid. Neurology 37:1392–1394
Gram LF, Hansen MG, Sindrup SH, Brosen K, Poulsen JH, Aaes-Jorgensen T, Overo KF (1993) Citalopram: interaction studies with levomepromazine, imipramine, and lithium. Ther Drug Monit 15:18–24
Greenblatt DJ, Friedman H, Burstein ES, Scarcne JM, Blyden GT, Ochs HR, Miller LG, Harmatz JS, Shader RI (1987) Trazodone kinetics: effect of age, gender, and obesity. Clin Pharmacol Ther 42:193–200
Greenblatt DJ, Ehrenberg BL, Gunderman JS, Locniskar A, Scavone JM, Harmatz JS, Shader RI (1989) Pharmacokinetic and electroencephalographic study of intravenous diazepam, midazolam and placebo. Clin Pharmacol Ther 45:356–365
Greenblatt DJ, von Moltke LL, Shader RI (1996) The importance of presystemic extraction in clinical pschopharmacology. J Clin Psychopharmacol 16:417–419
Greene DS, Salazar DE, Dockens RC, Kroboth P, Barbhaiya RH (1995) Coadministration of nefazodone and benzodiazepines: III. A pharmacokinetic interaction study with alprazolam. J Clin Psychopharmacol 15:399–408
Halper JP, Mann JJ (1988) Cardiovascular effects of antidepressant medications. Br J Psychiatry 153 (suppl 3):87–98
Hammer W, Sjoquist F (1967) Plasma levels of monomethylated tricyclic antidepressants during treatment with imipramine-like compounds. Life Sci 6:1895–1903
Harvey AT, Preskorn SH (1996a) Cytochrome P450 enzymes: interpretation of their interactions with selective serotonin reuptake inhibitors. Part I. J Clin Psychopharmacol 16:273–285
Harvey AT, Preskorn SH (1996b) Cytochrome P450 enzymes: interpretation of their interactions with selective serotonin reuptake inhibitors. Part II. J Clin Psychopharmacol 16:345–355
Harvey AT, Preskorn SH (2001). Fluoxetine pharmacokinetics and effect on CYP2C19 in young and elderly volunteers. J Clin Psychopharmacol 21:161–166.
Hegarty JE, Dundee JW (1977) Sequelae after the intravenous injection of three benzodiazepines-diazepam, lorazepam and flunitrazepam. Br Med J 22:1384–1385
Hemeryck A, Lefebvre RA, De Vriendt C, et al (2000) Paroxetine affects metoprolol pharmacokinetics and pharmacodynamics in healthy volunteers. Clin Pharmacol Ther 67:283–291
Henry JA (1991) Overdose safety with fluvoxamine. Int Clin Psychopharmacol 6(suppl 3):41–47
Hesse LM, Venkatakrishnan K, Court MH, Duan SX, von Moltke LL, Shader RI, Greenblatt DJ (2000) CYP2B6 mediates the invitro hydroxylation of bupropion: Potential drug interactions with other antidepressants. Drug Metabolism and Disposition 28:1176–1183
Hesse LM, von Moltke LL, Shader RI, Greenblatt DJ (2001) Ritonavir, efavirenz, and nelfinavir inhibit CYP2B6 activity in vitro: potential drug interactions with bupropion. Drug Metab Dispos 29:100–2
Holford NHG, Sheiner LB (1981) Understanding the dose-effect relationship: clinical application of pharmacokinetic-pharmacodynamic models. Clin Pharmacokinet 6:429–453
Hyttel J (1994) Pharmacological characterization of selective serotonin reuptake inhibitors (SSRIs). Int Clin Psychopharmacol 9(suppl 1):19–26
Jandhyala B, Steenberg M, Perel J, Manian AA, Buckley JP (1977) Effects of several tricyclic antidepressants on the hemodynamics and myocardial contractility of the anesthetized dogs. Eur J Pharmacol 42:403–410

Janicak PG, Davis JM (2000) Pharmacokinetics and drug interactions. In: Sadock BJ, Sadock V (eds) Kaplan & Sadock's comprehensive textbook of psychiatry, Vol. 2, 7th ed. Lippincott Williams & Wilkins, Philadelphia, pp 2250-2259

Janicak PG, Davis JM, Preskorn SH, Ayd Jr FJ (2001) Principles and practice of psychopharmacotherapy, 3rd edition. Lippincott Williams & Wilkins, Baltimore

Jann MW, Carson SW, Grimsley SR, Erikson SM, Kuman A, Carter JG (1995) Lack of effect of sertraline on the pharmacokinetics and pharmacodynamics of imipramine and its metabolites. Clin Pharmacol Ther 57:207 (abstract)

Jeppesen U, Gram LF, Vistisen K, Loft S, Poulsen HE, Brosen K (1996) Dose-dependent inhibition of CYP1A2, CYP2C19, and CYP2D6 by citalopram, fluoxetine, fluvoxamine and paroxetine. Eur J Clin Pharmacol 51:73-78

Kahn RS, Asiris GM, Wetzler S (1988) Neuroendocrine evidence for serotonin receptor hypersensitivity in panic disorder. Psychopharmacology 96:360-364

Kaul S, Shukla U, Barbhaiya RH (1995) Nonlinear pharmacokinetics of nefazodone after escalating single and multiple oral dose. J Clin Pharmacol 35:830-839

Kaye CM, Haddock RE, Langley PF Mellows G, Tasken TCG, Zussman BD, Greb WH (1989) A review of the metabolism of paroxetine in man. Acta Psychiatr Scand 80 (5 suppl 350):60-75

Ketter TA, Barnett J, Schroeder DH, Hinton ML, Chao J, Post RM (1992) Carbemazepine induces bupropion metabolism. Presented at the 145th Annual Meeting of the American Psychiatric Association, Washington, DC, May 2-7

Klamerus KJ, Maloney K, Rudolph RL, Sisenwine SF, Jusko WJ, Chiang ST (1992) Introduction of a composite parameter to the pharmacokinetics of venlafaxine and its active O-desmethyl metabolite. J Clin Pharmacol 32:716-724

Klotz U (1998) Effect of age on pharmacokinetics and pharmacodynamics in man. Int J Clin Pharmacol Ther 36:581-585

Kupfer A, Preisig R (1984) Pharmacogenetics of p-mephenytoin: a new drug hydroxylation polymorphism in man. Eur J Clin Pharmacol 26:753-759

Kurtz D, Bergstrom RF, Goldberg MJ, Cerimele BJ (1997) The effect of sertraline on the pharmacokinetics of desipramine and imipramine. Clin Pharmacol Ther 62:145-156

Lemberger L, Bergstrom RF, Wolen RL, Farid NA, Enas GG, Aronoff GR (1985) Fluoxetine: clinical pharmacology and physiologic disposition. J Clin Psychiatry 46(3 s 2):14-19

Levitt MA, Sullivan JB Jr, Owens SM (1986) Amitriptyline plasma protein binding: effect of plasma pH and relevance to clinical overdose. Am J Emerg Med 4:121-125

Lineberry CG, Johnston JA, Raymond RN, Samara B, Feighner JP, Harto NE, Granacher RP, Weisler RH, Carman JS, Boyer WF (1990) A fixed-dose (300 mg) efficacy study of bupropion and placebo in depressed outpatients. J Clin Psychiatry 51:194-199

Loonen AJ, Doorschot CH, Oostelbos MC, Sitsen JM (1999) Lack of drug interactions between mirtazapine and risperidone in psychiatric patients: a pilot study. Eur Neuropsychopharmacol 10:51-57

MacKichan JJ (1989) Protein binding drug displacement interactions: fact or fiction? Clin Pharmacokinet 16:65-73

Maes M, Westenberg H, Vandoolaeghe E, Demedts P, Wauters A, Neels H, Meltzer H (1997) Effects of trazodone and fluoxetine in the treatment of major depression: Therapeutic pharmacokinetic and pharmacodynamic interactions through formation of meta-chlorophenylpiperazine. J Clin Psychopharmacol 17:358-364

Marino MR, Langenbacher KM, Hammett JL, Nichola P, Uderman HD (1997) The effect of nefazodone on the single-dose pharmacokinetics of phenytoin in healthy male subjects. J Clin Psychopharmacol 17:27-33

Mayol RF, Cole CA, Luke GM, Colson KL, Kerns EH (1994) Characterization of the metabolites of the antidepressant drug nefazodone in human urine and plasma. Drug Metab Dispos 22:304-311

Mehta BR, Robinson BHB (1980) Lithium toxicity induced by triamterene-hydrochlorothiazide. Postgrad Med J 56:783–784
Montgomery SA (1995) Safety of mirtazapine: a review. Int Clin Psychopharmacology 10(Suppl 4):37–45
Naranjo CA, Sproule BA, Knoke DM (1999) Metabolic interactions of central nervous medications and selective serotonin reuptake inhibitors. Int Clin Psychopharmacol 14(suppl 2):S35-S47
Nelson JC, Jatlo PI (1987) Non linear desipramine kinetics: prevalence and importance. Clin Pharmacol Ther 41:666–670
Nemeroff CB (1993) Diagnosis and treatment of depression in medical practice. J Med Assoc Ga 82:461–464
Olyaei AJ, deMattos AM, Norman DJ, Bennett WM (1998) Interaction between tacrolimus and nefazodone in a stable renal transplant recipient. Pharmacotherapy 18:1356–1359
Otton SV, We D, Joffe RT, Cheung SW, Sellers EM (1993) Inhibition by fluoxetine of cytochrome P450 2D6 (CYP2D6) activity. Clin Pharmacol Ther 53:401–409
Otton SV, Ball SE, Cheung SW, Inaba T, Sellers EM (1994) Comparative inhibition of the polymorphic enzyme CYP2D6 by venlafaxine and other 5HT uptake inhibitors. Clin Pharmacol Ther 55:141
Ozdemir V, Naranjo CA, Herrmann N, Reed K, Sellers EM, Kalow W (1997) Paroxetine potentiates the central nervous system side effects of perphenazine: contribution of cytochrome P4502D6 inhibition in vivo. Clin Pharmacol Ther 62:334-47
Ozdemir V, Naranjo CA, Hermann N, et al (1998) The extent and determinants of changes in CYP2D6 and CYP1A2 activities with therapeutic doses of sertraline. J Clin Psychopharmacol 18:55–61
Palmer KJ, Benfield P (1994) Fluvoxamine: an overview of its pharmacological properties and review of its therapeutic potential in nondepressive disorders. CNS Drugs 1:57–87
Parker BM, Cusack BJ, Vestal RE (1995) Pharmacokinetic optimisation of drug therapy in elderly patients. Drugs Aging 7:10–18
Paxton JW (1980) Effects of aspirin on salivary and serum phenytoin kinetics in healthy subjects. Clin Pharmacol Ther 27:170–178
Pentel PR, Bullock ML, DeVane CL (1982) Hemoperfusion for imipramine overdose: elimination of active metabolites. J Toxicol Clin Toxicol 19:239–248
Physicians' Desk Reference (2003) Thomson PDR, Montvale
Pippenger CE (1987) Clinically significant carbamazepine drug interactions: an overview. Epilepsia 28:571–576
Pollock BG (1994) Recent developments in drug metabolism of relevance to psychiatrists. Harv Rev Psychiatry 2:204–213
Potter WZ, Manji HK (1990) Antidepressants, metabolites and apparent drug resistance. Clin Neuropharmacol 13(suppl 1):545–553
Preskorn SH (1982) Antidepressant response and plasma concentrations of bupropion. J Clin Psychiatry 43(12 sect 2):137–139
Preskorn SH (1989) Tricyclic antidepressants: the whys and hows of therapeutic drug monitoring, J Clin Psychiatry 50 (suppl 7):34–42
Preskorn SH (1991) Should bupropion dosage be adjusted based upon therapeutic drug monitoring? Psychopharmacol Bull 27:637–643
Preskorn SH (1993a) Pharmacokinetics of antidepressants: why and how they are relevant to treatment? J Clin Psychiatry 54(9, suppl):14–34
Preskorn SH (1993b) Recent pharmacologic advances in antidepressant therapy for the elderly. Am J Med 94 (suppl 5A):2–12
Preskorn SH (1996) Clinical pharmacology of selective serotonin reuptake inhibitors. Professional Communications, Caddo, Oklahoma

Preskorn SH (1997) Clinically relevant pharmacology of selective serotonin reuptake inhibitors: an overview with emphasis on pharmacokinetics and effects on oxidative drug metabolism. Clin Pharmacokinet 32(suppl 1):1–21

Preskorn SH (1998a) Debate resolved: there are differential effects of serotonin selective reuptake inhibitors on cytochrome P450 enzymes. J Psychopharmacology 12 (3 Suppl B):S89-S97

Preskorn SH (1998b) A Message from *Titanic*. J Pract Psychiatry Behav Health 4:236–242

Preskorn SH (1999) Outpatient management of depression. Professional Communications, Inc., Caddo, Oklahoma

Preskorn SH (2002) Drug approvals and withdrawals over the last 60 years. J Psychiatr Pract 8:41–50

Preskorn SH, Catterson M (1998) Antidepressants and the elderly: Focus on newer agents. Int J Geriatric Psychopharmacology 1:66–77

Preskorn SH, Fast GA (1991) Therapeutic drug monitoring for antidepressants: efficacy, safety and cost effectiveness. J Clin Psychiatry 52(suppl 8):23–33

Preskorn SH, Fast GA (1992) Tricyclic antidepressant-induced seizures and plasma drug concentration. J Clin Psychiatry 53:160–162

Preskorn SH, Irwin HA (1982) Toxicity of tricyclic antidepressants-kinetics, mechanism, intervention: a review. J Clin Psychiatry 43:151–156

Preskorn SH, Jerkovich GS (1990) Central nervous system toxicity of tricyclic antidepressants: phenomenology, course, risk factors, and role of therapeutic drug monitoring, J Clin Psychopharmacol 10:88–95

Preskorn SH, Mac DS (1985) Plasma levels of amitriptyline: effect of age and sex. J Clin Psychiatry 46:276–277

Preskorn SH, Othmer S (1984) Evaluation of bupropion hydrochloride: the first of a new class of atypical antidepressants. Pharmacotherapy 4:20–34

Preskorn SH, Burke MJ, Fast GA (1993) Therapeutic drug monitoring: principles and practice. In: Dunner DL (ed) The Psychiatric Clinics of North America. Saunders, Philadelphia, Pennsylvania, 16:611–641

Preskorn SH, Alderman J, Chung M, Harrison W, Messig M, Harris S (1994) Pharmacokinetics of desipramine coadministered with sertraline or fluoxetine. J Clin Psychopharmacol 14:90–98

Preskorn SH, Markowitz P, Stahl SM, Magnus R, Borian FE, Hamid S, Ieni JR, Jody DN (1997) Once daily dosing of nefazodone for the treatment of depression in patients previously stabilized on twice daily dosing. Presentation at the 37th Annual Meeting of New Clinical Drug Evaluation Units (NCDEU), Boca Raton, FL, May 27–30

Ragheb M, Ban TA, Buchanan D, Frolich JC (1980) Interaction of indomethacin and ibuprofen with lithium in manic patients under a steady-state lithium level. J Clin Psychiatry 41:397–398

Richelson E (1988) Synaptic pharmacology of antidepressants: an update. McLean Hosp J 13:67–88

Richelson E (1998) Pharmacokinetic interactions of antidepressants. J Clin Psychiatry 59(suppl 10):22–26

Rickels K, Schweizer E, Clary C, Fox I, Weise C (1994) Nefazodone and imipramine in major depression: a placebo-controlled trial. Br J Psychiatry 164:802–805

Robinson DS, Nies A, Cooper TB (1980) Relationship of plasma phenelzine levels to platelet monoamine oxidase inhibition, acetylator phenotype and clinical outcome in depressed outpatients. Clin Pharmacol Ther 27:280 (abstract)

Robinson D, Cooper T, Jindal S, et al. (1985) Metabolism and pharmacokinetics of phenelzine: lack of evidence for acetylation pathway in humans. J Clin Psychopharmacol 5:333–337

Rochat B, Amey M, Baumann P (1995) Analysis of enantiomers of citalopram and its demethylated metabolites in plasma of depressive patients using chiral reverse-phase liquid chromatography. Therap Drug Monit 17:273–279

Rosenblatt JE, Rosenblatt NC (1992) More about spontaneous postmarketing reports of bupropion related seizures. Curr Affect Illness 11:18–20
Routledge PA (1986) The plasma protein binding of basic drugs. Br J Clin Pharmacol 22:499–506
Salazar DE, Dockens RC, Milbrath RL, Raymond RH, Fulmor IE, Chaikin PC, Uderman HD (1995) Pharmacokinetic and pharmacodynamic evaluation of warfarin and nefazodone coadministration in healthy subjects. J Clin Pharmacol 35:730–738
Schmider J, Greenblatt DJ, von Moltke LL, Harmatz JS, Shader RI (1996) Inhibition of cytochrome P450 by nefazodone in vitro: studies of dextromethorphan O- and N-demethylation. Br J Clin Pharmacol 41:339–43
Schroeder DH (1983) Metabolism and kinetics of bupropion. J Clin Psychiatry 44(5 sect 2):79–81
Schwartz LL (1985) The debate over substitution policy: its evolution and scientific basis. Am J Med 79:38–44
Schweizer E, Weise C, Clary C, Fox I, Rickels K (1991) Placebo-controlled trial of venlafaxine for the treatment of major depression. J Clin Psychopharmacol 11:233–236
Seeman P (1995) Therapeutic receptor-blocking concentrations of neuroleptics. Int Clin Psychopharmacol 10(suppl 3):5–13
Shad MU, Preskorn SH (1997) A possible bupropion and imipramine interaction (letter. J Clin Psychopharmacol 17:118–119
Shad MU, Preskorn SH (2000) Antidepressants. In: Levy RH, Thummel KE, Trager WF, Hansten PD, Eichelbaum M (eds). Metabolic Drug Interactions. Lippincott, Williams & Wilkins, Philadelphia, pp 563–577
Sharma A, Goldberg MJ, Cerimele BJ (2000) Pharmacokinetics and safety of duloxetine, a dual-serotonin and norepinephrine reuptake inhibitor. J Clin Pharmacol 40:161-7.
Shen WW, Lin K-M (1991) Cytochrome P450 monooxygenases and interactions of psychotropic drugs. Int J Psychiatry 21:47–56
Sindrup SH, Brøsen K, Hansen MG, et al. (1993) Pharmacokinetics of citalopram in relation to the sparteine and the mephenytoin oxidation polymorphisms. Ther Drug Monit 15:11–17
Sitsen JMA, Zivkov M (1995) Mirtazapine: clinical profile. Central Nervous System Drugs 4(Suppl 1):39–48
Skjelbo E, Brosen K (1992) Inhibitors of imipramine metabolism by human liver microsomes, Br J Clin Pharmacol 34:256–261
Smith PG, Hinton ML, Fleck RJ (1980) Metabolism and excretion of the ^{14}C-bupropion in humans. Fed Proc 40:635 (abstract)
Solai LK, Mulsant BH, Pollock BG, Sweet RA, Rosen J, Yu K, Reynolds CF, III (1997) Effect of sertraline on plasma nortriptyline levels in depressed elderly. J Clin Psychiatry 58:440–443
Spina E, Pollicino AM, Avenoso A, Campo GM, Perucca E, Caputi AP (1993) Effect of fluvoxamine on the pharmacokinetics of imipramine and desipramine in healthy subjects. Ther Drug Monit 15:243–246
Spina E, Pisani F, Perucca E (1996) Clinically significant pharmacokinetic drug interactions with carbamazepine: an update. Clin Pharmacokinet 31:198–214
Sproule B, Otton SV, Cheung SW, Zhong XH, Romach MK, Sellers EM (1997) CYP2D6 inhibition in patients treated with sertraline. J Clin Psychopharmacol 17:102–106
Steiner E. Bertilsson L, Sawe J, Bertling I, Sjoqvist F (1988) Polymorphic debrisoquine hydroxylation in 757 Swedish subjects. Clin Pharm Ther 44:431–435
Sternbach H (1988) Danger of MAOI therapy after fluoxetine withdrawal. Lancet II 850–851
Stormer E, von Moltke LL, Shader RI, Greenblatt DJ (2000) Metabolism of the antidepressant mirtazapine in vitro: Contribution of cytochromes P-450, 1A2, 2D6 and 3A4. Drug Metabolism and Disposition 28:1168–1175

Stormer E, von Moltke LL, Perloff MD, Greenblatt DJ (2001) P-glycoprotein interactions of nefazodone and trazodone in cell culture. J Clin Pharmacol 41:708–14

Tanaka E, Hisawa S (1999) Clinically significant pharmacokinetic drug interactions with psychoactive drugs: antidepressants and antipsychotics and the cytochrome P450 system. J Clin Pharm Ther 24:7–16

Thomson AH, McGovern EM, Bennie P, Caldwell G, Smith M (1992) Interaction between fluvoxamine and theophylline. The Pharmaceutical Journal 249:137 (abstract)

Thurmann PA, Hompesch BC (1998) Influence of gender on the phamacokinetics and pharmacodynamics of drugs. Int J Clin Phamacol Ther 36:586–590

Timmer CJ, Lohmann AAM, Mink CPA (1995) Pharmacokinetic dose-proportionality study at steady state of mirtazapine from remeron tablets. Human Psychopharmacology 10(Suppl 2):97–107

Timmer CJ, Sitsen JM, Delbressine LP (2000) Clinical pharmacokinets of mirtazapine. Clinical Pharmacokinetics 38:461–474

Troy SM, Schultz RW, Parker VD, Chiang ST, Blum RA (1994) The effect of renal disease on the disposition of venlafaxine. Clin Pharmacol Ther 56:14–21

Vallner JJ (1977) Binding of drugs by albumin and plasma protein. J Pharm Sci 66:447–465

Vella JP, Sayegh MH (1998) Interactions between cyclosporine and newer antidepressant medications. Am J Kidney Dis 31:320–3

Von Moltke LL, Greenblatt DJ, Harmatz JS Shader RI (1994) Cytochromes in psychopharmacology. J Clin Psychopharmacol 14:1–4

Von Moltke LL, Greenblatt DJ, Court MH, Duan SX, Harmatz JS, Shader RI (1995) Inhibition of alprazolam and desipramine hydroxylation in vitro by paroxetine and fluvoxamine: comparison with other serotonin reuptake inhibitor antidepressants. J Clin Psychopharmacol 50:125–131

Voortman G, Paanakker JE (1995) Bioavailability of mirtazapine from remeron tablets after single and multiple oral dosing. Human Psychopharmacology 10(Suppl 2):83–97

Wang Q, Halpert JR (2002) Combined three-dimensional quantitative structure-activity relationship analysis of cytochrome P450 2B6 substrates and protein homology modeling. Drug Metab Dispos 30:86–95

Warrington SJ (1991) Clinical implications of the pharmacology of sertraline. Int Clin Psychopharmacol 6(suppl 2):11–21

Weintraub D (2001) Nortriptyline toxicity secondary to interaction with bupropion sustained-release. Depress Anxiety 13:50–2

Weller R, Preskorn SH (1984) Psychotropic drugs and alcohol: pharmacokinetic and pharmacodynamic interactions. Psychosomatics 25:301–309

Wright DH, Lake KD, Bruhn PS, Emery RW, Jr (1999) Nefazodone and cyclosporine drug–drug interaction. J Heart Lung Transplant 18:913–915

Wright JM, Stokes EF, Sweeney VP (1982) Isoniazid-induced carbamazepine toxicity and vice versa: a double drug interaction. N Engl J Med 307:1325–1327

Woosley RL, Yiwang C, Freiman JP, Gillis RA (1993) Mechanism of the cardiotoxic actions of terfenadine. JAMA 269:1532–1536

Yonkers KA, Kando JC, Hamilton J (1996) Gender issues in psychopharmacologic treatment. Essent Psychopharmacol 1:54–69

Zussman BD, Davie CC, Fowles SE, Kumar R, Lang U, Wargenau M, Sourgens (1995) Sertraline, like other SSRIs, is a significant inhibitor of desipramine metabolism in vivo. Br J Pharmacology 39:550–551

Therapeutic Drug Monitoring of Antidepressants

M. J. Burke[1] · S. H. Preskorn[2]

[1] Department of Psychiatry, University of Kansas School of Medicine,
1010 N. Kansas, Wichita, KS 67214, USA
e-mail: mjburke@kumc.edu

[2] Department of Psychiatry and Behavioral Sciences,
University of Kansas School of Medicine and Psychiatric Research Institute,
1010 N. Kansas, Wichita, KS 67214, USA

1	Introduction	88
2	Dose:Response Versus Concentration:Response Relationships	90
3	Basics Principles of TDM	91
3.1	Plasma:Tissue Concentration	91
3.2	Drug Concentration Variability	92
4	Utility of TDM with Antidepressant Pharmacotherapy	93
4.1	Avoiding Toxicity	94
4.2	Enhanced Therapeutic Response	95
4.3	General Application of TDM	97
5	Recommendations for TDM with Antidepressants	98
5.1	Tricyclic Antidepressants	98
5.1.1	Concentration:Response Relationship	99
5.1.2	Recommendation	100
5.2	Serotonin Selective Reuptake Inhibitors	100
5.2.1	Concentration:Response Relationship	101
5.2.2	Recommendation	103
5.3	Bupropion	103
5.3.1	Concentration:Response Relationship	104
5.3.2	Recommendations	104
5.4	Venlafaxine	105
5.5	Nefazodone and Trazodone	105
5.6	Mirtazapine	106
5.7	Monoamine Oxidase Inhibitors	106
5.7.1	Concentration:Response Relationship	107
5.7.2	Recommendations	107
6	Summary	107
	References	109

Abstract Therapeutic drug monitoring (TDM) can play an important role in use of a number of antidepressants and other psychotropic medications (e.g., tricyclic antidepressants). However, the field of psychiatry has been slow to adopt such monitoring as a routine part of care. Research supporting the clinical benefit and cost-effectiveness of TDM with newer antidepressants has also been limited. However, an increasing awareness of how significant inter- and intra-patient variability in pharmacokinetics may affect response to antidepressant therapy is creating growing interest in the use of TDM to optimize antidepressant treatment response. To interpret the results of TDM, it is important to understand the relationship between plasma and tissue concentrations of drugs and the factors that can cause variations in drug concentration. There is a correlation between plasma concentrations of psychotropic drugs and tissue concentrations. Thus, TDM provides an indirect measurement of drug concentration at effector sites in tissue compartments of interest (i.e., the central nervous system for psychotropic medications). Plasma drug concentrations of a given dose of drug may be affected by the pharmacokinetics of the individual patient (e.g., drug absorption, metabolism, elimination). TDM can be used to increase the safety of antidepressant pharmacotherapy with certain agents (e.g., tricyclic antidepressants) by enabling clinicians to avoid toxic levels of medication resulting from variations in patient pharmacokinetics. TDM can also be used to enhance therapeutic response. For those antidepressants with established concentration:response relationships, TDM has the potential to speed antidepressant response by providing a basis for more efficient dose adjustment and can also help clinicians monitor/confirm medication compliance. Recommendations for the use of TDM in clinical practice and directions for future research are provided.

Keywords Therapeutic drug monitoring · Drug dose · Plasma drug concentration · Drug toxicity · Therapeutic response

1
Introduction

The first chapter explained the basic potential pharmacodynamics of drugs such as antidepressants designed to affect brain function. The second chapter, in turn, laid out the basic principles of pharmacokinetics. These two chapters form the basis for understanding the clinical effects of psychiatric medications such as antidepressants as illustrated in the following equation:

$$\text{Response} = \text{pharmacodynamics} \times \text{pharmacokinetics} \times \text{biological variance} \quad (1)$$

This equation can also be stated as follows:

$$\text{Response} = \text{affinity for site of action} \times \text{drug concentration} \times \text{biological variance}$$
$$\begin{array}{ll} \text{absorption} & \text{genetics} \\ \text{distribution} & \text{age} \\ \text{metabolism} & \text{disease} \\ \text{elimination} & \text{internal environment} \end{array} \quad (2)$$

As can be readily seen in Eq. 2, pharmacokinetics is important because drug concentration determines which potential sites of action of the drug will be engaged and to what degree. At the lowest concentration, the drug will exert its effect only on the site(s) of action for which it has the greatest affinity. At higher concentrations, the drug will sequentially engage sites of action for which its affinity is lower. It is particularly important to understand this fact when dealing with drugs such as tertiary amine tricyclic antidepressants (TCAs) (e.g., amitriptyline) that can affect many different sites of action over a relatively narrow concentration range (see the chapter "Overview of Currently Available Antidepressants" by Preskorn and Ross, this volume).

Viewed from this perspective, clinical trials done for drug registration are in fact population pharmacokinetic studies in which the goal is to determine the usual dose needed in the usual patient enrolled in the clinical trial to achieve a concentration that engages the right site of action to the right degree to achieved the desired clinical effect. This fact also forms the basis for understanding the value of therapeutic drug monitoring (TDM) with psychiatric medications such as antidepressants, and particularly antidepressants such as TCAs, which can affect multiple sites of action over a relatively narrow dosing range. In essence, TDM is a refinement of the dose–response approach. The goal of TDM is to eliminate interindividual differences in drug clearance among patients (variable No. 3 in Eq. 2) as a factor affecting the outcome of treatment response.

To understand this statement, it is important to appreciate the following equation:

$$\text{Drug concentration} = \text{dosing rate}/\text{clearance} \tag{3}$$

If there were no interindividual differences in drug clearance, then TDM would provide no advantage over simply knowing the dose, because clearance would drop out as a variable in Eq. 3 and drug concentration would simply be a function of dose. However, the clearance of most drugs varies substantially among different patients due to inborn or genetic as well as acquired differences in the ability to clear a specific drug. The acquired differences are enumerated in Eq. 2 and include concomitant drug therapy that can change the pharmacokinetics of a co-prescribed drug by affecting its absorption, distribution, metabolism, or elimination. For example, the concomitant ingestion of a drug that inhibits the specific cytochrome P450 (CYP) enzymes mediating the biotransformation of a second drug leads to increased accumulation of the second drug and hence greater effect either on its highest affinity site of action or on a lower affinity site of action.

TDM then is essentially a means of detecting differences in drug clearance so that parallel adjustments can be made in the dose to produce a drug concentration that has the maximum chance of being therapeutic with a minimum risk of being toxic. That is the basic foundation underlining the information presented in this chapter.

The application of TDM in medical practice can have a dramatic affect on clinical response to pharmacotherapy. However, despite data supporting the benefit of TDM with certain antidepressants, namely TCAs, the field of psychiatry has been slow to adopt such monitoring for reasons that are not particularly clear. One factor that may have contributed to this problem is an under-appreciation among clinicians of the degree of interpatient variance in pharmacokinetics and the associated broad and unpredictable variability in plasma drug concentrations achieved at a given antidepressant dose.

Research to support the clinical benefit and cost-effectiveness of TDM with "newer" antidepressants (i.e., those antidepressants first approved for use in the 1990s) has been limited. In the absence of well-defined concentration–response relationships for these agents, it may not be surprising that many clinicians have perceived the use of TDM with antidepressants as an unnecessary, complicated, and costly procedure offering obscure clinical benefit. As such, demand for further research in this area has not been great. However, times are changing. Along with a burgeoning interest in pharmacogenetics, there is now an increasing appreciation of how significant inter- and intrapatient variability in pharmacokinetics may affect response to antidepressant pharmacotherapy. It seems reasonable to conclude that, at present, TDM is an underutilized procedure that has considerable untapped potential to enhance antidepressant drug therapy.

The goal of this chapter is to discuss the general principles underlying TDM and then to examine the present status of TDM as it applies to our current antidepressant armamentarium. The chapter concludes with recommendations for the use of TDM in routine clinical practice using antidepressants as well as directions for further research.

2
Dose:Response Versus Concentration:Response Relationships

At times, the boundary between the concepts of drug dose and drug concentration can become blurred with the two phenomena often considered synonymously. In fact, nothing could be further from the truth. Drug dose and the subsequently achieved drug concentration can have an erratic, unpredictable relationship. When technology did not permit accurate and precise routine determination of plasma drug concentration, dose was emphasized as the variable of singular importance both in clinical practice and pharmaceutical marketing. However, given present advances in technology that permit relatively low-cost and accessible determination of drug concentration, exclusive emphasis on drug dose as the critical treatment variable is naïve. It was, is, and will always be the concentration of drug at the effector sites that mediates both the beneficial and the adverse effects of drug treatment. Dose is only indirectly related to treatment response as an, at times, inaccurate estimate of drug concentration.

In the drug development process, the relationship between dose and concentration is straightforward at the in vitro phase of investigation. The investigational drug is added to a closed, controlled system and the amount of drug added (i.e., the dose) is the only variable determining drug concentration. A biological effect is then assayed (e.g., receptor occupancy) and a concentration-response relationship is determined which literally mirrors the dose:response relationship. In this case, dose and concentration *are* virtually synonymous.

Shifting to the in vivo phase of investigation, a drug dose is added to the system (i.e., given to a human subject), clinical response is assessed, and a dose-response relationship is determined, albeit with broader standard deviations around the mean response than seen in the controlled in vitro system. Under such conditions, the relationship between dose and drug concentration becomes considerably more complex due to pharmacokinetic variance among subjects. Dose shifts from the exclusive variable to one of many variables that determine drug concentration. A particular drug dose is, of course, associated with some mean plasma concentration in the population of study subjects, but there can be wide variance in the deviation from that mean (Burke and Preskorn 1999). Efforts are made to minimize deviation from the mean by reducing the variability in the study population (i.e., inclusion/exclusion criteria regarding characteristics such as age, concurrent illness, and concomitant medications). These data from clinical trials underestimate the substantial interpatient variability in pharmacokinetics that exists in a routine clinical population, as well as the considerable deviation that exists in the dose:concentration relationship. Hence, in clinical practice (i.e., in vivo) it is an error to tacitly assume that dose:response is equivalent to concentration:response. For those patients whose pharmacokinetics deviate from the population mean, the actual drug dose may have little predictive value, and knowledge of drug concentration can make the difference between a failed versus an optimal treatment response.

3
Basics Principles of TDM

As mentioned in the introduction, the cornerstone of pharmacotherapy, and the basic principle underlying TDM, is that clinical response to drug therapy is a function of a drug's biological activity (i.e., nature and potency of action at a biological target or site of action) and the concentration of drug available at that site of action to exert the effect (Eqs. 1 and 2).

The "working" extension of this relationship is that changes in drug concentration will change the magnitude, and perhaps quality, of the clinical response to drug therapy.

3.1
Plasma: Tissue Concentration

Another basic principle underlying TDM is that the plasma drug concentration reflects the concentration of the drug at the site of action that is responsible for the clinical effect. For antidepressant drugs, the site mediating the psychotropic effect is presumably in the brain. Sites mediating other effects can be located in other tissue compartments (e.g., orthostasis due to vascular α-adrenergic receptor blockade in the periphery). A critical point is that there is generally a correlation between plasma concentrations of psychotropic drugs and tissue concentration (Glotzbach and Preskorn 1982). Thus, TDM provides an indirect measurement of drug concentration at effector sites in the tissue compartments of interest (i.e., the central nervous system in the case of antidepressant effect). There are reports in which the relationship between plasma and tissue antidepressant drug concentration is obscure (Renshaw et al. 1992; Karson et al. 1993). These reports are not surprising when one considers the complex kinetic/dynamic relationship that must be taken into account to minimize error in these plasma/tissue correlational studies (e.g., delay in drug entry into the brain compartment due to protein binding, molecular size, lipophilicity) (Greenblatt and Harmatz 1993). More recent studies that have used fluorine magnetic resonance spectroscopy (MRS) to examine the kinetics of "newer" antidepressants have found that brain drug concentration correlates with plasma drug concentration and clinical response, although, as expected, changes in brain drug concentration lag behind changes in plasma drug concentration (Henry et al. 2000).

Advanced functional imaging technology (i.e., MRS, positron emission tomography [PET]) holds the promise of further advancing the cause of TDM (i.e., increasing awareness and understanding of the clinical utility of drug concentration monitoring). More recently the "reverse logic" of TDM has begun to be applied to the process of developing psychotropic treatments. Functional imaging is being used in this area to determine the amount of drug necessary at a site of action (e.g., receptor occupancy in particular brain regions) to produce a desired clinical response. Researchers can then determine the plasma drug concentration associated with the desired degree of biological activity and the dosing range needed to achieve this concentration (Martineza et al. 2001). From this type of research will come a refinement of dosage parameters and further application of TDM to guide dosing to achieve desired tissue drug concentrations.

3.2
Drug Concentration Variability

For an orally administered drug, the plasma concentration achieved is a function of the dose ingested and the pharmacokinetics of the individual patient (e.g., drug absorption, metabolism, elimination) (see Eq. 3). In this regard, the ingested drug dose has been and is used as a first approximation or estimate of the drug concentration achieved in an average individual. However, as seen in Eq. 3, dose is only one of the variables determining concentration. Drug dose, in and of itself, can be a strikingly misleading predictor of drug concentration achieved. Drug concentration can vary markedly due to patient-specific and environmental factors that affect the pharmacokinetic handling of the drug (Table 1). To the best of our understanding, the principal source of interpatient

Table 1 Factors affecting plasma drug concentration achieved at a given drug dose

Age/development (e.g., puberty, menopause)
Gender
Genetics (e.g., cytochrome P450 isoenzyme polymorphism)
Administration (e.g., time of day, food effect)
Body habitas (e.g., size, % body fat, nutritional status, hydration)
Physiological derangement (e.g., comorbid disease, impaired organ function)
Pharmacokinetic drug interactions (e.g., concomitant medications, environmental toxins)

variance in the pharmacokinetics of antidepressant agents is drug metabolism (Brosen 1996) (for further discussion of this issue, see chapters by Garlapati et al. and Preskorn and Catterson, this volume). The genetic polymorphism of CYP isoenzymes and drug interactions that can alter isoenzyme activity have received considerable attention in the last decade with regard to the effects they can have on antidepressant drug metabolism (Preskorn and Harvey 1998). Despite the attention to metabolism, variability in drug absorption and the effects of changes in cardiac and renal function on drug elimination should not be underestimated as factors that can contribute to the wide range of drug concentration achieved at a given dose (e.g., from near zero to an order of magnitude greater than predicted based on drug dose).

As mentioned in the introduction, TDM is essentially a refinement of the dose approximation of drug concentration, which accounts for, and eliminates, inter- and intrapatient variability in pharmacokinetics as a factor affecting treatment outcome. It provides the clinician with objective data to rationally guide dose adjustment. In this sense, it represents an advance over the time-consuming and often error-prone approach to dose titration based on clinical assessment of response. Clearly drug concentration, while a principal determinant of clinical response, is not the sole determinant, and clinical assessment remains the final arbiter of whether the treatment was successful or not. However, objective data, such as drug concentration, can assist the clinical decision-making process in a meaningful way (Yesavage 1986; Guthrie et al. 1987; Preskorn et al. 1993; Dahl and Sjoqvist 2000; Rasmussen and Brosen 2000; Tucker 2000). This type of data is particularly desirable in the treatment of depression, where (1) in the absence of biological markers, response is typically based on subjective assessment, and (2) the adverse effects of antidepressant drug therapy can mimic a worsening of depressive symptoms and lead the clinician to increase the drug dose when a reduction of dose would actually be more appropriate (see the discussion of "Avoiding Toxicity" below).

4
Utility of TDM with Antidepressant Pharmacotherapy

When the patient is taking a medication for which an assay is available, TDM differs from virtually all other laboratory tests in that it always yields useful in-

Table 2 Goals of therapeutic drug monitoring

Avoid toxicity	Identify pharmacokinetic variability and adjust dose to avoid concentration-dependent toxicity
Assess compliance	Obtain objective index of the amount of drug being consumed
Enhance response	Adjust dose to achieve optimal plasma concentration based on concentration:response data
Increase cost-efficiency	Obtain economic benefit of reducing titration time and optimizing clinical outcome of pharmacotherapy
Avoid medico-legal problems	Establish plasma level on prescribed dose; substantiate need for and safety of unusually high or low dosing

Table 3 Pharmacologic features of a drug consistent with the utility of therapeutic drug monitoring

Complex pharmacodynamics (drugs with multiple biological activities)
Substantial interpatient variability in pharmacokinetics affecting plasma drug concentration
Narrow therapeutic index
Difficult detection of early toxicity
Delayed onset of clinical benefit
Defined concentration:outcome relationships (effectiveness and toxicity)
Necessity of dose titration

formation. Many laboratory tests are ordered as screening procedures; hence, the results are likely to be normal for most patients. In the case of TDM, the test is directed at something that should be there: the drug being used to treat the patient. The drug concentration is either detectable or not. If it is not, that is useful information. If it is detectable, the question is whether the concentration is too low to expect a beneficial response, too high, so that toxicity may result or adverse effects will mask any therapeutic benefit, or at an appropriate level to optimize clinical outcome.

There is good reason to believe, and scientific support for the belief, that the use of TDM, although unlikely to be a panacea, can optimize antidepressant treatment response (see section "Recommendations for TDM with Antidepressants" later in this chapter). The general goals for TDM with any type of drug are all relevant to antidepressant pharmacotherapy (Table 2). Antidepressant agents also exhibit many of the pharmacological features that are considered predictive of TDM utility in clinical practice (Table 3). Some of these features will be reviewed below. The reader is referred to Burke and Preskorn (1999) for a more detailed discussion of these topics.

4.1
Avoiding Toxicity

The first goal in antidepressant, or for that matter any, pharmacotherapy is safety. TDM is a method that can be used to increase the safety of antidepressant pharmacotherapy. There is considerable interpatient variability in the clearance of virtually all antidepressant drugs (Preskorn et al 1993; Jerling 1995; Amsterdam et al. 1997). In controlled studies, plasma drug concentrations achieved with a given antidepressant at a given dose have been shown to vary more than 40-fold (Preskorn 1991; Rasmussen and Brosen 2000). This interpatient variability in antidepressant drug clearance can have serious, even life-threatening, consequences for those agents with a narrow therapeutic index, such as the TCAs; but even for agents with a wide therapeutic index, there can be significant consequences in terms of tolerability and a "failed treatment trial." The clinician can only determine a particular patient's rate of drug clearance by measuring the plasma drug level (i.e., TDM) and/or genotyping the patient (Preskorn 1993; Tucker 2000).

For several antidepressants, the early stages of toxicity may be clinically "silent" (e.g., slowing of intracardiac conduction by TCAs) or may mimic a worsening of the condition for which the medication was initially prescribed (e.g., insomnia and anorexia from a serotonin reuptake inhibitor). When early toxicity mimics worsening of the depressive syndrome (e.g., fatigue, sexual dysfunction, anxiety, anorexia), the clinician, based on a subjective assessment of "response," may increase the drug dose believing the problem to be lack of effectiveness due to inadequate levels rather than early toxicity due to excessive levels (Balant-Gorgia et al. 1989; Preskorn and Jerkovich 1990). By detecting excessively high plasma drug concentrations, TDM can provide the clinician with objective data to more accurately assess the cause of the adverse or inadequate treatment response and then quickly and safely optimize the dose. In complex patients who have multiple comorbid medical conditions and/or are taking multiple concomitant medications, TDM during the early stages of therapy can be useful in identifying patients at risk for toxicity at "therapeutic doses" (Rasmussen and Brosen 2000). In those patients who do experience an untoward or "idiosyncratic" response (e.g., syncope, seizure), TDM can help clarify the underlying cause (i.e., demonstrate that the patient achieves unusually high drug levels at a routine dose).

4.2
Enhanced Therapeutic Response

As a general rule, the faster the feedback, the more beneficial it is. Feedback from TDM is both faster and less ambiguous than feedback from clinical assessment of antidepressant response. For example, a clinician treating depression may have to wait several weeks to determine whether the selected antidepressant dose will be effective (Depression Guideline Panel 1994; Quitkin et al. 1996). This process of dose titration based exclusively on clinical response is slow, error prone, and costly, with the patient experiencing prolonged illness while the decision is being made to titrate the dose or change drugs. For those antidepressants with established concentration:response relationships, TDM can potentially hasten antidepressant response by providing a basis for more efficient dose adjustment (Preskorn and Fast 1991).

In psychiatry, identifying the relationship between drug concentration and effectiveness is fraught with difficulty. In a psychiatric illness like major depressive disorder, under routine practice conditions, the detection and quantification of improvement, especially early or partial response, is not reliable due to non-drug response (e.g., placebo response) and non-response. Hence, those studies that attempt to correlate concentration with response are more likely to yield a type II error (i.e., not finding a relationship when one exists) (Preskorn 1996a).

Despite the imprecision of response assessment, there have been a surprising number of studies demonstrating a plasma concentration:efficacy relationship for various antidepressant drugs (Preskorn et al. 1993; Pollock et al. 1996; Veefkind et al. 2000). In such cases, knowing the plasma drug level permits the clinician to adjust the drug dose rationally into a "therapeutic range" to hasten the onset and increase the likelihood of a therapeutic response. However, even in the absence of well-established concentration:response relationships, TDM may be useful. If a patient is not responding optimally to pharmacotherapy, TDM can provide data about whether the problem is likely to be too little drug (i.e., a concentration near zero) or too much drug (i.e., a concentration several times greater than the reference range associated with response in fixed-dose studies), which can guide further treatment decisions.

For those drugs for which concentration:response studies per se have not been done, clinical trial data on dose:response can provide a basis for estimating a concentration:response threshold or range. In those studies that employed a fixed-dose design, mean drug concentration determined at the "ineffective dose" and at the "usually effective minimum dose" can be used to estimate a minimum or threshold concentration range necessary for response. The concentrations achieved at higher doses, associated with either no additional benefit or with a response that is actually decreased (i.e., decreased efficacy or increased dropout rate due to adverse effects), are an estimate of the upper end of a therapeutic concentration range. In some cases, when a determination of plasma drug concentration is requested, clinical laboratories may report the concentration:response data from fixed-dose studies as the "reference range," which can be used as a target concentration range for clinical effectiveness. However, this is not always the case and it behooves the clinician to clarify this point with the reference laboratory when interpreting TDM results.

Pharmacokinetic differences aside, the issue of medication compliance is relevant to optimizing antidepressant treatment and particularly amenable to TDM. More than 40% of patients who receive psychotropic medications are noncompliant with their regimen (Ley 1981; Lin et al. 1995). Erratic compliance can lead to subtherapeutic antidepressant concentrations and thereby prolong the duration of the illness, or it can lead to excessive antidepressant concentrations and toxicity. Compliance may become even more of an issue if the patient's depressive syndrome includes psychotic features with impairment of concentration, motivation, and thought organization.

4.3
General Application of TDM

Ironically, one of the arguments mounted against the use of TDM in psychiatry has been that it is "not cost-effective." When the cost argument is made, generally only the cost of TDM is considered, not the costs such an approach can save. These potential savings include (1) the cost of failed antidepressant treatment trials due to a dose that was inadequate or too high; (2) the cost of manpower and other resources utilized during a prolonged dose titration based on clinical response; and (3) the cost of treatment for adverse effects related to excessive plasma drug concentration. What typically should be a one-time expenditure for a plasma drug assay must be balanced against TDM's potential usefulness and overall cost savings during an antidepressant treatment trial. As the availability and use of these drug assays have increased, their cost has gone down. In the case of newer antidepressant agents, the cost of the TDM is now typically less than the cost of a 1-month supply of drug.

With many of the more popular antidepressants (e.g., selective serotonin reuptake inhibitors[SSRIs]), the starting dose may be the effective antidepressant dose. When patients respond to the starting dose this is clearly advantageous. However, for those patients who do not respond to the "usually effective minimum dose," the appropriate next step in therapy has not been explicitly defined (Burke and Preskorn 1998; Depression Guideline Panel 1994; Quitkin et al. 1996). Should the dose be increased or decreased? Although most clinicians routinely increase the drug dose, the phenomenon of antidepressant "nonresponders" going on to become treatment "responders" after a dose reduction is established in the psychiatric literature (Dornseif et al. 1989; Cain 1992). Likewise, it is not clear how long the patient should be maintained at the "usually effective minimum dose" before determining that the dose is ineffective (Janicak et al. 1997; Burke and Preskorn 1998). Two weeks? Four weeks or longer? In such cases, the one-time use of TDM may be the most cost-effective way to assure that the patient receives a therapeutic treatment trial (Burke and Preskorn 1999).

TDM should usually be a one-time expenditure because assaying a plasma drug concentration at steady state measures the intrinsic ability of a particular patient to clear a specific drug (i.e., essentially phenotyping the patient as to their drug clearance capability). For most antidepressants, this intrinsic ability to clear drug is a trait phenomenon related to the patient's inherited complement of hepatic enzymes, body habitus, and organ function. Hence, under routine circumstances, the results of a single drug assay at steady-state concentration can be used to predict how a subsequent change in dose will correspondingly change the drug concentration in a particular patient. However, for those drugs with nonlinear kinetics, there may be a disproportionate increase in drug concentration as the dose increases beyond a certain level (i.e., a shift from first-order to zero-order elimination kinetics) (Janicak et al. 1997). Under such circumstances, it may be useful to repeat TDM to accurately determine how a dosing change affects plasma drug concentration.

A caveat to the predictive value of one-time TDM is that there can be substantial intrapatient changes in pharmacokinetics that can affect drug concentration (Table 1). A change in the intrinsic clearance capabilities of the individual can be caused by a change in organ function secondary to the onset of disease or by the addition of concomitant medications that directly alter hepatic enzyme activity (Preskorn 1993; Jerling 1995; Preskorn and Harvey 1998). In general, when TDM is repeated in an individual patient, it should be done for a specific reason such as (1) suspected noncompliance; (2) a significant change in general health status; (3) the addition of concomitant medications where there is risk of a pharmacokinetic drug–drug interaction; or (4) a change in treatment response (i.e., latent onset of adverse effects or the loss of treatment effectiveness).

In this era of litigation, a comment regarding the medico-legal relevance of TDM is appropriate. The lack of TDM in the case of a sudden death with substantially elevated postmortem plasma drug concentrations or in the case of a suicide with substantially low postmortem drug levels can be seen as negligence. TDM provides objective evidence to substantiate the need for and safety of unusually high doses of medications or the appropriateness of unusually low doses in specific patients (Goldman et al. 1989).

5
Recommendations for TDM with Antidepressants

At the present time, data suggest that TDM may offer a clinical benefit for a number of antidepressant agents, ranging from being mandatory with some types of antidepressants because of safety issues, to being a useful and cost-effective, but yet discretionary, means of enhancing response with others. This section will review information relevant to the use of TDM in routine clinical practice for the major classes of currently marketed antidepressant drugs. The goal is to help the reader decide, for which antidepressant agents, for which patients, and under what circumstances, TDM is a practical and efficient alternative to dose titration based on clinical assessment of response. Table 5 in the Appendix summarizes optimal plasma concentration levels.

5.1
Tricyclic Antidepressants

TCAs are the most extensively studied class of antidepressants with regard to TDM. The TCAs have several features consistent with the utility of TDM, including (1) complex pharmacodynamics with multiple biological activities, and (2) substantial interpatient variability in drug clearance due in large part to their dependence on the genetically polymorphic CYP isoenzyme 2D6 for their metabolism (Bolden-Watson and Richelson 1993; Preskorn 1993; Richelson 1994; Burke and Preskorn 1998). There are now genotype data demonstrating that TCA clearance decreases progressively, and hence TCA plasma concentration increases progressively, with the number of CYP 2D6 mutated alleles that a subject exhibits (Morita et al. 2000).

As a group, the TCAs exhibit a narrow therapeutic index, with the risk of cardiac and central nervous system toxicity occurring at dosages only four to five times higher than those recommended for antidepressant efficacy (Rudorfer and Young 1980; Veith et al. 1980; Preskorn et al. 1983; Preskorn and Jerkovich 1990; Preskorn and Fast 1992). Toxicity due to TCAs can be insidious, and early signs may either not be reported by patients or may be interpreted by the clinician as a worsening of the depression. Peripheral anticholinergic effects (e.g., blurred vision, dry mouth, constipation) are not well correlated with TCA-induced central nervous system toxicity and cannot be used as a marker to estimate plasma drug concentration reliably (Preskorn and Jerkovich 1990).

The usefulness of TDM and the potential lethality of TCAs have been underscored by numerous case reports of sudden death in which the autopsy ruled out acute over-ingestion and revealed no cause of death beyond sudden cardiac arrest and enormously elevated TCA plasma and tissue levels. In many of these cases, suits have been successfully brought to trial for failure to use TDM to avoid such fatal outcomes (Preskorn et al. 1989).

5.1.1
Concentration:Response Relationship

Despite the inherent difficulty in correlational studies, over the years there have been numerous successful attempts to identify a relationship between concentration and response for the TCAs. The strongest relationship between plasma concentration and therapeutic response has been demonstrated for nortriptyline (NT), desipramine (DMI), imipramine (IMI), and amitriptyline (AT). These data have been reviewed and summarized previously (Preskorn et al. 1993; Goodnick 1994; Burke and Preskorn 1999). In general, the optimal plasma concentration for therapeutic response is 100–300 ng/ml for tertiary-amine tricyclics and 50–150 ng/ml for secondary-amine tricyclics. Although specific target concentrations are not available, the lower end of the concentration range could be interpreted as the minimum effective concentration for most patients.

Studies of NT have demonstrated a curvilinear concentration:response relationship with an optimum range of 50-150 ng/ml (Goodnick 1994; Perry et al. 1987). Within this range, 70% of patients with primary major depressive disorder experience complete remission (e.g., a final score on the Hamilton Rating Scale for Depression [Hamilton 1960] equal to or less than 6) versus 29% of patients who develop TCA plasma concentrations outside this range. DMI studies demonstrate a curvilinear concentration:antidepressant response relationship similar to the one observed with NT. The "therapeutic window" for DMI derived from meta-analysis is 100-160 ng/ml (Perry et al. 1987; Goodnick 1994). There is a remission rate of 59% in this range versus 20% outside it. Of note, for the secondary-amine TCAs (i.e., NT, DMI) the response rate is generally higher at the lower end of this range than at the upper limit, suggesting that the common tendency to "push" the dose in the case of non-response may not be the best strategy with these agents.

The results for tertiary-amine tricyclic agents are less robust in terms of concentration:antidepressant response. However an optimal range can be discerned from a consideration of concentration:efficacy data and concentration:toxicity data. The optimum plasma level range for AT efficacy is 75-175 ng/ml. For a person taking AT, this concentration range refers to the sum of AT and its active metabolite NT. The remission rate is 48% within this range versus 29% outside it (Perry et al. 1987; Goodnick 1994). Based on meta-analysis, the threshold for optimum antidepressant response to IMI was close to the threshold for central nervous system and cardiac toxicity. Thus, the upper limit of the therapeutic range appears to be a function of toxicity rather than reduced efficacy as it is with the secondary amine TCAs (e.g., NT, DMI). At present, the threshold concentration proposed for patients taking IMI is 265 ng/ml with a remission rate of 42% above this threshold versus 15% below it. The concentration refers to the sum of IMI and its active metabolite DMI. Studies of clomipramine have identified a positive relationship between plasma concentration and clinical antidepressant response, with half reporting a linear relationship and the other half a curvilinear relationship. Clomipramine does have dose-dependent toxicity and this likely accounts for the upper limits of the "therapeutic" plasma range. Taken together, the data suggest an optimal antidepressant response when the combined clomipramine/desmethylclomipramine concentration is greater than 175 ng/ml and less than 400 ng/ml (Goodnick 1994).

5.1.2
Recommendation

The concentration:response data for the TCAs discussed above support the use of TDM at least once as a routine aspect of TCA therapy for major depressive disorder. TDM directed dosing of the TCAs can increase the likelihood of a successful antidepressant response two- to threefold, compared to use of a fixed dose, and can reduce the likelihood of experiencing serious toxicity (Perry et al. 1987). Because of their dependence on CYP 2D6 for metabolism, repeat TDM may be indicated when additional medications are initiated that may have affects on CYP 2D6 activity and hence alter plasma TCA concentration (Preskorn and Harvey 1998).

5.2
Serotonin Selective Reuptake Inhibitors

At the present time, most physicians consider the SSRIs to be the first-line pharmacotherapy for treating depression (Burke and Preskorn 1998; Olfson and

Klerman 1993). Any approach (e.g., TDM) that could improve the response rate for patients treated with these agents would therefore be of great interest. As a class, the SSRIs do exhibit a number of features that suggest the utility of TDM (Table 3). There is a delay in onset of the antidepressant response, a wide interpatient variability in the dose:concentration relationship, and data that suggest concentration:response relationships for drugs in the SSRI class.

Tolerability problems with the SSRIs are dose- and hence concentration-dependent. These include nausea, restlessness, tremors, and insomnia, all of which can lead to premature treatment discontinuation. Moreover, some of the adverse effects of the SSRIs can look like a worsening depressive disorder, which can complicate dose titration based on clinical assessment of response. This latter observation suggests that TDM of SSRIs could be a cost-effective strategy, even if used only in patients who do not respond to an initial trial of the usually effective minimum dose. However, TDM is not a standard of care issue with SSRIs due to their wide therapeutic index. No serious systemic toxicity occurs with the SSRIs even when taken in substantial overdose (Wernicke et al. 1987; Beasley et al. 1991; Dunner and Dunbar 1992; Carraci 1994).

5.2.1
Concentration:Response Relationship

A provocative feature of the SSRIs, based on the results of fixed-dose studies, is that the response rate, on average, does not increase at doses above the "minimum effective dose." This feature has been partly responsible for the popularity of the SSRIs, because of the suggestion that, unlike with other antidepressant agents, dose titration was unnecessary and the starting dose was also the effective dose. In double-blind, controlled studies of patients suffering from major depressive disorder, daily doses of 20, 40, and 60 mg of fluoxetine produced comparable remission (Altamura et al. 1988; Wernicke et al. 1988; Schweizer et al. 1990). Similarly, no difference was found in the overall remission rates in patients treated with 20, 30, or 40 mg/day of paroxetine or 50, 100, 150, or 200 mg/day of sertraline (Tasker et al. 1990; Dunner and Dunbar 1992; Fabre et al. 1995; Preskorn 1997). Despite these data that support a "flat dose:response" for SSRIs, there are clearly individual patients who only benefit from a dose other than the usually effective minimum dose. It is plausible that the need for a higher dose in a specific patient is related to pharmacokinetics (i.e., lower plasma concentrations of drug achieved at a given dose), but systematic TDM studies have not been carried out to explore this issue.

The degree of serotonin reuptake inhibition does correlate with SSRI concentration and, hence, the antidepressant effect related to this mechanism of action must be concentration-dependent as well (Preskorn 1993). However, it is likely that the "flat dose:response curve" of the SSRIs is due to the fact that 70%–80% of serotonin reuptake sites are inhibited at the "usually effective minimum dose" and little additional uptake is inhibited as the dose is increased (Wood et al. 1983; Lemberger et al. 1985; Marsden et al. 1987; Preskorn and Harvey 1996).

These data suggest the possibility of a threshold effect for the SSRIs, reminiscent of that of the monoamine oxidase inhibitors (MAOIs), for which inhibition of 70%–80% of monoamine oxidase (MAO) activity is necessary for the optimal antidepressant effect to occur, while further enzyme inhibition beyond that appears to be of limited additional value for most patients.

The reader is referred to Rasmussen and Brosen (2000) for a review of concentration:response studies of SSRIs. The "flat dose:response curve" of the SSRIs may partly explain why a number of attempts using "routine doses" have failed to identify a consistent relationship between SSRI plasma levels and antidepressant efficacy (Kelly et al. 1989; Preskorn et al. 1991; Goodnick 1994; Amsterdam et al. 1997). Prior, failed attempts to find a consistent relationship between SSRI concentration and antidepressant efficacy do not prove that one does not exist. Simple correlation is a very limited method of discovering a relationship between drug concentration and response, especially when it involves complex factors such as enantiomeric forms of the drug with differing biological activity, delayed time to achieve steady-state plasma/tissue drug concentration, delayed onset of clinical response, and interpatient variability in pharmacokinetic parameters (Koran et al. 1996).

Identifying concentration:response relationships for antidepressant agents is particularly problematic because there is an inherent signal-to-noise problem for antidepressant efficacy due to lack of objective measures of illness, a high placebo response rate, and diverse underlying psychopathology (e.g., substance abuse, Axis II disorders). If the study, like most clinical trials, is performed with outpatients, drug compliance and timing of plasma sampling can add considerable error to the interpretation of drug concentration determinations. In the case of fluoxetine and norfluoxetine, considerable time is required to achieve steady-state concentrations in the central nervous system, and failure to account for plasma:tissue equilibration time can obscure the relationship between plasma drug concentration and clinical response (Renshaw et al. 1992; Greenblatt and Harmatz 1993; Karson et al. 1993).

Even in the absence of an ideal correlational study, data from the double-blind, fixed-dose clinical trials provide an expected plasma drug concentration range for the usually effective dose of each SSRI. Since this dose on average separates drug treatment from placebo, the concentration that on average is achieved by this dose must also separate the drug from placebo. This concentration then is an estimate of the minimum concentration of the SSRI necessary to produce an antidepressant effect. Therapeutic plasma levels of three SSRIs derived in this way are 10–50 ng/ml of sertraline on 50 mg/day (Preskorn et al. 1994; Ronfeld et al. 1997), 70–120 ng/ml of paroxetine on 20 mg/day (Lund et al. 1979; Bayer et al. 1989; Hebenstreit et al. 1989; Lundmark et al. 1989; Sindrup et al. 1992), and 120-300 ng/ml of combined fluoxetine and norfluoxetine on 20 mg/day (Feighner and Cohn 1985; Lemberger et al. 1985; Goodnick 1991; Preskorn et al. 1991, 1994; Goodnick 1992). Mean plasma concentrations associated with the usually effective dose of the other SSRIs are 85 ng/ml for citalopram (40 mg/day) and 100 ng/ml for fluvoxamine (150 mg/day) (Kragh-

Sorenson et al. 1981; Fredricson 1982; Bjerkenstedt et al. 1985; Folgia et al. 1990; Nathan et al. 1990; Rochat et al. 1995).

5.2.2
Recommendation

Due to their wide therapeutic index, there is no compelling reason to monitor plasma levels of any of the SSRIs as a standard of care to prevent toxicity. However, because of the broad variability in plasma SSRI concentrations achieved at a routine dose, TDM is useful for individual dose optimization (Lundmark et al. 2000, 2001). Approximately 50% of patients on any single dose of an SSRI do not achieve an optimum response in terms of relief of their depressive episode (Preskorn 1996a). For these patients, TDM can provide important objective data to rule out noncompliance and identify those individuals who are pharmacokinetic outliers (i.e., have particularly slow or rapid drug clearance). In the "nonresponder" on the usually effective minimum dose of an SSRI, a determination of drug concentration can be used to direct treatment decisions (i.e., low drug concentration and lack of effectiveness, high drug concentration and poor tolerability) and reduce the delay in getting the patient on an optimal dose. Another unique use of TDM with the SSRI fluoxetine would be to determine if its metabolite, norfluoxetine, is still present after drug discontinuation. Such plasma monitoring would help the clinician decide when it is safe to start treatment with another agent and avoid a drug–drug interaction (e.g., when switching to an MAOI or to a substrate for one of the several drug-metabolizing enzymes inhibited by fluoxetine).

5.3
Bupropion

Bupropion undergoes extensive first-pass hepatic metabolism. The drug is converted into three metabolites that accumulate in concentrations several times that of the parent compound (Cooper et al. 1984; Perumal et al. 1986; Preskorn and Katz 1989; Goodnick 1991a). The three major known metabolites are: hydroxybupropion (HB), threobupropion (TB) and erythrobupropion (EB). The HB and TB metabolites are pharmacologically active, with half the potency of the parent compound, in animal model tests for antidepressant activity (Cooper et al. 1984; Perumal et al. 1986; Preskorn and Katz 1989; Goodnick 1991). There is considerable interindividual variability in plasma levels of bupropion and these metabolites (Preskorn et al. 1990).

A focus of concern with bupropion has been the risk of seizure, which is estimated at 4 per 1,000 at a dose of 450 mg/day and increases exponentially with higher doses (Davidson 1989). Several observations (Davidson 1989; Preskorn et al. 1990) suggest that the risk of seizures may be due to excessively high levels of bupropion or its metabolites: (1) that the incidence of seizures is dose related, and hence must be concentration related; (2) that seizures typically occur within days of a dose change and a few hours after the last dose, suggesting an effect due to peak plasma concentrations; and (3) that individuals with increased lean body mass such as anorexic patients are at an increased risk for seizures. Case report data link excessive levels of bupropion and its metabolites to other forms of CNS toxicity including psychosis, delirium, and extrapyramidal syndromes (Preskorn and Katz 1989; Preskorn et al. 1990).

5.3.1
Concentration:Response Relationship

In spite of the complicated pharmacokinetics of bupropion, a number of studies have examined the relationship between bupropion concentration and antidepressant efficacy. In most cases, the concentration:response studies of bupropion have assayed only plasma levels of the parent compound. These studies demonstrate an optimal antidepressant response occurring at plasma levels below 100 ng/ml of the parent compound, with patients appearing to do better in the 10–50 ng/ml range than at higher levels (Preskorn et al. 1990; Goodnick 1991a; Goodnick 1992). In a study by Golden et al. (1988), plasma concentrations of bupropion and all three of its metabolites were assayed, and it was found that the patients with higher metabolite levels did not do as well as those with lower levels. Taken together, the results of these concentration:response studies are surprisingly consistent: (1) there is a better response at lower (10–50 µg/ml) rather than higher plasma levels of the parent drug, bupropion; and (2) higher levels of the metabolites (HB >1200 ng/ml, TB >400 ng/ml, and EB >90 ng/ml) are associated with a poorer response. Clearance of the HB metabolite appears to be dependent on CYP 2D6. Patients phenotypically determined to be "poor metabolizers" (i.e., to have deficient CYP 2D6 activity) have been shown to accumulate high plasma concentrations of HB, which may increase the risk of toxicity and decrease antidepressant efficacy (Pollock et al.1996).

5.3.2
Recommendations

The available data support the conclusion that TDM would be likely to increase the safe and effective use of bupropion. TDM appears particularly applicable to bupropion since the risk of central nervous system toxicity is associated with the use of higher doses, and effectiveness occurs at lower rather than higher plasma levels of bupropion and its metabolites. However, TDM is not yet routinely used with bupropion, perhaps because of the limited database or the limited appreciation of its potential utility. In cases in which patients have other than the desired response to the usually effective doses of bupropion, either lack of efficacy or treatment-limiting adverse effects, plasma level monitoring would be a reasonable means to optimize dosing. TDM is particularly suited for preventing potential drug interactions and confirming safety in patients who are taking bupropion along with drugs that are known to inhibit CYP 2D6 activity.

5.4
Venlafaxine

Venlafaxine has linear pharmacokinetics over its clinical dosing range and is O-demethylated by CYP 2D6 to its major, active metabolite, O-desmethylvenlafaxine (Haskins et al. 1985; Klamerus et al. 1996). There are data that support an ascending dose:response and hence concentration:response relationship for venlafaxine (Harvey and Preskorn 2000). Patients with mutated CYP 2D6 alleles (i.e., poor metabolizers), or who are taking drugs that inhibit CYP 2D6 activity can develop significant increases in plasma venlafaxine concentration (i.e., 300%–400% increase) (Eap et al. 2000; Fukuda et al. 2000). As the plasma concentration of venlafaxine increases, associated norepinephrine reuptake inhibition becomes more prominent, and the risk of adverse signs of autonomic arousal (e.g., hypertension, tachycardia, diaphoresis, tremor) increases. These pharmacological and clinical features suggest the potential usefulness of TDM in optimizing the antidepressant response to venlafaxine.

Precise, selective assays for venlafaxine and its active metabolite have been successfully developed in recent years using high-performance liquid chromatography (Hicks et al. 1994; Vu et al. 1997). Preliminary data based on these assays suggest an optimal plasma concentration range from 195–400 ng/ml for therapeutic response with venlafaxine (Veefkind et al. 2000). The concentration range refers to the sum of venlafaxine and its active metabolite, O-desmethylvenlafaxine. It should be noted that, in this fixed-dose study by Veefkind et al (2000), "nonresponders" tended to have higher plasma drug concentrations, while the "responders" had higher ratios of metabolite to parent compound. Although O-desmethylvenlafaxine is believed to have comparable pharmacodynamics to the parent compound, this finding suggests that the metabolite or ratio of metabolite to parent compound may play some therapeutic role. More research is clearly needed; however, at present a case can be made in favor of using TDM for those patients who are not responding optimally to venlafaxine at the usually effective minimum doses.

5.5
Nefazodone and Trazodone

Nefazodone and trazodone have multiple effects on serotonin transmission in the central nervous system. Both drugs have a short half-life, nonlinear pharmacokinetics, and are biotransformed to several active metabolites (Shukla et al. 1992; Preskorn 1993; Goodnick 1994; Greene and Barbhaiya 1997). All these pharmacological features pose difficulties in trying to identify a meaningful relationship between plasma drug concentration and clinical outcome (e.g., modest differences in sampling time or compliance could substantially affect plasma drug concentration). Hence, it is not surprising that little is known about the relationship between plasma concentration and response for trazodone and nefazodone. There is no recommendation for the clinical application of TDM with these agents at this time. A case can be made that TDM research is particularly needed for drugs like nefazodone that have nonlinear kinetics and multiple active metabolites that make dose:concentration and concentration:response relationships clinically difficult to predict. Recently, plasma assays for nefazodone and its metabolites have been successfully developed that are sensitive, selective, and precise (Yao et al. 1998; Dodd et al. 1999; Yao and Srinivas 2000). Note that the average levels of bupropion, hydroxybupropion, erythrohydrobupropion, and threohydrobupropion achieved on 450 mg/day are 33, 1452, 138, and 671, respectively. Thus, the total for parent drug plus metabolites is over 2,000 ng/ml (2 ug/ml) (Preskorn and Katz 1989; Preskorn 2000). However, the use of these assays is at present confined to the research lab.

5.6
Mirtazapine

Mirtazapine is the most recent, novel addition to the antidepressant armamentarium. It has multiple metabolites, exhibits linear pharmacokinetics over its clinically relevant dosing range, and has a broad therapeutic index (Preskorn 1997). At present, because of the limited database (i.e., there are no published fixed-dose:response studies), there is no recommendation for the use of TDM with mirtazapine.

5.7
Monoamine Oxidase Inhibitors

Three MAOIs are marketed in the United States with an approved indication for the treatment of depression: phenelzine, isocarboxazid, and tranylcypromine. All three drugs irreversibly inhibit both subtypes of MAO (i.e., types A and B). These drugs are rapidly cleared from the plasma with a short half-life of 1–2 h (Goodnick 1994). They act as substrate for MAO, forming covalent bonds with the enzyme and thereby irreversibly inactivating it. Thus, the pharmacokinetics of this enzyme-inhibiting class of antidepressants is unique among psychotropic drugs in that the drug is rapidly cleared from the plasma but the biological effects persist. The most serious adverse effects of MAOIs are related to hypermetabolic states (i.e., hypertensive crisis, serotonin syndrome). These effects are related to the decreased MAO activity. A clear relationship between MAOI levels and the generally nuisance side effects (e.g., orthostatic hypotension, weight gain, sexual dysfunction, myoclonus) has not been established.

5.7.1
Concentration:Response Relationship

Given their rapid elimination and the fact that their presumed mechanism of action is irreversible, it is not surprising that a relationship between plasma concentration of the MAOIs and outcome has been difficult to demonstrate. In the case of the MAOIs, there is a dissociation between the effect and concentration (i.e., the magnitude of effect will persist even after the plasma drug concentration has fallen substantially, or even completely disappeared). For this reason, conventional TDM (i.e., measuring circulating concentrations of the drug) is not applicable to MAOIs. Yet, the pharmacodynamics of this class of drugs actually provides an opportunity to make a measurement that is potentially closer to an ideal measurement, namely measurement of the biological effect of the drugs: MAO inhibition. While the ideal measurement would be MAO activity in the brain, that measurement is currently impractical. Instead, platelet MAO activity has been used as a surrogate marker for central neuronal MAO activity. The use of this surrogate endpoint is based on the correlation between the degree of platelet and neuronal MAO inhibition. Optimal antidepressant efficacy has been correlated with approximately 80% inhibition of platelet MAO (Preskorn and Burke 1992).

5.7.2
Recommendations

TDM as it applies to MAOIs is not measuring plasma drug concentration but rather the drug effect (i.e., inhibition of platelet MAO activity) . The platelet assay for MAOI activity is cumbersome, expensive, and not always readily available. Two samples, rather than one, are needed, and to assure accuracy, the plasma sample must be handled in a more fastidious way than samples for measuring drug levels. The first sample determines basal enzyme activity, and the second sample is used to assess the degree of enzyme inhibition. These facts, coupled with the infrequent use of MAOIs, have limited the application of this approach despite the fact that monitoring enzyme inhibition may permit more optimal dosing.

6
Summary

Although the focus in administering antidepressant pharmacotherapy is dose, it is the concentration of a drug at the site of action that determines the clinical effect. Because of interpatient variance in pharmacokinetics, dose often falls short of predicting the concentration of drug that is achieved (i.e., higher than expected drug levels leading to toxicity, lower than expected drug levels that are without benefit). In treating a condition like depression, dose titration typically occurs without the benefit of objective measures of response. Interpreting treatment response can be confounded when adverse effects of antidepressant pharmacotherapy mimic a worsening of the depressive syndrome. TDM is a simple way to eliminate pharmacokinetic variability from the treatment equation. It provides rational, objective data to guide the clinician in optimizing the drug dose and the treatment trial. Despite the precedent found in other medical disciplines and decades of scientific support, TDM is underappreciated and underutilized in psychiatry.

The facts are (1) that there is broad variability in plasma drug concentrations achieved at a given antidepressant dose, (2) that this variability in drug concentration has clinical relevance to both efficacy and toxicity, and (3) that the antidepressant concentration achieved by a given patient taking a given drug dose cannot be reliably predicted by clinical assessment. The question is how best to account for pharmacokinetic variability so that antidepressant treatment can be optimized. With the recent increase in attention to how pharmacogenetic factors regulate drug metabolism, it has been suggested that general phenotyping or genotyping screens for hepatic enzyme activity be used to optimize antidepressant treatment (Tucker 2000). We shall see what the future will bring. For now, traditional TDM provides targeted, economical phenotyping that is directly relevant to the antidepressant treatment trial. In a sense, traditional TDM is actually a broader index of drug handling by the patient in that it takes into account drug absorption and elimination variables, whereas genotyping only assesses metabolic capabilities.

In the case of TCAs, a drug class characterized by a narrow therapeutic index, TDM has become a standard of care issue, guiding the dosing process to avoid serious toxicity and increasing the likelihood of an optimal response. With the other popular classes of antidepressants, which tend to be characterized by a broad therapeutic index, a case can be made that TDM can provide meaningful data to guide dose optimization. Does this mean that all persons started on antidepressant medication should have TDM? This is probably not necessary, since a good number of persons treated for depression respond well to the starting antidepressant dose. However, for those patients who do not respond to the usual effective dose of an antidepressant, TDM can suggest a reason for the nonresponse (e.g., too much drug, too little drug) and provide an alternative to a lengthy trial and error, dose titration process. As illustrated in Fig. 1, TDM is used principally as a means to guide dosing. However, when an outlier is identified (i.e., a patient with an unanticipated high or low drug concentration), the clinician has an objective basis for pursuing other issues related to treatment such as compliance or a suspected rare genotype (Dahl and Sjoqvist 2000).

Inherent in the clinical goals of TDM is an increase in the efficiency with which healthcare resources are utilized. For an illness like clinical depression, which represents a significant public and private economic burden, any measure that may hasten an optimal clinical outcome should be considered. In the treatment of clinical depression, TDM is often a one-time expenditure, the cost of which is typically less than a 1-month supply of medication. The low cost of TDM relative to (1) drug acquisition cost, (2) the cost of higher intensity clinical

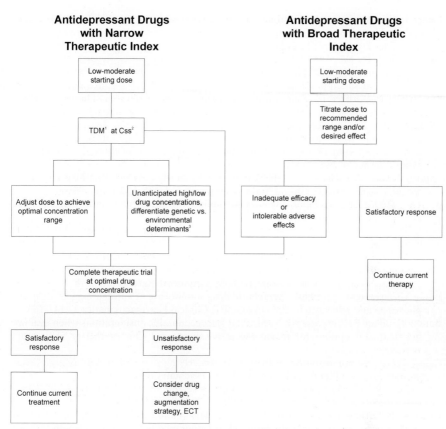

Fig. 1 Algorithm for recommended use of TDM in antidepressant pharmacotherapy. [1]TDM=therapeutic drug monitoring; [2]Css=steady-state plasma concentration; [3]genotype for CYP isoenzyme polymorphism, consider possible drug interaction, compliance issues, impaired organ function (copyright Burke and Preskorn 2003)

services during a prolonged dose titration, and (3) the potential consequences of a failed treatment trial (e.g., suicide attempts, accidents, unemployment) suggests that this clinical test more than pays for itself by providing objective data for rational dose optimization (Burke et al. 1994; Burke 1996).

References

Altamura A, Montgomery S, Wernicke J (1988) The evidence for 20 mg a day of fluoxetine as the optimal in the treatment of depression. Br J Psychiatry 153(suppl 3):109–112

Amsterdam JD, Fawcett J, Quitkin FM, Reimherr FW, Rosenbaum JF, Michelson D, Hornig-Rohan M, Beasley CM (1997) Fluoxetine and norfluoxetine plasma concentrations in major depression: a multicenter study. Am J Psychiatry 154:963–969

Balant-Gorgia AE, Balant LP, Garrone G (1989) High blood concentrations of imipramine and clomipramine and therapeutic failure: a case report study using drug monitoring data. Ther Drug Monit 11:415–20

Bayer AJ, Roberts NA, Allen EA, Horan M, Routledge PA, Swift CG, Byrne MM, Clarkson A, Zussman BD (1989) The pharmacokinetics of paroxetine in the elderly. Acta Psychiatr Scand 80(suppl 350):85–86

Beasley CM Jr, Dornseif BE, Pultz JA, Bosomworth JC, Sayler ME (1991) Fluoxetine versus trazodone: Efficacy and activating sedating effects. J Clin Psychiatry 52:294–9

Bjerkenstedt L, Flyckt L, Overo KF, Lingjaerde O (1985) Relationship between clinical effects, serum drug concentration and serotonin uptake inhibition in depressed patients treated with citalopram. a double-blind comparison of three dose levels. Eur J Clin Pharmacol 28:553–557

Bolden-Watson C, Richelson E (1993) Blockade by newly-developed antidepressants of biogenic amine uptake into rat brain synaptosomes. Life Sci 52:1023–1029

Brosen K (1996) Drug-metabolizing enzymes and therapeutic drug monitoring in psychiatry. Ther Drug Monit 18:393–396

Burke MJ (1996) The search for value: issues in the treatment of depression. J Pract Psychiatry Behav Health 2:2–13

Burke MJ, Preskorn SH (1998) Short-term treatment of mood disorders with standard antidepressants. In: Watson SJ (ed) Psychopharmacology: The Fourth Generation of Progress, Version 2, 1998 Edition CD-ROM. Lippincott-Raven, New York

Burke MJ, Preskorn SH (1999) Therapeutic drug monitoring of antidepressants: cost implications and relevance to clinical practice. Clinical Pharmacokinetics 37:147-165

Burke MJ, Silkey B, Preskorn SH (1994) Pharmacoeconomic considerations when evaluating treatment options for major depressive disorder. J Clin Psychiatry 55(9, suppl A):42–52

Cain JW (1992) Poor response to fluoxetine: underlying depression, serotonergic overstimulation, or a therapeutic window? J Clin Psychiatry 53:272–277

Carraci G (1994) Unsuccessful suicide attempt by sertraline overdose (letter). Am J Psychiatry 151:147

Cooper TB, Suckow RF, Glassman A (1984) Determination of bupropion and its major basic metabolites in plasma by liquid chromatography with dual wavelength U.V. detection. J Pham Sci 73:1104–1107

Dahl M, Sjoqvist F (2000) Pharmacogenetic methods as a complement to therapeutic monitoring of antidepressants and neuroleptics. Ther Drug Monit 22:114–117

Davidson J (1989) Seizures and bupropion: a review. J Clin Psychaitry 50:256–261

Depression Guideline Panel of the Agency for Health Care Policy and Research (1994) Synopsis of the Clinical Practice Guidelines for Diagnosis and Treatment of Depression in Primary Care. Arch Fam Med 3:85–92

Dodd S. Buist A, Burrows GD, Maguire KP, Norman TR (1999) Determination of nefazodone and its pharmacologically active metabolites in human blood plasma and breast milk by high-performance liquid chromatography. J Chromatogr B Biomed Sci Appl 730:249–255

Dornseif BE, Dunlop SR, Potvin JH, Wernicke JF (1989) Effect of dose escalation after low-dose fluoxetine therapy. Psychopharmacol Bull 25:71–79

Dunner DL, Dunbar GC (1992) Optimal dose regimen for paroxetine. J Clin Psychiatry 53(suppl 2):21–26

Eap CB, Bertel-Laubscher R, Zullino D, Amey M, Baumann P (2000) Marked increase of venlafaxine enantiomer concentration as a consequence of metabolic interactions: a case report. Pharmacopsychiatry 33:112–115

Fabre LF, Abuzzahab FS, Amin M, Claghorn JL, Mendels J, Petrie WM, Dube S, Small JG (1995) Sertraline safety and efficacy in major depression: a double-blind fixed dose comparison with placebo. Biol Psychiatry 38:592–602

Feighner JP, Cohn JB (1985) Double blind comparative trials of fluoxetine and doxepin in geriatric patients with major depressive disorder. J Clin Psychiatry 46:20–25

Folgia JP, Perel JM, Nathan RS, Pollock BG (1990) Therapeutic drug monitoring (TDM) of fluvoxamine, a selective antidepressant (abstract). Clin Chem 36:1043

Fredricson OK (1982) Kinetics of citalopram in man: plasma levels in patients. Prog Neuropsychopharmacol Biol Psychiatry 6:311–318

Fukuda T, Nishida Y, Zhou Q, Yamamoto I, Kondo S, Azuma J (2000) The impact of the CYP2D6 and CYP2C19 genotypes on venlafaxine pharmacokinetics in a Japanese population. Eur J Clin Pharmacol 56:175–178

Glotzbach RK, Preskorn SH (1982) Brain concentrations of tricyclic antidepressants: single dose kinetics and relationship to plasma concentration in chronically dosed rats. Psychopharmacology 78:25–27

Golden RN, DeVane CL, Laizure SC, Rudorfer MV, Sherer MA, Potter WZ (1988) Bupropion in depression: the roles of metabolites in clinical outcome. Arch Gen Psychiatry 45:145–149

Goldman DL, Katz SE, Preskorn SH (1989) What to do about extremely high plasma levels of tricyclics? Am J Psychiatry 146:401–402

Goodnick PJ (1991a) Pharmacokinetics of second generation antidepressants: bupropion. Psychopharmacol Bull 27:513–519

Goodnick PJ (1991b) Pharmacokinetics of second generation antidepressants: fluoxetine. Psychopharmacol Bull 27:503–512

Goodnick PJ (1992a) Blood levels and acute response to bupropion. Am J Psychiatry 149:399–400

Goodnick PJ (1992b) Fluoxetine blood levels and clinical response. Biol Psychiatry 31:186A

Goodnick PJ (1994) Pharmacokinetic optimization of therapy with newer antidepressants. Clin Pharmacokinet 27:307–330

Greenblatt DJ, Harmatz JS (1993) Kinetic-dynamic modeling in clinical psychopharmacology. J Clin Psychopharmacol 13:231–234

Greene D, Barbhaiya R (1997) Clinical pharmacokinetics of nefazodone. Clin Pharmacokinet 33:260–275

Guthrie S, Lane EA, Linnoila M (1987) Monitoring of plasma drug concentrations in clinical psychopharmacology. In: Meltzer HY (ed) Psychopharmacology: the third generation of progress. Raven, New York, pp 1323–1338

Hamilton M (1960) A rating scale for depression. J Neurol Neurosurg Psychiatry 23:56–62

Harvey AT, Rudolph RL, Preskorn SH (2000) Evidence of the dual mechanisms of action of venlafaxine in healthy male volunteers. Arch Gen Psychiatry 57:503–509

Haskins JT, Moyer JA, Muth EA, Sigg EB (1985) DMI, WY-45,030, WY-45,881 and ciramadol inhibit locus ceruleus neuronal activity. Eur J Pharmacol 115:139–146

Hebenstreit GF, Fellerer K, Zochling R, Zentz A, Dunbar GC (1989) A pharmacokinetic dose titration study in adult and elderly depressed patients. Acta Psychiatr Scand 80(suppl 350):81–84

Henry ME, Moore CM, Kaufman M, Michelson D, Schmidt ME, Stoddard E, Vuckevic AJ, Berreira PJ, Cohen BM, Renshaw PF (2000) Brain kinetics of paroxetine and fluoxetine on the third day of placebo substitution: A fluorine MRS study. Am J Psychiatry 157:1506–1508

Hicks DR, Wolaniuk D, Russell A, Cavanaugh N, Kraml M (1994) A high-performance liquid chromatographic method for the simultaneous determination of venlafaxine and O-desmethylvenlafaxine in biological fluids. Ther Drug Monit 16:100–107

Janicak PG, Davis JM, Preskorn SH, Ayd FJ (1997) Principles and practice of psychopharmacotherapy, 2nd edition. Williams & Wilkins, Baltimore, pp 61–82

Jerling M (1995) Dosing of antidepressants: the unknown art. J Clin Psychopharmacol 15:435–39

Karson CN, Newton JE, Livingston R, Jolly JB, Cooper TB, Sprigg J, Komoroski RA (1993) Human brain fluoxetine concentrations. J Neuropsychiatry Clin Neurosci 5:322–329

Kelly MW, Perry PJ, Holstad SG, Garvey MJ (1989) Serum fluoxetine and norfluoxetine concentration and antidepressant response. Ther Drug Monit 11:165–170

Klamerus KJ, Maloney K, Rudolph RL, Sisenwine SF, Jusko WJ, Chiang ST (1996) Introduction of a composite parameter to the pharmacokinetics of venlafaxine and its active O-desmethyl metabolite. J Clin Pharmacol 32:716–724

Koran LM, Cain JW, Dominguez RA, Rush AJ, Thiemann S(1996) Are fluoxetine plasma levels related to outcome in obsessive compulsive disorder? Am J Psychiatry 153:1450–1454

Kragh-Sorenson P, Overo KF, Peterson OL, Jensen K, Parnas W (1981) The kinetics of citalopram: single and multiple dose studies in man. Acta Pharmacol Toxicol 48:53–60

Lemberger L, Bergstrom RF, Wolen RL, Farid NA, Enas GG, Aronoff GR (1985) Fluoxetine: clinical pharmacology and physiologic disposition. J Clin Psychiatry 46(3, sec 2):14–19

Ley P (1981) Satisfaction, compliance and communication. Br J Clin Psychol 21:241–4

Lin EH, Von Korff M, Katon W, Bush T, Simon GE, Walker E, Robinson P (1995) The role of the primary care physician in patient's adherence to antidepressant therapy. Med Care 33:67–74

Lund J, Lomholt B, Fabricius J, Christensen JA, Bechgaard E (1979) Paroxetine: pharmacokinetics, tolerance and depletion of blood 5-HT in man. Acta Pharmacol Toxicol 44:289–295

Lundmark J, Scheel K, Thomsen I, Fjord-Larsen T, Manniche PM, Mengel H, Moller-Nielsen EM, Pauser H, Walinder J (1989) Paroxetine: pharmacokinetics and antidepressant effect in the elderly. Acta Psychiatr Scand 80(suppl 350):76–80

Lundmark J, Reis M, Begtsson F (2000) Therapeutic drug monitoring of sertraline: variability factors as displayed in a clinical setting. Ther Drug Monit 22:446–454

Lundmark J, Reis M, Begtsson F (2001) Serum concentration of fluoxetine in the clinical treatment setting. Ther Drug Monit 23:139–147

Marsden CA, Tyrer P, Casey P, Seivewright N (1987) Changes in human whole blood 5-hydroxytrypamine (5-HT) and platelet 5-HT uptake during treatment with paroxetine, a selective 5-HT uptake inhibitor. J Psychopharmacol 1:244–250

Martinez D, Hwang D, Mawlawi O, Slifstein M, Kent J, Simpson N, Parsey RV, Hashimoto T, Huang Y, Shinn A, Van Heertum R, Abi-Dargham A, Caltabiano S, Malizia A, Cowley H, Mann JJ, Laruelle M (2001) Differential occupancy of somatodendritic and postsynaptic 5HT1A receptors by pindolol: A dose-occupancy study with (11C)WAY 100635 and positron emission tomography in humans. Neuropsychopharmacology 24:209–229

Morita S, Shimoda K, Someya T, Yoshimura Y, Kamijima K, Kato N (2000) Steady-state plasma levels of nortriptyline and its hydroxylated metabolites in Japanese patients: Impact of CYP2D6 genotype on the hydroxylation of nortriptyline. J Clin Psychopharmacol 20:141–149

Nathan RS, Perel JM, Pollock BG, Kupfer DJ (1990) The role of neuropharmacologic selectivity in antidepressant action: fluvoxamine versus desipramine. J Clin Psychiatry 51:367–372

Olfson M, Klerman GL (1993) Trends in the prescription of antidepressants by office-based psychiatrists. Am J Psychiatry 150:571–577

Perry PJ, Pfohl BM, Holstad SG (1987) The relationship between antidepressant response and tricyclic antidepressant plasma concentrations. Clin Pharmacokinet 13:381–392

Perumal AS, Smith TM, Suckow RF, Cooper TB (1986) Effects of plasma from patients containing bupropion and its metabolites on the uptake of norepinephrine. Neuropharmacology 25:199–202

Pollock B, Sweet R, Kirshner M, Reynolds C (1996) Bupropion plasma levels and CYP2D6 phenotype. Ther Drug Monit 18:581-585
Preskorn SH (1991) Should bupropion dosage be adjusted based upon therapeutic drug monitoring? Psychopharmacol Bull 27(4):637-643
Preskorn SH (1993) Pharmacokinetics of antidepressants: why and how are they relevant to treatment. J Clin Psychiatry 54(9, suppl):14-34
Preskorn SH (1996a) Clinical pharmacology of selective serotonin reuptake inhibitors. Professional Communications, Caddo, OK, pp 99-105
Preskorn SH (1996b) Why did Terry fall off the dose-response curve? J Pract Psychiatry Behav Health 2:39-43
Preskorn SH (1997a) Clinically relevant pharmacology of the selective serotonin reuptake inhibitors: an overview with emphasis on pharmacokinetics and effects on oxidative drug metabolism. Clin Pharmacokinet 32(suppl 1):1-21
Preskorn SH (1997b) Selection of an antidepressant: Mirtazapine. J Clin Psychiatry 58(suppl 6):3-8
Preskorn SH (2000) Bupropion: what mechanism of action? J Psychiatr Pract 6:39-44
Preskorn SH, Burke MJ (1992) Somatic therapy for major depressive disorder: selection of an antidepressant. J Clin Psychiatry 53(9, suppl):5-1
Preskorn SH, Fast GA (1991) Therapeutic drug monitoring for antidepressants: efficacy, safety, and cost-effectiveness. J Clin Psychiatry 52(6, suppl):23-33
Preskorn SH, Fast GA (1992) Tricyclic antidepressant-induced seizures and plasma drug concentration. J Clin Psychiatry 53:160-162
Preskorn SH, Harvey A (1996) Biochemical and clinical dose-response curves with sertraline. Clin Pharmacol Ther 59:180 (abstract)
Preskorn S, Harvey A (1998) Cytochrome P450 enzymes and psychopharmacology. In Watson SJ (ed) Psychopharmacology: 1998 Edition CD-ROM. Lippincott-Raven, New York
Preskorn SH, Jerkovich GS (1990) Central nervous system toxicity of tricyclic antidepressants: phenomenology, course, risk factors, and role of therapeutic drug monitoring. J Clin Psychopharmacol 10:88-95
Preskorn S, Katz S (1989) Bupropion plasma levels: intraindividual and interindividual variability. Ann Clin Psychiatry 1:59-61
Preskorn SH, Weller EB, Weller RA, Glotzbach E (1983) Plasma levels of imipramine and adverse effects in children. Am J Psychiatry 140:1332-1335
Preskorn SH, Jerkovich GS, Beber JH, Widener J (1989) Therapeutic drug monitoring of tricyclic antidepressants: a standard of care issue. Psychopharmacol Bull 25:281-4
Preskorn SH, Fleck RJ, Schroeder DH (1990) Therapeutic drug monitoring of bupropion. Am J Psychiatry 147:1690-1691
Preskorn SH, Silkey B, Beber JH, Dorey C (1991) Antidepressant response and plasma concentration of fluoxetine. Ann Clin Psychiatry 3:147-151
Preskorn SH, Burke, MJ, Fast GA (1993) Therapeutic drug monitoring: principles and practice. Psych Clin N Am 16:611-645
Preskorn SH, Alderman J, Chung M, Harrison W, Messig M, Harris S (1994) Pharmacokinetics of desipramine coadministered with sertraline or fluoxetine. J Clin Psychopharmacol 14:90-98
Quitkin FM, McGrath PJ, Stewart JW, Ocepek-Welikson K, Taylor BP, Nunes E, Deliyannides D, Agosti V, Donovan SJ, Petkova E, Klein DF (1996) Chronological milestones to guide drug change: when should clinicians switch antidepressants? Arch Gen Psychiatry 53:785-792
Rasmussen B, Brosen K (2000) Is therapeutic drug monitoring a case for optimizing clinical outcome and avoiding interactions of the selective serotonin reuptake inhibitors? Ther Drug Monit 22:143-154

Renshaw PF, Guimaraes AR, Fava M, Rosenbaum JF, Pearlman JD, Flood JG, Puopolo PR, Clancy K, Gonzalez RG (1992) Accumulation of fluoxetine and norfluoxetine in human brain during therapeutic administration. Am J Psychiatry 149:1592–1594

Richelson E (1994) Pharmacology of antidepressants: characteristics of the ideal drug. Mayo Clin Proc 69:1069–81

Rochat B, Amey M, Bauman P (1995) Analysis of enantiomers of citalopram and its demethylated metabolites in plasma of depressive patients using chiral reverse-phase liquid chromatography. Therap Drug Monit 17:273–279

Ronfeld RA, Tremaine LM, Wilner KD, Henry EB (1997) Evaluation of the pharmacokinetic properties of sertraline and desmethylsertraline in elderly and young male and female volunteers. Clin Pharmacokinet 32(suppl 1):22–30

Rudorfer MB, Young RC (1980) Desipramine: cardiovascular effects and plasma levels. Am J Psychiatry 137:984–86

Schweizer E, Rickels K, Amsterdam JD, Fox I, Puzzuoli G, Weise C (1990) What constitutes an adequate antidepressant trial for fluoxetine. J Clin Psychiatry 1990; 51:8–11

Shukla U, Marathe P, Labudde J, Pittman KA, Barbhaiya RH (1992) Pharmacokinetics of nefazodone in the dog following single oral administration. Eur J Drug Metab Pharmacokinet 17:301–308

Sindrup SH, Brosen K, Gram LF (1992) Pharmacokinetics of the selective serotonin reuptake inhibitor paroxetine: nonlinearity and relation to the sparteine oxidation polymorphism. Clin Pharm Ther 51:288–295

Tasker TC, Kaye CM, Zussman BD, Link CG (1990) Paroxetine plasma levels: lack of correlation with efficacy or adverse events. Acta Psychiatr Scand 80(suppl 350):152–155

Tucker G (2000) Advances in understanding drug metabolism and its contribution to variability in patient response. Ther Drug Monit 22:110–113

Veefkind A. Haffmans P, Hoencamp E (2000) Venlafaxine serum levels and CYP2D6 genotype. Ther Drug Monit 22:202–208

Veith RC, Friedel RO, Bloom B, Bielski R (1980) Electrocardiogram changes and plasma desipramine levels during treatment. Clin Pharmacol Ther 27:796–802

Vu R, Helmeste D, Albers L, Reist C (1997) Rapid determination of venlafaxine and O-desmethylvenlafaxine in human plasma high-performance liquid chromatography with flourimetric detection. J Chromatogr B Biomed Sci Appl 703:195-201

Wernicke JF, Dunlop SR, Dornseif BE, Zerbe RL (1987) Fixed dose fluoxetine therapy for depression. Psychopharmacol Bull 23:164-168

Wernicke JF, Dunlop SR, Dornseif BE, Bosomworth JC, Humbert M (1988) Low-dose fluoxetine therapy for depression. Psychopharmacol Bull 24:183–188

Wood K, Swade C, Abou-Saeh M, Milln P, Coppen A (1983) Drug plasma levels and platelet 5-HT uptake inhibition during long-term treatment with fluvoxamine or lithium in patients with affective disorders. Br J Clin Pharmacol 15(suppl 3):365S–368S

Yao M, Srinivas N (2000) Simultaneous quantitation of d&-nefazodone, nefazodone, d7-hydroxy nefazodone, hydroxynefazodone, m-chlorophenylpiperazine and triazole-dione in human plasma by liquid chromatographic-mass spectrometry. Biomed Chromatogr 14(2):106-112

Yao M, Shah V, Shyu W, Srinivas N (1998) Sensitive liquid chromatographic-mass spectrometric assay for the simultaneous quantitation of nefazodone and its metabolites hydroxynefazodone, m-chlorophenylpiperazine and tiazole-dione in human plasma using single-ion monitoring. J Chromatogr Bbiomed Sci Appl 718:77-85

Yesavage, JA (1986) Psychotropic blood levels: a guide to clinical response. J Clin Psychiatry 47(9, suppl):16-19

Part 2
Basic Principles: Clinical Science

Biological Markers of Depression

T. J. Connor[1,2] · B. E. Leonard[1]

[1] Department of Pharmacology, National University of Ireland, Galway, Ireland
e-mail: connort@tcd.ie
[2] Trinity College Institute of Neuroscience, Department of Physiology, Trinity College, Dublin 2, Ireland

1	Introduction	118
2	Biological Markers of Depression	119
2.1	Why Do We Need Biological Markers of Depression?	119
2.2	State Versus Trait Markers of Depression	119
3	Neurotransmitter Changes in Depression: Data Generated from Analysis of Body Fluids and Postmortem Brain Tissue	120
3.1	Noradrenaline	120
3.2	Serotonin	121
3.3	Other Neurotransmitters That May Play a Role in the Etiology of Depressive Disorders	123
3.3.1	Dopamine	123
3.3.2	Excitatory Amino Acids	124
3.3.3	Neuropeptides: Corticotropin-Releasing Factor	125
4	Peripheral Markers of Neurotransmitter Receptors: Platelet and Lymphocyte Studies	127
4.1	Studies on the Noradrenergic System	127
4.1.1	Lymphocyte β-Adrenoceptors	127
4.1.2	Platelet α_2-Adrenoceptors	128
4.1.3	Platelet Imidazoline Receptors: A New Peripheral Marker of Depression?	128
4.2	Platelet Studies on the Serotonergic System	129
5	Immunological Markers	130
6	Neuroendocrine Markers of Depression	133
6.1	Hypothalamic-Pituitary-Adrenal Axis Responses	133
6.2	Growth Hormone Response to Clonidine	135
6.3	Prolactin Response to Serotonergic Enhancers: D-Fenfluramine, Clomipramine, L-Tryptophan, and 5-Hydroxytryptophan	135
6.4	Melatonin	136
7	Conclusion	137
References		138

Abstract Over the last four decades, a wealth of data has been generated concerning biological substrates of depression. In this chapter, the literature dealing with biological markers of depression is critically reviewed with respect to (1) the potential usefulness of such markers in the diagnosis of depressive disorders and (2) the relevance of such markers for elucidating the underlying biological basis of depression. Thus far a number of different hypotheses have emerged with respect to the underlying biological basis of depressive illness. The monoamine hypothesis suggests that depression results from a reduction in synaptic concentrations of serotonin and/or noradrenaline. The hypothalamic-pituitary-adrenal (HPA) axis hypothesis suggests that depression is secondary to hypersecretion of either corticotropin-releasing factor or glucocorticoids. In contrast, the macrophage theory suggests that depression occurs due a hyperactivity of macrophage arm of the immune response, resulting in increased production of pro-inflammatory cytokines such as interleukin (IL)-1β, IL-6 and interferon (IFN). While each hypothesis when examined in isolation looks impressive, there is a fundamental lack of integration of the various theories. In addition, the fact that depression is such a heterogeneous disorder with many subtypes makes a single "all-inclusive" biological marker of depression difficult to obtain. However, it is suggested that biological markers may be more predictive of certain subtypes of depression. For instance, hyperactivity of the HPA axis appears to be a particularly good marker of melancholic depression.

With regard to the future of biological markers, it is likely the psychiatric state of an individual may be diagnosed by a profile of different markers, as opposed to just the presence or absence of a single marker. Moreover, in vivo imaging techniques such as MRI, PET, and SPECT are likely to yield biological markers that will give new insights into the biological basis of depression, without having to use invasive procedures such as lumbar puncture in order to study brain function.

Keywords Depression · Biological markers · Neurotransmitter · Neuropeptide · Endocrine · Immune · Platelet

1
Introduction

Depressed mood is experienced by most people at one time or another. When mild, it is a passing feeling with no serious consequences. However, depressive illness involves an accentuation of the intensity of otherwise normal emotions and can cause severe distress and disruption of life and, if left untreated, can be fatal. In addition to the abnormal severity of the mood disturbance, the psychopathological state involves a combination of other features such as loss of motivation and an inability to experience pleasure (anhedonia), loss of self esteem, feeling of worthlessness and extreme pessimism, disturbances in sleep and appetite, loss of energy, psychomotor disturbances (retardation or agitation), autonomic nervous system and gastrointestinal disturbances, impairment of reality (hallucinations, delusions, or confusion), and suicidal tendencies. Depression is by no means a homogeneous disorder; rather it is a complex phenomenon that has many subtypes, and probably more than one etiology. However, in this chapter, the term depression, or depressive illness, is used throughout merely as a convention.

2
Biological Markers of Depression

2.1
Why Do We Need Biological Markers of Depression?

Because of the frequent difficulties that arise in making a precise diagnosis, specific markers of depression and its the underlying etiology would be particularly valuable. In recent decades, a search for such biological markers of depression has been undertaken. Much of the data on biological markers of depression described in this chapter were generated in an effort to understand the biological substrates underlying depressive illness. However, as noted above, depression is a complex phenomenon with many subtypes and probably more than one etiology. In addition, depressive symptoms are often comorbid with other psychiatric disorders such as anxiety, schizophrenia, and mania. Therefore, given the existence of many different subtypes of depression, and different comorbid states that may be caused by different biological substrates, one must ask whether it is realistic to try to identify an all-inclusive biological marker of depression. It seems more likely that biological markers may be an important tool for distinguishing different subtypes or co-morbidities of depression.

2.2
State Versus Trait Markers of Depression

A distinction is often made in biological psychiatry between so-called state and trait markers of depression. A trait marker is one that is present in the individual both during the active phase of the disease and also when the patient has been successfully treated and has gone into remission. Such a marker may therefore highlight an innate vulnerability of the individual to depressive illness. In contrast, a state marker is one that is evident or detectable only while depressive symptoms are present. Thus, such a marker may either be causally related to the depressive symptoms in some way, or alternatively, may occur as a direct result of the presence of depressive symptoms, thus being a secondary or coincidental change. In many studies, patients are assessed both clinically and biochemically during the active phase of the disease, following successful treatment, and occasionally following relapse. Such longitudinal studies give important insights into how biological changes correspond to psychiatric pathology and may yield important information regarding the biological basis of depression and possible diagnostic markers.

In this chapter, the literature dealing with biological markers of depression is critically reviewed with respect to (1) the potential usefulness of such markers in the diagnosis of depressive disorders and (2) the relevance of such markers for elucidating the underlying biological basis of depression.

3
Neurotransmitter Changes in Depression: Data Generated from Analysis of Body Fluids and Postmortem Brain Tissue

3.1
Noradrenaline

Schildkraut (1965) first proposed a role for the catecholamine neurotransmitter noradrenaline in depressive disorders. This hypothesis suggested that depression may be due to a reduced noradrenaline concentration in the synapse. After this, a great interest arose in examining biological markers of central noradrenergic activity in depressed patients and also in normal (non-depressed) control subjects. The initial studies on central noradrenergic function in depression were largely restricted to the analysis of the major central metabolite of noradrenaline, 3-methoxy-4-hydroxy phenyl glycol (MHPG), in urine and plasma, because there is evidence that under carefully controlled conditions of diet, exercise, and time of day at which the body fluid is collected, there is an equilibrium in the distribution of MHPG between the cerebrospinal fluid (CSF), blood, and urine (Potter et al. 1983). Thus it was hoped that analysis of MHPG concentrations in body fluids would provide an index of central noradrenergic activity. Early studies indicated that urinary MHPG concentrations were reduced in depressed patients (Maas et al. 1972), a finding that has subsequently been observed by other groups (Muscettola et al. 1984; Schatzberg et al. 1989). However, others have failed to detect changes in urinary MHPG in unipolar depressed patients and control subjects, but have noted that patients with bipolar depression had significantly lower urinary MHPG compared with unipolar depressed patients (Schildkraut et al. 1978; Beckmann and Goodwin 1980; Muscettola et al. 1984). In stark contrast to these findings, Potter et al. (1983) reported that urinary MHPG concentrations were higher in unipolar depressed patients than in control subjects. Thus, studies to date have reported either no change, an increase, or a decrease in urinary MHPG in depressed patients. Similarly, studies that have examined concentrations of MHPG in plasma or CSF in depression have yielded variable results (Redmond and Leonard 1997). In addition to examining MHPG concentrations, other approaches have been used to study the noradrenergic system in depression. For example, β-adrenoceptor density has been examined in brain tissue from suicide victims and a number of studies have reported a significant increase in β-adrenoceptor binding in suicide brain tissue (Zarko and Biegon 1983; Mann et al. 1986). However, others failed to replicate such findings in a group of suicide victims who had a history of depressive illness (De Paermantier et al. 1990). In addition to the β-adrenoceptor studies, an increased α_2-adrenoceptor density was observed in postmortem brain tissue from depressed suicide victims (Meana and Garcia-Seville 1987; Meana et al. 1992). This finding provided support for the hypothesis that major depression is related to supersensitive α_2-adrenoceptors that cause an inhibition of noradrenaline release (Meana and Garcia-Seville 1987).

Tyrosine hydroxylase (TH) has been the subject of a number of studies in depression research. Biegon and Fieldust (1992) reported that there was reduced TH immunoreactivity in the locus coeruleus of suicide victims. In contrast, another study reported increased TH levels in the locus coeruleus of antidepressant-free suicide victims compared to age-matched controls (Ordway et al. 1994). The results of the Ordway et al. (1994) conflict with the findings of Biegon and Fieldust (1992). However, in the Biegon and Fieldust study, only 50% of the patients had a confirmed history of depression. In addition, Ordway et al. quantified total TH in entire cross sections of the locus coeruleus, whereas Biegon and Fieldust examined TH concentrations within individual cell bodies. Thus, there are substantial methodological differences between these studies that may account for the divergent results observed.

3.2
Serotonin

A putative role for serotonin in depressive disorders was proposed based on the fact that reserpine-induced depletion of monoamine neurotransmitters provoked depressive symptoms in vulnerable individuals (Harris et al. 1957). However, as described above, much of the initial research on the biochemical basis of depression focused on the role of noradrenaline in the pathophysiology of depression (Schildkraut et al. 1965), and it wasn't until Coppen et al. (1963) showed that tryptophan potentiated the antidepressant effect of monoamine oxidase inhibitors (MAOIs) that a direct role for serotonin in the etiology of depression was proposed, and a serotonergic theory of depression was formulated (Coppen 1967). After this, there was a great interest in examining biological markers of central serotonergic activity in depressed patients in comparison to control subjects. CSF concentrations of the serotonin metabolite, 5-hydroxyindole acetic acid (5-HIAA), have been used as an index of brain serotonin turnover. However there is disagreement as to how far lumbar CSF 5-HIAA concentrations correlate with brain 5-HIAA concentrations and ultimately serotonergic activity within the central nervous system (CNS) (Stanley et al. 1985; Gjerris 1988). Nonetheless, it has been demonstrated that there was reduced CSF 5-HIAA concentrations in depressed patients (van Praag et al. 1972) and suicide victims (Asberg et al. 1976). These data added more support to the serotonergic hypothesis of depression. Later studies also reported reduced 5-HIAA in the CSF of depressed patients relative to healthy control subjects (Agren 1980; Asberg et al. 1984). However, others have reported that there was either no change, or even an increase, in CSF 5-HIAA in depressed patients (Vestergaard et al. 1978; Banki et al. 1981; Koslow et al. 1983; Roy et al. 1985). Thus, studies to date in depressed patients have yielded inconsistent findings, but overall it can be concluded that depressed patients as a group do not have reliably lower concentrations of CSF 5-HIAA (Cowen 1996). However, it has been suggested that reduced CSF 5-HIAA may be a marker for suicide as opposed to depression per se, since reduced CSF 5-HIAA concentrations have been observed in suicide attempters irrespective of whether they were suffering from depression (Arango et al. 1997). Thus, it has been suggested that reduced CSF 5-HIAA concentrations are more pronounced in a subset of depressed patients who attempted suicide, and particularly in those patients who were persistent suicide attempters with a preference for a violent method of suicide (Asberg et al. 1987). Therefore, it is not unreasonable to suggest that reduced CSF 5-HIAA may be a biological marker of suicidal intention in depressed patients.

In addition to measuring CSF 5-HIAA concentrations in depressed patients, many other studies have focused on the measurement of plasma concentrations of tryptophan, the amino acid precursor of serotonin. Because tryptophan hydroxylase (TPH), the enzyme catalyzing the rate-limiting step in serotonin biosynthesis, is not normally saturated, it was hypothesized that the reduced brain serotonergic activity observed in depression may be due to reduced availability of the essential amino acid tryptophan. The studies conducted to date have examined both total and free tryptophan, and also the ratio of tryptophan to large neutral amino acids that compete with tryptophan for an amino acid transporter that carries amino acids across the blood-brain barrier. There are several reports indicating that there is a reduction in plasma-free tryptophan concentrations in patients with major depression in comparison to healthy controls or subjects with minor depression (see Meltzer and Lowy 1987). However, a number of other studies have failed to detect such a reduction in depressed patients (see Meltzer and Lowy 1987). The data generated on plasma amino acid concentrations in depression suggest that a reduced free tryptophan: large neutral amino acid ratio in depressives is due to a reduction in plasma tryptophan concentrations, as opposed to an increase in the concentration of the competing amino acids (Maes and Meltzer 1995). With regard to the role of tryptophan availability in depression, it is of interest that dietary tryptophan depletion using a tryptophan-free diet and amino acid loading causes patients who had previously responded to antidepressant therapy with selective serotonin reuptake inhibitors (SSRIs) to relapse into a depressive state; this was quickly reversed following replenishing of tryptophan stores (Delgado et al. 1990, 1999).

Shaw et al. (1967) were the first to suggest that abnormalities in serotonin neurotransmission in the brainstem were present in the brains of suicide victims. Presynaptic and postsynaptic serotonin receptor alterations have been observed in the prefrontal cortex of suicide victims that are consistent with reduced serotonergic function. Binding to the serotonin transporter was found to be reduced in some but not all studies, particularly in the ventral and lateral prefrontal cortical regions (Arango et al. 1997). In this regard, a recent in vivo imaging study using single photon emission computed tomography (SPECT) revealed that there was a reduction in the number of serotonin uptake sites in the brain stem of depressed patients as indicated by reduced binding of $[^{123}I]$-2 β-carbomethoxy-3 β-(4-iodophenyl)tropane ($[^{123}I]$ β-CIT) (Malison et al. 1998). In this study by Malison et al. (1998), platelet $[^3H]$ paroxetine binding was not altered in depressed patients and was not significantly correlated with brainstem $[^{123}I]$ β-CIT binding, thus raising questions about the validity of the platelet serotonergic system in our understanding of brain serotonergic function. In addition to changes observed in serotonin transporter number, increased binding to postsynaptic 5-HT$_{1A}$ and 5-HT$_{2A}$ receptors have been noted in many studies (Arango et al. 1997). Such an increase in serotonin receptors may be compensatory in nature due to the reduction in synaptic serotonin concentrations.

3.3
Other Neurotransmitters That May Play a Role in the Etiology of Depressive Disorders

Noradrenaline and serotonin are the two neurotransmitters that have received the most attention with respect to depressive illness, and to date the pharmacological approaches used in the treatment of depressive disorders have involved agents that modulate serotonergic and/or noradrenergic neurotransmission. However, numerous groups have been investigating the possible role of other neurotransmitter systems in the biology of depression. Data that support a role for other neurotransmitters in the etiology of depression are discussed in the following sections.

3.3.1
Dopamine

Dopamine may be regarded as the forgotten monoamine with respect to depressive disorders. Depression is a disorder that is characterized by a disruption of normal reward processes. An inability to experience pleasure (anhedonia), or a loss of motivation (lack of interest) are key features of melancholic depression according to the diagnostic and statistical manual of mental disorders (DSM-IV) (American Psychiatric Association 1994). It has been suggested that hypoactivity of the mesolimbic dopaminergic system (MDS) may be responsible for these particular features of the disorder (Willner 1983), since dopamine is a neurotransmitter that is central to motivation and reward processes (Willner 1983). However, the evidence that dopamine has a role in depression has come almost entirely from preclinical research.

Numerous studies have tried to assess forebrain dopaminergic function in depressed patients by measuring concentrations of the dopamine metabolite homovanillic acid (HVA) in CSF. Measurement of CSF HVA concentrations has provided one of the most direct means for assessing brain dopaminergic function in clinical studies (Jimerson 1987). In some studies, the subjects were treated with probenecid to block the transport of HVA out of the CSF; this allows the accumulation of HVA in the CSF and is therefore thought to give a better approximation of dopamine turnover. The majority of studies tend to report reduced HVA concentrations in the CSF of depressed patients relative to control subjects; five out of six studies demonstrated that depressed subjects who were medication free for at least 10 days had reduced CSF HVA concentrations in comparison to control subjects (Jimerson 1987). It has been reported that such decreases in HVA are particularly pronounced in patients who display marked psychomotor retardation (Willner 1983). There are also reports of decreased HVA concentrations in the CSF of depressed patients who attempted suicide (Brown and Gershon 1993). Consistent with these findings, it has been reported that there was a decrease in 24-h urinary excretion of HVA and another dopamine metabolite, dihydroxyphenylacetic acid (DOPAC), in depressed suicide attempters (Roy et al. 1992). Thus, these studies provide further evidence for a decrease in dopamine turnover in depressed patients. However, the interpretation of these data is by no means straightforward. From an analysis of the various studies it appears that reduced CSF HVA concentrations are more closely related to motor activity than mood. For example, CSF HVA concentrations were reported to be lower in patients suffering from melancholic rather than non-melancholic depression, a relationship that may be explained by the fact that patients with melancholic depression display psychomotor retardation (Willner 1995). In fact, reduced CSF HVA concentrations have also been observed in non-depressed patients suffering from Parkinson's disease and Alzheimer's disease who display impaired motor function (Wolfe et al. 1990). Thus the specificity of such dopaminergic changes as a marker of depressive illness remains questionable. In addition to the dopamine metabolite studies in depressed patients, it has also been reported that there is lower binding of the selective D_1 receptor ligand SCH 23390 in the frontal cortex of patients with bipolar disorder (Suhara et al. 1992). In addition, an in vivo SPECT study indicated that there is an increase in the density of D_2 receptors in the basal ganglia of depressed subjects when compared to control subjects (D'haenen and Bossuyt 1994). Such an increase in D_2 density is compatible with reduced dopamine turnover as indicated by the low CSF HVA levels that have been reported in depressed patients. However, more recently Bowden et al. (1997) reported that there were no significant differences between antidepressant-free suicide victims and control subjects with respect to D_1 or D_2 density.

3.3.2
Excitatory Amino Acids

The idea that excitatory amino acids such as glutamate and aspartate may play a role in depressive disorders is quite a recent one. Most of the support for this hypothesis has come from preclinical studies which demonstrated that chronic treatment with a variety of antidepressant drugs alter the ligand binding properties of the *N*-methyl-D-aspartate (NMDA) subtype of glutamate receptors. In addition, functional antagonists of the NMDA receptor have been reported to display antidepressant-like activity in both behavioral and biochemical screening procedures used to predict antidepressant activity in rodents (Paul 1997). The hypothesis that the NMDA receptor is involved in the neurobiology of depression is given credence by the finding that the glutamate recognition site on the NMDA receptor and the allosteric regulation of this site by glycine differ in the frontal cortex of suicide victims compared with those from age- and postmortem interval-matched control subjects (Nowak et al. 1995). Specifically, the binding of the glutamate receptor antagonist [^{3}H]CGP-39653 was significantly (40%) lower in cortices from suicide victims than in controls (Nowak et al. 1995).

The high-affinity component of glycine displacement of [^{3}H]CGP-39653 is also significantly reduced in frontal cortical homogenates from suicide victims. However, other studies that examined the NMDA receptor in postmortem brain tissue found no significant differences either in the affinity (Kd) of [^{3}H] dizocilpine or receptor density (B_{max}) between suicide victims and control subjects or depressed suicide victims and controls (Holemans et al. 1993; Palmer et al. 1994). Thus, the postmortem data that support a role for the NMDA receptor in depressive illness are limited, and further studies are required to establish that NMDA has a definite role in depressive disorders.

3.3.3
Neuropeptides: Corticotropin-Releasing Factor

Corticotropin-releasing factor (CRF) is a 41 amino acid peptide that was first isolated, characterized, and synthesized in 1981 by Vale et al. (1981). CRF is released from the hypothalamus and stimulates release of adrenocorticotropin (ACTH) from the pituitary gland and is consequently a central regulator of glucocorticoid secretion from the adrenal glands. However, in the last decade, a wealth of data have accumulated to suggest that CRF can also act as a central neurotransmitter via mechanisms that are independent of its ACTH-releasing properties (Owens and Nemeroff 1991; De Souza and Grigoriadis 1995). To date, animal studies have demonstrated that many of the alterations in neurotransmission elicited by CRF are similar to those engendered by acute stressors, such as the increased release of noradrenaline and dopamine in a number of brain structures (Dunn and Berridge 1987; Lavicky and Dunn 1993). Chronic CRF administration also upregulates locus coeruleus tyrosine hydroxylase activity in a manner similar to chronic stress (Melia and Duman 1991). In addition, central CRF administration provokes many stress-like changes in behavior, such as anxiogenic behavior in laboratory animals (Koob et al. 1993). In terms of biological psychiatry, CRF has been a focus of investigation with respect to both depressive and anxiety disorders for a number of years. Nemeroff et al. (1984) were the first to observe elevated CSF concentrations of CRF in depressed subjects in comparison to control subjects. In addition, Nemeroff et al. (1984) examined patients suffering from Alzheimer's disease, schizophrenia, and alcoholism and failed to observe the increases in the CSF CRF concentrations that were seen in depressed patients. These findings were replicated in a study conducted by Banki et al. (1987), which examined larger patient numbers and demonstrated that depressed patients displayed CSF CRF concentrations that were almost twofold higher than those seen in either control subjects or non-depressed psychiatric patients. It is of interest that, in this study, four patients with bipolar disorder were examined and displayed CSF CRF concentrations that were no different from those of control subjects (Banki et al. 1987). Thus, elevated CSF CRF concentrations appear to be a selective marker for unipolar depressive illness. Moreover, elevated CSF CRF appears to be specific to patients suffering form primary depression as opposed to those who suffer from depression that is associated with an underlying physical complaint such as chronic back pain (France et al. 1988). Other studies have reported either no significant change (Roy et al. 1987; Molchan et al. 1993; Pitts et al. 1995) or even a decrease (Geracioti et al. 1992, 1997) in CSF CRF concentrations in depressed patients when compared to control subjects. The reason for these discrepancies is unclear. However, in many of these studies, the group of depressed patients examined was heterogeneous and often included patients with both unipolar and bipolar depression. The inclusion of bipolar subjects may well have skewed these analyses, as it is well established that CSF CRF concentrations are not altered in bipolar illness (Banki et al. 1987; Berrettini et al. 1987; Kling et al. 1991; Risch et al. 1992). Other difficulties include the use of small sample sizes and failure to consider confounding factors such as gender, age, and episode characteristics (Mitchell 1998). In addition to the clinical data, a number of studies have examined CRF and CRF receptor expression in tissue derived from suicide victims. Arato et al. (1989) found increased CRF in cisternal CSF samples from suicide victims in comparison to appropriate controls. Raadsheer et al. (1994) reported a fourfold increase in total CRF neuron count in the hypothalamic paraventricular nucleus and a threefold increase in neurons which co-express CRF and arginine vasopressin (AVP). Other studies that measured CRF content in cortical sites found no significant difference between suicide victims or depressed patients and control subjects (Charlton et al.1988; Leake et al. 1990). With respect to CRF receptor numbers in depressed patients, Nemeroff et al. (1988) reported decreased CRF receptor binding in the frontal cortex of suicide victims compared to control subjects. However Leake et al. (1990) failed to detect any changes in CRF receptor binding in cortical tissue obtained from depressed patients. In all, the data generated from both basic and clinical studies have led to proposals that hypersecretion of CRF is a core defect in psychiatric disorders such as anxiety and depression (Owens and Nemeroff 1993; Heim et al. 1997). Also, elevated CSF CRF concentrations appear to be a state-specific marker of depression that subsides following effective treatment either with antidepressant drugs or electroconvulsive therapy (Nemeroff et al. 1991; DeBellis et al. 1993). Moreover, failure of CRF to return to baseline following treatment appears to be indicative of relapse (Banki et al. 1992). In addition, it has been suggested that CRF antagonists may have an important therapeutic utility in treating depression and anxiety (Owens and Nemeroff 1993; Schulz et al. 1996).

4
Peripheral Markers of Neurotransmitter Receptors: Platelet and Lymphocyte Studies

Certain analogies exist between the neurotransmitter pharmacology of the brain and of the peripheral tissues. Since it is often not possible to study the brain directly in man, one window of opportunity would be to study these same neurotransmitters and neurotransmitter receptors in peripheral tissues whenever they behave in a manner analogous to those in the CNS. It is with this in mind that the platelet has been developed as a model of the serotonergic and noradrenergic nerve terminal (Stahl 1985). Some structural properties of platelets are similar to neurons. Both possess a limiting membrane rich in receptors and both contain mitochondria and dense-cored vesicles in which neurotransmitters are stored (for reviews, see Sneddon 1973; Stahl 1985). However, one of the most basic limitations of platelets is that they lack a synapse and are therefore a poor model of neuronal synaptic transmission. Nonetheless, numerous investigations have been conducted over the last two decades that have examined serotonin uptake and numerous serotonergic and noradrenergic receptors on platelets in depressed patients. In addition to the studies conducted on platelets, peripheral lymphocytes have been used to study β-adrenergic receptors in depressed patients.

4.1
Studies on the Noradrenergic System

4.1.1
Lymphocyte β-Adrenoceptors

β-Adrenoceptor downregulation in the frontal cortex has been commonly used as a biochemical marker of antidepressant efficacy. The human lymphocyte β-adrenoceptor has been used as a peripheral marker of central β-adrenoceptor function. A reduction in β-adrenoceptor binding on lymphocyte membranes had been identified in depressed patients using radioligand binding studies (Extein et al. 1979; Carstens et al. 1987; Pandey et al. 1987; Magliozzi et al. 1989). However, other studies have reported an increase (Healy et al. 1983; Butler and Leonard 1986) or no change (Cooper et al. 1985; Mann et al. 1985) in lymphocyte β-adrenoceptor density in depressed patients. Thus, studies conducted to date on lymphocyte β-adrenoceptor numbers have yielded equivocal results (Elliot 1991; Potter et al. 1993), possibly as a result of the heterogeneity of the patients studied, types of radioligands used in the analysis, and the different periods allowed for washout of psychotropic medication taken by patients. Therefore, the lymphocyte β-adrenoceptor does not appear to be a reliable marker of central β-adrenergic function in depression.

4.1.2
Platelet α_2-Adrenoceptors

Considerable attention has been paid to α_2-adrenoceptors in depressive illness, as presynaptic α_2-adrenoceptors control the release of noradrenaline from central noradrenergic neurons (Langer 1981). In addition, rodent studies indicated that chronic, but not acute, treatment with the tricyclic antidepressant desipramine results in decreased sensitivity of α_2-adrenoceptors, thus increasing synthesis and release of noradrenaline from neurons (Spyraki and Fibiger 1980). In addition, the finding of increased α_2-adrenoceptor density in postmortem brain tissue from depressed suicide victims provided strong support for the hypothesis that endogenous depression is related to supersensitive α_2-adrenoceptors (Meana and Garcia-Seville 1987). In an attempt to monitor human α_2-adrenoceptor activity in relation to affective disorders, a number of research groups have compared platelet α_2-adrenoceptor binding in depressed patients with that in matched healthy controls. However, the studies to date that have examined the α_2-adrenoceptor in depressed patients have yielded contradictory results. Studies of the α_2 agonists ([^{3}H]-clonidine or [^{3}H]-p-amino-clonidine) have found increased (Garcia-Sevilla et al. 1981, 1986; Takeda et al. 1989; Piletz et al. 1990), decreased (Doyle et al. 1985; Carstens et al. 1986), or no significant change (Georgotas et al. 1987; Werstiuk et al. 1992) in α_2-adrenoceptor number in depressed subjects. Similarly, using the selective α_2-adrenoceptor antagonists such as [^{3}H]-rauwolscine or [^{3}H]-yohimbine, which label a larger number of sites but may not label all the high-affinity sites, researchers have found an elevated (Healy et al. 1983), reduced (Wood and Coppen 1982; Wood et al. 1985; Maes et al. 1999), or unaltered (Lenox et al. 1983; Campbell et al. 1985) number of α_2-adrenoceptor sites in depressed subjects. Antidepressant treatment has been reported to either reduce or not to alter platelet α_2-adrenoceptor number (Garcia-Sevilla et al. 1981; Healy et al. 1983; Pimoule et al. 1983; Campbell et al. 1985; Pandey et al. 1989; Werstiuk et al. 1992). In a review by Piletz et al. (1986) it was suggested that the inconsistencies in the platelet α_2-adrenoceptor literature on depression are due to variations in the biochemical techniques used, such as different conditions for the isolation and preparation of platelets, age of the platelet sample, and whether agonists or antagonists were used in the binding assay.

4.1.3
Platelet Imidazoline Receptors: A New Peripheral Marker of Depression?

Imidazoline receptors are a newly discovered family of receptors, some of which, like α_2-adrenoceptors, have a presynaptic inhibitory effect on the release of noradrenaline. Two subtypes of imidazoline receptor have been identified (I_1 and I_2 receptors) based on their differential affinities for imidazoline compounds such as clonidine, idazoxan, guanabenz, and yohimbine. Because of the ability of imidazoline receptors to modulate noradrenaline release, these receptor are of interest when studying the neurobiology of depressive illness. To date, a number of studies have examined platelet imidazoline receptors in depressed patients (Garcia-Sevilla et al. 1996; Piletz et al. 1996), and one recent study examined imidazoline receptor binding in brain tissue from suicide victims (Garcia-Sevilla et al. 1996). It has been reported that there is an increase in I_1 receptor number in platelets from depressed patients in comparison to control subjects (Garcia-Sevilla et al. 1996; Piletz et al. 1996). In addition, the Kd for $[^{125}I]$-p-iodoclonidine binding at I_1 sites in depressed patients was two- to three-fold lower than in control subjects (Piletz et al. 1994). Thus, it appears that both I_1 receptor number and affinity are increased in depressed patients compared to healthy controls. In addition, it was reported that chronic treatment with both desipramine and fluoxetine reduced the number of I_1 binding sites, suggesting that an increased I_1 receptor number may be a state marker of depressive illness (Piletz et al. 1994). It is also of interest that this increase in I_1 receptor number appears to be quite specific to unipolar depression since euthymic patients with bipolar disorder or patients suffering from generalized anxiety disorder did not display increases in platelet I_1 receptor number (Garcia-Sevilla et al. 1996; Piletz et al. 1996). Thus, the data generated to date indicate that increased I_1 receptor expression may be a new, relatively specific state marker of depressive illness. Nonetheless, further studies are required to assess the validity of increased platelet I_1 binding as a biological marker of depression.

4.2
Platelet Studies on the Serotonergic System

The platelet has been widely studied as a peripheral model of serotonergic neurons (Pletscher and Laubscher 1980; Leonard 1991). The platelet shares with its CNS nerve terminal counterpart the ability to take up, store, and release 5-HT (Stahl 1977), thus representing an easily accessible marker of serotonergic function. In the absence of the sophisticated neuroimaging techniques that are now becoming available, platelet studies have provided much information regarding adaptations to the serotonergic system in major depression and other psychiatric disorders (for review, see Nugent and Leonard 1998). Various serotonergic abnormalities have been described in the platelets from depressed patients, including a deficit in the transport of 5-HT; it has been shown that the maximal rate of transport (V_{max}) of serotonin into the platelet is reduced in depressed patients in comparison to healthy controls (Tuomisto and Tukainen 1976; Tuomisto et al. 1979; Butler and Leonard 1988; for a review, see Meltzer and Lowy 1987). Attempts to differentiate between depressive subtypes according to 5-HT uptake indicate that no differences appear to exist (Stahl et al. 1983; Arora et al. 1984; Rausch et al. 1986). Clinical improvement in patients is frequently, but not always, accompanied by a normalization of platelet 5-HT uptake, suggesting that this is a state-dependent marker of depressive illness (Butler and Leonard 1988). With regard to the specificity of impaired 5-HT uptake to depressed patients, it has been reported that reduced platelet 5-HT uptake is also observed in other psychiatric disorders such as schizophrenia and panic disorder (Rotman et al. 1979; Arora and Meltzer 1982; Pecknold et al. 1988).

There is also a reduced number of [^{3}H]-imipramine binding sites on platelets from depressed patients when compared to healthy control subjects, without any significant alteration in the binding affinity (Briley et al. 1980). This finding was explained in terms of a reduced number of 5-HT uptake sites on the platelet membrane of depressed patients in comparison with control subjects. Subsequent to this initial finding, numerous research groups have examined imipramine binding in depressed patients, only half of whom have described a significant decrease in this parameter (for a review, see Elliot 1991). Therefore, the reliability of platelet [^{3}H]-imipramine binding as a peripheral marker of depression has come into question. It should be also emphasized that no change has been reported in the [^{3}H]-paroxetine binding site on the serotonin transporter of depressed patients (Stain-Malmgren et al. 1998).

In addition to carrying serotonin uptake sites on their membranes, platelets also carry functional 5-HT$_2$ receptors that, when stimulated, initiate phosphoinositol (PI) turnover. Platelet 5-HT$_2$ receptor activation provokes a change in platelet shape and, at higher concentrations, platelet aggregation. Studies conducted in depressed patients examining 5-HT-induced platelet aggregation responses have yielded variable results, with some groups observing no significant change between depressives and controls (Wood et al. 1984) and others finding a reduced aggregation response in depressed patients that normalizes following successful treatment (Healy et al. 1983; Butler and Leonard 1988). Based on 5-HT$_2$ receptor binding studies in depressed patients, it was reported that there were either no significant differences between depressed patients and control subjects (Cowen et al. 1987; McBride et al. 1987) or that depressed patients had increased receptor number compared to controls (Biegon et al. 1987; Butler and Leonard 1988). In the study by Butler and Leonard (1988), increased platelet 5-HT$_2$ receptor number was paradoxically accompanied by a decreased platelet aggregation response, which was interpreted as being due to an uncoupling of the 5-HT$_2$ receptor from the second messenger system in the platelet. A recent study reported increased protein kinase C activity in the cytosolic fraction of platelets from depressed patients, suggesting that a hyperactive PI signaling system may reflect the increased density of 5-HT$_2$ receptors on the platelet membrane of depressed patients (Pandey et al. 1998).

5
Immunological Markers

It is now widely accepted that psychological stress and psychiatric illness can compromise immune function (Leonard 1995). Furthermore, it is well established that soluble mediators released by immune cells (cytokines) can affect CNS function and produce alterations in behavior (see Anisman et al. 1993; Dantzer 1994). In addition to the behavioral changes that are evident in depressed patients, many endocrine and immune abnormalities have also been identified. Most of the initial research that examined the effect of depression on immunity indicated that there were suppressive effects on many aspects of immune function. Thus, some studies reported impaired zymosan-induced neutrophil phagocytosis (O'Neill and Leonard 1990), mitogen-stimulated lymphocyte proliferation (Kronfol and House 1989; Maes et al. 1989), and natural killer (NK) cell cell activity (Irwin et al. 1990, 1992) in depressed patients. Despite the numerous reports documenting a suppression of various indices of immune function in depression, contradictory studies have been reported in which researchers failed to detect any significant alteration in some of these immune parameters in depressed patients (Albrecht et al. 1985; Darko et al. 1989). It has been suggested that these inconsistencies were a result of, among other things, evaluation of varying forms of depression (unipolar patients displayed more immune alterations than bipolar patients) and varying severities of illness (suppression of immune function was positively correlated with illness severity). In addition, studies that examined immune correlates in relatively young outpatients with milder symptoms of depression found that lymphocyte function was not altered compared to control subjects (for a review, see Weisse 1992).

Despite these initial findings of immunosuppression in depressed patients, other studies have indicated that immune activation could also be present in depressed patients, and it has been suggested that such immune activation may play a role in the onset of depressive symptoms (Smith 1991; Maes et al. 1995a). Traditionally, major depression was viewed as a disorder that occurred as a result of abnormalities in central monoaminergic neurotransmitter systems (primarily the noradrenergic and serotonergic systems), and that such neurotransmitter changes gave rise to the behavioral sequelae and the reported alterations in endocrine and immune function. However, in recent times it has been suggested that the behavioral deficits, central monoamine abnormalities, and hypothalamic-pituitary-adrenal (HPA) axis activation seen in depressed patients may, in fact, be secondary to alterations in immune function, at least in some cases of depression (Smith 1991; Maes et al. 1995a).

Depression is associated with an increase in plasma concentrations of the complement proteins C3 and C4 and immunoglobulin (IgM) (Song et al. 1994). Song et al. (1994) also reported increases in plasma concentrations of the positive acute phase proteins (APPs) haptoglobin, α_1-antitrypsin, and α_1 and α_2 macroglobulin in depressed patients. These findings were consistent with earlier reports of increases in complement and positive acute APPs in depressed patients (Kronfol and House 1989; Maes et al. 1992), whereas negative APP concentrations were reduced in depressed patients (Maes et al. 1995a).

Recently, it has been reported that increased concentrations of interleukin (IL)-6, soluble IL-6 receptor (sIL-6R), soluble IL-2 receptor (sIL-2R), IL-1, IL-1 receptor antagonist (IL-1ra), and interferon (IFN)-γ occur in depressed patients (Maes et al. 1993, 1994, 1995b,c; Sluzewska et al. 1995a,b), and many of these alterations were accompanied by increased positive APP concentrations (see Maes et al. 1995a). In addition, increased monocyte phagocytosis has been observed in depressed patients, an effect that is reversed upon successful antidepressant treatment, suggesting that increased monocyte phagocytosis is a state marker of depression (McAdams and Leonard 1993). Similarly, Griffiths et al. (1996) reported that increased serum IL-1 concentrations in depressed patients returned to normal in those patients who responded to a 12-week course of sertraline. Also, elevated serum IL-6 and α_1-acid glycoprotein concentrations observed in a group depressed patients were normalized following chronic fluoxetine treatment (Sluzewska et al. 1995a), suggesting that, like increased macrophage phagocytosis, increased concentrations of IL-1, IL-6, and α_1-acid glycoprotein may be state markers of depression and may be causally involved in the onset of depressive symptomatology. In this regard, there is evidence suggesting that signs of immunological activation, such as increased serum IL-6 and positive APP concentrations, are more evident in patients suffering from treatment-resistant depression, compared to those showing a positive response to antidepressant treatment (Sluzewska et al. 1995b). Seidel et al. (1995) reported a significant increase in mitogen-stimulated γ-IFN and sIL-2R production from peripheral blood mononuclear cell (PBMC) cultures and elevated serum APP concentrations in depressed patients, which were maximal during the acute phase of the illness and returned to control levels over a 6-week hospitalization period during which time a concomitant decrease in Hamilton Depression Rating Scale (HAM-D) (Hamilton 1960) scores was apparent. In stark contrast to the findings already outlined, Weizman et al. (1994) reported that IL-1β, IL-2, and IL-3 production from mitogen-stimulated PBMC cultures was significantly reduced in depressed patients, when compared to age- and sex-matched controls. Furthermore, the reduced cytokine secretion in these depressed patients was normalized following 4 weeks of clomipramine treatment (Weizman et al. 1994). The reason for this discrepancy between studies is not apparent, and clearly further studies with larger sample sizes need to be conducted to evaluate both circulating cytokine concentrations and ex vivo mitogen-stimulated cytokine production in various subtypes of depressed patients.

Studies employing flow cytometric analysis have revealed that depressed patients have an increased number of T helper ($CD4^+$), T memory ($CD4^+,CD44RO^+$), activated T cells ($CD25^+$ T cells and $HLA-DR^+$ T cells), and B cell subsets, indicating the presence of immunological activation in these patients (see Maes et al. 1995a).

Although many discrepancies exist between the clinical studies conducted to date, a large volume of data has been generated that, for the most part, indicates that signs of immunological activation (elevated cytokine and positive APP concentrations) are evident in depressed patients. Nonetheless, relatively few of these studies have examined the effect of antidepressant treatment on the observed changes. However, the few studies that have evaluated the effect of antidepressant therapy suggest that signs of immunological activation are state, as opposed to trait, markers of depression (McAdams and Leonard 1993; Sluzewska et al. 1995a; Griffiths et al. 1996). It remains to be seen if altered concentrations of cytokines, soluble cytokine receptors and APPs are specific to particular subtypes of depression and other conditions in which stress plays a major role. In this regard, it is of interest that a recent study reported that increased serum IL-1 concentrations were present in patients suffering from typical depression, but absent from patients suffering from atypical depression with increased neurovegetative symptoms (Griffiths et al. 1996).

6
Neuroendocrine Markers of Depression

Since hypothalamic control of pituitary hormone release is primarily regulated by the same biogenic amine neurotransmitters as have been implicated in the etiology of depression, it is not unreasonable to suggest that alterations in endocrine functioning may be associated with depressive disorders. Because direct, invasive studies on the limbic system in humans are impossible, the neuroendocrine system can be studied as an indirect measure of limbic integrity or dysfunction. With this goal in mind, neuroendocrine challenge tests were developed in an attempt to determine which neurotransmitter receptors were dysfunctional in depression; it has been proposed that some of these tests could be used to aid the diagnosis of depression (Checkley 1980).

6.1
Hypothalamic-Pituitary-Adrenal Axis Responses

Because the HPA axis plays an integral role in the pathophysiology of stress, and because stress has long been thought to precipitate episodes of depression in vulnerable individuals (Anisman and Zacharko 1982), the HPA axis has been the most extensively studied of all endocrine axes with respect to psychiatric disorders. A large volume of data suggests that depression is associated with hyperactivity of the HPA axis. There have been reports of elevated urinary-free cortisol concentrations in depressed patients relative to healthy controls (Carroll 1986). Also, as mentioned earlier, depression is associated with elevated CRF concentrations in the CSF (Nemeroff et al. 1984), and recent studies using imaging techniques have demonstrated that adrenal size is increased in depressed patients but normalizes following clinical recovery from the disease (Nemeroff et al. 1992; Rubin et al. 1995). In addition to these baseline alterations in HPA-axis activity, depression is also associated with an inability to suppress ACTH and cortisol secretion in response to a 1- to 2-mg oral dose of the synthetic glucocorticoid dexamethasone (Carroll 1982). The dexamethasone suppression test (DST) is undoubtedly the most intensively studied single test of HPA-axis activity in depressed patients. However, despite its initial promise of being a diagnostic tool for depressive illness, its specificity must be called into question, in that dexamethasone nonsuppression is evident in a significant proportion of patients with alcoholism, anorexia nervosa, or Alzheimer's disease in the absence of depression. In addition, only about 50% of depressed patients display dexamethasone nonsuppression and, in the majority of cases, these are melancholic depressed patients (Dinan 1994). Thus, the use of the DST as a diagnostic tool in the diagnosis of depression per se is questionable. However, it may be useful as a marker in the diagnosis of depression of the melancholic subtype; in addition, the DST does seem to be a useful predictor of response to antidepressant therapy, in that depressed patients who persistently display dexamethasone nonsuppression following treatment tend to have a poor clinical outcome (Nemeroff and Evans 1984; Carroll and Haskett 1985).

Another challenge test that has been used to assess the status of the HPA axis in depressed patients is the ACTH and cortisol responses to intravenous CRF administration (Holsboer et al. 1999). Reports indicate that depressed patients display a blunted ACTH response to intravenous CRF administration in the presence of a normal cortisol response (Gold et al. 1988; Ur et al. 1992; Holsboer et al. 1999). Such a phenomenon has been explained in terms of a downregulation of CRF receptors on cells of the anterior pituitary. Such a downregulation of CRF receptors is likely to have occurred as a result of CRF hypersecretion in depressed patients as indicated by elevated CSF concentrations of CRF in depressed patients (Nemeroff et al. 1988). Nonetheless, the adrenal gland appears to be secreting cortisol independently of ACTH.

A combined dexamethasone suppression-CRF challenge test has also been employed to assess the status of the HPA axis in depressed patients (von Bardeleben and Holsboer 1989). In this test, dexamethasone-pretreated patients show enhanced ACTH and cortisol responses to an infusion of 100 μg of CRF. A comparison of the dose–response curves for depressed and normal individuals demonstrated that depressed patients required higher doses of dexamethasone to suppress ACTH and cortisol secretion following CRF administration (Holsboer and Barden 1996). It has been suggested that the sensitivity of the combined dexamethasone suppression-CRF challenge test greatly exceeds that of the standard DST (Heuser et al. 1994).

In addition to CRF, AVP is another hypothalamic peptide that stimulates the release of ACTH and ultimately cortisol. AVP alone is a weak stimulator of ACTH release. However, AVP and CRF act in a synergistic fashion with respect to ACTH release, so that, when both peptides are given together, ACTH is released far in excess of what it would be expected by either peptide given alone. For example, in man, concurrent administration of AVP and CRF produces a 30-fold increase in the release of ACTH in comparison to that achieved following administration of CRF alone (DeBold et al. 1984). Moreover, the ACTH-releasing potential of AVP is dependent on the ambient endogenous CRF level (DeBold et al. 1984). Although elevated levels of CSF AVP have not so far been reported in depression, it is of interest that fluoxetine treatment produces a reduction in CSF concentrations of both AVP and CRF in patients with major depression (DeBellis et al. 1993). It is also of interest that fluoxetine reduced hypothalamic AVP release in vitro (DeBellis et al. 1993). Also, it was reported that depressed patients exhibited an increase in the number of AVP-expressing CRF neurons in the hypothalamic paraventricular nucleus (Purba et al. 1996). Recently, Dinan et al. (1999) reported that depressed patients who were challenged with CRF displayed a blunted ACTH response. However, when challenged with a combination of CRF and the AVP analog desmopressin, the depressed patients displayed an ACTH response similar to that of healthy control subjects (Dinan et al. 1999). Thus, it was suggested that there may be a supersensitivity of the vasopressin V1b receptor in depressed patients, and that this may explain the continued adrenal overactivity in the presence of a blunted ACTH response to CRF (Scott and Dinan 1998; Dinan et al. 1999). However, a number of studies that have examined the effects of AVP or desmopressin on HPA activity in depression have yielded inconsistent results (Scott and Dinan 1998). Thus, further examination of the role of AVP in the HPA dysregulation observed in depression is warranted. In addition, the use of better selection criteria for patients in such studies may yield more conclusive results, since many of the studies conducted to date have examined groups of depressed patients who were clinically heterogeneous (Scott and Dinan 1998).

6.2
Growth Hormone Response to Clonidine

One of the most consistent neuroendocrine changes reported in depressed patients is a blunted growth hormone (GH) response to the α_2-adrenoceptor agonist, clonidine. The GH response to clonidine has been consistently found to be blunted in endogenously depressed patients in comparison to control subjects (Matussek et al. 1980; Checkley et al. 1981, 1984; Charney et al. 1982b; Siever et al. 1982a; Ansseau et al. 1984, 1988). However, there are also studies that have used low-dose oral or intravenous clonidine and found no significant difference in the GH response between depressed patients and controls (Dolan and Calloway 1986; Katona et al. 1986). Changes in GH release appear to be due to decreased sensitivity of α_2-adrenoceptors in depressed patients. However, for the most part it appears that antidepressant treatment has no effect on the GH response to clonidine (Charney et al. 1982a, 1984; Siever et al. 1982b; Price et al. 1986), suggesting that such a blunted GH response to clonidine is a trait marker of depression. Similarly, it has been reported that depressed patients display a blunted GH response to the noradrenaline reuptake inhibitor, desipramine (Meesters et al. 1985; Laakmann et al. 1986). In addition, it would appear that patients who were dexamethasone non-suppressors were far more likely to show a blunted GH response to desipramine than subjects who were suppressors, suggesting that this endocrine abnormality may be secondary to the hypercortisolemia observed in depressed patients (Dinan and Barry 1990).

6.3
Prolactin Response to Serotonergic Enhancers:
D-Fenfluramine, Clomipramine, L-Tryptophan, and 5-Hydroxytryptophan

Centrally acting serotonin enhancers such as fenfluramine, L-tryptophan, 5-hydroxytryptophan, and clomipramine increase the release of prolactin from the anterior pituitary. Because these compounds facilitate an increase in synaptic 5-HT concentrations, their physiological action is interpreted as providing an indirect index of central 5-HT responsivity (Quattorone et al. 1983; Cleare et al. 1996; Bauman et al. 1999), which is particularly useful when trying to understand the role of 5-HT in psychiatric disorders. Many studies using various serotonergic enhancers have observed a blunted prolactin response in depressed patients when compared with control subjects (O'Keane and Dinan 1991; Mann et al. 1995). However, Price et al. (1991) reported that a blunted prolactin response to L-tryptophan was only observed in non-melancholic depressives, and that prolactin secretion following tryptophan challenge was further elevated in both melancholic and psychotic depressives, indicating that a blunted prolactin response does not occur in all subtypes of depression. It is also noteworthy that a blunted prolactin response to serotonergic enhancers has been reported in patients who do not suffer from depression, but exhibit impulsive aggressive behavior (Coccaro et al. 1989). Thus, the specificity of such a marker for depressive disorders is questionable. It is also of interest that, in a recent study, it was reported that the prolactin response to D-fenfluramine was still abnormal in remitted patients in comparison to controls (Flory et al. 1998), suggesting that serotonergic changes in depression may be trait-dependent, rather than state-dependent.

6.4
Melatonin

A role for the pineal hormone melatonin in the neurobiology of depressive disorders was first suspected as depression is thought to be associated with alterations in circadian rhythms (Gillin 1993) and/or in noradrenergic function (Schildkraut 1965). With regard to the association between melatonin secretion and the noradrenergic system, it is noteworthy that β-adrenoceptors are located on pinealocytes and β-adrenoceptor stimulation by noradrenaline stimulates pinealocytes to synthesize melatonin from its precursor serotonin (Axelrod 1978). It has also been demonstrated that the β-adrenoceptor antagonist, propranolol, blocks the nocturnal rise in melatonin in human subjects (Hanssen et al. 1977). Thus, it has been suggested that, in man, melatonin secretion may serve as an index of both noradrenergic function and circadian rhythmicity (Lewy 1983). A blunting in the nocturnal rise of melatonin has been reported in patients suffering from major depression (Mendlewicz et al. 1979; Wetterberg et al. 1979; Brown et al. 1985, 1987; McIntyre et al. 1989). The reduced nocturnal melatonin concentrations observed in depressed patients appear to be particularly associated with melancholia, since melancholic depressed patients display significantly reduced nocturnal melatonin secretion compared to both nonmelancholic depressives and healthy control subjects (Brown et al. 1985). The mechanisms that underlie the suppression of nocturnal melatonin secretion in melancholic depressed patients still remain elusive. However, since a disturbance in noradrenergic activity may be involved in the altered pattern of nocturnal melatonin secretion, a recent study examined β-adrenoceptor binding in the pineal gland of suicide victims compared with control subjects (Little et al. 1997). No differences in pineal β-adrenoceptor binding were observed between depressed and control subjects (Little et al. 1997).

It is also of interest that there is an association between the suppression of nocturnal melatonin secretion, high serum cortisol, and DST nonsuppression in patients with an acute depressive episode (Wetterberg et al. 1979, 1981, 1984). However, it was reported that these individuals continued to display a suppression of melatonin secretion following remission of the depressive symptoms and normalization of cortisol levels (Branchey et al. 1982). Such data suggest that the blunted nocturnal secretion of melatonin observed in some depressed patients may be a trait, rather than state, marker of depressive illness. In addition to the alterations in nocturnal melatonin secretion that were observed in unipolar depressed patients, a number of reports indicate that mania is associated with increased melatonin secretion in comparison to control subjects (Lewy et al. 1979; Miles and Philbrick 1988). For example, Lewy et al. (1979) reported that manic patients displayed consistently elevated plasma melatonin concentrations both during the day and at night.

7
Conclusion

Over the last three decades, a wealth of data has been generated concerning biological substrates of depression. Different research groups have focused on a multitude of targets ranging from monoamine neurotransmitters to neuropeptides, and have even examined immunopeptides such as cytokines. Thus far, a number of different hypotheses have emerged, each of which explains the biological basis of depression quite eloquently. However, while each hypothesis when examined in isolation looks impressive, there is a fundamental lack of integration of all the various theories that have been proposed concerning biological substrates for depressive illness. In addition, the fact that depression is such a heterogeneous disorder, which can present itself in many different subtypes and exists in a number of comorbid states, makes a reliable all-inclusive biological marker of depression difficult to obtain. However, it is not unreasonable to suggest that correlations between biological markers and the psychiatric state may serve to give some indication of what biological substrates underlie different subtypes of depression. Ongoing studies are warranted in order to examine large patient sample sizes and a number of different biological markers in each patient. Such studies will allow the evaluation of a sufficient number of patients presenting with different subtypes of depression and different comorbidities. In the future, the psychiatric state of an individual may be diagnosed by a profile of different biological markers, as opposed to just the presence or absence of a single marker of depression. Moreover, the advent of in vivo imaging techniques such as MRI, PET and SPECT may yield biological markers that will give new insights into the biological basis of depression or new diagnostic tools for particular subtypes of depressive illness without having to use invasive procedures such as lumbar puncture in order to study brain function. However, until the time when reliable biological markers are identified, the psychiatrist's clinical judgment remains the most valid means of diagnosing and treating depressive disorders.

References

Agren H (1980) Symptom patterns in unipolar and bipolar depression correlating with monoamine metabolites in the cerebrospinal fluid: I. general patterns. Psychiatry Res 3:211–223

Albrecht J, Helderman JH, Schlesser MA, Rush AJ (1985) A controlled study of cellular immune function in affective disorders before and during somatic therapy. Psychiatry Res 15:185–193

American Psychiatric Association (1994) Diagnostic and statistical manual of mental disorders. (DSM-IV). American Psychiatric Association, Washington DC

Anisman H, Zacharko R (1982) Depression: the predisposing influence of stress. Behav Brain Sci 5:89–137

Anisman H, Zalcman S, Zacharko RM (1993) The impact of stressors on immune and central neurotransmitter activity: bidirectional communication. Rev Neurosci 4:147–180

Ansseau M, Scheyvaerts M, Doumont A, Poirrier R, Legros JJ, Franck G (1984) Concurrent use of REM latency, dexamethasone suppression, clonidine, and apomorphine tests as biological markers of endogenous depression: a pilot study. Psychiatry Res 12:261–272

Ansseau M, Von Frenckell R, Cerfontaine JL, Papart P, Franck G, Timsit-Berthier M, Geenen V, Legros JJ (1988) Blunted response of growth hormone to clonidine and apomorphine in endogenous depression. Br J Psychiatry 153:65–71

Arango V, Underwood MD, Mann JJ (1997) Biologic alterations in the brainstem of suicides. Suicide 20:581–593

Arato M, Banki CM, Bissette G, Nemeroff CB (1989) Elevated CSF CRF in suicide victims. Biol Psychiatry 25:355–359

Arora RC, Meltzer HY (1982) Serotonin uptake by blood platelets of schizophrenic patients. Psychiatry Res 6:327–333

Arora RC, Kregel L, Meltzer HY (1984) Seasonal variation in serotonin uptake in normal controls and depressed patients. Biol Psychiatry 19:795–804

Asberg M, Traskman L, Thoren P (1976) 5-HIAA in cerebrospinal fluid: a biochemical suicide predictor. Arch Gen Psychiatry 33:1193–1197

Asberg M, Bertilsson L, Martensson B, Scalla-Tomba G-P, Thoren P, Traskman-Bendz L (1984) CSF monoamine metabolites in melancholia. Acta Psychiat Scand 69:201–219

Asberg M, Schalling D, Traskman-Bendz L, Wagner A (1987) Psychobiology of suicide, impulsivity, and related phenomena. In: Meltzer HY (ed) Psychopharmacology: the third generation of progress. Raven Press, New York, pp 655–668

Axelrod J (1978) Introductory remarks on the regulation of pineal indolamine synthesis. J Neural Trans 13:73–79

Banki CM, Vojnik M, Molnar G (1981) Cerebrospinal fluid amine metabolites, tryptophan and clinical parameters in depression. Part I. Background variables. J Affect Disord 3:81–89

Banki CM, Bissette G, Arato M, O'Connor L, Nemeroff CB (1987) CSF corticotropin-releasing factor-like immunoreactivity in depression and schizophrenia. Am J Psychiatry 144:873–877

Banki CM, Karmacsi L, Bissette G, Nemeroff CB (1992) CSF corticotropin-releasing hormone and somatostatin in major depression: response to antidepressant treatment and relapse. Eur Neuropsychopharmacol 2:107–113

Baumann MH, Ayestas MA, Rothman RB (1999) In vivo correlates of central serotonin function after high dose fenfluramine administration. Ann NY Acad Sci 844:138–152

Beckmann H, Goodwin FK (1980) Urinary MHPG in subgroups of depressed patients and normal controls. Neuropsychobiology 6:91–100

Berrettini WH, Nurnberger JI, Zerbe RL, Gold PW, Chrousos GP, Tomai T (1987) CSF neuropeptides in euthymic bipolar patients and controls. Br J Psychiatry 150:208–212

Biegon A, Fieldust S (1992) Reduced tyrosine hydroxylase immunoreactivity in the locus ceruleus of suicide victims. Synapse 10:79–82

Biegon A, Weizman A, Karp L, Ram A, Tiano S, Wolff M (1987) Serotonin 5-HT2 receptor binding on blood platelets–a peripheral marker for depression? Life Sci 41:2485–2492

Bowden C, Theodorou AE, Cheetham SC, Lowther S, Katona CL, Crompton MR, Horton RW (1997) Dopamine D1 and D2 receptor binding sites in brain samples from depressed suicides and controls. Brain Res 752:227–233

Branchey L, Weinberg U, Branchey M, Linkowski P, Mendlewicz J (1982) Simultaneous study of 24hr patterns of melatonin and cortisol secretion in depressed patients. Neuropsychobiology 8:225–232

Briley MS, Langer SZ, Raisman R, Sechter D, Zarifian E (1980) Tritiated imipramine binding sites are decreased in platelets of untreated depressed patients. Science 209:303–305

Brown AS, Gershon S (1993) Dopamine and depression. J Neural Trans 91:75–109

Brown R, Kocsis JH, Caroff S, Amsterdam J, Winokur A, Stokes PE, Frazer A (1985) Differences in nocturnal melatonin secretion between melancholic depressed patients and control subjects. Am J Psychiatry 142:811–816

Brown RP, Kocsis JH, Caroff S, Amsterdam J, Winokur A, Stokes P, Frazer A (1987) Depressed mood and reality disturbance correlate with decreased nocturnal melatonin in depressed patients. Acta Psychiatr Scand 76:272–275

Butler J, Leonard BE (1986) Post-partum depression and the effect of nomifensine treatment. Int Clin Psychopharmacol 1:244–252

Butler J, Leonard BE (1988) The platelet serotonergic system in depression and following sertraline treatment. Int Clin Psychopharmacol 3:343–347

Campbell IC, McKernan RM, Checkley SA, Glass IB, Thompson C, Shur E (1985) Characterization of platelet alpha2 adrenoceptors and measurement in control and depressed subjects. Psychiatry Res 14:17–31

Carroll BJ (1982) The dexamethasone suppression test for melancholia. Br J Psychiatry 140:292–304

Carroll BJ (1986) Depression and urinary free cortisol. Br J Psychiatry 148:218

Carroll BJ, Haskett RF (1985) The DST in newly hospitalized patients. Am J Psychiatry 143:999–1000

Carstens ME, Engelbrecht AH, Russell VA, Aalberts C, Gagiano CA, Chalton DO, Taljaard JJ (1986) α2-adrenoceptor levels on platelets of patients with major depressive disorders. Psychiatry Res 18:321–331

Carstens ME, Engelbrecht AH, Russell VA, Aalbers C, Gagiano CA, Chalton DO, Taljaard JJ (1987) Beta-adrenoceptors on lymphocytes of patients with major depressive disorder. Psychiatry Res 20:239–248

Charlton BG, Cheetham SC, Horton RW, Katona CLE, Crompton MR, Ferrier IN (1988) Corticotropin-releasing factor immunoreactivity in post-mortem brain from depressed suicides. J Psychopharm 2:13–18

Charney DS, Heninger GR, Sternberg DE (1982a) Failure of chronic antidepressant treatment to alter growth hormone response to clonidine. Psychiatry Res 7:135–138

Charney DS, Heninger GR, Sternberg DE, Hafstead KM, Giddings S, Landis H (1982b) Adrenergic receptor sensitivity in depression. Effects of clonidine in depressed patients and healthy subjects. Arch Gen Psychiatry 39:290–294

Charney DS, Heninger GR, Sternberg DE (1984) The effect of mianserin on alpha2 adrenergic receptor function in depressed patients. Br J Psychiatry 144:407–416

Checkley SA (1980) Neuroendocrine tests of monoamine function in man: a review of basic theory and its application to the study of depressive illness. Psychol Med 10:35–53

Checkley SA, Slade AP, Shur E (1981) Growth hormone and other responses to clonidine in patients with endogenous depression. Br J Psychiatry 138:51–55

Checkley SA, Glass JB, Thompson C, Corn T, Robinson P (1984) The GH response to clonidine in endogenous as compared with reactive depression. Psychol Med 14:773–777

Cleare AJ, Murray RM, O'Keane V (1996) Reduced prolactin and cortisol responses to d-fenfluramine in depressed patients as compared to healthy matched controls. Neuropsychopharmacology 14:349–354

Coccaro EF, Siever LJ, Klar HM, Maurer G, Cochrane K (1989) Serotonergic studies in patients with affective and personality disorders. Arch Gen Psychiatry 46:587–599

Cooper SJ, Kelly JG, King DJ (1985) Adrenergic receptors in depression: effect of electroconvulsive therapy. Br J Psychiatry 147:23–29

Coppen AJ (1967) The biochemistry of affective disorders. Br J Psychiatry 113:1237–1264

Coppen AJ, Shaw D, Farrell JP (1963) The potentiation of the antidepressive effects of a monoamine-oxidase inhibitor by tryptophan. Lancet 24:61–64

Cowen PJ (1996) The serotonin hypothesis: necessary but not sufficient. In: Feighner JP, Boyer WF (eds) Selective serotonin re-uptake inhibitors: advances in basic research and clinical practice. Wiley, Chichester, pp 63–86

Cowen PJ, Charig EM, Fraser S, Elliott JM (1987) Platelet 5-HT receptor binding during depressive illness and tricyclic antidepressant treatment. J Affect Disord 13:45–50

Dantzer R (1994) How do cytokines say hello to the brain? Neural versus humoral mediation. Eur Cytokine Netw 5:271–273

Darko DF, Gillin JC, Risch SC, Bulloch K, Golshan S, Tasevska Z, Hamburger RN (1989) Mitogen-stimulated lymphocyte proliferation and pituitary hormones in major depression. Biol Psychiatry 26:145–155

D'haenen H, Bossuyt A (1994) Dopamine D2 receptors measured in the brain using SPECT. Biol Psychiatry 35:128–132

De Paermentier F, Cheetham SC, Crompton MR, Katona CLE, Horton RW (1990) Brain β-adrenoceptor binding sites in antidepressant-free depressed suicide victims. Brain Res 525:71–77

De Souza EB, Grigoriadis DE (1995) Corticotropin-releasing factor: physiology, pharmacology, and role in central nervous system and immune disorders. In: Bloom FE, Kupfer DJ (eds) Psychopharmacology: the fourth generation of progress. Raven Press, New York, pp 505–517

DeBellis MD, Gold PW, Geracioti TD Jr, Listwak SJMS, Kling MA (1993) Association of fluoxetine treatment with reduction in CSF concentration of corticotropin releasing hormone and arginine vasopressin in patients with major depression. Am J Psychiatry 15:656–657

DeBold CR, Sheldon WR, DeCherney GS, Jackson RV, Alexander AN, Vale W, Rivier J, Orth DN (1984) Arginine vasopressin potentiates adrenocorticotropin release induced by ovine corticotropin-releasing factor. J Clin Invest 73:533–538

Delgado PL, Charney DS, Price LH, Aghajanian GK, Landis H, Heninger GR (1990) Serotonin function and the mechanism of antidepressant action: reversal of antidepressant-induced remission by rapid depletion of plasma tryptophan. Arch Gen Psychiatry 47:411–418

Delgado PL, Miller HL, Salomon RM, Licinio J, Krystal JH, Moreno FA, Heninger GR, Charney DS. (1999) Tryptophan-depletion challenge in depressed patients treated with desipramine or fluoxetine: implications for the role of serotonin in the mechanism of antidepressant action. Biol Psychiatry 46:212–220.

Dinan TG (1994) Glucocorticoids in the genesis of depressive illness: a psychobiological model. Br J Psychiatry 164:365–371

Dinan TG, Barry S (1990) Responses of growth hormone to desipramine in endogenous and non-endogenous depression. Br J Psychiatry 156:680–684

Dinan TG, Lavelle E, Scott LV, Newell-Price J, Medbak S, Grossman AB (1999) Desmopressin normalizes the blunted adrenocorticotropin response to corticotropin-releasing hormone in melancholic depression: evidence of enhanced vasopressinergic responsivity. J Clin Endocrinol Metab 84:2238–2240

Dolan RJ, Calloway SP (1986) The human growth hormone response to clonidine: relationship to clinical and neuroendocrine profile in depression. Am J Psychiatry 143:772–774

Doyle MC, George AJ, Ravindran AV, Philpott R (1985) Platelet α2-adrenoceptor binding in elderly depressed patients. Am J Psychiatry 142:1489–1490

Dunn AJ, Berridge CW (1987) Corticotropin releasing factor administration elicits a stress-like activation of cerebral catecholaminergic systems. Pharmacol Biochem Behav 27:685–691

Elliot JM (1991) Peripheral markers in affective disorders. In: Horton R, Katona C (eds) Biological aspects of affective disorders. Academic Press, London, pp 96–144

Extein J, Tallman J, Smith CC, Goodwin FK (1979) Changes in lymphocyte beta-adrenergic receptors in depression and mania. Psychiatry Res 1:191–197

Flory JD, Mann JJ, Manuck SB, Muldoon MF (1998) Recovery from major depression is not associated with normalisation of serotonergic function. Biol Psychiatry 43:320–326

France RD, Urban B, Krishnan KRR, Bissette G, Banki CM, Nemeroff C, Speilman FJ (1988) CSF corticotropin-releasing factor-like immunoreactivity in chronic pain patients with and without major depression. Biol Psychiatry 23:86–88

Garcia-Sevilla JA, Zis AP, Hollingsworth PJ, Greden JF, Smith CB (1981) Platelet a2-adrenergic receptors in major depressive disorders: binding of tritiated clonidine before and after tricyclic antidepressant drug treatment. Arch Gen Psychiatry 38:1327–1333

Garcia-Sevilla JA, Guimon J, Garcia-Vallejo P, Fuster MJ (1986) Biochemical and functional evidence of supersensitive platelet α2-adrenoceptors in major depression. Arch Gen Psychiatry 43:51–57

Garcia-Sevilla JA, Escriba PV, Sastre M, Walzer C, Busquets X, Jaquet G, Reis DJ, Guimon J (1996) Immunodetection and quantitation of imidazoline receptor proteins in platelets of patients with major depression and in brains of suicide victims. Arch Gen Psychiatry 53:803–810

Georgotas A, Schweitzer J, McCue RE, Armour M, Friedhoff AJ (1987) Clinical and treatment effects on ^{3}H clonidine and ^{3}H imipramine binding in elderly depressed patients. Life Sci 40:2137–2143

Geracioti TD, Orth DN, Ekhator NN, Blumenkopf B, Loosen PT (1992) Serial cerebropinal fluid corticotropin-releasing hormone concentrations in healthy and depressed humans. J Clin Endocrinol Metab 74:1325–1330

Geracioti TD, Loosen PT, Orth DN (1997) Low cerebrospinal fluid corticotropin-releasing hormone concentrations in eucortisolaemic depression. Biol Psychiatry 42:166–174

Gillin JC (1993) Sleep studies in affective illness: diagnostic, therapeutic and pathophysiological implications. Psychiatry Annals 13:367–382

Gjerris A (1988) Do concentrations of neurotransmitters in lumbar CSF reflect cerebral dysfunction in depression? Acta Psychiat Scand 345:21–24

Gold P, Kling MA, Demitrack MA (1988) Clinical studies with corticotropin releasing hormone: Implications for hypothalamic pituitary adrenal dysfunction in depression and related disorders. In: Genten D, Pfaff D (eds) Current trends in neuroendocrinology. Springer Verlag, Berlin, pp 238–256

Griffiths J, Ravindran AV, Merali Z, Anisman H (1996) Immune and behavioural correlates of typical and atypical depression. Soc Neurosci Abstr 22:1350

Hamilton M (1960) A rating scale for depression. J Neurol Neurosurg Psychiatry 23:56–62

Hanssen T, Heyden T, Sundberg I, Wetterberg L (1977) Effect of propranolol on serum-melatonin. Lancet 2:309–310

Harris TH (1957) Depression induced by rauwolfia compounds. Am J Psychiat 950–963

Healy D, Carney PA, Leonard BE (1983) Monoamine-related markers of depression: changes following treatment. J Psychiatry Res 17:251–260

Heim C, Owens MJ, Plotsky PM, Nemeroff CB (1997) I. Endocrine factors in the pathophysiology of mental disorders. Psychopharmacol Bull 33:185–192

Heuser I, Yassouridis A, Holsboer F (1994) The combined dexamethasone/CRH-test: a refined laboratory test for psychiatric disorders. J Psychiatry Res 28:341–356

Holemans S, De Paermentier F, Horton RW, Crompton MR, Katona CLE, Maloteaux J-M (1993) NMDA glutamatergic receptors, labeled with [^{3}H] MK-801, in brain samples from drug free depressed suicides. Brain Res 616:138–143

Holsboer F, Barden N (1996) Antidepressants and hypothalamic-pituitary-adrenocortical regulation. Endo Rev 17:187–205

Holsboer F, Gerken A, Stalla GK (1999) ACTH, cortisol and corticosterone output after ovine corticotropin releasing factor challenge during depression and after recovery. Biol Psychiatry 20:276–286

Irwin M, Caldwell C, Smith TL, Brown S, Schuckit MA, Gillin JC (1990) Major depressive disorder, alcoholism, and reduced natural killer cell cytotoxicity. Arch Gen Psychiatry 47:713–719

Irwin M, Lacher U, Caldwell C (1992) Depression and reduced natural killer cytotoxicity: a longitudinal study of depressed patients and control subjects. Psychological Medicine 22:1045–1050

Jimerson DC (1987) Role of dopamine mechanisms in affective disorders. In: Meltzer HY (ed) Psychopharmacology: the third generation of progress. Raven Press, New York, pp 505–509

Katona CLE, Theodorou AE, Davies SL (1986) In: Deakin JWF (ed) The biology of depression. Gaskell, London, pp 121–136

Kling MA, Roy A, Doran AR, Calabrese JR, Rubinow DR, Whitfield HJ Jr, May C, Post RM, Chrousos GP (1991) Cerebrospinal fluid immunoreactive corticotropin-releasing hormone and adrencorticotropin secretion in Cushing's disease and major depression: potential clinical implications. J Clin Endocrinol Metab 72:260–271

Koob GF, Heinrichs SC, Pich EM, Menzaghi F, Baldwin H, Miczek K, Britton KT (1993) The role of corticotropin releasing factor in the behavioral response to stress, (Ciba foundation symposium, 172). 173–182.

Koslow SH, Maas JW, Bowden CL, Dairs JM, Hanin I, Javaid J (1983) CSF and urinary biogenic amines and metabolites in depression and mania. a controlled, univariate analysis. Arch Gen Psychiatry 40:999–1014

Kronfol Z, House JD (1989) Lymphocyte mitogenesis, immunoglobulin and complement levels in depressed patients and normal controls. Acta Psychiat Scand 80:142–147

Laakmann G, Zygan K, Schoen HW, Weiss A, Wittmann M, Meissner R, Blaschke D (1986) Effect of receptor blockers (methylsergide, propranolol, phentolamine, yohimbine, and prazosin) on desipramine-induced pituitary hormone stimulation in humans: I. growth hormone. Psychoneuroendocrinology 11:447–461

Langer SZ (1981) Presynaptic regulation of the release of catecholamines. Pharmacol Rev 32:337–362

Lavicky J, Dunn AJ (1993) Corticotropin-releasing factor stimulates catecholamine release in hypothalamus and prefrontal cortex in freely moving rats as assessed by microdialysis. J Neurochem 60:602–612

Leake A, Perry EK, Perry RH, Fairbairn AF, Ferrier IN (1990) Cortical concentrations of corticotropin-releasing hormone and its receptors in Alzheimer type dementia and major depression. Biol Psychiatry 28:603–608

Lenox RH, Ellis JE, Van Riper DA, et al (1983) Platelet 2-adrenergic receptor activity in clinical studies of depression.. In: Usdin E, Goldstein M, Friedhoff A (eds) Frontiers in neuropsychiatric research. MacMillan, London, pp 331–356

Leonard BE (1991) Blood cells as models of neurons: studies in the affective disorders. J Irish Coll Phys Surg 20:282–284

Leonard BE (1995) Stress and the immune system: immunological aspects of depressive illness. In: Leonard BE, Miller K (eds) Stress, the immune system and psychiatry. John Wiley and Sons, Oxford, pp 114–136

Lewy AJ (1983) Biochemistry and regulation of mammalian melatonin production. In: Relkin R (ed) The pineal gland. Elsevier, New York, pp 77–129.

Lewy AJ, Wehr TA, Gold PW (1979) Plasma melatonin in manic-depressive illness. In: Usdin E, Kopin IJ, Barchas J (eds) Catecholamines: basic and clinical frontiers. Pergamon Press, Oxford, pp 1173–1175

Little KY, Ranc J, Gilmore J, Patel A, Clark TB (1997) Lack of pineal beta-adrenergic receptor alterations in suicide victims with major depression. Psychoneuroendocrinology 1:53–62

Maas J, Fawcett J, Dekirmenjian H (1972) Catecholamine metabolism, depressive illness and drug responses. Arch Gen Psychiatry 26:252–262

Maes M, Meltzer HY (1995) The serotonin hypothesis of major depression. In: Bloom FE, Kupfer DJ (eds) Psychopharmacology: the fourth generation of progress. Raven Press, New York, pp 933–944

Maes M, Bosmans E, Suy E, Minner B, Raus J (1989) Impaired lymphocyte stimulation by mitogens in severely depressed patients: a complex interface with H.P.A.-axis hyperfunction, noradrenergic activity and the aging process. Br J Psychiatry 155:793–798

Maes M, Scharpe S, Van Grootel L, Uyttenbroeck W, Cooreman W, Cosyns P, Suy E (1992) Higher alpha 1-antitrypsin, haptoglobin, ceruloplasmin and lower retinol binding protein plasma levels during depression: Further evidence for the evidence of an acute phase response. J Affect Disord 24:183–192

Maes M, Bosmans E, Meltzer HY, Scharpe S, Suy E (1993) Il-1: a putative mediator of HPA-axis hyperactivity in major depression? Am J Psychiatry 150:1189–1193

Maes M, Scharpe S, Meltzer HY, Okayli G, Bosmans E, D'Hondt P, Vanden Bossche B, Cosyns P (1994) Increased neopterin and interferon-gamma secretion and lower availability of L-tryptophan in major depression: further evidence for an immune response. Psychiatry Res 54:143–160

Maes M, Smith R, Scharpe S (1995a) The monocyte-T-lymphocyte hypothesis of major depression. Psychoneuroendocrinology 20:111–116

Maes M, Meltzer HY, Bosmans E, Vandoolaeghe E, Ranjan R, Desnyder R (1995b) Increased plasma concentrations of interleukin-6, soluble interleukin-6, soluble interleukin-2, and transferrin receptor in major depression. J Affect Disord 34:301–309

Maes M, Vandoolaeghe E, Ranjan R, Bosmans E, Bergmans R, Desnyder R (1995c) Increased serum interleukin-1-receptor-antagonist concentrations in major depression. J Affect Disord 36:29–36

Maes M, Van Gastel A, Delmeire L, Meltzer HY (1999) Decreased platelet alpha-2 adrenoceptor density in major depression: effects of tricyclic antidepressants and fluoxetine. Biol Psychiatry 45:278–284

Magliozzi JR, Gietzen D, Maddock RJ, Haack D, Doran AR, Goodman T, Weiler PG (1989) Lymphocyte beta-adrenoceptor density in patients with unipolar depression and normal controls. Biol Psychiatry 26:15–25

Malison RT, Price LH, Berman R, van Dyck CH, Pelton GH, Carpenter L, Sanacora G, Owens MJ, Nemeroff CB, Rajeevan N, Baldwin RM, Seibyl JP, Innis RB, Charney DS (1998) Reduced brain serotonin transporter availability in major depression as measured by [123I]-2 beta-carbomethoxy-3-beta-(4-iodophenyl)tropane and single photon emission computed tomography. Biol Psychiatry 44:1090–1098

Mann JJ, Brown RP, Halper JP, Sweeney JA, Kocsis JH, Stocher PE, Bilezikian JP (1985) Reduced sensitivity of lymphocyte beta-adrenergic receptors in patients with endogenous depression and psychomotor agitation. N Engl J Med 313:715–720

Mann JJ, Stanley M, McBride PA, McEwen BS (1986) Increased serotonin 2 and beta-adrenergic receptor binding in the frontal cortices of suicide victims. Arch Gen Psychiatry 43:954–969

Mann JJ, McBride PA, Malone KM, DeMeo M, Kelip J (1995) Blunted serotonergic responsivity in depressed patients. Neuropsychopharmacology 13:53–64

Matussek N, Ackenheil M, Hippius H, Muller F, Schroder HT, Schultes H, Wasilewski B (1980) Effect of clonidine on growth hormone release in psychiatric patients and controls. Psychiatry Res 2:25–36

McAdams C, Leonard BE (1993) Neutrophil and monocyte phagocytosis in depressed patients. Prog Neuro-Psychopharmacol & Biol Psychiat 17:971–984

McBride PA, Mann JJ, Polley MJ, Wiley AJ, Sweeney JA (1987) Assessment of binding indices and physiological responsiveness of the 5-HT2 receptor on human platelets. Life Sci 40:1799–1809

McIntyre IM, Judd FK, Marriott PM, Burrows GD, Norman TR (1989) Plasma melatonin levels in affective states. Intern J Clin Pharmacol Res 9:159–164

Meana JJ, Garcia-Sevilla JA (1987) Increased β-adrenoceptor density in the frontal cortex of depressed suicide victims. J Neural Trans 70:377–381

Meana JJ, Barturen F, Garcia-Sevilla JA (1992) Alpha-2-adrenoceptors in the brain of suicide victims increased receptor density associated with depression. Biol Psychiatry 31:471–490

Meesters P, Kerkhofs M, Charles G, Decoster C, Vanderelst M, Mendlewicz J (1985) Growth hormone release after desipramine in depressive illness. Eur Arch Psychiat Neurol Sci 235:140–142

Melia KR, Duman RS (1991) Involvement of corticotropin-releasing factor in chronic stress regulation of the of the main noradrenergic system. Proc Natl Acad Sci USA 88:8382–8386

Meltzer HY, Lowy MT (1987) The serotonin hypothesis of depression. In: Meltzer HY (ed) Psychopharmacology: the third generation of progress. Raven Press, New York, pp 513–526

Mendlewicz J, Linkowski P, Branchey L (1979) Abnormal 24hr pattern of melatonin secretion in depression. Lancet 2:1362

Miles A, Philbrick DRS (1988) Melatonin and psychiatry. Biol Psychiatry 23:405–425

Mitchell AJ (1998) The role of corticotropin releasing factor in depressive illness: a critical review. Neurosci Biobehav Rev 22:635–651

Molchan SE, Hill JL, Martinez RA, Lawlor BA, Mellow AM, Rubinow DR, Bissette G, Nemeroff CB, Sunderland T (1993) CSF somatostatin in Alzheimer's disease and major depression: relationship to hypothalamic-pituitary-adrenal axis and clinical measures. Psychoneuroendocrinology 18:509–519

Muscettola G, Potter WZ, Pickar D, Goodwin FK (1984) Urinary 3-methoxy-4-hydroxyphenylglycol and major affective disorders. Arch Gen Psychiatry 41:337–342

Nemeroff CB, Evans DL (1984) Correlation between the dexamethasone suppression test in depressed patients and clinical response. Am J Psychiatry 141:247–249

Nemeroff CB, Widerlov E, Bisette G, Walleus H, Karlsson I, Eklund K, Kilts CD, Loosen PT, Vale W (1984) Elevated concentrations of CSF corticotropin-releasing factor-like immunoreactivity in depressed patients. Science 226:1342–1344

Nemeroff CB, Owens MJ, Bissette G, Andorn AC, Stanley M (1988) Reduced corticotropin-releasing factor binding sites in the frontal cortex of suicide victims. Arch Gen Psychiatry 45:577–579

Nemeroff CB, Bissette G, Akil H, Fink M (1991) Neuropeptide concentrations in the cerebrospinal fluid of depressed patients treated with electroconvulsive therapy: corticotropin-releasing factor, beta-endorphin, and somatostatin. Br J Psychiatry 158:59–63

Nemeroff CB, Krishnan KR, Reed D, Leder R, Beam C, Dunnick NR (1992) Adrenal gland enlargement in major depression. a computed tomographic study. Arch Gen Psychiatry 49:384–387

Nowak G, Ordway GA, Paul IA (1995) Alterations in the N-methyl-D-aspartate (NMDA) receptor complex in the frontal cortex of suicide victims. Brain Res 675:157–164

Nugent DF, Leonard BE (1998) Platelet abnormalities in depression. J Serotonin Res 4:251–267

O'Keane V, Dinan TG (1991) Prolactin and cortisol responses to d-fenfluramine in major depression: evidence for diminished responsivity of central serotonergic function. Am J Psychiatry 148:1009–1015

O'Neill B, Leonard BE (1990) Abnormal zymosan-induced neutrophil chemiluminescence as a marker of depression. J Affect Disord 19:265–272

Ordway GA, Smith KS, Haycock JW (1994) Elevated tyrosine hydroxylase in the locus coeruleus of suicide victims. J Neurochem 62:680–685

Owens MJ, Nemeroff CB (1991) Physiology and pharmacology of corticotropin releasing factor. Pharmacol Rev 4:425–473

Owens MJ, Nemeroff CB (1993) Corticotropin-releasing factor, (Ciba foundation symposium, 172). 296–316

Palmer AM, Burns MA, Arango V, Mann JJ (1994) Similar effects of glycine, zinc and an oxidizing agent on [^{3}H] dizocilpine binding to the N-methyl-D-aspartate receptor in neocortical tissue from suicide victims and controls. J Neural Transm Gen Sect 96:1–8

Pandey GN, Janicak PG, Davis JM (1987) Decreased beta-adrenergic receptors in the leucocytes of depressed patients. Psychiatry Res 22:265–273

Pandey GN, Janicak PG, Javaid JI, Davis JM (1989) Increased 3H-clonidine binding in the platelets of patients with depressive and schizophrenic disorders. Psychiatry Res 28:73–88

Pandey GN, Dwivedi Y, Kumari R, Janicak PG (1998) Protein kinase C in platelets of depressed patients. Biol Psychiatry 44:909–911

Paul IA (1997) NMDA receptors and affective disorders. In: Skolnick P (ed) Antidepressants: new pharmacological strategies. Humana Press, New Jersey, pp 145–158

Pecknold JC, Suranyi-Cadotte B, Chang H, Nair NP (1988) Serotonin uptake in panic disorder and agoraphobia. Neuropsychopharmacology 1:173–176

Piletz JE, Schubert DS, Halaris A (1986) Evaluation of studies on platelet alpha2-adrenoceptors in depressive illness. Life Sci 39:1589–1616

Piletz JE, Halaris A, Saran A, Marler MR (1990) Elevated ^{3}H-para-amino-clonidine binding to platelet purified plasma membranes from depressed patients. Neuropsychopharmacology 3:201–210

Piletz JE, Halaris A, Ernsberger PR (1994) Psychopharmacology of imidazoline and a2-adrenergic receptors: implications for depression. Crit Rev Neurobiol 91:29–66

Piletz JE, Halaris A, Nelson J, Qu Y, Bari M (1996) Platelet I1-imidazoline binding sites are elevated in depression but not generalized anxiety disorder. J Psychiatry Res 30:147–168

Pimoule C, Briley MS, Gay C, Loo H, Sechter D, Zarifian E, Raisman R, Langer SZ (1983) ^{3}H-rauwolscine binding in platelets from depressed patients and healthy volunteers. Psychopharmacology 79:308–312

Pitts AF, Samuelson SD, Meller WH, Bissette G, Nemeroff CB, Kathol RG (1995) Cerebrospinal fluid corticotropin-releasing hormone, vasopressin, and oxytocin concentrations in treated patients. Biol Psychiatry 38:330–335

Pletscher A, Laubscher A (1980) Blood platelets as models for neurons: uses and limitations. J Neural Trans 16:7–16

Potter WZ, Musclettola G, Goodwin FK (1983) Sources of variance in clinical studies of MHPG. In: Maas JW (ed) MHPG: basic mechanisms and basic psychopathology. Academic Press, New York, pp 145–165

Potter WZ, Grossman F, Rudorfer MV (1993) Noradrenergic function in depressive disorders. In: Mann JJ, Kupfer DJ (eds) Biology of depressive disorders, Part A: a systems perspective. Plenum Press, New York, pp 1–27

Price LH, Charney DS, Heninger GR (1986) Effects of trazodone treatment on alpha-2 adrenoceptor function in depressed patients. Psychopharmacology 89:38–44

Price LH, Charney DS, Delgado PL, Heninger GR. (1991) Serotonin function and depression: neuroendocrine and mood responses to intravenous L-tryptophan in depressed patients and healthy comparison subjects. Am J Psychiatry. 148:1518–1525.

Purba JS, Hoogendijk WJ, Hofman MA, Swaab DF (1996) Increased number of vasopressin- and oxytocin-expressing neurons in the paraventricular nucleus of the hypothalamus in depression. Arch Gen Psychiatry 53:137–143

Raadsheer FC, Hoogendijk WJG, Stam FC, Tilders FJH, Swaab DF (1994) Increased number of corticotropin-releasing hormone expressing neurons in the hypothalamic paraventricular nucleus of depressed patients. Neuroendocrinology 60:436–444

Rausch JL, Janowsky DS, Risch SC, Huey LY (1986) A kinetic analysis and replication of decreased platelet serotonin uptake in depressed patients. Psychiatry Res 19:105–112

Redmond AM, Leonard BE (1997) An evaluation of the role of the noradrenergic system in the neurobiology of depression: a review. Human Psychopharmacol 12:407–430

Risch SC, Lewine RJ, Kalin NH, Jewart RD, Risby ED, Caudle JM, Stipetic M, Turner J, Eccard MB, Pollard W (1992) Limbic-hypothalamic-pituitary-adrenal axis activity and ventricular-to-brain ratio studies in affective illness and schizophrenia. Neuropsychopharmacology 6:95–100

Rotman A, Modai I, Munitz H, Wijsenbeek H (1979) Active uptake of serotonin by blood platelets of acute schizophrenic patients. FEBS Lett 101:134–136

Roy A, Pickar D, Linnoila M, Doran AR, Ninan P, Paul SM (1985) Cerebrospinal fluid monoamine and monoamine metabolite concentrations in melancholia. Psychiatry Res 15:281–292

Roy A, Pickar D, Paul S, Doran A, Chrousos GP, Gold PW (1987) CSF corticotropin-releasing hormone in depressed patients and normal control subjects. Am J Psychiatry 144:641–645

Roy A, Karoum F, Pollack S (1992) Marked reductions in indices of dopamine transmission among patients with depression who attempt suicide. Arch Gen Psychiatry 49:447–450

Rubin RT, Phillips JJ, Sadow TF, McCracken JT (1995) Adrenal gland volume in major depression: increase during the depressive episode and decrease with successful treatment. Arch Gen Psychiatry 52:213–218

Schatzberg AF, Samson JA, Bloomingdale KL, Orsulak PJ, Gerson B, Kizuka PP, Cole JO, Schildkraut JJ (1989) Towards a biochemical classification of depressive disorders. X: urinary catecholamines, their metabolites, and D-type scores in subgroups of depressive disorders. Arch Gen Psychiatry 46:260–268

Schildkraut JJ (1965) The catecholamine hypothesis of affective disorders: a review of the supporting evidence. Am J Psychiatry 122:509–522

Schildkraut JJ, Orsulak PJ, Labrie RA, Schatzberg AF, Gudeman JE, Cole JO, Rohde WA (1978) Towards a biochemical classification of depressive disorders. II: application of multivariate discriminant function analysis to data on urinary catecholamines and metabolites. Arch Gen Psychiatry 46:260-268

Schulz DW, Mansbach RS, Sprouse J, Braselton JP, Collins J, Corman M, Dunaiskis A, Faraci S, Schmidt AW, Seeger T, Seymour P, Tingley FDI, Winston EN, Chen YL, Heym J (1996) CP-154,526: a potent and selective nonpeptide antagonist of corticotropin releasing factor receptors. Proc Natl Acad Sci USA 93:10477-10482

Scott LV, Dinan TG (1998) Vasopressin and the regulation of hypothalamic-pituitary-adrenal axis function: implications for the pathophysiology of depression. Life Sci 62:1985-1998

Seidel A, Arolt V, Hungstiger M, Rink L, Behnisch A, Kirschner H (1995) Cytokine production and serum proteins in depression. Scand J Immunol 41:534-538

Shaw DM, Camps FE, Eccleston EG (1967) 5-hydroxytrypatmine in the hind-brain of depressive suicides. Br J Psychiatry 113:1407-1411

Siever LJ, Uhde TW, Insel TR, Roy BF, Murphy DL (1982a) Growth hormone response to clonidine unchanged by chronic clorgyline treatment. Psychiatry Res 7:139-144

Siever LJ, Uhde TW, Silberman EK, Jimerson DC, Aloi JA, Post RM, Murphy DL (1982b) Growth hormone response to clonidine as a probe of noradrenergic responsiveness in affective disorder patients and controls. Psychiatry Res 6:171-183

Sluzewska A, Rybakowski JK, Laciak M, Mackiewicz A, Sobieska M, Wiktorowicz K (1995a) Interleukin-6 levels in depressed patients before and after antidepressant treatment. Ann N Y Acad Sci 762:474-477

Sluzewska A, Rybakowski JK, Sobieska M, Bosmans E, Pollet H, Wiktorowicz K (1995b) Increased levels of alpha-1acid glycoprotein and interleukin-6 in refractory depression. Depression 3:170-175

Smith RS (1991) The macrophage theory of depression. Med Hypoth 35:298-306

Sneddon JM (1973) Blood platelets as a model for monoamine-containing neurons. Prog Neurobiol 1:151-198

Song C, Dinan T, Leonard BE (1994) Changes in immunoglobulin, complement and acute phase protein levels in the depressed patients and in normal controls. J Affect Disord 30:283-288

Spyraki C, Fibiger HC (1980) Functional evidence for subsensitivity of noradrenergic alpha 2 receptors after chronic desipramine treatment. Life Sci 27:1863-1867

Stahl SM (1977) The human platelet: a diagnostic and research tool for the study of biogenic amines in psychiatric and neurologic disorders. Arch Gen Psychiatry 34:509-516

Stahl SM (1985) Platelets as pharmacological models for the receptors and biochemistry of monoaminergic neurons. In: Longenecker GL (ed) The platelets: Physiology and pharmacology. Academic Press, New York, pp 307-340

Stahl SM, Woo DJ, Mefford IN, Berger PA, Ciaranello RD (1983) Hyperserotonemia and platelet serotonin uptake and release in schizophrenia and affective disorders. Am J Psychiatry 140:26-30

Stain-Malongren R, Tham A, Asberg-Wistedt A (1998) Increased platelet 5-HT2 receptor binding after electroconvulsive therapy in depression. J ECT 14:15-24

Stanley M, Traskman-Bendz L, Dorovini-Zis K (1985) Correlations between aminergic metabolites simultaneously obtained from human CSF and brain. Life Sci 37:1279-1286

Suhara T, Nakayama K, Inoue O, Fukuda H, Shimizu M, Mori A, Tateno Y (1992) D1 dopamine receptor binding in mood disorders measured by positron emission tomography. Psychopharmacology 106:14-18

Takeda T, Harada T, Otsuki S (1989) Platelet ^{3}H-clonidine and ^{3}H-imipramine binding and plasma cortisol level in depression. Biol Psychiatry 26:52-60

Tuomisto J, Tukainen E (1976) Decreased uptake of 5-hydroxytryptamine in blood platelets from depressed patients. Nature 262:596–598

Tuomisto J, Tukainen E, Ahlfors UG (1979) Decreased uptake of 5-hydroxytryptamine in blood platelets from patients with endogenous depression. Psychopharmacology 65:141–147

Ur E, Dinan TG, O'Keane V, Clare AW, McLoughlin L, Rees LH, Turner TH, Grossman A, Besser GM (1992) Effect of metyrapone on the pituitary-adrenal axis in depression: relation to dexamethasone suppressor status. Neuroendocrinology 56:533–538

Vale WJ, Spiess J, Rivier C, Rivier J (1981) Characterization of a 41-residue ovine hypothalamic peptide that stimulates secretion of corticotropin and beta-endorphin. Science 213:1394–1396

Van Praag HM, Korf J, Dols LCW (1972) A pilot study of the predictive value of the probenecid test in the application of 5-hydroxytryptophan as an antidepressant. Psychopharmacology 25:14–21

Vestergaard P, Sorensen T, Hoppe E, Rafaelsen OJ, Yates CM, Nicolaaou N (1978) Biogenic amine metabolites in cerebrospinal fluid of patients with affective disorders. Acta Psychiat Scand 58:88–96

Von bardeleben U, Holsboer F (1989) Cortisol response to a combined dexamethasone-human corticotropin-releasing hormone (CRH) challenge in patients with depression. J Neuroendocrinol 1:485–488

Weisse CS (1992) Depression and immunocompetence: a review of the literature. Psychological Bulletin 111:475–489

Weizman R, Laor N, Podliszewski E, Notti I, Djaldetti M, Bessler H (1994) Cytokine production in major depressed patients before and after clomipramine treatment. Biol Psychiatry 35:42–47

Werstiuk ES, Auffarth SE, Coote M, Gupta RN, Steiner M (1992) Platelet alpha 2-adrenergic receptors in depressed patients and healthy volunteers: The effect of desipramine. Pharmacopsychiatry 25:199–206

Wetterberg L, Beck-Friis J, Aperia B, Kjellman BF, Ljunggren JG, Petterson U, Sjolin A (1979) Melatonin/cortisol ratio in depression. Lancet 2:1361

Wetterberg L, Aperia B, Beck-Friis J (1981) Pineal-hypothalamic-pituitary function in patients with depressive illness. In: Fuxe K, Gustafsson JA, Wetterberg L (eds) Steroid hormone regulation of the brain. Pergamon Press, Oxford, pp 397–403

Wetterberg L, Beck-Friis J, Kjellman BF (1984) Circadian rhythms in melatonin and cortisol secretion in depression. In: Usdin E, Asberg H, Bertilsson L Frontiers in biochemical and pharmacological research in depression. Raven Press, New York, pp 197–205

Willner P (1983) Dopamine and depression: a review of recent evidence. Brain Res Rev 6:211–246

Willner P (1995) Dopaminergic mechanisms in depression and mania. In: Bloom FE, Kupfer DJ (eds) Psychopharmacology: the fourth generation of progress. Raven Press, New York, pp 921–931

Wolfe N, Katz DI, Albert ML, Almozlino A, Durso R, Smith MC, Volicer L (1990) Neuropsychological profile linked to low dopamine: In Alzheimer's disease, depression and Parkinson's disease. J Neurol Neurosurg Psychiatry 53:915–917

Wood K, Coppen A (1982) Peripheral alpha 2-adrenergic activity in the affective disorders. Adv Biochem Psychopharmacol 32:13–19

Wood K, Swade C, Abou-Saleh M, Coppen A (1984) Peripheral serotonergic receptor sensitivity in depressive illness. J Affect Disord 7:59–65

Wood K, Swade C, Coppen A (1985) Platelet alpha 2-adrenergic receptors in depression: ligand binding and aggregation studies. Acta Pharmacol Toxicol 56 (Suppl. 1):203–211

Zarko M, Biegon A (1983) Increased beta-adrenergic receptor binding in human frontal cortex of suicide victims. Soc Neurosci Abstr Abstract 210.5.

Quality of Life and Rating Scales of Depression

P. Bech[1]

[1] Psychiatric Research Unit, Frederiksborg General Hospital, 3400 Hillerød, Denmark
e-mail: pebe@fa.dk

1	Introduction	150
2	Disorder-Oriented Symptom Rating Scales for Major Depressions	151
2.1	Interview Rating Scales for Depression	152
2.2	Self-Rating Scales for Depression	156
3	Ailment-Oriented Symptom Rating Scales	157
4	Stress Adaptation: Coping Skills and Compliance	158
5	Subjective Quality of Life: Positive Psychological Well-Being	159
5.1	Generic Psychological Well-Being Scales	160
5.2	Depression-Specific Quality of Life Scales	162
6	Conclusion	163
	References	163

Abstract Rating scales for depression can be considered as the quantification of the symptoms included in the DSM-IV diagnosis of major depression. Among the essential psychometric properties of depression rating scales is their sensitivity to discriminate between antidepressants and placebo. Among the interview-based rating scales, the Hamilton Depression Rating Scale (HAM-D) is still the one most widely used. It is, however, not a unidimensional scale and the factors or dimensions of depression, anxiety and sleep should be analysed separately in drug trials. Among the self-rating scales, the coverage of major depression is insufficient in the most widely used scale, the Beck Depression Inventory. The scales based on DSM-IV, such as the Major Depression Inventory, should therefore be considered for use in future trials with antidepressants.

Quality of life scales are essentially self-rating scales as quality of life is a subjective dimension. Positive well-being is the core dimension of quality of life and can be considered the opposite pole to major depression. The most widely used quality of life scales are the Medical Outcome Studies (MOS) SF-36 and the Psychological General Well-Being Scale (PGWB). As a short, non-intrusive quality of life scale, the WHO-Five (which covers positive well-being items of the PGWB) should be considered in trials with antidepressants. However, the assessment of quality of life is also to be considered an attempt to see the whole of the patient when evaluating the outcome of treatment.

Therefore, quality of life also covers coping skills and compliance in trials with antidepressants. The top-bottom approach of health-related quality of life in the psychopharmacological treatment of depressed patients encompasses the issues of impairments, disability, wanted and unwanted treatment effects, coping skills, compliance and subjective quality of life.

Keywords Depression rating scales · Quality of life · Unidimensionality · Sensitivity · Responsiveness · Applicability

1
Introduction

Figure 1 shows the top–bottom approach of health-related quality of life when used in the assessment of major depression. It is the holistic approach originally developed by the World Health Organization (WHO) to evaluate the sequence underlying chronic disorders (WHO 1980; Bech 1993): impairments, disability, treatment, stress adaptation and subjective quality of life. Thus, at the bottom is the patient's own assessment of his or her subjective quality of life.

At the top (Fig. 1) are listed impairments that refer to the neuropsychobiological structures causing disease (e.g., impairments in the serotonergic transmitters or receptors in the brain). Disabilities, then, reflect the behavioural dysfunctions (i.e., the clinical symptoms leading to restricted social performance).

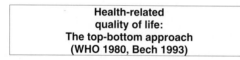

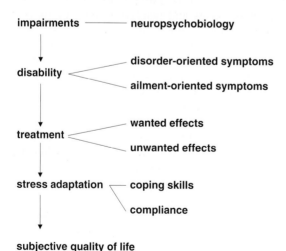

Fig. 1 Health-related quality of life: The top–bottom approach (WHO 1980; Bech 1993)

As discussed by Fayers et al. (1997), many quality of life rating scales have been developed without a clear model of validity, often combining items of disability and side-effects of treatment with subjective well-being.

This chapter discusses tools that have been developed to evaluate the various issues listed in Fig. 1, from disorder-oriented symptoms at the level of disability down to subjective well-being.

2
Disorder-Oriented Symptom Rating Scales for Major Depressions

Within the disability symptoms, Feinstein (1987) has differentiated between disorder-oriented and ailment-oriented symptoms to emphasize that clinicians should not only focus on the symptoms modified by treatment but should also measure to what extent the symptoms have a negative effect on the daily life of the patient.

However, in the criteria for major depressive episodes, DSM-IV (American Psychiatric Association 1994) has included both disorder-oriented and ailment-oriented symptoms. According to criterion B for major depression in DSM-IV, at least five of the clinical core symptoms should be present for at least 2 weeks. According to criterion C, these symptoms should cause clinically significant distress in social, occupational, and other important areas of functioning.

As pointed out by Winokur (1995), the modern systems of diagnosis in psychiatry deal mainly with a classification based on clinical symptomatology, implying that the definitions are referring to syndromes. Thus, the Feighner criteria (Feighner et al. 1972) are syndromatic diagnoses based on shared phenomenology of symptoms resulting in a high degree of inter-observer reliability. The Feighner criteria did lead subsequently to the development of the DSM-III (American Psychiatric Association 1980) and the ICD-10 (WHO 1993).

Figure 2 shows the psychometric triangle developed by Bech (1996) for the analysis of symptom rating scales for depression, including internal validity, inter-observer reliability, and external validity.

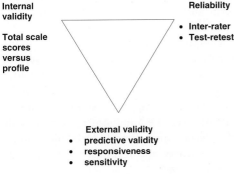

Fig. 2 Psychometric triangle for analysis of rating scales

Internal validity means both content validity ("face validity") and construct validity [i.e., the combination of items (symptoms) into a total score]. External validity means predictive validity for treatment outcome as well as responsiveness (i.e., the ability to measure changes during treatment). Sensitivity refers to the ability of a scale to discriminate between active treatment and placebo. For interview rating scales, inter-rater reliability is important. For self-rating scales, test–retest reliability is important.

2.1
Interview Rating Scales for Depression

The Hamilton Depression Rating Scale (HAM-D) (Hamilton 1960) was developed by Hamilton at the beginning of the antidepressant area (i.e., in the late 1950s). The inter-observer reliability of the scale was found to be high and acceptable in all settings in which it had been used.

In the randomized clinical trials by which antidepressants (e.g., imipramine) were shown to be superior to placebo, the HAM-D was found to be of acceptable responsiveness as well as sensitivity when compared to global improvement assessed by experienced psychiatrists. When using a global improvement response of "much" and "very much" improved, the short-term (typically 4–6 weeks) outcome of improvement with antidepressants is around 65%, whereas placebo shows 45%. When a 50% reduction on the HAM-D score from pre-treatment to endpoint (3–6 weeks later) is used as a response criterion, similar outcome rates of improvement are found, namely 65% versus 45% for antidepressants versus placebo. Thus, the responsiveness of the HAM-D and of the global improvement scales was very similar. However, the sensitivity of these scales was rather small. Thus, the advantage of antidepressants over placebo was 15%-20%.

Figure 3 shows the terminology of response, remission, relapse and recovery, as suggested by Frank et al. (1991) and Kupfer (1991). With reference to the HAM-D, response is defined as at least a 50% reduction in the pre-treatment score, and full remission is defined as a score of 7 or less. According to the European guidelines for antidepressants (1994), the treatment of an episode of major depression covers both a short-term and a medium-term period (i.e., the whole episode of major depression).

Figure 3 shows the definition of early improvement as a 25% reduction in the HAM-D score (Lauge et al. 1998). The delayed onset of action of the antidepressants, with early improvement seen after 2 weeks of therapy and response after 4 weeks, has also been investigated with depression rating scales other than the HAM-D. However, the response curve shown in Fig. 3 is similar for other depression rating scales (Lauge et al. 1998). Even the Montgomery-Åsberg Scale (MADRS) does not seem to be too different from the HAM-D, although it was originally intended to have better sensitivity and responsiveness than the HAM-D (Montgomery and Åsberg 1979; Bech 1989).

The internal validity of the HAM-D has historically been the most problematic feature of this scale. The original Hamilton manual (Hamilton 1960, 1967) is

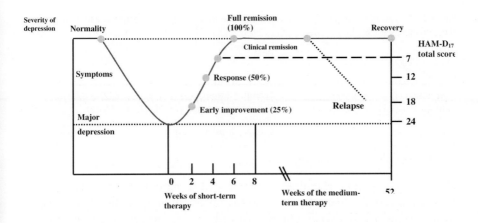

Fig. 3 Terminology of the treatment of a major depressive episode of which the short-term treatment is 8 weeks and the medium-term trial period is 44 weeks

Table 1 The universe symptoms in the HAM-D_{21}. The first 17 items constitute the HAM-D_{17}

Item no.	Symptoms
1	Depressed mood[a]
2	Feelings of guilt[a]
3	Suicide
4	Insomnia, early
5	Insomnia, middle
6	Insomnia, late
7	Work and activities[a]
8	Retardation[a]
9	Agitation
10	Anxiety, psychic[a]
11	Anxiety, somatic
12	Somatic symptoms, gastrointestinal
13	Somatic symptoms, general (loss of energy)[a]
14	Genital symptoms
15	Hypochondriasis
16	Insight
17	Loss of weight
18	Diurnal variation
19	Depersonalization and derealization
20	Paranoid symptoms
21	Obsessive and compulsive symptoms

[a] Items constitute the depression core factor.

rarely used today; rather, the American version published by Guy (1976) is now the most frequently used version (Zitman et al. 1990). However, Hamilton never accepted the definitions of the individual items in the American version (Hamilton and Shapiro 1990). Furthermore, the American version (Table 1) in-

Table 2 The two most frequently used versions of the HAM-D and their factor structure (in brackets the item nos., see Table 1)

HAM-D_{17}	HAM-D_{21}
Factors	Factors
Depression (1, 2, 7, 8, 10, 13)	Cognitive disturbances (2, 3, 9, 19, 20, 21)
Anxiety (9, 10, 11)	Anxiety/somatization (10, 11, 12, 13, 15, 16)
Sleep (4, 5, 6)	Sleep (4, 5, 6)
	Retardation (1, 7, 8, 14)

cludes 21 items while Hamilton recommended using only the first 17 items to measure severity of depressive states. In collaboration with Hamilton, Bech et al. (1986) published a version to be used with outpatients rather than inpatients. This version has been further modified for primary care physicians and family practitioners, who now treat more than 80% of depressed patients (Bech 1996).

Hamilton used factor analysis to evaluate the internal validity of the scale. He was, however, aware of the many problems involved in the use of factor analysis (e.g., the problem of generalization because the method is too sensitive to the distribution of the sample of patients selected). Table 2 compares the factors identified with the HAM-D_{17} and the HAM-D_{21}. Thus, the scale covers more than one factor. In the HAM-D_{21}, the item of depressed mood is included in the retardation factor. In the HAM-D_{17}, depressed mood is included in the depression factor. This factor, often referred to as the melancholia factor, has been identified by latent structure analysis (Bech et al. 1981). The melancholic factor includes six of the HAM-D items as indicated in Table 1 (HAM-D_6). On the basis of these 6 items, the Melancholia Scale (MES) was developed (Bech and Rafaelsen 1980). This scale consists of 11 items and has been shown by latent structure analysis to cover only one dimension (Bech 1996, 2002a). The Mania Scale (MAS) was constructed in a manner analogously to the MES (Bech et al. 1979; Bech 1996, 2002b; Smolka and Stieglitz 1999). In long-term trials with bipolar patients on lithium therapy, the MES and MAS have been shown to have a high applicability and validity in defining relapses or recurrence (Jensen et al. 1995).

Because the MES and the MAS had been developed before the release of the DSM-III, attempts have been made to modify the two scales to cover the DSM-IV and the ICD-10. Thus, the Major Depression Scale (Bech et al. 1997) and the DSM-IV Mania Scale (Bech 1996) were developed.

When the major depression symptoms are removed from the HAM-D, it is the factor of anxiety or anxiety/somatization that is most important. Therapeutically, anxiety is a substantial part of depressive states. Thus, in Europe, the Kielholz classification system of antidepressants includes a sedative-anxiolytic versus an activity profile (Bech 1993). In other words, the factor structure of the HAM-D seems important when classifying antidepressants.

The individual symptoms included in Table 1 are in accordance with Hamilton, measured on a scale of intensity. However, severity of symptoms can also be measured by frequency (i.e., how much of the time they occur). The Inventory of Depressive Symptomatology (IDS) was developed by Rush et al. by using frequency of symptoms, but with reference to the universe of the HAM-D, as shown in Table 1 (Rush et al. 1986). The final version of IDS contains 30 symptoms (Rush et al. 1996).

The IDS has both an interview version (clinician-rated, IDS-C) and a self-rating version (IDS-SR). Thus, when comparing clinician versus patient assessment of the severity of depression, the IDS measures the same content. Most other studies in this field compare not only clinicians' ratings with patients' ratings but also two different symptom contents [e.g., the HAM-D with the Beck Depression Inventory (BDI, Beck et al. 1961)]. A quick version of IDS (QIDS) has recently been developed (Rush et al. 2000) to cover the major depressive disorder items in the DSM-IV. The QIDS-C correlates 0.93 with the IDS-C_{30}, while the QIDS-SR correlates 0.94 with the IDS-SR_{30} (Rush et al. 2000).

The inter-rater reliability for the IDS-C is as high as for the HAM-D. The factor structure is also similar to the HAM-D, with the most important factors being cognitive/mood and anxiety (Rush et al. 1996).

In conclusion, the HAM-D is the most frequently used interview rating scale for measuring severity of depressive states. When most randomized clinical trials with antidepressants done over the last 40 years were reviewed (Bech 1999), it was found that the HAM-D was used in over 90% of the trials. Various attempts to develop scales that are psychometrically better than the HAM-D have not been successful. The Major Depression Scale has been developed to cover the DSM-IV concept of major depression better than the HAM-D. The most interesting attempt to modify the HAM-D has been made by Rush et al. with the IDS. The advantage of this scale is that it has both a clinician version (IDS-C) and a patient version (IDS-SR). The key issue in assessing depression based on symptoms is severity. Table 3 shows the concordance between the DSM-IV categories of a major depressive episode and total scores on the HAM-D_{17}.

Table 3 The key issue of major depression is severity. The concordance between the DSM-IV categories of major depression and the total severity score on HAM-D_{17}

DSM-IV categories of a major depressive episode (MDE)	HAM-D_{17} total scores
MDE with psychotic features	30 or higher
MDE with melancholic features	25–29
MDE without melancholic features	18–24
Less than major depression	13–17
Probably major depression	
Dysthymia	
Mixed anxiety depression	
Seasonal depression	

2.2
Self-Rating Scales for Depression

There are many more self-rating than interview scales for depression, although, in principle, disorder-oriented scales should be clinician-assisted. The most widely used self-rating scale for depression, especially in the psychotherapeutic literature, is the Beck Depression Inventory (BDI) (Beck et al. 1961). The universe of items includes 21 symptoms which mainly cover cognitive aspects according to Beck's therapy of depression. With the introduction of the DSM-IV criteria for major depressive disorder, the BDI is insufficient. However, the content of the scale has recently been modified to be closer to the DSM-IV criteria for major depression (Beck et al. 1996).

How much of the time during the past two weeks

		All of the time	Most of the time	Slightly more than half of the time	Slightly less than half of the time	Some of the time	At no time
1	Have you felt low in spirits or sad?	☐	☐	☐	☐	☐	☐
2	Have you lost interest in your daily activities?	☐	☐	☐	☐	☐	☐
3	Have you felt lacking in energy or strength?	☐	☐	☐	☐	☐	☐
4	Have you felt less self-confident?	☐	☐	☐	☐	☐	☐
5	Have you had a bad conscience or feelings of guilt?	☐	☐	☐	☐	☐	☐
6	Have you felt that life wasn't worth living?	☐	☐	☐	☐	☐	☐
7	Have you had difficulty in concentrating, e.g. when reading the newspaper or watching television?	☐	☐	☐	☐	☐	☐
8a	Have you felt very restless?	☐	☐	☐	☐	☐	☐
8b	Have you felt subdued?	☐	☐	☐	☐	☐	☐
9	Have you had trouble sleeping at night?	☐	☐	☐	☐	☐	☐
10a	Have you suffered from reduced appetite?	☐	☐	☐	☐	☐	☐
10b	Have you suffered from increased appetite?	☐	☐	☐	☐	☐	☐

Fig. 4 Major depressive inventory

In psychopharmacology, the Zung Self-Report Depression Scale (SRDS) (Zung 1965) has been used more than the BDI. However, the content of the SRDS is, like that of the BDI, not sufficient for measuring major depression in accordance with DSM-IV.

The Major Depression Inventory (MDI) (Bech 1998; Bech and Wermuth 1998) is a self-rating scale that covers major depression as defined in both the DSM-IV and the ICD-10. Psychometrically, the scale was found superior to the SRDS (Bech and Wermuth 1998). Furthermore, the MDI has shown adequate sensitivity and specificity when compared with the Present State Examination diagnosis of major depression (Bech et al. 2000). The MDI is shown in Fig. 4.

The Inventory of Depressive Symptomatology Self-Rating Version (IDS-SR) (Rush et al. 1996) has the advantage of covering the HAM-D universe of symptoms. The factor structure of the IDS-SR is similar to that of the IDS-C as well as of the HAM-D.

3
Ailment-Oriented Symptom Rating Scales

Ailment-oriented scales capturing patient-centred problems should, in principle, be self-report scales.

The Hopkins Symptom Checklist (HSCL) was developed by Parloff et al. (1954) to measure distress or discomfort in patients seeking psychotherapy help, independent of their psychiatric diagnosis. Frank (1974), who was among the co-authors of the Parloff paper, preferred the term "demoralization" to "distress", emphasizing that these patients are typically conscious of having failed to meet their own expectations or those of others, or are unable to cope with pressing problems. Parloff and Frank tried to go beyond symptoms to describe "distress" or "demoralization" but found that the symptoms are the language of distress.

The original Hopkins Symptom Checklist, which was published by Bech (1993), contained 41 items. Over the years, however, the HSCL has been enlarged. Thus, the final revision of the HSCL, the HSCL-90, has 90 items (Lipman 1989). Items have been added to cover more affective syndromes. Moreover, the scoring of the individual items was changed from "How much were you distressed?" to "How much were you bothered?", thereby focussing on psychopathological symptoms rather than on "distress" or "demoralization". Derogatis et al. (1974) identified many significant clinical factors apart from the original factor of interpersonal sensitivity. Among these factors are "retarded depression", "agitated depression", "phobic anxiety", "obsessive-compulsive", "anger-hostility", and "somatization". In other words, over the years, the HSCL has developed from an ailment-oriented rating scale to a disorder-oriented scale (i.e., covering various affective syndromes).

The Medical Outcomes Study (MOS) Short-Form (SF-36) is a questionnaire that asks about a mixture of ailment-oriented symptoms and well-being (Ware and Gandek 1994). SF-36 contains 36 items covering eight factors or dimen-

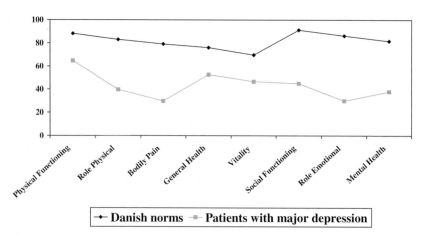

Fig. 5 The SF-36 profile of patients with major depression before treatment compared to the Danish norm means

sions: physical functioning; role limitations, physical; bodily pain; general health; vitality; social functioning; emotional health; and role limitations, emotional.

The internal validity of the eight dimensions has been found acceptable by Mcdowell and Newell (1996). The test–retest reliability has also been found adequate (Brazier et al. 1992).

The SF-36 has been widely used due to its capacity to discriminate between different diseases (e.g. Stewart et al. 1992). It has been shown that major depression has a negative impact on all eight dimensions of the SF-36 (Wells et al. 1992). The advantage of the SF-36 is that most countries now have national norms from general population studies. Figure 5 shows the Danish norms as well as the typical scores of a patient with major depression (Bjorner et al. 1998). Untreated cases of major depression are among the most disabling disorders when compared to diabetes, cancer or cardiovascular diseases.

The responsiveness of SF-36 during treatment has been evaluated only to a limited extent so far. However, when compared to the Sickness Impact Profile, the SF-36 was shown to be superior (Beaton et al. 1994)

4
Stress Adaptation: Coping Skills and Compliance

The Hopkins Symptom Checklist (HSCL) was mentioned earlier as an example of a questionnaire that was originally designed to measure distress, but which over the years has become a questionnaire for the assessment of various affective, psychopathological syndromes. On the other hand, the correspondence between the specific affective factors on the HSCL and types of coping skills has been investigated by Frank et al. in Baltimore (e.g., Frank et al. 1978) using the

Clark Personal and Social Adjustment Scale (Clark 1968) and the Locus of Control (Rotter 1966). A recent study has shown a close correspondence between the HSCL and coping strategies in patients with breast cancer (Schnoll et al. 1998).

Coping skills refer to a person's ability to change cognitive and behavioural efforts to manage specific demands that are appraised as taxing or exceeding the resources of this person (Lazarus and Folkman 1984). As discussed by Folkman et al. (1991), coping has two major functions: to manage or alter the problem that is causing distress (problem-focussed coping) and to regulate the emotional response to this problem (emotion-focussed coping).

In patients with chronic disorders, such as bipolar disorders, the changeable aspects for which problem-focussed coping is possible occur in the context of long-term prophylactic therapy with a mood stabilizer such as lithium.

In studies of the compliance of patients with bipolar disorder with long-term lithium treatment, it has been found that there are great discrepancies between psychiatrists' assessments and patients' self-reports (Cochran 1984). The main reasons for noncompliance with lithium treatment from the patients' point of view were a dislike of having medication controlling their moods and the side-effects of treatment.

Essentially, compliance means that patients have to do what doctors tell them to do. Quality of life is essentially an attempt to focus on patients' autonomy with regard to what they want to do with their lives and bodies (Brock 1988). The term "concordance" rather than "compliance" has been suggested to refer to the relationship between doctor and patient (Bech 1998).

5
Subjective Quality of Life: Positive Psychological Well-Being

Health-related quality of life assessments essentially help the patient communicate what is of importance to him or her during therapy, although this might not be of importance to other patients (Bech 1998). Self-reported scales or questionnaires are crucial in the field of subjective quality of life because only patients are able to report on how they feel.

In reviewing quality of life scales for chronic conditions such as cancer, hypertension and major depression, it became obvious that there are both disorder-specific quality of life scales and disease-anonymous (generic) quality of life scales (Bech 1995). The core components of all the generic quality of life scales include positive versus negative well-being, so that decreased positive well-being appears to be the most sensitive outcome measure (i.e., to have the greatest responsiveness). In a recent meta-analysis of nine quality of life scales (Pukrop 1997), it was concluded that a general quality of life dimension emerged, covering positive versus negative psychological well-being.

The SF-36 was discussed earlier in the chapter as a mixture of ailment-oriented symptoms and well-being. There are a total of five psychological well-being items on the SF-36, which measure both positive and negative well-being. These items are derived from the Psychological General Well-Being Scale (PGWB), which was originally developed by Dupuy (1984). The PGBW appears to be the most widely used psychological well-being scale. Another widely used well-being questionnaire is the General Health Questionnaire (GHQ), which was developed by Goldberg (1972) as a screening instrument to detect current affective disorders in general population surveys, in primary medical care settings or among general medical outpatients. Although there are many other scales in this field, in the following section, only PGWB and GHQ will be described as examples of generic quality of life scales. Among disease-specific quality of life scales, the Quality of Life in Depression Scale (QLDS) will be briefly described.

5.1
Generic Psychological Well-BeingScales

Research on quality of life questionnaires has led to improvements in two important psychometric aspects of these tools, namely standardization in normal population scores and in translation procedures. Both SF-36 and PGWB were originally developed in American English while GHQ was in British English. The SF-36 has been translated into several European languages through an International Quality of Life Assessment (IQOLA) Project (Ware and Gandek 1994). This project is unique in the literature on quality of life instruments. The *Journal of Clinical Epidemiology* has recently devoted a whole issue to the presentation of the results of these translations (Gandek and Ware 1998).

The PGWB and the GHQ have been through less extensive translation procedures; however, translations in several European languages exist. An English–English PGWB translation has been published by Hunt and McKenna (1992).

The PGBW consists of 22 items, 11 of which are positively keyed (measure positive well-being), while the rest are negatively keyed (negative well-being). Figure 6 shows the content validity of PGWB when compared to SF-36. Thus, the dimensions of general health, vitality and mental health overlap, while

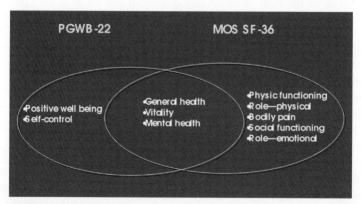

Fig. 6 Comparative content of scales

Table 4 Positive versus negative well-being: Quality of life and depression

PGWB/WHO	Major Depression (ICD-10) Inventory
Positive well-being	Negative well-being
1 In good spirits 2 Active and vigorous 3 Interested in things 4 Fresh and rested 5 Calm and relaxed	1 In low spirits 2 Lack of energy 3 Lack of interests 4 Sleep disturbances 5 Restless/subdued 6 Difficulty in concentration 7 Lacking self-confidence 8 Guilt feelings 9 Suicidal thoughts 10 Decreased/increased appetite

PGWB contains positive psychological well-being and self-control. Studies in Europe (WHO 1998; Rasmussen et al. 1999) have shown that the total scale score of PGWB is a sufficient statistic and that the score on the five items (WHO-Five) that measure positive psychological well-being is also a sufficient statistic (i.e., covers most of the variance). Table 4 shows the five items of the PGWB that measure positive well-being (Bech et al. 2003) and how they correspond to the items on the Major Depression Inventory (MDI) (Fig. 4). In other words, MDI measures negative well-being.

The test–retest reliability of PGWB has been found to be adequate (Ware et al. 1979). PGWB demonstrated the greatest responsiveness compared to ten other questionnaires in a study by Croog et al. (1986), in which the outcome of different antihypertensive therapies was investigated in terms of subjective quality of life.

In a study on patients with manic-depressive illness in prophylactic maintenance therapy, Thunedborg et al. (1995) showed that PGWB was among the best scales in predicting relapse of depressive episode.

In a trial with fluoxetine in the acute therapy of major depression, patients with low PGWB scores (meaning decreased quality of life) dropped out of treatment despite a reduction in symptoms measured by the treating physician (Hunt and McKenna 1993).

In a trial with patients suffering from both major depression and alcoholism (Borup and Undén 1994), it was found that combination therapy with disulfiram and fluoxetine improved patients' conditions to a degree equal to the national mean [which is 82; the theoretical score range of PGWB is from 0 (worst possible well-being) to 119 (best possible well-being)]. In another study performed by Speer (1998), it was shown that the patients (mainly elderly depressed patients) scored an average of 45.9 on the PGWB at admission to hospital. At the end of treatment, the average score had improved significantly to 68.5. However, the patients were still significantly below the national mean of 82. Thus, using national means as a reference, quality of life as measured by the PGWB can be used as a goal of treatment.

The PGWB includes items that measure intensity as well as frequency. Revicki et al. (1996) have suggested measuring all the PGWB items on frequency scales. WHO (1998) has also recommended using frequency items for the five items that measure positive psychological well-being (Bech et al. 2003).

The General Health Questionnaire (GHQ) included 60 items in its original version (Goldberg 1972). However, shorter forms (30, 28, and 12 items) have been recommended. The internal validity of the brief 12-item version shows that the total score is a sufficient statistic (Vieweg and Hedlund 1983). The test–retest reliability is also adequate (Goldberg 1978). The GHQ was developed primarily as a screening instrument for detecting patients with depressive illness. For this purpose, the 12-item version has a sensitivity of around 90% and a specificity of around 80% (Freeling and Tylee 1992). However, Henkel et al. (2003) have shown that the WHO-Five is superior to the GHQ in the screening of depression in the primary care setting.

Another version of the GHQ-QoL scale has been developed to measure positive psychological well-being (Bech 1993). The GHQ-QoL has recently been shown to have high applicability and validity in a Danish population study of persons aged 65 years or older who were being screened for mild dementia or major depression (Lauritzen 1998).

5.2
Depression-Specific Quality of Life Scales

The Quality of Life in Depression Scale (QLDS) was developed as a questionnaire to be used in patients with mild depression and even in the normal population (Tuynman-Qua et al. 1992). It was designed to be more easily administered than the Beck Depression Inventory (BDI) or the Zung Self-Report Scale. Although many of the items reflect depressive symptoms, the formulations of the individual items are more "soft" or "patient-friendly". The QLDS consists of 34 items which are formulated dichotomously (i.e., the response is either "yes" or "no"), so that the score ranges from 0 to 34. The QLDS has been found to have adequate internal validity (Dam et al. 1998; McKenna and Hunt 1992).

The original study with this scale, which was carried out in the Netherlands and in England (Tuynman-Qua et al. 1992), showed a correlation coefficient of around 0.60 when the QLDS was compared to the HAM-D. This is rather similar to the correlation coefficient found in many studies in which the BDI or the Zung Self-Report Scale have been compared to the HAM-D. When compared to the Psychological General Well-Being Scale (PGBW), the coefficient for the QLDS was around 0.80 (McKenna and Hunt 1992).

Studies comparing the responsiveness of the QLDS and the PGWB are still needed. In a recent pilot study comparing the antidepressive effect of fluoxetine plus placebo versus fluoxetine plus mianserin in hospitalized patients with major depression, the QLDS was found to be sensitive but not better than the HAM-D (Dam et al. 1998).

6
Conclusion

Both depression and quality of life are multi-dimensional concepts. While quality of life is a subjective concept, depression is both a subjective and behavioural concept. Hence, rating scales for depression include both interview or observer scales and self-rating scales, while scales measuring quality of life are essentially self-report scales or questionnaires.

The most widely used interview scale for depression is the HAM-D which is a multi-dimensional scale. Over the years, attempts have been made to modify the HAM-D to cover the symptoms of major depression according to DSM-IV as closely as possible. Among the self-rating scales for depression, the BDI and the Zung SRDS have been the most widely used. However, self-rating scales have also been developed to cover DSM-IV major depression, such as the MDI.

Among questionnaires that measure the zone between ailment-oriented symptoms and distress, the Hopkins SCL is the most widely used questionnaire. The Medical Outcomes Study, SF-36 can be considered a mixture of ailment-oriented symptoms and quality of life items. It is a multi-dimensional scale that shows that major depression is among the most disabling medical disorders.

Among the quality of life scales, the PGWB is the most widely used generic scale (i.e., like the SF-36, it is a scale that can be used across all medical disorders). The PGWB and the WHO-Five are essentially one-dimensional scales for measuring positive psychological well-being. In this sense, they are counterpoints to the MDI, which measures negative well-being.

The QLDS is a depression-specific quality of life scale. The advantage of this scale compared with the WHO-Five or the MDI is still to be proved.

References

American Psychiatric Association (1980) Diagnostic and statistical manual of mental disorders. 3rd Ed. American Psychiatric Association, Washington, DC
American Psychiatric Association (1994) Diagnostic and Statistical Manual of Mental Disorders. 4th Ed. (DSM-IV). American Psychiatric Association, Washington, DC
Beaton DE, Bombardier C, Hogg-Johnson S (1994) Choose your tool: A comparison of the psychometric properties of five generic health status instruments in workers with soft tissue injuries. Qual Life Res 3:50-56
Bech P (1989) Clinical effects of selective serotonin reuptake inhibitors. In: Dahl SG, Gram LF (eds) Clinical pharmacology in psychiatry. Springer, Berlin, pp 81-93.
Bech P (1993) Rating scales for psychopathology, health status and quality of life. Springer, Berlin
Bech P (1995) Quality-of-life measurements for patients taking which drugs? The clinical PCASEE perspective. PharmacoEconomics 7:141-151

Bech P (1996) The Bech, Hamilton and Zung Scales for mood disorders: screening and listening. Springer, Berlin

Bech P (1998) Quality of life in the psychiatric patient. Mosby-Wolfe, London

Bech P (1999) Pharmacological treatment of depressive disorders: a review. In: Maj M, Sartorius N (eds) Depressive disorders: WPA Series Evidence and Experience in Psychiatry. John Wiley & Sons, Chichester, pp 89-127

Bech P (2002a) The Bech-Rafaelsen Melancholia Scale (MES) in clinical trials of therapies in depressive disorders: a 20-year review of its use as outcome measure. Acta Psychiatr Scand 2002;106:252-264

Bech P (2002b) The Bech-Rafaelsen Mania Scale in clinical trials of therapies for bipolar disorder: 20-year review of its use as an outcome measure. CNS Drugs;16:47-63

Bech P, Rafaelsen OJ (1980) The use of rating scales exemplified by a comparison of the Hamilton and the Bech-Rafaelsen Melancholia Scale. Acta Psychiatr Scand 62(Suppl 285):128-132

Bech P, Wermuth L (1998) Applicability and validity of the Major Depression Inventory in patients with Parkinson's disease. Nord Psychiatr J 52:305–309

Bech P, Bolwig TG, Kramp P, Rafaelsen O J (1979) The Bech-Rafaelsen Mania Scale and the Hamilton Depression Scale: evaluation of homogeneity and inter-observer reliability. Acta Psychiatr Scand 59:420-430.

Bech P, Allerup P, Gram LF, Reisby N, Rosenberg R, Jacobsen O, Nagy A (1981) The Hamilton Depression Scale: evaluation of objectivity using logistic models. Acta Psychiatr Scand 63:290-299

Bech P, Kastrup M, Rafaelsen OJ (1986) Mini-compendium of rating scales for states of anxiety, depression, mania, schizophrenia with corresponding DSM-III syndromes. Acta Psychiatr Scand 73(suppl 326):1-37

Bech P, Stage KB, Nair NPV, Larsen JK, Kragh-Sørensen P, GjerrisA, DUAG (1997) The major depression rating Scale (MDS): inter-rater reliability and validity across different settings in randomized moclobemide trials. J Affect Dis 42:39-48

Bech P, Rasmussen NA, Olsen LR, Noerholm V, Abildgaard W (2001) The sensitivity and specificity of the Major Depression Inventory using the Present State Examination as the index of diagnostic validity. J Affect Dis 66:159-164

Bech P, Olsen LR, Kjoller M, Rasmussen NK (2003) Measuring well-being rather than absence of distress symptoms: a comparison of the SF-36 Mental Health subscale and the WHO-Five well-being scale. Int J Meth Psychiatr Res (in press)

Beck AT, Ward CH, Mendelson M, Mock J, Erbaugh J (1961) An inventory for measuring depression. Arch Gen Psychiatry 4:561–571

Beck AT, Steer RA, Ball R, Ranieri W (1996) Comparison of Beck Depression Inventories-IA and -II in psychiatric outpatients. J Pers Assess 67:588–597

Bjorner JB, Damsgaard MT, Watt T, Groenvold M (1998) Tests of data quality, scaling assumptions and reliability of the Danish SF-36. J Clin Epidemiol 51:1001-1011

Borup C, Undén M (1994) Combined fluoxetine and disulfiram treatment of alcoholism with comorbid affective disorders. A naturalistic outcome study, including quality of life measurements. Eur Psychiatry 9:83–89

Brazier JE, Harper R, Jones NMB (1992) Validating the SF-36 Health Survey questionnaire: new outcome measure for primary care. Br Med J 305:160–164

Brock DW (1988) Paternism and autonomy. Ethics 98:550–568

Clark AW (1968) The Personality and Social Network Adjustment Scale. Human Relations 21:85–96.

Cochran SD (1984) Preventing medical non-compliance in the outpatient treatment of bipolar affective disorders. J Consult Clin Psychol 5:873–878

Croog SH, Levine S, Testa M (1986) The effect of antihypertensive therapy on the quality of life. N Engl J Med 341:1657–1664

Dam J, Ryde L, Svejsø J, Lange N, Lauritsen B, Bech P (1998) Morning fluoxetine plus evening mianserin versus morning fluoxetine plus evening placebo in the acute treatment of major depression. Pharmacopsychiatry 73:393-400

Derogatis LR, Lipman RS, Rickels K, Uhlenhuth EH, Covi L (1974) The Hopkins Symptom Checklist (HSCL). In: Pichot P (ed) Modern problems of pharmacopsychiatry. Karger, Basel, pp 79-110

Dupuy HJ (1984) The Psychological General Well-Being (PGWB) index. In: Wenger NK, Mattson ME, Furberg CD, Elison J (eds) Assessment of quality of life in clinical trials of cardiovascular therapies. Le Jacq Publishing, New York, pp184-188

Fayers PM, Hand, DJ, Bjordal K and Groenvold M (1997) Causal indicators in quality of life research. Quality Life Res 6:393-406

Feighner JP, Robins E, Guze SB, Woodruff RA, Winokur G, Munoz R (1972) Diagnostic criteria for the use in research. Arch Gen Psychiatry 26:57-63

Feinstein AR (1987) Clinimetrics. Yale University Press, New Haven

Folkman S, Chesney M, McKusick L, Ironson G, Johnson DS, Coates TJ (1991) Translating coping theory into an intervention. In Eckenrode J (ed) The social context of coping. Plenum Press, New York, pp 239-260

Frank JD (1974) Persuasion and healing. Schocken, New York

Frank JD, Hoehn-Saric R, Imber SD, Liberman BL, Stone AR (1978) Effective ingredients of successful psychotherapy. Brunner Mazel, New York

Frank E, Prien RF, Jarrett RB (1991) Conceptualization and rationale for consensus definitions of terms in major depression. Arch Gen Psychiatry 48:851-855

Freeling PR, Tylee A (1992) Depression in general practice. In: Paykel ES (ed) Handbook of affective disorders. Churchill Livingstone, Edinburgh, pp 651-663

Gandek B, Ware JE (eds) (1998) Translating functional health and well-being: international quality of life assessment (IQOLA) project studies of the SF-36 health survey. J Clin Epidemiol 51/11(Special Issue)

Goldberg D (1972) The detection of psychiatric illness by questionnaire. Oxford University Press: Oxford

Goldberg D (1978) Manual of the General Health Questionnaire. NFER, London

Goldberg DP, Hillier VF (1979) A scaled version of the General Health Questionnaire. Psychol Med 9:139-145

Anonymous (1994) Guidelines on psychotropic drugs for the EC: antidepressant medical products. Eur Neuropsychopharmacol 4:62-65

Guy W (1976) Early Clinical Drug Evaluation (ECDEU) assessment manual for psychopharmacology. National Institute of Mental Health, Rockville, MD

Hamilton M (1960) A rating scale for depression. J Neurol Neurosurg Psychiat, 23:56-62

Hamilton M (1967) Development of a rating scale for primary depressive illness. Brit J Soc Clin Psychol 6:278-296

Hamilton M, Shapiro CM (1990) Depression. In: Peck DF, Shapiro CM (eds) Measuring human problems: a practical guide. Wiley, Chichester, pp 25-65

Henkel V, Mergl R, Kohnen R, Maier W, Möller H-J, Hegerl U (2003) Identifying depression in primary care: a comparison of different methods in a prospective cohort study. BMJ 326: 200-201

Hunt SM, McKenna SP (1992) A British adaptation of the General Well-Being Index: a new tool for clinical research. Br J Med Econ 2:49-60

Hunt SM, McKenna SP (1993) Measuring quality of life in psychiatry. In: Walker SR, Rosser RM (eds.). Quality of life assessment: key issues in the 1990's. Kluwer Academic Publishers, Dordrecht, pp 343-354

Jensen HV, Plenge P, Mellerup ET, Davidsen K, Toftegaard, L, Aggernaes H, Bjørum N (1995) Lithium prophylaxis of manic depressive disorder: daily lithium dosing schedule versus every second day. Acta Psychiatr Scand 92:69-74

Kupfer DJ (1991) Long-term treatment of depression. J Clin Psychiatr 52(Suppl 5):28-34

Lauge N, Behnke K, Søgaard J, Bahr B, Bech P (1998) Responsiveness of observer rating scales by analysis of number of days until improvement in patients with major depression. Eur Psychiatry 13:143–145

Lauritzen L (1998) Screening for dementia in a normal population study of persons of 65 years or older in Karlebo Kommune. Ph.D thesis, University of Copenhagen, Copenhagen

Lazarus RS, Folkman S (1984) Stress, appraisal, and coping. Springer, New York

Lipman RS (1989) Depression scales derived from Hopkins Symptoms Checklist. In: Sartorius N, Ban TA (eds) Assessment of depression. Springer, Berlin, pp 232–248

McDowell I, Newell C (1996) Measuring health: a guide to rating scales and questionnaires. Oxford University Press, New York

McKenna SP, Hunt SM (1992) A new measure of quality of life in depression: testing the reliability and construct validity of the QLDS. Health Policy 22:321–330

Montgomery SA, Åsberg M (1979) A new depression rating scale designed to be sensitive to change. Br J Psychiatry 134:382–389

Parloff MD, Kelman HC, Frank GD (1954) Comfort, effectiveness, and self-awareness as criteria for improvement in psychotherapy. Am J Psychiatry 3:343–351

Pukrop R (1997) Theoretische Explikation und empirische Validierung des Konstruktes Lebensqualität unter besonderer Berücksichtigung psychopathologischer Subgruppen und facettentheoretischer Methodik. Dissertation. Universität Bielefeld. Köln

Rasmussen NA, Nørholm V, Bech P (1999) Translation of the Psychological General Well-Being Schedule (PGWB). Quality of Life Newsletter 22:7

Revicki DA, Leidy NK, Howland L (1996) Evaluating the psychometric characteristics of the Psychological General Well-Being Index with a new response scale. Qual Life Res 5:419–425

Rotter JB (1966) Generalized expectancies for internal versus external control of reinforcement. Psychological Monographs 80:1–80

Rush AJ, Giles DE, Schlesser MA (1986) The Inventory of Depressive Symptomatology (IDS): preliminary findings. Psychiatry Res 18:65–87

Rush AJ, Gullion CM, Basco MR, Jarrett RB, Trivedi MH (1996) The Inventory of Depressive Symptomatology (IDS): psychometric properties. Psychol Med 26:477-86

Rush, AJ, Carmody T, Reimitz P-E (2000) The Inventory of Depressive Symptomatology (IDS): Clinician (IDS-C) and Self-Report (IDS-SR) ratings of depressive symptoms. International J Methods in Psychiatric Research 9:45–59

Schnoll RA, Harlow LL, Brandt U, Stolbach LL (1998) Using two factor structures of the Mental Adjustment to Cancer (MAC) scale for assessing adaptation to breast cancer. Psycho-Oncology 7:424–435

Smolka M, Stieglitz RD (1999) On the validity of the Bech-Raphaelsen Melancholic Scale. J Affect Disord 54:119–128

Speer DC (1998) Mental health outcome evaluation. Academic Press, San Diego

Stewart AL, Ware JE, Sherbourne CD, Wells KB (1992) Psychological distress/well-being and cognitive functioning measures. In: Stewart AL, Ware JE (eds.) Measuring functioning and well-being. Duke University Press, Durham, pp 102–142.

Thunedborg K, Black C, Bech P (1995) Beyond the Hamilton Depression Scale scores in manic-melancholic patients in long term relapse-prevention: a quality of life approach. Psychother Psychosom 64:131–140

Tuynman-Qua HG, De Jonghe F, McKenna S, Hunt S (1992) Quality of life in depression scale. Ibero, Houten The Netherlands

Vieweg BW, Hedlund JC (1983) The General Health Questionnaire (GHQ): a comprehensive review. J Operat Psychiat 14:74–85

Ware JE, Gandek B (1994) The SF-36 health survey: development and use in mental health research and the IQOLA Project. Int J Mental Health 23:49–73

Ware JE, Johnston SA, Davies-Avery A (1979) Conceptuation and measurement of health for adults in the Health Insurance Study. Publication no. R-1987/3. Rand Corporation, Santa Monica, California

Wells KB, Burman MA, Rogers W (1992) The course of depression in adult outpatients: results from the Medical Outcomes Study. Arch Gen Psychiatry 49:788–794

Winokur G (1995) Manic-depressive disease (bipolar): is it autonomous? Psychopathology 28(suppl 1):51–58

World Health Organization (1980) International classification of impairments, disabilities and handicaps (ICDIDH/WHO). World Health Organization, Geneva

World Health Organization (1993) International Classification of Diseases, tenth revision (ICD-10). Geneva: World Health Organization

World Health Organization (1998) Use of well-being measures in primary health care: the DEPCARE project. World Health Organization, Regional Office for Europe, Copenhagen

Zitman FG, Mennen MF, Griez E, Hooijer Cl (1990) The different versions of the Hamilton Depression Rating Scale. Psychopharmacol Ser 9:28–34

Zung WWK (1965) A self-rating depression scale. Arch Gen Psychiatry 12:63–70

Part 3
**Current Pharmacotherapy
of Major Depressive Disorder**

Part 1

Current Pharmacotherapy of Major Depressive Disorder

Overview of Currently Available Antidepressants

S. H. Preskorn[1] · R. Ross[2]

[1] Department of Psychiatry and Behavioral Sciences,
University of Kansas School of Medicine and Psychiatric Research Institute,
1010 N. Kansas, Wichita, KS 67214, USA
e-mail: spreskor@kumc.edu

[2] Ross Editorial, 228 Black Rock Mtn Ln, Independence, VA 24348, USA

1	Introduction	172
2	Mechanisms of Action of Antidepressants	174
3	Brief History of Antidepressant Development	177
References		181

Abstract Until the late 1980s when the first selective serotonin reuptake inhibitor (SSRI) was introduced, clinicians had limited options for treatment of clinical depression, with the only available antidepressants being the tricyclic antidepressants (TCAs), the monoamine oxidase inhibitors (MAOIs), and trazodone. In recent decades, antidepressant drug development has evolved from a process based on chance discovery to a rational development process that molecularly targets specific sites of action in the central nervous system. Such rational drug development has produced agents that are safer and better tolerated than the older agents such as the TCAs and the MAOIs. Currently available antidepressants have a variety of mechanism of actions. Some of them, such as the tertiary amine TCA amitriptyline, affect multiple sites of action (SOA) while others, such as the SSRIs, target a single SOA. Drugs such as venlafaxine and bupropion affect two SOAs. This chapter serves as an introduction to the chapters that follow, by reviewing the mechanisms of action of antidepressants and providing a brief overview of the history of antidepressant development. Each of the five chapters in the section that follows deals with different groups of antidepressants. The next chapter and the one following it cover the first classes of antidepressants that were introduced, the TCAs and the MAOIs, respectively. Newer antidepressants are then reviewed: the SSRIs and other recently developed antidepressants. Finally, Wheatley provides an overview of new findings concerning herbal agents in the treatment of depression.

Keywords Antidepressants · Mechanism of action · Site of action · Rational drug development

1
Introduction

The previous two sections of this book reviewed basic pharmacology and principles of neuroscience that are important to the safe and effective use of antidepressants. In this section, the different classes of currently marketed antidepressants are discussed in the order in which they were introduced. Lader (in the following chapter) and Kennedy et al. (in the chapter after that) cover the first classes of antidepressants that were introduced, the tricyclic antidepressants (TCAs) and the monoamine oxidase inhibitors (MAOIs). Newer antidepressants—selective serotonin reuptake inhibitors (SSRIs) and other recently developed antidepressants—are reviewed in the next two chapters by Preskorn et al. Finally, Wheatley provides an overview of new findings concerning herbal agents in the treatment of depression. Tables summarizing dosing recommendations and adverse events profiles of commonly used antidepressants as well as tables giving information on pharmacokinetic parameters and plasma drug concentrations are provided in the Appendix. This introductory chapter is intended to provide a transition from the first two sections and to establish a conceptual frame of reference to help readers understand the following chapters on the various classes of available antidepressants. For more information on the pharmacology of antidepressants and copies of other publications by the first author of this chapter, readers are referred to his *Applied Clinical Pharmacology* website (www.preskorn.com).

The modern era of antidepressant pharmacotherapy began in the second half of the twentieth century with the chance discovery of the TCAs and the MAOIs. These drugs were the mainstay of antidepressant pharmacotherapy for most of the remainder of that century and are still widely used around the world, particularly in developing countries where cost is a critical issue. They are also used in more developed countries but on a more limited basis, having given up substantial market share to newer antidepressants that were developed as a result of rational attempts to improve on the pharmacology of the TCAs and MAOIs. Based on an understanding of this history and development process, it is possible to rationally place all the currently available antidepressants into eight mechanistically different classes from which clinicians can select the best treatment for their patients (Table 1).

Over the last 30 years, antidepressant drug development evolved from a process based on chance to one based on rational research in which molecules were designed to selectively target specific sites of action in the central nervous system, such as specific neuroreceptors or neuronal transport pumps for specific neurotransmitters. The goal has been to move from medications such as the TCAs, which have multiple desired and undesired effects, to drugs that affect only one or two sites of action capable of mediating antidepressant efficacy while avoiding effects on other sites of action that mediate tolerability and/or safety problems. The overall goal of recent antidepressant drug development has been to produce medications that are safer, better tolerated, and less likely

Table 1 Classification of currently available antidepressants by putative mechanism(s) of action responsible for antidepressant efficacy (copyright Preskorn 2003)

Class	Antidepressants	Comments
Combined NE and 5-HT uptake inhibitors, which also have effects on multiple other neuroreceptors and fast sodium channels	Amitriptyline Amoxapine Clomipramine Doxepin Imipramine Lofepramine Trimipramine	All are tertiary amine TCAs except amoxapine
NE selective reuptake inhibitors (NSRIs)	Desipramine Maprotiline Nortriptyline Protriptyline Reboxetine	All are tertiary amine TCAs except maprotiline and the nontricyclic NSRIs such as reboxetine (not available in United States)
Monoamine oxidase inhibitors (MAOIs)	Phenelzine Tranylcypromine Selegiline Moclobemide Brofaromine	Only irreversible and nonselective MAOIs are available in the United States, but selective and reversible agents (RIMAs) such as moclobemide are available elsewhere
5HT$_{2A}$ receptor blockers and weak 5-HT uptake inhibitors	Trazodone Nefazodone	
Selective serotonin reuptake inhibitors (SSRIs)	Citalopram Escitalopram Fluoxetine Fluvoxamine Paroxetine Sertraline	
Combined 5-HT and NE uptake inhibitors (SNRIs)	Venlafaxine	
DA and NE uptake inhibitors	Bupropion	
5-HT (2A and 2C) and α_2-NE receptor blockers	Mirtazapine	

5-HT, 5-hydroxytryptamine (serotonin); DA, dopamine; NE, norepinephrine.

to interact with other co-prescribed drugs than the older agents such as the TCAs, MAOIs, and trazodone (Tallman and Dahl 1994; Preskorn 1995b). The success of this approach is evident in the great expansion in the use of newer generation antidepressants, such as the selective serotonin reuptake inhibitors, to treat even mild to moderate depression (Pincus et al. 1998). Due to the toxicity and tolerability problems of the older antidepressants (e.g., TCAs and MAOIs), physicians shied away from using these antidepressants to treat more mild but still clinically significant episodes of depression. Instead, such patients were referred for counseling approaches, which were not always successful but were nevertheless felt to be safer.

2
Mechanisms of Action of Antidepressants

Before tracing the history of antidepressant development, it may be helpful to briefly review how antidepressant drugs produce their clinical effects. For a more detailed discussion of these issues, readers are referred to the first two chapters of this volume. To produce a desired effect (e.g., antidepressant efficacy), a drug must act on a site of action (SOA) (e.g., an uptake pump, a receptor, an enzyme) that is physiologically relevant to that effect. The drug "recognizes" and binds to that SOA. The activation or inhibition of the SOA is called the drug's "mechanism of action." For example, a drug may be an agonist or antagonist at a specific serotonin receptor. The nature and magnitude of a drug's effect is determined by its:

- SOA
- Binding affinity for that SOA
- Concentration at the SOA (Preskorn 1996)
- Action on the site (i.e., agonism, antagonism, or inverse agonism)

These four factors determine the "usual" effect of the drug in the "usual" patient at the "usual" dose (as established in clinical trials). However, all patients are not "usual" because of individual variations due to factors such as age, genetics, intercurrent illnesses that affect organ functioning, concomitant medications, and social habits (e.g., smoking) (Lin and Poland 1994). Clinicians need to take such interindividual differences into account in selecting and dosing a drug. The three important variables that determine the effect of a drug in a specific patient are summarized in the following equation:

Effect =
Affinity for site of action × Concentration at site of action × Interindividual variability
(pharmacodynamics)   (pharmacokinetics)

The first two variables in this equation explain the relationship between pharmacodynamics and pharmacokinetics. The first variable determines the potential effect of the drug and how much drug is needed to engage the SOA to a clinically meaningful extent in the usual patient. The second variable determines the magnitude of the drug's effect by determining how much drug reaches the SOA. Concentration is determined by the dose the patient is taking relative to the patient's ability to clear the drug from the body (concentration=dosing rate/clearance). The third term explains how interindividual variability in patients can shift the dose–response curve to produce a greater or lesser effect than would usually be expected relative to the dose prescribed.

Knowledge of the SOA(s) of antidepressants will help clinicians understand their effects (Table 2). As shown in Table 2, some antidepressants, such as the SSRIs, directly affect only one SOA at the usual concentration achieved under therapeutic dosing conditions. [Note that some SSRIs also affect cytochrome

Table 2 Comparison of the mechanisms of action of antidepressants at clinically relevant doses[a] (adapted from de Boer et al. 1988, 1995; Bolden-Watson and Richelson 1993; Cusack et al. 1994; Frazer 1997) (copyright Preskorn 2003)

Mechanism of Action[b]	Ami-triptyline	Desi-pramine	Ser-traline	Ven-lafaxine	Nefa-zo-done	Mir-tazap-ine	Bu-propion	Tranyl-cypro-mine
Histamine-1 receptor blockade	Yes[c]	No	No	No	No	Yes[c]	No	No
Acetylcholine receptor blockade	Yes	No	No	No	No	No	No	No
NE uptake inhibition	Yes	Yes	No	Yes	No	No	Yes	No
5-HT$_{2A}$ receptor blockade	Yes	No	No	No	Yes[c]	Yes	No	No
α_1-NE receptor blockade	Yes	No	No	No	Yes	No	No	No
5-HT uptake inhibition	Yes	No	Yes	Yes[c]	Yes	No	No	No
α_2-NE receptor blockade	No	No	No	No	No	Yes	No	No
5-HT$_{2C}$ receptor blockade	No	No	No	No	No	Yes	No	No
5-HT$_3$ receptor blockade	No	No	No	No	No	Yes	No	No
Fast Na$^+$ channels inhibition	No	No	No	No	No	No	No	No
Dopamine uptake inhibition	No	No	No	No	No	No	Yes	No
Monoamine oxidase inhibition	No	No	No	No	No	No	No	Yes

5-HT, 5-hydroxytryptamine (serotonin); Na$^+$, sodium; NE, norepinephrine.

[a] Amitriptyline represents the mixed reuptake and neuroreceptor blocking class; desipramine—norepinephrine selective reuptake inhibitors; sertraline—serotonin selective reuptake inhibitors; venlafaxine—serotonin and norepinephrine reuptake inhibitors; nefazodone—5-HT$_{2A}$ and weak serotonin uptake inhibitors; mirtazapine—specific serotonin and norepinephrine receptor blockers; bupropion—dopamine and norepinephrine uptake inhibitor. Monoamine oxidase inhibitors (MAOIs) do not directly share any mechanism of action with other classes of antidepressants, although they affect dopamine, norepinephrine, and serotonin neurotransmission via their effects on monoamine oxidase.

[b] The effects of these various antidepressants are listed using a binary (yes/no) approach for simplicity and clinical relevance. The issue for clinician and patient is whether the effect is expected under usual dosing conditions. A "yes" means that the usual patient on the usually effective antidepressant dose of the drug achieves concentrations of parent drug and/or metabolites that should engage that specific target to a physiologically/clinically significant extent given the in vitro affinity of the parent drug and/or metabolites for that target. If the affinity for another target is within an order of magnitude of desired target, then that target is likely also affected to a physiologically relevant degree. For example, under usual dosing conditions, amitriptyline achieves concentrations that engage the norepinephrine uptake pump. At such concentrations, it also substantially blocks histamine-1 and muscarinic acetylcholine receptors, since it has even more affinity for those targets than it does for the norepinephrine uptake pump. Since the binding affinity of amitriptyline for the 5-HT$_{2A}$ and α_1-norepinephrine receptors and the serotonin uptake pump are within an order of magnitude of its affinity for the norepinephrine uptake pump, amitriptyline at usual therapeutic concentrations for antidepressant efficacy will also affect those targets. On the other hand, amitriptyline will not typically affect Na$^+$ fast channels at usual therapeutic concentrations because there is more than an order of magnitude (i.e., >tenfold) separation between its effects on this target versus norepinephrine uptake inhibition. Nevertheless, an amitriptyline overdose can result in concentrations which engage this target. That fact accounts for the narrow therapeutic index of the tricyclic antidepressants (TCAs) and is the reason therapeutic drug monitoring to detect unusually slow clearance is a standard of care when using such drugs.

[c] Most potent effect (i.e., effect that occurs at lowest concentration). See previous footnote for further explanation.

The binding affinities of all drugs listed in the table (except mirtazapine) are based on the work of Cusack et al. (1994) and Bolden-Watson and Richelson (1993). Information on the binding affinities of mirtazapine (including affinities for 5-HT$_{2A}$ and 5-HT$_3$ receptors) is based on the work of de Boer et al. (1988, 1995). Although the publications by the Richelson group did not include values for the 5-HT$_{2A}$ and 5-HT$_3$ receptors or the other antidepressants, Elliot Richelson of the Mayo Clinic in Jacksonville, Florida (personal communication) confirmed that the other antidepressants would be unlikely to affect these receptors under usual dosing conditions.

P450 (CYP) enzymes, which are non-neural sites of action, a fact that can have important consequences for drug–drug interactions.] However, other antidepressants affect more than one site: For example, venlafaxine and bupropion affect two SOAs and nefazodone affects three SOAs. Tertiary amine TCAs, such as amitriptyline, affect six different targets at usual therapeutic concentrations. Drugs that affect more than one site of action often act sequentially: i.e., as the concentration increases, they affect additional targets, so that their clinical effects are different at different doses. The multiple actions of the tertiary amine TCAs, like the effects of some newer antidepressants on CYP enzymes, would generally be considered a clinical disadvantage since they increase the risk of side effects or drug–drug interactions (Preskorn 1995a, 1996). The older classes of antidepressants, either because they affect a site of action with broad effects on organ function (e.g., the MAOIs, which affect an enzyme responsible for four major neurotransmitters) or because they affect multiple sites of action (e.g., the TCAs) typically have narrow therapeutic indices, poor tolerability profiles, and the potential to cause multiple types of pharmacodynamic interactions with concomitantly prescribed medications.

As a rule, most drugs including antidepressants act as antagonists at their SOA(s). Table 3 shows the effects produced by blocking a specific neural mechanism. If the drug were to act as an agonist at that site, it would have the opposite effect to what is shown in Table 3. For example, antidepressants that block neuronal uptake pumps for serotonin, norepinephrine, and dopamine act as indirect agonists by increasing the concentration of these neurotransmitters at their respective receptors. Therefore, the clinical effects of uptake inhibitors will be

Table 3 Sites of action and clinical and physiologic consequences of blockade or antagonism (adapted from with permission from Preskorn 1996, pp 48–49) (copyright Preskorn 2003)

Site of action	Consequence of blockade
Histamine-1 receptor	Sedation, antipruritic effect
Muscarinic acetylcholine receptor	Dry mouth, constipation, sinus tachycardia, memory impairment
NE uptake pump	Antidepressant efficacy, increased blood pressure, tremors, diaphoresis
5-HT$_{2A}$ receptor	Antidepressant efficacy, increased rapid eye movement sleep, antianxiety efficacy, anti-extrapyramidal symptoms
α_1-NE receptor	Orthostatic hypotension, sedation
5-HT$_2$ uptake pump	Antidepressant efficacy, nausea, loose stools, insomnia, anorgasmia
α_2-NE receptor	Antidepressant efficacy, arousal, increased libido
5-HT$_{2C}$ receptor	Antianxiety efficacy, increased appetite, decreased motor restlessness
5-HT$_3$ receptor	Antinauseant
Fast Na$^+$ channels	Delayed repolarization leading to arrhythmia, seizures, delirium
Dopamine uptake pump	Antidepressant efficacy, euphoria, abuse potential, antiparkinsonian activity, aggravation of psychosis
Monoamine oxidase	Antidepressant activity, decreased blood pressure[a]

5-HT, 5-hydroxytryptamine (serotonin); Na$^+$, sodium; NE, norepinephrine.
[a] Hypertensive crisis (i.e., markedly elevated blood pressure) and serotonin syndrome can occur when monoamine oxidase inhibitors are combined with noradrenergic and serotonin agonists, respectively.

the opposite of those shown in Table 3. By increasing the availability of serotonin receptors, SRIs (the SSRIs and the SNRI venlafaxine) act as indirect agonists at the receptors 5-HT$_{2A}$, 5-HT$_{2C}$, and 5-HT$_3$ (Glennon and Dukat 1994). This explains why all SRIs can (1) decrease rapid eye movement (REM) and stage IV sleep due to indirect stimulation of the 5-HT$_{2A}$ receptor (Oswald and Adam 1986; Saletu et al. 1991; Armitage et al. 1995; Staner et al. 1995; Sharpley et al. 1996), (2) decrease appetite and cause motor restlessness due to indirect stimulation of the 5-HT$_{2C}$ receptor (Tecott et al. 1995), and (3) cause nausea due to indirect stimulation of the 5-HT$_3$ receptor (Glennon and Dukat 1994). Indirect stimulation of one or more of these 5-HT receptors likely mediates the sexual dysfunction caused by all serotonin reuptake inhibitors (SSRIs and also dual reuptake inhibitors such as venlafaxine) (Modell et al. 1997; Montejo-Gonzales et al. 1997).

Blocking, rather than stimulation, of certain receptors can explain the activities of certain other antidepressants. For example, nefazodone and mirtazapine most likely increase stage IV sleep by blocking the 5-HT$_{2A}$ receptor (Rush et al. 1998). Mirtazapine most likely increases appetite and decreases motor restlessness by blocking the 5-HT$_{2C}$ receptor and can treat nausea by blocking the 5-HT$_3$ receptor (Glennon and Dukat 1994). Trazodone is a 5-HT$_{2A}$ blocker that is sometimes used as an antidote for the sleep disturbances caused by SRIs (Nierenberg et al. 1992). The effects of mirtazapine and trazodone on sleep are also amplified by their blockade of central histaminic receptors (Bolden-Watson and Richelson 1993; Cusack et al. 1994; Frazer 1997; De Boer et al. 1988, 1995). The fact that nefazodone and mirtazapine cause minimal, if any, sexual dysfunction is consistent with their minimal inhibition of serotonin uptake, perhaps coupled with their blockade of the 5-HT$_{2A}$ receptor (De Boer et al. 1988, 1995; Bolden-Watson and Richelson 1993; Cusack et al. 1994; Frazer 1997).

3
Brief History of Antidepressant Development

Before the introduction of the SSRIs, all psychotropic medications were the result of chance observation. For example, lithium came from studies looking for putative endogenous psychomimetic substances excreted in the urine of psychotic patients (Cade 1949). The phenothiazines came from a search for better preanesthetic agents (Laborit et al. 1952). The TCAs were the results of an unsuccessful attempt to improve on the antipsychotic effectiveness of phenothiazines (Kuhn 1958). The MAOIs came from a failed attempt to develop effective antitubercular medications (Crane 1957). The first studies of benzodiazepines were unsuccessful attempts to treat schizophrenia. Despite the initial failure of these drugs to be therapeutic in the areas for which they were being tested, astute clinical investigators recognized their therapeutic value for other conditions: lithium for bipolar disorder, phenothiazines for psychotic disorders, TCAs and MAOIs for major depression, and benzodiazepines for anxiety disorders.

This movement from chance to rational drug discovery has been the rule not only in psychiatry but also in all areas of drug development (Preskorn 1994). That change occurs in concert with a movement from a situation in which there is too little knowledge of the underlying biological causes of illnesses to permit a more rational approach to drug development to a stage where there is sufficient knowledge to begin rationally targeting mechanisms important in the pathophysiology of a disease. Ironically, many of the early chance discovery drugs have played an important role in providing insights into pharmacological mechanisms that mediate antidepressant efficacy.

The TCAs and MAOIs, the antidepressant properties of which were discovered by chance, were the first scientifically proven treatments for major depression. They demonstrated that major depression was amenable to medical interventions just like other medical conditions such as hypertension or diabetes. Unfortunately, like most chance discovery drugs, their beneficial effects were associated with numerous undesirable effects since the TCAs affect multiple SOAs over a relatively narrow range of concentrations (Preskorn and Irwin 1982; Preskorn and Fast 1991; Bolden-Richelson and Watson 1993; Cusack et al. 1994). The inhibition of fast sodium channels by the TCAs can cause potentially serious effects on cardiac conduction that occur at concentrations that are only an order of magnitude higher than those needed to inhibit the neuronal uptake pumps for NE and 5-HT, which are the putative sites of action mediating the antidepressant efficacy of the TCAs. This explains why an overdose of TCAs that is only 5–10 times the therapeutic dose can cause serious toxicity.

Table 4 illustrates the cocktail of drugs, each having only one predominant mechanism of action, that a patient would have to take to reproduce the effects that occur in a patient taking a tertiary amine TCA such as amitriptyline. Obviously, the problem with amitriptyline is that the patient has to experience a wide range of effects, some of them undesirable, to receive the benefit of the mechanism that medicates an antidepressant response. The problem is further complicated by the interindividual variability in the clearance of the TCAs that can result from a patient being a slow or fast metabolizer or from the effects of age,

Table 4 Amitriptyline: polydrug therapy in a single pill

Drug	Action
Chlorpheniramine	Histamine-1 receptor blockade
Cimetidine	Histamine-2 receptor blockade
Benztropine	Acetylcholine receptor blockade
Desipramine	NE uptake inhibition
Nefazodone	5-HT$_{2A}$ receptor blockade
Sertraline	Serotonin uptake inhibition
Prazosin	α_1-NE receptor blockade
Yohimbine	α_2-NE receptor blockade
Quinidine	Direct membrane stabilization

5-HT, 5-hydroxytryptamine (serotonin); NE, norepinephrine.

medical illness, or taking a concomitant medication that either induces or inhibits the clearance of the TCA (Preskorn 1993). Thus, certain patients can develop serious adverse effects while taking routine doses due to the accumulation of toxic concentrations (Preskorn et al. 1989). Although therapeutic drug monitoring may help clinicians adjust the dose to avoid these problems, adverse effects that are difficult to tolerate may often occur at concentrations of the TCAs that are actually subtherapeutic. This problem with the TCAs highlighted the need for new drugs with a much wider gap between the concentration at which they have potency for depression and the concentration at which they cause undesirable side effects.

Despite the problems associated with the TCAs and MAOs, these early classes of antidepressants served as roadmaps to improve our understanding of the mechanisms of action that medicated their desirable and undesirable effects. They implicated the potentiation of neurotransmission in one or more central biogenic amine neural systems as a potential mechanism of action responsible for their antidepressant efficacy. This information, coupled with improved means of isolating and studying the effects of drugs on specific neural mechanisms, was essential in the rational drug development efforts that led to the development of the newer generation antidepressants covered in chapters entitled "Selective Serotonin Reuptake Inhibitors" and "Other Antidepressants."

The goal of rational psychiatric drug development is to develop drugs that target only specific brain mechanisms believed to be important in the pathophysiology of a psychiatric syndrome (e.g., major depression), while avoiding effects on other targets that mediate unwanted effects (e.g., peripheral anticholinergic effects) (Preskorn 1990, 1995b; Tallman and Dahl 1994). This concept is also relevant to the last section of this book, which begins with a discussion of cutting-edge theories of mechanisms mediating antidepressant response and then proceeds to a discussion of investigational antidepressants either in development or soon to be in development and ends with a discussion of the human genome project and its role in identifying previously unknown targets that might be capable of mediating antidepressant response.

Figure 1 illustrates the evolution of antidepressants over the past three decades. Based on the knowledge gained from studying the pharmacology of the TCAs and the MAOIs, biogenic amine neurotransmission (e.g., dopamine, norepinephrine, and dopamine) became the target for rational psychiatric drug development during the last 30 years. This focus was further reinforced by basic neuroscience studies demonstrating that these neurotransmitters systems subserve a wide range of basic brain functions including the perception of pain, sleep, thermal regulation, appetite, gut regulation, balance, reproductive functioning, motor functioning, higher cognitive functioning, and sensory interpretation. Many of these basic functions are disturbed in patients suffering from major depression.

While newer antidepressants are more selective and thus produce a narrower range of effects in comparison to TCAs and MAOIs, they still have their limitations. For example, the SRIs (SSRIs and venlafaxine) are selective in terms of af-

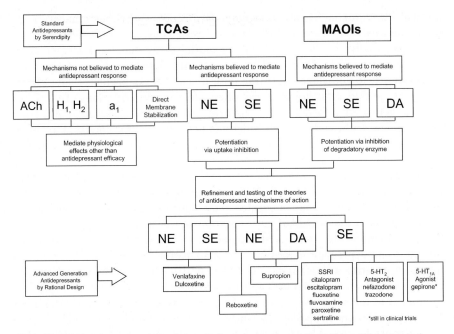

Fig. 1 Mechanism of action and development of standard and new generation antidepressants (copyright Preskorn 2003)

fecting the neuronal uptake pump for serotonin. Nevertheless, they are not selective in terms of their effects on the multitude of postsynaptic serotonin receptors such as 5-HT_{1A}, 5-HT_{1D}, 5-HT_{2A}, and 5-HT_3 (Hoyer et al. 1994). As a result, they can produce many different serotonin-mediated effects, some of which are likely unnecessary for antidepressant efficacy. For example, SRIs can cause nausea, a feeling of lack of coordination, decreased libido, suppression of REM sleep, and akathisia as well as being useful in the treatment of such apparently disparate disorders as major depression, anxiety disorders, pain disorders, and premature ejaculation. Nevertheless, in contrast to the TCAs, SRIs do not affect other neuroreceptors such as histamine, acetylcholine, and α-adrenergic receptors or fast sodium channels; and it is effects on those sites of action that are responsible for many of the safety and tolerability problems with the TCAs. This targeted development explains the similarities between the different SRIs.

While the introduction of the SSRIs and other newer antidepressants represented a significant advance in the treatment of major depression and in large measure account for the substantial expansion in the antidepressant market seen over the last decade, they also demonstrate the adage that nature sides with the "hidden flaw." Although intended to be selective, three of the SSRIs (i.e., fluoxetine, fluvoxamine, and paroxetine) have subsequently been found to be more potent as inhibitors of specific drug metabolizing CYP enzymes than as antidepressants (Preskorn and Catterson, this volume). That is because these antidepressants were developed before the isolation of the first CYP enzyme in 1988. As a result, these three SSRIs are prone to specific types of drug–drug interactions (for a more detailed discussion of this issue, see the chapter entitled "Selective Serotonin Reuptake Inhibitors", this volume). In fact, their effect on the CYP enzyme system has now been recognized as the principal characteristic that distinguishes among the different SSRIs. While fluoxetine, fluvoxamine, and paroxetine all inhibit one of more CYP enzymes to a clinically significant degree at usually effective antidepressant doses, citalopram and sertraline do not (Preskorn 1996). As a result, those two SSRIs have become the rational choice among the SSRIs, particularly when one considers that patients on an antidepressant are frequently taking multiple other medications. In essence, fluoxetine, fluvoxamine, and paroxetine would likely not be developed today. Instead, their propensity for inhibiting human drug metabolizing CYP enzymes would be recognized during the preclinical stage of development (Preskorn 2000a,b). Following that finding, the drug developers would most likely decide against advancing those drugs into human testing. Instead, they would tweak the structure of the molecule to produce a new candidate agent which, in this case, would be capable of blocking serotonin uptake without substantially inhibiting any human drug metabolizing CYP enzyme (e.g., as is the case with citalopram and sertraline). This brief discussion illustrates that drug development is a continually evolving process limited only by available knowledge and technology, as will be discussed in the last section of this volume.

References

Armitage R, Trivedi M, Rush AJ (1995) Fluoxetine and oculomotor activity during sleep in depressed patients. Neuropsychopharmacology 12:159-165
Bolden-Watson C, Richelson E (1993) Blockade by newly-developed antidepressants of biogenic amine uptake into rate brain synaptosomes. Life Sci 52:1023-1029
Cade JFJ (1949) Lithium salts in the treatment of psychotic excitement. Med J Aust 2:349-352
Crane GE (1957) Iproniazid (Marsilid) phosphate, a therapeutic agent for mental disorders and debilitating disease. Psychiatry Res Rep 8:142-152
Cusack B, Nelson A, Richelson E (1994) Binding of antidepressants to human brain receptors: focus on newer generation compounds. Psychopharmacology 114:559-565
de Boer TH, Maura G, Raiteri M, de Vos CJ, Wieringa J, Pinder RM (1988) Neurochemical and autonomic pharmacological profiles of the 6-aza-analogue of mianserin, Org 3770 and its enantiomers. Neuropharmacology 27:399-408
de Boer T, Ruigt G, Berendsen H (1995) The alpha2-selective adrenoceptor antagonist Org 3770 (mirtazapine remeron) enhances noradrenergic and serotonergic transmission. Hum Psychopharmacol 10:107S-118S
Frazer A (1997) Antidepressants. J Clin Psychiatry 58(suppl 6):9-25
Glennon RA, Dukat M (1994) Serotonin receptor subtypes. In: Bloom FE, Kupfer DJ (eds) Psychopharmacology: the fourth generation of progress. Raven Press, New York, pp 1861-1874
Hoyer D, Clarke DE, Fozard JR, Hartig PR, Martin GR, Mylecharane EJ, Saxena PR, Humphrey PP (1994) International Union of Pharmacology classification of receptors for 5-hydroxytryptamine (serotonin). Pharmacol Rev 46:157-203

Kuhn R (1958) The treatment of depressive states with G-22355 (imipramine hydrochloride). Am J Psychiatry 115:459-464

Laborit H, Huguenard P, Alluaume R (1952) Un nouveau stabil-isateur vegetatif, le 4560 RP. Presse Med 60:206-208

Lin KM, Poland RE (1994) Ethnicity, culture and psychopharmacology. In : Bloom FE, Kupfer DF (eds) Psychopharmacology: the fourth generation of progress. Raven Press, New York, pp 1907-1918

Modell JG, Katholi CR, Modell JD, DePalma RL (1997) Comparative sexual side effects of bupropion, fluoxetine, paroxetine, and sertraline. Clin Pharmacol Ther 61:476-487

Montejo-Gonzalez AL, Llorca G, Izquierdo JA, Ledesma A, Bousono M, Calcedo A, Carrasco JL, Ciudad J, Daniel E, De la Gandara J, Derecho J, Franco M, Gomez MJ, Macias JA, Martin T, Perez V, Sanchez JM, Sanchez S, Vicens E (1997) SSRI-induced sexual dysfunction: fluoxetine, paroxetine, sertraline, and fluvoxamine in a prospective, multicenter, and descriptive clinical study of 344 patients. J Sex Marital Ther 23:176-194

Nierenberg AA, Cole JO, Glass L (1992) Possible trazodone potentiation of fluoxetine: a case series. J Clin Psychiatry 53:83-85

Oswald I, Adam K (1986) Effects of paroxetine on human sleep. Br J Clin Pharmacol 22:97-99

Pincus HA, Tanielian TL, Marcus SC, Olfson M, Zarin DA, Thompson J, Zito JM (1998) Prescribing trends in psychotropic medications: primary care, psychiatry, and other medical specialties. JAMA 279:526-531

Preskorn SH (1990) The future and psychopharmacology: potentials and needs. Psychiatr Ann 20(suppl 11):625-633

Preskorn SH (1993) Pharmacokinetics of antidepressants: why and how they are relevant to treatment. J Clin Psychiatry 54(suppl 9):14-34

Preskorn SH (1994) Antidepressant drug selection: criteria and options. J Clin Psychiatry 55(suppl A):6-221

Preskorn SH (1995a) Comparison of the tolerability of bupropion, fluoxetine, imipramine, nefazodone, paroxetine, sertraline, and venlafaxine. J Clin Psychiatry 56(suppl 6):12-21

Preskorn SH (1995b) Should rational drug development in psychiatry target more than one mechanism of action in a single molecule? Intl Rev Psychiatry7:17-28

Preskorn SH (1996) Clinical pharmacology of selective serotonin reuptake inhibitors. Professional Communications, Caddo, OK

Preskorn SH (2000a) The human genome project and modern drug development in psychiatry. J Pract Psychiatry Behav Health 6:272-276 (available at www.preskorn.com)

Preskorn SH (2000b) The stages of drug development and the human genome project: drug discovery. J Pract Psychiatry Behav Health 6:341-344 (available at www.preskorn.com)

Preskorn SH, Fast GA (1991) Therapeutic drug monitoring for antidepressants: efficacy, safety and cost effectiveness. J Clin Psychiatry 52(suppl 6):23-33

Preskorn SH, Irwin HA (1982) Toxicity of tricyclic antidepressants-kinetics, mechanism, intervention: a review. J Clin Psychiatry 43:151-156

Preskorn SH, Jerkovich GS, Beber JH, Widener P (1989) Therapeutic drug monitoring of tricyclic antidepressants: a standard of care issue. Psychopharmacol Bull 25:281-284

Rush AJ, Armitage R, Gillin JC, Yonkers KA, Winokur A, Moldofsky H, Vogel GW, Kaplita SB, Fleming JB, Montplaisir J, Erman MK, Albala BJ, McQuade RD (1998) Comparative effects of nefazodone and fluoxetine on sleep in outpatients with major depressive disorder. Biol Psychiatry 44:3-14

Saletu B, Frey R, Krupka M, Anderer P, Grünberger J, See WR (1991) Sleep laboratory studies on the single-dose effects of serotonin reuptake inhibitors paroxetine and fluoxetine on human sleep and awakening qualities. Sleep 14:439-447

Sharpley AL, Williamson DJ, Attenburrow ME, Pearson G, Sargent P, Cowen PJ (1996) The effects of paroxetine and nefazodone on sleep: a placebo controlled trial. Psychopharmacology 126:50-54

Staner L, Kerkhofs M, Detroux D, Leyman S, Linkowski P, Mendlewicz J (1995) Acute, subchronic and withdrawal sleep EEG changes during treatment with paroxetine and amitriptyline: a double-blind randomized trial in major depression. Sleep 18:470-477

Tallman JF, Dahl SG (1994) New drug design in psychopharmacology: the impact of molecular biology. In: Bloom FE, Kupfer DF (eds) Psychopharmacology: the fourth generation of progress. Raven Press, New York, pp 1861-1874

Tecott LH, Sun LM, Akana SF, Strack AM, Lowenstein DH, Dallman MF, Julius D (1995) Eating disorder and epilepsy in mice lacking 5-HT2c serotonin receptors. Nature 374:542-546

Tricyclic Antidepressants

M. Lader[1]

[1] Institute of Psychiatry, King's College, Denmark Hill, London, SE5 9AF, UK
e-mail: spklmhl@iop.kcl.ac.uk

1	Introduction and History	186
2	Chemistry, Pharmacology and Clinical Effects	188
2.1	Imipramine as an Exemplar TCA	188
2.1.1	Basic Pharmacology	188
2.1.2	Clinical Pharmacology	188
2.1.3	Clinical Uses	189
2.1.4	Efficacy in Acute Depression	189
2.1.5	Rate of Onset of Efficacy	190
2.2	Amitriptyline	191
2.2.1	Nortriptyline	191
2.3	Other Tricyclic Antidepressants	192
3	Metabolism/Plasma Concentrations and Clinical Response	193
4	Safety and Adverse Effects	195
4.1	Drug–Drug Interactions	195
4.1.1	CNS Drugs	195
4.1.2	Cardiovascular Drugs	195
4.2	Adverse Effects	196
4.2.1	Central Nervous System	196
4.2.2	Cardiovascular	197
4.2.3	Gastrointestinal	198
4.2.4	Genitourinary	198
4.2.5	Metabolic-Endocrine	199
4.2.6	Miscellaneous	199
4.2.7	Drop-Out Rates as an Indicator of Tolerability	199
4.3	Overdose	200
4.4	Teratogenicity	201
5	Clinical Use and Therapeutic Indications: Major Depression	202
5.1	Acute Phase	202
5.2	Continuation Phase	203
5.3	Maintenance Therapy	203
6	Special Considerations: Use in Primary Care	204
References		205

Abstract The first truly effective antidepressants were the tricyclics (TCAs) such as imipramine and amitriptyline. Many more followed in the 1960s and 1970s, before they were partly superseded by newer drugs such as the selective serotonin reuptake inhibitors (SSRIs). TCAs were discovered serendipitously, their clinical potential discerned, and their pharmacology was later elucidated. The first members of the class were three-ringed in chemical structure, but later compounds were more varied chemically. They act to increase biogenic amines in the brain, especially serotonin, norepinephrine and sometimes dopamine. The mechanism is to block reuptake back into the pre-synaptic neuron. They have a gradual onset of action and help about 60%–80% of depressed patients. However, they seem most effective in patients of moderate degrees of severity of illness. Some correlation between plasma concentration and clinical response has been established but this is of limited clinical application.

The TCAs have a wide range of unwanted effects, involving many bodily systems. Some induce drowsiness, some alertness. The convulsive threshold is lowered. The most troublesome side effects are autonomic, with dry mouth, constipation, blurring of vision, hypotension and tachycardia. Urinary retention occasionally occurs in elderly males. Adherence to treatment may be disappointingly low because of intolerability. A further major disadvantage is toxicity in overdose. Drug interactions involve other CNS drugs, especially sedatives and alcohol.

In clinical usage, the TCAs are given in the acute phases of depression and continued into the maintenance phase. They are also indicated as prophylactic therapy in patients prone to recurrent episodes. Patients with depression secondary to a dementing process often respond well to small doses.

The TCAs are still widely used, particularly in primary care, but mainly for reasons of cost. In most countries, secondary care practitioners prefer SSRIs because of their better side-effect profile and superior safety.

Keywords Tricyclic antidepressants · Reuptake inhibition · Brain biogenic amines · Depressive disorder · Relapse · Recurrence

1
Introduction and History

Tricyclic antidepressants (TCAs) such as amitriptyline and imipramine along with the monoamine oxidase inhibitors (Kennedy et al., this volume) were the first truly effective antidepressants. They, like most of the first drugs in any therapeutic class at the time, were chance discovery medications. Ironically, the TCAs were the result of a failed attempt to produce a better antipsychotic by modifying the structure of chlorpromazine. Although the TCAs were chance discovery drugs, they had three major effects on psychiatry and clinical psychopharmacology. First, they were the first medications that could successfully treat clinical depression, a serious psychiatric medical condition. Second, they showed that such psychiatric illnesses could be treated like other medical conditions and thus provided an impetus for the remedicalisation of psychiatry. Third, they provided the first useful clues as to what might underlie clinical depression and thus gave rise to the biogenic amine theories and served as blueprints for the development of subsequent generations of antidepressants (see the two chapters by Preskorn et al., "Selective Serotonin Reuptake Inhibitors" and "Other Antidepressants" in this volume).

The nomenclature of this class of antidepressants is problematic because the term "tricyclic" refers to the chemical structure and, as such, is descriptive but not specific, nor does it convey any information about the pharmacology of these drugs. For example, phenothiazines also have a three-ringed core structure. Also, some newer drugs of this pharmacological type considered to be "tricyclics" contain one, two or four rings. The term "monoamine re-uptake inhibitors" has been alternatively proposed instead of "tricyclic" to parallel the description of mechanism of action used with other classes of antidepressants [e.g., monoamine oxidase inhibitors, selective serotonin re-uptake inhibitors (SSRIs), non-serotonin re-uptake inhibitors]. However, having made this point, by convention, in this chapter I will use the term "tricyclic antidepressant" or TCA.

The first tricyclic compound produced, imipramine, is an iminodibenzyl compound. In the later 1940s, such chemicals were evaluated as possible antihistaminic and antiparkinsonian agents, but proved relatively ineffective for these purposes. However, the discovery of chlorpromazine as an antipsychotic caused renewed interest in these compounds. Although imipramine was ineffective as an antipsychotic agent, Roland Kuhn tried imipramine in depressed patients as well as in patients with schizophrenia and discovered its antidepressant properties.

Imipramine was first marketed in 1959 and gradually became established in the treatment of depression. Several companies then synthesised other tricyclic compounds, which were evaluated in the treatment of depression and introduced. Amitriptyline and trimipramine also have marked sedative properties. Many other compounds have subsequently been marketed, most of which have few if any advantages over these prototypal compounds. Over the years, other indications for the TCAs in addition to depression have been studied. These include nocturnal enuresis, bulimia nervosa and obsessive-compulsive disorder (OCD) (Orsulak and Waller 1989).

2
Chemistry, Pharmacology and Clinical Effects

2.1
Imipramine as an Exemplar TCA

2.1.1
Basic Pharmacology

An extensive range of pharmacological actions have been described for imipramine, and its pharmacology will be described as the exemplar for the TCAs as a class (Stahl and Palazidou 1986). Imipramine affects several neurotransmitters, including 5-hydroxytryptamine, norepinephrine, and acetylcholine. Imipramine has sympathomimetic effects due to its ability to inhibit the re-uptake of norepinephrine from the synaptic cleft into the presynaptic neuron. The active re-uptake transport system is blocked, as is that related to 5-HT uptake. In vivo, imipramine is more potent than desipramine in blocking 5-HT re-uptake. Conversely, desipramine is more potent than imipramine in blocking norepinephrine uptake. Consequently, with chronic administration, the uptakes of both amines are blocked. Adaptive processes occur within the neuron and in receptors to overcome the amine re-uptake blockade (Stahl and Palazidou 1986): the synthesis of the neurotransmitters is reduced, as is the consequent turnover. The effects on brain amines and mechanisms are therefore complex.

Imipramine's agonistic effect on biogenic amines, postulated to be involved in the pathophysiology of depression, can be demonstrated by its prevention, and partial reversal, of the depressant effects of reserpine and tetrabenazine. In addition, amphetamine, cocaine and other central stimulants are potentiated by imipramine, and the uptake of indirectly acting sympathomimetic amines, such as tyramine, is blocked.

Imipramine binds to cholinoceptors and has anticholinergic properties similar to those of atropine. Thus, it can inhibit acetylcholine-induced spasm of the guinea-pig ileum and physostigmine-induced arousal of the cat's electroencephalogram (EEG). However, the antihistaminic effects of imipramine are weak. (Readers are referred to the earlier chapter in this volume by Horst for a more detailed discussion of the synaptic pharmacology underlying such effects).

2.1.2
Clinical Pharmacology

In man, imipramine induces slow-wave (theta) activity in the EEG with generalised desynchronisation. It may activate epilepsy. REM sleep is suppressed. Cognitive, psychomotor and affective functions in normal subjects are impaired (Seppala and Linnoila 1983), with memory being particularly affected. Positive mood changes can be detected in normal subjects but only those selected for their high scores on ratings of depressive affect.

In the periphery, the anticholinergic actions of imipramine are the most easily demonstrable. Potentiation of norepinephrine can also be shown. The blockade of amine uptake is demonstrable by injecting tyramine intravenously (i.e. more tyramine is needed to increase the blood pressure after imipramine).

2.1.3
Clinical Uses

In his original account of the clinical effects of imipramine in over 500 patients with various types of depressive illness, Kuhn (1958) concluded that "The main indication is without doubt a simple endogenous depression," and that "every complication of the depression impairs the chances of success of the treatment." Although these statements are still generally true, some cases of "reactive depression" seem to respond to tricyclic antidepressants.

Patients with depression associated with early dementia are also responsive to TCAs (Raskin and Crook 1976). However, depressed schizophrenics and depressive reactions in the context of personality disorders are much less amenable to treatment, as are adolescent depressives (Hazell et al. 1995). In this review, the researchers performed a meta-analysis of 12 randomised controlled trials and concluded that the TCAs were no more effective than placebo in depressed subjects aged 16–18 years. The pooled-effect size was only 0.35 standard deviations. The pooled odds ratio was only 1.08, again indicating no significant benefit of treatment.

2.1.4
Efficacy in Acute Depression

Many controlled clinical trials have shown imipramine to be superior to placebo, but there have also been failed trials (Klein and Davis 1969). By and large, 60%–70% of patients with major depressive disorder respond adequately to imipramine as compared with one-third of patients given placebo. However, some trials suggest that imipramine may not be as effective in more severely depressed patients. In these patients, who are usually admitted to the hospital, imipramine has been found to be less effective than electroconvulsive therapy. At the other end of the continuum of severity, the efficacy of imipramine as compared with placebo in mildly depressed patients has not been established. In one trial, patients with a moderate degree of depression improved significantly, whereas those who only just met inclusion criteria for severity did not (Stewart et al. 1983). These findings may reflect the nature of the illness rather than the treatment (i.e. mild depression may be more responsive to placebo and more severe depression may be unresponsive to any treatment).

Nelson et al. (1984b) identified symptoms that improve significantly with antidepressant treatment and correlated them with those symptoms that were found to be responsive in two previous studies (Montgomery and Asberg 1979; Bech et al. 1981). A fair degree of concordance is apparent: across all three stud-

Table 1 Comparison of symptoms useful for measuring change in severity of depression

Nelson et al. 1984b	Montgomery and Asberg 1979	Bech et al. 1981
Depressed mood (includes apparent and reported sadness); hopelessness	Apparent sadness; reported sadness (includes hopelessness)	Depressed mood (includes apparent and reported sadness and hopelessness)
Guilt, worthlessness	Pessimistic thoughts (includes guilt and worthlessness)	Guilt (includes worthlessness)
Early morning awakening	Reduced sleep	–
Decreased interest in work and activities	Lassitude (difficulty initiating and performing work)	Decrease in work and interests
Decreased appetite	Reduced appetite	–
Loss of energy	–	Somatic symptoms: general (includes loss of energy)
Anhedonia	Inability to feel	–
Somatic anxiety	–	–
–	Inner tension	Psychic anxiety
–	Concentration difficulties	–
–	Suicidal thoughts	–
–	–	Retardation

ies, the symptoms susceptible to improvement are depressed mood (sadness), guilt and worthlessness, and decreased involvement in work and interests. A comparison of symptoms that have been found useful for measuring change in severity of depression are presented in Table 1.

2.1.5
Rate of Onset of Efficacy

Multiple explanations for the apparent delay in the onset of the action of antidepressant drugs have been proposed. Pharmacokinetics could be one factor involved in the delay, especially with regard to TCAs, which are begun at lower doses and then increased as the patient tolerates the side-effects. Stassen et al. (1996) looked at this time delay in some detail by taking the data from two multicentre double-blind studies comparing amitriptyline, oxaprotiline, imipramine, moclobemide and placebo. Their meta-analyses showed that, among improvers, the time to improvement was the same whether or not active drug or placebo had been given. This effect was constant, despite the eventual difference in efficacy between active drug and placebo. In addition, 70% of patients who showed improvement in the first 14 days of treatment became responders. The authors concluded from this that the therapeutic qualities of antidepressants do not lie in the suppression of symptoms but rather in their ability to "elicit and maintain certain conditions which enable recovery in a sub-group of patients who would otherwise remain non-responders".

With respect to the time course of response, improvement in somatic symptoms is often the first change to be noted in patients treated with TCAs. Patients remark that they sleep better and longer, and that they feel less tired during the day. Their ability to do things also improves, with patients becoming more alert and attentive. Retardation becomes less marked. Only later does the depressive affect begin to lift: this effect may be delayed for up to 4 weeks from the initiation of drug therapy. In patients with anxious depression, anxiety may be slow to respond or refractory to treatment (Lader et al. 1987). In bipolar patients, a hypomanic swing may occur. An upswing in mood induced by the TCA may be the first indication that a patient previously believed to be unipolar is probably bipolar.

Attempts to identify those patients who are particularly likely to respond to drug therapy (Brown and Khan 1994) have not been entirely successful. Age and educational background do not seem to affect outcome. A favourable outcome is associated with few previous episodes, insidious onset, brief duration and lack of precipitating causes. Conversely, neurotic, hypochondriacal and histrionic personality traits predict poor response. Delusional patients are also less responsive to TCAs.

In depressed schizophrenic patients, imipramine may increase hostility, belligerence, agitation, delusions and thought disorder. The effect of imipramine by itself on the constellation of symptoms associated with schizophrenia is usually disappointing, so that clinicians must resort to using a combination of an antidepressant and an antipsychotic drug.

2.2
Amitriptyline

Amitriptyline closely resembles imipramine in its pharmacology and clinical effects, but is more sedative and has more pronounced anticholinergic effects. Like imipramine, amitriptyline is mainly metabolised by demethylation (to nortriptyline) and by hydroxylation and glucuronide formation (see Preskorn and Catterson, this volume, for a more detailed discussion of the metabolism of the TCAs). It also has a fairly long plasma half-life. Many controlled trials have established its effectiveness in depression, especially when anxiety or agitation coexists. Amitriptyline became generally established in Europe as the archetypal tricyclic antidepressant with which newly developed putative antidepressants were compared.

2.2.1
Nortriptyline

The pharmacokinetics of nortriptyline have been extensively studied as adequate methods for its estimation became available early (see Preskorn and Catterson's chapter in this volume for a more detailed discussion). Treatment with standard dosage results in an up to 30-fold variation in steady-state concentrations. Twin and family studies suggest that genetic factors strongly influence metabolism. Plasma concentrations correlate with biological effects, such as inhibition of noradrenaline uptake into rat iris, or into cerebral cortex, or as shown by the reduced pressor effects of intravenous tyramine in man. Side-effects increase with plasma concentrations as judged by subjective complaints or by measurements of visual accommodation.

In clinical practice, nortriptyline is about as effective as amitriptyline, but it is less sedating.

2.3
Other Tricyclic Antidepressants

More than 20 TCAs have been introduced throughout the world, most with rather similar clinical characteristics. However, the inhibition of uptake of brain amines varies greatly among these drugs (Table 2) (Rudorfer and Potter 1989).

Desipramine, available in the U.S. but unavailable in Europe, was believed to be the active principle of imipramine with a quicker onset of action. This has not proven to be the case.

Clomipramine, the dihydrobenzazepine analogue of chlorpromazine, is strongly sedative. Although a powerful inhibitor of 5-HT uptake, on repeated administration its metabolite, desmethylclomipramine, accumulates and preferentially blocks noradrenaline uptake (Balant-Gorgia et al. 1991). Claims that it is

Table 2 The relative inhibitory potencies of antidepressant and related drugs on the uptake of noradrenaline, 5-HT, and dopamine in the rat brain in vitro (synaptosomes), in vivo (various techniques), and 5-HT uptakes in human platelets under clinical conditions

Drug	Rat brain In vitro			In vivo			Human platelets
	NA	5-HT	DA	NA	5-HT	DA	5-HT
Amitriptyline	++	++	–	+	(+)	–	+(+)
Citalopram	–	+++	–	–	+++	–	++
Clomipramine	++	+++	–	++	+(+)–	+++	
Desipramine	+++	(+)	–	+++	(+)	–	+
Doxepin	+	+	–	+	+	–	+
Fluoxetine	+	++	–	–	++	–	++
Fluvoxamine	–	++	–	–	+	–	
Imipramine	++	+(+)	–	+++	+	–	++
Iprindole	–	–	–	–	–	–	–
Lofepramine	+++	(+)	–	+++	(+)	–	+
Maprotiline	++	–	–	++	–	–	–
Mianserin	+	–	–	–	–	–	
Nomifensine	++	–	+	++	–	+	
Nortriptyline	++	(+)	–	++	(+)	–	
Viloxazine	+	–	–	(+)	–	–	

+++, Very high potency; ++, high potency; +, moderate potency; –, low potency; parentheses denote possible effects; DA, dopamine; NA, noradrenaline.
The data are compiled from numerous reports.

rapidly therapeutic following intravenous infusion have not been adequately tested. It has proven efficacy in patients with obsessional and panic disorders (McTavish and Benfield 1990). Its efficacy in OCD appears to be independent of its antidepressant activity. It is also favoured by some experienced clinicians in the management of refractory depression. *Trimipramine* has a branched side chain and is strongly sedating (Settle and Ayd 1980).

Lofepramine is a tricyclic compound that resembles imipramine except that it has a more complex side chain. It binds only weakly to cholinergic receptors and therefore has fewer anticholinergic side-effects (Feighner et al. 1982; Rickels et al. 1982). Despite being partly metabolised to desipramine, it has less cardiotoxicity and seems safer in overdosage than typical TCAs (Lancaster and Gonzalez 1989a). It has been popular in the United Kingdom, especially in the treatment of elderly patients with depression.

Protriptyline is strongly stimulant and may produce restlessness and insomnia if taken during the late afternoon or evening.

Doxepin may have fewer side-effects than amitriptyline (Lancaster and Gonzalez 1989b) whereas *dothiepin* is very similar to amitriptyline. Both drugs are sedative antidepressants and have been recommended for the treatment of anxious depression.

Maprotiline is chemically a tricyclic compound with a bridge structure across the central ring. It selectively blocks norepinephrine re-uptake.

Amoxapine is chemically related to the antipsychotic compound, loxapine (Jue et al. 1982). It appears fairly similar in efficacy to the standard TCAs with a similar profile of unwanted effects.

3
Metabolism/Plasma Concentrations and Clinical Response

This is a complex subject that has excited much discussion in the past (Furlanut et al. 1993). Nevertheless, some discussion of this topic is necessary because plasma concentration estimations are available in many centres, the use of such measures can sometimes be helpful in optimising clinical dosage (see Connor and Leonard, this volume), and therapeutic drug monitoring can play a role in preventing adverse drug effects.

It was believed that a range of plasma concentrations could be established for each TCA, within which a clinical response was predictable (Risch et al. 1979a,b). Of the 20%–30% of depressed patients who fail to respond adequately, some are undoubtedly failing to take their medication as prescribed; however, others are believed to metabolise the drug rapidly, resulting in sub-therapeutic levels.

As noted earlier, nortriptyline has been the most widely studied TCA in this respect. Some researchers have felt that *routine* monitoring of nortriptyline plasma concentrations is not justified (Risch et al. 1981). However, enough of a relationship has been established to make it worthwhile taking a standardised plasma sample (i.e. just before the morning dose) in non-responders and adjusting the dosage of nortriptyline if the plasma drug concentration seems excessively high or inadequately low (between 50–200 ng/ml).

Clinical response is related to plasma amitriptyline concentrations in an even more complex way. Both amitriptyline and its active metabolite, nortriptyline, have to be considered, and addition of the two concentrations is too simplistic. Unfortunately, the relevant studies are somewhat contradictory. Again, it is difficult to discern any clear guidelines, but it is unlikely that the amitriptyline plus nortriptyline levels have a relationship to response totally at variance with that of nortriptyline alone. Accordingly, it is sensible to lower dosage if the levels of the two together exceed 250 ng/ml or if that of nortriptyline exceeds 150 ng/ml (Vandel et al. 1978). The risk of serious adverse effects begins to appear at levels above 350 ng/ml and the risk and seriousness of these adverse effects increases as the concentration increases (Preskorn 1986). The reader is referred to the chapters by Preskorn and Catterson ("General Principles of Pharmacokinetics") and Burke and Preskorn ("Therapeutic Drug Monitoring of Antidepressants") earlier in this volume for more details.

Studies with imipramine are perhaps the clearest (Task Force 1985). When the combined levels of imipramine and desipramine in patients exceed 200 ng/ml, clinical response is better than in patients with lower levels. Higher levels (over 250 ng/ml) are associated with no better clinical response than 200–250 ng/ml, but side-effects are increased. Clomipramine studies yield essentially similar data. The threshold value for satisfactory antidepressant effect is a combined clomipramine and desmethylclomipramine concentration of 160–200 ng/ml (Faravelli et al. 1984).

Monitoring of plasma levels can also play an important role in preventing adverse drug effects. For example, individuals who are genetically deficient in the hepatic isoenzyme CYP 2D6 (approximately 7% of Caucasians) (Eichelbaum 1990) may attain plasma levels of TCAs that are fourfold higher than those of non-affected individuals (Nelson and Jatlo 1987; Preskorn et al. 1993). In such patients, conventional doses of a TCA such as imipramine (150–250 mg/day) may lead to toxic side-effects such as delirium and seizures (Preskorn and Jerkovich 1990; Preskorn and Fast 1992). Therapeutic drug monitoring can help prevent such adverse outcomes. For a more detailed discussion of pharmacokinetic issues and therapeutic drug monitoring with TCAs, readers are referred to the chapters in this volume entitled "General Principles of Pharmacokinetics" and "Therapeutic Drug Monitoring of Antidepressants". A table summarising information on plasma concentrations of commonly used antidepressants in provided in the Appendix (Table 5).

4
Safety and Adverse Effects

4.1
Drug–Drug Interactions

For a discussion of drug–drug interactions mediated by pharmacokinetic factors, readers are referred to the chapter by Preskorn and Catterson in this volume.

4.1.1
CNS Drugs

The sedative actions of many TCAs are potentiated by central depressant drugs such as alcohol, barbiturates and benzodiazepines. Thus, patients taking antidepressants must be warned about drinking alcohol and exhorted never to drink and drive. The commonly prescribed combination of a TCA and a benzodiazepine results in definite but usually mild potentiation of central sedation.

The stimulant effects of amphetamine, as with other sympathomimetics, are potentiated.

Potentiation of phenothiazines is only a problem at high doses. Anticholinergic effects include confusion, constipation, dry mouth and urinary retention. Toxic effects on the heart may develop. Potentiation occurs both at the receptor level and in the liver where drugs compete for the microsomal liver enzymes.

4.1.2
Cardiovascular Drugs

TCAs prevent some obsolescent antihypertensive agents (e.g. guanethidine, bethanidine, debrisoquine and perhaps methyldopa) from being taken up by the amine pump to their site of action in the presynaptic noradrenergic neuron (Cocco and Ague 1977). Consequently, if a TCA is given to a depressed hypertensive patient maintained on one of these antihypertensive agents, control of the blood pressure may be lost. The rise in blood pressure is sometimes marked and difficult to control.

Clonidine is more complex. It is transported by the norepinephrine pump, but also acts as a presynaptic agonist, the stimulation of these receptors producing a feedback inhibition of norepinephrine synthesis.

In managing the depressed hypertensive patient, the clinician should first review whether specific antihypertensive therapy is necessary or whether salt restriction and a diuretic might suffice. If drugs are judged necessary, a beta-adrenoceptor antagonist should be used. Otherwise, an SSRI should be the antidepressant of choice.

TCAs markedly potentiate some directly acting sympathomimetic amines, especially norepinephrine. Attacks of severe hypertension may be caused. Patients receiving injections of local anaesthetics formulated with epinephrine are at risk, so that dentists and others who use such injections should routinely ask their patients whether they are taking TCAs.

TCAs may potentiate anti-arrhythmics such as quinidine and cardiac drugs of the digoxin type. Coumarin anticoagulants are occasionally potentiated by TCAs. Thyroid hormones are also mildly potentiated. Conversely, thyroid medication has been advocated as an adjunct to TCAs in the treatment of refractory depression.

4.2
Adverse Effects

The unwanted effects of the TCAs are often more than a minor or a transient nuisance (Blackwell 1981a,b). The patient may find them so unpleasant that they are unable to attain an adequate therapeutic dose level. Refusal to continue with medication after the first dose or two is commonly encountered in specialist practice. Because patients with depression have many subjective bodily complaints anyway, it is often difficult to disentangle drug-related effects from depressive symptoms (Nelson et al. 1984a). The unwanted effects of the TCAs are discussed in the following sections by bodily system. For more information on adverse effects, refer to the adverse effects tables in the Appendix (Tables 2 and 3).

4.2.1
Central Nervous System

Many of the antidepressants induce drowsiness and torpor and a feeling of heaviness. This comes on rapidly, reaches its peak in the first few days, and then wanes over the next week or two. Amitriptyline, doxepin, dothiepin, trimipramine, clomipramine and mianserin are the most sedating; imipramine and nortriptyline have less effect, whereas desipramine and protriptyline are actually mild stimulants. As well as drowsiness, the patient may complain of feeling light-headed, almost to the point of psychological detachment and even depersonalisation. Thinking, attention, concentration and especially memory are impaired at high doses (Thompson 1991). However, these functions are typically impaired anyway in depressed patients and TCAs may improve cognitive performance after a week or so as a harbinger of general clinical improvement (Glass et al. 1981; Thompson and Trimble 1982).

A persistent tremor may be induced in the elderly by the TCAs. High doses, especially of amitriptyline, may precipitate extrapyramidal reactions. A few instances of syndromes resembling tardive dyskinesia have even been reported. Rare neurotoxic reactions have also been documented, comprising ataxia, nystagmus, dysarthria, agitation, tremor and hyperreflexia.

Almost all TCAs and related drugs lower the convulsive threshold. Patients with epilepsy may experience an increase in seizure frequency, and a hitherto non-epileptic subject may have a seizure for the first time (Burley et al. 1978). Of the TCAs, maprotiline is particularly likely to induce seizures (Edwards et al. 1986). As would be expected, the risk of seizures increases with high dose and polypharmacy (Rosenstein et al. 1993).

A consequence of all antidepressant therapy is that the patient may erupt into hypomania. This is true with TCAs (Sultzer and Cummings 1989), and patients with documented bipolar illnesses and with a pronounced diurnal variation in mood are most at risk. It has been claimed that TCAs cannot only induce a hypomanic reaction but can also induce rapid cycling.

The anticholinergic properties of the TCAs, particularly tertiary amine TCAs (e.g. amitriptyline, doxepin and imipramine) may lead to confusional reactions, especially in the elderly. The initial symptom is heightened visual awareness, which typically progresses to visual illusions such as shimmering and moiré effects, and flashes of colour. Eventually visual hallucinations, usually of a fairly unformed type, supervene. Rarely, a full-blown delirium is induced, with fully organised hallucinations. Confusional reactions, as with other anticholinergic reactions, are more likely when the TCA is combined with other anticholinergic drugs. The triple combination of a TCA, an anticholinergic antipsychotic (e.g. thioridazine) and an antiparkinsonian agent is even more likely to result in this unfortunate outcome.

The abrupt withdrawal of TCAs is occasionally followed by an akathisia-like syndrome or behavioural activation. Other symptoms include abdominal pain, nausea, vomiting, diarrhoea and flu-like symptoms such as fatigue, anxiety and agitation, nightmares, and sleep disturbances (Disalver and Greden 1984; Disalver et al. 1987; Disalver 1989, 1994). Such symptoms have been reported in up to half of patients who abruptly stop high-dose imipramine, and in a third of those withdrawing from clomipramine (Coupland et al. 1996). In such cases, the TCA should be re-instituted and then withdrawn more gradually. Alternatively, a modest dose of atropine can be given to combat the cholinergic overactivity rebound (Disalver et al. 1983). Despite the misgivings of some researchers (e.g. Medawar 1997), this does not represent true dependence but only a physical withdrawal syndrome (Drug and Therapeutics Bulletin 1999).

4.2.2
Cardiovascular

Because of the α-adrenergic blocking effects, postural hypotension may occur and is particularly common in the very young, the very old, and the physically debilitated. Tachycardia, found in many patients on TCAs, is a reflex mechanism to compensate for the hypotension. The effect is most pronounced during the first few weeks of medication but it may persist. It can be minimised by increasing the dose gradually at the onset of treatment.

In patients taking TCAs, changes in cardiac conduction resemble those seen in patients on phenothiazine therapy (Taylor and Braithwaite 1978). Repolarisation is prolonged, as shown in the ECG by a lengthened Q-T interval, flattened or inverted T-waves, and prominent U-waves. In addition, TCAs have a quinidine-like effect with prolongation of A-V conduction and even A-V block. More serious conduction defects include bundle branch and complete heart block. Bradycardia or tachycardia, ventricular extrasystoles, and atrial and ventricular arrhythmias may occur. The pharmacological mechanisms are complex and include cholinergic blockade of the vagus, a direct toxic effect on the myocardium, and a heightened sensitivity of the heart to circulating catecholamines (Glassman and Stage 1994).

The cardiac arrhythmias with TCAs have been blamed for an excess of sudden deaths in depressed patients with a history of cardiac disease. Amitriptyline seems the most suspect TCA but this may reflect the extent of its usage. Nevertheless, its use should be avoided in cardiac patients. In the elderly, TCAs should be administered in divided doses rather than as one large dose at night. Congestive cardiac failure may be precipitated or aggravated by TCAs in patients with recent myocardial infarction.

Lofepramine may be less cardiotoxic. This is also true for many of the newer antidepressants (see the two chapters by Preskorn et al., this volume: "Selective Serotonin Reuptake Inhibitors" and "Other Antidepressants"). Therefore, one of these drugs is preferred in the depressed patient with cardiovascular pathology.

Therapeutic drug monitoring can also be useful in monitoring for plasma drug levels that may lead to cardiotoxicity (for a more detailed discussion of this issue, see Preskorn and Catterson, "General Principles of Pharmacokinetics" and Burke and Preskorn "Therapeutic Drug Monitoring of Antidepressants", this volume).

4.2.3
Gastrointestinal

Anticholinergic effects comprise reduced gut motility with constipation which may become severe or even culminate in paralytic ileus. Achalasia of the oesophagus and relaxation of the oesophageal sphincter may result in a hiatus hernia or exacerbate a pre-existing one. The best treatment is to lower the dose rather than to administer cholinomimetic drugs such as neostigmine or bethanechol. However, acid formation in the stomach is decreased by TCAs, which can therefore help depressed patients with symptomatic peptic ulcers.

4.2.4
Genitourinary

The cholinergic blockade of the TCAs results in increased bladder sphincter tonus. Some degree of urinary retention may ensue but this is usually of clinical significance only in patients with pre-existing pathology such as prostatic enlargement. Consequently, elderly males should be asked about difficulties with micturition before being prescribed TCAs.

A common sexual effect of the TCAs is delayed orgasm with slow ejaculation. This effect can be exploited therapeutically in patients with premature ejaculation. Anorgasmia may occur, but the incidence of this side-effect is higher with the SSRIs than with the TCAs. Erectile dysfunctions and changes in libido may also occur (Mitchell and Popkin 1983; Segraves 1989).

4.2.5
Metabolic-Endocrine

Effects on the pituitary–gonadal axis can produce menstrual irregularities, breast enlargement and galactorrhea in women and impotence, gynaecomastia and testicular swelling in men. Libido may be increased or decreased but such changes are difficult to interpret because of alterations in sexual function associated with the depressive illness itself.

Weight gain is often marked in patients treated with TCAs. Part of the weight gain is a reversal of the weight loss earlier in the illness. Nonetheless, many patients report a craving for carbohydrates, and TCAs are known to have complex effects on blood glucose levels. Weight gain can be upsetting for many patients. Obesity is also a risk factor for hypertension and diabetes and cardiovascular problems.

4.2.6
Miscellaneous

Other anticholinergic effects include pupillary dilatation and loss of accommodation with blurring of vision. Because intraocular pressure rises, these drugs should be used with the utmost caution and with appropriate ophthalmic monitoring in patients with narrow angle glaucoma. One of the newer antidepressants with few anticholinergic actions such as an SSRI is to be preferred. Paradoxically, hyperhidrosis has been described, especially with imipramine, and was believed to be a harbinger of a good therapeutic response: the mechanism is unclear.

Minor symptoms include headache, fatigue, nausea, anorexia, stomatitis, peculiar taste sensations and nightmares. Liver function tests may show minor impairment, especially with high doses of TCAs, but usually return to normal even while the drug is being continued. Mild cholestatic jaundice occasionally develops. Allergic reactions include rashes, urticaria and oedema (Warnock and Knesevick 1988); agranulocytosis, thrombocytopenia and eosinophilia have been reported.

4.2.7
Drop-Out Rates as an Indicator of Tolerability

Drop-out rates have been used as an operational indicator of the tolerability of various antidepressant drugs. In one study (Schmidt et al. 1986) that used a multi-centred hospital-based drug monitoring system, the drop-out rate in patients on TCAs was 7.4%, as compared with 3.1% in patients on newer antidepressants such as maprotiline. A comparison of published trials of TCAs and SSRIs found that 18.8% of patients on TCAs and 15.4% of patients on SSRIs dropped out because of side-effects, a difference that did not quite reach significance (Song et al. 1993). Parenthetically, this study categorised trazodone, mianserin, bupropion and nomifensine as TCAs. A meta-analysis of 42 controlled trials carefully compared TCAs with SSRIs, omitting antidepressants not falling unequivocally into one or other category and found that drop-out due to side-effects averaged 27% with TCAs and 19% with SSRIs ($p<0.01$) (Montgomery et al. 1994).

4.3
Overdose

TCAs constitute a major problem in overdose (Proudfoot 1986). First, they are prescribed for depressive disorders, a salient feature of which can be suicidal ideation or intentions. Next, most of these drugs are dangerous in overdose, producing a complex clinical picture that is difficult to treat. For this reason, prescriptions should be for conservative amounts; however, even a week's supply of a TCA at full dosage can be lethal, and the determined patient can hoard his supplies. Entrusting antidepressants to a trustworthy relative or friend is a wise precaution, but this strategy should be carefully negotiated with the patient. The problem of accidental overdosage among children is also a concern: even 250 mg has proven fatal on occasion. A toddler might take a parent's TCAs or those prescribed to an elder sibling for enuresis.

The standard TCAs of the imipramine and amitriptyline family are all dangerous in overdose. Doses above 600 mg are likely to produce serious effects in adults. Doses over 2.5 g (i.e. one hundred 25-mg tablets) are likely to prove fatal. [Contrary to popular belief among general practitioners in the United Kingdom, dothiepin appears to be at least as toxic as other TCAs (Buckley et al. 1994).] Other antidepressants such as the SSRIs, mianserin, trazodone, nefazodone and bupropion may be significantly less toxic. They are preferred when prescribing for a patient who is socially isolated and whom one suspects of harbouring suicidal intent. Lofepramine also appears to be safer (Dorman 1985; Malmvik et al. 1994).

The seriousness of a TCA overdose is correlated both with the dose ingested and the plasma drug level achieved (see the chapter in this volume by Preskorn and Catterson "General Principles of Pharmacokinetics" and Burke and Preskorn "Therapeutic Drug Monitoring of Antidepressants").

Symptoms of overdose with a TCA usually appear within 4 h (Crome 1982). Common features include dry mouth, blurred vision, dilated pupils, sinus tachycardia, extrapyramidal signs and either drowsiness or excitement.

The cardiovascular effects of overdosage with the TCAs are the most life-threatening and range from sinus tachycardia to major arrhythmias. Serious arrhythmias are more likely in the elderly, those with pre-existing cardiac problems, and in patients with pre-existing respiratory difficulties. The ECG may show prolongation of the QRS complex and ST segment, and T-waves are often abnormal. The ECG changes may resemble ventricular or supraventricular tachycardia or bundle branch block; atrioventricular block and bradycardia sometimes occur. The more serious effects probably stem from a quinidine-like myocardial depressant action of the TCAs. The EEG of all patients who have taken an overdose of a TCA should be monitored until they have fully recovered.

Central effects comprise excitement or coma with shock. Coma with shock suggests that a sedative drug such as a benzodiazepine may have been taken as well. Metabolic acidosis is sometimes present. Convulsions may occur, especially in children; maprotiline is the TCA most likely to induce seizures. Respiratory depression may develop with sudden apnoea. Other effects that have been reported include hypotension, agitation or delirium, hyperreflexia, myoclonus and chorea, and bowel and bladder paralysis. Total recovery is still possible despite a flat EEG and dilated non-reactive pupils.

Because the anticholinergic actions of the drug may delay gastric emptying, it is worth washing out the stomach within the first 12 h. Activated charcoal may also help reduce absorption. Because of the high tissue and plasma binding, forced diuresis, haemodialysis and haemoperfusion are ineffective.

Respiration must be maintained and electrolyte and blood gas disturbances rectified. The correction of the metabolic acidosis by intravenous infusion of bicarbonate may of itself lessen the cardiac abnormalities. Intravenous physostigmine salicylate 2 mg over 5 min may reverse both peripheral and central anticholinergic effects; however, because it tends to cause convulsions, on balance it is best avoided. Diazepam 10 mg may be needed to combat any seizures which develop. Some centres prefer intravenous phenytoin because it also has an antiarrhythmic effect; however, phenytoin may not be a good choice because it can add to the delay in intracardiac conduction caused by the TCA. Otherwise treatment is generally supportive, with bodily temperature and vital functions being monitored and maintained.

4.4
Teratogenicity

Imipramine has been implicated in teratogenic effects in animals but reports of congenital limb deformities in babies born of mothers who had taken this drug during pregnancy have not been confirmed. Nevertheless, unless there is a compelling need to use these drugs in pregnant women, they should be avoided, especially in the first trimester. If a depressed woman taking a TCA becomes pregnant, the possibility of recommending a therapeutic abortion should be considered. At the least, the fetus should be carefully monitored. Based on a recent review of the use of antidepressants in pregnancy and lactation, it appears that the key factor is whether the likely benefits to the mother outweigh the risks to the fetus (McElhatton 1999). Among TCAs, the drugs of choice for use during pregnancy are amitriptyline and imipramine. Neonatal withdrawal symptoms may also occur and this should be monitored for.

5
Clinical Use and Therapeutic Indications: Major Depression

Current recommendations for the use of antidepressant drug therapy distinguish between the acute phase, the continuation phase and maintenance therapy (Drug and Therapeutics Bulletin 1999). Depression is now recognised as a relapsing and recurring condition (Angst 1992), especially in the more severe forms requiring referral to a specialist. About 1 in 3 such patients who are not maintained on antidepressant therapy will relapse within 6 months of the index episode. Even more will relapse or enter a new depressive phase within 2 years (Belsher and Costello 1988). Predictors of relapse include a history of previous episodes, severe acute illness, persistent symptoms, especially somatic symptoms, ongoing medical and/or social problems, and lack of social support. For a summary of U.S. package insert dosing guidelines for clinical depression, refer to the dosing table in the Appendix (Table 1).

5.1
Acute Phase

The initial treatment phase, referred to as the acute phase, typically involves the first 2–4 months of treatment, during which a significant clinical response will occur in about two-thirds of patients. The rest of the patients remain depressed despite apparently adequate drug dosage and compliance (Quitkin et al. 1996). If no sign of improvement can be detected, treatment with a TCA at full dose should not be continued beyond 4 weeks. A longer time is justified if some response can be discerned, and an increase in dosage in partial responders is a worthwhile strategy but should be guided by therapeutic drug monitoring to ensure that the patient is not developing toxic levels (see the chapter entitled "Therapeutic Drug Monitoring of Antidepressants", this volume). A standardised estimate of plasma drug concentration may be helpful in detecting the clandestine non-complier, as well as in detecting slow metabolisers and confirming ultra-rapid metabolisers.

In many countries, newer antidepressants, especially the SSRIs, have become the treatment of choice for depression; however, in other countries the TCAs are still considered a first choice treatment for depression. Patients who have responded well to a TCA during a previous episode will probably respond again. Conversely, TCAs are best avoided in the old and the physically ill. Unfortunately, the medico-scientific issues have been overshadowed by pharmaco-economic ones. In my experience, most patients can just tolerate full doses of a TCA but are more comfortable taking an SSRI or another newer compound.

There is no rationale for using combinations of TCAs. Combinations of a TCA and some SSRIs, principally fluoxetine, fluvoxamine and paroxetine, can result in high TCA levels in the body because of competition for microsomal enzymes. An increase in the dosage of the TCA is a more logical and less hazardous manoeuvre.

5.2
Continuation Phase

Several placebo-controlled trials indicate that if patients are maintained on a TCA for 4 to 6 months until they are symptom free, this halves the rate of relapse (Reimherr et al. 1998). Contrary to earlier practice, it is now accepted that the TCA should be continued at full, rather than half, dosage (Paykel and Priest 1992). A question that is still unanswered is whether continued treatment for months after apparent recovery is needed in more mildly ill patients seen in primary care. In whatever context, it is probably wise to continue medication if any symptoms (e.g. insomnia, anhedonia) persist. When TCAs are withdrawn in due course, the dose should be tapered and abrupt discontinuation avoided.

5.3
Maintenance Therapy

The decision as to whether to recommend long-term maintenance therapy to lessen the risk of recurrence depends on several factors, the most important of which being the number, frequency and severity of previous episodes. Another issue is how disruptive the episodes have been and if they are increasing in severity and/or duration. A general criterion is that long-term therapy should be considered if a patient has had two or more previous episodes of depression, requiring specialist advice or care over the preceding 5 years.

Maintenance therapy is usually quite successful (e.g. Frank et al. 1990; Kupfer et al. 1992; Reynolds et al. 1999). Again, full dosage is usually recommended. Attempts to lower the dose may result in the return of symptoms such as lack of motivation or insomnia, or a full-blown episode may supervene and prove rather treatment-resistant. Sometimes, despite full dosage, efficacy in preventing recurrence appears to wane (Byrne et al. 1998). A change of medication may then be needed.

Little is known of the efficacy of TCAs in primary care in preventing recurrence. However, one encounters numerous patients who have been taking low doses of TCAs for years and believe they are deriving continuing benefit.

The decision to stop long-term therapy is a difficult one and depends on a variety of factors. It is important to try to discern by a careful symptomatic enquiry whether subclinical episodes are still occurring. For example, some patients describe a few weeks in which some insomnia recurs or they feel under the weather. Withdrawal may result in recurrence in these patients. As a rule, the older the patient, the more important it is to continue TCA medication.

6
Special Considerations: Use in Primary Care

In many countries, the administration of antidepressants is very much within the purview of general practitioners. In one large-scale study in the United Kingdom (Martin et al. 1997), a representative sample of 250 doctors recorded their prescribing activity every 4 weeks, for a total of 4,000 general practitioner weeks of recording per year. Included in the record were any patients who began a new course of an antidepressant or whose treatment was stopped or changed by the GP. During the course of the study, there were nearly 14,000 inceptions and 4,000 discontinuations of various antidepressants. The use of antidepressants increased between 1990 and 1995, due mostly to an increase in the prescribing of SSRIs. Furthermore, the ratio of discontinuations to inceptions was significantly lower for SSRIs (22%) than for TCAs (33%). However, there was more switching away from SSRIs when they failed than from TCAs. This study raises interesting questions about the different ways in which TCAs and SSRIs medications are used in primary care.

A large study constituted the first pan-European survey of depression in the community (Lepine et al. 1997). A total of 13,359 of the 78,463 adults who were screened by interview were identified as suffering from depression. This translates into a 6-month prevalence rate of 17%. Major depression accounted for about 7% of these and 1.8% for the rest. A significant proportion of the sufferers from depression (43%) failed to seek treatment for their depressive symptoms. Of those who did seek help, most consulted a primary care physician. Sufferers from major depression posed the greatest demand and made frequent visits to their general practitioners. However, more than two thirds of depressed patients were not prescribed any treatment and, when drug therapy was prescribed, only a quarter of these subjects were given antidepressant drugs. There was widespread use of tranquillisers, which it was thought might perhaps reflect the high frequency of insomnia or concomitant anxiety in these depressed patients. Indeed, with the exception of Germany and the United Kingdom, more depressed subjects were given tranquillisers than antidepressant drugs. The only country in which antidepressants were prescribed significantly more frequently than tranquillisers was the United Kingdom (31% versus 8%), which may reflect all the publicity over the past 20 years concerning problems with the benzodiazepines.

Another problem in the primary care use of TCAs that has been remarked on in many countries for many years is the low and supposedly ineffective level of the TCA dosages used. The duration of treatment is also believed to be less than optimal. In a study by MacDonald et al. (1996), the TCAs seemed safe in the dosages used but they may not have been exerting any real therapeutic activity.

In the United States, the advent of managed care has highlighted problems with the use of antidepressants. The SSRI manufacturers are well aware that the TCAs are less expensive and hence may be preferred by the insurers in health management organisations on cost grounds. However, the question is whether the total cost of treating a patient with a TCA rather than an SSRI is less. There are studies that suggest that treatment with SSRIs may be less costly than treatment with TCAs when the total cost of providing a course of treatment is considered. In this regard, Hylan et al. (1998) reported that patients who initiate therapy with SSRIs consistently achieve minimum therapeutic doses for the duration of therapy whereas those on TCAs are less likely to do so.

References

Angst J (1992) How recurrent and predictable is depressive illness? In: Montgomery SA, Rouillon F (eds) Long-term treatment of depression. John Wiley, Chichester

Balant-Gorgia AE, Gex-Fabry M, Balant LP (1991) Clinical pharmacokinetics of clomipramine. Clin Pharmacokinet 20:447–462

Bech P, Allerup P, Gram L, Reisby N, Rosenberg R, Jacobsen O, Nagy A. (1981) The Hamilton Rating Scale: evaluation of objectivity using logistic models. Acta Psychiatr Scand 63:290-299

Belsher G, Costello CG (1998) Relapse after recovery from unipolar depression: a critical review. Psychol Bull 104:84–96

Blackwell B (1981a) Adverse effects of antidepressant drugs. Part I: monoamine oxidase inhibitors and tricyclics. Drugs 21:201–219

Blackwell B (1981b) Adverse effects of antidepressant drugs. Part 2: 'second generation' antidepressants and rational decision making in antidepressant therapy. Drugs 21:273–282

Brown WA, Khan A (1994) Which depressed patients should receive antidepressants? CNS Drugs 1:341–347

Buckley NA, Dawson AH, Whyte IM, Henry DA (1994) Greater toxicity in overdose of dothiepin than of other tricyclic antidepressants. Lancet 343:159–162

Burley D, Jukes A, Steen J (1978) Maprotiline hydrochloride and grand-mal seizures. BMJ 2:1230

Byrne SE, Rothschild AJ (1998) Loss of antidepressant efficacy during maintenance therapy: possible mechanisms and treatments. J Clin Psychiatry 59:279–288

Cocco G, Ague C (1977) Interactions between cardioactive drugs and antidepressants. Eur J Clin Pharmacol 11:389–393

Coupland NJ, Bell CJ, Potokar JP (1996) Serotonin reuptake inhibitor withdrawal. J Clin Psychopharmacol 16:356–362

Crome P (1982) Antidepressant overdosage. Drugs 23:431–461

Disalver SC (1989) Antidepressant withdrawal syndromes: phenomenology and pathophysiology. Acta Psychiatr Scand 79:113–117

Disalver SC (1994) Withdrawal phenomena associated with antidepressant and antipsychotic agents. Drug Saf 10:103–114

Disalver SC, Greden JF (1984) Antidepressant withdrawal phenomena. Biol Psychiatry 19:237–256

Disalver SC, Feinberg M, Greden JF (1983) Antidepressant withdrawal symptoms treated with anticholinergic agents. Am J Psychiatry 140:249–251

Disalver SC, Greden JF, Snider RM (1987) Antidepressant withdrawal syndromes: phenomenology and pathophysiology. Int Clin Psychopharmacol 2:1–19

Dorman T (1985) Toxicity of tricyclic antidepressants: are there important differences? J Int Med Res 13:77–83

Drug and Therapeutics Bulletin (1999) Withdrawing patients from antidepressants. DTB 37:49–51

Edwards JG, Long SK, Sedgwick EM, Wheal HV (1986) Antidepressants and convulsive seizures: clinical, electroencephalographic, and pharmacological aspects. Clin Neuropharmacol 9:329-60

Eichelbaum M, Gross AS (1990) The genetic polymorphism of debrisoquine/sparteine metabolism: clinical aspects. Pharmacol Ther 46:377-394

Faravelli C, Ballerini A, Ambonetti A, Broadhurst AD, Das M (1984) Plasma levels and clinical response during treatment with clomipramine. J Affect Disord 6:95-107

Feighner JP, Meredith CH, Dutt JE, Hendrickson GG (1982) A double blind comparison of lofepramine, imipramine and placebo in patients with primary depression. Acta Psychiatr Scand 66:100-108

Frank E, Kupfer DJ, Perel JM, Cornes C, Jarrett DB, Mallinger AG, Thase ME, McEachran AB, Grochocinski VJ (1990) Three-year outcomes for maintenance therapies in recurrent depression. Arch Gen Psychiatry 47:1093-1099

Furlanut M, Benetello P, Spina E (1993) Pharmacokinetic optimisation of tricyclic antidepressant therapy. Clin Pharmacokinet 24:301-318

Glass RM, Uhlenhuth EH, Hartel FW, Matuzas W, Fischman MW (1981) Cognitive dysfunction and imipramine in outpatient depressives. Arch Gen Psychiatry 38:1048-1051

Glassman AH, Stage KB (1994) Depressed patients with cardiovascular disease: treatment considerations. CNS Drugs 1:435-440

Hazell P, O'Connell D, Heathcote D, Robertson J, Henry D (1995) Efficacy of tricyclic drugs in treating child and adolescent depression: a meta-analysis. BMJ 310:897-901

Hylan TR, Buesching DP, Tollefson GD (1998) Health economic evaluations of antidepressants: a review. Depress Anxiety 7:53-64

Jue SG, Dawson GW, Brogden RN (1982) Amoxapine: a review of its pharmacology and efficacy in depressed states. Drugs 24:1-23

Klein DF, Davis JM (1969) Diagnosis and drug treatment of psychiatric disorders. Williams & Wilkins, Baltimore, pp 193-194

Kuhn R (1958) The treatment of depressive states with G22355 (imipramine hydrochloride). Am J Psychiatry 115:459-463

Kupfer DJ, Frank E, Perel JM, Cornes C, Mallinger AG, Thase ME, McEachran AB, Grochocinski VJ (1992). Five-year outcome for maintenance therapies in recurrent depression. Arch Gen Psychiatry 49:769-773

Lader M, Lang RA, Wilson GD (1987) Patterns of improvement in depressed in-patients. Oxford University Press, Oxford

Lancaster SG, Gonzalez JP (1989a) Lofepramine: a review of its pharmacodynamic and pharmacokinetic properties, and therapeutic efficacy in depressive illness. Drugs 37:123-140

Lancaster SG, Gonzalez JP (1989b) Dothiepin: a review of its pharmacodynamic and pharmacokinetic properties, and therapeutic efficacy in depressive illness. Drugs 38:123-147

Lépine J-P, Gastpar M, Mendlewicz J, Tylee A, on behalf of the DEPRES Steering Committee (1997) Int Clin Psychopharmacol 12:19-29

MacDonald TM, McMahon AD, Reid IC, Fenton GW, McDevitt DG (1996) Antidepressant drug use in primary care: a record linkage study in Tayside, Scotland. BMJ 313:860-861

Malmvik J, Löwenhielm CGP, Melander A (1994) Antidepressants in suicide: differences in fatality and drug utilisation. Eur J Clin Pharmacol 46:291-294

Martin RM, Hilton SR, Kerry SM, Richards NM (1997) General practitioners' perceptions of the tolerability of antidepressant drugs: a comparison of selective serotonin reuptake inhibitors and tricyclic antidepressants. BMJ 314:646-651

McElhatton P (1999) Use of antidepressants in pregnancy and lactation. Prescriber April:101-111

McTavish D, Benfield P (1990) Clomipramine: an overview of its pharmacological properties and a review of its therapeutic use in obsessive compulsive disorder and panic disorder. Drugs 39:136–153

Medawar C (1997) The antidepressant web: marketing depression and making medicines work. International Journal of Risk and Safety Medicine 10:75–126

Mitchell JE, Popkin MK (1983) Antidepressant drug therapy and sexual dysfunction in men: a review. J Clin Psychopharmacol 3:76–77

Montgomery SA, Asberg M (1979) A new depression scale designed to be sensitive to change. Br J Psychiatry 134:382–389

Montgomery SA, Henry J, McDonald G, Dinan T, Lader M, Hindmarch I, Clare A, Nutt D (1994) Selective serotonin reuptake inhibitors: meta-analysis of discontinuation rates. Int Clin Psychopharmacol 9:47–53

Nelson JC, Jatlow PI, Quinlan DM (1984a) Subjective complaints during desipramine treatment: relative importance of plasma drug concentrations and the severity of depression. Arch Gen Psychiatry 41:55–59

Nelson JC, Mazure C, Quinlan DM, Jatlow PI (1984b) Drug-responsive symptoms in melancholia. Arch Gen Psychiatry 41:663–668

Nelson JC, Jatlo PI (1987) Non linear desipramine kinetics: prevalence and importance. Clinical Pharmacol Ther 41:666–670

Orsulak PJ, Waller D (1989) Antidepressant drugs: additional clinical uses. J Fam Pract 28:209–216

Paykel ES, Priest RG (1992) Recognition and management of depression in general practice: consensus statement. BMJ 305:1198–1202

Preskorn SH (1986) Tricyclic antidepressant plasma level monitoring: an improvement over the dose-response approach. J Clin Psychiatry 47 (suppl.1):24–30

Preskorn SH, Fast GA (1992) Tricyclic antidepressant-induced seizures and plasma drug concentration. J Clin Psychiatry 53:160–162

Preskorn SH, Jerkovich GS (1990) Central nervous system toxicity of tricyclic antidepressants: phenomenology, course, risk factors, and role of therapeutic drug monitoring, J Clin Psychopharmacol 10:88–95

Preskorn SH, Burke MJ, Fast GA (1993) Therapeutic drug monitoring: principles and practice. In: Dunner DL (ed) The Psychiatric Clinics of North America. Saunders, Philadelphia, Pennsylvania, 16:611–641

Proudfoot AT (1986) Acute poisoning with antidepressants and lithium. Prescriber's Journal 26:97–106

Quitkin FM, McGrath PJ, Stewart JW, Ocepek-Welikson K, Taylor BP, Nunes E, Deliyannides D, Agosti V, Donovan SJ, Petkova E, Klein DF (1996) Chronological milestones to guide drug change: when should clinicians switch antidepressants? Arch Gen Psychiatry 53:785–792

Raskin A, Crook TH (1976) The endogenous-neurotic distinction as a predictor of response to antidepressant drugs. Psychol Med 6:59–70

Reimherr FW, Amsterdam JD, Quitkin FM, Rosenbaum JF, Fava M, Zajecka J, Beasley CM Jr, Michelson D, Roback P, Sundell K (1998). Optimal length of continuation therapy in depression: a prospective assessment during long-term fluoxetine treatment. Am J Psychiatry 155:1247–1253

Reynolds CF III, Frank E, Perel JM, Imber SD, Cornes C, Miller MD, Mazumdar S, Houck PR, Dew MA, Stack JA, Pollock BG, Kupfer DJ (1999) Nortriptyline and interpersonal psychotherapy as maintenance therapies for recurrent major depression: a randomized controlled trial in patients older than 59 years. JAMA 281:39–45

Rickels K, Weise CC, Zal HM, Csanalosi I, Werblowsky J (1982) Lofepramine and imipramine in unipolar depressed outpatients: a placebo controlled study. Acta Psychiatr Scand 66:109–120

Risch SC, Huey LY, Janowksy DS (1979a) Plasma levels of tricyclic antidepressants and clinical efficacy: review of the literature, Part I. J Clin Psychiatry 40:4–16

Risch SC, Huey LY, Janowsky DS (1979b) Plasma levels of tricyclic antidepressants and clinical efficacy: a review of the literature, Part II. J Clin Psychiatry 40:58–69

Risch SC, Kalin NH, Janowksy DS, Huey LY (1981) Indications and guidelines for plasma tricyclic antidepressant concentration monitoring. J Clin Psychopharmacol 1:59–63

Rosenstein DL, Nelson JC, Jacobs SC (1993) Seizures associated with antidepressants: a review. J Clin Psychiatry 54:289–299

Rudorfer MV, Potter WZ (1989) Antidepressants: a comparative review of the clinical pharmacology and therapeutic use of the 'newer' versus the 'older' drugs. Drugs 37:713–738

Schmidt LG, Grohmann R, Müller-Oerlinghausen B, Ochsenfahrt H, Schönhöfer PS (1986) Adverse drug reactions to first- and second-generation antidepressants: a critical evaluation of drug surveillance data. Br J Psychiatry 148:38–43

Segraves RT (1989) Effects of psychotropic drugs on human erection and ejaculation. Arch Gen Psychiatry 46:275–284

Seppala T, Linnoila M (1983) Effects of zimeldine and other antidepressants on skilled performance: a comprehensive review. Acta Psychiatr Scand 68:135–140

Settle EC, Ayd FJ (1980) Trimipramine: twenty years worldwide clinical experience. J Clin Psychiatry 41:266–274

Song F, Freemantle N, Sheldon TA, House A, Watson P, Long A, Mason J (1993) Selective serotonin reuptake inhibitors: meta-analysis of efficacy and acceptability. BMJ 306:683–686

Stahl SM, Palazidou L (1986) The pharmacology of depression: studies of neurotransmitter receptors lead the search for biochemical lesions and new drug therapies. TIPS 7:349–354

Stassen HH, Angst J, Delini-Stula A (1996) Delayed onset of action of antidepressant Drugs? Survey of results of Zurich meta-analyses. Pharmacopsychiatry 29:87–96

Stewart JW, Quitkin FM, Liebowitz MR, McGrath PJ, Harrison WM, Klein DF (1983) Efficacy of desipramine in depressed outpatients: response according to research diagnostic criteria diagnoses and severity of illness. Arch Gen Psychiatry 40:202–207

Sultzer DL, Cummings JL (1989) Drug-induced mania–causative agents, clinical characteristics and management: a retrospective analysis of the literature. Med Toxicol Adverse Drug Exp 4:127–143

Task Force on the use of laboratory tests in psychiatry (1985) Tricyclic antidepressants: blood level measurements and clinical outcome: Am J Psychiatry 142:155–162

Taylor DJE, Braithwaite RA (1978) Cardiac effects of tricyclic antidepressant medication: a preliminary study of nortriptyline. Br Heart J 40:1005–1008

Thompson PJ (1991) Antidepressants and memory: a review. Human Psychopharmacology 6:79–90

Thompson PJ, Trimble MR (1982) Non-MAOI antidepressant drugs and cognitive functions: a review. Psychol Med 12:539–548

Vandel S, Vandel B, Sandoz M, Allers G, Bechtel P, Volmat R (1978) Clinical response and plasma concentration of amitriptyline and its metabolite nortriptyline. Eur J Clin Pharmacol 14:185–190

Warnock JK, Knesevich JW (1988) Adverse cutaneous reactions to antidepressants. Am J Psychiatry 145:425–430

Monoamine Oxidase Inhibitors

S. H. Kennedy[1] · A. Holt[2] · G. B. Baker[3]

[1] University Health Network, University of Toronto, 200 Elizabeth St.,
  Eaton North, 8th Floor, Room 222, Toronto, ON, M5G 2C4, Canada
  e-mail: sidney.kennedy@uhn.on.ca
[2] Alviva, Suite 218, 111 Research Drive, Saskatoon, SK, S7N 3R2, Canada
[3] Department of Psychiatry, University of Alberta, 1E7.44 W,
  Mackenzie Health Sciences Centre, Edmonton, AB, T6G 2B7, Canada

1	**Introduction and History**	210
2	**Chemistry**	211
2.1	Hydrazine Status	212
2.2	Selectivity and Reversibility	212
3	**Pharmacology**	212
3.1	Characteristics of the Monoamine Oxidase Enzyme System	212
3.2	Platelet MAO	214
3.3	Effects of MAOIs on Brain Amine and Amino Acid Levels	214
4	**Metabolism/Plasma Concentration and Clinical Response**	217
4.1	Phenelzine	217
4.2	Tranylcypromine	218
4.3	Selegiline	218
4.4	Moclobemide and Brofaromine	219
5	**Safety and Adverse Effects**	219
5.1	Drug–Drug Interactions	219
5.2	Adverse Effects	221
6	**Clinical Use/Therapeutic Indications**	222
6.1	Atypical Depression	223
6.2	Anergic Bipolar Depression	224
6.3	Anxiety and Eating Disorders	224
6.4	Depression in the Medically Ill	225
6.5	MAOIs in Combination Therapy	225
7	**Beyond MAO Inhibition: Binding of MAOIs to Other Amine Oxidases and Imidazoline Binding Sites**	226
	References	229

Abstract The monoamine oxidase inhibitors (MAOIs) have been in clinical use for five decades. The coincidental discovery that inhibiting brain monoamine oxidase resulted in antidepressant benefits indirectly led to the norepinephrine (NE) and serotonin hypotheses for depression. Phenelzine (PLZ) and tranylcypromine (TCP) typify the classical, nonselective and irreversible inhibitors; selegiline (SEL) selectively but irreversibly inhibits MAO-B and is an established adjunct therapy for Parkinson's disease; while moclobemide and befloxatone represent examples of selective and reversible inhibitors. These agents provide opportunities to examine brain amine and amino acid levels during treatment and have also contributed to emerging aware of neurogenesis and neuronal rescue as potential antidepressant properties. Also of emerging interest are the extended sites of action beyond MAO for these agents including γ-aminobutyric acid (GABA) and imidazoline binding sites.

Despite the well-known clinical concerns about food and drug interactions and the high side-effect burden associated with classical MAOIs, they continue to be third line agents for treatment-resistant depression. PLZ is also a second line agent for atypical depression. Moclobemide is the only generally available reversible inhibitor of MAO-A and has obtained limited acceptance as a first line treatment in some countries for major depressive disorder (MDD), particularly in patients with prominent anxiety symptoms. It also has one of the lowest side-effect burdens of the MAOIs and, unlike the SSRIs, rarely produces sexual dysfunction. A transdermal form of SEL has been investigated for treating MDD. With appropriate cautions, the MAOIs continue to provide an important alternative class of antidepressants for the treatment of various forms of depressive illnesses.

Keywords Monoamine oxidase inhibitors (MAOI) · Phenelzine (PLZ) · Tranylcypromine (TCP) · Selegiline (SEL) · Moclobemide · GABA · Neuronal rescue · Imidazoline

1
Introduction and History

"The best news for psychiatry in 1957 was the discovery that iproniazid protects serotonin from monoamine oxidase. Serotonin, allowed free activity in the brain, is perhaps the most energetic releaser of reserve power in the human machine..." (Robie 1958).

This enthusiastic comment heralded the advent of antidepressant pharmacotherapy, based on the preliminary experience of patients with tuberculosis and subsequently schizophrenia who were the first recipients of iproniazid.

The monoamine oxidase inhibitor (MAOI) "honeymoon" proved to be relatively brief. Iproniazid was ultimately withdrawn due to its toxic effects on the liver. Shortly after tranylcypromine (TCP) became available as a less toxic alternative, the "cheese reaction" was attributed to this drug's interaction with tyramine in mature cheese (Blackwell 1963), and one of the first multicentre trials in the UK involving phenelzine (PLZ), imipramine, electroconvulsive therapy and placebo failed to demonstrate any clinical superiority for PLZ over placebo (MRC 1965).

Nevertheless, a small but convinced group of psychopharmacologists in both the United States and the United Kingdom continued to evaluate the safety and efficacy of TCP, PLZ and isocarboxazid (ISO) as the three surviving representatives of what have come to be known as the "classical" MAOIs.

Continuing interest in antidepressant specificity for "atypical depression" and "anxious" subtypes of major depression (Quitkin et al. 1994), as well as the emergence of a subclass of reversible and selective agents within the MAOIs (Waldmeier et al. 1994) has confirmed the continuing importance of the MAOI family of drugs in the treatment of depression and related disorders. There are also encouraging findings regarding the efficacy and safety of a transdermal delivery system for selegiline (SEL) (also called l-deprenyl) in clinical populations.

2
Chemistry

As of 2002, only irreversible MAOIs have approval from the Food and Drug Administration in the United States. In addition to PLZ and TCP, SEL has approval for the treatment of Parkinson's disease. Pargyline was approved for treatment of severe hypertension but is no longer manufactured. Although manufacturing of the antidepressant ISO was discontinued in 1990 in the United States, it was

Table 1 Some MAO inhibitors and their actions

Generic name	Brand name or company code	Currently marketed in North America	Reversible/irreversible, selectivity
Iproniazid	Marsilid	No	Irreversible, nonselective
Isoniazid, INH	Nydrazid	Yes, as antitubercular	Irreversible, nonselective
Phenelzine	Nardil	Yes	Irreversible, nonselective
Isocarboxazid	Marplan	Yes (relaunched in U.S.)	Irreversible, nonselective
Tranylcypromine	Parnate	Yes	Irreversible, nonselective
Clorgyline	–	No	Irreversible, MAO-A-selective
Pargyline	Eutonyl	Recently withdrawn as antihypertensive	Irreversible, MAO-B-selective
Selegiline	Eldepryl, l-Deprenyl	Yes, trials with transdermal system completed	Irreversible, MAO-B-selective
Moclobemide	Aurorix, Manerix	Available in Canada and elsewhere (not in U.S.)	Reversible, MAO-A-selective
Brofaromine	Consonar	No	Reversible, MAO-A-selective.
Lazabemide	Ro 196327	No	Reversible, MAO-B- selective
Mofegiline	–	No	Irreversible, MAO-B-selective
Milacemide	–	No	Partially reversible, relatively nonselective
Toloxatone	–	No	Reversible, MAO-A-selective
Befloxatone	–	No	Reversible, MAO-A-selective

relaunched in 1998. Moclobemide, a reversible inhibitor of MAO-A (RIMA), has been available in Europe, Canada, Australia and elsewhere since the early 1990 s but is not available in the United States (Fulton and Benfield 1996). Brofaromine, another RIMA with demonstrated antidepressant efficacy (Volz 1997), is not currently under active investigation. However, a new RIMA compound, befloxatone, is undergoing clinical trial evaluation in the United States and Europe (Curet et al. 1996).

The MAOIs may be classified according to chemical structure (hydrazine versus nonhydrazine), by affinity for substrate site (reversible versus irreversible), or by degree of selectivity for the A or B isoenzyme (A or B inhibitor; see Table 1).

2.1
Hydrazine Status

Both PLZ and ISO are hydrazine derivatives, with a nitrogen–nitrogen bond in their side chains. The first generation of hydrazine compounds (iproniazid and isoniazid) was withdrawn due to hepatotoxicity. TCP and SEL represent nonhydrazine arylalkylamines. These compounds are structurally related to amphetamine, and in a single case report of TCP overdose (Youdim et al. 1979) detectable levels of amphetamine were found, although this finding has not been replicated by other researchers (Sherry et al. 2000). However, methamphetamine and amphetamine are prominent metabolites of SEL (Heinonen et al. 1994).

2.2
Selectivity and Reversibility

Pargyline was the first selective MAOI to receive FDA approval. Although it proved selective for the B isoenzyme and did not require dietary restriction of tyramine intake (the gastrointestinal tract contains predominantly MAO-A), it was not an effective antidepressant. Although SEL is also a B inhibitor at low doses, at therapeutic doses in antidepressant trials, its selectivity is diminished (Lipper et al. 1979). Both pargyline and SEL are examples of *irreversible* selective inhibitors, in contrast to moclobemide, which is both reversible *and* selective (Table 1).

3
Pharmacology

3.1
Characteristics of the Monoamine Oxidase Enzyme System

The monoamine oxidases are flavoproteins found on the outer membrane of mitochondria. They catalyse the oxidative deamination of a variety of alkylamines, resulting in the formation of hydrogen peroxide and an aldehyde derivative of the amine. The aldehyde is then further reduced or oxidized to the corresponding alcohol or carboxylic acid, respectively. Free cytoplasmic and extraneuronal neurotransmitter amines would be most susceptible to MAO metabolism. The presence of high concentrations of MAO in the blood–gut and blood–brain barriers supports the idea that MAO serves a protective or detoxifying role (Murphy and Kalin 1980).

Irreversible inhibitors attach to the flavin–adenine dinucleotide group, resulting in a decreased breakdown of the substrate neurotransmitters and a denaturing of the original MAO. This results in a deficiency of MAO to denature dietary tyramine or amine medications. In contrast, the reversible inhibitors such as moclobemide have a direct inhibitory action on MAO-A and may be competitively displaced by tyramine, substantially reducing the likelihood of food and drug interactions (Bieck and Antonin 1989).

The enzyme is expressed widely through eukaryotic organisms and in all mammals studied to date. Many cells express both forms of MAO, in differing proportions, but some tissues contain only one form of the enzyme, which has facilitated purification of the individual. MAO-A is found in the placenta while MAO-B is found in platelets. Because platelets are relatively easy to procure, much of the research on MAO in humans has examined MAO-B. Although the human brain expresses both enzymes, MAO-B predominates (80%–95%); in contrast, in rat brain, MAO-A predominates over MAO-B. Immunohistochemical studies in postmortem human brain have reported that serotonergic neurons contain predominantly MAO-B, while catecholaminergic neurons contain MAO-A (Levitt et al. 1982; Westlund et al. 1988). These findings suggest that the two forms of MAO are independently regulated and perform different functions. The presence of MAO-B in serotonergic neurons is surprising since serotonin (5-HT) is a preferential substrate of MAO-A; this finding suggests that MAO-B may function to metabolize dietary and endogenous amines such as phenylethylamine (PEA) or tyramine (TA), but may have little effect on 5-HT unless levels of this amine are elevated. Normally, the classical neurotransmitter amines metabolized by MAO [norepinephrine (NE), dopamine (DA), 5-HT] are preferentially stored in vesicles where they are protected from MAO.

Homology between the amino acid sequences of both A and B isoforms is 70%, and both are encoded by separate genes located on the short arm of the X chromosome. During development, the MAO-A form is expressed first, followed by MAO-B, which increases in proportion through life. The complete absence of MAO has been described (Sims et al. 1989) in a family of male cousins with a deletion of Xp21-p11, an area that includes the gene for Norrie disease. Norrie disease is a rare X-linked recessive neurologic disorder. Characteristics of Norrie disease include retinal dysplasia with blindness, mental retardation and progressive hearing loss. Complete absence of MAO is thought to be related to features such as somatic growth failure, abnormal sexual maturation, autonomic nervous system dysfunction with hypotension, sleep disturbances with a marked reduction in the amount of rapid eye movement (REM) sleep, flushing, atonic seizures, motoric hyperactivity and hyperreflexia.

Numerous attempts have been made to correlate hereditary variations in MAO activity with human disease. Wide interindividual variations in platelet and fibroblast MAO activity, as much as 50-fold, have been described in studies of normal volunteers and those with psychiatric disorders. Low MAO-B activity has been associated with stimulus-seeking and suicidal behaviour, alcoholism and bipolar affective disorder. MAO-B activity increases with age in animals, elderly healthy humans and, to a greater extent, in patients with neurodegenerative disease (Manuck et al. 2000). The increased activity of MAO-B with age may reflect the glial cell proliferation that is linked to neuronal loss during ageing, since the concentration of the enzyme is thought to be higher in glial than in neuronal cells.

3.2
Platelet MAO

There are reports (Ravaris et al. 1976) that patients with greater than 80% inhibition of platelet MAO after 2 weeks of treatment have a better antidepressant response than patients with less than 80% inhibition. However, other studies (Sharma et al. 1990) have not demonstrated a significant relationship between clinical response and percentage of platelet MAO inhibition with PLZ, TCP or SEL. A problem with measuring platelet MAO is, of course, that it is exclusively MAO-B and does not necessarily reflect the central effect of the drugs on MAO-A. Hence, there is no role for measurement of platelet MAO inhibition with the RIMAs.

3.3
Effects of MAOIs on Brain Amine and Amino Acid Levels

The inhibition of MAO by drugs such as PLZ and TCP results in a marked elevation of levels of a number of brain amines termed "trace amines" (e.g. PEA, meta- and para-tyramine, tryptamine and octopamine). In addition, PEA is a metabolite of PLZ. These trace amines can have marked effects on uptake and/or release of both catecholamines and 5-HT at nerve endings. They may also act as neuromodulators through direct actions on receptors for the catecholamines and/or 5-HT (Baker et al. 1992).

Several groups of researchers have noted that administering PLZ to rats results in an elevation of brain levels of γ-aminobutyric acid (GABA). Acute and chronic studies suggest that this elevation is the result, at least in part, of an inhibition of the catabolic enzyme GABA-transaminase (GABA-T) (Popov and Matthies 1969; Baker et al. 1991; McManus et al. 1992; McKenna et al. 1994; Parent et al. 2000). Elevation of brain levels of the amino acid alanine and inhibition of alanine transaminase has also been observed (Wong et al. 1990; Tanay et al. 2001). It is of interest that there is increasing evidence that GABA is involved in the action of antipanic drugs, and PLZ is commonly used to treat panic disorder. Although the N^2-acetyl analogue of PLZ is also an effective MAO inhibitor, it neither alters GABA levels nor produces anxiolytic effects in rat (Paslawski et al. 1996). The inhibition of GABA-T and elevation of GABA by PLZ can be dramatically reduced by pre-treating rats with another MAO inhibitor (Popov and Matthies 1969; Todd and Baker 1995). This finding suggests that a metabolite of PLZ produced by the action of MAO on PLZ is ultimately responsible for the elevation of brain GABA (Paslawski et al. 2001). In theory, this would make PLZ a preferred antidepressant in the treatment of depressed patients with seizure disorders.

The structures of PLZ and TCP are similar to those of PEA and amphetamine, and, not surprisingly, they have effects on the uptake and/or release of DA, NA and, to a lesser extent, 5-HT. In animal studies that used therapeutically equivalent doses of TCP, sufficiently high levels of this drug were attained in the brain to affect the uptake and release of these neurotransmitters. High dosages of TCP (1.3 to 3.0 mg/kg/day) are effective in treating patients who suffer from refractory depression (Amsterdam 1991). Since these doses are well above those that have been reported to inhibit MAO by more than 90%, effects of TCP other than the inhibition of MAO may contribute to the antidepressant effects of high-dose TCP. Levels of free TCP may also contribute to the side-effects of this drug, including a mean orthostatic drop in systolic blood pressure and a rise in pulse rate. Elevations in blood pressure have also been significantly correlated with the dose of TCP (Keck et al. 1991). One group of investigators hypothesized that the initial hypertensive response to TCP is mediated by NE and that the orthostatic hypotensive effect is mediated by a direct interaction between TCP and α-adrenergic receptors (Keck et al. 1991).

Of the RIMA compounds that have been investigated so far, moclobemide appears to have few effects beyond MAO inhibition. In contrast, brofaromine is also a relatively potent inhibitor of 5-HT reuptake, which may contribute to its therapeutic efficacy (Waldmeier et al. 1994).

The selective, irreversible inhibitor of MAO-B, SEL, has been used extensively in the treatment of Parkinson's disease, primarily in early stages before treatment with L-dopa is required, or, in later stages, in combination with L-dopa (Parkinson Study Group 1993; Olanow et al. 1998). Although it was originally assumed that SEL was effective in this condition because it slowed the inhibition of the metabolism of DA, this theory has been disputed (Paterson et al. 1990). Some researchers feel that the bioactive amine 2-phenylethylamine may play an important role in the actions of this drug. SEL causes a marked increase in striatal levels of the 2-phenylethylamine that occurs before increases in DA levels become apparent with this drug. 2-Phenylethylamine has been reported to potentiate the postsynaptic effects of DA. SEL also increases expression of the mRNA coding for L-aromatic amino acid decarboxylase, the enzyme primarily responsible for synthesis of 2-phenylethylamine (Paterson et al. 1990).

There is now also a reasonably large body of evidence indicating that SEL may exert its therapeutic effects through actions on "oxidative stress" (Bentué-Ferrer et al. 1996; Gerlach et al. 1996). Several laboratories have reported that SEL increases neostriatal superoxide dismutase (SOD) activity in rodents (Knoll

1988; Carrillo et al. 1991; Clow et al. 1991). In addition, Wu et al. (1993) reported that SEL can suppress formation of the hydroxyl radical formation following intrastriatal administration of the neurotoxin MPP$^+$, and Chieuh et al. (1994) demonstrated that this drug can also reduce hydroxyl radical formation by inhibiting DA autoxidation and subsequent melanin formation. It has also been demonstrated that SEL can protect noradrenergic neurons from the neurotoxic effects of DSP-4 (Zhang et al. 1995) and dopaminergic neurons from the neurotoxic actions of 6-hydroxydopamine (Knoll 1988; Salonen et al. 1996). It is of interest that both SEL and its metabolite, N-desmethyldeprenyl (N-propargylamphetamine) have protective effects against excitotoxicity induced by glutamate or glutamate receptor agonists (Semkova et al. 1996; Mytilineou et al. 1997a,b). SEL has been reported to reduce apoptotic death and to provide neuroprotective/neurorescue effects in a remarkably large number of diverse models of neurotoxicity (Magyar et al. 1998; Paterson and Tatton 1998; Boulton 1999; Xu et al. 1999).

Recent reports in the literature suggest that the second messenger nitric oxide (NO) may also be involved in the actions of SEL. Zhang and Yu (1995) found that the neurotoxin DSP-4, in addition to depleting noradrenaline levels in hippocampus, also reduced NO synthase (NOS) activity in neurons of the dentate gyrus and that SEL prevented the noradrenaline depletion and the loss of NOS activity. In this regard, it is also of interest that N^w-nitro-L-arginine (LNNA), a NOS inhibitor, has been reported to reduce delayed neuronal death in gerbil brain after transient global ischemia (Kohno et al. 1996) and that SEL has been reported to induce rapid increases in NO production in brain tissue and cerebral vessels (Thomas et al. 1998). Surprisingly, SEL has also been reported to protect human dopaminergic neuroblastoma SH-SY5Y cells from apoptosis induced by excess NO. Obviously the relationship between SEL and NO is an interesting one that warrants further research.

It is also of interest that SEL has been reported to have neurorescue properties (Tatton and Greenwood 1991) in addition to neuroprotective actions. Tatton and Chalmers-Redman (1996) have proposed that these neurorescue properties as well as some of the neuroprotective effects of SEL are independent of inhibition of MAO. It has recently been suggested that SEL has neurotrophic actions and can also alter expression of genes for SOD-1 and SOD-2 and for factors involved in neuronal apoptosis (Tatton and Chalmers-Redman 1996; Paterson and Tatton 1998; Gelowitz and Paterson 1999). In this regard, it is also of interest that several studies have now suggested that SEL has beneficial effects in Alzheimer's disease (Filip and Kolibas 1999 and references contained in that article). Changes in several pre-synaptic and post-synaptic receptors may occur subsequent to the increased levels of the amines and/or amino acid neurotransmitters produced by the MAOIs. These delayed effects may be associated with the lag between administration of the MAO inhibitors and onset of antidepressant effect. Down-regulation of β- and α_2-adrenergic, 5-HT$_2$, and tryptamine receptors have been shown to occur after chronic administration of PLZ or TCP.

In addition to inhibiting GABA-T and alanine-T, PLZ has been reported to inhibit tyrosine amino transferase, aromatic amino acid decarboxylase, and dopamine β-hydroxylase (Yu and Bolton 1992). Aromatic amino acid decarboxylase, GABA-T and tyrosine amino transferase are pyridoxal-dependent enzymes. Malcolm et al. (1990) reported that PLZ depletes blood pyridoxal-5-phosphate levels in humans. In vitro studies of rats indicate that PLZ increases phenylethanolamine N-methyltransferase and catechol O-methyltransferase activities. It has also been reported that chronic administration of TCP results in an increase in activity of aromatic amino acid decarboxylase (Robinson et al. 1979).

Numerous reports indicate that both PLZ and TCP interact with enzymes involved in drug metabolism. Patients receiving PLZ or TCP may be taking other drugs concomitantly that have the potential for metabolic drug–drug interactions. PLZ and TCP have been reported to inhibit the degradation of drugs such as hexobarbital, ethylmorphine, aminopyrine, meperidine and antipyrine (Eade and Renton 1970; Clark et al. 1972; Smith et al. 1980; McDaniel 1986; Dupont et al. 1987). TCP is also a relatively potent inhibitor of cytochrome P450 2C19 (Parkinson 1996).

Another factor that must be considered in the actions of TCP is the presence of chiral centres. TCP is used clinically as the racemate but the (+)enantiomer has been shown to be more potent than (−)TCP at inhibiting MAO, whereas (−)TCP has been demonstrated to be more effective than (+)TCP as an inhibitor of uptake of catecholamines (Smith 1980). The two enantiomers also differ in their interaction with 5-HT_1 receptors in human post-mortem frontal cortex, with (−)TCP displaying a higher affinity than (+)TCP. Results to date, from studies on laboratory animals and humans (Fuentes et al. 1976) indicate marked pharmacokinetic differences between the two enantiomers in brain and plasma.

4
Metabolism/Plasma Concentration and Clinical Response

4.1
Phenelzine

Both PLZ and TCP are absorbed rapidly after oral administration and have short elimination half-lives. Despite the fact that both drugs have been commercially available for many years, much is still unknown about the metabolism of these two drugs and the contribution of metabolites to their overall pharmacological profiles. Numerous studies have been conducted on the relationship between acetylator status and response of patients to treatment with PLZ. Because isoniazid has structural similarity to PLZ and is a known substrate for acetylation, it has been assumed that PLZ is also acetylated, although the existence of N-acetyl-PLZ as a metabolite of PLZ was not demonstrated until relatively recently, and it appears to be only a minor metabolite (Robinson et al. 1985; Mozayani et al. 1988). PLZ is a unique drug since it is a substrate for, as well as an auto-inhibitor of, MAO (Clineschmidt and Horita 1969a,b).

There is also evidence that phenylacetic acid (PAA) and *p*-hydroxyphenylacetic acid (*p*-OH-PAA) are important metabolites of PLZ (Clineschmidt and Horita 1969a,b; Robinson et al. 1985). It is of interest that *p*-OH-PAA is a metabolite of the endogenous amine *p*-TA (*p*-OH-β-phenylethylamine) and that 2-phenylethylamine is also a known metabolite of PLZ (Dyck et al. 1985). There is also indirect evidence for the formation of *p*-hydroxyphenelzine (*p*-OH-PLZ) from PLZ (McKenna et al. 1990). Thus, possible routes for the formation of *p*-OH-PAA are as follows (Baker et al. 2000):

- PLZ→PAA→*p*-OH-PAA or
- PLZ→PEA→*p*-TA (and/or PAA)→*p*-OH-PAA; or
- PLZ→*p*-OH-PLZ→*p*-OH-PAA (with or without *p*-TA intermediate).

These routes have not been studied in detail nor is there extensive information on the pharmacological activity of metabolites such as *p*-OH-PLZ. Phenylethylidene hydrazine (PhCH$_2$CH = N-NH$_2$) or 1-(2-phenylethyldiazene) (PhCH$_2$CH$_2$N= NH) may be intermediates in going from PLZ to PAA, and hydrazine (H$_2$NNH$_2$) itself is a possible end-product of PLZ metabolism (Tipton and Spires 1971; Yu and Tipton 1989). *N*-methylation is another possible route of PLZ metabolism (Yu et al. 1991). To our knowledge there is no information available about the involvement of cytochrome P450 (CYP) isozymes in the various steps of PLZ metabolism.

4.2
Tranylcypromine

Although several popular textbooks mention that amphetamine is a metabolite of TCP, the bulk of evidence in the literature indicates that this is not the case. Despite a case report on the presence of amphetamine in the plasma of a patient who had overdosed on TCP, other studies conducted in humans and rats have not revealed amphetamine in human urine or rat brain after the administration of pharmacologically relevant doses of TCP (Sherry et al. 2000). The presence of *N*-acetyl and ring hydroxylated metabolites of TCP has been demonstrated in rats (Baker et al. 1986; Calverley et al. 1981; Foster et al. 1991) and microbes. There is little information available about the involvement of specific CYP isozymes in the metabolism of TCP, but TCP itself is a relatively potent inhibitor of CYP 2C19 (Goldstein et al. 1994; Parkinson 1996).

4.3
Selegiline

SEL is extensively metabolized to (−)methamphetamine, (−)amphetamine and (−)*N*-desmethyldeprenyl (*N*-propargylamphetamine) (Heinonen et al. 1994). CYP isozymes are involved in these metabolic pathways (Grace et al. 1994; Bach et al. 2000). The relative contribution of these metabolites to the therapeutic actions and side-effects of SEL continues to be a matter of debate.

4.4
Moclobemide and Brofaromine

The two RIMAs, moclobemide and brofaromine, demonstrate similar T-max values, but brofaromine is more highly protein- bound and has a longer elimination half-life than moclobemide (Waldmeier et al. 1994). Both agents are extensively metabolized. There is very little information available on the involvement of CYP isozymes in these metabolic routes, although it has been reported that moclobemide appears to be a substrate for CYP 2C19 and an inhibitor of CYP 2C19, CYP 2D6, and CYP 1A2, and that CYP 2D6 contributes to O-demethylation of brofaromine (Gram and Brossen 1993; Jedrychowski et al. 1993; Hartter et al. 1996). It has also been suggested that one-half the usual dose of moclobemide should be used when it is given in combination with the potent CYP 2D6 inhibitor cimetidine. It has been suspected for some time, but not yet definitely shown, that an active metabolite may contribute to the MAO-inhibiting activity of moclobemide (Da Prada et al. 1989; Waldmeier et al. 1994).

5
Safety and Adverse Effects

5.1
Drug–Drug Interactions

The major drug–drug and food–drug interactions involving MAOIs are summarized in Table 2. The so-called cheese effect has been discussed in the preceding section on "Precautions and Adverse Reactions."

Although coadministration of certain MAOIs with TCAs, selective serotonin reuptake inhibitors (SSRIs) and psychomotor stimulants such as amphetamine are listed in this table, all three of these combinations have been used successfully in the treatment of refractory depression (Amsterdam 1991); however, these combinations should be used only with extreme caution. Although MAOIs have been used effectively in combination with amitriptyline, imipramine, doxepin and trazodone, their use with SSRIs or clomipramine may lead to the potentially fatal serotonin syndrome. Combinations involving TCP and clomipramine are among the most dangerous and should be avoided. When switching from irreversible MAOIs to TCAs or SSRIs, a minimum washout of 2 weeks is required to allow complete recovery of MAO activity (American Psychiatric Association 2000; Kennedy et al. 2001). Combinations of moclobemide and an SSRI or a TCA have been reported to be safe and efficacious, although this work is preliminary and close observation of patients is important (Joffe and Bakish 1994). When moclobemide or brofaromine is withdrawn, only 24–48 h is required for complete recovery of MAO activity. When switching from a TCA or

Table 2 Interactions of MAOIs with drugs and food (reprinted with permission from Hansten and Horn 1995)

Type of interacting drug	Examples	Outcome	MAOI expected to interact
Foods containing tyramine (releases norepinephrine)	Cheese, red wine, beer, yeast extracts, aged meats	Hypertension	Nonselective MAO-A selective[a] MAO-B selective[a]
Drugs that release norepinephrine from sympathetic neurons	Amphetamines, ephedrine, phenylpropanolamine, pseudoephedrine	Hypertension	Nonselective MAO-A selective MAO-B selective?[b]
Drugs metabolized by MAO	Phenylephrine (oral)	Hypertension	Nonselective MAO-B selective[c]
	Sumatriptan (Imitrex)	Increased serum levels of sumatriptan	Nonselective[d] MAO-A selective
Drugs that can inhibit serotonin reuptake at synapses	SSRIs, clomipramine, imipramine, meperidine, dextromethorphan, propoxyphene?, venlafaxine	Serotonin syndrome (confusion, agitation, hypomania, sweating, myoclonus, fever, coma). Can be fatal	Nonselective MAO-A selective[e] MAO-B selective[f]
Serotonin agonists	Sumatriptan	Sertonin syndrome?[d]	Nonselective[d] MAO-A selective[d]

MAO, monoamine oxidase; MAOIs, monoamine oxidase inhibitors; SSRIs, selective serotonin reuptake inhibitors.

[a] If doses of MAOIs and/or tyramine are large enough, tyramine-induced hypertension can occur.

[b] Recommended doses of selegiline ($\leq$10 mg/day) are unlikely to inhibit MAO-A; thus, theoretically, selegiline should not interact with indirect-acting sympathomimetics. However, doses of selegiline >10 mg/day may result in inhibition of MAO-A.

[c] Initial evidence suggests that moclobemide produces only a minor increase in the pressor response of IV phenylephrine, but the risk should be greater with oral phenylephrine.

[d] Based upon theoretical considerations.

[e] It is recommended that moclobemide not be used with meperidine or dextromethorphan. Initial evidence suggests that moclobemide plus an SSRI does not result in a serotonin syndrome, but more data are needed to establish whether moclobemide and an SSRI can be used safely together.

[f] A case report suggests that selegiline may interact with meperidine, but the patient also was on imipramine 700 mg/day. Little is known about the use of selegiline with other inhibitors of serotonin uptake.

an SSRI to an MAOI, a washout period of 10–14 days is sufficient, except in the case of fluoxetine, where a washout period of 5 weeks is required, primarily because of the long elimination half-life of its active metabolite, norfluoxetine.

Effective lithium augmentation of MAOIs was first reported in 1972 (Himmelhoch et al. 1972). In countries where it is available, L-tryptophan may also be used in augmentation strategies, but should be started at low doses (250–500 mg/day) and gradually titrated upward to reduce the risk of developing the serotonin syndrome (Barker et al. 1987). This syndrome, which results from excessive increases in serotonergic tone, is characterized by tremor at rest, hypertonicity, myoclonus and autonomic signs. Hallucinations may also result and a life-threatening hyperthermia may occur.

PLZ and TCP have been reported to inhibit the activity of a number of enzymes involved in drug metabolism (Baker et al. 1992). TCP has been identified as a relatively potent inhibitor of CYP 2C19 (Parkinson 1996) and moclobemide has been reported to be a substrate for CYP 2C19 and to inhibit CYP 2D6, CYP 2C19 and CYP 1A2 (Gram et al. 1995; Hartter et al. 1996). Thus, the potential for metabolic drug–drug interactions between these MAOIs and coadministered drugs that are substrates for and/or inhibitors of these CYP isozymes should be considered.

It is also possible that interactions with prodrugs of amino acids or of valproic acid, which are converted in the body by MAO-B to amino acids or to valproic acid, will occur. Several such drugs are under investigation as anticonvulsants, and the potential for a serious drug–drug interaction with MAOIs must be considered because the MAOIs will interfere with metabolism of the prodrug to the active drug of interest.

5.2
Adverse Effects

The most feared side-effect of irreversible MAOIs is the TA-induced hypertensive crisis. Once it is decided to initiate a trial of an irreversible MAOI, the patient should be instructed to avoid foods with substantial TA content (other sympathomimetic amines such as PEA may also be culprits in this regard). TA, a substrate of MAO present in significant concentrations in certain fermented food stuffs including red wine, tap beer, cheese, yeast extracts and pickled fish, causes a vasopressor response, an effect that is dramatically accentuated in patients taking an MAOI (Shulman et al. 1997; Shulman and Walker 1999). This response is due in part to the MAO-inhibited patient's inability to deaminate TA (which is normally broken down by MAO-A in the gut) resulting in displacement of intracellular stores of NA. When dietary restrictions are followed, the risk of a serious hypertensive problem is curtailed. Although lists of foods that should be avoided by patients taking MAOIs are available, there is a general consensus that many of these are too extensive, and there has been a trend in recent years to develop more "user-friendly" MAOI diets (Gardner et al. 1996). There are also case reports of patients who develop idiosyncratic, spontaneous hypertensive episodes shortly after taking initial doses of TCP and, more rarely, PLZ. Reports of minimal increases in blood pressure in response to intravenous and oral TA administration in patients taking RIMAs (moclobemide and brofaromine) suggest that these drugs are safer and can be used without stringent dietary control (Bieck and Boulton 1994). However, it is suggested that they be given after meals to reduce the likelihood of causing any inhibition of TA metabolism. At usual therapeutic doses, the cheese effect should not be a problem with the selective MAO-B inhibitor SEL, but at higher doses of this drug, there is some inhibition of MAO-A and dietary restrictions should be followed.

Because of its α-adrenergic blocking properties, phentolamine (5 mg IM or IV) has traditionally been the recommended treatment for an MAOI-induced hypertensive crisis. Onset of action is rapid, usually within 5 min, and a single dose is generally effective. A diuretic such as furosemide may be given intravenously to avoid fluid retention and, if necessary, a beta-blocker such as propranolol may be given to control tachycardia. Additional doses of phentolamine can be given over several hours if needed. A recent alternative treatment is sublingual nifedipine (10 mg with a repeat dose after 20 min if required). As with phentolamine, the onset of action occurs within 5 min.

Although a hypertensive crisis is the most feared reaction with classical MAOIs, paradoxically, hypotension is a more common cardiovascular side-effect and can lead to dizziness and fainting after sudden postural changes (Thase et al. 1995). In fact, postural hypotension may be a useful clinical sign for determining therapeutic dose (i.e. the dose may be titrated upwards cautiously until some postural hypotension is observed). If postural hypotension does become a problem in otherwise responsive patients, mineralocorticoids may be given while monitoring serum potassium levels to avoid hypokalaemia.

Insomnia can also be problematic with MAOIs, ultimately necessitating discontinuation of the drug in a small number of people; this side-effect can often be managed by lowering the dose or by adding a sedative drug (e.g. a benzodiazepine or trazodone) to the therapeutic regimen at bedtime. Paradoxically, some patients are sedated on PLZ and take the medication prior to sleep. The irreversible MAOIs may cause a marked suppression of REM sleep, which is less likely to be experienced with RIMAs (Flanigan and Shapiro 1994). Other common side-effects of the MAOIs include weight gain, peripheral oedema and sexual dysfunction. A peripheral neuropathy resulting from pyridoxine deficiency produced by PLZ may occur in some patients. There is a reduced incidence of sleep disorders, sexual dysfunction and weight gain with moclobemide compared with the classical MAOIs. There are fewer reports of sexual dysfunction associated with moclobemide than with most other antidepressants (Kennedy et al. 2000).

A discontinuation syndrome consisting of arousal, mood disturbance and somatic symptoms has been reported with sudden cessation of PLZ or TCP administration (Dilsaver 1988). As with other classes of antidepressants, it is prudent to taper the dose of MAOIs and RIMAs over several weeks. With MAOIs but not with RIMAs, it is necessary to avoid the administration of other antidepressants for at least 10–14 days after drug discontinuation to allow a full return to normal MAO activity before starting a new drug.

6
Clinical Use/Therapeutic Indications

The MAOIs have found an important niche in the therapeutic armamentarium of clinicians, particularly for atypical depression and depression associated with anxiety, panic and phobias. PLZ continues to be used as a second-line therapy for atypical depression and TCP may be effective in previously "treatment-resistant depression".

Demonstrated antidepressant efficacy with SEL appears to require oral doses (30–60 mg/day) that result in a loss of selective MAO-B inhibition and the need for dietary precautions (Bodkin and Kwon 2001). A promising alternative delivery system for SEL involves the Selegiline Transdermal System (STS) that allows direct absorption into the systemic circulation with consequent central MAO-B inhibition and less inhibitory effects on gut or other peripheral MAO sites. This STS "patch" has been found to be significantly better than placebo in the treatment of major depression in both rates of response and remission. The only side-effect that occurred significantly more frequently with the SEL group was a local skin reaction at the patch site (Bodkin and Amsterdam 1999). If approved, this novel delivery system for antidepressant medication could serve to increase clinician confidence in prescribing MAOI therapy.

A summary of dosing guidelines for clinical depression is presented in the dosing table in the Appendix (Table 1).

6.1
Atypical Depression

The concept of "atypical depression" has been applied to different subtypes of depression; in the DSM-IV (American Psychiatric Association 1994), the term is used to refer to a depression characterized by preserved mood reactivity accompanied by at least two of the following symptoms: extreme sensitivity to interpersonal loss or rejection, prominent anergia, increased appetite, weight gain or hypersomnia. Atypical depression was previously subclassified into anxious (type A) and reversed vegetative (type V) types (Davidson et al. 1982). The Columbia University definition of atypical depression combined reversed vegetative symptoms with mood reactivity and intense rejection sensitivity. This group has been able to demonstrate a superior response to PLZ over imipramine (IMI) in patients with "Columbia University Atypical Depression" (Quitkin et al. 1994). There also appears to be considerable overlap between the predisposition towards "rejection sensitivity" and Axis II personality disorders, especially borderline personality disorder (BPD). TCP was found to be superior to alprazolam, carbamazepine and trifluoperazine in BPD (Cowdry and Gardner 1988)

Recent investigations have compared moclobemide and sertraline in outpatients with a diagnosis of atypical depression (Sogaard et al. 1999). Although both sertraline and moclobemide improved the symptoms of atypical depression equally, the low dose of moclobemide (300–450 mg) limits conclusions about comparative effectiveness of both agents in clinical practice.

6.2
Anergic Bipolar Depression

A group at the University of Pittsburgh has emphasized the presence of type V (reversed vegetative) symptoms in what has been described as "anergic bipolar depression". Although their results have not been replicated elsewhere, this group has demonstrated the superiority of TCP over IMI in bipolar depression (Himmelhoch et al. 1991). The extent to which findings from the Columbia group are specific to PLZ and those of the Pittsburgh group are specific to TCP remains unknown.

In a review of pharmacological strategies for the treatment of bipolar depression, Thase and Sachs (2000) recommended TCP and venlafaxine as preferred agents in treatment-resistant cases of bipolar depression. A lower rate of switching from bipolar depression to hypomania or mania was also observed with lamotrigine, paroxetine and moclobemide in three large-scale, double-blind studies of bipolar depression (Calabrese et al. 1999), although other antidepressants including bupropion have not been systematically evaluated in similar large multicentre studies.

6.3
Anxiety and Eating Disorders

Many patients with atypical depression also suffer from panic attacks and anxiety symptoms (Quitkin et al. 1990). PLZ has been reported by some authors to be the treatment of choice for panic disorder (Westenberg 1996). In other investigations, moclobemide was found to be comparable in efficacy to clomipramine (Kruger and Dahl 1999) and fluoxetine (Tiller et al. 1999), with approximately 60% of patients achieving panic-free status at the end of both studies. Patients with moclobemide also reported significantly fewer side-effects. However, moclobemide was less effective than CBT and conferred no extra benefit when added to CBT (Loerch et al. 1999).

Patients with social phobia also respond favourably to MAOI and RIMA therapy. Studies involving PLZ provided support for the 60-mg dose (Gelerenter et al. 1991; Liebowitz 1992), while, in one of the few direct comparisons of PLZ and moclobemide, Versiani et al. (1992) confirmed the comparable efficacy of moclobemide at a dose of 600 mg/day. PLZ also produced a superior outcome to cognitive–behavioural group therapy during the acute phase of treatment (Liebowitz et al. 1999).

Although dietary concerns have limited their use in patients with bulimia nervosa, controlled trials support the efficacy of PLZ (Walsh et al. 1988) isocarboxazid (Kennedy et al. 1988) and brofaromine (Kennedy et al. 1993).

6.4
Depression in the Medically Ill

There is limited evidence to support the use of moclobemide as a treatment for comorbid depression in multiple sclerosis patients (Barak et al. 1999) and for comorbid dementia and depression (Amrein et al.1999). Moclobemide also appears to be well tolerated and effective in treating depression associated with traumatic brain injury (Newburn et al. 1999).

MAOIs have also been employed in the therapeutic management of various conditions that include anginal pain, atypical facial pain, refractory thought disorders, neurodermatitis, treatment-resistant narcoleptic states, migraine, narcolepsy, attention-deficit disorder and idiopathic orthostatic hypotension.

6.5
MAOIs in Combination Therapy

The practice of combining MAOI or RIMA antidepressants with other antidepressant agents, including TCAs and SSRIs, to treat refractory depression has been in existence for decades, despite the potential for serious adverse drug interactions (White and Simpson 1981; Joffe and Bakish 1994; Berlanga and Ortega-Soto 1995; Ebert et al. 1995).

There is also anecdotal evidence to suggest tranylcypromine may be effective in combination with bupropion (Pierre and Gitlin 2000) or risperidone (Stoll and Haura 2000) in the management of treatment-refractory depression.

Following the publication of case series reports concerning the use of a combination of moclobemide with SSRIs (Joffe and Bakish 1994) in previously non-responsive depressed patients, a triple therapy of moclobemide, trazodone and lithium (Magder et al. 2000) was also found to be effective and safe. These potentially hazardous combination therapies should be used only by specialists in pharmacology and with full patient awareness of the potential for adverse drug interactions.

Continuation and maintenance treatments with MAOI and RIMA therapies have not been adequately studied. Findings from three randomized, placebo-controlled trials involving long-term MAOI therapy have been published. All three studies involved PLZ, which proved superior to nortriptyline in a maintenance trial in elderly depressives (Georgotas et al. 1989) and superior to placebo in adult depressed populations (Harrison et al. 1986; Robinson et al. 1991).

Although acute treatment outcomes were comparable following fluoxetine or PLZ treatment for atypical depression (Pande et al. 1996), none of the PLZ-treated group had chosen to remain on this medication at a 2-year follow-up (Zubieta et al. 1999).

7
Beyond MAO Inhibition: Binding of MAOIs to Other Amine Oxidases and Imidazoline Binding Sites

Not all experimental or clinical effects elicited by MAOIs might be explained through binding to, and inhibition of, MAO enzymes. Hydrazine-based agents in particular are notorious for their lack of specificity, inhibiting many enzymes with carbonyl-containing cofactors such as pyridoxal phosphate. Thus, PLZ and isocarboxazid might reasonably be expected to have rather widespread effects on, for example, decarboxylase and aminotransferase enzymes such as those involved in GABA metabolism and the urea cycle. While a full discussion of the implications of such interactions is beyond the scope of this chapter, some mention of the effects of MAOIs on other amine oxidases and related proteins is both pertinent and revealing.

MAO substrates such as dopamine, tyramine, β-phenylethylamine, tryptamine and benzylamine are also substrates, to a greater or lesser degree and in a tissue and species-dependent manner, for a family of enzymes known collectively as the semicarbazide-sensitive amine oxidase (SSAO), or benzylamine oxidase enzymes (Lyles 1996). These enzymes of the EC 1.4.3.6 group exist as cell surface glycosylated type II ectoenzymes, mainly on vascular and non-vascular smooth muscle (Lewinsohn 1981; Lyles and Singh 1985) and on white and brown adipocytes (Barrand and Callingham 1982; Barrand and Fox 1982; Barrand et al. 1984; Raimondi et al. 1991). Soluble SSAO enzymes present in the plasma of humans (Lyles et al. 1990) and other species are more usually referred to as plasma amine oxidases (PAO), and sometimes erroneously as serum amine oxidases, with the SSAO abbreviation generally limited to the tissue-bound enzymes. The soluble PAO enzymes have substrate preferences distinct from the tissue-bound SSAO enzymes, most notably in that they metabolize the polyamines, spermine and spermidine (Morgan 1985). That the EC 1.5.3.11 enzyme, polyamine oxidase, is an integral player in the polyamine cycle and is also often abbreviated to PAO continues to be a source of some confusion (Seiler 1995). In the present review, PAO refers to plasma SSAO enzymes.

Both the soluble and tissue-bound SSAO enzymes contain a trihydroxyphenylalanine quinone (TPQ) cofactor (Janes et al. 1990, 1992; Holt et al. 1998) as well as active site copper (Klinman and Williams 2000). These cofactors facilitate an aminotransferase-type ping-pong oxidation of substrate amines, resulting in release of aldehyde, hydrogen peroxide and ammonia. Thus, the products of deamination are identical to those from MAO, although the reaction mechanism is somewhat different. While MAO and SSAO/PAO enzymes share several substrates, a few endogenous SSAO and/or PAO substrates are not turned over by MAO, including histamine (also a diamine oxidase substrate) (Buffoni 1995), methylamine (Precious et al. 1988; Lyles et al. 1990), aminoacetone (Lyles and Chalmers 1992) and the polyamines, spermidine and spermine.

Several clinically used MAOIs are also able to inhibit SSAO enzymes. The hydrazine derivative, PLZ, can be considered a tight-binding suicide-type SSAO inhibitor, with enzyme recovery partly as a result of slow metabolism of the inhibitor by the enzyme (Andree and Clarke 1982). It is likely that doses of PLZ sufficiently high to have a substantial effect on MAO will also inhibit SSAO to a similar or greater degree. Isocarboxazid has not been thoroughly examined as an SSAO inhibitor, but it is a hydrazine derivative and is structurally related to iproniazid, a potent SSAO inhibitor (Lyles et al. 1983). At a concentration of 1 mM, TCP caused complete inhibition of rat aorta SSAO in vitro, although inhibition was readily reversible by dialysis (Lyles 1996). A comprehensive review by Lyles (1996) provides numerous examples of other therapeutics, such as hydralazine, carbidopa, benserazide, procarbazine and mexiletine, which also inhibit SSAO and PAO enzymes.

In the experimental setting, SSAO inhibition by established or novel MAOI under investigation must be accounted for in situations where, for example, amine or metabolite concentrations are being measured following MAOI administration and the amine of interest is a combined MAO/SSAO (or PAO) substrate. In the clinical setting, the consequences of SSAO/PAO inhibition are far less easy to predict, as a result of our lack of understanding of the physiological functions of these enzymes. In fact, although several pathophysiological functions for SSAO and PAO enzymes have been proposed (Callingham et al. 1995; Yu 2000), one of only two physiological functions described to date for tissue-bound SSAO is that of a vascular adhesion protein (VAP)-1 involved in lymphocyte trafficking (Salminen et al. 1998) and it is not yet clear if the enzymatic capability of SSAO is even required for lymphocyte trafficking to occur. In adipocytes, SSAO appears to be important in the regulation of glucose transport, with hydrogen peroxide precipitating glucose uptake through GLUT4 translocation to the plasmalemma (Enrique-Tarancón et al. 2000).

Deamination of the endogenous amines, methylamine and aminoacetone, by SSAO/PAO yields formaldehyde and methylglyoxal respectively, as well as ammonia and hydrogen peroxide. Both of these aldehydes are toxic and induce cross-linking and damage in proteins and DNA (Callingham et al. 1995). Elevated SSAO activities are seen in diabetes, atherosclerosis, obesity and chronic heart failure, and it has recently been confirmed that mortality rates in chronic heart failure patients are significantly higher in those patients with elevated PAO activity (Boomsma et al. 2000; Yu 2000). In combination with the increased methylamine or aminoacetone levels that are seen in diabetes, pregnancy, pregnancy toxaemia and uraemia, formation of formaldehyde and methylglyoxal is likely to be greatly accelerated in these conditions and it seems that some benefit may be gained from inhibiting SSAO and reducing the rate of formation of toxic metabolites. However, one should caution against the enthusiastic administration of SSAO inhibitors, or MAOI with the capacity to inhibit SSAO/PAO, since there are many situations in which active SSAO would seem to be desirable. Langford et al. (1999) showed that weanling rats treated chronically with an experimental SSAO inhibitor failed to develop functional aortae, particularly with respect to vessel wall and elastin integrity. One might argue that this could be related to a disruption of polyamine metabolism, given the integral role played by spermine and spermidine and their metabolites in cell growth, differentiation and survival (Pegg and McCann 1982; Bonneau and Poulin 2000). In fact, polyamines have been shown to possess both cytotoxic and protective capabilities under a variety of conditions (Gilad and Gilad 1992; Poulin and Bonneau 2000; Segal and Skolnick 2000) and inhibition of SSAO/PAO may not be without risk.

The past decade has seen the emergence of a new family of receptors or binding sites, the imidazoline receptors (IR or IBS). These are reported to be involved in the control of blood pressure and renal function, to modulate the conductance of some ion channels, and perhaps to exert some control over MAO activities (Renouard et al. 1993; Carpéné et al. 1995a,b; Raddatz et al. 1995; Tesson et al. 1995; Alemanay et al. 1997; Gargalidis-Moudanos et al. 1997; Molderings 1997; Ozaita et al. 1997; Raddatz and Lanier 1997; Molderings and Gothert 1999; Olmos et al. 1999). At least three binding site subtypes have been identified, and some of these can be further subdivided based on ligand affinity data (Eglen et al. 1998). While strong evidence now exists to support the contention that I_1BS does correspond to functional receptors, this may not be the case for I_2BS and I_3BS. In fact, the presence of a high-affinity I_2BS on MAO has been demonstrated (Tesson et al. 1995), although clearly this site is not a receptor. While imidazoline ligands are able to inhibit MAO, the high-affinity I_2BS appears to be distinct from the enzyme-active site, and ligand concentrations required to inhibit MAO are much higher than their corresponding K_D values at the I_2BS (Ozaita et al. 1997). Inhibition may thus be through an initially competitive interaction with the active site, and a function has not yet been ascribed to the high-affinity I_2BS on MAO.

Numerous imidazoline ligands have been shown to inhibit diamine oxidase, a rather specialized member of the SSAO family (A. Holt, unpublished data; see Holt and Baker 1995) and an imidazoline binding site has also been identified on plasma amine oxidase enzymes (Carpéné et al. 1995a). Thus, it would seem reasonable to conclude that at least some IBS are actually the active sites of different amine oxidase enzymes, and this would contribute to the apparent multiplicity of IBS. But is it merely coincidence that such a high degree of ligand cross-recognition exists? Agmatine is an endogenous IBS ligand, and is also a substrate for both porcine kidney diamine oxidase (Holt and Baker 1995) and porcine vascular SSAO (A. Holt, unpublished). Polyamines are endogenous imidazoline ligands, as well as substrates for plasma amine oxidases. Tryptamine has micromolar affinity for I_2BS in rabbit brain (Price et al. 1999), and is a substrate for both MAO and SSAO. Interestingly, several β-carbolines, which can be formed endogenously from tryptamine, had nanomolar affinities for I_2BS in rabbit brain (Price et al. 1999). One such compound, harmane, produced hypotension when injected into the rostral ventrolateral medulla of the rat, most likely through an action at I_1 receptors (Musgrave and Badoer 1999).

While it is likely that more than one endogenous imidazoline receptor ligand exists, current evidence suggests that one or more of these may be substrates for one or more amine oxidase enzymes. This being the case, it would seem reasonable to suggest that amine oxidase inhibition may increase tissue levels of the ligand(s), presumably increasing imidazolinergic transmission and consequently decreasing imidazoline receptor density. In fact, ongoing studies in our laboratories have provided preliminary evidence that rat brain I_1 receptors are down-regulated following chronic treatment with selective amine oxidase inhibitors (A. Holt and G.B. Baker, unpublished). Platelet I_1 receptors are up-regulated in depression and several related disorders, while treatment with desipramine or fluoxetine reduced receptor density (Piletz et al. 1994, 1996a,b). Taken together with the well-documented (although arguable) role played by imidazoline receptors in blood pressure control (Bousquet and Feldman 1999) and the fact that several imidazolines, including agmatine, appear to be neuroprotective (Olmos et al. 1999), it seems that the use of amine oxidase inhibitors to manipulate imidazolinergic systems might prove to be therapeutically beneficial. However, until the roles of imidazoline receptors are more clearly established, and the effects of individual amine oxidase inhibitors, including MAOI, on this system have been defined, it would seem prudent for users of MAOI to bear in mind that the potential exists for clinical or experimental anomalies that might not be attributable to MAO inhibition.

Acknowledgements. The authors are grateful to the Canadian Institutes of Health Research for ongoing funding support.

References

Alemanay R, Olmos G, García-Sevilla J (1997) Labelling of I_{2B}-imidazoline receptors by [^{3}H]2-(2-benzofuranyl)-2-imidazoline (2-BFI) in rat brain and liver: characterization, regulation and relation to monoamine oxidase enzymes. Naunyn-Schmiedeberg's Arch Pharmacol 356:39-47
American Psychiatric Association (1994) Diagnostic and statistical manual of mental disorders, 4th edition. American Psychiatric Association, Washington, DC
American Psychiatric Association. Practice guidelines for major depressive disorder, second edition.Washington, DC: American Psychiatric Association; 2000: suppl. p 11.
Amrein R, Martin JR, Cameron AM (1999) Moclobemide in patients with dementia and depression. Adv Neuro 80:509-519
Amsterdam JD. (1991) Use of high dose tranylcypromine in resistant depression. In: Amsterdam JD (ed) Advances in neuropsychiatry and psychopharmacology, volume 2. Refractory depression. Raven Press, New York
Andree TH, Clarke DE (1982) Characteristics and specificity of phenelzine and benserazide as inhibitors of benzylamine oxidase and monoamine oxidase. Biochem Pharmacol 31:825-830
Bach MV, Coutts RT, Baker GB (2000) Metabolism of N,N-dialkylated amphetamines, including deprenyl, by CYP2D6 expressed in a human cell line. Xenobiotica 30:297-306
Baker GB, Hampson DR, Coutts RT, Micetich RG, Hall TW, Rao TS (1986) Detection and quantitation of a ring-hydroxylated metabolite of the antidepressant drug tranylcypromine.J Neural Transm 1986;65:233-243
Baker GB, Wong JTF, Yeung JM, Coutts RT (1991) Effects of the antidepressant phenelzine on brain levels of γ-aminobutyric acid (GABA). J Affect Disord 21:207-211

Baker GB, Coutts RT, McKenna KF, Sherry-McKenna RL (1992) Insights into the mechanisms of action of the MAO inhibitors phenelzine and tranylcypromine: a review. J Psychiatry Neurosci 17:206–214

Baker GB, Coutts RT, Greenshaw AJ (2000) Neurochemical and metabolic aspects of antidepressants: An overview. J Psychiatr Neurosci 25:481–496

Barak Y, Ur E, Achiron A (1999) Moclobemide treatment in multiple sclerosis patients with comorbid depression: an open-label safety trial. J Neuropsychiatry Clin Neurosci 11:271–273

Barker WA, Scott J, Eccleston D (1987) The Newcastle chronic depression study: results of a treatment regime. Int Clin Psychopharmacol 2:261–272

Barrand MA, Callingham BA (1982) Monoamine oxidase activities in brown adipose tissue of the rat: some properties and subcellular distribution. Biochem Pharmacol 31:2177–2184

Barrand MA, Fox SA (1982) Amine oxidase activities in brown adipose tissue of the rat: identification of semicarbazide-sensitive (clorgyline-resistant) activity at the fat cell membrane. J Pharm Pharmacol 36:652–658

Barrand MA, Fox SA, Callingham BA (1984) Amine oxidase activities in brown adipose tissue of the rat: identification of semicarbazide-sensitive (clorgyline-resistant) activity at the fat cell membrane. J Pharm Pharmacol 38:288–293

Bentué-Ferrer D, Ménard G, Allain H (1996) Monoamine oxidase B inhibitors: Current status and future potential. CNS Drugs 6:217–236

Berlanga C, Ortega-Soto HA (1995) A 3-year follow-up of a group of treatment-resistant depressed patients with a MAOI/tricyclic combination. J Affect Disord 34:187–92

Bieck PR, Antonin KH (1989) Tyramine potentiation during treatment with MAO inhibitors: brofaromine and moclobemide vs irreversible inhibitors. J Neural Transm Suppl 28:21–31

Bieck P, Boulton A (1994) Tyramine potentiation during treatment with MAOIs. In: Kennedy S (ed) Clinical advances in monoamine oxidase inhibitors therapies, vol. 43. American Psychiatric Press, Washington, DC, pp 83–110.

Blackwell B. (1963) Hypertensive crisis due to monamine-oxidase inhibition. Lancet 2:849–851.

Bodkin JA, Amsterdam JD (1999) Transdermal selegiline in the treatment of patients with major depression: a double-blind placebo-controlled trial. Presented at the annual meeting of American College of Neuropsychopharmacology, December 1999, Acapulco, Mexico

Bodkin JA, Kwon AE (2001) Selegiline and other atypical monoamine oxides inhibitors in depression. Psychiatric Annals 31 385–391

Bonneau MJ, Poulin R (2000) Spermine oxidation leads to necrosis with plasma membrane phosphatidylserine redistribution in mouse leukemia cells. Exp Cell Res 259:23–34

Boomsma F, de Kam PJ, Tjeerdsma G, van den Meiracker AH, van Veldhuisen DJ (2000) Plasma semicarbazide-sensitive amine oxidase (SSAO) is an independent prognostic marker for mortality in chronic heart failure. Eur Heart J 21:1859–63

Boulton AA (1999) Antiapoptotic drugs: a progression of research. Can J Cont Med Educ March:195–202

Bousquet P, Feldman J (1999) Drugs acting on imidazoline receptors: a review of their pharmacology, their use in blood pressure control and their potential interest in cardioprotection. Drugs 58:799–812

Buffoni F (1995) Semicarbazide-sensitive amine oxidases: some biochemical properties and general considerations. Prog Brain Res 106:323–331

Calabrese JR, Rapport DJ, Kimmel SE, Shelton MD (1999) Controlled trials in bipolar I depression: focus on switch rates and efficacy. Eur Neuropsychopharmacol Suppl 4:S109–112

Callingham BA, Crosbie AE, Rous BA (1995) Some aspects of the pathophysiology of semicarbazide-sensitive amine oxidase enzymes. Prog Brain Res 106:305–321

Calverley DG, Baker GB, Coutts RT, Dewhurst WG (1981) A method for measurement of tranylcypromine in rat brain regions using gas chromatography with electron capture detection. Biochem Pharmacol 30:861–867

Carpéné C, Collon P, Remaury A, Cordi A, Hudson A, Nutt D, Lafontan M (1995a) Inhibition of amine oxidase activity by derivatives that recognize imidazoline I_2 sites. J Pharm Exp Ther 272:681–688

Carpéné C, Marti L, Hudson A, Lafontan M (1995b) Nonadrenergic imidazoline binding sites and amine oxidase activities in fat cells. Ann N Y Acad Sci 763:380–397

Carrillo MC, Kanai S, Nokubo M, Kitani K (1991) L-deprenyl induces activities of both superoxide dismutase and catalase but not of glutathione peroxidase in the striatum of young male rats. Life Sci 48:517–521

Chiueh CC, Huang SJ, Murphy DL (1994) Suppression of hydroxyl radical formation by MAO inhibitors: a novel possible neuroprotective mechanism in dopaminergic neurotoxicity. J Neural Transm Suppl 41:189–196

Clark B, Thompson JW, Widdington G (1972) Analysis of the inhibition of pethidine N-demethylation by monoamine oxidase inhibitors and some other drugs with special reference to drug interactions in man. Br J Pharmacol 44:89–99

Clineschmidt BV, Horita A (1969a) The monoamine oxidase catalyzed degradation of phenelzine-1-^{14}C, an irreversible inhibitor of monoamine oxidase: I. Studies in vitro. Biochem Pharmacol 18:1011–1020

Clineschmidt BV, Horita A (1969b) The monoamine oxidase catalyzed degradation of phenelzine-1-^{14}C, an irreversible inhibitor of monoamine oxidase: II. Studies in vivo. Biochem Pharmacol 18:1021–1028

Clow A, Hussain T, Glover V, Sandler M, Dexter DT, Walker M (1991) (-)-Deprenyl can induce soluble superoxide dismutase in rat striata. J Neural Transm Gen Sect 86:77–80

Cowdry RW, Gardner DL (1988) Pharmacotherapy of borderline personality disorder. Alprazolam, carbamazepine, trifluoperazine, and tranylcypromine. Arch Gen Psychiatry 45:111–119

Curet O, Damoiseau G, Aubin N, Sontag N, Rovei V, Jarreau FX (1996) Befloxatone, a new reversible and selective monoamine oxidase-A inhibitor. I. Biochemical profile. J Pharmacol Exp Ther 277:253–264

Da Prada M, Kettler R, Keller HH, Burkard WP, Muggli-Maniglio D, Haefely WE (1989) Neurochemical profile of moclobemide, a short-acting and reversible inhibitor of monoamine oxidase type A. J Pharmacol Exp Ther 248:400–414

Davidson JR, Miller RD, Turnbull CD, Sullivan JL (1982) Atypical depression. Arch Gen Psychiatry 39:527–534

Dilsaver SC (1988) Monoamine oxidase inhibitor withdrawal phenomena: Symptoms and pathophysiology. Acta Psychiatr Scand 78:1–7

Dupont H, Davies DS, Strolin-Benedetti M (1987) Inhibition of cytochrome P-450-dependent oxidation reactions by MAO inhibitors in rat liver microsomes. Bichem Pharmacol 36:1651–1657

Dyck LE, Durden DA, Boulton M (1985) Formation of β-phenylethylamine from the antidepressant, β-phenylethythydrazine. Biochem Pharmacol 34:1925–1929

Eade NR, Renton KW (1970) Effect of monoamine oxidase inhibitors on the N-demethylation and hydrolysis of meperidone. Biochem Pharmacol 19:2243–2250

Ebert D, Albert R, May A, Stosiek I, Kaschka W (1995) Combined SSRI-RIMA treatment in refractory depression. Safety data and efficacy. Psychopharmacology (Berl) 119:342–344

Eglen RM, Hudson AL, Kendall DA, Nutt DJ, Morgan NG, Wilson VG, Dillon MP (1998) 'Seeing through a glass darkly': casting light on imidazoline 'I' sites. Trends Pharmacol Sci 19:381–390

Enrique-Tarancon G, Castan I, Morin N, Marti L, Abella A, Camps M, Casamitjana R, Palacin M, Testar X, Degerman E, Carpene C, Zorzano A (2000) Substrates of semi-carbazide-sensitive amine oxidase co-operate with vanadate to stimulate tyrosine phosphorylation of insulin-receptor- substrate proteins, phosphoinositide 3-kinase activity and GLUT4 translocation in adipose cells. Biochem J 350(Pt 1):171–180

Filip V, Kolibas E (1999) Selegiline in the treatment of Alzheimer's disease: A long-term randomized placebo-controlled trial. Czech and Slovak Senile Dementia of Alzheimer Type Study Group. J Psychiatry Neurosci 24:234–243

Flanigan M, Shapiro C (1994) MAOIs and sleep. In: Kennedy S (ed) Clinical advances in monoamine oxidase inhibitor therapies. American Psychiatric Press, Washington, DC, pp 125-145

Foster BC, Litster DL, Zamecnik J, Coutts RT (1991) The biotransformation of tranylcypromine by Cunninghamella echinulata. Can J Microbiol 37:791–795

Fuentes JA, Oleshansky MA, Nef NH (1976) Comparison of the antidepressant activity of (-) and (+) tranylcypromine in an animal model. Biochem Pharmacol 25:801–804

Fulton B, Benfield P (1996) Moclobemide. An update of its pharmacological properties and therapeutic use. Drugs 52:450–474

Gardner DM, Shulman KI, Walker SE, Tailor SA (1996) The making of a user friendly MAOI diet. J Clin Psychiatry 57:99–104

Gargalidid-Moudanos C, Remaury A, Pizzinat N, Parini A (1997) Predominant expression of monoamine oxidase B isoform in rabbit renal proximal tubule: regulation by I_2 imidazoline ligands in intact cells. Mol Pharmacol 51:637–643

Gelerenter CS, Uhde TW, Cimbolic P, Arnkoff DB, Vittone BJ, Tancer ME, Bartko JJ (1991) Cognitive-behavioral and pharmacological treatments of social phobia. A controlled study. Arch Gen Psychiatry 48:938–945

Gelowitz DL, Paterson IA (1999) Neuronal sparing and behavioral effects of the antiapoptotic drug, l-deprenyl, following kainic acid administration. Pharmacol Biochem Behav 62:255–262

Georgotas A, McCue RE, Cooper TB (1989) A placebo-controlled comparison of nortriptyline and phenelzine in maintenance therapy of elderly depressed patients. Arch Gen Psychiatry 46:783–786

Gerlach M, Youdim MB, Riederer P (1996) Pharmacology of selegiline. Neurology 47(Suppl 3):S137–145

Gilad GM, Gilad VH (1992) Polyamines in neurotrauma. Ubiquitous molecules in search of a function. Biochem Pharmacol 44:401–407

Goldstein JA, Faletto MB, Romkes-Sparks M, Sullivan T, Kitareewan S, Raucy JL, Lasker JM, Ghanayem BI (1994) Evidence that CYP2C19 is the major (S)-mephenytoin 4'-hydroxylase in humans. Biochemistry 1994;33:1743–1752

Grace JM, Kinter MT, Macdonald TL (1994) Atypical metabolism of deprenyl and its enantiomer, (S)-(+)-N,alpha-dimethyl-N-propynylphenethylamine, by cytochrome P450 2D6. Chem Res Toxicol. 7:286–290

Gram LF, Brosen K. (1993) Moclobemide treatment causes a substantial rise in the sparteine metabolic ratio. Danish University Antidepressant Group. Br J Clin Pharmacol. 35:649–652

Gram LF, Guentert TW, Grange S, Vistisen K, Brosen K (1995) Moclobemide, a substrate of CYP2C19 and an inhibitor of CYP2C19, CYP2D6, and CYP1A2: A panel study. Clin Pharmacol Ther 57:670–677

Hansten PD, Horn JR (1995) Drug interactions and updates quarterly. Applied Therapeutics, Vancouver, WA

Harrison W, Rabkin J, Stewart JW, McGrath PJ, Tricamo E, Quitkin F (1986) Phenelzine for chronic depression: a study of continuation treatment. J Clin Psychiatry 47:346–349

Hartter S, Dingemanse J, Baier D, Ziegler G, Hiemke C (1996) The role of cytochrome P450 2D6 in the metabolism of moclobemide. Eur Neuropsychopharmacol 6:225–230

Heinonen EH, Anntila MI, Lammintausta RAS (1994) Pharmacokinetic aspects of l-deprenyl (selegiline) and its metabolites. Clin Pharmacol Ther 56:742–749

Himmelhoch JM, Detre T, Kupfer DJ, Swartzburg M, Byck R (1972) Treatment of previously intractable depressions with tranylcypromine and lithium. J Nerv Ment Dis 155:216–220

Himmelhoch JM, Thase ME, Mallinger AG, Houck P (1991) Tranylcypromine versus imipramine in anergic bipolar depression. Am J Psychiatry 148:910–916

Holt A, Baker GB (1995) Metabolism of agmatine (clonidine-displacing substance) by diamine oxidase and the possible implications for studies of imidazoline receptors. Prog Brain Res 106:187–197

Holt A, Alton G, Scaman CH, Loppnow GR, Szpacenko A, Svendsen I, Palcic MM (1998) Identification of the quinone cofactor in mammalian semicarbazide-sensitive amine oxidase. Biochemistry 37:4946–4957

Janes SM, Mu D, Wemmer D, Smith AJ, Kaur S, Maltby D, Burlingame AL, Klinman JP (1990) A new redox cofactor in eukaryotic enzymes: 6-hydroxydopa at the active site of bovine serum amine oxidase. Science 248:981–987

Janes SM, Palcic MM, Scaman CH, Smith AJ, Brown DE, Dooley DM, Mure M, Klinman JP (1992) Identification of topaquinone and its consensus sequence in copper amine oxidases. Biochemistry 31:12147–12154

Jedrychowski M, Feifel N, Bieck PR, Schmidt EK (1993) Metabolism of the new MAO-A inhibitor brofaromine in poor and extensive metabolizers of debrisoquine. J Pharm Biomed Anal 11:251–255

Joffe RT, Bakish D (1994) Combined SSRI-moclobemide treatment of psychiatric illness. J Clin Psychiatry 55:24–25

Keck PE, Carter WP, Nierenberg M, Cooper TB, Potter WZ, Rothschild AJ (1991) Acute cardiovascular effects of tranylcypromine: correlation with plasma drug, metabolite, norepinephrine and MHPG levels. J Clin Psychiatry 52:250–254

Kennedy SH, Piran N, Warsh JJ, Prendergast P, Mainprize E, Whynot C, Garfinkel PE (1988) A trial of isocarboxazid in the treatment of bulimia nervosa. J Clin Psychopharmacol 8:391–396

Kennedy SH, Goldbloom DS, Ralevski E, Davis C, D'Souza JD, Lofchy J (1993) Is there a role for selective monoamine oxidase inhibitor therapy in bulimia nervosa? A placebo-controlled trial of brofaromine. J Clin Psychopharmacol 13:415–422

Kennedy SH, Eisfeld BS, Dickens SE, Bacchiochi JR, Bagby RM (2000) Antidepressant-induced sexual dysfunction during treatment with moclobemide, paroxetine, sertraline, and venlafaxine. J Clin Psychiatry 61:276–281

Kennedy SH, Lam RW, Cohen NL, Ravindran AV (2001) Clinical guidelines for the treatment of depressive disorders. IV. Medications and other biological treatments. Can J Psychiatry 46(Suppl 1):38S–58S

Klinman JP, Williams NK (2000) Whence topa? Models for the biogenesis of topaquinone in copper amine oxidases. J Mol Cat B 8:95–101

Knoll J (1988) The striatal dopamine dependency of life span in male rats. Longevity study with l-deprenyl. Mech Ageing Dev 46:237–262

Kohno K, Ohta S, Kohno K, Kumon Y, Mitani A, Sakaki S, Kataoka K (1996) Nitric oxide synthase inhibitor reduces delayed neuronal death in gerbil hippocampal CA1 neurons after transient global ischemia without reduction of brain temperature or extracellular glutamate concentration. Brain Res 738:275–280

Kruger MB, Dahl AA (1999) The efficacy and safety of moclobemide compared to clomipramine in the treatment of panic disorder. Eur Arch Psychiatry Clin Neurosci 249(Suppl 1):S19–24

Langford SD, Trent MB, Balakumaran A, Boor PJ (1999) Developmental vasculotoxicity associated with inhibition of semicarbazide-sensitive amine oxidase. Toxicol Appl Pharmacol 155:237–244

Levitt P, Pintar JE, Breakefield XO (1982) Immunocytochemical demonstration of monoamine oxidase B in brain astrocytes and serotonergic neurons. Proc Nat Acad Sci USA 79:6385–6389

Lewinsohn R (1981) Amine oxidase in human blood vessels and non-vascular smooth muscle. J Pharm Pharmacol 33:569–575

Liebowitz MR (1992) Reversible MAO inhibitors in social phobia, bulimia, and other disorders. Clin Neuropharmacol 15:434A–435A

Liebowitz MR, Heimberg RG, Schneier FR, Hope DA, Davies S, Holt CS, Goetz D, Juster HR, Lin SH, Bruch MA, Marshall RD, Klein DF (1999) Cognitive-behavioral group therapy versus phenelzine in social phobia: Long-term outcome. Depress Anxiety 10:89–98

Lipper S, Murphy DL, Slater S, Buchsbaum MS (1979) Comparative behavioral effects of clorgyline and pargyline in man: a preliminary evaluation. Psychopharmacology 62:123–128

Loerch B, Graf-Morgenstern M, Hautzinger M, Schlegel S, Hain C, Sandmann J, Benkert O (1999) Randomised placebo-controlled trial of moclobemide, cognitive-behavioural therapy and their combination in panic disorder with agoraphobia. Br J Psychiatry 174:205–212

Lyles GA (1996) Mammalian plasma and tissue-bound semicarbazide-sensitive amine oxidases: Biochemical, pharmacological and toxicological aspects. Int J Biochem Cell Biol 28:259–274

Lyles GA, Chalmers J (1992) The metabolism of aminoacetone to methylglyoxal by semicarbazide-sensitive amine oxidase in human umbilical artery. Biochem Pharmacol 43:1409–1414

Lyles GA, Singh I (1985) Vascular smooth muscle cells: a major source of the semicarbazide-sensitive amine oxidase of the rat aorta. J Pharm Pharmacol 37:637–643

Lyles GA, Garcia-Rodriguez J, Callingham BA (1983) Inhibitory actions of hydralazine upon monoamine oxidizing enzymes in the rat. Biochem Pharmacol 32:2515–2521

Lyles GA, Holt A, Marshall CM (1990) Further studies on the metabolism of methylamine by semicarbazide-sensitive amine oxidase activities in human plasma, umbilical artery and rat aorta. J Pharm Pharmacol 42:332–338

Magyar K, Szenda B, Lengyel J, Tarezali J, Szatmary I (1998) The neuroprotective and neuronal rescue effects of (-)-deprenyl. J Neurol Transm Suppl 52:109–123

Magder DM, Aleksic I, Kennedy SH (2000) Tolerability and efficacy of high-dose moclobemide alone and in combination with lithium and trazodone. J Clin Psychopharmacol 20:394–395

Malcolm DE, Yu PE, Bowen RC, O'Donovan C, Hawkes J (1990) Phenelzine and plasma vitamin B6 (pyridoxine) levels. Proceedings of the Annual Meeting of the Canadian Psychiatric Association, Toronto, Canada

Manuck SB, Flory JD, Ferrell RE, Mann JJ, Muldoon MF (2000) A regulatory polymorphism of the monoamine oxidase-A gene may be associated with variability in aggression, impulsivity, and central nervous system serotonergic responsivity. Psychiatry Res 95:9–23

McDaniel KD (1986) Clinical pharmacology of monoamine oxidase inhibitors. Clin Neuropharmacol 9:207–234

McKenna KF, Baker GB, Coutts RT, Rauw G, Mozayani A, Danielson TJ (1990) Recent studies on the MAO inhibitor phenelzine and its possible metabolites. J Neural Transm 32:113–118

McKenna KF, McManus DJ, Baker GB, Coutts RT (1994) Chronic administration of the antidepressant phenelzine and its N-acetyl analogue: effects on GABAergic function. J Neural Transm Suppl 41:115–122

McManus DJ, Baker GB, Martin IL, Greenshaw AJ, McKenna KF (1992) Effects of the antidepressant/antipanic drug phenelzine on GABA concentrations and GABA-T activity in rat brain. Biochem Pharmacol 43:2486–2489

Molderings G (1997) Imidazoline receptors: Basic knowledge, recent advances and future prospects for therapy and diagnosis. Drug Fut 22:757–772

Molderings GJ, Gothert M (1999) Imidazoline binding sites and receptors in cardiovascular tissue. Gen Pharmacol 32:17–22

Morgan DM (1985) Polyamine oxidases. Biochem Soc Trans 13:322–326

Mozayani A, Coutts RT, Danielson TJ, Baker GB (1988) Metabolic acetylation of phenelzine in rats. Res Commun Chem Pathol Pharmacol 62:397–406

Medical Research Council by its Clinical Psychiatry Committee (1965) Clinical Trial of the treatment of depressive illness. Br Med J 1:881–886

Murphy DL, Kalin NH (1980) Biological and behavioral consequences of alterations in monoamine oxidase activity. Schizophr Bull 6:355–367

Musgrave IF, Badoer E (1999) Harmane produces hypotension following microinjection into the RVLM: possible role of I_1-imidazoline receptors. Br J Pharmacol 129:1057–1059

Mytilineou C, Radcliffe P, Leonardi EK, Werner P, Olanow CW (1997a) L-deprenyl protects mesencephalic dopamine neurons from glutamate receptor-mediated toxicity in vitro. J Neurochem 68:33–39

Mytilineou C, Radcliffe PM, Olanow CW (1997b) L-(-)-desmethylselegiline, a metabolite of selegiline [L-(-)-deprenyl], protects mesencephalic dopamine neurons from excitotoxicity in vitro. J Neurochem. 68:434–436

Newburn G, Edwards R, Thomas H, Collier J, Fox K, Collins C (1999) Moclobemide in the treatment of major depressive disorder (DSM-3) following traumatic brain injury. Brain Inj 13:637–642

Olanow CW, Myllyla VV, Sotaniemi KA, Larsen JP, Palhagen S, Przuntek H, Heinonen EH, Kilkku O, Lammintausta R, Maki-Ikola O, Rinne UK (1998) Effect of selegiline on mortality in patients with Parkinson's disease: A meta-analysis. Neurology 51:825–830

Olmos G, DeGregorio-Rocasolano N, Paz Regalado M, Gasull T, Assumpcio Boronat M, Trullas R, Villarroel A, Lerma J, Garcia-Sevilla JA (1999) Protection by imidazol(ine) drugs and agmatine of glutamate-induced neurotoxicity in cultured cerebellar granule cells through blockade of NMDA receptor. Br J Pharmacol 127:1317–1326

Ozaita A, Olmos G, Boronat MA, Lizcano JM, Unzeta M, Garcia-Sevilla JA (1997) Inhibition of monoamine oxidase A and B activities by imidazol(ine)/guanidine drugs, nature of the interaction and distinction from I2-imidazoline receptors in rat liver. Br J Pharmacol 121:901–912

Pande AC, Birkett M, Fechner-Bates S, Haskett RF, Greden JF (1996) Fluoxetine versus phenelzine in atypical depression. Biol Psychiatry 40:1017–1020

Parent M, Habib MK, Baker GB (2000) Time-dependent changes in brain monoamine oxidase activity and in brain levels of monoamines and amino acids following acute administration of the antidepressant/antipanic drug phenelzine. Biochem Pharmacol 59:1253–1263

Parkinson A (1996) Biotransformation of xenobiotics. In: Klassen CD (ed) Casarett and Doull's toxicology: the basic science of poisons, fifth edition. McGraw-Hill, New York, pp 119–186

Parkinson Study Group (1993) Effects of tocopherol and deprenyl on the progression of disability in early Parkinson's disease. N Engl J Med 328:176–183

Paslawski T, Treit D, Baker GB, George M, Coutts RT (1996) The antidepressant drug phenelzine produces antianxiety effects in the plus-maze and increases in rat brain GABA. Psychopharmacol 127:19–24

Paslawski T, Knaus E, Iqbal N, Coutts RT, Baker GB (2001) β-Phenylethylidenehydrazine, a novel inhibitor of GABA transaminase. Drug Devel Res 54:35–39

Paterson IA, Tatton WG (1998) Antiapoptotic actions of monoamine oxidase B inhibitors. Adv Pharmacol 42:312–315

Paterson IA, Juorio AV, Boulton AA (1990) 2-Phenylethylamine: a modulator of catecholamine transmission in the mammalian central nervous system? J Neurochem 55:1827–1837

Pegg AE, McCann PP (1982) Polyamine metabolism and function. Am J Physiol 243: C212–C221

Pierre JM, Gitlin MJ. (2000) Bupropion-tranylcypromine combination for treatment-refractory depression. J Clin Psychiatry 61:450–451

Piletz JE, Halaris A, Ernsberger PR (1994) Psychopharmacology of imidazoline and α_2-adrenergic receptors: implications for depression. Crit Rev Neurobiol 9:29–66

Piletz JE, Halaris A, Nelson J, Qu Y, Bari M (1996a) Platelet I_1-imidazoline binding sites are elevated in depression but not generalized anxiety disorder. J Psychiatr Res 30:147–168

Piletz JE, Halaris A, Chikkala D, Qu Y (1996b) Platelet I_1-imidazoline binding sites are decreased by two dissimilar antidepressant agents in depressed patients. J Psychiatr Res 30:169–184

Popov N, Matthies H (1969) Some effects of monoamine oxidase inhibitors on the metabolism of γ-aminobutyric acid in rat brain. J Neurochem 16:899–907

Poulin R, Bonneau M-J (2000) Spermine oxidation leads to necrosis with plasma membrane phosphatidylserine redistribution in mouse leukemia cells. Exp Cell Res 259:23–24

Precious E, Gunn CE, Lyles GA (1988) Deamination of methylamine by semicarbazide-sensitive amine oxidase in human umbilical artery and rat aorta. Biochem Pharmacol 37:707–713

Price RE, Luscombe S, Tyacke RJ, Nutt DJ, Hudson AL (1999) Affinities of β-carbolines for I_2-binding sites in rabbit brain membranes. Br J Pharmacol 127(Suppl): 57P

Quitkin FM, McGrath PJ, Stewart JW, Harrison W, Tricamo E, Wager SG, Ocepek-Welikson K, Nunes E, Rabkin JG, Klein DF (1990) Atypical depression, panic attacks, and response to imipramine and phenelzine. A replication. Arch Gen Psychiatry 47:935–941

Quitkin F, Rothschild R, Stewart JW, McGrath PJ, Harrison WM (1994) Atypical depression: A unipolar depressive subtype with preferential response to MAOIs (Columbia University Depressive Studies). In: Kennedy SH (ed) Clinical advances in monoamine oxidase inhibitor therapies, vol 43. American Psychiatric Press, Washington, DC, pp 181–203

Raddatz R, Lanier SM (1997) Relationship between imidazoline/guanidinium receptive sites and monoamine oxidase A and B. Neurochem Int 30:109–117

Raddatz R, Parini A, Lanier SM (1995) Imidazoline/guanidinium binding domains on monoamine oxidases. Relationship to subtypes of imidazoline-binding proteins and tissue-specific interaction of imidazoline ligands with monoamine oxidase B. J Biol Chem 270:27961–27968

Raimondi L, Pirisino R, Ignesti G, Capecchi S, Banchelli G, Buffoni F (1991) Semicarbazide-sensitive amine oxidase activity (SSAO) of rat epididymal white adipose tissue. Biochem Pharmacol 41:467–470

Ravaris CL, Nies A, Robinson DS, Ives JO, Lamborn KR, Korson L (1976) A multiple-dose, controlled study of phenelzine in depression-anxiety states. Arch Gen Psychiatry 33:347–350

Renouard A, Widdowson PS, Cordi A (1993) [^{3}H]-idazoxan binding to rabbit cerebral cortex recognizes multiple imidazoline I_2-type receptors: pharmacological characterization and relationship to monoamine oxidase. Br J Pharmacol 109:625–631

Robie T (1958) Marsilid in depression. Am J Psychiatry 114:936

Robinson DS, Campbell IC, Walker M, Statham NH, Lovenberg W, Murphy DL (1979) Effects of chronic monoamine oxidase inhibitor treatment on biogenic amine metabolism in rat brain. Neuropharmacol 18:771–776

Robinson DS, Cooper TB, Jindal SP, Corcella J, Lutz T (1985) Metabolism and pharmacokinetics of phenelzine: Lack of evidence for acetylation pathway in humans. J Clin Psychopharmacol 5:333–337

Robinson DS, Lerfald SC, Bennett B, Laux D, Devereaux E, Kayser A, Corcella J, Albright D (1991) Continuation and maintenance treatment of major depression with the monoamine oxidase inhibitor phenelzine: A double-blind placebo-controlled discontinuation study. Psychopharmacol Bull 27:31–39

Salminen TA, Smith DJ, Jalkanen S, Johnson MS (1998) Structural model of the catalytic domain of an enzyme with cell adhesion activity: Human vascular adhesion protein-1 (HVAP-1) D4 domain is an amine oxidase. Protein Eng 11:1195–1204

Salonen T, Haapalinna A, Heinonen E, Suhonen J, Hervonen A (1996) Monoamine oxidase B inhibitor selegiline protects young and aged rat peripheral sympathetic neurons against 6-hydroxydopamine-induced neurotoxicity. Acta Neuropathol 91:466–474

Segal JA, Skolnick P (2000) Spermine-induced toxicity in cerebellar granule neurons is independent of its actions at NMDA receptors. J Neurochem 74:60–69

Seiler N (1995) Polyamine oxidase, properties and functions. Prog Brain Res 106:333–344

Semkova I, Wolz P, Schilling M, Krieglstein J (1996) Selegiline enhances NGF synthesis and protects central nervous system neurons from excitotoxic and ischemic damage. Eur J Pharmacol 315:19–30

Sharma RP, Janicak PG, Javaid JI, Pandey GN, Gierl B, Davis JM (1990) Platelet MAO inhibition, urinary MHPG, and leukocyte beta-adrenergic receptors in depressed patients treated with phenelzine. Am J Psychiatry 147:1318–1321

Sherry RL, Rauw G, McKenna KF, Paetsch PR, Coutts RT, Baker GB (2000) Failure to detect amphetamine or 1-amino-3-phenylpropane in humans or rats receiving tranylcypromine. J Affect Disord 61:23–29

Shulman KI, Tailor SA, Walker SE, Gardner DM (1997) Tap (draft) beer and monoamine oxidase inhibitor dietary restrictions. Can J Psychiatry 42:310–312

Shulman KI, Walker SE (1999) Refining the MAOI diet: Tyramine content of pizzas and soy products. J Clin Psychiatry 60:191–193

Sims KB, Ozelius L, Corey T, Rinehart WB, Liberfarb R, Haines J, Chen WJ, Norio R, Sankila E, de la Chapelle A, et al. (1989) Norrie disease gene is distinct from the monoamine oxidase genes. Am J Hum Genet 45:424–434

Smith DF (1980) Tranylcypromine stereoisomers, monoaminergic transmission and behaviour: a review. Pharmacopsychiatry 13:130–136

Smith SE, Lambourn J, Tyrer PF (1980) Antipyrine elimination by patients under treatment with monoamine oxidase inhibitors. Br J Clin Pharmacol 9:21–25

Sogaard J, Lane R, Latimer P, Behnke K, Christiansen PE, Nielsen B, Ravindran AV, Reesal RT, Goodwin DP (1999) A 12-week study comparing moclobemide and sertraline in the treatment of outpatients with atypical depression. J Psychopharmacol 13:406–414

Stoll AL, Haura G (2000) Tranylcypromine plus risperidone for treatment-refractory major depression. J Clin Psychopharmacol 20:495–496

Tanay VA-MI, Parent MB, Wong JTF, Paslawski T, Martin IL, Baker GB (2001) Effects of the antidepressant/antipanic drug phenelzine on alanine and alanine transaminase in rat brain. Cell Mol Neurobiol 21:325–339

Tatton WG, Chalmers-Redman RM (1996) Modulation of gene expression rather than monoamine oxidase inhibition: (−)-deprenyl-related compounds in controlling neurodegeneration. Neurology 47:S171–183

Tatton WG, Greenwood CE (1991) Rescue of dying neurons: A new action for deprenyl in MPTP parkinsonism. J Neurosci Res 30:666–672

Tesson F, Limon-Boulez I, Urban P, Puype M, Vandekerckhove J, Coupry I, Pompon D, Parini A (1995) Localization of I_2-imidazoline binding sites on monoamine oxidase. J Biol Chem 270:9856–9861

Thase ME, Sachs GS (2000) Bipolar depression: pharmacotherapy and related therapeutic strategies. Biol Psychiatry 48:558–572

Thase ME, Trivedi MH, Rush AJ (1995) MAOIs in the contemporary treatment of depression. Neuropsychopharmacology 12:185–219

Thomas T, McLendon C, Thomas G (1998) L-deprenyl: nitric oxide production and dilation of cerebral blood vessels. Neuroreport 9:2595–2600

Tiller JW, Bouwer C, Behnke K (1999) Moclobemide and fluoxetine for panic disorder. International Panic Disorder Study Group. Eur Arch Psychiatry Clin Neurosci 249:S7–10

Tipton KF, Spires IPC (1971) Oxidation of 2-phenylethylhydrazine by monoamine oxidase. Biochem Pharmacol 21:268–270

Todd KG, Baker GB (1995) GABA-elevating effects of the antidepressant/antipanic drug phenelzine in brain: effects of pretreatment with tranylcypromine, (-)-deprenyl and clorgyline. J Affect Disord 35:125–129

Versiani M, Nardi AE, Mundim FD, Alves AB, Liebowitz MR, Amrein R (1992) Pharmacotherapy of social phobia. A controlled study with moclobemide and phenelzine. Br J Psychiatry 161:353–360

Volz HP, Gleiter CH, Moller HJ (1997) Brofaromine versus imipramine in-patients with major depression: A controlled trial. J Affect Disord 44:91–99

Waldmeier P, Amrein R, Schmid-Burgk W (1994) Pharmacology and pharmacokinetics of brofaromine and moclobemide in animals and humans. In: Kennedy SH (ed) Clinical advances in monoamine oxidase inhibitor therapies, vol 43. American Psychiatric Press, Washington, DC, pp 33–59

Walsh BT, Gladis M, Roose SP, Stewart JW, Stetner F, Glassman AH (1988) Phenelzine vs placebo in 50 patients with bulimia. Arch Gen Psychiatry 45:471–475

Westenberg HG (1996) Developments in the drug treatment of panic disorder: What is the place of the selective serotonin reuptake inhibitors? J Affect Disord 40:85–93

Westlund KN, Denney RM, Rose RM, Abell CW (1988) Localization of distinct monoamine oxidase A and monoamine oxidase B cell populations in human brainstem. Neurosci 25:439–456

White K, Simpson G (1981) Combined MAOI-tricyclic antidepressant treatment: A reevaluation. J Clin Psychopharmacol 1:264–282

Wong JTF, Baker GB, Coutts RT, Dewhurst WG (1990) Long-lasting elevation of alanine in brain produce by the antidepressant phenelzine. Brain Res Bull 25:179–181

Wu RM, Chiueh CC, Pert A, Murphy DL (1993) Apparent antioxidant effect of l-deprenyl on hydroxyl radical formation and nigral injury elicited by MPP+ in vivo. Eur J Pharmacol 243:241–247

Xu L, Ma J, Seigel GM, Ma J (1999) L-deprenyl, blocking apoptosis and regulating gene expression in cultured retinal neurons. Biochem Pharmacol 58:1183–1190

Youdim MB, Aronson JK, Blau K, Green AR, Grahame-Smith DG (1979) Tranylcypromine ('Parnate') overdose: measurement of tranylcypromine concentrations and MAO inhibitory activity and identification of amphetamines in plasma. Psychol Med 9:377–382

Yu PH (2000) Semicarbazide-sensitive amine oxidase and mortality in chronic heart failure. Eur Heart J 21:1812–1814

Yu PH, Boulton M (1992) A comparison of the effect of brofaromine, phenelzine and tranylcypromine on the activities of some enzymes involved in the metabolism of different neurotransmitters, Res Commun Chem Path Pharmacol 16:141–153

Yu PH, Tipton KF (1989) Deuterium isotope effect of phenelzine on the inhibition of rat liver mitochondrial monoamine oxidase activity. Biochem Pharmacol 38:4245–4251

Yu PH, Davis BA, Durden DA (1991) Enzymatic N-methylation of phenelzine catalyzed by methyltransferases from adrenal and other tissues. Drug Metab Dispos 19:830–834

Zhang X, Yu PH (1995) Depletion of NOS activity in the rat dentate gyrus neurons by DSP-4 and protection by deprenyl. Brain Res Bull 38:307–311

Zhang X, Zuo DM, Yu PH (1995) Neuroprotection by R(-)-deprenyl and N-2-hexyl-N-methylpropargylamine on DSP-4, a neurotoxin, induced degeneration of noradrenergic neurons in the rat locus coeruleus. Neurosci Lett 186:45–48

Zubieta JK, Pande AC, Demitrack MA (1999) Two year follow-up of atypical depression. J Psychiatr Res 33:23–9

Selective Serotonin Reuptake Inhibitors

S. H. Preskorn[1] · R. Ross[2] · C. Y. Stanga[3]

[1] Department of Psychiatry and Behavioral Sciences,
University of Kansas School of Medicine and Psychiatric Research Institute,
1010 N. Kansas, Wichita, KS 67214, USA
e-mail: spreskor@kumc.edu
[2] Ross Editorial, 228 Black Rock Mtn Ln, Independence, VA 24348, USA
[3] W8057 Moonlite Road, Neillsville, WI 54456, USA

1	Introduction and History	242
1.1	Organization	243
1.2	Currently Available SSRIs and Indications for Their Use	243
1.3	Prescription Data	244
1.4	Rational Drug Development	244
2	Chemistry, Pharmacology, and Metabolism	247
3	Safety and Tolerability	249
3.1	Drug–Drug Interactions	249
3.1.1	Pharmacodynamic Interactions	249
3.1.2	Pharmacokinetic Interactions	249
3.2	Adverse Effects	250
3.3	Overdose	252
3.4	Teratogenicity	252
3.5	Long-Term Safety	253
4	Clinical Indications/Uses	253
4.1	Acute Treatment	254
4.2	Dosing Recommendations	256
4.3	Onset of Action	256
4.4	Maintenance Treatment	257
	References	258

Abstract The selective serotonin reuptake inhibitors (SSRIs) (citalopram, escitalopram, fluoxetine, fluvoxamine, paroxetine, and sertraline) are the most widely prescribed class of antidepressants in the United States and a number of other countries. All the SSRIs except fluvoxamine have an indication for major depressive disorder (MDD) in the United States. The SSRIs were the first rationally developed class of psychiatric medications and thus share many similarities: equivalent acute and maintenance antidepressant efficacy, flat dose-response curve for antidepressant efficacy, ascending dose–response curve for adverse effects, adverse effect profile consistent with excessive serotonin agonism,

60%–80% inhibition of serotonin uptake at their lowest, usually effective antidepressant dose, and efficacy for both depressive and anxiety disorders. The SSRIs have good tolerability and do not cause severe side effects seen with certain other antidepressants (e.g., intracardiac conduction delays, seizures, postural hypotension). Because of their wide therapeutic index, the SSRIs have demonstrated good safety in overdose and there is no evidence of long-term safety problems. However, the different SSRIs do differ considerably in pharmacokinetics and effects on cytochrome P450 enzymes. Three of the SSRIs—fluoxetine, fluvoxamine, and paroxetine—inhibit one or more CYP enzymes to a substantial degree and have the potential to cause clinically meaningful drug–drug interactions. There is some debate about the efficacy of SSRIs in patients with more severe depression. If a patient has not been able to tolerate one SSRI, many clinicians will try a second SSRI; however, most psychiatrists prefer to switch to a drug with a different mechanism of action if a patient has had an inadequate response to an adequate trial of one SSRI, since there is no compelling evidence showing that nonresponders to one SSRI will respond to a trial of a different SSRI. All the SSRIs have a flat dose–response curve, so that there is usually no advantage in using doses above the usually effective minimum dose.

Keywords Selective serotonin reuptake inhibitors · Rational drug development · Side effects · Tolerability · Citalopram · Escitalopram · Fluoxetine · Fluvoxamine · Paroxetine · Sertraline

Portions of this chapter have been adapted with permission from Janicak et al. (2001).

1
Introduction and History

As reviewed earlier in this volume ("Overview of Currently Available Antidepressants"), drug development in psychiatry has evolved from a process based on chance to one based on molecularly targeting specific sites of action in the central nervous system (e.g., specific neuroreceptors or neuronal uptake pumps for specific neurotransmitters). This chapter will review the first class of psychiatric medications rationally designed using in vitro receptor binding affinity technology, the selective serotonin reuptake inhibitors (SSRIs) (Preskorn 1996).

The objective of such rational drug development was to create agents that would affect only mechanisms that mediate the desired effect, in this case antidepressant efficacy, while avoiding effects on mechanisms that did not contribute to such efficacy but instead mediated the tolerability or safety problems that plagued the use of older chance-discovery antidepressants, specifically the tricyclic antidepressants (see Lader, this volume; Preskorn 1990, 1995b; Tallman and Dahl 1994).

1.1
Organization

This chapter first describes the development of the SSRIs and then reviews their chemistry, pharmacology, metabolism, safety and adverse effects, and clinical use in the treatment of depression. For a more detailed discussion of the pharmacokinetics of these agents, see Preskorn and Catterson (this volume), and for a discussion of the use of therapeutic drug monitoring (TDM) of antidepressants, see Burke and Preskorn (also in this volume). Summary tables on dosing, adverse effects, pharmacokinetic parameters, and plasma concentrations are provided in the Appendix. For more information on the pharmacology of the SSRIs and copies of other publications by the first author of this chapter, readers are referred to his "Applied Clinical Pharmacology" website (www.preskorn.com).

1.2
Currently Available SSRIs and Indications for Their Use

The following SSRIs are currently available in the United States and many other countries (the product names in use in the United States are given in parentheses): citalopram (Celexa), escitalopram (Lexapro), fluoxetine (Prozac), fluvoxamine (Luvox), paroxetine (Paxil), and sertraline (Zoloft). All the SSRIs, except fluvoxamine, have an indication for major depressive disorder in the United States. Although fluvoxamine is marketed as an antidepressant in many countries, in the United States it is only labeled for the treatment of obsessive-compulsive disorder (OCD). In the United States, fluoxetine is also indicated for the treatment of OCD, panic disorder, premenstrual dysphoric disorder, and bulimia nervosa; paroxetine is indicated for the treatment of OCD, panic disorder, social anxiety disorder (social phobia), generalized anxiety disorder, and posttraumatic stress disorder (PTSD); and sertraline is indicated for the treatment of OCD, panic disorder, social anxiety disorder, PTSD, and premenstrual dysphoric

Table 1 Approved FDA indications for the SSRIs in the United States

SSRI	Depression	Panic disorder	OCD	Social anxiety disorder	GAD	PTSD	Bulimia nervosa	PMDD
Citalopram	X							
Escitalopram	X							
Fluoxetine	X	X	X				X	X
Fluvoxamine			X					
Paroxetine	X	X	X	X	X	X		
Sertraline	X	X	X*	X		X		X

OCD, obsessive-compulsive disorder; GAD, generalized anxiety disorder; PTSD, posttraumatic stress disorder; PMDD, premenstrual dysphoric disorder.
* Also approach for pediatric OCD

disorder. The indications for the SSRIs that are currently approved by the U.S. Food and Drug Administration are summarized in Table 1.

1.3
Prescription Data

The SSRIs are the most widely prescribed class of antidepressants in the United States as well as in a number of other countries. The National Ambulatory Medical Care Survey done in 1993–1994 found that half of the prescriptions for antidepressants written by psychiatrists were for SSRIs (Olfson et al. 1998). An analysis of data from two representative surveys of the general population of the United States, the 1987 National Medical Expenditure Survey and the 1997 Medical Expenditure Panel Survey, found that individuals being treated with depression were 4.5 times more likely to be treated with a psychotropic medication in 1997 than in 1987 and that this increase in antidepressant use was primarily attributable to the introduction of the SSRIs, which were unavailable in 1987; SSRIs were prescribed to 58.3% of individuals receiving outpatient treatment for depression in 1997 (Olfson et al. 2002). A similar trend has been seen in other countries, where the prescribing of SSRIs has also increased significantly [e.g., in New Zealand, where SSRIs accounted for 20.9% of antidepressant prescriptions in 1993 but 44.8% in 1997 (Roberts and Norris 2001)]. Based on data from the national prescription audit done in the United States by IMS Health, the SSRIs accounted for approximately 55% of prescriptions for antidepressant products in the first 11 months of 2002 (IMS Health 2002).

1.4
Rational Drug Development

As noted above, the SSRIs were the first class of psychotropic drugs to be rationally designed, and their introduction launched a new era in psychotropic drug development. In the 1960s and 1970s, researchers began to investigate the pharmacology of the tricyclic antidepressants (TCAs) and monoamine oxidase inhibitors (MAOIs). Although the antidepressant activity of these drugs was discovered serendipitously, they provided the first evidence that central serotonin agonism might be a means of producing antidepressant response (Preskorn 1994). Study of the TCAs demonstrated that drugs that inhibit the serotonin transport protein can act as indirect central serotonin agonists. With this information available, pharmaceutical developers in the 1970s began a research effort to rationally develop drugs that would inhibit the neuronal uptake pump for serotonin without affecting the various other neuroreceptors or fast sodium channels that are affected by the TCAs and are responsible for many of their side effects (Preskorn and Fast 1991; Preskorn and Burke 1992). In other words, the goal of this rational drug development process was to find agents that would be efficacious in treating depression without causing the side effects associated with the older agents. These efforts resulted in the development of the SSRIs.

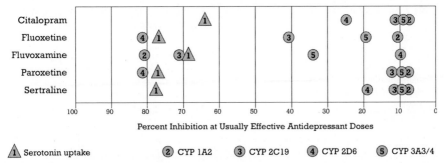

Fig. 1 In vivo profile of SSRIs: serotonin uptake inhibition versus CYP enzyme inhibition. (Based on data from Shad and Preskorn 2000) (copyright Preskorn 2003)

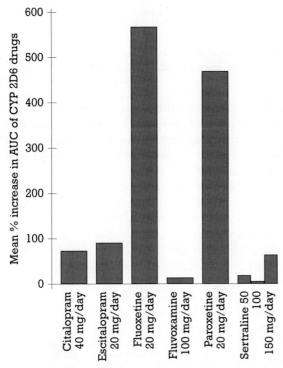

Fig. 2 Mean percentage increase in the concentration of CYP 2D6 model substrate/drugs as a function of co-administration of different SSRIs. *AUC*, area under the plasma concentration–time curve (copyright Preskorn 2003)

All of the SSRIs were the product of rational drug development. The major goal was to develop drugs with a precise, limited (or focused) range of pharmaceutical actions. Although this goal was achieved in terms of the neural mechanisms of action of the SSRIs, it was not achieved in all respects. One problem is

Table 2 Summary of the effect of the various SSRIs on the major human drug metabolizing cytochrome P450 enzymes (from Preskorn 1996, 1999b; Harvey and Preskorn 1996a,b; Madsen et al. 2001) (copyright Preskorn 2003)[a]

	1A2	2D6	2C9/10	2C19	3A3/4
Citalopram		++			
Escitalopram		++			
Fluoxetine		+++	+++	++	+
Fluvoxamine	+++		+++[b]	+++	++
Paroxetine		+++			
Sertraline		+			

Blank, no or minimal effect (<20%); +, mild effect (20%–50%); ++, moderate effect (50%–150%); +++, substantial effect (>150%).

[a] The information in the table is based on the results of formal in vivo studies in humans. The ratings are based on the following doses because the effects are dose dependent and the studies were principally conducted at these doses: 40 mg/day of citalopram; 20 mg/day of escitalopram, fluoxetine, and paroxetine; 150 mg/day of fluvoxamine; and 100 mg/day of sertraline, with some studies testing doses as high as 200 mg/day

[b] This table has been updated to reflect the results of a recent in vivo human study showing that fluvoxamine at a dose of 150 mg/day caused a 300% increase in tolbutamide levels (Madsen et al. 2001).

that several, although not all, of the SSRIs have substantial, unintended, inhibitory effects on cytochrome P450 enzymes that mediate the bulk of human oxidative drug metabolism (see Figs. 1 and 2 and Table 2). Another problem is that two SSRIs, citalopram and fluoxetine, were marketed as racemates containing both the S- and the R-enantiomeric forms of these drugs.

If a molecule has an asymmetrical carbon (i.e., is "chiral"), then there are two enantiomeric forms of the compound. All of the SSRIs except fluvoxamine have an asymmetrical carbon. However, only the neuropharmacologically active enantiomeric forms of paroxetine and sertraline were marketed. In contrast, citalopram and fluoxetine were marketed as the racemates of their two enantiomers, meaning that patients taking citalopram or fluoxetine were receiving both the S- and the R-forms of the molecule.

The existence of two enantiomeric forms of a molecule such as citalopram or fluoxetine raises questions about whether there are substantial differences in the pharmacodynamics and pharmacokinetics of the different enantiomeric forms and whether these differences could contribute in a meaningful way to variance in drug response and side effects in different patients.

The presence of enantiomeric forms of a molecule can also complicate the use of TDM (Preskorn 1996, 1999b) because it typically takes more sophisticated assays to separately quantify the concentration of each enantiomer than are typically used in clinical assays for TDM. If the enantiomers have substantially different pharmacology, these differences can complicate the ability to reliably and accurately interpret the TDM results.

To resolve these chirality issues with citalopram and fluoxetine, the R- and S-enantiomers of each of these SSRIs were identified, synthesized, and studied.

Both the R- and S-enantiomers of fluoxetine were found to be pharmacologically active in ways that suggested the potential to be clinically beneficial as antidepressants. For a variety of reasons, the company that developed fluoxetine chose the R-enantiomer for subsequent development as an antidepressant. That would have given them a patented product even after generic fluoxetine became available. However, this development program was abandoned when the R-enantiomer was found to be capable of slowing intracardiac conduction as witnessed by prolongation of the QTc interval.

In contrast to fluoxetine, the S-enantiomer of citalopram appeared to have all of the beneficial properties of citalopram, while the R-enantiomer was found to be generally pharmacologically inactive. This development program was successful and the S-enantiomer of citalopram is now available as escitalopram. The pivotal question awaiting clarification is whether R-citalopram has any negative properties that are not present in escitalopram or whether escitalopram is simply a more potent form of citalopram. Some countries, such as Sweden, have declined to give a favorable price to escitalopram until the manufacturer, Lundbeck, demonstrates a value in removing R-citalopram from the racemic product.

2
Chemistry, Pharmacology, and Metabolism

The goal in developing the SSRIs was to produce antidepressants that both *potently* and *selectively* inhibit the serotonin uptake pump. Potency and selectivity are key concepts in understanding why the SSRIs are so different from the TCAs and so similar to one another in terms of their central nervous systems (CNS) effects. The SSRIs all selectively block the neuronal uptake pumps for serotonin, thus increasing the availability of serotonin, and act as indirect agonists at the following receptors: $5-HT_{2A}$, $5-HT_{2C}$, and $5-HT_3$ (Glennon and Dukat 1994). Selectivity is defined as a separation of at least one order of magnitude (a tenfold separation) between the effects of a drug on its most potent site of action and its next most potent site of action; such a drug is capable of having pharmacologically and clinically meaningful effects on its most potent target while having no effects on any other target (i.e., the drug is "selective") (Preskorn 1995b, 1999a).

As a result of their selective action on the serotonin transporter protein, SSRIs can achieve concentrations that produce virtually complete inhibition of the serotonin uptake pump without producing effects on other neural mechanisms of action such as inhibition of the norepinephrine or dopamine transporter protein (i.e., "uptake pumps"). Given this fact, the SSRIs share many similarities (Preskorn 1999b) as follows:

1. Equivalent acute and maintenance antidepressant efficacy
2. A flat dose–response curve for antidepressant efficacy
3. An ascending dose–response curve for adverse effects

4. An adverse effect profile consistent with excessive serotonin agonism
5. 60%–80% Inhibition of serotonin uptake at their lowest, usually effective antidepressant dose
6. Efficacy in both depressive and anxiety disorder

However, while the SSRIs are very similar in terms of their efficacy and adverse effect profiles, the different SSRIs do differ considerably in their pharmacokinetics and their effects on cytochrome P450 (CYP) drug metabolizing enzymes (Figs. 1 and 2 and Table 2) (Harvey and Preskorn 1996a,b; Preskorn 1996, 1999b; Shad and Preskorn 2000). Despite their rational development, the SSRIs were not developed to avoid inhibitory effects on CYP enzymes, since these agents were developed before the isolation of the first human CYP enzyme, CYP 2D6. For this reason, the technology for screening for such unwanted effects on these important drug metabolizing enzymes did not exist in the 1970s when the SSRIs were in drug discovery. Hence, three of the SSRIs—fluoxetine, fluvoxamine, and paroxetine—inhibit one or more CYP enzymes to a substantial degree and thus have the potential to cause clinically meaningful, specific CYP enzyme-mediated pharmacokinetic drug–drug interactions (Preskorn 1999b). This fact needs to be taken into consideration in choosing between the different SSRIs in clinical practice. (For a more detailed discussion of the pharmacokinetics of antidepressants, see the chapter by Preskorn and Catterson, this volume.)

The technology to screen for effects on CYP does exist now. As a result of this technology, fluoxetine, fluvoxamine, and paroxetine ironically would probably not enter human testing today, much less be marketed (Preskorn 1995a, 1996). Instead, inhibitory effects on CYP enzymes would now be detected in preclinical testing. If found, the companies developing these drugs would probably go back to the design board and tweak their structures to produce SSRIs which, like citalopram, escitalopram, and sertraline, do not inhibit any CYP enzymes to a substantial degree under usual antidepressant dosing conditions (Preskorn 1996, 1999b; Shad and Preskorn 2000). Interestingly, data on market shares of the various SSRIs show that the most prescribed agent in 2002 was sertraline, followed by paroxetine, fluoxetine, citalopram, and fluvoxamine (with the newly marketed escitalopram more widely prescribed than fluvoxamine by the end of 2002) (IMS Health 2002). This information on market share commanded by fluoxetine and paroxetine suggests that many prescribers do not understand the significant problems posed by the fact that these SSRIs are prone to cause clinically significant drug–drug interactions as a result of their inhibition of specific CYP enzymes and have no offsetting advantage when compared with the other SSRIs (e.g., citalopram, escitalopram, and sertraline) that do not have this problem (Preskorn 1999b).

With the exception of fluoxetine, the half-lives of the SSRIs under steady-state conditions are all close to 24 h: citalopram and escitalopram 1.5 days (Milne and Goa 1991), fluvoxamine 0.5–1 day (de Vries et al. 1992), paroxetine 1 day (at 20 mg/day) (Preskorn 1993, 1996), and sertraline 1 day (Preskorn 1993,

1996). For this reason, it is recommended that all but fluvoxamine be dosed once a day (Janicak et al. 2001; Preskorn 1993). In addition, steady-state and virtually complete washout are achieved within approximately 5 or 6 days of starting and stopping these antidepressants (Preskorn 1996).

The only antidepressant with a half-life that is substantially longer than 24 h is fluoxetine (Preskorn 1999b). The parent drug has a half-life of 2–4 days and its active metabolite, norfluoxetine, has a half-life of 7–15 days (Preskorn 1996). The half-life of fluoxetine and norfluoxetine is even longer at doses above 20 mg/day and in physically healthy older patients (Preskorn et al. 1998). In the physically healthy elderly over 65 years of age, the average half-life of norfluoxetine is 3 weeks (Harvey and Preskorn 2001) meaning that it takes 4 months to reach steady-state and 4 months to fully clear the drug in the physically healthy elderly over 65 years of age. These long half-lives mean that fluoxetine is essentially an oral depot antipsychotic.

In beginning treatment with fluoxetine, clinicians need to keep in mind that it requires an extended period of time to reach steady-state and to completely clear this drug. For this reason, adverse effects can develop gradually, sometimes making detection of the cause difficult, and can persist for a sustained period after this drug is discontinued. Likewise, the clinician who is switching a patient from fluoxetine to another antidepressant must keep this prolonged accumulation and washout in mind when starting or stopping fluoxetine (Preskorn 1999b).

3
Safety and Tolerability

3.1
Drug–Drug Interactions

3.1.1
Pharmacodynamic Interactions

The SSRIs demonstrate good safety in terms of pharmacodynamically mediated drug–drug interactions, which are limited to those mediated by serotonin reuptake inhibition. For example, all the SSRIs can interact with monoamine oxidase inhibitors (MAOIs) to cause the serotonin syndrome (Sternbach 1991; Beasley et al. 1993b; Bhatara and Bandettini 1993; Coplan and Gorman 1993; Nierenberg and Semprebon 1993; Graber et al. 1994; Mills 1995; Sporer 1995).

3.1.2
Pharmacokinetic Interactions

As discussed above, the risk of pharmacokinetic interactions differs among the different SSRIs. Three of the SSRIs, fluoxetine, fluvoxamine, and paroxetine, inhibit one or more CYP enzymes to a substantial degree and thus have the potential to cause clinically significant pharmacokinetic drug–drug interactions (Figs. 1 and 2 and Table 2) (Harvey and Preskorn 1996a,b). Fluoxetine and fluvoxamine cause the most types of CYP enzyme-mediated drug interactions because they inhibit more than one CYP enzyme to a substantial degree at their usually effective antidepressant doses. Substantial inhibition of CYP enzymes means that the coadministration of these SSRIs causes a several fold increase in levels of co-prescribed drugs whose clearance is dependent on the CYP enzyme inhibited by the SSRI. Such an increase in levels of the affected drug can cause a variety of negative effects, including decreased tolerability, withdrawal syndromes, or increased adverse effects and toxicity (Preskorn 1999b). Prescribers may also attribute the adverse outcomes to a problem of "sensitivity" or "resistance" and inappropriately adjust the dose or add additional medications, compounding the problem (Preskorn 1998a). Such interactions are a concern especially with fluoxetine because its inhibition of CYP enzymes can persist for weeks after fluoxetine is discontinued due the long half-life of this drug and its active metabolite, norfluoxetine (Preskorn et al. 1994). Clinicians need to keep the possibility of drug–drug interactions in mind when prescribing for any patient who has recently been taking fluoxetine (Preskorn 1999b). It should be noted that the three SSRIs that carry a risk of causing such pharmacokinetic drug–drug interactions do not have any known therapeutic advantages over the other SSRIs (i.e., citalopram, escitalopram, and sertraline) that do not cause such problems.

3.2
Adverse Effects

The SSRIs as a group have good tolerability. An important benefit of the SSRIs is the absence of the severe side effects, such as intracardiac conduction delays, seizures, and postural hypotension, seen with certain other types of antidepressants.

Because of the similarity in their mechanism of action, the side effect profile of all members of this class is very similar (Preskorn 1996, 1999b) (for a comparison of the placebo-subtracted incidence rate (%) of frequent adverse effects for citalopram, fluoxetine, fluvoxamine, paroxetine, and sertraline, see Table 2 in the Appendix to this volume). The adverse effects associated with the SSRIs are all consistent with excessive serotonin agonism and many of them are dose dependent. Since the SSRIs all have a flat dose–response curve and treatment can be begun at a clinically effective dose, clinicians should increase the dose slowly if at all in the first 2–4 weeks of treatment, since there is no improvement on average in response gained by using higher than the usually effective dose. Parenthetically, that does not mean that some individuals will not benefit from a higher (or even a lower dose) as a result of interindividual variability in their ability to clear the drug.

Nausea and loose stools can occur early in treatment; these side effects are dose dependent and tolerance typically develops with continued treatment. Nausea may be lessened by starting with a lower dose or by a dose reduction. It may also be helpful to take the medication with food or to add an antacid or a medication that blocks serotonin receptors. For example, cisapride was given to counteract the nausea caused by SSRIs until its removal from the market, which occurred when it was recognized that an SSRI such as fluvoxamine can cause dangerous and even lethal elevations in the levels of co-administered cisapride as result of substantial inhibition of the CYP responsible for cisapride clearance, CYP 3A3/4 (Janicak et al. 2001). Other side effects that may appear early in treatment are headache, dizziness, somnolence or insomnia, sweating, tremor, dry mouth, anxiety, and restlessness. A dose reduction or the addition of a benzodiazepine early in treatment may be helpful with anxiety, nervousness, or tremor; a beta-blocker may also be added for tremor. Less frequent side effects include weight gain, sexual dysfunction (inhibition of ejaculation or orgasm), bruxism, myoclonus, and paraesthesia (Janicak et al. 2001).

All antidepressants that produce substantial serotonin uptake inhibition can cause anorgasmia, decreased libido, and delayed ejaculation (Modell et al. 1997; Montejo-Gonzales et al. 1997). Sexual dysfunction occurs in approximately 30% of patients treated with SSRIs. While most adverse effects caused by antidepressants occur shortly after the first dose or at least within the first week of treatment, sexual dysfunction is an exception (Modell et al. 1997; Montejo-Gonzales et al. 1997). It frequently does not occur for several weeks or even months after initiation of treatment and tolerance often does not develop. Unfortunately, some patients will decide to discontinue treatment with an SSRI during the maintenance phase because of this side effect and thus put themselves at risk for a recurrence. A number of possible antidotes to the sexual dysfunction that occurs with the SSRIs have been tried with varying degrees of success, including treatment with bupropion, cyproheptadine, yohimbine, sildenafil, and topically applied 1% testosterone creams for anorgasmia in women. Although most of these treatments are supported only by case reports and open label studies, sildenafil has been found to be superior to placebo in a double-blind study (Nurnberg et al. 2000; Nurnberg 2001).

The SSRIs can also produce a withdrawal syndrome when they are stopped (Lazowick and Levin 1995; Coupland et al. 1996; Price et al. 1996; Landry and Roy 1997; Lejoyeux and Ades 1997; Schatzberg et al. 1997; Stahl et al. 1997; Zajecka et al. 1997; Rosenbaum et al. 1998). The withdrawal syndrome may include flu-like symptoms (fatigue, myalgia, loose stools, nausea), lightheadedness/dizziness, uneasiness/restlessness, sleep and sensory disturbances, and headache, which can be remembered using the mnemonic FLUSH (Preskorn 1996, 1999b). The SSRI withdrawal syndrome can mimic worsening of the underlying depression or even emergence of mania and such misdiagnosis may lead to inappropriate treatment (Lazowick and Levin 1995; Landry and Roy 1997; Lejoyeux and Ades 1997; Rosenbaum and Zajecka 1997; Rosenbaum et al. 1998). The diagnosis of a withdrawal syndrome is confirmed when symptoms remit, usually within 12 to 24 h, after the SSRI is restarted. The SSRI can then be tapered more gradually. Risk factors for an SSRI withdrawal syndrome are duration of time on the antidepressant, the potency of the antidepressant, and the half-life of the drug (Coupland et al. 1996; Price et al. 1996; Rosenbaum et al. 1998). The shorter the half-life of the drug, the more likely it is that the drug will wash out of the system before the brain has had an opportunity to re-equilibrate (e.g., for upregulation of receptors) and that a withdrawal syndrome will occur. Thus, following abrupt discontinuation, fluvoxamine and paroxetine are the most likely to cause a withdrawal syndrome, followed by citalopram, escitalopram, and sertraline. The SSRI withdrawal syndrome is least likely to occur following the discontinuation of fluoxetine due to the long half-life of the parent drug and its active metabolite, norfluoxetine (Coupland et al. 1996; Price et al. 1996; Schatzberg et al. 1997; Rosenbaum et al. 1998).

The increased risk of a withdrawal syndrome with paroxetine may initially seem to be inconsistent with its half-life, but that is because of its nonlinear pharmacokinetics: at concentrations lower than those usually achieved under steady-state conditions on 20 mg/day, paroxetine is preferentially cleared by CYP 2D6 which is a high-affinity but low-capacity enzyme for this biotransformation. That is consistent with the fact that the half-life of paroxetine is 10 h at a dose of 10 mg/day. At higher concentrations, paroxetine saturates this enzyme and hence inhibits its own metabolism via this pathway. For this reason, paroxetine's level increases disproportionately relative to the dose increase and the half-life is extended to 20 h under steady-state conditions achieved on 20 mg/day. When paroxetine is discontinued, the same phenomena operate in reverse and paroxetine is cleared more quickly than expected as its levels fall and CYP 2D6 becomes the principal mechanism of biotransformation and elimination. That is the likely explanation for the increased risk of the SSRI discontinuation syndrome with paroxetine compared with citalopram, escitalopram, or sertraline. For further details, the reader is referred to the earlier chapter by Preskorn and Catterson ("General Principles of Pharmacokinetics"), which reviews the pharmacokinetics of antidepressants, including paroxetine, in more detail.

3.3
Overdose

The SSRIs have demonstrated good safety in overdose because of their wide therapeutic index (Preskorn 1996, 1999b). No serious systemic toxicity has been demonstrated, even after substantial overdoses of SSRIs.

3.4
Teratogenicity

Most studies concerning the use of SSRIs in pregnancy have focused on fluoxetine, because, given its *long* half-life, even if patients stop this medication when they know they are pregnant, the medication will persist in the body for many weeks into the first trimester. In a large-scale study of fluoxetine exposure during pregnancy, Chambers et al. did not find any increase in the rate of spontaneous abortions or major structural anomalies, but did find more than a twofold increase in the incidence of three or more minor structural anomalies in neonates exposed to fluoxetine in utero (Chambers et al. 1996). They also found a higher rate of premature deliveries, lower birth weight, admission to special care nurseries, cyanosis on feeding, and jitteriness compared with controls, particularly in those exposed during the last trimester. These findings are similar to those of Pastusak et al. (1993): they found no difference in rates of major structural anomalies in neonates exposed to fluoxetine in utero compared with those exposed to TCAs or non-teratogens, but did find double the rate of miscarriage. The findings of Chambers et al. and Pastusak et al. are similar to those of Goldstein (1995), who found postnatal complications in 13% of neonates exposed to fluoxetine during the last trimester.

These findings support the recommendations in *The Expert Consensus Guidelines on the Treatment of Depression in Women 2001* which recommend the use of antidepressant medication in pregnancy as a first line option only if the current depression is severe or if the patient has a history of severe depression so that there is a high likelihood of relapse of recurrence if medication is discontinued. If it is decided that an antidepressant is needed during pregnancy, the *Expert Consensus* guidelines recommend the SSRIs as the safest choice, followed by the TCAs, recommendations that are consistent with the published literature on these agents (Altshuler et al. 2001).

3.5
Long-Term Safety

Antidepressants are often taken on a long-term basis to prevent recurrence of depression. Studies have not found any evidence of long-term safety problems with the SSRIs (Eric 1991; Doogan and Caillard 1992; Janicak et al. 2001).

4
Clinical Indications/Uses

Empirical data support the effectiveness of the SSRIs in a number of areas. The SSRIs have been shown to be effective in both major depression and in anxiety disorders; this is especially important because of the high incidence of comorbid anxiety disorders in depressed patients. SSRIs have been shown to be effective in both acute and long-term treatment (i.e., during the period when patients are vulnerable to relapse). Data also support the efficacy of the SSRIs across the lifespan in both younger and older patients, across different clinical settings ranging from specialty psychiatric to primary care settings, in patients with and without significant comorbid medical conditions, in outcome domains (e.g., improvement of life functioning) that go beyond just reduction in symptoms, and in producing full remission as well as simply an acute response in the depressive syndrome.

4.1
Acute Treatment

The major indication for the SSRIs is the treatment of major depression. In general, the overall efficacy of the different SSRIs for the acute treatment of MDD is similar, as would be expected given their common mechanism of action. The SSRIs demonstrate the same relative superiority over placebo in terms of acute efficacy. They are also similar in preventing relapse, based on double-blind, randomized clinical trials (Montgomery et al. 1988; Eric 1991; Doogan and Caillard 1992; Montgomery and Dunbar 1993; Janicak et al. 2001).

Two meta-analyses presented in Janicak et al. 2001 have shown that the SSRIs are clearly superior to placebo and have efficacy comparable to other standard antidepressants. The results of the first meta-analysis, which examined double-blind, placebo-controlled studies, are shown in Table 3; results of the second meta-analysis, which involved studies that compared the SSRIs with standard antidepressants, are shown in Table 4. However, in all studies comparing the SSRIs with the TCAs, the SSRIs have demonstrated a better tolerability profile in comparison to tertiary amine TCAs as demonstrated by both lower dropout rates and fewer adverse events.

One limitation that should be noted is that the majority of the studies listed in Table 4 were done in depressed outpatients. Consequently, there is some debate as to how efficacious the SSRIs are in patients with more severe depression, in particular in patients who have been hospitalized.

Table 3 SSRIs versus placebo: acute treatment (reprinted with permission from Janicak et al. 2001)

Drug	Number of studies	Number of subjects	Responders (%)			p Value
			SSRI (%)	Placebo (%)	Difference (%)	
Fluoxetine	9	1,365	65	41	24	10^{-22}
Fluvoxamine	3	125	67	42	25	0.008
Paroxetine	9	649	65	36	29	10^{-14}
Sertraline	3	575	78	48	30	10^{-12}

Table 4 SSRIs versus standard antidepressants: acute treatment (reprinted with permission from Janicak et al. 2001)

Drug	Number of studies	Number of subjects	Responders (%)		Difference (%)
			SSRI (%)	Standard antidepressant (%)	
Citalopram	6	347	73	74	−1
Fluoxetine	16	1,549	63	64	−1
Fluvoxamine	4	137	70	66	+4
Paroxetine	16	1,322	62	60	+2
Sertraline	3	682	68	65	+3

Some studies have suggested that the SSRIs may be less effective than certain other antidepressants, in particular those that have dual effects on both serotonin and norepinephrine. Two double-blind, active controlled studies found that clomipramine produced a better response than paroxetine or citalopram in patients hospitalized for major depression (Danish University Antidepressant Group 1986, 1990). Two other double-blind studies found venlafaxine and mirtazapine to be more effective than fluoxetine in patients hospitalized with depression (Wheatley et al. 1998; Clerc et al. 1994). Finally, several studies have found that adding desipramine (a selective norepinephrine reuptake inhibitor) to an SSRI can change nonresponders or partial responders to full responders (Weilburg et al. 1989; Nelson et al. 1991; Blier 2001).

Other studies support the effectiveness of the SSRIs in severe depression; these studies were done in patients hospitalized for depression, in patients with severe depression (HAM-D score of 25 or higher), and in patients with melancholia (Janicak et al. 2001). One caveat concerning these studies is that some included only active controls and simply found no difference between, for example, an SSRI and a TCA, which does not prove that the SSRI is as efficacious as the TCA but may only reflect the fact that the sample was too small to separate the treatments (Janicak et al. 2001). Five double-blind studies have compared various SSRIs with different TCAs in patients with HAM-D scores of 25 or higher (Reimherr et al. 1988; Bowden et al. 1993; Feighner et al. 1993; Pande and Sayler 1993; Kasper et al. 1995), three of which included both inpatients and outpatients (Bowden et al. 1993; Pande and Sayler 1993; Kasper et al. 1995) and two of which were outpatient studies (Reimherr et al. 1988; Feighner et al. 1993). In the three studies that were placebo-controlled (Reimherr et al. 1988; Feighner et al. 1993; Kasper et al. 1995), the SSRI tested (fluvoxamine, paroxetine, or sertraline) was either found to be superior to both the TCA and placebo or comparable to the TCA and superior to placebo. In the other two studies, which did not include a placebo control, no difference was found between TCA and SSRI. Four studies and one meta-analysis of European antidepressant trials also found no difference in efficacy between various SSRIs and several tertiary amine TCAs in patients hospitalized for major depression (Ginestet 1989; Tignol et al. 1992; Beasley et al. 1993a; Arminen et al. 1994; Ottevanger 1995). Finally, two small studies have found that fluoxetine and fluvoxamine had antidepressant efficacy superior to placebo in patients with melancholia (Feighner et al. 1989; Heiligenstein et al. 1993), while another larger study did not find a difference between paroxetine and amitriptyline in such patients (Stuppaeck et al. 1994).

There is controversy over whether it makes sense to switch to a second SSRI if a patient does not respond to a trial of one SSRI. If the patient has not been able to tolerate the first SSRI, many clinicians will try a second SSRI (Janicak et al. 2001). However, most psychiatrists prefer to switch to a drug with a different mechanism of action if a patient has achieved inadequate efficacy after an adequate trial of one SSRI (Fredman et al. 2000). There is no compelling evidence showing that nonresponders to one SSRI will respond to a trial of a different

SSRI (Janicak et al. 2001). For a detailed discussion of treatment-refractory depression, see the chapter by Trivedi et al. (later in this volume).

4.2
Dosing Recommendations

All the SSRIs have a flat dose–response curve, so that there is usually no advantage in using doses above the usually effective minimum dose, except in the instance of either unusually slow or rapid metabolism, leading to unusually high or low accumulation of drug, respectively. The former might be clinically suspected because of increased frequency or severity of serotonin-mediated adverse effects while the latter might be suspected because of the absence of therapeutic benefit on usually therapeutic doses. These suspicions can be confirmed by use of TDM even though that is rarely done in clinical practice for somewhat inexplicable reasons. The interested reader is referred to the earlier chapters by Preskorn and Catterson ("General Principles of Pharmacokinetics") and Burke and Preskorn ("Therapeutic Drug Monitoring of Antidepressants") for more detailed discussion of genetically determined differences in pharmacokinetics and the use of TDM.

Based on the labeling in the FDA package inserts for the various SSRIs, their respective usually effective, lowest antidepressant dose for the treatment of major depressive disorder are as follows:

1. Citalopram, 40 mg/day
2. Escitalopram, 20 mg/day
3. Fluoxetine, 20 mg/day
4. Fluvoxamine, 150–200 mg/day (note that this dose range has not been confirmed in definitive clinical trials)
5. Paroxetine, 20 mg/day
6. Sertraline, 50 mg/day

These usually effective doses produce comparable effects on plasma serotonin levels and the serotonin uptake pump regardless of the SSRI involved (Preskorn 1996). More detailed information on the dosing of the SSRIs is provided in Table 1 of the Appendix to this volume.

4.3
Onset of Action

Like most of the antidepressants reviewed in this book, there is a delay of at least 1–2 weeks between starting an SSRI and seeing a clinically meaningful antidepressant response. To date, all data regarding the rapidity of the SSRIs' onset of action have been drawn from the retrospective subanalysis of clinical trials that were not designed to evaluate onset of action (Blier 2001), and there is need for prospective studies to examine comparative onset of action using sensitive designs and adequate sample sizes (Thase 2001) to clearly differentiate between the different SSRIs. However, several randomized, double-blind studies have suggested that fluoxetine may have a slower onset of action than other antidepressants, including citalopram (Patris et al. 1996), moclobemide (Gattaz et al. 1995), paroxetine (De Wilde et al. 1993), and venlafaxine (Clerc et al. 1994). The delayed onset of antidepressant action with fluoxetine may be due to the long half-lives of fluoxetine and norfluoxetine and the fact that it therefore takes 35–75 days to reach steady state in healthy young individuals and an average of 105 days in healthy individuals over the age of 65. Complete clearance of the drug after the achievement of steady state takes a comparable period of time after discontinuation. This pharmacokinetic profile, coupled with fluoxetine's substantial effects on multiple CYP enzymes, is a compelling reason to avoid the use of fluoxetine in general and in particular in elderly patients and those on multiple other medications.

The fact that it takes several weeks for onset of antidepressant action with the other SSRIs raises important questions about the mechanism of their antidepressant effect. While the SSRIs were developed based on the theory that central serotonin uptake inhibition could confer antidepressant activity, there is a temporal disconnect between the onset of that action and the onset of their antidepressant activity. In fact, substantial serotonin uptake inhibition occurs within hours of the first dose of most SSRIs. If that effect alone was sufficient to mediate an antidepressant response, then these medications should work within hours rather than weeks. It now appears that, while the blockade of serotonin uptake is a necessary action for the antidepressant efficacy of the SSRIs, by itself it is not sufficient. Instead, this blockade sets in motion a cascade of more complex adaptive responses in the brain which downregulate various pre- and postsynaptic serotonin receptors as well as other neurotransmitter systems. It may be that these adaptive responses more directly mediate the antidepressant response than the blockade of serotonin uptake inhibition per se, since the time course for these adaptive changes fits better with the time course of SSRI antidepressant response.

4.4
Maintenance Treatment

Six random-assignment, double-blind, placebo-controlled, crossover studies have found that all the SSRIs (except fluvoxamine) can prevent relapse after remission has been achieved (Janicak et al. 2001). These six studies involved 910 adults with major depression who had responded to acute treatment with an SSRI and were randomly assigned in a double-blind fashion to either continue on the SSRI or switch to placebo: 10% of those who were continued on the SSRI relapsed over 24 weeks versus 35% of those who were switched to placebo (Janicak et al. 2001). The findings from the three longest studies are summarized in Table 5. In those studies, the advantage of remaining on the SSRI was still apparent at 44 and 52 weeks; over the course of a year, there was approximately a 30%

Table 5 Relapse prevention continuation studies: double-blind, random assignment to remain on SSRIs or crossover to placebo (adapted with permission from Preskorn 1996, p. 92)

SSRI (study)	Duration (weeks)	Relapse rate (%) outcome			p Value
		Placebo	SSRI	Difference	
Citalopram (Montgomery et al. 1993)	24	31	11	20	<0.05
Fluoxetine (Montgomery et al. 1988)	52	57	26	31	<0.001
Paroxetine (Montgomery and Dunbar 1993)	52	43	16	27	<0.01
Sertraline (Doogan and Caillard 1992)	44	46	13	33	<0.001

greater relapse rate in patients who were switched to placebo than in those who remained on an SSRI. A study comparing the efficacy of fluvoxamine and sertraline in relapse prevention found no difference between the agents (Franchini et al. 1997).

References

Altshuler LL, Cohen LS, Moline ML, et al. (2001) The expert consensus guideline series: treatment of depression in women 2001. Postgrad Med Special Report, March:1–115

Arminen SL, Ikonen U, Pulkkinen P, et al. (1994) A 12-week, double-blind, multicentre study of paroxetine and imipramine in hospitalized depressed patients. Acta Psychiatr Scand 89:382–389

Beasley CM, Holman SL, Potvin JH (1993a) Fluoxetine compared with imipramine in the treatment of inpatient depression: a multicenter trial. Ann Clin Psychiatry 5:199–208

Beasley CM, Masica DN, Heiligenstein JH, et al (1993b) Possible monoamine oxidate inhibitor-serotonin uptake inhibitor interaction: fluoxetine clinical data and preclinical findings. J Clin Psychopharmacol 13:312–320

Bhatara VS, Bandettini FC (1993) Possible interaction between sertraline and tranylcypromine. Clin Pharmacol 12:222–225

Blier P (2001) Possible neurobiological mechanisms underlying faster onset of antidepressant action. J Clin Psychiatry 62:7–11

Bonomo V, Fogliani AM (2000) Citalopram and haloperidol for psychotic depression. Am J Psychiatry 157:1706–7

Bowden CL, Schatzberg AF, Rosenbaum A, et al.(1993) Fluoxetine and desipramine in major depressive disorder. J Clin Psychopharmacol 13:305–311

Chambers CD, Johnson KA, Dick LM, et al. (1996) Birth outcomes in pregnant women taking fluoxetine. N Engl J Med 335:1010–1015

Clerc GE, Ruimy P, Verdeau-Pallès J (1994) A double-blind comparison of venlafaxine and fluoxetine in patients hospitalized for major depression and melancholia. The Venlafaxine French Inpatient Study Group. Int Clin Psychopharmacol. 9:139–143

Coplan JD, Gorman JM (1993) Detectable levels of fluoxetine metabolites after discontinuation: an unexpected serotonin syndrome. Am J Psychiatry 150:837

Coupland NJ, Bell CJ, Potokar JP (1996) Serotonin reuptake inhibitor withdrawal. J Clin Psychopharmacol 16:356–362

Danish University Antidepressant Group (1986) Citalopram: clinical effect profile in comparison with clomipramine: a controlled multicenter study. Psychopharmacology 90:131–138

Danish University Antidepressant Group (1990) Paroxetine: a selective serotonin reuptake inhibitor showing better tolerance, but weaker antidepressant effect than clomipramine in a controlled multicenter study. J Affect Disord 18:289-299

de Vries MH, Raghoebar M, Mathlener IS, van Harten J (1992) Single and multiple oral dose fluvoxamine kinetics in young and elderly subjects. Ther Drug Monit 14:493-498

De Wilde J, Spiers R, Mertens C, et al.(1993) A double-blind, comparative, multicentre study comparing paroxetine with fluoxetine in depressed patients. Acta Psychiatr Scand 87:141-5

Doogan D, Caillard V (1992) Sertraline in the prevention of depression. Br J Psychiatry 160:217-222

Eric L (1991) A prospective double-blind comparative multicentre study of paroxetine and placebo in preventing recurrent major depressive episodes. Biol Psychiatry 29(suppl 11):254S-255S

Feighner JP. Boyer WF, Meredith CH, et al. (1989) A placebo-controlled inpatient comparison of fluvoxamine maleate and imipramine in major depression. Int Clin Psychopharmacol 4:239-244

Feighner JP, Cohn JB, Fabre LF, et al. (1993) A study comparing paroxetine, placebo, and imipramine in depressed patients. J Affect Disord 28:71-79

Franchini L, Gasperini M, Perez J, et al. (1997) A double-blind study of long-term treatment with sertraline or fluvoxamine for prevention of highly recurrent unipolar depression. J Clin Psychiatry 58:104-107

Fredman SJ, Fava M, Kienke AS, et al.(2000) Partial response, nonresponse, and relapse on SSRIs in major depression: a survey of current "next-step" practices. J Clin Psychiatry 61:403-407

Gattaz WF, Vogel P, Kick H, et al.(1995) Moclobemide versus fluoxetine in the treatment of inpatients with major depression. J Clin Psychopharmacol 15(suppl 2):35S-40S

Ginestet D (1989) Fluoxetine in endogenous depression and melancholia versus clomipramine. Int Clin Psychopharmacol 4:37-40

Glennon RA, Dukat M (1994). Serotonin receptor subtypes. In: Bloom FE, Kupfer DJ, eds. Psychopharmacology: The fourth generation of progress. Raven Press, New York, NY, pp. 415-430

Goldstein DJ (1995) Effects of third trimester fluoxetine exposure on the newborn. J Clin Psychopharmacol 15:417-420

Graber MA, Hoehns TB, Perry PJ (1994) Sertraline-phenelzine drug interaction: a serotonin syndrome reaction. Ann Pharmacother 28:732-735

Harvey AT. Preskorn SH (1996a) Cytochrome P450 enzymes: Interpretation of their interactions with selective serotonin reuptake inhibitors. Part I. J Clin Psychopharmacol 16:273-285

Harvey AT. Preskorn SH (1996b) Cytochrome P450 enzymes: Interpretation of their interactions with selective serotonin reuptake inhibitors. Part II. J Clin Psychopharmacol 16:345-355

Harvey AT, Preskorn SH (2001) Fluoxetine pharmacokinetics and effect on CYP2C19 in young and elderly volunteers. J Clin Psychopharmacol 21:161-166

Heiligenstein JH, Tollefson GD, Faries DE (1993) Response patterns of depressed outpatients with and without melancholia: a double-blind, placebo-controlled trial of fluoxetine versus placebo. J Affect Disord 30:163-173

IMS Health. Source Prescription Audit. January-November 2002

Janicak PG, Davis JM, Preskorn SH, Ayd FJ, Jr (2001) Principles and practice of psychopharmacotherapy. Lippincott Williams & Wilkins, Philadelphia

Kasper S. Moller H-J, Montgomery SA, et al. (1995) Antidepressant efficacy in relation to item analysis and severity of depression: a placebo-controlled trial of fluvoxamine versus imipramine. Int Clin Psychopharmacol 9:3-12:

Landry P, Roy L (1997) Withdrawal hypomania associated with paroxetine. J Clin Psychopharmacol. 17:60–61

Lazowick AL, Levin GM (1995) Potential withdrawal syndrome associated with SSRI discontinuation. Ann Pharmacother 29:1284–1285

Lejoyeux M, Ades J (1997) Antidepressant discontinuation: a review of the literature. J Clin Psychiatry 58(suppl 7):11–16

Madsen H, Enggaard TP, Hansen LL, et al. (2001) Fluvoxamine inhibits the CYP2C9 catalyzed biotransformation of tolbutamide. J Clin Pharmacol Ther 69:41–47

Mills KC (1995) Serotonin syndrome. Am Fam Physician 52:1475–1482

Milne RJ, Goa KL (1991) Citalopram: a review of its pharmacodynamic and pharmacokinetic properties, and therapeutic potential in depressive illness. Drugs 41:450–477

Modell JG, Katholi CR, Modell JD, et al. (1997) Comparative sexual side effects of bupropion, fluoxetine, paroxetine, and sertraline. Clin Pharmacol Ther 61:476–487

Montejo-Gonzales AL, Llorca G, Izquierdo JA, et al. (1997) SSRI-induced sexual dysfunction: fluoxetine, paroxetine, sertraline, and fluvoxamine in a prospective, multicenter, and descriptive clinical study of 344 patients. J Sex Marital Ther 23:176–194

Montgomery SA, Dunbar G. Paroxetine is better than placebo in relapse prevention and the prophylaxis of recurrent depression. Int Clin Psychopharmacol 1993 Fall; 8(3):189–95

Montgomery SA, Dufour H, Brion S, et al.(1988) The prophylactic efficacy of fluoxetine in unipolar depression. Br J Psychiatry 153(suppl 3):69–76

Montgomery SA, Rasmussen JG, Tanghoj P (1993) A 24-week study of 20 mg citalopram, 40 mg citalopram, and placebo in the prevention of relapse of major depression. Int Clin Psychopharmacol 8:181–188

Nelson JC, Mazure CM, Bowers MB Jr, et al. (1991) A preliminary, open study of the combination of fluoxetine and desipramine for rapid treatment of major depression. Arch Gen Psychiatry 48:303–307

Nierenberg DW, Semprebon M (1993) The central nervous system serotonin syndrome. Clin Pharmacol Ther 53:84–88

Nurnberg HG (2001) Managing treatment-emergent sexual dysfunction associated with serotonergic antidepressants: before and after sildenafil. Journal of Psychiatric Practice 7:92–108

Nurnberg HG, Gelenberg AJ, Fava M, et al.(2000) Sildenafil for serotonin reuptake inhibitor associated sexual dysfunction: A 3-center, 6-week double-blind placebo-controlled study in 90 men. Program and abstracts from the 153rd Annual American Psychiatric Association Meeting, Chicago, IL, May 13–18 (Abstract NR419)

Olfson M, Marcus SC, Pincus HA, Zito JM, Thompson JW, Zarin DA (1998) Antidepressant prescribing practices of outpatient psychiatrists. Arch Gen Psychiatry 55:310–316

Olfson M, Marcus SC, Druss B, Elinson L, Tanielian T, Pincus HA (2002) National trends in the outpatient treatment of depression. JAMA 287:203–209

Ottevanger EA (1995) Fluvoxamine and clomipramine in depressed hospitalized patients; results from a randomized, double-blind study. Encephale 21:317–321

Pande AC, Sayler ME (1993) Severity of depression and response to fluoxetine. Int Clin Psychopharmacol 8:243–245

Pastuszak A, Schick-Boschetto B, Zuber C, et al. Pregnancy outcome following first trimester exposure to fluoxetine (Prozac). JAMA 1993;269:2246–2248

Patris M, Bouchard J-M, Bougerol T, et al. (1996) Citalopram versus fluoxetine: a double-blind, controlled, multicentre, phase II trial in patients with unipolar major depression treated in general practice. Int Clin Psychopharmacol 11:129–136

Preskorn SH (1990) The future and psychopharmacology: potentials and needs. Psychiatr Ann 1990;20(suppl 11):625–633

Preskorn SH (1993) Pharmacokinetics of antidepressants: why and how they are relevant to treatment. J Clin Psychiatry 54(suppl 9):14–34

Preskorn SH (1994) Antidepressant drug selection: criteria and options. J Clin Psychiatry 55(suppl A): 6–22

Preskorn SH (1995a) Comparison of the tolerability of bupropion, fluoxetine, imipramine, nefazodone, paroxetine, sertraline, and venlafaxine. J Clin Psychiatry 56(suppl 6):12–21

Preskorn SH (1995b) Should rational drug development in psychiatry target more than one mechanism of action in a single molecule? Int Rev Psychiatry 1995;7:17–28

Preskorn SH (1996) Clinical pharmacology of selective serotonin reuptake inhibitors. Professional Communications, Caddo, OK (Available on the website www.preskorn.com)

Preskorn SH. (1998a) A message from Titanic. J Prac Psych Behav Hlth. 4:236–242

Preskorn SH (1998b) Sweetness! J Pract Psychiatry Behav Health 4:304–308

Preskorn SH (1999a) Defining "Is." Journal of Practical Psychiatry and Behavioral Health 5:224–228 (Available on the website www.preskorn.com)

Preskorn SH (1999b) Outpatient management of depression. Professional Communications, Caddo, OK (Available on the website www.preskorn.com)

Preskorn SH, Burke M (1992) Somatic therapy for major depressive disorder: selection of an antidepressant. J Clin Psychiatry 53(suppl 9):5–18

Preskorn SH, Fast GA (1991) Therapeutic drug monitoring for antidepressants: efficacy, safety and cost effectiveness. J Clin Psychiatry 52(suppl 6):23–33

Preskorn SH, Alderman J, Chung M, Harrison W, Messig M, Harris S (1994) Pharmacokinetics of desipramine coadministered with sertraline or fluoxetine. J Clin Psychopharmacol 14:90–98

Preskorn SH, Shad MU, Alderman J, Lane R (1998) Fluoxetine: age and dose dependent pharmacokinetics and CYP 2C19 inhibition. Am Soc Clin Pharmacol Ther 63:166. Abstract

Price JS, Waller PC, Wood SM, MacKay AV (1996) A comparison of the post-marketing safety of four selective serotonin re-uptake inhibitors including the investigation of symptoms occurring on withdrawal. Br J Clin Pharmacol 42:757–763

Reimherr FW, Byerley WF, Ward MF, et al.(1988) Sertraline, a selective inhibitor of serotonin uptake, for the treatment of outpatients with major depressive disorder. Psychopharmacol Bull 24:200–205

Roberts E, Norris P (2001) Growth and change in the prescribing of anti-depressants in New Zealand: 1993–1997. NZ Med J 114:25–27

Rosenbaum JF, Zajecka J (1997) Clinical management of antidepressant discontinuation. J Clin Psychiatry 58(suppl 7):37–40

Rosenbaum JF, Fava M, Hoog SL, Ascroft RC, Krebs WB (1998) Selective serotonin reuptake inhibitor discontinuation syndrome: A randomized clinical trial. Biol Psychiatry 44:77–87

Schatzberg AF, Haddad P, Kaplan EM, et al. (1997) Possible biological mechanisms of the serotonin reuptake inhibitor discontinuation syndrome. Discontinuation Consensus Panel. J Clin Psychiatry 58(suppl 7):23–27

Shad MU, Preskorn SH (2000) Antidepressants. In: Levy RH, Thummel KE, Trager WF, Hansten PD, Eichelbaum M (eds) Metabolic Drug Interactions. Lippincott, Williams & Wilkins, Philadelphia, pp. 563–577

Sporer KA (1995) The serotonin syndrome. Implicated drugs, pathophysiology and management. Drug Saf 13:94–104

Stahl MM, Lindquist M, Pettersson M, et al.(1997) Withdrawal reactions with selective serotonin re-uptake inhibitors as reported to the WHO system. Eur J Clin Pharmacol 53:163–169

Sternbach H (1991) The serotonin syndrome. Am J Psychiatry 148:705–713

Stuppaeck CH, Geretsegger C, Whitworth AB, et al. (1994) A multicenter double-blind trial of paroxetine versus amitriptyline in depressed inpatients. J Clin Psychopharmacol 14:241–246

Tallman JF, Dahl SG, New drug design in psychopharmacology: the impact of molecular biology. In: Bloom FE, Kupfer DJ, eds. Psychopharmacology: The Fourth Generation of Progress. Raven Press, New York; NY, 1994:1861–1874

Thase ME (2001) Methodology to measure onset of action. J Clin Psychiatry 62(Suppl 15):18–21

Tignol J, Stoker MJ, Dunbar GC (1992) Paroxetine in the treatment of melancholia and severe depression. Int Clin Psychopharmacol 7:91–94

Weilburg JB, Rosenbaum JF, Biederman J, et al. (1989) Fluoxetine added to non-MAOI antidepressants converts nonresponders to responders: a preliminary report. J Clin Psychiatry 50:447–449

Wheatley DP, van Moffaert M, Timmerman L, et al.(1998) Mirtazapine: efficacy and tolerability in comparison with fluoxetine in patients with moderate to severe major depressive disorder. J Clin Psychiatry 59:306–312

Zajecka J, Tracy KA, Mitchell S (1997) Discontinuation symptoms after treatment with serotonin reuptake inhibitors: a literature review. J Clin Psychiatry 58:291–297

Other Antidepressants

S. H. Preskorn[1] · R. Ross[2]

[1] Department of Psychiatry and Behavioral Sciences,
  University of Kansas School of Medicine, and Psychiatric Research Institute,
  1010 N. Kansas, Wichita, KA 67214, USA
  e-mail: spreskor@kumc.edu
[2] Ross Editorial, 228 Black Rock Mtn Ln, Independence, VA 24348, USA
  e-mail: ross@ls.net

1	Introduction	268
1.1	Dose–Response Curves	268
1.2	Organization of the Chapter	269
2	**Bupropion**	269
2.1	Chemistry, Pharmacology, and Metabolism	269
2.2	Safety and Adverse Effects	270
2.2.1	Drug–Drug Interactions	270
2.2.2	Adverse Effects	270
2.2.3	Overdose	271
2.3	Clinical Use/Therapeutic Indications	272
2.3.1	Efficacy for Acute Treatment	272
2.3.2	Dosing Recommendations	273
2.3.3	Onset of Action	273
3	**Mirtazapine and Mianserin**	273
3.1	Chemistry, Pharmacology, and Metabolism	273
3.2	Safety and Adverse Effects	274
3.2.1	Drug–Drug Interactions	274
3.2.2	Adverse Effects	275
3.2.3	Overdose	276
3.3	Clinical Use/Therapeutic Indications	276
3.3.1	Efficacy for Acute Treatment	276
3.3.2	Dosing Recommendations	276
3.3.3	Onset of Action	277
4	**Nefazodone and Trazodone**	277
4.1	Chemistry, Pharmacology, and Metabolism	277
4.2	Safety and Adverse Effects	277
4.2.1	Drug–Drug Interactions	277
4.3	Adverse Effects	278
4.3.1	Overdose	278
4.4	Clinical Use/Therapeutic Indications	278
4.4.1	Efficacy for Acute Treatment	278
4.4.2	Dosing Recommendations	279
4.4.3	Onset of Action	280

5	**Reboxetine**	280
5.1	Chemistry, Pharmacology, and Metabolism	280
5.2	Safety and Adverse Effects	283
5.2.1	Drug-Drug Interactions	283
5.2.2	Adverse Effects	283
5.2.3	Overdose	284
5.3	Clinical Use/Therapeutic Indications	284
5.3.1	Efficacy for Acute Treatment	284
5.3.2	Dosing Recommendations	285
5.3.3	Onset of Action	285
6	**Venlafaxine**	286
6.1	Chemistry, Pharmacology, and Metabolism	286
6.2	Safety and Adverse Effects	286
6.2.1	Drug-Drug interactions	286
6.2.2	Adverse Effects	287
6.2.3	Overdose	288
6.3	Clinical Use/Therapeutic Indications	289
6.3.1	Efficacy for Acute Treatment	289
6.4	Dosing Recommendations	290
6.4.1	Onset of Action	291
7	**Duloxetine**	291
7.1	Chemistry, Pharmacology, and Metabolism	291
7.2	Safety and Adverse Effects	294
7.2.1	Drug-Drug Interactions	294
7.2.2	Adverse Effects	294
7.2.3	Overdose	295
7.3	Clinical Use/Therapeutic Indications	295
7.3.1	Efficacy for Acute Treatment	295
7.3.2	Dosing Recommendations	295
7.3.3	Onset of Action	295
8	**Milnacipran**	296
8.1	Chemistry, Pharmacology, and Metabolism	296
8.2	Safety and Adverse Effects	299
8.2.1	Drug-Drug Interactions	299
8.2.2	Adverse Effects	299
8.2.3	Overdose	299
8.3	Clinical Use/Therapeutic Indications	299
8.3.1	Efficacy for Acute Treatment	299
8.3.2	Dosing Recommendations	301
8.3.3	Onset of Action	301
9	**Tianeptine**	301
9.1	Chemistry, Pharmacology, and Metabolism	301
9.2	Safety and Adverse Effects	306
9.2.1	Drug-Drug Interactions	306
9.2.2	Adverse Effects	306
9.2.3	Overdose	307
9.3	Clinical Use/Therapeutic Indications	307
9.3.1	Efficacy for Acute Treatment	307

9.3.2 Dosing Recommendations 309
9.3.3 Onset of Action. 309

10 Conclusion 309

References 311

Abstract This chapter covers the following antidepressants, which do not belong to one of the major classes described in the three preceding chapters: bupropion, mirtazapine and mianserin, nefazodone and trazodone, reboxetine, venlafaxine, duloxetine, milnacipran, and tianeptine. Unlike the selective serotonin reuptake inhibitors (SSRIs), many of these antidepressants have an ascending rather than a flat dose–response curve. The chapter provides a brief review of the chemistry, pharmacology, metabolism, safety and adverse effects, clinical use, and therapeutic indications of each antidepressant. *Bupropion* is a weak dual uptake inhibitor of both dopamine and norepinephrine (NE). Information concerning its pharmacodynamic and pharmacokinetic properties is limited, primarily because of the age of the drug (clinical trials begun in the mid-1970s). Bupropion's most serious side effect is dose-dependent seizures, so that the highest recommended doses are 450 and 400 mg/day for the immediate release (IR) and sustained release (SR) formulations, respectively. Other adverse effects include restlessness, activation, tremors, insomnia, and nausea. Bupropion was found to be an effective antidepressant in several double-blind studies that generally used doses higher than the maximum recommended dose of 450 mg/day. Despite fairly modest evidence of antidepressant efficacy, bupropion may be useful in a number of clinical situations, including for patients with prominent psychomotor retardation, Parkinson's disease, or attention-deficit/hyperactivity disorder; for patients who have failed to respond to other antidepressants; and for patients who cannot tolerate sexual side effects of other antidepressants. Bupropion has also been approved as an aid in smoking cessation. *Mirtazapine* and its forerunner mianserin, are tetracyclic compounds with a unique mechanism of action. Mirtazapine is an α_2-antagonist that increases noradrenergic and serotonergic neurotransmission, the primary mechanism thought to underlie its antidepressant activity. Mirtazapine does not cause many of the side effects associated with the SSRIs (e.g., nausea, loose stools, disturbed sleep pathology, sexual dysfunction) and causes minimal anticholinergic effects, no quinidine-like effects, and no effects on blood pressure. Sedation can be a problem, especially early in treatment, although this may be an advantage for patients with prominent insomnia, anxiety, or agitation. Mirtazapine can cause increased appetite and weight gain, transient neutropenia, and transient mild elevations of liver function tests. Three cases of agranulocytosis were reported out of 3,000 patients in the mirtazapine clinical trial program, an incidence too low to draw any conclusion about cause and effect. Although postmarketing experience has not found an unusual number of cases of agranulocytosis, the package insert in the United States contains a warning that a white blood cell count should be done if a patient taking mirtazapine develops signs of fever or infection. Because of its unique mechanism of action, mirtazapine may be efficacious for patients who have not benefited from other types of antidepressants. *Nefazodone* and *trazodone* have chemically related structures that incorporate 5-HT$_{2A}$ receptor blockade plus weak 5-HT uptake blockade and possibly NE uptake blockade. Trazodone is widely used as a nonhabit-forming sleep aid rather than as an antidepressant. The antihistaminergic properties of trazodone are partly responsible for its popularity as a sleep aid, but can cause significant problems with daytime sedation when it is used as an antidepressant; however, it may be useful for treating agitation in geriatric patients. Nefazodone, a more potent serotonin reuptake inhibitor (SRI) than trazodone, was designed with the goal of producing a better antidepressant than trazodone, although it is a much weaker SRI than the SSRIs or venlafaxine. Because it substantially inhibits CYP 3A3/4, nefazodone can elevate levels of coprescribed drugs metabolized via CYP 3A/3/4. Nefazodone produces less activation and sexual dysfunction than the SSRIs and venlafaxine; it does not cause blood pressure elevation or disturb sleep physiology; it improves subjective sleep quality. The incidence and severity of the following adverse effects increase in a dose-dependent fashion as a function of the starting dose of nefazodone: dizziness/lightheadedness, confusion, sedation, gastrointestinal side effects. Nefazodone appears to have efficacy in patients with clinical depression and prominent anxiety. Although there appears to be greater interpatient variability in response to nefazodone than to many of the other newer antidepressants, nefazodone can be a useful option for patients who are unable to tolerate the adverse effects of the SSRIs. *Reboxetine* is a selective NE reuptake pump inhibitor. Its most common adverse effects are insomnia, sweating, constipation, dry mouth, and urinary hesitancy. Most of the published trials of reboxetine have been active rather than placebo-controlled and results were not published in full with rigorous peer review, compromising the ability to make an assessment of efficacy. Sufficient evidence of the efficacy of reboxetine in major depression has not been presented to receive approval for marketing in the United States, but reboxetine is available in several other countries. *Venlafaxine*, a phenylethylamine, first inhibits the neuronal uptake pump for serotonin (SE) and then at higher concentrations inhibits the uptake pump for NE. Unlike tertiary amine tricyclic antidepressants (TCAs), which also inhibit the SE and NE uptake pumps, venlafaxine has low affinity for most other neural receptors and does not inhibit sodium fast channels, making it relatively safe in overdose. Its adverse effects change qualitatively as the dose increases because of progressively greater blockade of NE uptake with increasing doses. At low doses, the adverse-effect profile is similar to an SSRI with nausea, loose stools, sexual dysfunction, while venlafaxine at higher doses can produce generally mild increases in blood pressure, diaphoresis, tachycardia, tremors, and anxiety. A disadvantage of venlafaxine relative to the SSRIs is the potential for dose-dependent blood pressure elevation, most likely due to the NE uptake inhibition caused by higher doses; however, this adverse effect is infrequently observed at doses below 225 mg/day. Venlafaxine and the SSRIs have similar advantages over the TCAs and monoamine oxidase inhibitors. Venlafaxine also has a number of potential advantages over the SSRIs, including an ascending dose–antidepressant response curve, with possible greater overall efficacy at higher doses, evidence of more rapid onset of antidepressant action, evidence of superior efficacy in hospitalized patients with major depressive disorder compared with placebo or fluoxetine, and minimal effects on CYP enzymes in contrast to fluoxetine, fluvoxamine, paroxetine, and the non-SSRI, bupropion. *Duloxetine* is a SE–NE re-uptake pump inhibitor, which is pending approval in the United States and other countries in late 2003. It will be the third member of this pharmacological class, which also contains venlafaxine and milnacipran (sibutramine, the fourth member of this class, is marketed for obesity rather than major depression). Only a limited number of articles have been published on this compound but more should be expected shortly after its market introduction. The manufacturer is initially seeking indications for both major depression and urinary incontinence. Due to its inhibition of the SE and NE uptake pumps, duloxetine will undoubtedly carry a warning against use in combination with monoamine oxidase inhibitors. It is also a moderate inhibitor of CYP 2D6, so that modest dose reductions and careful monitoring will be needed when prescribing duloxetine in combination with drugs that are preferentially metabolized by CYP 2D6, particularly those with narrow therapeutic indexes. The most common side effects identified in clinical trials to date appear to be nausea, dry mouth, dizziness, constipation, insomnia, asthenia, and hypertension, consistent with its mechanisms of action. Clinical trials to date have demonstrated rates of response and remission in patients with major depression that are comparable to other marketed antidepressants reviewed in this book. Although *milnacipran* is marketed in France, Japan, and a few other countries, its development in the United States was discontinued. It is an SE and NE reuptake inhibitor in the same class as venlafaxine, duloxetine, and the anti-obesity drug, sibutramine. Milnacipran would be predicted to be susceptible to the same pharmacodynamic drug–drug interactions as other drugs in this class, but would not be expected to be involved in any CYP enzyme-mediated drug–drug interactions. Milnacipran at doses of 50–200 mg/day has a favorable adverse-effect profile when compared with tertiary amine TCAs, including a lower incidence of abnormal liver function tests. At doses of 50 or 100 mg twice a day but not 100 mg once a day, it caused a lower incidence of nausea and anxiety but a higher incidence of headache, dry mouth, and dysuria than did fluoxetine, 20 mg/day, or fluvoxamine, 100 mg twice a day. As with other drugs in this class, dysuria is the most common troublesome and dose-dependent adverse effect (occurring in up to 7% of patients). High-dose milnacipran has been reported to cause blood pressure elevation. Like reboxetine, most of the published trials of milnacipran have been active rather than placebo-controlled and results were not published in full with rigorous peer review, compromising the ability to make an assessment of efficacy; however, findings to date suggest that milnacipran produces a superior antidepressant response compared with placebo at doses of 50 and 100 mg twice a day. *Tianeptine* is marketed in France but few other countries around the world and there is little knowledge of this drug in the United States and other English-speaking countries. However, a surprising amount of research, particularly preclinical, has been done with tianeptine, in part because of its apparent novel mechanism of action: Tianeptine, in contrast to most other antidepressants, increases SE uptake into neurons rather than blocking it. However, its side-effect profile is similar to that of other newer antidepressants, with low abuse potential and a low risk of adverse effects on the cardiovascular system, the cholinergic systems, sleep/arousal, cognition, psychomotor functioning, and weight. The most common adverse effects include nausea, constipation, abdominal pain, headache, dizziness, and altered dreaming. Hepatotoxicity has been reported but is rare. Like reboxetine and milnacipran, most of the published trials of tianeptine have been active rather than placebo-controlled and results were not published in full with rigorous peer review, compromising the ability to assess efficacy; however, trials to date do suggest efficacy in patients with major depression.

Keywords Bupropion · Mirtazapine · Mianserin · Nefazodone · Trazodone · Reboxetine · Venlafaxine · Duloxetine · Milnacipran · Tianeptine

Portions of this chapter are adapted from Janicak et al. (2001) with permission from Lippincott Williams & Wilkins, and from four columns on reboxetine, duloxetine, milnacipran, and tianeptine that are in press in the *Journal of Psychiatric Practice*, with permission from Sheldon H. Preskorn.

1
Introduction

This chapter focuses on the following antidepressants, which do not belong to one of the major classes described in the three preceding chapters: bupropion, mirtazapine and mianserin, nefazodone and trazodone, reboxetine, venlafaxine, duloxetine, milnacipran, and tianeptine. The pharmacodynamic properties of these antidepressants differ substantially from those of the tricyclic antidepressants (TCAs), the monoamine oxidase inhibitors, the selective SE reuptake inhibitors (SSRIs), and from each other. For example, while each of the SSRIs has at least one order of magnitude separation between its binding affinity for the serotonin (SE) reuptake pump and its affinity for its next most potent potential site of action, this is not true for the antidepressants covered in this chapter. Among the antidepressants described in this chapter, venlafaxine most closely approximates the binding affinity profile of an SSRI. However, it is five times more potent (i.e., has a higher binding affinity) as an inhibitor of the SE compared with the norepinephrine (NE) uptake pump, although it is not as "selective" as the SSRIs (Bolden-Watson and Richelson 1993).

1.1
Dose–Response Curves

Unlike the SSRIs, many of the antidepressants in this chapter have an ascending as opposed to a flat dose–response curve. That is consistent with their in vitro pharmacology. For example, studies indicate that bupropion and its metabolites and mirtazapine have curvilinear dose–response curves, at least for some of their effects. Drugs with curvilinear dose–response curves can produce an effect at one dose that they do not produce at a higher dose, because at the lower dose the drug preferentially engages a site of action capable of mediating an effect, while at higher concentrations (doses) the drug engages a second site of action that blocks or antagonizes that effect. For example, mirtazapine may cause sedation at low doses, which may then dissipate at higher doses (Preskorn 2000b). Optimum response to bupropion occurs at steady-state plasma levels of 50–100 ng/ml, while higher levels of both parent compound and its metabolites are associated with a poorer outcome (Golden et al. 1985). Nefazodone and venlafaxine have ascending dose–antidepressant response curves (Preskorn 1995, 1999). However, the short half-lives of both nefazodone and trazodone (approximately 4 h) mean that even small differences in the sample timing result in substantial changes in observed plasma levels of the parent compounds. Thus, the short half-life of these drugs makes it methodologically difficult to test for a concentration–response relationship.

1.2
Organization of the Chapter

For each antidepressant discussed in this chapter, we briefly review its chemistry, pharmacology, metabolism, safety and adverse effects, clinical use, and therapeutic indications. For a more detailed discussion of the pharmacokinetics of these agents, see the chapter on "General Principles of Pharmacokinetics" by Preskorn and Catterson (this volume). For a discussion of the use of therapeutic drug monitoring of antidepressants, see the chapter by Burke and Preskorn. Summary tables on dosing, adverse effects, pharmacokinetic parameters, and plasma concentrations are provided in the Appendix.

2
Bupropion

2.1
Chemistry, Pharmacology, and Metabolism

Information concerning the pharmacodynamics and pharmacokinetics of bupropion is fairly limited, primarily because of the age of this antidepressant. Clinical trials of bupropion were begun in the mid-1970s and the drug was approved before fluoxetine, although marketing was delayed because of the risk of seizures (Soroko et al. 1977; Preskorn and Katz 1989; Preskorn 1991, 2000a; Silkey et al. 2000). Consequently, when bupropion was in clinical trials, much of the technology needed to study the pharmacodynamics and pharmacokinetics of central nervous system drugs did not exist, nor were such studies required for drug approval at that time.

Bupropion is a weak dual uptake inhibitor of both dopamine (DA) and NE. Bupropion is structurally related to catechols and hence to psychostimulants such as diethylpropion (Tenuate) and amphetamine. Consistent with its structure, in animals bupropion blocks the uptake of both DA and NE. Animals will self-administer bupropion, as they will other amphetamine-like drugs.

Bupropion's weak affinity for the DA and NE uptake pumps has raised questions about whether either of these mechanism of action is relevant to its antidepressant efficacy. However, the parent drug, which has a half-life of 8–10 h, is biotransformed by oxidative metabolism into three active metabolites: hydroxybupropion, threohydrobupropion, and erythrohydrobupropion, all of which have half-lives of 24 h or more and thus accumulate to a greater extent than the parent drug (Schroeder 1983; Golden et al. 1985; Preskorn and Katz 1989; Martin et al. 1990; Ascher et al. 1995). When the combined plasma concentrations of bupropion plus its three active metabolites are considered, it is likely that clinically relevant inhibition of both of the DA and NE uptake pumps occurs under usual antidepressant dosing conditions. Consistent with its weak effects on DA and NE uptake, the combined levels of bupropion and metabolites needed for antidepressant efficacy are in the microgram/ml range vs the nanogram/ml range for almost all other antidepressants (Preskorn 2000a). The total concentration of bupropion plus its metabolites far exceeds the concentration of any other antidepressant except nefazodone and its metabolites. The high concentration is consistent with its relatively weak in vitro affinity for 5-HT_{2A} receptors and the SE uptake pumps. It should also be noted that there are substantial interindividual variability in plasma levels of bupropion so that patients on the same dose of bupropion may have quite different plasma levels.

Cytochrome P450 (CYP) 2B6 is responsible for the conversion of bupropion to hydroxybupropion (Ekins et al. 1999). The mechanisms responsible for converting bupropion to its two other metabolites are unknown.

2.2
Safety and Adverse Effects

2.2.1
Drug–Drug Interactions

As noted above, because of the age of bupropion, knowledge of its potential to cause pharmacokinetically mediated drug–drug interactions is limited. There are some data from case reports that coadministration of fluoxetine may increase levels of two of bupropion's metabolites (Preskorn 1991). However, there have also been case reports that bupropion can substantially affect the metabolism of imipramine (Shad and Preskorn 1997).

2.2.2
Adverse Effects

The most serious side effect associated with bupropion is seizures. The high concentrations of bupropion and its metabolites needed for antidepressant efficacy mean that bupropion has a narrow therapeutic index in terms of the dose needed for antidepressant efficacy vs the dose that causes seizures. The minimum recommended dose for antidepressant efficacy is 300 mg/day, and some patients will need the maximum dose of 450 mg/day to achieve a response. Davidson (1989) found that the risk of seizures with bupropion was higher at doses greater than the maximum recommended dose of 450 mg/day; doubling the dose to 900 mg/day produces a fivefold increase in the seizure risk. An increased risk of seizures was found in patients with eating disorders (i.e., bulimia), which led to a delay in the marketing of the drug in the 1980s. The risk of seizures with the immediate release (IR) formulation is estimated to be 4/1,000 at doses of 450 mg/day, while the risk of seizures with the sustained release formulation may be as low as 1/1,000 in patients with no pre-existing history or risk factors at doses of 400 mg/day. The occurrence of seizures appears to be concentration dependent, given that (1) the occurrence of seizures is dose dependent, (2) seizures tend to occur within a few days of a dose increase and within a few hours of a dose, and (3) that patients with a lean body mass (e.g., with bulimia) appear to be at higher risk for seizures. Given the dose-dependent nature of seizures, patients who seize on lower doses are probably slow metabolizers who accumulate unusually high levels of bupropion or one or more of its three active metabolites (Preskorn 1991). This suggests that therapeutic drug monitoring (TDM) might play a role in minimizing the risk of seizures with bupropion but TDM with bupropion has not been adequately studied (see the chapter "Therapeutic Drug Monitoring of Antidepressants" by Burke and Preskorn, this volume). Particular care should be taken when switching from an antidepressant that substantially inhibits specific drug-metabolizing CYP enzymes (e.g., fluoxetine, fluvoxamine, nefazodone, or paroxetine) to bupropion or when adding bupropion to the treatment regimen of patients taking one of these antidepressants or other drugs that substantially inhibit CYP enzymes (e.g., macrolides, fluoroquinoles, antifungals, protease inhibitors).

The other main adverse effects that occur with bupropion are associated with its DA and NE uptake inhibition: restlessness, activation, tremors, insomnia, and nausea (Preskorn and Othmer 1984; Preskorn 1995).

Bupropion causes few, if any, anticholinergic, antihistaminic, or orthostatic hypotensive effects. It is nonsedating and does not cause weight gain, nor does it have any effect on intracardiac conduction. In contrast to the SSRIs and venlafaxine, bupropion causes minimal or no sexual dysfunction, so that it may be a useful alternative or adjunct for patients who experience sexual side effects with other antidepressants (Labbhate et al. 1997; Rowland et al. 1997; Ashton and Rosen 1998).

2.2.3
Overdose

Bupropion is safer in overdose than the TCAs; death is a rare outcome with an overdose of bupropion alone since the drug has no adverse effects on the cardiovascular or respiratory systems (Preskorn and Othmer 1984). The main adverse effect associated with an overdose of bupropion is an increased risk of seizures (Davidson 1989).

2.3
Clinical Use/Therapeutic Indications

2.3.1
Efficacy for Acute Treatment

Bupropion has been found to be an effective antidepressant in several double-blind studies; however, most of the studies used doses higher than the maximum recommended dose of 450 mg/day (Fabre et al. 1983; Meredith and Feighner 1983; Musso et al. 1993).

Bupropion was originally approved based on clinical efficacy studies that used doses as high as 900 mg/day. However, after its approval, the maximum recommended daily dose was capped at 450 mg/day because of concern about seizures. Reanalysis of the data from only those patients who had received no more than 450 mg/day in the clinical trials indicated that the response was not as robust in this subgroup of patients, but the FDA permitted marketing of the product based on this information.

Approval of the sustained release (SR) version of bupropion was based on three double-blind studies. The FDA concluded that the three studies all failed to demonstrate that the sustained release version of bupropion in doses of 400 mg/day or less was superior to placebo in the treatment of outpatients with major depression (bupropion summary basis of approval). Nevertheless, the SR formulation was approved based on bioequivalence with the IR formulation, possibly because it was believed that the SR formulation would be less likely to cause seizures than the IR formulation at comparable doses.

Despite the fairly modest evidence of bupropion's antidepressant efficacy, it may be a useful option in a number of clinical situations. Because of its activating properties, bupropion may be particularly useful for patients with prominent psychomotor retardation. Given its DA agonistic properties, bupropion may be uniquely helpful for patients with Parkinson's disease as well as those with attention-deficit/hyperactivity disorder (Wender 1998). In the original clinical trials with bupropion, it was found to be efficacious in patients who historically had not responded to TCAs (Stern et al. 1983) and may also be helpful for patients who do not respond to SE reuptake inhibitors (SRIs). Bupropion may also be a good option of patients who cannot tolerate the sexual side effects of other antidepressants (Hirschfeld 1999).

Finally, bupropion has also been approved as an aid in smoking cessation (Package insert for Zyban, Physicians' Desk Reference, 2003). Given that smokers have an increased incidence of depression and that clinical depression impairs the ability to stop smoking, bupropion may serve a dual role both to treat clinical depression and as an aid to stop smoking.

2.3.2
Dosing Recommendations

The maximum recommended doses of the IR and SR formulations are 450 and 400 mg/day, respectively. The difference is due to the strengths of the IR and SR dosage forms. The IR formulation should be given three times a day with doses administered at least 4 h apart to minimize the risk of seizures (Davidson 1989). The SR formulation should be given twice a day, with doses at least 4 h apart. In selecting the dose, the goal is to aim for the lowest possible dose to minimize the risk of seizures without compromising therapeutic efficacy.

2.3.3
Onset of Action

Based on the published double-blind studies, the time to onset of bupropion's antidepressant efficacy (as defined by statistical separation from a parallel group treated with placebo) is approximately 2 weeks, which is comparable with most other antidepressants.

3
Mirtazapine and Mianserin

3.1
Chemistry, Pharmacology, and Metabolism

Mirtazapine (6-azo-mianserin) and its forerunner, mianserin, are tetracyclic compounds that differ from other antidepressant classes in terms of the mechanism thought to be responsible for their antidepressant efficacy (Preskorn 1997). Mianserin, the older drug of the two, is marketed in several countries around the world but not in the United States; mirtazapine has been marketed in the United States since August 1996, and is also available in several other countries.

The mechanism of action of mirtazapine is unique. It is an α_2-antagonist that produces increases in noradrenergic and serotonergic neurotransmission, which is the primary mechanism thought to underlie its antidepressant activity (deBoer et al. 1995). It does not block the uptake pump for SE, NE, or DA. Mirtazapine's most potent site of action is the histamine-1 receptor. At higher concentrations, the drug first blocks the 5-HT$_{2A}$ receptor and then the α_2-adrenergic receptor. Blockade at the α_2-receptor increases NE release by a direct effect on the presynaptic α_2-adrenergic autoreceptor, indirectly increasing SE release due to increasing noradrenergic input to raphe neurons. Mirtazapine also binds the 5-HT$_{2C}$ and 5-HT$_3$ receptors nearly as avidly as it binds the 5-HT$_{2A}$ receptor (deBoer et al. 1995). Doses of mirtazapine required to treat clinical depression block all of these receptors (i.e., one or more of these mechanisms may mediate its antidepressant efficacy). Unlike the tertiary amine TCAs (e.g., amitriptyline), mirtazapine has minimal affinity for muscarinic cholinergic receptors and α_1-adrenergic receptors. Unlike the SSRIs, venlafaxine, and duloxetine, mirtazapine does not directly affect the uptake of SE or NE.

The half-life of mirtazapine is 20–40 h, which means that steady-state levels can be achieved with once-daily dosing at bedtime, thus taking advantage of the sedative and sleep enhancing qualities of the drug (Voortman and Paanakker 1995).

Mirtazapine is eliminated primarily through biotransformation that is mediated by CYP 1A2, CYP 2D6, and CYP 3A3/4 (Montgomery 1995; Verhoeven et al. 1996; Stormer et al. 2000; Timmer et al. 2000); about 25% of the elimination of mirtazapine is dependent on a phase II conjugation reaction with glucuronic acid.

Mirtazapine has linear kinetics across its clinically relevant dose range (15–80 mg/day; Timmer et al. 1995), suggesting that mirtazapine does not inhibit the three CYP enzymes responsible for its biotransformation. This conclusion is further supported by in vitro studies showing that mirtazapine is a weak inhibitor of these CYP enzymes (Delbressin et al. 1997).

Mirtazapine may have a curvilinear dose–response curve in terms of its sedative effect. This effect may be more pronounced at lower (15 mg) than at higher (30 mg/day or more) doses. This claim is based principally on comparisons of the adverse effect rates between European and American trials with the drug. Specifically, there was a higher incidence of sedation in American trials, which used lower doses of mirtazapine, than in European trials, which used higher doses. There is also a theoretical basis for this claim: the sedative effect of mirtazapine at low doses is consistent with its high affinity for the histamine-1 (H-1) receptor. In fact, histamine binding to a physiologically relevant degree occurs at lower doses of mirtazapine (i.e., <15 mg/day) than the doses need for the binding of the receptors responsible for its antidepressant actions. At higher doses, however, blockade of the α_2-adrenergic receptors occurs and can produce arousal effects like those seen with yohimbine. Thus, as a result of α_2-adrenergic antagonism, higher doses would be expected to counteract the sedation produced by lower doses of mirtazapine. To date, no fixed dose trials have been published that rigorously test this concept.

3.2
Safety and Adverse Effects

3.2.1
Drug–Drug Interactions

Given the multiple and approximately equal pathways for the biotransformation and elimination of mirtazapine (Preskorn 1997), this drug is relatively protected against being the victim of a clinically meaningful CYP enzyme-mediated drug–drug interaction. Nevertheless, fluvoxamine has been reported to cause a three- to fourfold elevation in mirtazapine plasma levels (Antilla et al. 2001). Of interest, a case of SE syndrome has been reported following the coadministration of fluvoxamine and mirtazapine (Demers and Malone 2001).

In one study (Preskorn et al., in press), the potential for drug–drug interactions was examined following a crossover from fluoxetine to mirtazapine. Fluoxetine was the SSRI of choice due to its long half-life and inhibition of multiple CYP isoenzymes. In this study, patients ($n=40$) who had not benefited from or tolerated fluoxetine were crossed over to open-label treatment with mirtazapine. As expected, in addition to measurable mirtazapine concentrations, measurable and pharmacologically relevant plasma concentrations of fluoxetine, and norfluoxetine were present for several weeks after fluoxetine had been discontinued. Despite these findings, none of the patients developed symptoms of the SE syndrome, and their adverse effects were qualitatively and quantitatively the same as would be expected for patients on mirtazapine alone, 15 mg/day. Fluoxetine did not significantly alter the plasma concentration of mirtazapine, consistent with the multiple pathways responsible for its elimination.

3.2.2
Adverse Effects

Mirtazapine does not cause many of the side effects associated with the SSRIs. It does not cause nausea or loose stools, disturb sleep pathology [it actually increases sleep efficiency (Frazer 1997)], and does not cause sexual dysfunction (Preskorn 1999). Mirtazapine causes minimal anticholinergic effects, no quinidine-like effects, and no effects on blood pressure, including orthostatic hypotension (Janicak et al. 2001).

Consistent with its blockade of the H-1 receptor, sedation can be a problem, especially in early treatment with mirtazapine (Preskorn 1997); some clinicians therefore start with a lower dose (e.g., 7.5 mg) but this strategy would be predicted to aggravate rather than avoid sedation and compromise efficacy (Nelson 1997). Sedation was listed as the reason for early termination in 10% of patients in the double-blind registration studies for the drug (Mirtazapine Package Insert, Thompson PDR, 2003). However, the sedation associated with mirtazapine may also be an advantage for patients with prominent insomnia, anxiety, or agitation (Wheatley et al. 1998). As noted above, higher doses may theoretically decrease the sedative effects of mirtazapine.

Mirtazapine can cause increased appetite and weight gain, transient neutropenia (which typically presents and resolves within the first 6 weeks of treatment), and transient mild elevations of liver function tests, particularly serum glutamic pyruvic transaminase (SGPT), (which typically presents and resolves within the first 6 weeks of treatment) (Janicak et al. 2001). Three cases of agranulocytosis out of 3,000 patients were reported in the clinical trial program for mirtazapine (Nelson 1997; Preskorn 1997). This incidence was too low to draw any conclusion about cause and effect, and postmarketing experience with mirtazapine has not indicated an unusual number of cases of agranulocytosis with this drug. Nevertheless, the package insert contains a warning that a white blood cell count should be done if a patient taking mirtazapine develops signs of fever or infection (Preskorn 1999).

3.2.3
Overdose

Mirtazapine, like the other newer antidepressants, has a wide therapeutic index and is generally safe in overdose (Nelson 1997; Bremner et al. 1998).

3.3
Clinical Use/Therapeutic Indications

3.3.1
Efficacy for Acute Treatment

The approval of mirtazapine in the United States was based on six double-blind, placebo- and amitriptyline-controlled studies, which found mirtazapine to be superior to placebo and comparable in efficacy to amitriptyline in the treatment of depression (Zivkov et al. 1995; Fawcett and Barkin 1997). A double-blind crossover study found that 63% of patients who were nonresponders to 6 weeks of double-blind treatment with amitriptyline responded to mirtazapine (Catterson and Preskorn 1996). Mirtazapine was found to be efficacious in the treatment of patients hospitalized for major depressive disorder in two studies; in the first study, it was found to be comparable to amitriptyline and superior to placebo (Montgomery et al. 1998), and in the second study, it was found to be superior to fluoxetine (Wheatley et al. 1998).

Three double-blind studies have found mianserin to be superior to placebo, but 15 studies have found it slightly less effective than tertiary amine TCAs (Janicak et al. 2001).

Because of its unique mechanism of action, which is not shared with any other types of antidepressant, mirtazapine may be efficacious for patients who have not benefited from other types of antidepressants, and there is some evidence supporting this possibility (Catterson and Preskorn 1996; Wheatley et al. 1998).

3.3.2
Dosing Recommendations

Mirtazapine can be started at a clinically effective dose and can be taken once a day. However, no double-blind fixed-dose studies have been published to establish the optimal dose of mirtazapine.

Liver failure can lead to a 30% decrease and renal failure to a 50% decrease in the clearance of mirtazapine (Organon, data on file). Accordingly, dosage adjustments may be needed for populations of depressed patients with liver or renal failure.

3.3.3
Onset of Action

Based on the published double-blind studies, the time to onset of mirtazapine's antidepressant efficacy (as defined by statistical separation from a parallel group treated with placebo) is approximately 2 weeks, which is comparable with most other antidepressants.

4
Nefazodone and Trazodone

4.1
Chemistry, Pharmacology, and Metabolism

Nefazodone and trazodone have chemically related structures; both incorporate 5-HT$_{2A}$ receptor blockade plus weak 5-HT uptake blockade and possibly NE uptake blockade (Yocca et al. 1985; Feighner et al. 1989; Fontaine et al. 1991; Weise et al. 1991).

More prescriptions are written for trazodone than for nefazodone, which most likely reflects the widespread use of trazodone as a nonhabit-forming sleep aid rather than as an antidepressant. Nefazodone is a structural analog of trazodone and was designed with the goal of producing a better antidepressant than trazodone (Yocca et al. 1985). Nefazodone is a more potent SRI than trazodone but is a much weaker SRI than the SSRIs or venlafaxine (Bolden-Watson and Richelson 1993; Frazer 1997). It is likely that nefazodone only produces SE (5HT$_{2A}$) inhibition at doses of 300 mg/day or above, and even at 500 mg/day it does not appear to produce the same degree of SE reuptake inhibition as the SSRIs and venlafaxine at their usual starting doses (Narayan et al. 1998). The pharmacology of nefazodone is consistent with its clinical advantages and disadvantages as described below.

4.2
Safety and Adverse Effects

4.2.1
Drug–Drug Interactions

Because of its ability to substantially inhibit CYP 3A3/4, nefazodone, like bupropion, fluoxetine, fluvoxamine, and paroxetine, has the potential to cause clinically meaningful CYP enzyme mediated adverse drug–drug interactions. Nefazodone-induced inhibition of CYP 3A3/4 can elevate levels of co-prescribed drugs that are dependent on this CYP enzyme for their oxidative metabolism (Preskorn 1999; Shad and Preskorn 2000).

4.3
Adverse Effects

Nefazodone produces less activation and sexual dysfunction than the SSRIs and venlafaxine, and it does not cause blood pressure elevation (Janicak et al. 2001). Nefazodone does not disturb sleep physiology and improves the subjective quality of sleep (Armitage et al. 1994; Gillin et al. 1997).

The incidence and severity of the following adverse effects increase as a function of the starting dose of nefazodone: dizziness/lightheadedness, confusion, sedation, and gastrointestinal side effects (Preskorn 1999).

The antihistaminergic properties of trazodone are partly responsible for its popularity as a sleep aid, but cause significant problems with daytime sedation when it is used as an antidepressant (Preskorn 1999). Because it is sedating, it may also be useful for treating agitation in geriatric patients (Janicak et al. 2001). Trazodone is not antiarrhythmic, it may induce or exacerbate ventricular arrhythmia (not proved), may cause priapism, and postural hypotension. However, it causes no anticholinergic effects and it causes no quinidine-like effects (Janicak et al. 2001).

4.3.1
Overdose

Nefazodone has good safety in overdose due to its wide therapeutic index (Robinson et al. 1996a). Trazodone is also relatively safe in overdose.

4.4
Clinical Use/Therapeutic Indications

4.4.1
Efficacy for Acute Treatment

Unlike the clinical trials of the SSRIs and venlafaxine, a true fixed-dose study of nefazodone was done but never published. This study had major limitations: the doses used were low (50–300 mg/day) compared with doses that were found to be optimally effective in later studies and the number of subjects was small.

Later studies used a targeted fixed "dose-range" design [i.e., low dose=50–300 mg/day and high dose=300–600 mg/day (Robinson et al. 1996b; Janicak et al. 2001)]. Patients were randomly assigned to the lowest dose in either range and the investigator could then titrate the patient within the assigned ranges to maximize treatment outcome. This design, while reasonable and valid, does not permit as straightforward an assessment of the dose–response curve as do conventional fixed-dose studies. This difference should be kept in mind, particularly when comparing the antidepressant effect of nefazodone with that of other antidepressants that were tested in strict fixed-dose designs. The absence of true fixed-dose studies probably reflects the need for multi-step dose titration, which has limited the clinical acceptance of nefazodone.

In a series of double-blind trials comparing nefazodone with placebo and other antidepressants using such fixed dose-range designs (D'Amico et al. 1990; Presentation to the FDA Psychopharmacology Advisory Council July 1993; Fontaine et al. 1994; Rickels et al. 1994; Mendels et al. 1995; Robinson et al. 1996b), nefazodone was found to be superior to placebo and equivalent to imipramine and fluoxetine in efficacy. A meta-analysis of efficacy studies with nefazodone examined the magnitude of the antidepressant response as a function of the average daily dose in an attempt to establish an antidepressant response curve. This meta-analysis indicated that the magnitude of antidepressant effect at doses of 300–400 mg/day is approximately comparable to that seen with the SSRIs and lower doses of venlafaxine, indicating that nefazodone used at these doses is effective in the same percentage of patients as other antidepressants. However, at doses of 500 mg/day, the magnitude of the response exceeded that seen with the SSRIs, but was less than that seen with high-dose venlafaxine (i.e., 225–375 mg/day). Response to nefazodone at 600 mg/day was virtually the same as to placebo. This finding is likely to be an artifact of the study design in which the clinician could titrate nonresponders to the highest tolerated dose producing a curvilinear dose–response curve with a diminution in response and increase in dropouts due to adverse effects at higher doses. Nevertheless, these data are consistent with the conclusion that there is generally no advantage in exceeding 500 mg/day of nefazodone.

Nefazodone appears to have efficacy in patients with clinical depression and prominent anxiety (Fawcett et al. 1995).

There appears to be a greater degree of interpatient variability in response to nefazodone than occurs with many of the other newer antidepressants. Nevertheless, nefazodone can be a useful option for patients who are unable to tolerate the adverse effects of the SSRIs.

Several random-assignment, well-controlled clinical trials have found that trazodone is 32% more effective than placebo and 4% more effective than tertiary TCAs (Janicak et al. 2001).

4.4.2
Dosing Recommendations

Tolerability problems limit the starting dose of nefazodone (Preskorn 1995). Stepwise dose titration reduces the frequency and severity of side effects by allowing patients to develop some tolerance. The package insert (Physician's Desk Reference 2003) recommends starting nefazodone at 100 mg twice a day for 1 week before increasing the dose in increments of 100–200 mg/day at intervals of no less than 1 week. Using this strategy, the dose of nefazodone can be increased as needed to a maximum of 600 mg/day in equally divided doses administered two or three times a day.

While the therapeutic dose of nefazodone is listed as 200–600 mg/day, the daily dose that is usually needed for most patients to achieve an adequate response appears to be in the 400–500 mg/day range given in equally divided doses two to four times a day. As noted above, nefazodone requires at least twice-a-day dosing and careful dose adjustment. The development program for a sustained release formulation was discontinued. There appears to be more variability between patients in the optimal dose of nefazodone for efficacy and tolerability compared with other antidepressants (Janicak et al. 2001). Antidepressant efficacy at the initial starting dose is less robust than with other antidepressants, and dose titration is needed to improve tolerability and increase efficacy (Janicak et al. 2001).

While trazodone has fewer anticholinergic adverse effects compared with the tricyclic antidepressants, excessive sedation can limit dose titration. Generally, the use of trazodone is now limited to single bedtime doses of 50–200 mg as a soporific agent to counteract the insomnia that can be produced by 5-HT uptake inhibitors and/or as an augmentation strategy with these medications.

4.4.3
Onset of Action

Nefazodone has been found to produce a statistically significant improvement in sleep and anxiety symptoms in patients with major depression within 1 week (Fawcett et al. 1995). However, full antidepressant response usually takes a period comparable to that needed with the SSRIs. There is no evidence that higher doses produce more rapid onset.

5
Reboxetine

5.1
Chemistry, Pharmacology, and Metabolism

Reboxetine, (RS)-2-[(RS)-α-(2-ethoxyphenox) benzyl] morpholine methanesulfonate, is a chiral compound and is marketed as a racemic mixture of (R,R)- and (S,S)-reboxetine in a number of countries but not in the United States (Frigerio et al. 1997). It is not a tricyclic antidepressant. Nevertheless, pharmacologically reboxetine is a selective NE reuptake pump inhibitor (Raggi et al. 2002). (S,S)-reboxetine is the more potent inhibitor of the NE uptake pump (Fleishaker 2000). There are pharmacokinetic and toxicological differences between the two enantiomers as described latter in this section (Raggi et al. 2002).

There is considerable preclinical animal evidence supporting an in vivo selective effect of reboxetine on the NE uptake pump with secondarily mediated effects on other biogenic amine systems, particularly DA. Osmotic pump infusion of reboxetine produced a 600% and 342% increase in extracellular NE and DA, respectively, in the prefrontal cortex of rats (Invernizzi et al. 2001). Administration of clonidine reduced this effect but this reduction was attenuated after 14 days compared to 2 days of continuous reboxetine infusion consistent with a reboxetine-induced desensitization of terminal α_2-adrenergic receptors (Invernizzi et al. 2001). Intravenous administration of reboxetine produced a dose-dependent decrease in the firing of NE in the locus coeruleus but not SE neurons in dorsal raphe, and this effect persisted and was enhanced by chronic treatment up to 21 days (Szabo and Blier 2001a). This result is consistent with a selective effect on NE but not SE uptake pumps. The dose-dependent magnitude of the reduction in firing rate following 1.25, 2.5, 5.0, and 10.0 mg/kg was 52%, 68%, 81%, and 83%, respectively. Cumulative intravenous injections of reboxetine increased the recovery time of CA3 pyramidal neurons in the hippocampus following iontophoretic application of NE but not SE consistent with a selective effect on the NE uptake pump (Szabo and Blier 2001b). However, administration of the selective 5-HT_{1S} receptor antagonist WAY 100,635 induced a 140% increase in basal pyramidal neuronal firing in rats treated with reboxetine compared with those treated with saline, suggesting a possible tonic activation of the postsynaptic 5-HT_{1A} receptors that is possibly attributable to the known effect of reboxetine of desensitizing α_2-adrenergic heteroreceptors (Szabo and Blier 2001b). These results suggest that a selective NE uptake pump inhibitor can nevertheless have secondary effects on SE neurotransmission. Despite being a selective NE uptake pump inhibitor, systemic administration of reboxetine in rats produced an increase in the burst firing but not the average firing frequency of DA neurons in the ventral tegmental area (VTA) and enhanced DA output in the medial prefrontal cortex but not the nucleus accumbens (Linner et al. 2001). These results suggest that reboxetine can facilitate both prefrontal DA output and excitability of VTA DA neurons, which might explain its effects on drive and motivation observed in clinical trials. Consistent with NE uptake pump inhibition as a mechanism mediating its antidepressant effects, reboxetine was ineffective in the forced swim test in mice genetically bred to be unable to synthesis NE and epinephrine, even though it was effective in normal mice (Cryan et al. 2001). Chronic administration of reboxetine also causes a net deamplification of the NE-mediated signal transduction cascade through a desensitization of the β-adrenergic receptor-coupled adenylate cyclase system and decreases nuclear cAMP-response element binding protein (CREB)-P, indicating a direct effect beyond the β-adrenergic receptors themselves (Manier et al. 2002). Systemic administration of reboxetine in rats has been shown to inhibit nicotinic acetylcholine receptors through an as-yet-unexplained mechanism other than an apparent direct effect on these receptors (Miller et al. 2002).

In man, reboxetine at doses of 4 and 8 mg has been shown to enhance cortical excitability as assessed by transcranial stimulation in the absence of changes in motor threshold, intracortical inhibition, M-response, and F-wave or H-reflexes (Anonymous 2002). In a placebo-controlled trial, reboxetine at doses of 4 and 8 mg/day in healthy adult humans caused a dose-dependent decrease in recall and recognition memory performance for emotional elements of a series of slides accompanied by a narrative (Papps et al. 2002). This effect was similar to that seen with β-adrenergic receptor antagonists and the opposite of that seen with yohimbine and the opposite of that predicted by the NE uptake pump inhibition produced by reboxetine. The results of this study, if replicated, require further work to understand the mechanism underlying this effect. In an interesting, double-blind, random assignment study of the differential social behavioral effects of reboxetine, a NE selective reuptake inhibitor, vs citalopram, an SSRI, in 60 normal volunteers it was found that subjects treated with reboxetine had reduced hand fiddling and more cooperative communication during a mixed-motive game and were more likely to be classified as cooperative players in comparison to subjects treated with citalopram, who conversely showed more evidence of protection of self from the negative consequences of social interaction (Tse and Bond 2002).

Based on in vitro studies using human liver microsomes, the metabolism of both reboxetine enantiomers is principally mediated by CYP 3A (Wienkers et al. 1999). At concentrations achieved on clinically used doses, reboxetine has no effect in vitro on CYP 1A2, 2C9, 2D6, 2E1, or 3A activity (Fleishaker 2000). Based on the in vivo study in man of the effect of chronic administration of reboxetine on the excretion rate of 6-β-hydroxycortisol (a model CYP 3A substrate), there was no evidence for significant induction of CYP 3A (Pellizzoni et al. 1996).

In vivo, the ratio of the area under the curve concentration–time curves (AUC) for the (S,S)- to (R,R)-enantiomers in man is approximately 0.5 and there is no evidence of chiral inversion (Fleishaker 2000). Differences in the clearance of the two enantiomers may be explained by differences in protein binding. Racemic reboxetine is greater than 97% protein bound principally to α_1-acid glycoprotein (Edwards et al. 1995). Reboxetine is well absorbed after oral administration with absolute bioavailability of approximately 95% (Fleishaker et al. 1999). Maximal concentrations are generally achieved 2–4 h after oral ingestion. Food affects the rate but not the extent of absorption. Reboxetine in man has linear pharmacokinetics over its clinically relevant dosing range (Dostert et al. 1997; Rey et al. 1999). It is primarily biotransformed prior to renal elimination with less than 10% excreted unchanged. Reboxetine is the primary circulating species in plasma, but a number of metabolites are formed by hepatic biotransformation. Clearance is principally via the kidneys with less than 10% excreted unchanged. Gender does not affect the pharmacokinetics of either of the reboxetine enantiomers (Fleishaker et al. 1999). The unbound clearance of reboxetine was statistically significantly less for Asians compared with blacks or Caucasians [3,742+/−1,468; 5,187+/−2,027; and 5,294+/−1,163 ml/min, respectively (Hendershot et al. 2001)]. However, the differences in AUCs were modest and unlikely to require dosage adjustment for ethnicity. Reboxetine plasma levels (AUCs) were four times higher and $t_{1/2}$ twice as long (24 vs 12 h) in elderly but otherwise healthy volunteers (mean age 81+/−9 years) compared to young volunteers (Bergmann et al. 2000). Similar results were found in elderly (80+/−4 years) females patients hospitalized for major depression or dysthymic disorder (Poggesi et al. 2000). Comparable changes in reboxetine clearance were observed in plasma levels of volunteers with severe renal impairment (9–19 ml/min) (Coulomb et al. 2000) and in volunteers with moderate to sever hepatic dysfunction (Fleishaker 2000).

5.2
Safety and Adverse Effects

5.2.1
Drug–Drug Interactions

Pharmacodynamically, reboxetine would be predicted to interact with other drugs capable of promoting NE function. Due to its ability to block the neuronal uptake of tyramine, reboxetine theoretically might protect against the risk of a hypertensive crisis when patients on monoamine oxidase inhibitors consumed a tyramine-rich meal (Dostert et al. 1997). However, the hypertensive crisis, if it did occur, would be predicted to be more severe and last longer in the presence of reboxetine. For this reason, it would be prudent to avoid such a combination until there are formal studies to asses the relative risk.

The in vivo human pharmacokinetic drug–drug interaction studies are consistent with the predictions based on the in vitro metabolism studies. In man, coadministration of 200 mg of ketoconazole increased the AUC of (R,R)- and (S,S)-enantiomers of reboxetine by 58% and 43%, respectively, and decreased their oral clearance by 34% and 24%, respectively (Herman et al. 1999). Conversely, neither quinidine nor fluoxetine, both substantial CYP 2D6 inhibitors, affected the clearance of either enantiomer (Fleishaker 2000). Conversely, coadministration of reboxetine did not affect the clearance of either alprazolam or dextromethorphan, model substrate/drugs for CYP 3A and 2D6, respectively (Fleishaker 2000) nor clozapine or risperidone (Spina et al. 2001).

5.2.2
Adverse Effects

Consistent with its pharmacology, the most common adverse effects of reboxetine compared with placebo were insomnia, sweating, constipation, dry mouth, and urinary hesitancy; compared with fluoxetine, reboxetine had lower rates of nausea, diarrhea, and somnolence (Scates and Doraiswamy 2000). In short-term (4- to 6-week) studies, the discontinuation rate in patients on reboxetine up to 8 mg/day was comparable to placebo (Tanum 2000). In an open-label, forced titration study to 8 mg/day in 12 elderly patients (75–87 years of age), one patient discontinued prematurely and two were held at 6 mg/day because of changes in cardiac rhythm. A total of 5 out of 11 (including the 2 held at 6 mg/day) experienced tachycardia (Andreoli et al. 1999). In contrast to SE uptake drug inhibitors, there is no evidence of a discontinuation syndrome following abrupt discontinuation of reboxetine and minimal spontaneous reports of adverse effects on decreased libido or anorgasmia (Tanum 2000). Consistent with its effect as a NE uptake pump inhibitor, there have been case reports of prolonged orgasm of reduced intensity, seminal emission after defecation accompanied by pain, spontaneous and painful ejaculation, and urinary hesitancy and retention (Andreoli et al. 1999; Benazzi 2000; Kasper 2000; O'Flynn and Michael 2000; Demyttenaere and Huygens 2002). These effects can also be seen with other antidepressants capable of inhibiting the NE uptake pump.

5.2.3
Overdose

There is limited published data on the safety of reboxetine in overdose. A placebo-controlled, crossover study in 20 healthy young volunteers revealed no statistically significant prolongation of the QTc (Fridericia correction) after treatment with 2, 4, and 6 mg of reboxetine alone and 6 mg in combination with 200 mg of ketoconazole twice daily for 7 days (Fleishaker et al. 2001). That result is consistent with no effect on cardiac repolarization and suggests cardiac safety in overdose. Consistent with this finding, a 71-year-old woman recovered uneventfully following an acute ingestion of 48 mg of reboxetine in a suicide attempt (Agelink et al. 2002). Nevertheless, more data on safety in overdose would be reassuring.

5.3
Clinical Use/Therapeutic Indications

5.3.1
Efficacy for Acute Treatment

Sufficient evidence of the efficacy of reboxetine in major depression has not been presented to receive approval for marketing in the United States, but reboxetine is available in several other countries around the world. Consistent with the fact that it was principally developed in Europe, most of the clinical trials with reboxetine are active but not placebo-controlled. To be a positive study (i.e., disprove the null hypothesis), the investigational drug in such studies must produce superior efficacy to the comparator. That has not been in the case in the majority of such studies with reboxetine and likely contributed to its failure to date to receive approval in the United States.

The publications with reboxetine are somewhat perplexing. Papers reporting on meta-analyses indicate that there have been four short-term (4- to 8-week) trials of reboxetine vs placebo totaling 703 patients (Ferguson et al. 2002); however, a PubMed search yielded only two published trials involving approximately 300 patients (Versiani et al. 1999; Andreoli et al. 2002). The meta-analysis reported improvements in psychomotor retardation, cognitive disturbance, and anxiety but did not mention improved overall antidepressant response, suggesting that perhaps there was no difference between this presumably positive efficacy outcome. In a published 8-week, double-blind, random assignment, multicenter trial involving 381 patients assigned to reboxetine, 8–10 mg/day, fluoxetine, 20–40 mg/day, or placebo, reboxetine and fluoxetine produced a statistically significant higher response and remission rate (Andreoli et al. 2002). A 6-week, double-blind, random assignment, placebo-controlled trial involving 52 hospitalized patients found a response rate of 74% in patients treated with reboxetine vs 20% in patients treated with placebo ($p<0.001$), with 64% of placebo patients vs 15% of reboxetine-treated patients withdrawn prematurely due to lack of efficacy (Versiani et al. 2000). In a double-blind relapse prevention study, patients who responded to 6 weeks of open label treatment with reboxetine were randomly assigned to either stay on reboxetine or be switched to placebo (Versiani et al. 1999). At the end of 6 months, there was a statistically significant lower relapse rate in patients treated with reboxetine vs placebo: 22% vs 56%, respectively ($p<0.001$).

The other published studies with reboxetine were not placebo-controlled. In an 8-week double-blind, random assignment trial in 347 elderly depressed patients, 68% of patients treated with reboxetine vs 71% of patients treated with imipramine experienced adverse events (not statistically different) (Katona et al. 1999). There have also been a number of double-blind, random assignment, antidepressant trials that have not shown an overall statistically superior response rate for reboxetine vs fluoxetine (Massana 1998; Massana et al. 1999; Keller 2001).

Post hoc analyses have proposed superior efficacy for reboxetine vs fluoxetine in more severely depressed patients (Massana et al. 1999) and in social functioning (Massana 1998; Keller 2001). The latter result is consistent with the previously mentioned prospective study of the effect of reboxetine on social functioning in normal volunteers compared with subjects receiving placebo. Nevertheless, post hoc analyses of clinical trials should be considered as hypothesis generating rather than hypothesis confirming.

5.3.2
Dosing Recommendations

The effective dose of reboxetine in the positive clinical trials discussed above has mainly been between 8–10 mg/day. As mentioned above, pharmacokinetic studies suggest that dose reductions are needed to achieve comparable plasma reboxetine levels in the elderly and in those with significant hepatic and renal impairment.

5.3.3
Onset of Action

There is no evidence of a faster onset of action with reboxetine.

6
Venlafaxine

6.1
Chemistry, Pharmacology, and Metabolism

Venlafaxine, a phenylethylamine, first inhibits the neuronal uptake pumps for SE and then the uptake pump for NE (Harvey et al. 2000). Venlafaxine is approximately five times more potent in vitro as an inhibitor of the SE than the NE uptake pump (Bolden-Watson and Richelson 1993). However, unlike tertiary amine TCAs, which also inhibit the SE and NE uptake pumps, venlafaxine has low affinity for most other neural receptors and does not inhibit sodium fast channels, making it relatively safe in overdose. It has been reported that the sequential effects of venlafaxine on the SE and then the NE uptake pumps are dose- and concentration-dependent (Harvey et al. 2000). At 75 mg/day, venlafaxine is predominantly an SRI, like the SSRIs (Preskorn 1999), whereas at 375 mg/day it produces NE uptake inhibition comparable to that of desipramine.

Venlafaxine is biotransformed by CYP 2D6 to its active metabolite, O-desmethylvenlafaxine (ODV). ODV has pharmacological activity comparable to the parent drug and is believed to contribute equally with the parent drug to efficacy and adverse effects; ODV is then metabolized via CYP 3A3/4 (Klamerus et al. 1992; Preskorn and Burke 1992; Troy et al. 1992). The half-life of venlafaxine is approximately 5 h, while the half-life of ODV is approximately 12 h.

In contrast to the SSRIs, which have a flat dose–response curve, venlafaxine has an ascending dose–response curve. This ascending dose–response curve is consistent with venlafaxine's sequential, concentration-dependent effects on the SE and NE uptake pumps. Clinically, this ascending dose–response curve provides a stronger rationale for using higher doses of venlafaxine in patients who have partially responded or not responded to an initial lower starting dose (e.g., 75 mg/day) (Janicak et al. 2001); this is in contrast to the SSRIs, which have a flat dose–response curve and for which, consequently, there is a weaker rationale for increasing the dose above the usually clinically effective dose.

6.2
Safety and Adverse Effects

6.2.1
Drug–Drug interactions

Venlafaxine appears to have a good profile in terms of pharmacodynamically mediated drug–drug interactions (Preskorn 1999). At lower doses, these are mediated by SE reuptake inhibition, whereas at higher doses, venlafaxine is susceptible to the same potential interactions as the secondary amine TCAs, which are NE selective reuptake inhibitors (NSRIs) (for example, hypertensive crisis when combined with other NE agonists, particularly monoamine oxidase inhibitors) (Feighner 1995; Nelson 1997).

Venlafaxine also appears to have a good profile in terms of pharmacokinetically mediated drug–drug interactions (Preskorn 1999). Venlafaxine does not appear to produce substantial inhibition of any drug metabolizing CYP enzyme at its usually effective, minimum antidepressant dose.

Because the effect of venlafaxine appears to be dependent on the combined concentrations of venlafaxine (the biotransformation of which is mediated by CYP 2D6) and its principal metabolite ODV (which is metabolized via CYP 3A3/4), CYP 2D6 deficiency, which occurs in approximately 7% of Caucasians, may have fewer clinical implications for venlafaxine than for other medications (such as paroxetine) that are metabolized by CYP 2D6. However, the co-administration of an agent that is a substantial inhibitor of CYP 3A3/4 could result in a clinically meaningful increase in venlafaxine and ODV levels, particularly in patients who are CYP 2D6 deficient, which could lead to an increase in the incidence or severity of adverse effects.

6.2.2
Adverse Effects

Whereas the adverse effect profile of an SSRI generally stays the same qualitatively, but increases quantitatively over its clinically relevant dosing range, the adverse effects of venlafaxine change qualitatively as the dose increases because of progressively greater blockade of NE uptake with increasing dose and hence increasing concentration over its clinically relevant dosing range.

Venlafaxine has a generally benign tolerability profile. At low doses, consistent with its action on the SE system, the adverse effect profile is similar to an SSRI: nausea, loose stools, sexual dysfunction. The IR formulation of venlafaxine appears to cause a somewhat higher incidence of gastrointestinal adverse effects, in particular nausea, compared with the SSRIs (Preskorn 1999). Venlafaxine appears to have the same liability as the SSRIs for causing sexual dysfunction (Janicak et al. 2001). At higher doses, venlafaxine can produce the same adverse effects as an NSRI: generally mild increases in blood pressure, diaphoresis, tachycardia, tremors, and anxiety (Preskorn 1995; Nelson 1997). A disadvantage of venlafaxine relative to the SSRIs is the potential for dose-dependent blood pressure elevation, most likely due to the NE uptake inhibition caused by higher doses of venlafaxine. However, elevated blood pressure is a dose-dependent effect and is infrequently observed at doses below 225 mg/day (Presentation to the U.S. Food and Drug Administration April 1993; Grunder et al. 1993). Since venlafaxine has been marketed, patients who have developed clinically meaningful elevations in blood pressure but who had not responded to other antidepressants but did respond to venlafaxine have been able to remain on the drug while their blood pressure was managed with a hypertensive (Janicak et al. 2001). Other management strategies, when feasible, include a dose reduction or a switch to another class of antidepressant. Elevated blood pressure developed within

2 months of patients being stabilized on a given dose, so patients should be carefully monitored during this period (Grunder et al. 1993; Presentation to the U.S. Food and Drug Administration April 1993; Feighner 1994). It is recommended that blood pressure be monitored in patients receiving doses of 225 mg/day or more at every visit during the first 2 months after dose stabilization (Janicak et al. 2001). As noted above, another dose-dependent adverse effect of venlafaxine is tremors. This effect might be due to NE uptake inhibition; however, this effect also occurs with the SSRIs (Preskorn 2000c). It therefore appears more likely that these tremors are mediated through an SE effect on either the NE or DA systems.

Because of its relatively short half-life, venlafaxine has a risk of causing an SE withdrawal syndrome (for a description, see the preceding chapter on the SSRIs) comparable to that of fluvoxamine and paroxetine (Schatzberg et al. 1997). The incidence and severity of the SE withdrawal syndrome is partly a function of the half-life of the agent involved—the half-life of venlafaxine, including the half-life of its active metabolite ODV, is approximately 12 h (Klamerus et al. 1992). The extended release formulation dose not change the half-life of the drug and therefore does not affect the risk of SE withdrawal syndrome upon abrupt discontinuation. Although the withdrawal syndrome that can occur with venlafaxine is not as medically serious as the sedative hypnotic withdrawal syndrome, it can be unpleasant for patients, who may mistake the withdrawal symptoms for a relapse of their depressive illness, and may even become so dysphoric or agitated that they experience suicidal ideation (Rosenbaum and Zajecka 1997; Rosenbaum et al. 1998). It can also mimic mania, which may lead to a misdiagnosis and inappropriate treatment (Landry and Roy 1997).

6.2.3
Overdose

Venlafaxine inhibits the SE and NE uptake pumps, but has low affinity for most other neural receptors and does not inhibit sodium fast channels or potassium rectifying channels (as do the tertiary amine TCAs), making it relatively safe in overdose due to its wide therapeutic index (Muth et al. 1991; Feighner 1995; Janicak et al. 2001). During clinical trials, patients survived acute ingestion of over 6,750 mg of venlafaxine without serious sequelae (Presentation to the U.S. Food and Drug Administration April 1993). These acute overdoses required no specific therapeutic interventions beyond general nursing care with the most common effects being nausea and vomiting, which further reduced the severity of the overdose.

6.3
Clinical Use/Therapeutic Indications

6.3.1
Efficacy for Acute Treatment

Venlafaxine and the SSRIs have similar advantages over the TCAs and monoamine oxidase inhibitors. Venlafaxine also has a number of potential advantages over the SSRIs, including an ascending dose–antidepressant response curve, with possible greater overall efficacy at higher doses, evidence of more rapid onset of antidepressant action, and evidence of superior efficacy in hospitalized patients with major depressive disorder (MDD) compared with placebo or fluoxetine, and minimal effects on CYP enzymes in contrast to fluoxetine, fluvoxamine, paroxetine, and the non-SSRI, bupropion.

At a dose of 75 mg/day, venlafaxine demonstrates efficacy comparable to any other type of antidepressant in outpatients with clinical depression (Preskorn 1994). Venlafaxine has an ascending dose–response curve in terms of antidepressant efficacy in contrast to the SSRIs, consistent with its sequential effects on SE and NE uptake inhibition. In essence, a dose increase with venlafaxine is pharmacologically and clinically a built-in augmentation strategy comparable to adding an NSRI such as desipramine to an SSRI such as sertraline (Preskorn 1998).

Ferrier (2001) reviewed findings from studies that reported rates of remission [defined as a score of 10 or less on the 17-item Hamilton Depression Rating Scale (HAM-D) or a clinicians' global impression (CGI) improvement score of 1, very much improved] for MDD and found remission rates of approximately 20%–30% with SSRIs during short-term therapy and remission rates with venlafaxine that ranged from 37% to 62%. In these studies, the remission rates for venlafaxine were greater than those seen with fluoxetine, sertraline, or paroxetine, especially at doses of 75–300 mg/day. While data on remission rates are clinically important, several caveats concerning this analysis should be kept in mind. First, these studies were designed to collect data on rates of response, not remission. Second, they used different dosing approaches ranging from conservative to more aggressive. Third, length of time to endpoint of study varied from one study to another. Nevertheless, some evidence exists in support of the idea that antidepressants that combine noradrenergic with serotonergic mechanisms of action, such as the TCAs and venlafaxine, may have a more robust effect than purely serotonergic antidepressants.

In addition to these meta-analyses, there have been a few head-to-head comparison with other antidepressant examining overall efficacy. These studies are briefly reviewed below.

Rudolph and Feiger (1999), in a double-blind, placebo-controlled, parallel-group design study, compared the efficacy and tolerability of venlafaxine XR (75–225 mg/day) and fluoxetine (20–60 mg/day) in 232 outpatients with MDD. Remission rates, defined as a HAM-D$\leq$7, in this study were 37%, 22%, and 18%

for venlafaxine XR, fluoxetine, and placebo, respectively, which were statistically significant differences. Discontinuation rates secondary to adverse events were low for both drug treatment groups: 6% for venlafaxine ER and 9% for fluoxetine. One possible explanation for the difference in remission rates could be the short duration of treatment, which could have biased the results against fluoxetine due to its long half-life and the longer time needed to achieve a steady-state concentration (C_{ss}) of both parent drug and active metabolite, norfluoxetine. Another possibility is that venlafaxine could be more efficacious than fluoxetine, perhaps due to its dual mechanism of action.

Clerc et al. (1994) compared the efficacy of venlafaxine IR (200 mg/day) and fluoxetine (40 mg/day) in a double-blind study of 68 inpatients hospitalized with melancholic depression, with initial minimum Montgomery-Asberg Depression Rating Scale (MADRS) scores of 25. Dropout rates for venlafaxine and fluoxetine were 18% and 35%, respectively. There were significantly more responders at weeks 4 and 6 in the venlafaxine group: 76% vs 47% at week 4 and 76% vs 41% at week 6 for venlafaxine and fluoxetine, respectively. One obvious but understandable limitation of this study is the lack of a placebo group.

An examination of the fixed-dose efficacy data on venlafaxine illustrates how study design can affect results. In clinical trials of venlafaxine, the magnitude of the antidepressant effect of venlafaxine was smaller at 375 mg/day than at the lower dose of 225 mg/day (Preskorn 1994). This finding is consistent with the higher early drop-out rate seen in those patients who were assigned to the highest dose of venlafaxine compared to the 225 mg/day dose. Such dropouts occurred before the patients could have reasonably been expected to experience an antidepressant response. In fixed-dose studies, patients are randomly assigned to predetermined doses regardless of need and titrated rapidly to that predetermined dose. This approach to dosing is in stark contrast to what occurs in clinical practice, where patients are typically started on the lowest, usually effective, dose and titrated upward according to response and side effects. Titration in clinical practice is frequently done much more gradually than in a fixed-dose study and, as a result, the patient treated in clinical practice has more time to adjust to any dose-dependent adverse effects. For example, patients in clinical practice typically develop tolerance to many of the acute adverse effects (e.g., nausea) produced by drugs that potentiate 5-HT, such as the SSRIs and venlafaxine.

6.4
Dosing Recommendations

The dosing of venlafaxine is simplified because clinicians can prescribe a usually effective dose (75 mg/day) from the start without dose titration. As a result of the development of an extended release formulation, venlafaxine can be taken once a day. The use of the immediate release formulation is now principally limited to rapid dose titration in hospitalized patients.

6.4.1
Onset of Action

There is evidence that venlafaxine has a more rapid onset of antidepressant action than the SSRIs. One study indicated that high-dose venlafaxine (>300 mg/day) can produce a response in 20% of patients within 1 week and can statistically separate from placebo as soon as 4 days (Preskorn 1994), a finding that is consistent with the combined inhibition of both SE and NE uptake pumps at such doses (Janicak et al. 2001).

7
Duloxetine

7.1
Chemistry, Pharmacology, and Metabolism

Duloxetine is a chiral compound [(+/−)-N-methyl-3-(1-napthalenyloxy)-3-(2-thiophene) propanamine] which can be formed from the building blocks of (S)-3-chloro-1-(2-thienyl)-1-propanil and the corresponding (R)-butanoate (Wong et al. 1988; Liu 2000). More importantly for the prescriber, it is an SE–NE reuptake pump inhibitor which is pending approval in the United States and other countries in late 2003. Thus, it will be a third member of this pharmacological class, joining venlafaxine and milnacipran; sibutramine is a fourth member of this class but is marketed for obesity rather than major depression (Jackson et al. 1997). Given that it has not yet been marketed, there are only a limited number of published articles available on this compound, but more should be expected shortly after its market introduction. The manufacturer (Lilly) is initially seeking indications for both major depression and urinary incontinence, illustrating the fact that the brain and the body are in fact inseparable.

The positive enantiomer (LY 248686) is a slightly more potent than the negative enantiomer (LY 248685) as an inhibitor of SE uptake; neither enantiomer is a substantial inhibitor of the DA uptake pump (Wong et al. 1993). There is both in vitro and in vivo preclinical pharmacological evidence suggesting that duloxetine may be a more balanced inhibitor of SE and NE reuptake pumps than is venlafaxine, meaning that the ratio of its affinity constants for these two pumps is closer to one (Beique et al. 1998; Bymaster et al. 2001). Like venlafaxine and milnacipran, duloxetine has low affinity for muscarinic, histamine-1, adrenergic, DA, and SE receptors (Wong et al. 1988). That suggests a low potential for causing adverse effects mediated by actions on these mechanisms, which has been confirmed in the clinical trials discussed below. The ability of duloxetine to block both uptake pumps in vivo has been demonstrated in a number of animal studies, including the usual animal models of depression (Wong et al. 1993; Fuller et al. 1994; Engleman et al. 1995, 1996; Katoh et al. 1995; Kihara and Ikeda 1995; Kasamo et al. 1996; Gobert et al. 1997; Gongora-Alfaro et al. 1997; Smith and Lakoski 1997, 1998; Rueter et al. 1998a,b; Gobert and Millan 1999). A brief summary of this in vivo preclinical work follows.

Duloxetine at 15 mg/kg administered intraperitoneally to rats produced a large increase in extracellular levels of SE (250%) and NE (1,100%) in both the hypothalamus and cerebral cortex (Engleman et al. 1995). In rats and mice, duloxetine prevented tetrabenazine-induced ptosis, inhibited reserpine-induced hypothermia, and potentiated the effects of 5-hydroxytrytophan (a precursor of SE) but did not affect oxotremorine-induced (oxotremorine is a cholinergic agonist) lacrimation or salivation (Katoh et al. 1995). These results support inhibition of both SE and NE uptake at doses that do not block muscarinic receptors, which is consistent with the in vitro data and with the adverse effect profile seen in clinical trials. Oral administration of duloxetine produced a dose-dependent increase in the output of both NE and SE from the frontal cortex and DA output from the nucleus accumbens. The latter is most likely an indirect effect mediated via its direct effects on the central SE and/or NE circuits (Kihara and Ikeda 1995) and is consistent with the anatomical connections of these systems and other electrophysiological studies, including the observation that acute administration of duloxetine and other SE uptake pump inhibitors increases the firing frequency of DA neurons in the substantia nigra at the same time that they suppress the spontaneous firing of SE neurons in the dorsal raphe. Duloxetine antagonized the SE- and NE-depleting effect of p-chloroamphetamine and 6-hydroxydopamine but not DA depletion in both mice and rats and decreased brain 5-hydroxyindoleacetic acid (5HIAA) and SE turnover consistent with in vivo blockade of the SE and NE uptake pumps, but had no direct effect on the DA uptake pump at the doses used (Fuller et al. 1994). The recovery times of dorsal hippocampal CA3 pyramidal neurons in rats were significantly prolonged after microiontophoretic applications of SE and NE following treatment with duloxetine, consistent with inhibition of both the SE and NE uptake pumps (Rueter et al. 1998a). In the same study, electrically evoked release of SE was enhanced in the midbrain and hippocampus presumably due to desensitization of the 5-HT$_{1D}$ and 5-HT$_{1A}$ autoreceptors and α_2-adrenergic heteroreceptors, respectively. Acute intravenous administration of duloxetine in rats suppressed spontaneous firing activity of both SE and NE neurons in the dorsal raphe and locus coeruleus, respectively, with ED$_{50}$ values of 99 and 475 µg/kg, respectively (Bruno et al. 1999). The firing rate of SE neurons in the dorsal raphe nucleus was decreased after 2 days of duloxetine administration but returned to control levels after 21 days consistent with acute SE uptake inhibition followed by compensatory desensitization of the somatodendritic 5-HT$_{1A}$ autoreceptors (Rueter et al. 1998b). That conclusion was further supported in this same study by the finding that administration of the 5-HT$_{1A}$ antagonist WAY 100635 increased hippocampal firing rates in rats treated for 21 days to a greater extent that in either rats treated for 2 days or control rats.

The ability of duloxetine to inhibit SE uptake has also been demonstrated in healthy human volunteers (Kasahara et al. 1996; Turcotte et al. 2001). In one study, duloxetine at doses up to 60 mg/day administered for 14 days did not impede the usual increase in blood pressure that follows an intravenous tyramine infusion, whereas clomipramine (a tertiary amine TCA) did (Turcotte et al. 2001). However, duloxetine at the 60-mg dose in this study did increase supine systolic blood pressure. The results from this single study need replication but suggest that 60 mg/day may be near the threshold dose for NE reuptake pump inhibition.

Duloxetine at doses of 20–40 mg twice a day in 12 healthy male volunteers exhibits linear pharmacokinetics with a mean oral clearance, apparent volume of distribution, and half-life of 114 l/h (range:44–218 l/h), 1943 l (range: 803–3,531 l), and 12.5 h (range: 9.2–19.1 h), respectively (Sharma et al. 2000).

There are several lines of in vitro, ex vivo, and in vivo evidence suggesting that duloxetine dissociates slowly from (i.e., binds tightly to) the SE uptake pump (Kasahara et al. 1996; Ishigooka et al. 1998). This finding, along with duloxetine's large volume of distribution and concentration in the brain, suggests that the plasma half-life underestimates the half-life in the brain and hence likely accounts for the fact that it has been efficacious for the treatment of depressed patients when given once a day.

Preclinical animal studies with duloxetine also provide some evidence of potential pharmacodynamic drug–drug interactions that may occur in man and may have either beneficial or adverse consequences depending on the magnitude of the effect. 1-(2-pyrimidinyl) piperazine (1-PP) is the major metabolite of buspirone and is an α_2-adrenergic antagonist. 1-PP potentiated the duloxetine-induced increase in SE, NE, and DA levels in the frontal cortex by 290%, 1,320%, and 600%, respectively, consistent with the fact that α_2-adrenergic receptors tonically inhibit NE and DA and phasically inhibit SE release in the frontal cortex (Gobert et al. 1997). (−)-Pindolol, a 5-HT$_{1A}$, 5-HT$_{1B}$, and β_1-/β_2-adrenergic receptor antagonist potentiated the increase in frontal cortex of levels of both SE and DA but not NE produced by administration of duloxetine and fluoxetine (Gobert et al. 1999). Systemic administration of buspirone, a 5-HT$_{1A}$ receptor partial agonist, transiently inhibited the duloxetine- and fluoxetine-induced increases in SE levels but markedly and synergistically increased duloxetine- and fluoxetine-induced increases in DA levels (550% and 240%, respectively) and in NE levels (750% and 350%, respectively) in the frontal cortex (Gobert et al. 1997). Consistent with this study was the finding that coadministration of the 5-HT$_{1A}$ antagonist, LY 206130, produced a 570%, 480%, and 300% increase in duloxetine-induced increases in SE, NE, and DA levels, respectively, in the hypothalamus of conscious, freely moving rats consistent with antagonism of the normal auto-inhibitory feedback loop mediated by somatodendritic 5-HT$_{1A}$ autoreceptors (Engleman et al. 1996) with duloxetine.

7.2
Safety and Adverse Effects

7.2.1
Drug–Drug Interactions

Preclinical evidence of pharmacodynamic drug–drug interactions with duloxetine was reviewed above. These interactions are mediated by duloxetine-induced inhibition of the SE and NE uptake pumps and thus would be common to this class. The SE- and NE-mediated interactions would also be common to other antidepressants sharing either of these specific mechanism (e.g., tertiary and secondary amine TCAs reviewed in the chapter by Lader in this volume, and SSRIs reviewed in the chapter by Preskorn, Ross, and Stanga). For these reasons, duloxetine will undoubtedly carry a warning against use in combination with monoamine oxidase inhibitors.

Duloxetine at a dose of 120 mg/day is also a moderate inhibitor of CYP 2D6 (i.e., less than fluoxetine or paroxetine at their lowest usually effective dose) but greater than escitalopram, citalopram, or sertraline at their lowest usually effective dose) (Skinner et al. 2003). Thus, modest dose reductions and careful monitoring will be needed when prescribing duloxetine in combination with drugs that are preferentially metabolized by CYP 2D6, particularly those with narrow therapeutic indexes.

7.2.2
Adverse Effects

A PubMed search yielded only four published clinical trials with duloxetine in major depression (Berk et al. 1997; Detke et al. 2002a,b; Goldstein et al. 2002) and one trial in stress urinary incontinence (Norton et al. 2002). Thus, the published data on adverse effects are limited but consistent with the preclinical pharmacology of the drug and the adverse effect profiles of other drugs with similar mechanisms of action. In the stress urinary incontinence study, the discontinuation rate for adverse effects was dose dependent as follows: placebo– 5%, 20 mg/day–9%, 40 mg/day–12%, and 80 mg/day–15%, with nausea being the most common symptom leading to discontinuation (Norton et al. 2002). Similar results but with fewer doses were seen in the depression trials. In one such trial, the discontinuation rate on placebo was 4.3% vs 12.5% with duloxetine, 60 mg/day, with the most common adverse effects being nausea, dry mouth, dizziness, and constipation (Detke et al. 2002a). In a forced titration trial to 120 mg/day, the only adverse effects that were reported to a statistically greater degree ($p<0.05$) on duloxetine compared with placebo were insomnia and asthenia (Goldstein et al. 2002). The adverse effects in these trials are consistent with indirect SE and NE agonism mediated by uptake pump inhibition. There was a significant incidence of hypertension reported in the published trials as has been reported with some antidepressants that inhibit the NE reuptake pump.

7.2.3
Overdose

There are no published data on overdoses with duloxetine. However, the preclinical pharmacology of the drug suggests that it should have a wide therapeutic index.

7.3
Clinical Use/Therapeutic Indications

7.3.1
Efficacy for Acute Treatment

At this time, approval for marketing is being sought for the treatment of both major depression and stress urinary incontinence. There are published random assignment, double-blind, placebo-controlled studies demonstrating statistically better reduction in symptoms in both conditions on duloxetine vs placebo (Berk et al. 1997; Detke et al. 2002a,b; Goldstein et al. 2002; Norton et al. 2002). The duloxetine specific response rates (i.e., rate on drug–rate of placebo) at doses of 60 and 120 mg/day were 23% and 16%, respectively (Detke et al. 2002a; Goldstein et al. 2002). The duloxetine specific remission rate (i.e., rate on drug–rate on placebo) at doses of 60 and 120 mg/day were 15%, 28%, and 24% (Detke et al. 2002a,b; Goldstein et al. 2002). These rates are comparable to drug-specific response and remission rates for other marketed antidepressants reviewed in this book. In addition to higher antidepressant response and remission rates (e.g., reduction in rating scale scores) compared to placebo, patients treated with duloxetine also experienced a greater reduction in associated painful physical symptoms that are commonly seen in depressed patients, including overall pain, back pain, shoulder pain, and time in pain while awake (Detke et al. 2002a; Goldstein et al. 2002).

7.3.2
Dosing Recommendations

The dosing guidelines at this time have not been formalized, but the published positive trials have used doses of 60–120 mg/day.

7.3.3
Onset of Action

There is no published evidence to date of a faster onset of action with duloxetine vs other marketed antidepressants.

8
Milnacipran

8.1
Chemistry, Pharmacology, and Metabolism

Milnacipran (also earlier called midalcipran) is chemically the cyclopropane derivative, 1-phenyl-1-diethyl-aminocarbonyl-cyclopropane (z) hydrochloride. It is a racemic mixture composed of two enantiomers which are both active based on preclinical pharmacology (Deprez et al. 1998; Atmaca et al. 2003). Milnacipran is marketed in France, Japan, and a few other countries (Tajima 2002). It was in development in the United States in the late 1980s and early 1990s but that development was discontinued.

Pharmacologically, milnacipran is a SE and NE reuptake inhibitor and thus is in the same class as venlafaxine and duloxetine as well as the anti-obesity drug sibutramine. It is virtually equipotent at the inhibition of both of these uptake pumps, produces several-fold and long-lasting elevations in the extracellular levels of monoamines in mammalian brain, and is active in animal models of depression (Briley et al. 1996; Delini-Stula 2000). It has no direct effect on the DA. Other than its SE and NE transporter binding, milnacipran has low affinity for receptors for acetylcholine, DA, histamine NE, and SE and does not affect ion channels nor does it inhibit monoamine oxidase (Briley et al. 1996; Atmaca et al. 2003). These findings qualify milnacipran as a member of the selective SE and NE uptake pump inhibitors along with venlafaxine and duloxetine and is consistent with its clinical adverse effect profile discussed later.

In rat brains, specific binding of milnacipran has been found in structures dense in SE innervation including the dorsal raphe, basal ganglia, colliculi, and cerebral cortex (Barone et al. 1994). Selective lesioning of SE neurons caused large decreases in milnacipran binding in septal nuclei, caudate, hippocampus, thalamus, and ventral and dorsal hypothalamus, but in other brain regions had only partial (putamen) or no effect (amygdala, lateral hypothalamus). Milnacipran was differentially displaceable in these various areas by SSRIs (e.g., paroxetine) and NSRIs (e.g., desipramine). These binding data are consistent with milnacipran binding to both the SE and NE uptake pumps. In contrast to desipramine, milnacipran, 50 mg/day for 21 days, did not modify basal or induced release of NE, tissue concentrations of NE, or α_2-adrenergic receptor sensitivity in the rat (Moret and Briley 1994). These results suggest that milnacipran under these dosing conditions in this species did not meaningfully inhibit the NE uptake pump; however, neither plasma nor tissue concentrations of milnacipran were obtained in this study, and thus the relevance of this experimental dosing condition to clinical dosing is open to question.

Consistent with its in vitro profile and its anatomical distribution, local administration of milnacipran increased SE output in the rat frontal cortex and the dorsal raphe by seven- and tenfold, respectively, via a calcium- and tetrodotoxin-dependent mechanism (Bel and Artigas 1999). A smaller magnitude increase was seen in these regions following systemic administration and was modestly potentiated by coadministration of the 5-HT$_{1A}$ antagonist WAY 100635, consistent with the inhibition of the 5-HT$_{1A}$ autoreceptor regulating synaptic SE concentration. Systemic administration of milnacipran for 2 days in rats reduced the firing rate of both NE neurons in the locus coeruleus and SE neurons in the dorsal raphe (Mongeau et al. 1998). The reduction in NE but not SE neuronal firing persisted following 14 days of continuous milnacipran administration. Chronic administration (14 days continuous administration) of milnacipran substantially attenuated the ability of the α_2-adrenergic agonist clonidine to suppress both NE and SE neuronal firing but had no effect on the ability of the 5-HT$_{1A}$ agonist to suppress SE neuronal firing (Mongeau et al. 1998). Finally, milnacipran was able to suppress SE neuronal firing in intact but not in NE-denervated rats. Taken together, this set of studies confirms that milnacipran has acute effects on both SE and NE uptake inhibition but differential long-term effects on α_2-adrenergic and 5-HT$_{1A}$ autoreceptors and suggests the possibility that the mechanism by which 5-HT neurons regain their normal firing during chronic milnacipran treatment may be mediated through its effects on the NE system, specifically the α_2-adrenergic heteroreceptor. These results were consistent with findings in several behavioral studies in rats, including its effects on the forced swimming test, clonidine-induced aggression, and methoxamine-induced exploratory hyperactivity (Maj et al. 2000; Reneric et al. 2002).

Despite these results, milnacipran differs from a number of (but not all) established antidepressants in that chronic administration, including osmotic mini-pump infusion for 27 days, did not downregulate β-adrenergic receptors or β-adrenergic-stimulated adenylate cyclase activity (i.e., β-adrenergic second messenger function) in rats (Assie et al. 1992; Neliat et al. 1996; Atmaca et al. 2003). Such dosing also did not alter α_1- or α_2-adrenergic, or 5-HT$_1$ or 5-HT$_2$ receptors, and did not modify the uptake or accumulation of SE or NE. In contrast to citalopram, chronic administration of milnacipran also did not downregulate SE autoreceptors, which is the apparent mechanism permitting an increase in SE neurotransmission with SSRIs (Moret and Briley 1990). However, other work suggests that chronic administration of milnacipran desensitizes somatodendritic but not postsynaptic 5-HT$_{1A}$ receptors in rats (Mochizuki et al. 2002a). Like the results for duloxetine reviewed earlier, chronic administration of milnacipran does affect DA neurotransmission in the brain as witnessed by an increased density of DA D$_2$ receptors in the striatum (Rogoz et al. 2000).

Nevertheless, milnacipran is active in several animal models of depression including the forced swim test in both mice and rats, learned helplessness in rats, conditioned fear stress test, and the olfactory bulbectomized rat (Briley et al. 1996; Rogoz et al. 1999; Mochizuki et al. 2002b; Reneric et al. 2002). Milnacipran also antagonized the depressant effect of tetrabenazine, yohimbine-induced mortality, and p-chloroamphetamine-induced hyperthermia in mice and enhanced L-tryptophan-induced behavioral changes in rats, but produced no anticholinergic, sedative, or stimulant effects (Stenger et al. 1987). These latter results are consistent with its in vitro pharmacological profile: SE and NE uptake inhibition without effects on neuroreceptors. Finally, milnacipran was able to induce rapid and sustained translocation of the glucocorticoid receptor into the nucleus of human lymphocytes (Okuyama-Tamura et al. 2003). The same was true for a variety of mechanistically different antidepressants, including desipramine, clomipramine, fluoxetine, and clorgyline but was not true for CNS active medications from other therapeutic classes including antipsychotics, benzodiazepines, lithium, and verapamil.

In anesthetized guinea pigs, intravenous milnacipran caused ventricular arrhythmias and cardiac arrests. However, milnacipran had a 22 times wider therapeutic index compared with imipramine.

Some studies have extended these preclinical studies to man. A study in 12 healthy volunteers demonstrated that milnacipran at clinically used doses produced concentrations which are capable ex vivo of inhibiting SE and NE uptake into human platelets and rat hypothalamic homogenates, respectively (Palmier et al. 1989). These results support milnacipran's ability to inhibit the uptake pumps for both SE and NE with similar potency and under clinically used dosing conditions. Also consistent with its in vitro and preclinical pharmacology, milnacipran in comparison to placebo and amitriptyline had a benign effect profile on cognitive function in both young and elderly (>65 years) volunteers as measured by a psychometric test battery consisting of critical flicker fusion, choice reaction time, compensatory tracking, short-term memory, and subjective sedation and sleep (Hindmarch et al. 2000). Intravenous infusion of milnacipran to normal volunteers at doses up to 0.8 mg/kg produced an average 18% increase in heart rate, 22% increase in systolic blood pressure, decreases in the functional refractory period of the atrium and atrioventricular node and in the effective refractory period of the right ventricle, as well as transient nausea in 50% of these volunteers (Caron et al. 1993).

Milnacipran demonstrates linear pharmacokinetics over a dose range of 25–200 mg/day, is rapidly and extensively absorbed (>85%), and has a half-life of 12 h with steady-state being reached within 48 h when administered twice a day (Delini-Stula 2000). Its metabolism does not require CYP enzyme-mediated biotransformation. Milnacipran is principally eliminated by renal excretion. Consistent with that fact, the clearance of milnacipran was significantly prolonged in patients with renal failure (creatinine clearance=9–84.5 ml/min) with its half-life being three times longer in renal failure vs normal volunteers (Puozzo et al. 1998b). Conversely, its pharmacokinetics were essentially unchanged in volunteers with even severe liver impairment (Puozzo et al. 1998a).

8.2
Safety and Adverse Effects

8.2.1
Drug–Drug Interactions

Milnacipran would be predicted to be susceptible to the same pharmacodynamic drug–drug interactions as other SE and NE uptake inhibitors. However, milnacipran would not be expected to be either the victim or perpetrator of any CYP enzyme-mediated drug–drug interactions.

8.2.2
Adverse Effects

Consistent with its absence of effects on muscarinic, adrenergic, and histamine receptors, milnacipran at doses of 50–200 mg/day has a favorable adverse effect profile when compared with tertiary amine TCAs such as amitriptyline and imipramine, including a lower incidence of abnormal liver function tests based on analysis of a database of over 3,300 patients (Montgomery et al. 1996; Puech et al. 1997). Milnacipran at doses of 50 or 100 mg twice a day but not 100 mg once a day caused a lower incidence of nausea and anxiety but a higher incidence of headache, dry mouth, and dysuria than did fluoxetine, 20 mg/day, or fluvoxamine, 100 mg twice a day, in 4- to 12-week trials (Lopez-Ibor et al. 1996). As with other drugs in this class, dysuria is the most common troublesome and dose-dependent adverse effect of milnacipran occurring in up to 7% of patients (Spencer and Wilde 1998; Delini-Stula 2000). Consistent with its NE uptake inhibition, high dose milnacipran has been reported to cause blood pressure elevation (Yoshida et al. 2002).

8.2.3
Overdose

There have been over 15 substantial overdoses of milnacipran with no fatalities and each having a uneventful course and favorable outcome (Montgomery et al. 1996).

8.3
Clinical Use/Therapeutic Indications

8.3.1
Efficacy for Acute Treatment

Like reboxetine and tianeptine and consistent with its primarily European development, most of the published trials of milnacipran have been active but not placebo controlled and results were not published in full with rigorous peer review (Spencer and Wilde 1998). That limits the ability to make definitive statements about the drug's efficacy. Given this important caveat, a meta-analysis of three multicenter, double-blind, short (4- to 8-week), acute efficacy trials in inpatients and outpatients with moderate to severe depression found milnacipran produced a superior antidepressant response compared with placebo at doses of 50 and 100 mg twice a day but not at a dose of 25 mg twice a day (Macher et al. 1989; Lecrubier et al. 1996). Another meta-analysis of seven randomized, double-blind studies of milnacipran, 50 mg twice a day, found comparable antidepressant efficacy to imipramine, clomipramine, and amitriptyline, 150 mg/day (Ansseau et al. 1989; Kasper et al. 1996; Van Ameringen et al. 2002). However, one randomized, double-blind, parallel-group study found milnacipran at a dose of 200 mg/day produced superior antidepressant response compared to milnacipran 50 and 100 mg/day and a comparable antidepressant response in reference to amitriptyline 150 mg/day (von Frenckell et al. 1990). In an 8-week, double-blind, random assignment trial involving 219 depressed elderly patients, milnacipran and imipramine at doses of 50 mg twice a day had comparable antidepressant efficacy but imipramine produced a greater number of adverse effects, particularly those attributable to muscarinic acetylcholine receptor blockade (Tignol et al. 1998). In a 6-week, double-blind, random assignment study involving treatment-refractory patients, milnacipran, 200 mg/day, and clomipramine, 150 mg/day, produced comparable but low antidepressant response rates (Steen and Den Boer 1997)

In 4- to 12-week acute efficacy trials, milnacipran at doses of 50 or 100 mg twice a day but not 100 mg once a day was as effective as fluoxetine, 20 mg/day, and possibly more effective than fluvoxamine, 100 mg twice a day, particularly in patients with higher depressive rating scale scores (Ansseau et al.1994; Lopez-Ibor et al. 1996; Clerc 2001; Fukuchi and Kanemoto 2002). A double-blind, random assignment study of milnacipran vs fluvoxamine did not support the usefulness of a loading dose strategy involving the administration of 300 mg/day for the first 2 weeks followed by a maintenance dose of 150 mg/day (Ansseau et al. 1991).

In addition to these conventional acute efficacy trials, small studies have also been done testing the effectiveness of milnacipran in patients with post-stroke depression and interferon-α-induced depression (Higuchi et al. 2002; Yoshida et al. 2003).

In a 4-month continuation study, relapse rates were 16% for patients treated with milnacipran (50 mg twice a day) vs 24% for placebo ($p<0.05$) (Rouillon et al. 2000). This finding was replicated in a 6-month continuation study in which there was not only a lower relapse rate in patients treated with milnacipran compared with those receiving placebo but also the patients treated with milnacipran had higher quality of life scores at the end of the continuation phase (Rouillon et al. 2000). However, clomipramine (75–150 mg/day) in a 6-month continuation study produced a statistically significant lower relapse rate than did milnacipran (100–200 mg/day): 63% vs 45%, respectively (Leinonen et al. 1997).

8.3.2
Dosing Recommendations

As reviewed above, milnacipran has been tested at doses of 50–200 mg/day, given on a twice-a-day schedule. In general, doses greater than 50 mg/day have been needed to demonstrate superior efficacy to placebo and comparable efficacy to TCAs and fluoxetine. There is some evidence that 200 mg/day may produce superior response compared to 100 mg/day. Based on one trial, divided doses are needed to demonstrate efficacy.

8.3.3
Onset of Action

When doses were titrated (not a requirement) from 50 to 100 mg/day, milnacipran had a slower onset of action than amitriptyline, 150 mg/day, in European but not Japanese patients and a comparable onset when European patients were started on 200 mg/day (Spencer and Wilde 1998).

9
Tianeptine

9.1
Chemistry, Pharmacology, and Metabolism

Tianeptine is a 3-chlorodibenzothiazepine nucleus with an aminoheptanoic side chain (Kroeze et al. 2002). It can also be chemically described as a substituted dibenzothiazepine nucleus with a long lateral chain (Labrid et al. 1992). There are highly specific structural requirements for molecules in the tianeptine series to be active, including the requirement for an aminocarboxylic chain with an optimal length of six methylene links, a tricyclic nucleus with an electron donor heteroatom in position 5, and an aromatic substitution with a moderate electron-acceptor atom in position 3. These highly specific requirements for the tianeptine series are in marked contrast with the lack of specific requirements for the classical tricyclic series (Kroeze et al. 2002).

Tianeptine is chiral with (+) and (−) enantiomers (Oluyomi et al. 1997). The drug is marketed as a racemic mixture. Both of the enantiomers are active but the (−) enantiomer is more active than the (+) enantiomer.

Like milnacipran, tianeptine and amineptine are marketed in France but few other countries and there is little knowledge of these drugs in the United States and other English-speaking countries around the world (Mitchell 1995). There has, however, been a surprising amount of research done with tianeptine, particularly preclinical research. That is in part because of its apparent novel mechanism of action: tianeptine, in contrast to most other antidepressants, increases SE uptake into neurons rather than blocking it (Wagstaff et al. 2001). In man, tianeptine increases platelet SE uptake, which is consistent with its in vitro pharmacology and opposite to that of the effect of SE reuptake inhibitors (Kamoun et al. 1994).

While the acute putative antidepressant mechanism of action of tianeptine differs radically from other antidepressants, its side-effect profile is similar to that of other newer antidepressants, particularly in terms of having low abuse potential and a low risk of adverse effects on the cardiovascular system, the cholinergic systems, sleep/arousal, cognition, psychomotor functioning, and weight (Loo and Deniker 1988; Wagstaff et al. 2001). Consistent with that fact, tianeptine does not bind to α- or β-adrenergic, DA, SE, GABA, glutamate, benzodiazepine, muscarinic, or histamine receptors nor did it affect calcium channels (Kato and Weitsh 1988). It also directly blocks the neuronal DA transport (Vaugeois et al. 1993).

The major effect in rat platelets and synaptosomes is a small increase in SE uptake after subchronic administration.

While its acute effects are quite different from that of most other antidepressants, chronic tianeptine treatment (14 days) reduced the expression of the SE transporter mRNA and the number of SE transporter binding sites (Watanabe et al. 1993; Kuroda et al. 1994). This latter effect is more in line with the effect of classical SE uptake pump inhibitors that are antidepressants and suggests a difference between the acute effects of tianeptine and the chronic adaptive changes, which may be more relevant to its antidepressant properties in man. In addition, chronic administration of tianeptine did not alter the concentration or the affinity of α_2, β_1, SE-1, SE-2, or GABA receptors or imipramine or benzodiazepine binding sites (Kato and Weitsh 1988).

Tianeptine in vivo, after both acute and repeated treatment, enhances SE uptake in the cortex and hippocampus but not in the mesencephalon and has no direct effect on NE or DA uptake (Mennini et al. 1987). Short-term treatment with tianeptine also increases NE levels in the preoptic area, the parietal sensory cortex, and dorsal raphe and decreases NE turnover in the parietal sensory cortex, dorsal raphe, and parietal motor cortex, indicating that tianeptine can also alter the central NE system, perhaps indirectly through its action on the SE system (Frankfurt et al. 1995). In addition, short-term tianeptine treatment increased extracellular DA concentration in the nucleus accumbens and, at higher doses, in the frontal cortex but not in the striatum, and the effect on DA was not diminished by marked depletion of SE by intracerebroventricular administration of 5,7-dihyrdoxytryptamine (Invernizzi et al. 1992; Sacchetti et al. 1993). Acute treatment with tianeptine also reduced acetylcholine release from the dorsal hippocampi by 40% and from the frontal cortices by 30% (Bertorelli et al. 1992). This effect of tianeptine on acetylcholine release is mediated by its SE effects, as witnessed by the fact that it is abolished by chemical lesion of the median raphe nucleus or by coadministration of metergoline, a SE receptor blocker.

These findings were confirmed in a study in which tianeptine increased but sertraline decreased the concentration of 5-HIAA, the major metabolite of SE formed by oxidation mediated by monoamine oxidase (MAO)-A, particularly in the hippocampus and hypothalamus (De Simoni et al. 1992; Marinesco et al. 1996). Tianeptine also decreases corticotropin-releasing factor (CRF) and adrenocorticotropic hormone (ACTH) levels, abolishing the stress-induced reduction of hypothalamic CRF concentration and markedly reducing the stress-induced increase in plasma ACTH and corticosterone levels (Delbende et al 1991, 1994).

Microiontophoretic application of tianeptine onto dorsal hippocampus CA3 pyramidal neurons increased their firing frequency. This effect was unchanged by the coadministration of 5,7-dihydroxytryptamine, indicating that the SE transporter is not involved in mediating this effect of tianeptine (Kamoun et al. 1989; Pineyro et al. 1995b).

Sustained treatment with tianeptine did not modify the firing of SE neurons in the dorsal raphe or their responsiveness to intravenous injections of lysergic acid diethylamide (LSD) or 8-OH-DPAT, which are agonists at the somatodendritic SE autoreceptor (Pineyro et al. 1995a). The firing of hippocampal CA3 pyramidal neurons in response to microiontophoretic application of SE also remained unchanged. Taken together, these findings indicate that sustained treatment with tianeptine in contrast to a number of other antidepressants does not modify the efficacy of SE synaptic transmission in the rat hippocampus.

In addition to studying the effect of tianeptine alone, there have been a number of preclinical drug–drug interaction studies with tianeptine alone and in combination with another drug. In rats, both tianeptine and citalopram reduced in a concentration-dependent manner the inhibitory effect of a $5HT_{1B}$ receptor agonist on stimulation-induced release of acetylcholine in the hippocampus (Bolanos-Jimenez et al. 1993). $5HT_{1B}$ presynaptic heteroreceptors are located on cholinergic terminals in this and other brain structures. Conversely, tianeptine did not antagonize the inhibitory effect of the muscarinic receptor agonist carbachol on K(+)-evoked release, indicating that the effect is most likely indirect through the SE system.

As mentioned above, despite its unusual and counterintuitive acute mechanism of action, tianeptine shares with the classical antidepressants the ability to reduce the expression of the SE transporter mRNA and the number of SE transporter binding sites. Like the classical antidepressants, tianeptine is also active in several animal models of depression, including (1) stress-induced spatial memory impairment in the rat (Conrad et al. 1996), (2) isolation-induced aggression in mice and behavioral despair in rats (Kamoun et al. 1994), (3) hyperactivity in the olfactory bulbectomized rat (Kelly and Leonard 1994), and (4) the rat pup ultrasonic vocalization (USV) model, which may be even more a model for anxiety disorders (Olivier et al. 1998). Tianeptine also:

- Reverses stress-induced deficits in exploratory activity (Borqua et al. 1992).
- Attenuates the behavioral signs of sickness behavior in rats induced by peripherally but centrally administered cytokine inducer lipopolysaccharide (LPS) or the prototypical proinflammatory cytokine interleukin (IL)-1β (Castanon et al. 2001).

- Shares with several TCAs and fluoxetine the ability to inhibit corticosterone-induced gene transcription (Budziszewska et al. 2000).

Additionally, tianeptine prevents atrophy of apical dendrites in the CA3 region of the hippocampus caused by chronic daily restraint stress or by daily administration of corticosterone (Kuroda and McEwen 1998; Conrad et al. 1999; Watanabe et al. 2003). This effect may not generalize to all antidepressants, since tianeptine in several studies produced this effect, but desipramine, fluoxetine, and fluvoxamine did not (McEwen et al. 1997; Magarinos et al. 1999). This neuroprotective effect was not mediated through neurotropic factors such as brain-derived neurotropic factor (BDNG), neurotrphin-3 (NT-3), or basic fibroblast growth factor (bFGF), nor through effects on neuronal developmental factors, GAP-43, or MAP2 (Kuroda and McEwen 1998). Perhaps this finding is relevant to the ability of tianeptine to improve working and reference memories in rodents (Bassant et al. 1991; Nowakowska et al. 2000). It has been postulated that tianeptine has a beneficial effect on memory by counteracting the SE-induced inhibition of medial septal neurons (Bassant et al. 1991). These findings suggest a potential role for tianeptine in the treatment of cognitive impairment in the elderly, particularly in those with dementing illness.

Taken together, the extensive preclinical pharmacology described above suggests that tianeptine is a unique and interesting CNS medication which may have many implications and uses beyond being just an antidepressant. Unfortunately, as discussed later in this section, tianeptine has not been studied as rigorously in phase I–III as one would like so that the full human pharmacology of the drug could be understood and potentially exploited.

Tianeptine is converted by hamster, mouse, and rat CYP enzymes into a reactive metabolite (Letteron et al. 1998). That finding has been extended to human liver microsomes and involves glucocorticoid inducible CYP enzymes, possibly CYP 3A but not CYP 2D6 or CYP 1A2 (Larrey et al. 1990). Based on further studies in animals, high doses of tianeptine (360 times the human therapeutic dose) produced covalent binding to liver, lung, and kidney proteins, depleted hepatic glutathione by 60%, and increased SGPT levels fivefold (Letteron et al. 1989). The significance of this finding to humans has not been determined including its possible predictive value for the potential for idiosyncratic and immunoallergic reactions. Tianeptine inhibits both β-oxidation and tricarboxylic acid cycle in mice products. At even higher doses (600 times the human oral dose), tianeptine administered to mice inhibits the oxidation of medium- and short-chain fatty acids and causes microvesicular steatosis in the liver (Fromenty et al. 1989). Again, the implications for humans are uncertain; however, severe and prolonged impairment of mitochondrial β-oxidation can cause microvesicular steatosis and may in severe cases cause liver failure, coma, and death (Fromenty and Pessayre 1995). As discussed in the section on adverse affects below, there have been cases of hepatotoxicity with microvesicular steatosis reported in patients treated with tianeptine.

Tianeptine does not undergo first pass metabolism during absorption from the gastrointestinal (GI) tract, and has high bioavailability and a low volume distribution (Royer et al. 1988). There is a modest food effect with increased time to the maximum peak (by 0.5 h) and lower peak concentrations by 25%; however, the magnitude of these effects are of doubtful clinical significance (Dresse et al. 1988). Tianeptine is highly bound to plasma protein (approximately 95%), particularly to human serum albumin and has saturable binding to α_1-acid glycoprotein (Zini et al. 1990). High non-esterified fatty acids (NEFA) which are seen in patients with chronic renal failure increased the unbound fraction of tianeptine threefold (Zini et al. 1990).

Tianeptine is rapidly eliminated from the body and is primarily cleared via the kidneys (Grislain et al. 1990; Wagstaff et al. 2001). It has a short half-life of 2.5 h (Royer et al. 1988). The major metabolites are analogs of tianeptine with a C5 and C3 lateral chain and a *N*-demethylated derivative (Royer et al. 1988).

Given the importance of the kidneys in the clearance of tianeptine, a number of studies have been done in volunteers with various degrees of impaired renal function. There was a 1-h prolongation of the half-life in elderly subjects and in patients with renal failure (Royer et al. 1988, 1989). However, the MC5 metabolite terminal half-life was almost tripled in patients with chronic renal failure [i.e., creatinine clearance less than 19 ml/min (Zini et al. 1991)]. Based on preclinical evidence that the MC5 metabolite is pharmacologically active, the dose of tianeptine should be reduced in such patients. In addition, tianeptine and its MC5 metabolite show low dialyzability so that hemodialysis is unlikely to be an effective method to speed the elimination of tianeptine after an overdose (Zini et al. 1991).

Given that cirrhosis secondary to alcoholism is a significant health concern in France as well as in the rest of the world, a number of studies have been done examining the interactions between tianeptine and alcohol and between tianeptine and various hepatic conditions. Acute alcohol administration decreased the absorption rate of tianeptine and lowered its plasma levels by approximately 30% but did not affect the pharmacokinetics of the MC5 metabolites (Salvadori et al. 1990). There was only a modest effect of even significant alcohol-induced hepatitis.

The pharmacokinetics of tianeptine are similar in elderly (72–81 years old) vs young volunteers, but the MC5 metabolite levels were higher, suggesting the need for a possible dose reduction (Demotes-Mainard et al. 1991). Women have a modestly lower (31%) volume of distribution compared to males, but the magnitude of this effect is of dubious clinical significance (Grasela et al. 1993).

9.2
Safety and Adverse Effects

9.2.1
Drug–Drug Interactions

In contrast to antipsychotics, benzodiazepines, and diclofenac, salicylic acid at high plasma concentrations can displace tianeptine from its plasma binding site and thus may potentiate its effects (Zini et al. 1991). Only limited drug–drug interaction studies have been published with tianeptine. One reports no interaction either way between tianeptine and oxazepam (Toon et al. 1990).

Tianeptine significantly reduced the wet dog shakes induced by 5-hyroxytryptophan, which would reduce the likelihood of the SE syndrome in the event that tianeptine is inadvertently used with other drugs capable of causing the SE syndrome (Datla and Curzon 1993).

9.2.2
Adverse Effects

The most common adverse effects include nausea, constipation, abdominal pain, headache, dizziness, and altered dreaming (Wagstaff et al. 2001). The following adverse effects are more common with amitriptyline than tianeptine: dry mouth (38% vs 20%), constipation (19% vs 15%), dizziness/syncope (23% vs 13%), drowsiness (17% vs 10%), and postural hypotension (8% vs 3%). However, insomnia and nightmares were more common with tianeptine than amitriptyline (20% vs 7%) (Wilde and Benfield 1995). Another report based on 1,458 outpatients on 37.5 mg/day of tianeptine who were followed for 3 months by 392 general practitioners in an open label study found that fewer than 5% of patients stopped prematurely due to adverse effects and that none of the adverse effects were clinically severe and/or serious. No cardiovascular, hematological, hepatic, or other biochemical abnormalities were found nor was there any evidence of abuse, dependence, or withdrawal symptoms.

Hepatotoxicity has been reported but is rare. In one such patient, there were hypersensitivity manifestations suggestive of an immuno-allergic mechanism and histological evidence of microvesicular steatosis. Discontinuation of tianeptine was followed by complete recovery (Lebricquir et al. 1994). This case is possibly due to oxidation of tianeptine to reactive metabolites and the inhibition of mitochondrial β-oxidation of fatty acids.

The cardiovascular effects of tianeptine have been assessed in specific placebo-controlled trials in healthy volunteers and by heart rate, blood pressure, and ECG data analysis in 5 trials involving 3,300 depressed patients (Juvent et al. 1990; Kasahara et al. 1996). Based on these studies, tianeptine does not affect heart rate, blood pressure, ventricular function (i.e., cardiac output), or intracardiac conduction, including in elderly or alcoholic patients or in those with pre-existing cardiac abnormalities. There were few instances of orthostatic hypotension. Suicide attempts did not result in death due to cardiovascular complications.

In a placebo-controlled and mianserin-controlled crossover study in healthy volunteers, tianeptine, in contrast to mianserin, did not affect a brake-reaction time "on-the-road" measure, choice reaction time, critical flicker fusion, or self assessed ratings of sedation (Ridout and Hindmarch 2001).

9.2.3
Overdose

There are limited published data on the tianeptine overdose risk. In a long-term maintenance study, seven patients attempted suicide by tianeptine overdose and all had uneventful outcomes (Loo et al. 1992). In addition to this experience, there is also the report of a patient who abused tianeptine for stimulant effect, taking 150 tablets daily (i.e., 50 times the recommended dose) for an unspecified period of time (Vandel et al. 1999). No severe toxicity, including hepatotoxicity, was observed. This patient initially experienced nausea, vomiting, abdominal pain, anorexia with weight loss, and constipation, but these adverse effects progressively disappeared despite continued and even escalating drug use. She was abruptly discontinued off tianeptine and experienced a mild withdrawal syndrome consisting of myalgia and cold sensation.

9.3
Clinical Use/Therapeutic Indications

9.3.1
Efficacy for Acute Treatment

Like reboxetine and milnacipran and consistent with its primarily European development, most of the published trials of tianeptine have been active rather than placebo-controlled and results were not published in full with rigorous peer review, compromising the ability to make an assessment of efficacy.

A summary report of three multicenter, randomized, double-blind, placebo-controlled trials involving a total of 556 patients has been published (Ginstet 1997). One of these studies conducted in Brazil has also been published separately (Costa e Silva et al. 1997). In the Brazilian study, which involved 126 depressed outpatients meeting DSM-III-R for major depression or bipolar disorder, tianeptine, 25–50 mg/day, produced a 58% response rate compared with 41% for placebo and statistically significantly superior ($p<0.01$) reduction in the final MADRS. In a second study in 186 patients with major depression or bipolar disorder, depressed phase, there was a statistically significantly greater reduction ($p<0.05$) in the final MADRS scores in both tianeptine (37.5 mg/day) and imipramine (150 mg/day) treated patients in comparison to placebo. The response rates were 56% for tianeptine, 48% for imipramine, and 32% for placebo. In the largest study involving 244 patients with major depression, tianeptine at doses of 37.5 and 75 mg/day did not produce superior response compared with placebo due to a greater than 65% response rate on placebo. The other published studies in major depression have been active- rather than placebo-controlled with amitriptyline being the most common comparator used.

An overview report that briefly summarized five double-blind studies found that tianeptine was comparable in efficacy and better tolerated than reference antidepressants, mainly amitriptyline, in patients with either DSM-III major depression, single or recurrent (without melancholia and psychotic features) or dysthymic disorder with or without an additional diagnosis of alcohol abuse or dependence (Guelfi 1992). One of these was a 6-week, multicenter trial conducted in Italy involving 300 inpatients or outpatients with DSM-III major depression. There was no difference in efficacy, but tianeptine produced fewer adverse effects (Invernizzi et al. 1994). There was also a comparison study vs fluoxetine in 387 patients with major depression, recurrent depressive disorder, or bipolar affective disorder as diagnosed using ICD-10 criteria. There was no difference on any efficacy parameter and the response rate was 58% and 56% for tianeptine and fluoxetine, respectively.

There have also been some trials of tianeptine in patient populations not typically studied in U.S. registration trials. A placebo-controlled trial reported greater improvement with tianeptine in patients with psychasthenia (Grivois et al. 1992). A maprotiline-controlled trial in 83 menopausal or premenopausal women with anxiodepressive disorder reported statistically superior ($p<0.01$) response and better tolerability in the tianeptine-treated patients (Chaby et al. 1993). In a 4- to 8-week, random assignment, double-blind study involving 129 chronic alcoholic patients with comorbid major depression or dysthymia, tianeptine (37.5 mg/day) produced comparable efficacy but was better tolerated in comparison to amitriptyline (75 mg/day), particularly in terms of better vigilance as well as better overall adverse effect profile (Loo et al. 1988). In a 6-week, double-blind, random assignment study in 265 outpatients with DSM-III dysthymic disorder with manifest anxiety, tianeptine (37.5 mg/day) produced comparable efficacy to amitriptyline (75 mg/day) with treatment response rates of 78% vs 83%, respectively (Guelfi et al. 1989).

There have also been some studies in elderly patients. In a 12-week, double-blind, random assignment study in 237 elderly patients (>65 years of age) with DSM-III-R major depression, fluoxetine (20 mg/day) produced a higher remission rate (MADRS<or=10) compared with tianeptine (37.5 mg/day) ($p<0.01$) (Guelfi et al. 1999). There have been two other open-label trials in such patients, with one trial having been published twice (Saiz-Ruiz et al. 1997, 1998; Andrusenko et al. 1999). In these studies, tianeptine was well tolerated.

In addition to acute efficacy trials, there have also been relapse prevention studies. While most were open label, a double-blind, placebo-controlled study has been reported (Dalery et al. 1997). In this study, 286 patients meeting DSM-III-R criteria for recurrent major depression (i.e., had to have at least one previous episode in the last 5 years) were initially treated open label with tianeptine, 37.5 mg/day for 6 weeks; 185 patients who responded were randomly assigned in a double-blind fashion to either continue of tianeptine or be switched to placebo and followed for 18 months. By 18 months, the relapse rate in the placebo-treated patients was 36% vs 16% in the patients treated with tianeptine. In addition to this study, there have been several open-label studies of several patients maintained on tianeptine for more than 1 year with maintenance of response in over 80% and no late emergent unexpected adverse effects (Loo et al. 1991, 1992). Finally, there was an interesting, albeit open-label, study of 130 patients treated with tianeptine for both major depression and comorbid alcoholism (Malka et al. 1992). Only one patient dropped out because of late-emergent adverse effects and 5% because of relapse of their alcoholism.

9.3.2
Dosing Recommendations

Doses of tianeptine used in most of the clinical trials have been between 25 and 50 mg/day with 37.5 mg/day appearing to be the most efficacious for most patients.

9.3.3
Onset of Action

There have been no published studies demonstrating a faster onset of action for tianeptine.

10
Conclusion

One useful method of assessing for clinically meaningful differences in the pharmacology of antidepressants is to examine their relative risk of causing specific adverse effects. Tables 2 and 3 in the Appendix to this volume present comparisons of the placebo-subtracted incidence rate (percentage) of frequent adverse effects for a number of commonly used antidepressants based on data from the double-blind trials presented to the FDA for drug approval. The placebo-subtracted incidence can assist with attributing adverse events to the specific effects of a drug, as opposed to other factors such as the underlying illness (e.g., major depressive disorder). The data in this table are a function of the average dose used in the clinical trials program that led to the drug's approval. Unlike the SSRIs, the adverse effect profiles of the non-SSRI antidepressants differ substantially from one another in a way that is consistent with the differences in their in vitro binding affinities. In other words, whereas the relative adverse effect profiles of the various SSRIs differ quantitatively (see the chapter by Preskorn et al. on the SSRIs), the adverse effect profiles of the non-SSRIs differ qualitatively.

In examining the adverse effects of the non-SSRIs, from most frequent to least frequent, their qualitative differences become apparent. Drowsiness, the most common side effect of mirtazapine, is consistent with blockade of H-1 receptors, its most potent action. Consistent with its SE uptake blockade, venlafaxine's most common side effect is nausea. The dizziness reported with nefazodone is similar to that caused by buspirone, due to its 5-HT$_{1A}$ agonism. By contrast, dizziness seen with imipramine is accompanied by an orthostatic drop in blood pressure attributable to α_1-adrenergic blockade (Preskorn 2000d).

Imipramine, mirtazapine, and venlafaxine all cause dry mouth; however, the mechanism of action for this adverse event differs for each agent, as evidenced by each drug's in vitro binding data. While imipramine causes this effect due to blockade of muscarinic acetylcholine receptor blockade, both mirtazapine and venlafaxine are only weak blockers of this receptor. Instead, both of these drugs likely cause this side effect by potentiation of NE. Venlafaxine does this by direct inhibition of the NE uptake pump. Mirtazapine does this via α_2-adrenergic receptor blockade which can increase NE release (Preskorn 2000d).

The point of these examples is that the adverse effects of antidepressants can be understood, and even anticipated, through knowledge of the medication's in vitro pharmacology coupled with a knowledge of the concentration of the drug achieved under clinically relevant dosing conditions.

In examining these data, keep in mind that the data come from product labeling as opposed to head-to-head trials. In other words, such data may not reflect the actual rate of these adverse effects in clinical practice or the actual differences between drugs. The ideal comparison would be based on large prospective double-blind studies in which every drug was compared at an equivalent antidepressant dose. Realistically, the concept of "equivalent antidepressant dose" has remained elusive, given the nature of the drug registration process and the vagaries of clinical trials in psychiatry. To truly compare the relative efficacy, and incidence and severity of side effects, of all the available antidepressants in a single study with adequate power to detect clinically meaningful differences would be cost-prohibitive—hence such a study is unlikely.

In addition, compilations and meta-analyses of adverse event data can come from a number of different types of databases that clearly lack methodological homogeneity. For example, in the registration trials for bupropion, a standardized checklist was used to elicit adverse event data, while data for other antidepressants came from other sources. These caveats should be factored into a critical examination of analyses of comparative adverse events across antidepressants.

Given these limitations of the available critical analyses, it is helpful to be knowledgeable about the binding affinities and different mechanisms of action of the available antidepressants at clinically relevant dosing ranges. As discussed in the chapter on therapeutic drug monitoring (TDM) by Burke and Preskorn (this volume), TDM, although currently underutilized, can be helpful in the use of some of the antidepressants discussed in this chapter.

References

Agelink MW, Ullrich H, Passenberg P, Sayar K, Brockmeyer NH (2002) Superior safety of reboxetine over amitriptyline in the elderly. Eur J Med Res 7:415–416

Andreoli V, Carbognin G, Abati A, Vantini G (1999) Reboxetine in the treatment of depression in the elderly: pilot study. J Geriatr Psychiatry Neurol 12:206–210

Andreoli V, Caillard V, Deo RS, Rybakowski JK, Versiani M (2002) Reboxetine, a new noradrenaline selective antidepressant, is at least as effective as fluoxetine in the treatment of depression. J Clin Psychopharmacol 22:393–399

Andrusenko MP, Sheshenin VS, Iakovleva OB (1999) Use of tianeptine (coaxil) in the treatment of late-life depression. Zh Nevrol Psikhiatr Im S S Korsakova 99:25–30

Anonymous (2002) Enhancement of human cortico-motoneuronal excitability by the selective norepinephrine reuptake inhibitor reboxetine. Neurosci Lett 330:231–234

Ansseau M, von Frenckell R, Mertens C, De Wilde J, Botte L, Deviotille JM, Evard JL, De Nayer A, Dairmont P, Dejaiffe G (1989) Controlled comparison of two doses of milnacipran (F 2207) and amitriptyline in major depressive inpatients. Psychopharmacology 98:163–168

Ansseau M, von Frenckell R, Gerard MA, Mertens C, De Wilde J, Botte L, Devoitille JM, Evard JL, De Nayer A, Dairmont P (1991) Interest of a loading dose of milnacipran in endogenous depressive inpatients. Comparison with the standard regimen and with fluvoxamine. European Neuropsychopharmacology 1:113–121

Ansseau M, Papart P, Troisfontaines B, Bartholome F, Bataille M, Charles G, Schittecatte M, Dairmont P, Devoitille JM, De Wilde J (1994) controlled comparison of milnacipran and fluoxetine in major depression. Psychopharmacology 114:137

Anttila S, Rasanen I, Leinonen E (2001) Fluvoxamine augmentation increases serum mirtazapine concentrations three to fourfold. Ann Pharmacother 35:1221–1223

Armitage R, Rush AJ, Trivedi M, et al (1994) The effects of nefazodone on sleep architecture in depression. Neuropsychopharmacology 10:123–127

Ascher JA, Cole JO, Colin J-N, et al (1995) Bupropion: a review of its mechanism of antidepressant activity. J Clin Psychiatry 56:395–401

Ashton AK, Rosen RC (1998) Bupropion as an antidote for serotonin reuptake inhibitor-induced sexual dysfunction. J Clin Psychiatry 59;112–115

Assie MB, Charveron M, Palmier C, Puozzo C, Moret C, Briley M (1992) Effects of prolonged administration of milnacipran, a new antidepressant, on receptors and monoamine uptake in the brain of the rat. Neuropharmacology 31:149–155

Atmaca M, Kuloglu M, Tezcan E, Buyukbayram A (2003) Switching to tianeptine in patients with antidepressant-induced sexual dysfunction. Hum Psychopharmacol 18:277–280

Barone P, Moret C, Briley M, Fillion G (1994) Autoradiographic characterization of binding sites for [3H] milnacipran, a new antidepressant drug, and their relationship to the serotonin transporter in rat brain. Brain Research 30:129–143

Bassant MH, Lee BH, Jazat F, Lamour Y (1991) Comparative study of the effects of tianeptine and other antidepressants on the activity of medial septal neurons in rats anesthetized with urethane. Naunyn Schmiedebergs Arch Pharmacol 344:568–573

Beique JC, Lavoie N, de Montigny C, Debonnel G (1998) Affinities of venlafaxine and various reuptake inhibitors for the serotonin and norepinephrine transporters. Eur J Pharmacol 349:129–132

Bel N, Artigas F (1999) Modulation of the extracellular 5-hydroxytryptamine brain concentrations by the serotonin and noradrenaline reuptake inhibitor, milnacipran. Microdialysis studies in rats. Neuropsychopharmacology 21:745–754

Benazzi F (2000) Urinary retention with reboxetine-fluoxetine combination in a young man. Can J Psychiatry 45:936

Bergmann JF, Laneury JP, Duchene P, Fleishaker JC, Houin G, Segrestaa JM (2000) Pharmacokinetics of reboxetine in healthy, elderly volunteers. Eur J Drug Metab Pharmacokinet 25:195–198

Berk M, du PA, Birkett M, Richardt D (1997) An open-label study of duloxetine hydrochloride, a mixed serotonin and noradrenaline reuptake inhibitor, in patients with DSM-III-R major depressive disorder. Lilly Duloxetine Depression Study Group. Int Clin Psychopharmacol 12:137–140

Bertorelli R, Amoroso D, Girotti P, Consolo S (1992) Effect of tianeptine on the central cholinergic system: involvement of serotonin. Naunyn Schmiedebergs Arch Pharmacol 345:276–281

Bolanos-Jimenez F, de Castro RM, Fillion G (1993) Antagonism by citalopram and tianeptine of presynaptic 5-HT1B heteroreceptors inhibiting acetylcholine release. Eur J Pharmacol 242:1–6

Bolden-Watson C, Richelson E (1993) Blockade by newly-developed antidepressants of biogenic amine uptake into rat brain synaptosomes. Life Sci 52:1023–1029

Borqua P, Baudrie V, Laude D, Chaouloff F (1992) Influence of the novel antidepressant tianeptine on neurochemical neuroendocrinological and behavioral effects of stress in rats. Biol Psychiatry 31:391–400

Bremner JD, Wingard P. Walshe TA (1998) Safety of mirtazapine in overdose. J Clin Psychiatry 59:233–235

Briley M, Prost JF, Moret C (1996) Preclinical pharmacology of milnacipran. International Clinical Psychopharmacology 11(Suppl 4):9–14

Bruno R, Debiaggi M, Sacchi P, et al (1999) Daily interferon regimen for chronic hepatitis C: A prospective randomised study. Clin Drug Invest 18:11–16

Budziszewska B, Jaworska-Feil L, Kajta M, Lason W (2000) Antidepressant drugs inhibit glucocorticoid receptor-mediated gene transcription-a possible mechanism. Br J Pharmacol 130:1385–1393

Bymaster FP, Dreshfield-Ahmad LJ, Threlkeld PG, et al (2001) Comparative affinity of duloxetine and venlafaxine for serotonin and norepinephrine transporters in vitro and in vivo, human serotonin receptor subtypes, and other neuronal receptors. Neuropsychopharmacology 25:871–880

Caron J, Libersa C, Hazard JR, Lacroix D, Facq E, Guedon-Moreau L, Fautrez V, Solles A, Puozzo C, Kacet,S (1993) Acute electrophysiological effects of intravenous milnacipran, a new antidepressant agent. Eur Neuropsychopharmacol 3:493–500

Castanon N, Bluthe RM, Dantzer R (2001) Chronic treatment with the atypical antidepressant tianeptine attenuates sickness behavior induced by peripheral but not central lipopolysaccharide and interleukin-1beta in the rat. Psychopharmacology 154:50–60

Catterson M, Preskorn SH (1996) Double-blind crossover study of mirtazapine, amitriptyline, and placebo in patients with major depression. In: New research program and abstracts of the 149th annual meeting of the American Psychiatric Association, New York, May 6:NR 157

Chaby L, Grinsztein A, Weitzman JJ, de Bodinat C, Dagens V (1993) Anxiety-related and depressive disorders in women during the premenopausal and menopausal period. Study of the efficacy and acceptability of tianeptine versus maprotiline. Presse Med 22:1133–8

Clerc GE, Ruimy P, Verdeau-Pailles J (1994) A double-blind comparison of venlafaxine and fluoxetine in patients hospitalized for major depression and melancholia. Int Clin Psychopharmacol 9:139–143

Clerc G (2001) Antidepressant efficacy and tolerability of milnacipran, a dual serotonin and noradrenaline reuptake inhibitor: a comparison with fluvoxamine. Int Clin Psychopharmacol 16:145–151

Conrad C, Galea LA, Kuroda Y, McEwen BS (1996) Chronic stress impairs rat spatial memory on the Y maze, and this effect is blocked by tianeptine pretreatment. Behav Neurosci 110:1321–1334

Conrad C, LeDoux JE, Magarinos AM, McEwen BS (1999) Repeated restraint stress facilitates fear conditioning independently of causing hippocampal CA3 dendritic atrophy. Behav Neurosci 113:902–913

Costa e Silva JA, Ruschel SI, Caetano D, Rocha FL, da Silva Lippi JR, Arruda S (1997) Placebo-controlled study of tianeptine in major depressive episodes. Neuropsychobiology 35:24–29

Coulomb F, Ducret F, Laneury JP, et al (2000) Pharmacokinetics of single-dose reboxetine in volunteers with renal insufficiency. J Clin Pharmacol 40:482–487

Cryan JF, Dalvi A, Jin SH, Hirsch BR, Luchi I, Thomas SA (2001) Use of dopamine-betahydroxylase-deficient mice to determine the role of norepinephrine in the mechanism of action of antidepressant drugs. J Pharmacol Exp Ther 298:651–657

Dalery J, Dagens-Lafant V, De Bodinat C (1997) Value of tianeptine in treating major recurrent unipolar depression. Study versus placebo for 16 1/2 months of treatment. Encephale 23:56–64

D'Amico M, Roberts D, Robinson D, et al (1990) Placebo-controlled dose-ranging trial designs in phase II development of nefazodone. Psychopharmacol Bull 26:147–150

Datla KP, Curzon G (1993) Behavioural and neurochemical evidence for the decrease of brain extracellular 5-HT by the antidepressant drug tianeptine. Neuropharmacology 32:839–845

Davidson J (1989) Seizures and bupropion: a review. J Clin Psychiatry 50:256–61

deBoer T, Ruight GSF, Berendsen HHG (1995) The alpha-2 selective adrenoceptor antagonist Org 3770 (mirtazapine, Remeron) enhances noradrenergic and serotonergic transmission. Human Psychopharmacology 10:S107–S118

Delbende C, Contesse V, Mocaer E, Kamoun A, Vaundry H (1991) The novel antidepressant tianeptine reduces stress-evoked stimulation of the hypothalamo-pituitary-adrenal axis. Eur J Clin Pharmacol 202:391–396

Delbende C, Tranchand Bunel D, Tarozzo G, et al (1994) Effect of chronic treatment with the antidepressant tianeptine on the hypothalamo-pituitary-adrenal axis. Eur J Pharmacol 251:245–251

Delbressin LP, Preskorn S, Horst D (1997) Characterization and inhibition of P450 enzymes involved in the metabolism of mirtazapine. In: New research program and abstracts of the 150th annual meeting of the American Psychiatric Association. San Diego, May 17–22

Delini-Stula A (2000) Milnacipran: an antidepressant with dual selectivity of action on noradrenaline and serotonin uptake. Hum Psychopharmacol 15:255–260

Demers JC, Malone M (2001) Serotonin syndrome induced by fluvoxamine and mirtazapine. Ann Pharmacother 35:1217–20

Demotes-Mainard F, Galley P, Manciet G, Vinson G, Salvadori C (1991) Pharmacokinetics of the antidepressant tianeptine at steady state in the elderly. J Clin Pharmacol 31:174–178

Demyttenaere K, Huygens R (2002) Painful ejaculation and urinary hesitancy in association with antidepressant therapy: relief with tamsulosin. Eur Neuropsychopharmacol 12:337–341

Deprez D, Chassard D, Baille P, Mignot A, Ung HL, Puozzo C (1998) Which bioequivalence study for a racemic drug? Application to milnacipran. Eur J Drug Metab Pharmacokinet 23:166–171

De Simoni MG, De Luigi A, Clavenna A, Manfridi A (1992) In vivo studies on the enhancement of serotonin reuptake by tianeptine. Brain Res 574:93–97

Detke MJ, Lu Y, Goldstein DJ, McNamara RK, Demitrack MA (2002a) Duloxetine 60 mg once daily dosing versus placebo in the acute treatment of major depression. J Psychiatr Res 36:383

Detke MJ, Lu Y, Goldstein DJ, Hayes JR, Demitrack MA (2002b) Duloxetine, 60 mg once daily, for major depressive disorder: a randomized double-blind placebo-controlled trial. J Clin Psychiatry 63:308–315

Dostert P, Benedetti MS, Poggesi I (1997) Review of the pharmacokinetics and metabolism of reboxetine, a selective noradrenaline reuptake inhibitor. Eur Neuropsychopharmacol 7(Suppl 1):S23-S35

Dresse A, Rosen JM, Brems H, Masset H, Defrance R, Salvadori C (1988) Influence of food on tianeptine and its main metabolic kinetics. J Clin Pharmacol 28:1115–1119

Edwards DM, Pellizzoni C, Breuel HP, et al (1995) Pharmacokinetics of reboxetine in healthy volunteers. Single oral doses, linearity and plasma protein binding. Biopharm Drug Dispos 16:443–460

Ekins S, Bravi G, Ring BJ, et al (1999) Three-dimensional quantitative structure activity relationship analyses of substrates for CYP2B6. J Pharmacol Exp Ther 288:21-29

Engleman EA, Perry KW, Mayle DA, Wong DT (1995) Simultaneous increases of extracellular monoamines in microdialysates from hypothalamus of conscious rats by duloxetine, a dual serotonin and norepinephrine uptake inhibitor. Neuropsychopharmacology 12:287–295

Engleman EA, Robertson DW, Thompson DC, Perry KW, Wong DT (1996) Antagonism of serotonin 5-HT1A receptors potentiates the increase in extracellular monoamines induced by duloxetine in rat hypothalamus. J Neurochem 66:599–603

Fabre LF, Brodie HK, Garver D, et al (1983) A multicenter evaluation of bupropion versus placebo in hospitalized depressed patients. J Clin Psychiatry 44:88–94

Fawcett JA, Barkin R (1997) Efficacy issues with antidepressants. J Clin Psychiatry 58(suppl 6):32–39

Fawcett J, Marcus R, Anton S, et al (1995) Response of anxiety and agitation symptoms during nefazodone treatment of major depression. J Clin Psychiatry 56(suppl 6):37–42

Feighner JP (1994) The role of venlafaxine in rational antidepressant therapy. J Clin Psychiatry 55 (suppl 9A):62–68

Feighner JP (1995) Cardiovascular safety in depressed patients: focus on venlafaxine. J Clin Psychiatry 56:574–579

Feighner JP, Pambakian R, Fowler RC, et al (1989) A comparison of nefazodone, imipramine, and placebo in patients with moderate to severe depression. Psychopharmacol Bull 25:219–221

Ferguson JM, Mendels J, Schwart GE (2002) Effects of reboxetine on Hamilton Depression Rating Scale factors from randomized, placebo-controlled trials in major depression. Int.Clin Psychopharmacol 17:45–51

Ferrier IN (2001) Characterizing the ideal antidepressant therapy to achieve remission. J Clin Psychiatry 62(Suppl 26):10–5

Fleishaker JC (2000) Clinical pharmacokinetics of reboxetine, a selective norepinephrine reuptake inhibitor for the treatment of patients with depression. Clin Pharmacokinet 39:413–427

Fleishaker JC, Mucci M, Pellizzoni C, Poggesi I (1999) Absolute bioavailability of reboxetine enantiomers and effect of gender on pharmacokinetics. Biopharm Drug Dispos 20:53–57

Fleishaker JC, Francom SF, Herman BD, Knuth DW, Azie NE (2001) Lack of effect of reboxetine on cardiac repolarization. Clin Pharmacol Ther 70:261–269

Fontaine R, Ontiveros A, Faludi G, et al (1991) A study of nefazodone, imipramine, and placebo in depressed outpatients. Biol Psychiatry 29:118A

Fontaine R, Ontiveros A, Elie R, et al (1994) A double-blind comparison of nefazodone, imipramine, and placebo in major depression. J Clin Psychiatry 55(suppl 6):234–241

Frankfurt M, McKittrick CR, McEwen BS, Luine VN (1995) Tianeptine treatment induces regionally specific changes in monoamines. Brain Res 696:1–6

Frazer A (1997) Antidepressants. J Clin Psychiatry 58(suppl 6):9–25

Frigerio E, Benecchi A, Brianceschi G, et al (1997) Pharmacokinetics of reboxetine enantiomers in the dog. Chirality 9:303-306

Fromenty B, Freneaux E, Labbe G, et al (1989) Tianeptine, a new tricyclic antidepressant metabolized by beta-oxidation of its heptamoic side chain, inhibits the mitochondrial oxidation of medium and short chain fatty acids in mice. Biochem Pharmacol 38:3743-3751

Fromenty B, Pessayre D (1995) Inhibition of mitochondrial beta-oxidation as a mechanism of hepatotoxicity. Pharmacol Ther 67:101-154

Fukuchi T, Kanemoto K (2002) Differential effects of milnacipran and fluvoxamine, especially in patients with severe depression and agitated depression: a case-control study. International Clinical Psychopharmacology 17:53-58

Fuller RW, Hemrick-Luecke SK, Snoddy HD (1994) Effects of duloxetine an antidepressant drug candidate on concentrations of monoamines and their metabolites in rats and mice. J Pharmacol Exp Ther 269:132-136

Gillin JC, Rapaport M, Erman MK, et al (1997) A comparison of nefazodone and fluoxetine on mood and on objective, subjective, clinician-rated measures of sleep in depressed patients: a double-blind, 8-week, clinical trial. J Clin Psychiatry 58:185-192 (for erratum, see J Clin Psychiatry 1997;58:275)

Ginstet D (1997) Efficacy of tianeptine in major depressive disorders with or without melancholia. Eur Neuropsychopharmacol 7:S341-S45

Gobert A, Millan MJ (1999) Modulation of dialysate levels of dopamine, noradrenaline, and serotonin (5-HT) in the frontal cortex of freely-moving rats by (−)-pindolol alone and in association with 5-HT reuptake inhibitors: comparative roles of B-adrenergic, 5-HT$_{1A}$, and 5-HT$_{1B}$ receptors. Neuropsychopharmacology 21:269-284

Gobert A, Rivet JM, Cistarelli L, Melon C, Millan MJ (1997) Alpha2-adrenergic receptor blockade markedly potentiates duloxetine- and fluoxetine-induced increases in noradrenaline, dopamine, and serotonin levels in the frontal cortex of freely moving rats. J Neurochem 69:2616-2619

Golden, RN, DeVane CL, Laizure SC, Rudorfer MD, Sherer MA, Potter WZ (1985) Bupropion in depression: The role of metabolites in clinical outcome. Arch Gen Psychiatry 45:145-49

Goldstein DJ, Mallinckrodt C, Lu Y, Demitrack MA (2002) Duloxetine in the treatment of major depressive disorder: a double-blind clinical trial. J Clin Psychiatry 63:225-231

Gongora-Alfaro JL, Hernandez-Lopez S, Flores-Hernandez J, Galarraga E (1997) Firing frequency modulation of substantia nigra reticulata neurons by 5-hydroxytryptamine. Neurosci Res 29:225-231

Grasela TH, Fiedler-Kelly JB, Salvadori C, Marey C, Jochemson R (1993) Development of a population pharmacokinetic database for tianeptine. Eur J Clin Pharmacol 45:173-179

Grislain L, Gele P, Bertrand M, et al (1990) The metabolic pathways of tianeptine, a new antidepressant, in healthy volunteers. Drug Metab Dispos 18:804-808

Grivois H, Deniker P, Ganry H (1992) Efficacy of tianeptine in the treatment of psychasthenia. A study versus placebo. Encephale 18:591-598

Grunder G, Wetzel H, Schloer R, et al (1993) Subchronic antidepressant treatment with venlafaxine or imipramine and effects on blood pressure: assessment by automatic 24 hour monitoring. Pharmacopsychiatry 26:155

Guelfi JD (1992) Efficacy of tianeptine in comparative trials versus reference antidepressants. An overview. Br J Psychiatry Suppl 15:72-75

Guelfi JD, Pichot P, Dreyfus JF (1989) Efficacy of tianeptine in anxious-depresses patients: results of a controlled multicenter trial versus amitriptyline. Neuropsychobiology 22:41-48

Guelfi JD, Bouhassira M, Bonett-Perrin E, Lancrenon S (1999) The study of the efficacy of fluoxetine versus tianeptine in the treatment of elderly depressed patients followed in general practice. Encephale 25:265-270

Harvey AT, Rudolph RL, Preskorn SH (2000) Evidence of the dual mechanisms of action of venlafaxine. Arch Gen Psychiatry 57:503–509,

Hendershot PE, Fleishaker JC, Lin KM, Nuccio ID, Poland RE (2001) Pharmacokinetics of reboxetine in healthy volunteers with different ethnic descents. Psychopharmacology (Berl) 155:148–153

Herman BD, Fleishaker JC, Brown MT (1999) Ketoconazole inhibits the clearance of the enantiomers of the antidepressant reboxetine in humans. Clin Pharmacol Ther 66:374–379

Higuchi H, Yoshida K, Takahashi H, Naito S, Tsukamoto K, Kamata M, Ito K, Sato R, Shimizu T (2002) Remarkable effect of milnacipran in the treatment of Japanese depressive patients. Hum Psychopharmacol 17:195–196

Hindmarch I, Rigney U, Stanley N, Briley M (2000) Pharmacodynamics of milnacipran in young and elderly volunteers. British Journal of Clinical Pharmacology 49:118–125

Hirschfeld RM (1999) Care of the sexually active depressed patient. J Clin Psychiatry 60:32–35

Invernizzi G, Pozzi L, Garattini S, Samanin R (1992) Tianeptine increases the extracellular concentration of dopamine in the nucleus. Neuropharmacology 31:221–227

Invernizzi G, Aguglia E, Bertolino A, et al (1994) The efficacy and safety of tianeptine in the treatment of depressve disorder: results of a controlled double-blind multicenter study vs. amitriptyline. Neuropsychbiology 30:85–93

Invernizzi RW, Parini S, Sacchetti G, et al (2001) Chronic treatment with reboxetine by osmotic pumps facilitates its effect on the extracellular noradrenaline and may desensitize alpha (2) -adrenoceptors in the prefrontal cortex. Br J Pharmacol 2001;132:183–188

Ishigooka J, Kasahara T, Nagata E, Murasaki M, Miura S (1998) Effects of washing procedure on platelets pretreated with serotonin uptake inhibitors in vitro: low Ki values predict long-lasting inhibition of serotonin uptake in vivo. Nihon Shinkei Seishin Yakurigaku Zasshi 18:19–21

Jackson HC, Needham AM, Hutchins LJ, Mazurkiewicz SE, Heal DJ (1997) Comparison of the effects of sibutramine and other monoamine reuptake inhibitors on food intake in the rat. Br J Pharmacol 121:1758–1762

Janicak PG, Davis JM, Preskorn SH, Ayd FJ, Jr (2001) Principles and practice of psychopharmacotherapy. Lippincott Williams & Wilkins, Philadelphia

Juvent M, Douchamps J, Delcourt E, et al (1990) Lack of cardiovascular side effects of the new tricyclic antidepressant tianeptine. A double-blind, placebo-controled study in young healthy volunteers. Clin Neuropharmacol 13:48–57

Kamoun A, Labrid C, Mocaer E, Perret L, Poirier JP (1989) Tianeptine an uncommon psychotropic drug. Encephale 15:419–422

Kamoun A, Delalleau B, Ozun M (1994) Can a serotonin uptake agonist be an authentic antidepressant? Results of a multicenter, multinational therapeutic trial. Encephale 20:521–525

Kasahara T, Ishigooka J, Nagata E, Murasaki M, Miura S (1996) Long-lasting inhibition of 5-HT uptake of platelets in subjects treated by duloxetine, a potential antidepressant. Nihon Shinkei Seishin Yakurigaku Zasshi 16:25–31

Kasamo K, Blier P, de Montigny C (1996) Blockade of the serotonin and norepinephrine uptake processes by duloxetine: in vitro and in vivo studies in the rat brain. J Pharmacol Exp Ther 277:278–286

Kasper S (2000) Managing reboxetine-associated urinary hesitancy in a patient with major depressive disorder: a case study. Psychopharmacology (Berl) 159:445–446

Kasper S, Pletan Y, Solles A, Tournoux A (1996) Comparative studies with milnacipran and tricyclic antidepressants in the treatment of patients with major depression: a summary of clinical trial results. International Clinical Psychopharmacology 11(Suppl 4):35–39

Kato G, Weitsh AF (1988) Neurochemical profile of tianeptine a new antidepressant drug. Clin Neuropharmacol 11:S43-S50

Katoh A, Eigyo M, Ishibashi C, et al (1995) Behavioral and electroencephalographic properties of duloxetine (LY248686) a reuptake inhibitor of norepinephrine and serotonin, in mice and rats. J Pharmacol Exp Ther 272:1067-1075

Katona C, Bercoff E, Chiu E, Tack P, Versiani M, Woelk H (1999) Reboxetine versus imipramine in the treatment of elderly patients with depressive disorders: a double-blind randomised trial. J Affect Disord 55:203-13

Keller M (2001) Role of serotonin and noradrenaline in social dysfunction: A review of data on reboxetine and the Social Adaptation Self-evaluation Scale (SASS). Gen Hosp Psychiatry 23:15-19

Kelly JP, Leonard BE (1994) The effect of tianeptine and sertraline in three models of depression. Neuropharmacology 33:1011-1016

Kihara T, Ikeda M (1995) Effects of duloxetine a new serotonin and norepinephrine uptake inhibitor on extracellular monoamine levels in rat frontal cortex. J Pharmacol Exp Ther 272:177-183

Klamerus KJ, Maloney K, Rudolph RL, et al (1992) Introduction of a composite parameter to the pharmacokinetics of venlafaxine and its active O-desmethyl metabolite. J Clin Pharmacol 32:716-724

Kroeze WK, Kristiansen K, Roth BL (2002) Molecular biology of serotonin receptors structure and function at the molecular level. Curr Top Med Chem 2:507-528

Kuroda Y, Watanabe A, McEwen B (1994) Tianeptine decreases both serotonin transporter MRNA and binding sites in rat brain. Eur J Pharmacol 268:R3-R5

Kuroda Y, McEwen BS (1998) Effect of chronic restraint stress and tianeptine on growth factors, growth-associated protein-43 and microtubule-associated protein 2 mRNA expression in the rat hippocampus. Brain Res Mol Brain Res 59:35-39

Labbate LA, Grimes JB, Hines A, et al (1997). Bupropion treatment of serotonin reuptake antidepressant-associated sexual dysfunction. Ann Clin Psychiatry 9:241-245

Labrid C, Mocaer E, Kamoun A (1992) Neurochemical and pharmacological properties for tianeptine a novel antidepressant. Br J Psychiatry Suppl 15:56-60

Landry P, Roy L (1997) Withdrawal hypomania associated with paroxetine. J Clin Psychopharmacol 17:60-61

Larrey D, Tinel M, Letteron P, Maurel P, Loeper J, Belghiti J, Pessayre D (1990) Metabolic activation of the new tricyclic antidepressant tianeptine by human liver cytochrome P450. Biochem Pharmacol 40:545-550

Lebricquir Y, Larrey D, Blanc P, Pageaux GP, Michel H (1994) Tianeptine—an instance of drug-induced hepatotoxicity predicted by prospective experimental studies. J Hepatol 21:771-773

Lecrubier Y, Pletan Y, Solles A, Tournoux A, Magne V (1996) Clinical efficacy of milnacipran: placebo-controlled trials. International Clinical Psychopharmacology 11(suppl 4): 29-33

Leinonen E, Lepola U, Koponen H, Mehtonen OP, Rimon R (1997) Long-term efficacy and safety of milnacipran compared to clomipramine in patients with major depression. Acta Psychiatrica Scandinavica 96:497-504

Letteron P, Labbe G, Descatoire V, et al (1989) Metabolic activation of the antidepressant tianeptine. II. In vivo covalent binding and toxicological studies at sublethal doses. Biochem Pharmacol 38:3247-3251

Letteron P, Descatoire V, Tinel M, Maurel P, Labbe G, Loeper J, Larrey D, Freneaux E, Pessayre D (1998) Metabolic activation of the antidepressant tianeptine. I. Cytochrome P-450-mediated in vitro covalent binding. Biochem Pharmacol 38:3241-3246

Linner L, Endersz H, Ohman D, Bengtssonm F, Schalling M, Svensson TH (2001): Reboxetine modulates the firing pattern of dopamine cells in the ventral tegmental area and selctively increases dopamine availability in the prefrontal cortex. J Pharmacol Exp Ther 297:540-546

Liu H, Hoff BH, Anthonsen T (2000) Chemo-enzymatic synthesis of the antidepressant duloxetine and its enantiomer. Chirality 12:26–29

Loo H, Deniker P (1988) Position of tianeptine among antidepressive chemotherapies. Clin Neuropharmacol 11:S97–S102

Loo H, Malka R, Defrance R, et al (1988) Tianeptine and amitriptyline. Controlled double-blind trial in depressed alcoholic patients. Neuropsychobiology 19:79–85

Loo H, Ganry H, Dufour H, et al (1991) Role of tianeptine in the prevention of depression relapse and recurrence. First estimations. Presse Med 20:1864–1868

Loo H, Ganry H, Dufour H, et al (1992) Long-term use of tianeptine in 380 depressed patients. Br J Psychiatry Suppl 15:61–65

Lopez-Ibor J, Guelfi JD, Pletan Y, Tournoux A, Prost JF (1996) Milnacipran and selective serotonin reuptake inhibitors in major depression. International Clinical Psychopharmacology 11(Suppl 4):41–46

Macher JP, Sichel JP, Serre C, von Frenckell R, Huck JC, Demarez JP (1989) Double-blind placebo-controlled study of milnacipran in hospitalized patients with major depressive disorders. Neuropsychobiology 22:77–82

Magarinos AM, Deslandes A, McEwen BS (1999) Effects of antidepressants and benzodiazepine treatments on the dendritic structure of CA3 pyramidal neurons after chroni stress. Eur J Pharmacol 371:113–122

Maj J, Rogoz Z, Dlaboga D, Dziedzicka-Wasylewska M (2000) Pharmacological effects of milnacipran, a new antidepressant, given repeatedly on the alpha1-adrenergic and serotonergic 5-HT2A systems. Journal of Neural Transmission 107:1345–1359

Malka R, Loo H, Ganry H, Souche A, Marey C, Kamoun A (1992) Long-term administration of tianeptine in depressed patients after alcohol withdrawal. Br J Psychiatry Suppl 15:66–71

Manier DH, Shelton RC, Sulser F (2002) Noradrenergic antidepressants: does chronic treatment increase or decrease nuclear CREB-P? J Neural Transm 109:91–99

Marinesco S, Poncet L, Debilly G, Jouvet M, Cespuglio R (1996) Effects of tianeptine, sertraline and clomipramine on brain serotonin metabolism: a voltammetric approach in the rat. Brain Res 736:82–90

Martin P, Massol J, Colin JN, Lacomblez L, Puech AJ (1990) Antidepressant profile of bupropion and three metabolites in mice. Pharmacopsychiatry 23:187–94

Massana J (1998) Reboxetine versus fluoxetine: An overview of efficacy and tolerability. J Clin Psychiatry 59:8–10

Massana J, Moller HJ, Burrows GD, Montenegro RM (1999) Reboxetine: a double-blind comparison with fluoxetine in major depressive disorder. Int Clin Psychopharmacol 14:73–80

McEwen BS, Conrad C, Kuroda Y, Frankfurt M, Magarinos AM, McKittrick C (1997) Prevention of stress-induced morphological and cognitive consequences. Eur Neuropsychopharmacol 7:S323-S328

Mendels J, Reimherr F, Marcus R, et al (1995) A double-blind, placebo-controlled trial of two dose ranges of nefazodone in the treatment of depressed outpatients. J Clin Psychiatry 56 (suppl 6):30–36

Mennini T, Mocaer E, Garattini S (1987) Tianeptine, a selective enhancer of serotonin uptake in rat brain. Naunyn Schmiedebergs Arch Pharmacol 336:478–482

Meredith CH, Feighner JP (1983) The use of bupropion in hospitalized depressed patients. J Clin Psychiatry 44:85-87

Miller DK, Wong EH, Chesnut MD, Dwoskin LP (2002) Reboxetine: Functional inhibition of monoamine transporters and nicotinic acetylcholine receptors. J Pharmacol Exp Ther 302:687–695

Mitchell PB (1995) Novel French antidepressants not available in the United States. Psychopharmacol Bull 31:509–519

Mochizuki D, Hokonohara T, Kawasaki K, Miki N (2002a) Repeated administration of milnacipran induces rapid desensitization of somatodendritic 5-HT1A auto receptors but not postsynaptic 5-HT1A receptors. J Psychopharmacol 16:253–260

Mochizuki D, Tsujita R, Yamada S, Kawasaki K, Otsuka Y, Hashimoto S, Hattori T, Kitamura Y, Miki N (2002b) Neurochemical and behavioural characterization of milnacipran a serotonin and noradrenaline reuptake inhibitor in rats. Psychopharmacol 162:323–332

Mongeau R., Weiss M, de Montigny C, Blier P (1998) Effect of acute, short and long-term milnacipran administration on rat locus coeruleus noradrenergic and dorsal rashe serotonergic neurons. Neuropharmacology 37:905–918

Montgomery SA (1995) Safety of mirtazapine: a review. Int Clin Psychopharmacology 10(Suppl 4):37–45

Montgomery SA, Prost JF, Solles A, Briley M (1996) Efficacy and tolerability of milnacipran: an overview. Int Clin Psychopharmacol 11:47–51

Montgomery SA, Reimitz PE, Zivkov M (1998) Mirtazapine versus amitriptyline in the long-term treatment of depression: a double-blind placebo-controlled study. Int Clin Psychopharmacol 13:63–73

Moret C, Briley M (1990) Serotonin autoreceptor subsensitivity and antidepressant activity. European Journal of Pharmacology 180:351–356

Moret C, Briley M (1994) Effect of milnacipran and desipramine on noradrenergic alpha 2-autoreceptor sensitivity. Prog Neuropsychopharmacol Biol Psychiat 18:1063–1072

Musso DL, Mehta NB, Soroko FE, et al (1993) Synthesis and evaluation of the antidepressant activity of the enantiomers of bupropion. Chirality 5:495–500

Muth EA, Moyer JA, Haskins JT, et al (1991) Biochemical, neurophysiological, and behavioral effects of WY-45,233 and other identified metabolites of the antidepressant venlafaxine. Drug Dev Res 23:191–199

Narayan M, Anderson G, Cellar J, et al (1998) Serotonin transporter-blocking properties of nefazodone assessed by measurement of platelet serotonin. J Clin Psychopharmacol 18:67–71

Neliat G, Bodinier MC, Panconi E, Briley M (1996) Lack of effect of repeated administration of milnacipran, a double noradrenaline and serotonin reuptake inhibitor, on the beta-adrenoceptor-linked adenylate cyclase system in the rat cerebral cortex. Neuropharmacology 35:589–593

Nelson JC (1997) Safety and tolerability of the new antidepressants. J Clin Psychiatry 58(suppl 6):26–31

Norton PA, Zinner NR, Yalcin I, Bump RC (2002) Duloxetine versus placebo in the treatment of stress urinary incontinence. Am J Obstet Gynecol 187:40–48

Nowakowska E, Kus K, Chodera A, Rybakowski J (2000) Behavioural effects of fluoxetine and tianeptine, two antidepressants with opposite action mechanisms, in rats. Arzneimittelforschung 50:5–10

O'Flynn R, Michael A (2000) Reboxetine-induced spontaneous ejaculation. Br J Psychiatry 177:567–568

Okuyama-Tamura M, Mikuni M, Kojima I (2003) Modulation of the human glucocorticoid receptor function by antidepressive compounds. Neurosci Lett 342:206–210

Olivier B, Molewijk HE, van der Heyden JA, et al (1998) Ultrasonic vocalizations in rat pups: effects of serotonergic ligands. Neurosci Biobehav Rev 23:215–227

Oluyomi AO, Datla KP, Curzon G (1997) Effects of the (+) and (−) enantiomers of the antidepressant drug tianeptine on 5-HTP-induced behavior. Neuropharmacology 36:383–387

Palmier C, Puozzo C, Lenehan T, Briley M (1989) Monoamine uptake inhibition by plasma from healthy volunteers after single oral doses of the antidepressant milnacipran. European Journal of Clinical Pharmacology 37:235–238

Papps BP, Shajahan PM, Ebmeier KP, O'Carroll RE (2002) The effects of noradrenergic re-uptake inhibition on memory encoding in man. Psychopharmacology 159:311-318

Pellizzoni C, Poggesi I, Jorgensen NP, Edwards DM, Paus E, Strolin BM (1996) Pharmacokinetics of reboxetine in healthy volunteers. Single against repeated oral doses and lack of enzymatic alterations. Biopharm Drug Dispos 17:623-633

Physicians' Desk Reference, 57th Edition (2003) Thomson PDR, Montvale, NJ

Pineyro G, Deveault L, de Montigny C, Blier P (1995a) Effect of prolonged administration of tianeptine on 5-HT neurotransmission: an electrophysiological study in the rat hippocampus and dorsal raphe. Naunyn Schmiedebergs Arch Pharmacol 351:119-125

Pineyro G, Deveault L, Blier P, Dennis T, de Montigny C (1995b) Effect of acute and prolonged tianeptine administration on the 5-HT transporter: electrophysiological, biochemical and radioligand binding studies in the rat brain. Naunyn Schmiedebergs Arch Pharmacol 351:111-118

Poggesi I, Pellizzoni C, Fleishaker JC (2000) Pharmacokinetics of reboxetine in elderly patients with depressive disorders. Int J Clin Pharmacol Ther 38:254-259

Presentation to the Food and Drug Administration Psychopharmacology Advisory Committee on nefazodone. Washington, DC, July 1993

Presentation to the Food and Drug Administration on venlafaxine. Psychopharmacology Advisory Committee. Washington DC, April 1993

Preskorn SH (1991) Should bupropion dosage be adjusted based on therapeutic drug monitoring? Psychopharmacol Bull 27:637-43

Preskorn SH (1994) Antidepressant drug selection: criteria and options. J Clin Psychiatry 55(suppl 9A):6-22

Preskorn SH (1995) Comparison of the tolerability of bupropion, fluoxetine, imipramine, nefazodone, paroxetine, sertraline, and venlafaxine. J Clin Psychiatry 56(suppl 6):12-21

Preskorn SH (1997) Selection of an antidepressant: mirtazapine. J Clin Psychiatry 58(suppl 6):3-8

Preskorn SH (1998) Recent dose-effect studies regarding antidepressants. In: Balant LP, Benitez J, Dahl SG, et al (eds) European cooperation in the field of scientific and technical research. European Commission, Belgium, pp 45-61

Preskorn SH (1999) Outpatient management of depression: A guide for the primary-care practitioner. Professional Communications Inc, Caddo, OK

Preskorn SH (2000a) Bupropion: What mechanism of action? J Psych Pract 6:39-44

Preskorn SH (2000b) Imipramine, mirtazapine, and nefazodone: multiple targets. Journal of Psychiatric Practice 6:97-102

Preskorn SH (2000c) The adverse effect profiles of the selective serotonin reuptake inhibitors: Relationship to in vitro pharmacology. J Psych Pract 6:153-157

Preskorn SH (2000d) The relative adverse effect profile of non-SSRI antidepressants: Relationship to in vitro pharmacology. J Psych Pract 6:218-223

Preskorn SH, Katz SE (1989) Bupropion plasma levels: intraindividual and interindividual variability. Ann Clin Psychiatry 1:59-61

Preskorn S, Othmer S (1984). Evaluation of bupropion hydrochloride: the first of a new class of atypical antidepressants. Pharmacotherapy 4:20-34

Preskorn SH, Burke M (1992) Somatic therapy for major depressive disorder: selection of an antidepressant. J Clin Psychiatry 53[Suppl 9]:5-18

Preskorn SH, Omo K, Magnus R, Shad MU (in press) A study of the safety, tolerability, and efficacy of mirtazapam in patients immediately switched from fluoxetine. Clin Therapeut

Puech A, Montgomery SA, Prost JF, Solles A, Briley M (1997) Milnacipran a new serotonin and noradrenaline reuptake inhibitor: an overview of its antidepressant activity and clinical tolerability. Int Clin Psychopharmacol 12:99-108

Puozzo C, Albin H, Vincon G, Deprez D, Raymond JM, Amouretti M (1998a) Pharmacokinetics of milnacipran in liver impairment. European Journal of Drug Metabolism and Pharmacokinetics 23:273–279

Puozzo C, Pozet N, Deprez D, Baille P, Ung HL, Zedkova L (1998b) Pharmacokinetics of milnacipran in renal impairment. Eur J Drug Metab Pharmacokinet 23:280–286

Raggi MA, Mandrioli R, Sabbioni C, et al (2002) Separation of reboxetine enantiomers by means of capillary electrophoresis. Electrophoresis 23:1870–1877

Reneric JP, Bouvard M, Stinus L (2002) In the rat forced swimming test, chronic but not subacute administration of dual 5-HT/NA antidepressant treatments may produce greater effects than selective drugs. Behav Brain Res 136:521–32

Rey E, Dostert P, D'Athis P, Jannuzzo MG, Poggesi I, Olive G (1999) Dose proportionality of reboxetine enantiomers in healthy male volunteers. Biopharm Drug Dispos 20:177–181

Rickels K, Schweizer E. Clary C, et al (1994) Nefazodone and imipramine in major depression: a placebo-controlled trial. Br J Psychiatry 164:802–805

Ridout F, Hindmarch I (2001) Effects of tianeptine and mianserin on car driving skills. Psychopharmacology 154:356–361

Robinson DS. Roberts DL. Smith JM, et al (1996a) The safety profile of nefazodone. J Clin Psychiatry 57(suppl 2):31–38

Robinson D, Roberts D, Archibald D, et al (1996b) Therapeutic dose range of nefazodone for the treatment of major depression. J Clin Psychiatry 57(suppl 2):6–9

Rogoz Z, Skuza G, Maj J (1999) Pharmacological profile of milnacipran, a new antidepressant, given acutely. Pol J Pharmacol 51:317–322

Rogoz Z, Margas W, Dlaboga D, Goralska M, Dziedzicka-Wasylewska M, Maj J (2000) Effect of repeated treatment with milnacipran on the central dopaminergic system. Pol J Pharmacol 52:83–92

Rosenbaum JF, Zajecka J (1997) Clinical management of antidepressant discontinuation. J Clin Psychiatry 58(suppl 7):37–40

Rosenbaum JF, Fava M, Hoog SL, et al (1998) Selective serotonin reuptake inhibitor discontinuation syndrome: a randomized clinical trial. Biol Psychiatry 44:77–87

Rouillon F, Berdeax G, Bisserbe JC, Warner B, Mesbah M, Smadja C, Chwalow J (2000) Prevention of recurrent episodes with milnacipran: consequences on quality of life. J Affect Disord 58:171–180

Rowland DL, Myers L, Culver A, Davidson JM (1997) Bupropion and sexual function: a placebo-controlled prospective study on diabetic men with erectile dysfunction. J Clin Psychopharmacol 17:350–357

Royer RJ, Albin H, Barrucand D, Salvadori-Failler C, Kamoun A (1988) Pharmacokinetic and metabolic parameters of tianeptine in healthy volunteers and in populations with risk factors. Clin Neuropharmacol 11:S90–S96

Royer RJ, Royer-Morrot MJ, Paille F, et al (1989) Tianeptine and its main metabolite pharmacokinetics in chronic alcoholism and cirrhosis. Clin Pharmacokinet 16:186–191

Rudolph RL, Feiger AD (1999) A double-blind, randomized, placebo-controlled trial of once-daily venlafaxine extended release (XR) and fluoxetine for the treatment of depression. J Affect Disord 56:171–81

Rueter LE, Kasamo K, de Montigny C, Blier P (1998a) Effect of long-term administration of duloxetine on the function of serotonin and noradrenaline terminals in the rat brain. Naunyn Schmiedebergs Arch Pharmacol 357:600–610

Rueter LE, de Montigny C, Blier P (1998b) Electrophysiological characterization of the effect of long-term duloxetine administration on the rat serotonergic and noradrenergic systems. J Pharmacol Exp Ther 285:404–412

Sacchetti G, Bonini I, Waeterloos GC, Samanin R (1993) Tianeptine raises dopamine and blocks stress-induced noradrenaline release in the rat frontal cortex. Eur J Pharm 236:171–175

Saiz-Ruiz J, Montes JM, Alvaraz E, et al (1997) Treatment with tianeptine for depressive disorders in the elderly. Actas Luso Esp Neurol Psiquiatr Cienc Afines 25:79–83

Saiz-Ruiz J, Montes JM, Alvarez E, et al (1998) Tianeptine therapy for depression in the elderly. Prog Neuropsychopharmacol Biol Psychiat 22:319–329

Salvadori C, Ward C, Defrance R, Hopkins R (1990) The pharmacokinetics of the antidepressant tianeptine and its main metabolite in healthy humans—influence of alcohol co-administration. Fundam Clin Pharmacol 4:115–125

Scates AC, Doraiswamy PM (2000) Reboxetine: a selective norepinephrine reuptake inhibitor for the treatment of depression. Ann Pharmacother 34:1302–1312

Schatzberg AF, Haddad P, Kaplan EM, et al (1997) Possible biological mechanisms of the serotonin reuptake inhibitor discontinuation syndrome. Discontinuation Consensus Panel. J Clin Psychiatry 58 (suppl 7):23–27

Schroeder DH (1983) Metabolism and kinetics of bupropion. J Clin Psychiatry 44(Suppl 5):79–81

Shad MU, Preskorn SH (1997) A possible bupropion and imipramine interaction. J Clin Psychopharmacol 17:118–119 (letter)

Shad MU, Preskorn SH (2000) Antidepressants. In: Levy RH, Thummel KE, Trager WF, Hansten PD, Eichelbaum M (eds). Metabolic Drug Interactions. Lippincott, Williams & Wilkins, Philadelphia, pp 563–577

Sharma A, Goldberg MJ, Cerimele BJ (2000) Pharmacokinetics and safety of duloxetine, a dual-serotonin and norepinephrine reuptake inhibitor. J Clin Pharmacol 40:161–167

Silkey B, Preskorn SH, Golbech A (2000) Interindividual variability in steady-state plasma levels of bupropion and its major metabolites. 40th Annual Meeting NCDEU, Boca Raton, FL

Skinner MH, Kuan HY, Sathirakul K, Gonzales CR, Yeo KP, Reddy S, Lim M, Ayan-Oshodi M, Wise SD (2003) Duloxetine is both an inhibitor and a substrate of cytochrome P4502D6 in healthy volunteers. Clin Pharmacol Ther 73:170–177

Smith JE, Lakoski JM (1997) Electrophysiological study of the effects of the reuptake inhibitor duloxetine on serotonergic responses in the aging hippocampus. Pharmacology 55:66–77

Smith JE, Lakoski JM (1998) Cellular electrophysiological effects of chronic fluoxetine and duloxetine administration on serotonergic responses in the aging hippocampus. Synapse 30:318–328

Soroko FE, Mehta NB, Maxwell RA, Ferris RM, Schroeder DH (1977) Bupropion hydrochloride ((+/-) alpha-t-butylamino-3-chloropropriophenone HCI): a novel antidepressant agent. J Pharm Pharmacol 29:767–70

Spencer CM, Wilde MI (1998) Milnacipran. A review of its use in depression. Drugs 56:405–427

Spina E, Avenoso A, Scordo MG, Ancione M, Madia A, Levita A (2001) No effect of reboxetine on plasma concentrations of clozapine, risperidone, and their active metabolites. Ther Drug Monit 23:675–678

Steen A, Den Boer JA (1997) A double-blind six months comparative study of milnacipran and clomipramine in major depressive disorder. International Clinical Psychopharmacology 12:269–281

Stenger A, Couzinier JP, Briley M (1987) Psychopharmacology of midalcipran, 1-phenyl-1-diethyl-amino-carbonyl-2-aminomethylcyclopropane hydrochloride (F2207), a new potential antidepressant. Psychopharmacology 91:147–153

Stern WC, Harto-Truax N, Bauer N (1983) Efficacy of bupropion in tricyclic-resistant or intolerant patients. J Clin Psychiatry 44(sec 2): 148-152

Stormer E, von Moltke LL, Shader RI, Greenblatt DJ (2000) Metabolism of the antidepressant mirtazapine in vitro: Contribution of cytochromes P-450, 1A2, 2D6 and 3A4. Drug Metab Disposition 28:1168–1175

Szabo ST, Blier P (2001a) Effect of the selective noradrenergic reuptake inhibitor reboxetineon the firing activity of noradrenaline and serotonin neurons. Eur J Neurosci 13:2077–2087

Szabo ST, Blier P (2001b) Effects of the selective norepinephrine reuptake inhibitor reboxetine on norepinephrine and serotonin transmission in the rat hippocampus. Neuropsychopharmacology 25:845–857

Tajima O (2002) Japanese experience with dual-action antidepressants. Int Clin Psychopharmacol 17:S37–S42

Tanum L (2000) Reboxetine: tolerability and safety profile in patients with major depression. Acta Psychiatr Scand Suppl 402:37–40

Thomson PDR. Physician's Desk Reference. Thomas PDF, Montvale, NJ, 2003

Tignol J, Pujol-Domenech J, Chartres JP, Leger JM, Pletan Y, Tonelli I, Tournoux A, Pezous N (1998) Double-blind study of the efficacy and safety of milnacipran and imipramine in elderly patients with major depressive episode. Acta Psychiatrica Scandinavica 97:157–165

Timmer CJ, Lohmann AAM, Mink CPA (1995) Pharmacokinetic dose-proportionality study at steady state of mirtazapine from remeron tablets. Hum Psychopharmacol 10(Suppl 2):97–107

Timmer C, Voortman G, Delbressin L (1996) Pharmacokinetic profile of mirtazapine. Eur Neuropharmacol 6(suppl 3):41

Timmer CJ, Sitsen JM, Delbressine LP (2000) Clinical pharmacokinets of mirtazapine. Clin Pharmacokinet 38:461–474

Toon S, Holt BL, Langley SJ, et al (1990) Pharmacokinetic and pharmacodynamic interaction between the antidepressant tianeptine and oxazepam at steady-state. Psychopharmacology 101:226–232

Troy S. Piergies A, Lucki I, et al (1992) Venlafaxine pharmacokinetics and pharmacodynamics. Clin Neuropharmacol 15 (suppl 1):324B

Tse WS, Bond AJ (2002) Difference in serotonergic and noradrenergic regulation of human social behaviors. Psychopharmacol 159:216–221

Turcotte JE, Debonnel G, de Montigny C, Hebert C, Blier P (2001) Assessment of the serotonin and norepinephrine reuptake blocking properties of duloxetine in healthy subjects. Neuropsychopharmacology 24:511–521

Van Ameringen M, Ferrey G, Tournoux A (2002) A randomized double-blind comparison of milnacipran and imipramine in the treatment of depression. J Affect Disord 72:21–31

Vandel P, Regina W, Bonin B, Sechter D, Bizourd P (1999) Abuse of tianeptine. A case report. Encephale 25:672–673

Vaugeois JM, Corera AT, Deslandes A, Costentin J (1993) Although chemically related to amineptine, the antidepressant tianeptine is not a dopamine uptake inhibitor. Pharmacol Biochem Behav 63:285–290

Verhoeven CHJ, Vos RME, Bogaards JJP (1996) Characterization and inhibition of human cytochrome P450 enzymes involved in the in vitro metabolism of mirtazapine. Presentation at International Symposium on Microsomes and Drug Oxidations. Los Angeles, July 21–24

Versiani M, Mehilane L, Gaszner P, Arnaud-Castiglioni R (1999) Reboxetine, a unique selective NRI, prevents relapse and recurrence in long-term treatment of major depressive disorder. J Clin Psychiatry 60:400–406

Versiani M, Amin M, Chouinard G (2000) Double-blind, placebo-controlled study with reboxetine in inpatients with severe major depressive disorder. J Clin Psychopharmacol 20:28–34

von Frenckell R, Ansseau M, Serre C, Sutet P (1990) Pooling two controlled comparisons of milnacipran (F2207) and amitriptyline in endogenous inpatients. A new approach in dose ranging studies. International Clinical Psychopharmacology 5:49–56

Voortman G, Paanakker JE (1995) Bioavailability of mirtazapine from remeron tablets after single and multiple oral dosing. Human Psychopharmacology 10(Suppl 2):83-97

Wagstaff AJ, Ormrod D, Spencer CM (2001) Tianeptine: a review of its use in depressive disorders. CNS Drugs 15:231-259

Watanabe Y, Sakai RR, McEwen BS, Mendelson S (1993) Stress and antidepressant effects on hippocampal and cortical 5-HT1A and 5-HT2 receptors and transport sites for serotonin. Brain Res 615:87-94

Watanabe Y, Gould E, Daniels DC, Cameron H, McEwen BS (2003) Tianeptine attenuates stress-induced morphological changes in the hippocampus. Eur J Pharmacol 222:157-162

Weise C, Fox I, Clary C, et al (1991) Nefazodone in the treatment of outpatient major depression. Biol Psychiatry 29:363S

Wender PH (1998) Pharmacotherapy of attention-deficit/hyperactivity disorder in adults. J Clin Psychiatry 59(suppl 7):76-79

Wheatley DP, van Moffaert M, Timmerman L, Kremer CM (1998) Mirtazapine: efficacy and tolerability in comparison with fluoxetine in patients with moderate to severe major depressive disorder. Mirtazapine-Fluoxetine Study Group. J Clin Psychiatry 59:306-312

Wienkers LC, Allievi C, Hauer MJ, Wynalda MA (1999) Cytochrome P-450-mediated metabolism of the individual enantiomers of the antidepressant agent reboxetine in human liver microsomes. Drug Metab Dispos 27:1334-1340

Wilde MI, Benfield P (1995) Tianeptine. A review of its pharmacodynamic and pharmacokinetic properties, and therapeutic efficacy in depression and coexisting anxiety and depression. Erratum 50:156

Wong DT, Bymaster FP, Mayle DA, Reid LR, Krushinski JH, Robertson DW (1993) LY248686 a new inhibitor of serotonin and norepinephrine uptake. Neuropsychopharmacology 8:23-33

Wong, DT, Robertson DW, Bymaster FP, Krushinski JH, Reid LR (1988) LY227942 an inhibitor of serotonin and norepinephrine uptake: Biochemical pharmacology of a potential antidepressant drug. Life Sciences 43:2049-2057

Yocca FD, Hyslop DK, Taylor DP (1985) Nefazodone: a potential broad spectrum antidepressant. Trans Am Soc Neurochem 16:115

Yoshida K, Higuchi H, Shimizu T (2002) Elevation of blood pressure induced by high-dose milnacipran. Hum Psychopharmacol 17:431-431

Yoshida K, Higuchi H, Takahashi H, Shimizu T (2003) Favorable effect of milnacipran on depression induced by interferon-alpha. J Neuropsychiatry. Clin Neurosci 15:242-243

Zini R, Morin D, Salvadori C, Tillement JP (1990) Tianeptine binding to human plasma proteins and plasma from patients with hepatic cirrhosis or renal failure. Br J Clin Pharmacol 29:9-18

Zini R, Morin D, Salvadori C, Tillement JP (1991) the influence of various drugs on the binding of tianeptine to human plasma proteins. Int J Clin Pharmacol 29:64-6

Zivkov M, Roes KCB, Pols AB (1995) Efficacy of Org 3770 (mirtazapine) vs. amitriptyline in patients with major depressive disorder: a meta-analysis. Hum Psychopharmacol 10:S135-S145

Current Role of Herbal Preparations

D. Wheatley[1, 2]

[1] 10 Harley Street, London, W1G 2PF, UK
e-mail: wheatley@ukgateway.net
[2] 15 Elizabeth Court, Lower Kings Rd., Kingston-on-Thames, Surrey, KT2 5HP, UK

1	Introduction	326
2	Hypericum in the Treatment of Depression	327
2.1	Chemical Constitution	327
2.2	Mode of Action	328
2.3	Pharmacokinetics	330
2.4	Major Depression	330
2.4.1	Placebo Comparisons in Major Depression	331
2.4.2	Drug Comparisons in Major Depression	331
2.5	Adverse Events	333
2.6	Photosensitivity	334
2.7	Toxicology	335
2.8	Drug Interactions	335
2.9	Seasonal Affective Disorder	336
2.10	Antidepressant Potential	336
3	Kava and the Anxiety Element	338
3.1	Background	339
3.2	Anxiolytic Properties	339
3.3	Anxiolytics in Depression	341
4	Valerian and Sleep Disturbance	342
4.1	Rationale	342
4.2	Polysomnography	342
4.3	Stress-Induced Insomnia	343
4.4	Sleep in Depression	343
5	Ginkgo Biloba, Memory and Sexual Dysfunction	344
5.1	Memory Enhancement	344
5.2	Ginkgo and Sexual Dysfunction	345
6	Comment	346
References		347

Abstract For many herbal remedies, precise information concerning composition, dose standardisation, clinical activity, interactions with other drugs and side effects is lacking. One of the best-researched herbal agents is St John's wort (hypericum). It has been shown to be an effective and safe antidepressant in more than 50 double-blind trials that involved comparisons with both placebo and standard antidepressants including selective serotonin reuptake inhibitors (SSRIs). Its mode of action, which involves reuptake inhibition of serotonin, noradrenaline and dopamine, is comparable to those of chemical antidepressants. Like the chemical antidepressants, it suffers from a "lag-period" of 3–4 weeks before it becomes fully effective and must be taken for at least 6 months to effect a lasting remission. It has been used in seasonal affective disorder (SAD) with good effect. Rarely, photosensitivity may occur, but only in cases of pathological sun allergy. Like chemical antidepressants, hypericum interacts in the metabolism of a number of other drugs, notably anticoagulants, theophylline preparations, oral contraceptives, antimigraine drugs, and HIV drugs. This may result in reduced effectiveness of both drugs and such combinations are therefore best avoided. A number of clinical trials have shown kava kava to be an effective anxiolytic and sleep-inducer, particularly in patients under stress. However, following reports of liver toxicity, it has been withdrawn from sale pending further investigation of these reports. On the other hand, valerian would appear to be a safe hypnotic for long-term use, although not for an immediate effect. Valerian increases deep (delta) sleep strikingly and beneficially and, by so doing, may augment the immune system. Ginkgo biloba enhances memory and the peripheral circulation. In double-blind trials, its efficacy has been demonstrated even in Alzheimer's disease, and it does not appear to have any appreciable side effects. Herbal remedies have much to offer in the treatment of psychiatric disorders and merit further scientific investigation.

Keywords St. John's wort · Kava kava · Valerian · Ginkgo biloba · Major depressive disorder · Anxiety · Sleep · Seasonal affective disorder · Stress · Immune system

1
Introduction

Not surprisingly, there exists a prejudice among physicians against herbal extracts because the preparations are often of dubious composition, the nature of the active principle may be unknown and dose standardisation has not been accurately determined. In consequence, clinical activity may vary widely from the extremes of too little (inactivity) to too much (toxicity), both between brands and between batches. The seeming lack of properly controlled clinical trials may indeed make the clinician somewhat wary of using herbal extracts. In addition, since most are sold "over the counter" (OTC), they are not subject to the usual regulatory controls that apply to synthetic chemical drugs (Drew and Myers 1997). Yet there is a wealth of relevant information available, although it is unfortunately often archived in relatively obscure publications. The aim of this chapter is to bring some of this information to the attention of the readers of this book, so that they can better form their own conclusions about the potential value of some of these preparations in the management of depression.

The main impetus for research on phytopharmaceuticals has come from Germany, where such compounds enjoy great popularity and, in some cases, are even prescribable under that country's health service. In 1996, the 2nd International Congress on Phytomedicine was held in Munich, during which 268 presentations were given on a wide variety of these compounds (Farnsworth and Wagner 1996/97), the most notable of which dealt with the use of paclitaxel, which is still partially derived from plant cells, in the treatment of breast cancer (DeFuria 1996).

Four plant products that may be of use in the management of depression are: hypericum, derived from St. John's wort (antidepressant), kava kava from *Piper methysticum* (anxiolytic), valerian from the flower *Valeriana officinalis* (hypnotic/anxiolytic), and ginkgo biloba from the tree of the same name (memory enhancer/aphrodisiac).

2
Hypericum in the Treatment of Depression

Hypericum perforatum is a member of the Hypericaceae family. Extracts of the plant have been used in herbal medicine since ancient times. The Greek name was *hupereikon* and plants were hung over religious images or pictures to protect against evil at the midsummer festival when the plant is in flower. St. John's wort was one of the herbs of St. John the Baptist and was collected and burned on St. John's Day (June 26th) to ward off "goblins, devils and witches" (Grigson 1958). In 1863, Porcher wrote that it was "greatly in vogue at one time, and was thought to cure demoniacs" (Porcher 1863). It was renowned as a healer of wounds and, among other appellations, was known as "balm-of-warrior's-wound", "God's-wonder-plant" and "touch-and-heal" (Coffey 1993).

Hypericum has been in use in Germany for the past 15 years, where it is licensed as a prescription drug and is also sold OTC. In 1993, over 2.7 million prescriptions were analysed in Germany and it was found that hypericum preparations were among the seven most popular antidepressant drugs (ADs) being used to treat depression (Lohse and Muller-Oerlinghausen 1994). Subsequently, hypericum has achieved the distinction in Germany of outselling (by volume) the world's most widely used AD, fluoxetine (Prozac) (IMS monthly marketing data 1994).

2.1
Chemical Constitution

Hypericum contains a number of compounds (Nahrstedt and Butterweck 1997), including phenylpropanes, flavonol glycosides, biflavones, tannins and proanthocyanidins, xanthones, phloroglucinols, essential oils, amino acids, and naphthodianthrones. The first four groups of compounds are biogenetically related and together are the main constituents of dry crude *H. perforatum*. Tannins and proanthocyanidins constitute 15% and flavonol glycosides (quercetin, isoquercetin, hyperoside, and rutin) 2%–4% of extracts of hypericum. Xanthones are present in very low concentrations. Phloroglucinols are a main constituent (4%) of the buds and flowers of the plant. The naphthodianthrones, hypericin and pseudo-hypericin, are considered to be important compounds for standardisation.

A number of extracts of hypericum are available commercially, including LI 160 (Jarsin, Kira, USA; Kira and Kira one-a-day, UK), Hyperforat, Psychotonin, Psychotonin M, Neuropas, and Esbericum. Most clinical trials have been performed using LI 160 (Jarsin). With the exception of Kira (UK), these preparations contain 300 mg of dry extract, standardised to contain a total of 900 μg of hypericin. Care is needed not to confuse hyper*icum* with hyper*icin*, as 300 mg of hypericum equals 900 μg hypericin. For example: Kira (USA) and Kira-one-a-day (UK) contain 300 mg hyper*icum* (900 μg hyper*icin*), whereas Kira (UK) contains 100 mg hyper*icum*. So, to treat depression, the daily doses would be Kira (USA) and Kira one-a-day (UK) 3 tablets, but Kira (UK) 9 tablets. Confusing? And since there are many unbranded products of herbal preparations available, this simply emphasises the care necessary when evaluating reports on their use.

2.2
Mode of Action

A number of the compounds present in hypericum have putative psychotropic effects; these have been reviewed by Nahrstedt and Butterweck (1997). Flavonols and xanthones exert some inhibitory activity against monoamine oxidase (MAO)-A, while pure hypericin does not inhibit MAO, and little is known about the effects of procyanidines. Cott (1997) undertook studies on in vitro receptor binding and enzyme inhibition by hypericum. In receptor binding assays, hyper*icin* had affinity only for N-methyl-D-aspartate receptors, whereas the crude extract showed significant affinity for adenosine (non-specific to subtypes), serotonin (5-HT$_1$), γ-aminobutyric acid (GABA)$_A$ and GABA$_B$, and benzodiazepine receptors, as well as for other neurotransmitters such as forskolin, inositol triphosphate and MAO. Cott (1997) speculated on the putative importance of GABA receptor binding and suggested that inhibition of MAO was not of significance in the pharmacology of hypericum. Experiments in rats have demonstrated that serotonin (5-HT), dopamine (DA), and norepinephrine (NE) reuptake are all inhibited by hypericum in vitro (Cott 1997; Muller et al. 1997) Furthermore, hypericum appears to have a similar potency for inhibiting the uptake of each of these transmitters. Oral administration of hypericum 240 mg/kg to rats for 14 days resulted in a significant reduction in B-adrenoceptor density in the frontal cortex (Muller et al. 1997), an effect similar to that produced by imipramine. Teufel-Mayer and Gleitz (1997) investigated the effects of long-term administration (26 weeks) of hypericum on serotonin 5-HT_{1A} and 5-HT_{2A} receptors in rats. While there was no effect on receptor affinity, the densities of both receptor subtypes were increased. The authors commented that such effects are similar to those seen with other antidepressants, indicating upregulation of both receptors. Ozturk et al. (1996/97) demonstrated that hypericum exerted effects similar to other antidepressants in animal tests of antidepressant activity, including antagonising the behavioural effects of reserpine and ketamine, potentiating the behavioural effects of amphetamine and serotonin, and potentiating yohimbine toxicity. They concluded that the antidepressant effects of hypericum were not due to hypericin, and that other hypericum species that do not contain this compound might be better suited to clinical use, since hypericin causes photosensitivity. More recently, Chatterjee et al. (1998) postulated that the active compound in hypericum extracts is, in fact, the phloroglucinol derivative, hyperforin, a potent re-uptake inhibitor of 5-HT, DA, NE, GABA, and L-glutamate. Muller et al. (1998) go further, commenting: "It is also remarkable that hyperforin inhibits all neuro-uptake systems with similar potency, since a similar phenomenon has not been reported yet for any other drug known." On the other hand, Bennett et al. (1998) comment that "hypericum extracts have only weak activity in assays related to mechanisms of the synthetic antidepressants—the clinical efficacy of St. John's wort could be attributable to the combined contribution of several mechanisms, each one too weak by itself to account for the overall effect."

Hypericum has been tested in a number of standard animal models that are used to indicate antidepressant activity. Thus, it decreases swimming time of mice in the forced swimming test, an effect that seems to be resistant to blockade with opioid antagonists (reviewed by Ozturk 1997) and decreases immobility time in the tail-suspension test in mice (Butterweck et al. 1997). Spontaneous motor activity in rats is decreased (Winterhoff et al. 1995), as is exploratory behaviour expressed as head-dips in mice (reviewed by Ozturk 1997). Hypericum causes a dose-dependent analgesia in "tail-clip" experiments in mice, an effect that is inhibited by naloxone (reviewed by Ozturk 1997). Further animal experiments were undertaken by Butterweck et al. (1997), who pointed out that neither the mode of action nor the active constituent(s) of hypericum are known. They noted that the dopaminergic, rather than the noradrenergic, system seems to be involved in the mode of action, since the effect of apomorphine (a DA agonist) on temperature is accentuated by hypericum, in contrast to other antidepressants that antagonise this effect. Also, hypericum causes a decrease in ketamine-induced sleep time, an effect that is similar to that of the dopaminergic antidepressant bupropion, which is abolished by haloperidol or sulpiride (DA antagonists).

These experimental data do not clarify the mode of action of hypericum, but do indicate that it has significant effects on central biochemical systems that may be involved in depression.

2.3
Pharmacokinetics

Brockmollor et al. (1997) assessed the pharmacokinetics of hypericum in 13 volunteers who were treated with single doses of 900, 1,800 or 3,600 mg hypericum or placebo in a fourfold, double-blind, crossover design. Maximum total hypericin concentrations were observed about 6 h after drug administration. The elimination half-life of the dose regimens was an average of 28.1 h for hypericin and 17.7 h for pseudohypericin. In an extension of the study, 50 volunteers were treated with hypericum 600 mg 3 times daily for 15 days. The terminal half-life during steady state was 41.7 h for hypericin and 22.8 h for pseudohypericin. This suggests that once daily medication should suffice for effective treatment of depression.

2.4
Major Depression

A considerable number of clinical trials of hypericum in major depression have been undertaken. These studies, which included both uncontrolled and double-blind studies comparing hypericum with placebo and a number of standard ADs, have been masterfully reviewed by Linde et al. (1996), who concluded that: "There is evidence that extracts of hypericum are more effective than placebo for the treatment of mild to moderately severe depressive disorders." This assessment was based on 23 randomised or possibly randomised trials, published between 1984 and 1994: 14 against placebo, 6 against other drugs, and 3 involving hypericum combinations. Trials that measured only physiological parameters were excluded, as were those on human volunteers.

However, there were a number of methodological problems in these reports. Different preparations of hypericum were used in different studies and the classification of depression was by no means consistent. Furthermore, the definition of response to treatment varied from trial to trial. Excluding the studies which examined combinations, most of the remainder involved 1–6 investigators treating 30–120 patients, although there were two "outriders" in which 20 investigators treated 135 patients and 50 investigators treated 112, respectively. Most of the studies quoted in this review were of 4–8 weeks duration, adequate to demonstrate some antidepressant effect but not to measure the full potential of a drug. Most studies used being a "responder" as the main measure of effect, with response defined as an improvement of 50% or more over baseline at the end of the trial. This was measured either using rating scales [usually the Hamilton Rating Scale for Depression (HAM-D)] (Hamilton 1960) or global changes, with the measures used varying from trial to trial. Despite these pitfalls of meta-analyses, this form of analysis provides the main evidence available to determine the role of hypericum as an antidepressant. This role has been reviewed by the author (Wheatley 1998a).

2.4.1
Placebo Comparisons in Major Depression

When the results from the 14 papers quoted in the review article were pooled, using the definition of response given above, there were 225 responders (55.1%) in the hypericum groups, as compared to 94 (22.3%) in the placebo groups, a highly significant difference. For example, Sommer and Harrer (1994) treated 105 patients with mild depression of short duration with hypericum 900 mg or placebo for 4 weeks and found a 50% or greater reduction in mean HAM-D score in 67% of patients on hypericum as compared to 28% on placebo at the end of the 4-week trial period ($p<0.01$). Furthermore, the difference was also significant at 2 weeks ($p<0.05$). However, these results were presented as graphs without exact figures for the mean HAM-D, which detracts from their usefulness in assessing the comparison.

In another study by Hansgen et al. (1994), 72 patients diagnosed as having major depression by DSM-III-R criteria (American Psychiatric Association 1987) were randomly allocated to treatment with either hypericum (LI 160) or placebo for a period of 4 weeks, followed by another 2 weeks when all patients received hypericum as an ethical requirement. At the end of 4 weeks, the mean HAM-D score fell from 21.8 to 9.3 in the hypericum group and from 20.4 to 14.7 in the placebo group ($p< 0.001$ between groups). At the end of 6 weeks, there were further reductions to 6.3 in the hypericum group and, more strikingly, to 8.5 in the original placebo group who were now receiving hypericum (significance not stated). The HAM-D responder rate at 4 weeks amounted to 81% in the hypericum group and 26% in the placebo group. The incidence of adverse events was trivial: one case of sleep disturbance in the hypericum group and two of gastrointestinal upsets in the placebo group. Similar results were recorded by Schrader et al. (1998).

Despite the omission of precise information on a number of aspects of these reports, they provide reasonable evidence that hypericum does exert an antidepressant effect over and above that of a placebo in mild to moderate depression.

2.4.2
Drug Comparisons in Major Depression

Double-blind comparative trials have been undertaken against the following standard tricyclic antidepressants (TCAs): imipramine (Vorbach et al. 1994, 1997), maprotiline (Harrer et al. 1994) and amitriptyline (Wheatley 1997a). In the first study against imipramine (Vorbach et al. 1994), 135 patients were treated for 6 weeks with either 900 mg hypericum or 75 mg imipramine. In the hypericum group, the mean HAM-D score fell from 20.2 to 8.8 ($p<0.001$), compared to a fall from 19.4 to 10.7 in the imipramine group ($p<0.001$), with no significant between-group differences. In a subgroup of 51 patients with initial HAM-D scores greater than 21, the response was actually significantly better with hypericum than with imipramine ($p<0.05$). The second study with this control drug (Vorbach et al. 1997) was notable for the use of higher doses of both drugs—hypericum 1800 mg and imipramine 150 mg—in 200 patients with *severe* major depressive disorders. There were no significant differences between the treatment groups, with both treatments found to be equally effective. Similar effectiveness for hypericum and the respective control ADs was reported in the other comparative trials listed above.

I was involved in a double-blind comparison of hypericum (LI 160) 900 mg and amitriptyline 75 mg (Wheatley 1997a). Entry was limited to those with HAM-D scores of 17–24 to ensure that only mild to moderate cases were included; following 3–7 days on placebo, active treatment was given for 6 weeks. The main measure of clinical effect was the HAM-D, which was administered at the beginning and end of the control period and at 2, 4, and 6 weeks of treatment. The Montgomery-Asberg depression scale (MADRS) (Montgomery and Asberg 1979) and the Clinical Global Impressions (CGI) Scale (Guy 1976) were also completed. Routine haematology and biochemistry parameters were measured before and after treatment. A total of 196 patients were screened and, when drop-outs in the first week were excluded, 156 were suitable for analysis: 83 on hypericum and 73 on amitriptyline. Females exceeded males by 4.2:1, with a mean age of 40.1 years (range 20–65) and an initial mean HAM-D score of 20.7. The mean number of previous attacks was two and the mean duration of the present attack was 1 year. Distribution of cases according to patient data was very similar, with no significant between-group differences. The mean changes in the HAM-D scores (±SEM) are shown in Fig. 1.

There were no significant differences in either group from start to end of the control period, during which patients were receiving placebo, with the initial mean scores being virtually identical on both occasions. Subsequently there was

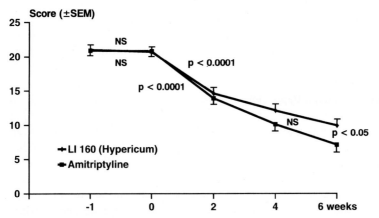

Fig. 1 Mean Hamilton Depression (HAM-D) scores during a 6-week, double-blind trial comparing 87 depressed patients treated with hypericum (LI 160) 900 mg/day with 78 patients treated with amitriptyline 75 mg/day (Wheatley 1997a)

a progressive reduction in the mean scores in both groups, which became highly significant at week 2 and continued to be significant to the end of the 6-week treatment period. There were no significant between-group differences until week 6, at which point the mean change score on the HAM-D was significantly better with amitriptyline ($p<0.05$). Similar results were recorded on the MADRS. Patients were further classified into "responders" (final HAM-D score of <10 *or* >50% reduction from baseline) and "non-responders." In the hypericum group, there were 60% responders, and 78% (NS) in the amitriptyline group. On the CGI severity scale, there was a very marked and highly significant shift to the right in favour of improvement in both groups ($p<0.001$), with no significant between-group differences. Similarly, on the CGI improvement scale, there was very marked improvement in both groups ($p<0.001$), with no significant between-group differences.

Satisfactory relief of depression was achieved with both hypericum and the control drug amitriptyline, the only significant between-drug difference being found in mean change scores on the HAM-D at the end of the trial in favour of amitriptyline, although there were no such differences on final HAM-D improvement scores or on the CGI. This was not considered to be clinically significant.

In view of the suggestion that hyperforin is the active constituent of hypericum, a trial by Laakman et al. (1998) is particularly relevant. In a double-blind, randomised trial in 147 depressed patients, they compared two preparations of hypericum extract, one containing hyperforin 0.5% and one containing hyperforin 5.0%, with placebo. Reduction on the HAM-D score in the hyperforin 5.0% group was significantly greater than in either the 0.5% or placebo groups ($p<0.01$). There were no significant differences between the 0.5% and placebo groups.

2.5
Adverse Events

In their meta-analysis of 1,757 cases, Linde et al. (1996) found that there were only 2 (0.8%) dropouts for adverse events with hypericum, as compared to 7 (3%) with the comparator ADs. Furthermore, side-effects occurred in 19.8% of the patients on hypericum compared with 52.8% of the patients on the other ADs, both differences being highly significant. Woelk et al. (1994) reported an investigation by 663 private practitioners, who treated 3,250 patients with hypericum for 4 weeks. Undesired drug effects were reported by 79 patients (2.4%), and 48 (1.5%) discontinued treatment. The most frequently reported events were gastrointestinal (0.6%), allergic (0.5%), tiredness (0.4%) and restlessness (0.3%). Serious adverse events have only been very rarely reported, such as two accounts of hypomania involving 3 patients (O'Breasail and Arguarch 1998; Schneck 1998). This is a well-recognised occurrence with ADs and may well represent a swing-over to the manic phase of a previously undiagnosed bipolar variant of the disorder. In my own study (Wheatley 1997a), ad-

Table 1 Number (%) of adverse events occurring more than once in a double-blind trial comparing 87 patients treated with hypericum (LI 160) 900 mg/day with 78 patients treated with amitriptyline 75 mg/day (Wheatley 1997a)

	LI 160 ($n=87$)	Amitriptyline ($n=78$)
Dry mouth	4 (5%)	32 (41%)
Drowsiness	1 (1%)	11 (14%)
Sleepiness	2 (2%)	8 (10%)
Dizziness	1 (1%)	6 (8%)
Lethargy	1 (1%)	3 (4%)
Nausea/vomiting	6 (7%)	6 (7%)
Headache	6 (7%)	2 (3%)
Constipation	4 (5%)	1 (1%)
Pruritus	2 (2%)	1 (1%)

verse events were recorded in response to direct questioning; baseline records of symptoms were also made and these were subtracted from those events occurring during the trial. Table 1 shows the numbers of patients who did not record any adverse events as well as the numbers of adverse events (cases) that occurred more than only once in either group. Early dropouts have been included, giving 87 patients on hypericum and 78 on amitriptyline. Compared to baseline, the overall occurrence of adverse events was considerably lower with hypericum (32 cases, 37%) than with amitriptyline (50 cases, 64%) ($p<0.05$). Furthermore, the incidence of all the individual adverse events was less with hypericum, with the differences being highly significant in the case of dry mouth and drowsiness ($p<0.001$ for each one).

2.6
Photosensitivity

Cows like to gorge themselves on St. John's wort when it is in flower, and some have developed photosensitivity as a result. According to Leuschner (1996/7), there is no risk of photosensitivity until very high levels of hypericum have been taken; however, in a review of the literature, Stevinson and Ernst (1999) quoted four reported cases of sensitisation rashes that resolved when treatment was discontinued. Furthermore, Bove (1998) reported the case of a 35-year-old woman who took hypericum 500 mg/day for 4 weeks and developed subacute polyneuropathy and a photosensitive rash, both of which also resolved on stopping the treatment. Clearly, individuals who are known to be sensitive to sunlight should not take hypericum, but it is reassuring that the incidence of this complication is extremely low and that it is reversible.

2.7
Toxicology

The dose of 900 mg that has mainly been used in man is equivalent to 13 mg in animals; in the rat, no signs of toxicity have been seen at doses up to 9,000 mg/kg (Leuschner 1996/7). When doses of 300, 900 and 2,700 mg/kg were administered to rats and dogs for 26 weeks, the lowest "toxic" dose was found to be 900 mg/kg, at which dose "very non-specific" signs of toxicity developed. Furthermore, all such changes reverted to normal when treatment was omitted for 4 weeks. There were no effects on reproduction and no mutagenic effects. A 2-year study in mice and rats has now been concluded, and showed no evidence of carcinogenic potential for hypericum (Leuschner 1996/7).

2.8
Drug Interactions

Because the metabolism of hypericum involves the cytochrome P450 enzyme system, interactions could occur between hypericum and the many other drugs metabolised via this pathway. The implicated compounds are diverse and include cyclosporin, warfarin, digoxin, theophylline, anticonvulsants, drugs used to treat HIV, oral contraceptives, other antidepressants and anti-migraine drugs. But, provided that such drugs are not indicated, there would seem to be little reason to prohibit the use of hypericum, since most other synthetic antidepressants, including many selective serotonin reuptake inhibitors (SSRIs) and tricyclics, exert similar effects. The possibility of such interactions is no reason to stop using these agents, but simply calls on the physician to be more discriminating in prescribing them. It would appear that the whole question of drug involvement in the cytochrome P450 enzyme system is far more complex than was originally supposed. At a recent meeting in the United States (Ereshelisky 2000), it emerged that the same drug can both antagonise or stimulate this system, depending on such diverse factors as dose, mode of administration, ethnicity of the individual and whether the person is a fast or slow drug metaboliser. There are also other systems that may be relevant to the problem that have not as yet been studied.

Perhaps the putative interaction of most concern is with oral contraceptives, since this involves a group of patients who are very susceptible to depressive illness. Theoretically oestrogens might be involved, again because of the common P450 metabolic pathway, although pharmacokinetic studies have not demonstrated any effect of hypericum on blood oestrogen levels (Lichtwer Pharma, personal communication). Also, it should be noted that intermenstrual bleeding, of which there have been a few reports with hypericum, also not infrequently occurs with low-dose contraceptives anyway. So, it would not seem that this interaction, even if confirmed, is of any clinical significance. However, patients should be warned of possible "pill failure", so that if they wish to take hypericum, they can use additional barrier methods to avoid conception.

The majority of drugs listed in pharmacopoeias carry drug-interaction warnings, which are very necessary to avoid dangerous combinations. However, such warnings do not constitute an embargo against any one drug, either used singly or in combination with other drugs with which there are no such interactions. Then too, the clinician must balance the benefits of pharmacological interventions against any possible harm that they may cause, remote though such a likelihood may be. For example, in untreated resistant depression, there is a high probability of a fatal outcome (suicide) and it is accepted practice to use combinations of different antidepressants, though there is a risk of interactions between them.

2.9
Seasonal Affective Disorder

Seasonal affective disorder (SAD) is a well-established variant of major depression (Rosenthal et al. 1984), in which depressive episodes are confined to the winter months, with resolution of symptoms in the summertime. Light therapy is an effective treatment (Terman et al. 1989), but it is also time-consuming, since it involves looking into a light-box for a minimum of 2 h daily. The SAD Association, a patient support group in the United Kingdom, organised a postal survey of its members before and after 8 weeks of treatment with hypericum plus light therapy ($n=133$) or with hypericum alone ($n=168$) (Wheatley 1999a). When response was evaluated using an 11-item rating scale (with maximum score=44), the mean score fell from 20.6 to 11.8 ($p<0.001$) with combined therapy, and from 21.3 to 13.0 ($p<0.001$) with hypericum alone. There were no significant between-group differences except in sleep disturbance, which was significantly better with combined therapy. Limitations are the non-blind nature of the study and the use of a low dose of hypericum (equivalent to only 300 mg/day of the extract). However, in view of these positive results, a more formal study is undoubtedly warranted, since hypericum alone would seem to obviate the need for light-box therapy.

2.10
Antidepressant Potential

Criticism of the reported studies on the use of hypericum in depression has been based on the fact that, with four exceptions (Vorbach et al. 1997; Phillipp et al. 1999; Brenner et al. 2000; Schrader 2000), comparisons were made to the lowest recommended doses of the control drugs. Doubtless this reflects the European custom of generally employing such lower doses, as compared with the practice in the United States. In my trial described earlier in the chapter (Wheatley 1997a), 6 weeks of treatment with 75 mg/day of amitriptyline resulted in a 54% reduction in mean HAM-D score. Furthermore, a mean end score of 9.5 matches anything that can be achieved with the SSRIs, as exemplified in three contemporary studies (Dunbar et al. 1991; Mendells et al. 1993; Rosenberg et al. 1994). Extensive research on a number of the characteristics of depressed patients (Moller 1996) failed to find any factors that accounted for the differences in trans-Atlantic prescribing practices, with one notable exception: patients in the United States were significantly more overweight than their European counterparts. Since the effective dose of any drug depends on the body weight of the patient, one could speculate that this might provide the explanation.

So what of these four higher-dose trials? The first was a double-blind comparison in 209 patients who were randomly assigned to 6 weeks of treatment with either hypericum, 1,800 mg/day, or imipramine, 150 mg/day (Vorbach et al. 1994). Treatment was equally effective in both groups, with no problems with side-effects in the patients treated with hypericum. The second trial (Phillipp et al. 1999) was another randomised double-blind study which compared hypericum, imipramine and placebo, with no differences between the two active drugs, both being significantly better than placebo. Next came a randomised double-blind study by Brenner et al. (2000), which used an SSRI, sertraline, at a dose of 75 mg/day as the control agent. Once again, results were equally good with both agents. Finally, Schrader et al. (2000) used another SSRI, fluoxetine, as the control drug in 240 patients in a 6-week, double-blind, random allocation study. They concluded that "hypericum and fluoxetine are equipotent with respect to all parameters used to investigate antidepressants in this population" and that "hypericum safety was substantially better than fluoxetine."

Hypericum would appear to be a well-tolerated and effective alternative to the synthetic ADs in the treatment of mild to moderate depression, particularly when adverse events of the synthetic ADs become intolerable to the patient. This might be of particular benefit to the elderly, many of whom do become depressed and are usually more sensitive to adverse events than younger patients (Smith 1997). Further research is clearly needed to better elucidate the mode of action of hypericum and to define its optimal dose range and further indications for clinical usage. Additional studies comparing hypericum with the SSRIs and other recently introduced ADs are needed, as well as studies examining higher doses in severe depression. Investigation of other indications for hypericum are also needed, such as for the treatment of dysthymia, general anxiety disorder, panic disorder, social and other phobias, obsessive-compulsive disorder, premenstrual and peri-menopausal syndromes, and stress and posttraumatic stress disorders. Most of these conditions are accompanied by depressive symptoms of varying degree, but often of mild to moderate severity.

Such is the nature of depressive illness that patients often give up hope that they can be cured and may stop treatment before a therapeutic effect becomes apparent. This problem is considerably aggravated when adverse events occur, since these symptoms may be perceived by patient and caregiver alike as indicating worsening of the illness caused by the treatment. It is hardly surprising then that it is difficult to ensure compliance with that treatment. The difficulty is compounded by the slow onset of effect that is characteristic of all AD drugs (including hypericum), so that the patient experiences little benefit during the first 2–3 weeks of treatment until the therapeutic effect becomes established. The introduction of the SSRIs has undoubtedly reduced the incidence of side-effects, but these agents are not without problems of their own, most notably gastric irritation, insomnia, anxiety and hypomania (Smith 1997). Furthermore, impairment of libido and other sexual functions and problems with male erection and ejaculation are integral components of the illness. Even when the depression responds to treatment with ADs, sexual function often does not improve, since sexual problems can be perpetuated by the majority of ADs themselves (Wheatley 1998a). Thus, AD-induced sexual problems may occur in as many as 67% of patients (Segraves 1998). Therefore the advent of an AD that is virtually free of adverse effects, even if not therapeutically superior to those ADs that are already available, heralds a notable advance in the treatment of depression.

3
Kava and the Anxiety Element

Anxiety is an inescapable component of depression, and the symptoms of the latter are often hidden by the overlay of those of the former, a condition aptly termed "masked depression." Nowhere is this more relevant than in the psychiatric response to stress (Wheatley 1993a). The most immediate reaction to a

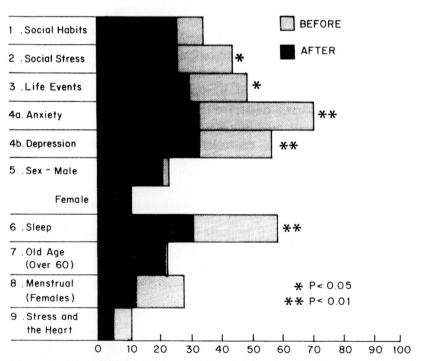

Fig. 2 Mean Wheatley Stress Profiles (WSP), before and after 12 weeks of treatment with antidepressant drugs in 21 patients suffering from stressful situations (Wheatley 1993a)

stressful situation is embodied in the primitive "fight or flight" response involving the production of a number of anxiogenic symptoms, the importance of which is emphasised in the original HAM-D rating scale (Hamilton 1960). With the resolution of the acute stress situation, the anxiety that it has engendered may also resolve. However, if there is a situation of continuing mental stress, as is often the case in the social milieu of today, then depression may develop insidiously under the cloak of the much more prominent anxiety symptoms. This is illustrated by an analysis of the first 100 patients seen at the Maudsley stress clinic in London, of whom no fewer than 49 (49%) were suffering from major depression (Wheatley 1997b). After treatment with various ADs, these patients achieved significant relief from anxiety as well as depressive symptoms, as illustrated in Fig. 2. However, modern antidepressant drugs, particularly the SSRIs, may induce anxiety symptoms as an adverse effect, thus requiring concomitant anxiolytic therapy until these anxiety symptoms resolve as the depression is relieved. Since there is no acceptable non-habit-forming drug of this type, kava, if effective, might have a secondary role to play in the treatment of depressive illness.

3.1
Background

Kava (or kava kava) is an extract of the roots of the Polynesian plant *Piper methysticum* and is used in the South Pacific for its sedative, aphrodisiac and stimulatory effects, both recreationally and in religious ceremonies (Singh 1992). It contains a number of active compounds, among which are the kava pyrones, kawain, dihydrokavain, methysticin, dihydromethysticin, yangonin and desmethoxyyangonin, although it is not known which of these may be responsible for any anxiolytic actions it may have (Hansel and Haas 1984; Walden et al. 1997). These authors also showed evidence for serotonergic and calcium antagonistic potencies of the kava pyrone kawain, but later Grunze and Walden (1999), discussing effects on GABAergic transmission, concluded that "the cellular actions of the kawain appear heterogeneous, but all of them counter excitation." This led them to suggest that it might prove to be a useful treatment for epilepsies, bipolar disorder, depression and anxiety, all conditions characterised by increased cellular excitability.

3.2
Anxiolytic Properties

There have been relatively fewer clinical trials of kava than of hypericum. Thus, in the book *Rational Phytotherapy* by Schulz et al. (1998), six such trials of the extract are described, two in the climacteric (Warnecke et al. 1995), one in perioperative (surgical) mood (Bhate et al. 1989) and three in anxiety (Kinzler et al. 1991; Woelk et al. 1993; Volz 1995). The dose in four of the extract trials was 210 mg/day, using a standardised preparation containing 70% kavapyrones, and

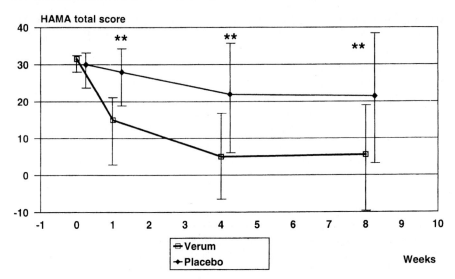

Fig. 3 Mean Hamilton Anxiety (HAM-A) scores during an 8-week trial in 40 anxious patients comparing kava (verum) 210 mg/day with placebo. Reprinted with permission from Warnecke et al. (1995)

all were double-blind comparisons, three against placebo and one against standard benzodiazepines. There were significant differences from placebo in each of the relevant studies and no significant differences between kava, bromazepam and oxazepam in the study that used benzodiazepines (Woelk et al. 1993). The results on the Hamilton Anxiety Rating Scale (HAM-A) (Hamilton 1959) in the study by Warnecke et al. (1995) are shown in Fig. 3.

There have also been nine studies using isolated DL-kawain. Two of the studies compared DL-kawain with placebo, while the others compared it with reference anxiolytics. All nine studies showed results similar to those of the extract studies (Volz and Hansel 1994).

These results have been confirmed in two more recent double-blind, placebo-controlled studies, both of which used a daily dose of 300 mg of the extract WS1490; in these studies, the HAM-A was used to assess results. Lehmann et al. (1996) treated 58 patients; with kava, the HAM-A fell from 25.6 initially to 12.6 at endpoint, while with placebo scores went from 24.5 to 21.0. The results were significantly in favour of kava from week 1 onwards ($p<0.02$). The second study by Volz and Kieser (1997) is notable for a longer duration of treatment, namely 25 weeks, but similar results were recorded. Thus, with kava, the HAM-A fell from 30.7 to 9.7, while with placebo, HAM-A scores went from 31.4 to 15.2. The between-group difference was significantly in favor of kava from week 8 ($p=0.02$) onwards to week 24 ($p=0.001$).

Two very recent studies have been undertaken in patients with a formal diagnosis of Generalized Anxiety Disorder (GAD) by DSM-IV criteria (American Psychiatric Association 1994).

The first study was my uncontrolled pilot trial in 24 patients treated for 4 weeks (Wheatley 2001a). Two dosage schedules were compared: 120 mg qid and 45 mg t.i.d. in a randomised-order crossover open trial. There were highly significant reductions in mean scores on the HAM-A ($p<0.001$), irrespective of dosage schedule, treatment order or sex of the patients. The impact of side-effects was relatively low and only 1 patient on t.i.d. dosing had to omit treatment because of nausea. No side-effects were experienced by 9 patients (37%) on t.i.d. dosing or 5 (22%) on qid dosing; daytime drowsiness occurred in 8 (33%) patients in the t.i.d. dosing group and in 2 (9%) in the qid group. Thirteen patients preferred qid dosing and 8 preferred t.i.d. doses, with no preferences in the remaining 3 cases. The author considered that the trial had shown kava at both schedules to be highly effective in relieving anxiety symptoms in these patients suffering from GAD.

A double-blind comparative trial was required to verify these results and this has been provided by Connor et al. (2000), who treated 35 patients with kava 280 mg/day for 4 weeks in a double-blind, placebo-controlled trial. Results were significantly better with the active drug, which was "well-tolerated and not associated with withdrawal at the doses administered".

However, there has been a case report from Switzerland of hepatitis associated with kava kava (Escher and Desmeules 2001) and additional cases have been reported in Germany (data not yet published). Should this reported finding be confirmed then it may be necessary to reconsider the role of kava kava in therapeutics. Indeed in the UK kava has been withdrawn from the market.

Otherwise, the reported incidence of adverse events has been low, overall varying between 1.5% in 4,049 patients to 2.3% in 3,029 patients (Hoffmann and Winter 1993; Hansel et al. 1994). The adverse events that occurred were described as "mild and reversible", but did include 31 cases of gastrointestinal disturbance and 31 cases of allergic reactions.

3.3
Anxiolytics in Depression

The degree and nature of the anxiety symptoms accompanying the depression and the choice of AD will determine whether concomitant anxiolytic therapy is required. The most relevant stage of treatment is during the first 3–4 weeks, before the antidepressant effect begins to be established, but when adverse events exert their greatest impact and may weaken the patient's resolve to persevere with treatment. Were it not for the reported hepatotoxicity, kava might be useful in an ancillary role under these circumstances, since it does not appear to cause any other notable adverse events or drug interactions per se. However, its dependence potential remains unknown, although Hansel et al. (1994) concluded that there was no evidence of any potential for physical or psychological dependency. Further rigorously controlled trials are required to confirm kava's potential as a safe anxiolytic. In the context of this chapter, which is concerned only with the treatment of depression, if an anxiolytic is required, then this is best confined to the first 3–4 weeks of treatment. When the patient is receiving hypericum or a sedative AD, such use is unlikely to be needed (Wheatley 1999a).

4
Valerian and Sleep Disturbance

Although a small proportion of depressed patients oversleep, sleep disturbance is an integral component of the illness in most (Wheatley 1993b). Not only are the duration and quality of sleep reduced, but there is selective reduction in deep sleep (short-wave or delta sleep, SWS) as measured by stages 3 and 4 on the sleep electroencephalogram (EEG), with resultant adverse effects on many body mechanisms, notably the immune system (Wheatley 1992). The problem of depression-induced insomnia is aggravated if stimulant ADs are used in treatment; this applies particularly to the SSRIs, which also selectively interfere with SWS. Additional hypnotic drugs may therefore be required to counteract these effects; benzodiazepine hypnotics are often used for this purpose. Unfortunately, although effective in prolonging the light stages of sleep, this is accomplished at the expense of further reduction in SWS and there is a potent hazard that the patient may develop dependence. Therefore, there is a need for a safer hypnotic drug for use with antidepressants. Can valerian fulfill this function?

4.1
Rationale

This time-honoured remedy is noted as: "a soother of troubled nerves and an inducer of untroubled sleep, mild in effect but safe in use" (Schulz et al. 1998). There are a multitude of valerian species worldwide, but medicinal valerian is derived from *Valeriana officinalis* (Schulz et al. 1998), either as aqueous or ethanol extracts, which do not necessarily yield equivalent doses. A number of clinical and polysomnographic trials have been undertaken. Lindahl and Lindwall (1989) undertook a double-blind crossover trial of valerian versus placebo for 2 nights only in 27 patients, using a simple subjective sleep assessment scale. The results showed significantly better sleep on the valerian nights and the dose used was equivalent to 400 mg of the root of the plant. Schulz et al. (1998) reviewed a number of other studies and concluded that: "valerian is not a suitable agent for the acute treatment of insomnia". The essential value of valerian may lie in its ability to promote natural sleep after several weeks of use, with no risk of dependence or adverse health effects. The polysomnographic studies are of more interest.

4.2
Polysomnography

When Leathwood and Chauffard (1983) administered single doses of valerian to 29 normal volunteers, EEG recordings in a sleep laboratory showed no significant differences from placebo (quoted by Schulz et al. 1998). On the other hand, Donath et al. (1996/97) undertook a double-blind comparison of 800 mg valerian extract (LI 156) and placebo, in a crossover design involving two 2-week periods of treatment with an intervening 2-week washout period, in 16 patients with insomnia. There were significant differences in favour of valerian in the following areas: shorter sleep latency and SWS onset, increase in duration of SWS and reduction in rapid eye-movement sleep (REM). However, a further report from the same authors (Donath et al. 2000), although confirming the SWS changes, found an *increase* in REM sleep with both valerian *and* placebo.

4.3
Stress-Induced Insomnia

I have undertaken an unblinded pilot study to compare kava, valerian and the combination of both in 24 patients with stress-induced insomnia, with patients acting as their own controls (Wheatley 2001b). Patients were first treated for 6 weeks with kava 120 mg/day, followed by 2 weeks off treatment. Then, 5 having dropped out, 19 patients received valerian, 600 mg/day for another 6 weeks. Then there was a further 2-week period off treatment, and a final 6 weeks of treatment with both drugs combined. Stress was measured in three areas: social, personal and life-events. Insomnia was also measured in three areas: time to fall asleep, hours slept and waking mood. Total severity of stress as well as insomnia was significantly relieved by both compounds ($p<0.01$) with no significant differences between them; and there was also further improvement with the combination, which was significant in the case of insomnia ($p<0.05$). On direct questioning, 16 patients (67%) did not report any side-effects from kava, 10 (53%) did not report any side-effects from valerian and 10 (53%) did not report any side-effects from the combination. The commonest effect was vivid dreams [kava+valerian, 4 cases (21%); and valerian alone, 3 cases (16%)], followed by gastric discomfort and dizziness with kava [3 cases of each (12% each)]. These results were considered to be very promising, but further studies may be required to determine the relative roles of valerian in the long-term.

4.4
Sleep in Depression

Personal experience with valerian would seem to confirm that it has a mild hypnotic action of slow onset. Of considerable theoretical interest (if confirmed) are the EEG findings showing an increase in SWS, with all the accompanying physiological advantages involved. These effects on sleep might be particularly advantageous for a role as adjuvant therapy in depression, in which illness similar sleep impairments are present. On the other hand, hypericum, although not a notable hypnotic per se, also selectively increases SWS.

Thus, Schulz and Jobert (1994) undertook a 2-week × 2-week, double-blind crossover study of hypericum (LI 160) versus placebo, in 12 female volunteers.

The mean duration of SWS on the sleep EEG increased from 1.5% to 6.0% in the subjects receiving hypericum, while it fell from 4.1% to 2.5% in those receiving placebo. However, there were no improvements in either onset or total duration of sleep, nor in awakenings during the night. As far as I am aware, no studies have been undertaken on the use of valerian in insomnia accompanying depression, and so it is difficult to assess any role that it may have to play. Certainly, if continuing sleep disturbance is a persistent problem, adjuvant use of valerian might be considered, but the extract would not appear to offer any help in the critical first few weeks of treatment.

5
Ginkgo Biloba, Memory and Sexual Dysfunction

Two prominent symptoms of major depression are impairment of memory and loss of libido and other sexual functions. Although these are likely to improve as the depression is relieved by AD treatment, such improvement is often delayed. Therefore, these problems may persist until and even beyond completion of the appropriate course of treatment. Ginkgo biloba is extracted from the leaves of the ginkgo tree that grows in far eastern countries and the United States and is cultivated in Europe (Cott 1995). Some 50 original papers are available on the pharmacologic actions of ginkgo, mostly using the extract EGb761. The main effects of the extract seem to be related to its anti-oxidant properties, which result in increased tolerance to hypoxia, especially in brain tissue (Oyama et al. 1994). A number of studies measuring pain-free walking distances, in patients with peripheral vascular disorders provide evidence for the beneficial circulatory effects of the extract. Thus, in a meta-analysis of such studies, Schneider (1992) recorded improvements of between 30 and 161 metres with gingko compared with placebo. Such is the putative rationale for the reputation of ginkgo biloba as a memory enhancer, since these effects are not confined to the corporal circulation but also embrace the cerebral component (Oyama et al. 1994).

5.1
Memory Enhancement

Currently, the most notable clinical evidence for the efficacy of ginkgo is the placebo-controlled double-blind clinical trial undertaken by Le Bars et al. (1997), in mild to severe cases of Alzheimer's disease. The product used was EGb 761 at a dose of 120 mg/day; 202 patients were treated for 52 weeks, at the end of which time, ginkgo demonstrated significant advantages over placebo. The results were significantly better for the active compound on the following measures: Alzheimer's Disease Assessment Scale (ADAS-Cog) (Mohs et al. 1983) ($p=0.04$) and the Geriatric by Relatives Rating Instrument (GERRI) (Schwartz and Loew 1983) ($p=0.005$), although this was not so in the case of the Clinical Global Impression of Change (CGIC). There were no significant differences from placebo in relation to either incidence or severity of adverse events. The authors concluded: "Although modest, the changes induced by EGb were objectively measured by the ADAS-Cog and were of sufficient magnitude to be recognised by the caregivers in the GERRI."

Might ginkgo therefore be useful for depressed patients with memory impairment? As far as I know, no such trials have been undertaken. However, in view of the social distress that this symptom may engender, there might be a limited role for the extract when the problem fails to resolve despite AD treatment.

5.2
Ginkgo and Sexual Dysfunction

As previously outlined, reduced or absent libido is a common effect of depressive illness, with significant adverse consequences for patient and partner alike. Inevitably, the stress that this generates constitutes an adverse prognostic influence on the response to treatment. Also, as already noted, many of the ADs, far from alleviating the problem, actually contribute to it through the sexual adverse effects that most of them can cause. A study by Baldwin et al. (1997) has drawn attention to these problems. The well-established TCAs, as well as the monoamine oxidase inhibitors (MAOIs), are all implicated. Thus, sexual dysfunction has been reported by 80% of men on phenelzine (Baldwin et al. 1997), and clomipramine has been associated with anorgasmia in 42%–96% of patients (Stein and Hollander 1994). The more recently introduced SSRIs can have just as detrimental an effect on sexual functioning; for example, fluoxetine has been reported to induce sexual dysfunction in as many as 75% of patients (Segraves 1998). In addition to effects on libido, these antidepressant drugs can also inhibit orgasm in both sexes, sometimes to the extent of complete inhibition. This catalogue of adverse effects does not end there, however, since in males there may be inability to obtain or maintain an erection and lack of ejaculation, while in females vaginal lubrication may be inhibited. Sedative ADs, such as nefazodone and mirtazapine, are far less likely to cause such problems (Wheatley 1998b,c,d).

There have been isolated anecdotal reports of correction of AD-induced sexual dysfunction by ginkgo biloba. These reports prompted Cohen and Bartlik (1998) to undertake an open trial of ginkgo in 63 patients with this problem; the average dosage was 207 mg/day for 4 weeks. These researchers reported improvement in desire, excitement, orgasm and afterglow, but there was no statistical analysis of these results.

In order to investigate any potential ginkgo may have in such cases, I undertook a small pilot study in 12 patients (Wheatley 1999b) who were suffering from recent sexual dysfunction as a direct consequence of treatment with ADs for either depression or social phobia. The trial period was 6 weeks and the dose of ginkgo 240 mg. The sexual stress questionnaire from the Wheatley Stress Profile (WSP) (Wheatley 1993a) was used to measure loss of libido, resultant stress, effects on sexual relationship, physical problems, guilt, masturbation and, in fe-

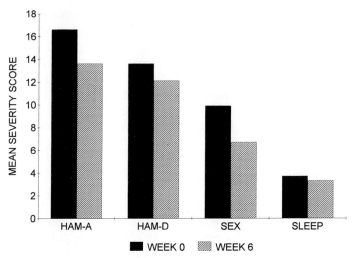

Fig. 4 Mean severity scores before and after 6 weeks concomitant treatment with ginkgo biloba 240 mg/day in 12 depressed patients suffering from sexual dysfunction induced by antidepressant drugs (Wheatley 1999b)

males, dyspareunia and fear of pregnancy. Sleep impairment was also recorded using the WSP sleep questionnaire, which measures sleep onset, times waking, early morning waking, duration of sleep and waking mood. The results for the HAM-A, HAM-D, Sex (max=24) and Sleep (max=10) are shown in Fig. 4.

There was improvement on all items, and this was significant in the case of anxiety ($p<0.05$) and the sex total ($p<0.01$). The incidence of gastric irritation was 14% but no other adverse events were recorded. However, a subsequent double-blind placebo-controlled 12-week study (not yet published) has shown similar beneficial effects with both active and placebo medications. Again, further work is needed to confirm or refute these findings.

6
Comment

It would be an irrational sceptic indeed who could ignore the abundance of evidence confirming that hypericum is a safe and effective antidepressant, at least in mild to moderate cases of major depression. The psychiatrist's dream of an instant antidepressant it is not, but for cases such as these, hypericum has a number of intrinsic advantages over current synthetic antidepressant drugs. These include infinitesimal incidence of true adverse events, resultant good patient compliance, psychological appeal as a herbal preparation, no dependence potential, and foreseeable and preventable drug interactions. Should hyperforin be confirmed as the active principle, subsequent isolation, purification and standardisation *may* result in an even more potent therapeutic tool with which to combat the illness. On the other hand, it has been postulated that the efficacy of hypericum may be due to a summation of the actions of a number of its constituents, each individually of too low a potency to exert an appreciable clinical effect (Bennett et al. 1998). Either way it is difficult to escape the conclusion that hypericum might well become the premier treatment of choice for mild to moderate depression; its role in severe depression still remains to be determined.

Not nearly so much research has been undertaken on the other herbal remedies discussed, but it would certainly appear that kava kava is an effective anxiolytic and also may be useful in the induction and maintenance of sleep. However, should the reports of hepatotoxicity be confirmed, this will considerably reduce its therapeutic potential, particularly for self-medication. On the other hand, valerian would appear to be perfectly safe, although with a slow onset of effect, so unsuitable for treating acute insomnia. Because of its effects in improving the quality of sleep, as evidenced by sleep EEG studies, it may well have a role to play in management of chronic insomnia, particularly in the elderly and as adjuvant treatment for insomnia caused by other ailments, notably depression. Thus, the stimulation of the immune system that occurs during the sleep phase of SWS can only be beneficial in many illnesses, and a compound that promotes this phase of the sleep cycle is to be welcomed.

Ginkgo biloba would also appear to be a very safe and relatively inexpensive alternative to the undoubtedly effective but expensive chemical memory enhancers that have recently been introduced for the treatment of Alzheimer's disease and allied conditions. This may well be due to its effects on the circulation generally and on the cerebral circulation specifically. Whether or not these may embrace other systems, such as sexual functioning, must await the results of further studies. Again, improvement in these parameters would be beneficial in the management of depression, thus suggesting an ancillary role for ginkgo biloba in selected cases of the illness.

These other psychoactive herbal preparations constitute important sources for ancillary management of depression, with its multi-faceted spectrum of symptoms. All-in-all, herbal medicine represents a relatively untapped treatment source for the potentially fatal illness of depression.

References

American Psychiatric Association (1987) Diagnostic and statistical manual of mental disorders, 3rd edition, revised. American Psychiatric Association, Washington, DC
American Psychiatric Association (1994) Diagnostic and statistical manual of mental disorders, 4th edition. American Psychiatric Association, Washington, DC
Baldwin D, Thomas SC, Birtwhistle J (1997) Effects of antidepressant drugs on sexual function. Int J Psych Clin Prac 1:47–58
Bennett DA Jr, Phun L, Polk JF, Voglino SA, Zlotnik V, Raffa RB (1998) Neuropharmacology of St John's Wort (Hypericum). Ann Pharmacother 32:1201–1208
Bhate H, Gerster G, Graczca E (1989) Orale Pramedikation mit zubereitungan aus piper methysticum bei operativen eingriffen in epiduralanasthesie. Erfahrungsheilkunde 6:339–345

Bove GM (1998) Acute neuropathy after exposure to sun in a patient treated with John's Wort. Lancet 352:1121-1122

Brenner R, Azbel V, Madhausoodanan S, Pawlowska M (2000) Comparisons of an extract of hypericum (Li 160) and sertraline in the treatment of depression: A double-blind, randomized pilot study. Clin Ther 22:411-419

Brockmoller J, Reum T, Bauer S, Kerb R, Hubner WD, Roots I (1997) Hypericin and pseudohypericin: pharmacokinetics and effects on photosensitivity in humans. Pharmacopsychiatry 30(Suppl 2):94-101

Butterweck V, Wall A, Lieflander-Wulf U, Winterhoff H, Nahrstedt A (1997) Effects of the total extract and fractions of *Hypericum perforatum* in animal assays for antidepressant activity. Pharmacopsychiatry 30(Suppl 2):117-124

Chatterjee SS, Bhattacharya SK, Wonnemann M, Singer A, Muller WE (1998) Hyperforin as a possible antidepressant component of hypericum extracts. Life Sci 63:499-510

Coffey T (1993) The history and folklore of North American wild flowers. Houghton Miflin, Boston, pp 64-66

Cohen AJ, Bartlik (1998) Ginkgo biloba for antidepressant-induced sexual dysfunction. J Sex Marital Ther 24:139-145

Connor KM, Watkins LL, Davidson JRT (2000) A study of an herbal anxiolytic. Presented at the 39th annual Meeting of the American College of Neuropsychopharmacology, 2000

Cott JM (1995) Natural product formulations available in Europe for psychotropic medication. Psychopharmacol Bull 31:745-751

Cott JM (1997) In vitro receptor binding and enzyme inhibition by hypericum perforatum extract. Pharmacopsychiatry 30(Suppl 2):108-112

DeFuria MD (1996) Taxol: Clinical development and commercializaton. Phytomedicine 3 (Suppl.1): 4

Donath F, Quispe Bravo S, Diefenbach K, Fietz I, Roots I (1996/97) Polysomnographic and subjective findings in insomniacs under treatment with placebo and valerian extract (LI 156). Phytomedicine 3(Suppl 1):75

Donath F, Quispe Bravo K, Diefenbach I, Maurer A, Fietze I, Roots I (2000) Critical evaluation of valerian extract on sleep structure and sleep quality. Pharmacopsychiatry 33:47-53

Drew AK, Myers SP (1997) Safety issues in herbal medicine. Med J Aust 166:538-541

Dunbar GC, Cohn IF, Feighner JP (1991) A comparison of paroxetine, imipramine and placebo in depressed outpatients. Human Psychopharmacology 11:283-291

Ereshefsky L (2000) Advance in psychotropic CYP interactions. New Clinical Drug Evaluation Unit (NCDEU) annual meeting Boca Raton, National Institute of Mental Health USA

Escher M, Desmeules J (2001) Hepatitis associated with kava, a herbal remedy for anxiety. BMJ 322:139

Farnsworth NR, Wagner H (1996/97) 2nd International congress on phytomedicine. Phytomedicine 3(Suppl1):1-304

Grigson G (1958) The Englishmans flora. Hart-Davis McGibbbon, London, pp 83-89

Grunze H, Walden J (1999) Kawain limits excitation in CA 1 pyramidal neurons of rats by modulating ionic currents and attenuating excitatory synaptic transmission. Hum Psychopharmacol Clin Exp 14:63-66

Guy W. ECDEU Assessment Manual for Psychopharmacology-3/4Revised (DHEW Publ No ADM 76-338). Rockville, MD, U.S. Department of Health, Education, and Welfare, Public Health Service, Alcohol, Drug Abuse, and Mental Health Administration, NIMH Psychopharmacology Research Branch, Division of Extramural Research Programs, pp. 218-222

Hamilton M (1959) The assessment of anxiety states by rating. Br J Med Psychol 32:50-55

Hamilton M (1960) A rating scale for depression. J Neurol Neurosurg Psychiatry 23:56-62

Hansel R, Haas H (1984) Therapie mit phytopharmaka. Springer Verlag, Heidelberg & New York

Hansel R, Keller K, Rimpler H, Schneider D (eds) (1994) Hagers handbuch der pharmazeutischen praxis 6th ed, Drogen E-O. Springer-Verlag, Berlin Heidelberg New York, pp 201–221

Hansgen KD, Vesper J, Ploch M (1994) Multi-center double blind study examining the antidepressant effectiveness of hypericum extract LI 160. J Geriatr Psychiatry Neurol 7(Suppl 1):15–18

Harrer G, Hubner WD, Podzuweit H (1994) Effectiveness and tolerance of the hypericum extract LI 160 compared to maprotiline: a multi-center double-blind study, J Geriatr Psychiatry Neurol 7(Suppl 1):24–28

Hofmann R, Winter U (1993) Therapeutische moglichkeiten mit einem hochdosierten standardisierten kava-kava-paparat (antares 120) bei angsterkrangungen. V. Phytotherapiekongress, Bonn, Nov 3–5

Kinsler E, Kromer J, Lehmann (1991) Wirksamkeit eines kava-spezial-extraktes bei patienten mit angst-, spannungs-, und erregungszustanden nicht-psychotischer genese. Arzneim Forsch/Drug Res 41:584–588

Laakmann G, Schule C, Baghai T, Kieser M (1998) St John's Wort in mild to moderate depression: The relevance of hyperforin for the clinical efficacy. Pharmacopsychiatry 31:54–59

Le Bars PL, Katz MM, Berman N, Itil TM, Freedman AM, Schatzberg AF (1997) A placebo-controlled, double-blind, randomized trial of an extract of ginkgo biloba for dementia. JAMA 278:1327–1332

Leathwood PD, Chaufford F (1983) Quantifying the effects of mild sedatives. J Psychiatr Res 17:115–122

Lehmann E, Kinsler E, Friedemann J (1996) Efficacy of a special kava extract (piper methysticum) in patients with states of anxiety, tension and excitedness of non-mental origin: A double-blind placebo-controlled study of four weeks treatment. Phytomedicine 3:113–119

Leuschner J (1996/7) Preclinical toxicological profile of hypericum extract LI 160 (abstract) Int J Phytother Phytopharmacol 3(Suppl 1):34–38

Lindahl O, Lindwall L (1989) Double blind study of a valerian preparation. Pharmacol Biochem Behav 32:1065–1066

Linde K, Ramirez G, Mulrow CD, Pauls A, Weidenhammer W, Melchart D (1996) St John's wort for depression—An overview and meta-analysis of randomised clinical trials. BMJ 313:253–258

Lohse MJ, Muller-Oerlinghausen B (1994) In: Schwabe U, Paffrath D (eds) Arznieverordnungsreport, Stuttgart: 354–370

Mendels J, Johnston R, Mattes J, Riesenberg R (1993) Efficacy and safety of b.i.d. doses of venlafaxine in a dose-response study. Psychopharmacol Bulletin 29:169–174

Mohs RC. Rosen WG, Davis KL (1983) The Alzheimer's Disease Assessment Scale: an instrument for assessing treatment efficacy. Psychopharmacol Bull 19:448–450

Moller HJ (1996) Differences observed between Europe and USA with comparators, ACNP 35th Annual meeting, Puerto Rico

Montgomery SA, Asberg M (1979) A new depression scale designed to be sensitive to change. Br J Psychiatry 134:382–389

Monthly marketing data institute for medical statistics (1997) IMS, Frankfurt: December

Muller WE, Rolli M, Schafer C, Hafner U (1997) Effects of hypericum extract (LI 160) in biochemical models of antidepressant activity. Pharmacopsychiatry 30(Suppl 2):102–107

Muller WE, Singer A, Wonnermann M (1998) Hyperforin represents the major re-uptake inhibiting constituent of St John's Wort, Poster American College of Psychopharmacology annual meeting, Puerto Rico.

Nahrstedt A, Butterweck V (1997) Biologically active and other constituents of the herb of *Hypericum perforatum*. Pharmacopsychiatry 30(Suppl 2):129–134

Oyama Y, Fuchs PA, Katayama N, Noda K (1994) Myricetin and quercetin, the flavonoid constituents of ginkgo biloba extract, greatly reduce oxidative metabolism in both resting and Ca-loaded brain neurons. Brain Res 635:125–129

O'Breasail AM, Arguarch S (1998) Hypomania and St John.s Wort Can J Psychiatry 43:746–747

Ozturk Y, Aydin S, Ozturk N, Bayer K. (1996/97) Effects of certain hypericum species on the central nervous system of mice (abstract), Int J Phytother Phytopharmacol 3(Suppl 2):102

Ozturk Y (1997) Testing of antidepressant effects of hypericum species on animal models Pharmacopsychiatry 30(Suppl 2):125–128

Philipp M. Kohnen R, Hiller IO (1999) Hypericum extract versus imipramine or placebo in patients with moderate depression: Randomized multicenter study of treatment for eight weeks. BMJ 319:1534–9

Porcher FP (1863) Resources of the fields and forests: Medical, economical and agricultural: prepared and published by order of the surgeon-general. Richmond VA. Reprint (1970), Arno Press, New York

Rosenberg C, Damsbo N, Fuglum E, Jacobsen LV, Horsgard S (1994) Citalopram and imipramine in the treatment of depressive patients in general practice: A Nordic multicentre clinical study. Int Clin Psychopharmacol 9(Suppl 1):41–48

Rosenthal NE, Sack DA, Gillin JC, Lewy AJ, Goodwin FK, Davenport Y, Mueller PS, Newsome DA, Wehr TA (1984) Seasonal affective disorder: A description of the syndrome and preliminary findings with light therapy. Arch Gen Psychiatry 41:72–80

Schneck C (1998) St John's Wort and hypomania. J Clin Psychiatry 59:12

Schneider B (1992) Ginkgo-biloba-extrakt bei peripheren arteriellen verschlusskrankheiten. Arzneim Forsch/Drug Res 42:428–436

Schrader E (2000) Equivalence of St. Johns wort extract (Ze 117) and fluoxetine: A randomized, controlled study in mild-moderate depression Int Clin Psychopharmacology 15:61–8.

Schrader E, Meier B, Brattstrom A (1998) Hypericum treatment of mild-moderate depression in a placebo-controlled study. A prospective, double-blind, randomized, placebo-controlled, multicentre study. Human Psychopharmacology 13:163–169

Schulz V, Hansel R, Tyler VE (1998) Rational phytotherapy. Springer-Verlag, Berlin, Heidelberg

Schulz V, Jobert M (1994) Effects of hypericum extract on the sleep EEG in older volunteers. J Geriatr Psychiatry Neurol 7(Suppl 1):S49–S43

Schwartz GE, Loew DM (1983) Assessment of psychopathology by significant others. In: Crook T, Ferrio S, Bartus R (eds) Assessment in geriatric psychopharmacology. Mark Powley Associates, New Canaan CT, pp 111–117

Segraves RT (1998) Antidepressant-induced sexual dysfunction. J Clin Psychiatry 59(Suppl 4):48–54

Singh YN (1992) Kava: An overview. J Ethnopharmacol 37:13–45

Sommer H, Harrer G (1994) Placebo-controlled double-blind study examining the antidepressant effectiveness of a hypericum preparation in 105 mildly depressed patients J Geriatr Psychiatry Neurol 7(Suppl):9–11

Smith DA (1997) Side-effects of psychotropic drugs. In: Wheatley D, Smith DA (eds) Psychopharmacology of cognitive and psychiatric disorders in the elderly. Chapman & Hall, London, pp 36–54

Stein JD, Hollander E (1994) Sexual dysfunction associated with the drug treatment of psychiatric disorders. CNS Drugs 2:78–86

Stevinson C, Ernst E (1999) Safety of hypericum in patients with depression. CNS Drugs 2:125–132

Terman M, Terman JS, Quitkin FM, McGrath PJ, Stewart JW, Rafferty B (1989) Light therapy for seasonal affective disorder: A review of efficacy. Neuropsychopharmacolgy 2:1–22

Teufel-Mayer R, Gleitz J (1997) Effects of long-term administration of hypericum extracts on the affinity and density of the central serotinergic 5-HT1A and HT2A receptors. Pharmacopsychiatry 30(Suppl 2):113–116

Volz HP (1995) Die anziolytische wirksamkeit von kava-spezialextrakt WS 1490 unter langzeithetherapie–eine randomisierte doppelblindstudie. Z Phytother Abstractbande: 9

Volz HP, Hansel R (1994) Kava-kava und kavain in der psychopharmakotherapie. Psychopharmakotherapie 1:33–39

Volz HP, Kieser M (1997) Kava-kava extract WS1490 versus placebo in anxiety disorders– a randomised placebo-controlled 25-week outpatient trial. Pharmacopsychiatrt 30:1–5

Vorbach EU, Arnoldt KH, Hubner WD (1997) Efficacy and tolerability of St John's Wort extract LI 160 versus imipramine in patients with severe depressive episodes according to ICD-10. Pharmacopsychiatry 30(suppl):81–85

Vorbach EU, Hubner WD, Arnoldt KH (1994) Effectiveness and tolerance of the hypericum extract LI 160 in comparison with imipramine: Randomised double-blind study in 135 outpatients. J Geriatr Psychiatry Neurol 7(Suppl):19–23

Walden J, von Wegerer J, Winter U, Berger M, Grunze H (1997) Effects of kawain and dihydromethisticin on field potential changes in the hippocampus. Prog Neuropsychopharmacol Biol Psychiatry 21:697–706

Warnecke G, Pfaender H, Gerster G, Gracza E (1995) Wirksamkeit von kava-spezialextrakt beim klimakterischen syndrom. Z Phytother 11:81–86

Wheatley D (1992) Prescribing short-acting hypnosedatives. Drug Safety 7:106–115

Wheatley D (1993a) The Wheatley Stress Profile. Stress Medicine 9:5–9

Wheatley D (1993b) Sleep in anxiety and depression associated with stress. Stress Medicine 9:127–129

Wheatley D (1997a) LI 160, an extract of St John's Wort, versus amitriptyline in mildly to moderately depressed outpatients—A controlled 6-week clinical trial. Pharmacopsychiatry 30(Suppl):77–80

Wheatley D (1997b) Stress, anxiety and depression. Stress Medicine 13:173–177

Wheatley D (1998a) Hypericum extract: Potential in the treatment of depression. CNS Drugs 6:431–440

Wheatley D (1998b) Sex, stress and sleep. Stress Medicine 14:245–248

Wheatley D (1998c) Mirtazapine, a new dual action antidepressant. Prescriber 9:97–99

Wheatley D (1998d) Sexual dysfunction and sleep disturbance. Psychiatry in Practice 17:10–12

Wheatley D (1999a) Hypericum in seasonal affective disorder (SAD). Curr Med Res Opin 15:33–37

Wheatley D (1999b) Ginkgo biloba in the treatment of sexual dysfunction due to antidepressant drugs. Human Psychopharmacology 14:511–513

Wheatley D (2001a). Kava-kava in the treatment of generalized anxiety disorder. Primary Care Psychiatry 7:97–100

Wheatley D (2001b) Stress-induced insomnia treated with kava and valerian singly and in combination. Human Psychopharmacology 16:353–356

Winterhoff H, Butterweck V, Nahrsted A, Schulz V, Wall A (1995) Pharmacologische untersuchungen zur antidepressiven virkung von hypericum perforatum L. In: Loew D, Rietbrook N (eds) Phytopharmaca im forschund Klinischer anwendung. Steinkopf, Darmstadt, pp 39–56

Woelk H, Burkhard G, Grunwald J (1994) Hypericum extract LI 160: Drug monitoring with 3250 patients. J Geriatr Psychiatry Neurol 7(Suppl 1):34–38

Woelk H, Kapoula O, Lehrl S (1993) Berhandlung von angst-patienten. Z Allg Med 69:271–277

Part 4
Use of Antidepressants in Special Populations

Children and Adolescents

J. F. Bober[1] · S. H. Preskorn[2]

[1] University of Kansas School of Medicine-Wichita, 1010 N. Kansas, Wichita,
KS 67214-3199, USA
e-mail: jbober@kumc.edu

[2] Department of Psychiatry and Behavioral Sciences,
University of Kansas School of Medicine and Psychiatric Research Institute,
1010 N. Kansas, Wichita, KS 67214, USA

1	Introduction	357
2	Pediatric Psychopharmacology	357
3	Treatment Issues	359
4	Pharmacokinetic Issues	359
5	Major Depressive Disorder in Children and Adolescents	362
5.1	Selective Serotonin Reuptake Inhibitors	363
5.2	Tricyclic Antidepressants	365
5.3	Other Antidepressants	366
5.4	Continuation Treatment	366
5.5	Adverse Effects of Antidepressants	366
6	Mania	367
7	Anxiety Disorders in Children and Adolescents	368
7.1	Obsessive-Compulsive Disorder	368
7.1.1	Clomipramine	368
7.1.2	Selective Serotonin Reuptake Inhibitors	370
7.2	Other Anxiety Disorders	371
8	Attention Deficit Disorder in Children and Adolescents	371
9	Enuresis	373
10	Conclusion	373
References		373

Abstract Special considerations arise in treating children and adolescents with antidepressants. Empirical data on antidepressants (and other pharmacological agents) in young patients are quite limited. Psychiatrists, faced with depriving children of potentially effective medication or prescribing medications "off label," need information on which to base treatment decisions, and efforts are underway (e.g., by the National Institutes of Health, the American Academy of

Pediatrics, and the Food and Drug Administration) to promote research in this area. Clinically significant differences in pharmacokinetics and possibly pharmacodynamics between adults and younger patients can also complicate treatment (e.g., younger patients may need higher doses on a milligram-per-kilogram basis to achieve the same drug concentration as an adult on a usually effective adult dose). Younger patients may also be more sensitive to adverse effects of medications.

The selective serotonin reuptake inhibitors (SSRIs) have superceded tricyclic antidepressants (TCAs) as first-choice pharmacotherapy based on studies demonstrating their superior safety and efficacy in children with major depressive disorder (MDD). TCAs are now usually reserved for children or adolescents with at least moderate depression who have not responded to at least one newer antidepressant; it is recommended that therapeutic drug monitoring (TDM) of the TCA be done at least once to ensure that the patient does not develop toxic plasma levels. The safety, pharmacokinetics, and tolerability of venlafaxine and nefazodone have been tested in children, but data on efficacy are not yet available. The adverse effect profiles of the SSRIs, the TCAs, venlafaxine, and nefazodone are similar to those in adults.

The TCA clomipramine and the SSRIs fluvoxamine and sertraline have indications for obsessive-compulsive disorder in pediatric patients. A number of TCAs and SSRIs have been studied in the treatment of other anxiety disorders (e.g., separation anxiety disorder, school phobia, elective mutism, generalized anxiety disorder) but none has received labeling for those indications. Antidepressants have been studied in the treatment of attention-deficit/hyperactivity disorder (ADHD). The TCA desipramine and bupropion have been found efficacious in ADHD, although desipramine causes higher rates of adverse effects than stimulant medications. Current treatment algorithms generally recommend trying an antidepressant after failed trials of several different stimulant medications. Atomoxetine, a nonstimulant medication, was recently approved for the treatment of ADHD in children, adolescents, and adults. Although behavioral management is preferred for treatment of enuresis, the TCA imipramine has also been found effective, although the relapse rate is as high as 50% upon discontinuation.

Given the paucity of data on antidepressants in pediatric patients and the clinically significant pharmacokinetic differences between younger patients and adults, clinicians should carefully consider and cautiously monitor any treatment plan involving antidepressant medications in order to maintain the risk to benefit ratio in favor of the child or adolescent patient.

Keywords Children · Adolescents · Selective serotonin reuptake inhibitors · Tricyclic antidepressants · Venlafaxine · Nefazodone · Clomipramine · Bupropion · Major depressive disorder · Obsessive-compulsive disorder · Anxiety disorders · Attention-deficit/hyperactivity disorder · Enuresis

1
Introduction

Special issues arise in treating children and adolescents with antidepressants. First, clinical data concerning the use of these agents in young patients are quite limited. In addition, there are clinically significant pharmacokinetic (and perhaps) pharmacodynamic differences between adults and younger patients that can complicate treatment management. This chapter first reviews some general issues related to pediatric psychopharmacology and then reviews the use of antidepressants in the treatment of children and adolescents with the following types of disorders:

- Major depressive disorder (MDD)
- Mania
- Anxiety disorders (obsessive-compulsive disorder and other anxiety disorders)
- Attention-deficit/hyperactivity disorder (ADHD)
- Enuresis

2
Pediatric Psychopharmacology

The benefits of effectively treating children and adolescents are potentially greater than those that can be achieved in the rest of general psychiatry. Yet the evidence from clinical studies upon which treatment decisions can be based is quite limited in virtually all areas of pediatric clinical pharmacology. Of the prescription drugs currently marketed in the United States, 80% have not been approved by the Food and Drug Administration (FDA) for use in children (Kauffman 1998). The situation has changed little over the past 20 years. In 1973, the labeling of 78% of the 2,000 prescription medications listed in the Physician's Desk Reference included a proscription against their use in children; the same was true in 1992. Of 53 new drugs approved in 1996, 37 were approved for the treatment of conditions that occur in children as well as adults—yet 30 of these drugs were approved only for adults.

Psychiatrists who are prescribing psychiatric medications for children and adolescents will likely take little comfort from the fact that they are no worse off than their colleagues in other areas of pediatric medicine. These psychiatrists are faced with the conundrum of either depriving children of potentially effective medication treatment or prescribing such medication "off label" and without optimal information on dosing, efficacy, and safety in children and adolescents.

Pediatric psychiatric disorders may lead to the following negative outcomes:

- Persistent poor self-image
- Poor interpersonal relationships

- Chronic underachievement
- Poor psychosocial development
- Dropping out of school
- Poor work performance
- Substance abuse
- Legal problems
- Frequent psychiatric and medical hospitalizations
- Death through suicide or accidents

Such negative consequences are likely to persist even after the resolution of the acute episode and can profoundly affect further psychosocial adjustment (Tosyali and Greenhill 1998). The irony is that early and effective treatment intervention in a disease process may often lessen its long-term sequelae.

Over the last decade, there has been a growing awareness on the part of the government, the medical profession, the pharmaceutical industry, and the public concerning the importance of having empirical data upon which to base medication treatment decisions for children and adolescents. Both the National Institute of Mental Health (NIMH) and the FDA have taken steps to increase the amount of information available on the optimal treatment of children and adolescents with medications. At the request of the Director of NIMH, the Institute of Medicine (IOM) formed a committee to assess the status of research in mental disorders affecting children and adolescents. The resulting report, *Research on Children and Adolescents with Mental, Behavioral, and Developmental Disorders*, led to a 5-year plan to stimulate a wide range of clinical research, including clinical psychopharmacology studies, and to develop young investigators in this area (NIMH 1991, 1993). During the same timeframe, the American Academy of Pediatrics issued its *Guidelines for the Ethical Conduct of Studies to Evaluate Drugs in Pediatric Populations* (Kauffman et al. 1995). To encourage research, in the late 1990s, the FDA offered 6 months of patent extension to manufacturers of selected approved drugs if they conducted appropriate studies in children and adolescents. This approach meant that a company could more than recoup the cost of the study by the additional revenue generated during the extended period of patent protection. To put this matter in perspective, 6 months of additional patent protection is worth 100 million dollars if the drug has annual sales of 200 million dollars.

As a result of these efforts, more research is underway in child and adolescent clinical psychopharmacology than ever before. The full impact of this increased research activity will take several years to be realized due to the lag time between initiating and completing a research program in clinical psychopharmacology. The first information will be on the pharmacokinetics of newly approved drugs in children and adolescents, simply because such studies take less time to complete than clinical trials evaluating efficacy, safety, and tolerability. Such pharmacokinetic data can aid researchers in determining the optimal dose and dosing schedule for use in efficacy/safety trials (Preskorn et al. 1983, 1989; Derivan et al. 1995; Findling et al. 2000).

3
Treatment Issues

There are some potentially important differences in the physician–patient relationship when the patient is a child or unemancipated adolescent: First, the child psychiatric patient often does not present for treatment, but instead is brought to the clinician because someone else (e.g., a parent or teacher) is concerned or annoyed by the child's behavior. Thus, the patient may be a passive, if not reluctant, participant in treatment. Second, there is a greater possibility with children than with adults that a beneficial medication may have deleterious effects on growth and development. Therefore, clinicians need to consider the following questions carefully prior to initiating treatment:

- Does the child or adolescent have a disorder or syndrome of a type and of sufficient severity to warrant medication?
- Do the needs and the desires of the child or adolescent and the parent conflict or are they in synchrony?
- What is the patient's social and family situation and how will it influence treatment outcome?
- Will the parent or guardian be able to assist with the administration and monitoring of the medication?
- What other forms of treatment may be needed (e.g., education about the condition and about better behavioral management techniques, family therapy, and/or individual psychotherapy)?
- What does the patient think about his or her condition, the need for treatment, and the specific treatment that is being recommended?
- How will the treatment affect the patient's self-concept and relations with others?
- How is the patient doing in school and why? (If there have been any significant changes in level of functioning, then the time course and magnitude of the changes should be carefully assessed.)
- Have all the options been reasonably discussed and their relative merits and liabilities weighed?
- What outcome parameters will be used to document the potential beneficial and adverse effects of the medication?
- Is the addition of a second or third medication necessary (e.g., if the first treatment fails, should it be discontinued rather than resorting to polypharmacy)?

4
Pharmacokinetic Issues

Ideally, drug doses in children and adolescents should be based on systematic studies in this age group. As outlined above, more work is now underway in this area than ever before. While the pharmacokinetic data will be an important step forward, optimal dosing should also be based on efficacy and safety studies in these populations. In addition to differences in pharmacokinetics, children may also be more sensitive to the beneficial and/or adverse effects of specific medications. Until such data are available, determining optimal dosing in children and adolescents will be difficult, especially when extrapolating from adult data. Hence, the guiding principle remains to start low and go slow and to aim for the lowest effective dose.

Although the concentration-response curves for efficacy and safety may be different in children and/or adolescents than in adults, the prescriber can use therapeutic drug monitoring (TDM) with at least some of the newer medications to determine whether the patient is achieving a concentration that has proven to be both effective and safe in adults. TDM can be used to assess the ability of a patient to clear a drug. Using these results, the prescriber can adjust the dose to achieve adult concentrations. Still, this approach is only an approximation of what might be optimal for the child and adolescent patient. These patients may be either more or less sensitive to the beneficial or adverse effects of the drug and thus might need a concentration that is either higher or lower than that needed in adults to achieve an optimal response. Nevertheless, TDM results can serve as a reasonable reference point in the absence of any more definitive efficacy and safety data in children and adolescents. Using TDM as a frame of reference, the prescriber should carefully titrate the dose based on clinical assessment of safety and efficacy. TDM can also be used to assess treatment adherence. (See the chapter by Burke and Preskorn in this volume for a more detailed discussion of the use of TDM.)

With regard to specific antidepressants, the AACAP Practice Parameter for Major Depressive Disorder (American Academy of Child and Adolescent Psychiatry 1998) has advocated monitoring specific blood levels of the tricyclic antidepressants (TCAs). There are no clear data on the use of TDM with selective serotonin reuptake inhibitors (SSRIs) in children and adolescents. In a recent presentation, Emslie indicated that the practice at the University of Texas Southwest Medical Center, if the patient's depression is not responding to 20 mg/day of fluoxetine, is to check the combined fluoxetine and norfluoxetine level. If the level is less than 250 ng/ml, the dose is raised to 40 mg/day (Emslie 2001).

There are a number of pharmacokinetic differences between children and adolescents compared with adults. These differences frequently lead to the need for higher doses on a milligram-per-kilogram basis to achieve the same drug concentrations as are achieved in adults on the usually effective adult doses. These differences are summarized below.

Most psychotropic medications are highly lipophilic. The percentage of total body fat, which is a reservoir for these lipid-soluble compounds, increases during the first year of life and then decreases until the prepubertal increase (Briant 1978; Tosyali and Greenhill 1998). Thus, children at different ages have different volumes of deep storage, which can affect the overall residual time a drug remains in the body after its discontinuation.

Table 1 Developmental patterns for specific drug-metabolizing enzymes

Cytochrome P450	Development pattern
Phase I enzymes	
1A2	Adult level reached by 4 months but exceeded by age 1–2 years. Declines to adult levels by the conclusion of puberty. Gender differences are possible during puberty
2C9, 2C19	Adult activity reached by 6 months but exceeded by 1.5–1.8 times by age 3–4. Declines to adult levels by conclusion of puberty
2D6	Adult levels obtained by 3–5 years of age
3A4	Adult levels by 6–12 months but then exceeded by 1–4 years of age. Declines to adult levels by the conclusion of puberty
Phase II enzymes	
NAT2	Adult activity present by 1–3 years of age
TPMT	Adult activity achieved by 7–9 years of age
UGT	Adult activity by 6–18 months of age
ST	May exceed adult levels during early childhood

NAT2, N-acetyltransferase-2; ST, sulfotransferase; TPMT, thiopurine methyltransferase; UGT, glucuronosyltransferase.

The acquisition of adult levels of both cytochrome P450 (CYP) and phase II drug-metabolizing enzymes is enzyme-specific and isoform-specific (Leeder and Kearns 1997). Although recent research has shown that the traditional view of a locked step progression of drug-metabolizing capacity is overly simplistic, some generalizations can still be made. The activity of most drug-metabolizing enzymes is absent in the fetus but rapidly increases over the first years of life, so that toddlers and older children have levels of several, but not all, drug-metabolizing enzymes that exceed those of adults. These levels decline from that point until "usual" adult levels are achieved by the conclusion of puberty. Developmental changes over the first two decades of life in the activity of specific CYP enzymes is reflected in the increase and then fall in theophylline clearance (1A2), the decline in the clearance of phenytoin (2C9/10, 2C19), and the fall in the ratio of carbamazepine-10, 11 epoxide to carbamazepine (3A3/4) (Milavetz et al. 1986; Korinthenberg et al. 1994). Further details are provided in Table 1.

The rate of drug metabolism is also partly dependent on liver mass. Relative to body weight, the mass of the liver of a toddler is 40%–50% greater, and that of a 6-year-old is 30% greater than that of an adult (Briant 1978; Tosyali and Greenhill 1998). That is another reason why children tend to clear drugs more rapidly than adults and frequently need higher doses on a milligram-per-kilogram basis to achieve the same plasma levels and clinical effect.

By 1 year of age, glomerular filtration rate and renal tubular mechanisms for secretion have reached adult levels; however, fluid intake may be greater in children. Thus, lithium has a shorter half-life and more rapid renal clearance in children than in adults (Carlson 1990).

Nevertheless, the interindividual differences in the clearance of psychiatric medications are as great in children and adolescents as in adults. This should not be surprising, since the same interindividual differences exist in children and adolescents as in adults. For example, genetically determined differences in CYP 2D6 function are expressed at birth. Hence, 5%–10% of children and adolescents of northern European origin are deficient in CYP 2D6 activity and will develop 4–6 times higher levels of drugs that are predominantly metabolized by CYP 2D6 than individuals with a functional copy of this enzyme.

Among the psychiatric medications, the pharmacokinetics of the psychostimulants have been the best studied in children and adolescents. The next best pharmacokinetic studies in children and adolescents have been done with antidepressants, while few studies have been done on the pharmacokinetics of antipsychotics and anxiolytics in this age range.

It may be prudent to divide the daily doses more frequently than is done in adults to avoid excessively high peak plasma drug concentrations that may be associated with increased tolerability and safety problems.

5
Major Depressive Disorder in Children and Adolescents

MDD can occur in children as young as 6 years of age. The diagnosis is based on the same criteria as in adults, with the exception that irritable mood can be substituted for depressed mood. Children and adolescents with MDD typically have high familial loading for psychiatric disorders (Hughes et al. 1989). Over 70% of mothers of children and adolescents with MDD also have MDD, either pure or complicated by the presence of other psychiatric syndromes. However, the fathers are more likely to have alcohol abuse or dependence, as opposed to MDD. Given this familial pattern, it is not surprising that many children and adolescents with MDD frequently also meet criteria for other psychiatric syndromes, in particular conduct disorder and oppositional defiant disorder (Hughes et al. 1989).

Despite the diagnostic challenges that remain in trying to understand the nature of MDD in children and adolescents, considerable progress has been made in its treatment in recent years. SSRIs have superceded the TCAs as the treatment of first choice based on considerations of both efficacy and safety. As in adults, specific psychotherapies (cognitive therapy, cognitive-behavioral therapy, and interpersonal therapy) may be as effective as antidepressant medication, at least in mild to moderate depression in children and adolescents (Brent et al. 1997). There is also evidence that depression in children and adolescents may be more influenced by psychosocial variables, such as peers and family, as well as other environmental factors than is depression in adults (Hammen et al. 1999).

This finding may partly account for the higher placebo response rate seen in children and adolescents, particularly in many of the earlier antidepressant trials. Puig-Antich et al. (1979) reported a placebo response rate of over 68%, and others have found similar rates (Hughes et al. 1990). One reason for the high "placebo" response rate in these earlier studies was the degree of psychosocial treatment that these patients received. Most of these studies involved several weeks of hospitalization, which was more feasible in the 1980s than now. These hospitalizations removed the patient from what was often a chaotic home life. The patients also received intensive individual, group, and milieu psychotherapy. Generally, such intensive psychotherapy services were not available in more recent trials of the newer antidepressants—which is consistent with the fact that these studies have generally had lower placebo response rates. Thus, the newer antidepressants may not actually be more effective than the TCAs. Instead, the difference may be that the "placebo" treatment in the more recent studies was less effective than the earlier "placebo" treatment in the older studies, and that difference in placebo effect alone could permit the study to detect a drug–placebo difference in efficacy.

The high placebo response rate found in children and adolescents with MDD may also be due to the heterogeneous nature of the disorder mentioned above. Hughes et al. (1990) found that children and adolescents with major depression plus concomitant conduct or oppositional defiant disorder had a higher response rate to placebo than to imipramine. Nevertheless, an empirical trial of an antidepressant, particularly the safer newer antidepressants, may still be warranted in such patients.

5.1
Selective Serotonin Reuptake Inhibitors

From 1996 to 1997, 792,000 prescriptions for SSRIs were written to treat depression in children and adolescents between 6 and 18 years of age (Hoar 1998). This increase was due to the safety of these medications compared with the TCAs and to growing evidence of their efficacy. However, the first double-blind study of fluoxetine failed to show a difference in response rates, although it suffered from a high "placebo" or nonspecific treatment response rate (Simeon et al. 1990).

Emslie et al. (1997b) were the first to demonstrate the superiority of an SSRI over placebo in both children and adolescents aged 7–17 years old. In their study, 48 outpatients were randomized to drug and 48 to placebo at a single academic medical center. Patients with bipolar disorder (or a family history of bipolar disorder), previous psychosis, or substance abuse were excluded. Using the "intent to treat" sample, 27 (56%) of those receiving 20 mg/day of fluoxetine for 8 weeks responded versus 16 (33%) who received placebo, with response defined as a Clinical Global Improvement (CGI) rating (Guy 1976) of "much" or "very much" improved ($p<0.05$). There was also significant improvement on the Children's Depression Rating Scale-Revised (CDRS-R) (Poznanski et al. 1985). However, there were no significant differences in terms of complete symptom remission defined as a score of 28 or less on the CDRS-R. Remission occurred in 31% of patients treated with fluoxetine versus 23% of patients treated with placebo. There was also no difference in scores on the Children's Global Assessment Scale (CGAS) (Shaffer et al. 1983) between the two groups at completion, with both groups showing improvement in functioning. There was no difference in response and remission rates between subjects 12 years of age and younger ($n=48$) and subjects who were 13 years of age or older ($n=48$). Interestingly, of the patients who responded, only 20% remained on the medication after 6 months.

A multisite trial of fluoxetine in depressed children and adolescents has now been completed (Emslie 2001). In this study, 219 outpatients ages 8–17 were enrolled, with 109 randomized to fluoxetine (10 mg for 1 week and then 20 mg for 8 weeks) and 110 to placebo; 52% of those receiving fluoxetine responded versus 37% of those on placebo, with response defined as a CGI rating of "much" or "very much" improved ($p=0.028$). There was also significant improvement ($\geq 50\%$ reduction) on the CDRS-R (58% versus 41%, $p=0.014$). There was a significant difference in complete symptom remission, defined as a score of less than or equal to 28 on the CDRS-R. Remission occurred in 41% of patients treated with fluoxetine versus 20% of those receiving placebo ($p<0.01$).

Keller et al. (2001) were the first to demonstrate the superiority of paroxetine over imipramine and placebo in adolescents aged 12–18 years. In their study, 275 outpatient subjects with MDD of at least 8 weeks duration were randomized to paroxetine (20–40 mg, average 28 mg), while 95 were randomized to imipramine (200–300 mg, average 206 mg), and 87 to placebo at 10 centers in the United States and 2 in Canada. Patients with bipolar disorder, posttraumatic stress disorder (PTSD), current suicidal ideation or suicide attempt by overdose, previous psychosis, or substance abuse were excluded. Of those receiving paroxetine for 8 weeks, 59 (66%) responded versus 49 (52%) on imipramine and 42 (48%) on placebo, with response defined as a CGI rating of "much" or "very much" improved ($p<0.05$). There was also significant improvement on the HAM-D depressed mood item and on the Schedule for Affective Disorders and Schizophrenia for School-Age Children: Lifetime Version (K-SADS-L) (Kaufman et al. 1997) depressed mood item. There was a significant difference in complete symptom remission defined as a score of 8 or less on the 17-item Hamilton Rating Scale for Depression (HAM-D) (Hamilton 1967). Remission occurred in 63% of patients treated with paroxetine compared with 50% of those treated with imipramine and 46% of those receiving placebo. There was no difference in parent- or self-rating measures between the three groups at completion. CGAS scores were improved from baseline (average 43) in all groups, but differences between the groups were not significant. Patients on paroxetine withdrew from the study due to adverse effects at a rate similar to that of patients in the placebo group (10% versus 7%), while there was a much higher rate of withdrawals (32%) due to adverse effects in the group receiving imipramine.

Sertraline has also shown promising effects in an open-label trial in adolescents with major depression (Ambrosini et al. 1999). Industry-sponsored, randomized controlled trials have been completed with sertraline, citalopram, and venlafaxine. The data are currently being analyzed and no results are yet available yet. Two identical multisite randomized controlled trials have been completed using mirtazapine (15–45 mg), involving a total of 250 outpatients

(7–17 years old, average age 12 years), with 165 assigned to medication and 85 to placebo. The study failed to show efficacy, perhaps due to a high placebo response rate (Emslie 2001).

NIMH is currently supporting a six-site nationwide study called the Treatment of Resistant Depression in Adolescents (TORDIA). Data from this study, which will be available in a few years, should answer questions about the role of the SSRIs as a treatment option for adolescents who have not fully responded to a previous adequate trial of an SSRI.

5.2
Tricyclic Antidepressants

As mentioned above, the clinical trials with TCAs in children and adolescent with MDD have generally been disappointing (Birmaher et al. 1996a,b; Keller et al. 2001). A meta-analysis by Hazell et al. (1995) looked at 12 studies from 1981–1992. These studies were generally hindered by small treatment groups and small sample sizes. They found a small treatment effect but it was not clinically significant when compared to the placebo group. In addition, the TCAs have a less favorable adverse effect profile and thus higher patient attrition rates during the acute treatment phase compared to the newer antidepressants. They also have a lower therapeutic index (i.e., difference between therapeutic and toxic dose, see the chapter by Lader, this volume).

Despite these challenges, there have been two positive, double-blind, placebo-controlled studies showing TCAs to be superior to placebo in children suffering from major depression (Kashani et al. 1984; Preskorn et al. 1987). The number of subjects in each study was small. In the study by Kashani et al., there were only nine subjects in each treatment group, the maximum dose of amitriptyline used was low (1.5 mg/kg/day), and the statistical results were modest ($p<0.05$, based on a one-tail t-test, presuming drug would be superior to placebo). In the double-blind, placebo-controlled study by Preskorn et al., the group sample sizes were somewhat larger ($n=15$ per cell). The imipramine dose was adjusted based on TDM to ensure that the patient achieved plasma levels that had previously been found to be safe and effective (Preskorn et al. 1982, 1983). Imipramine was superior to placebo through the first 3 weeks of treatment ($p<0.05$, based on a two-tail t-test); however, the response rates in the imipramine and placebo groups were not different by the end of the 6-week study.

Kramer and Feiguine (1981) conducted the first double-blind study of amitriptyline in adolescents with major depression. At the end of 6 weeks, all subjects had improved and the only difference was a lower score on the Depression Adjective Checklist (Lubin and Himelstein 1976; Lubin and Levitt 1979) in the group receiving the TCA versus the controls. Geller et al. (1992) were unable to find a difference in efficacy between nortriptyline and placebo. Kutcher et al. (1994) randomly assigned 60 adolescent subjects to either desipramine (up to 200 mg/day) or placebo and found a 48% response rate in patients on drug versus 35% in those on placebo.

5.3
Other Antidepressants

The first studies of venlafaxine and nefazodone in both children and adolescents examined their pharmacokinetics, safety, and tolerability (Derivan et al. 1995; Findling et al. 2000). As expected, the clearance of both of these medications was modestly more rapid in children and adolescents than in adults, but the difference was not sufficient to warrant a significant change in the mg/kg daily dose. These studies now need to be followed by appropriately designed efficacy studies.

5.4
Continuation Treatment

There are no studies that provide guidance concerning how long antidepressants should be continued in children and adolescents once a response has been achieved. Emslie et al. (1997a) performed a naturalistic study of 70 children and adolescents with MDD and found that 98% recovered from their index episodes of MDD within 1 year of their initial evaluation. Over 80% received antidepressants but the nature of the treatment was determined by the individual clinician rather than being dictated by a treatment protocol. Over 60% of these patients had at least one recurrence during a 1- to 5-year follow-up period. Of those who experienced a recurrence, 47% had one within 1 year of their recovery and 70% within 2 years. These results are consistent with earlier studies indicating that 54%–72% of children and adolescents with MDD have a recurrent episode when followed for 3–8 years (McCauley et al. 1993; Rao et al. 1995). Until further studies are done, it would seem prudent to follow the same continuation and maintenance treatment guidelines for children and adolescents with MDD as for adults with MDD.

5.5
Adverse Effects of Antidepressants

The adverse effects of the SSRIs, venlafaxine, and nefazodone in children and adolescents are comparable to those in adults (see the two chapters by Preskorn et al., this volume) and have been documented in both clinical trials and clinical practice (Derivan et al. 1995; Tierney et al. 1995; Emslie et al. 1997b; Ambrosini et al. 1999; Findling et al. 2000; Keller et al. 2001). As in adults, there have been isolated case reports of behavioral activation in children and adolescents treated with SSRIs (Guiles 1996; Go et al. 1998). The significance of such reports in terms of a causal link to the drug is difficult to determine due to their rare and anecdotal nature and the fact that the patients are at increased risk for such behavioral disturbances relative to the general population as a result of their underlying psychiatric disorder.

The adverse effects of TCAs are also similar to those reported in adults (see chapter by Lader, this volume). The secondary amine TCAs (e.g., desipramine, nortriptyline) are generally as well tolerated as newer antidepressants. Increased blood pressure may be more likely to occur in children than adults but hypertension per se is rare (Lake et al. 1979; Preskorn et al. 1983). The most common cardiovascular effect is mild tachycardia. Despite their generally favorable adverse effect profile, secondary amine TCAs can cause serious toxicity in children and adolescent just as in adults when taken in an overdose or when high TCA plasma levels occur due to slow metabolism (Preskorn 1998). For that reason, most clinicians will reserve TCAs for the child or adolescent who has at least a moderate depressive disorder that has not responded to a trial of at least one newer antidepressant. In such instances, TDM should be done at least once to assure that the patient does not develop plasma concentrations above 450 ng/ml (Preskorn and Fast 1991). Such levels are associated with an increased risk of:

- Delirium
- Seizures
- Slowing of intracardiac conduction, which can lead to heart blocks, arrhythmias, and sudden death (Preskorn et al. 1983, 1988)

There have been several cases of sudden death in children and adolescents taking desipramine for a variety of indications (Biederman 1991; Riddle et al. 1991). These cases have raised significant concern among child psychiatrists, even though the drug was barely detectable at autopsy, indicating that it was unlikely to have been a contributor to the sudden death. These cases have led some clinicians to recommend frequent electrocardiogram (ECG) monitoring (baseline and at every dose increase) without evidence that such monitoring will achieve early detection of a problem and avoid an untoward outcome. In fact, this recommendation may lead some clinicians not to use TCAs at all in children and adolescents and thus may deprive them of a potentially therapeutic option.

6
Mania

It can be quite difficult to differentiate between the emotional vicissitudes of adolescence and mild episodes of bipolar disorder. Nonetheless, a sizable minority (30%) of adult patients with bipolar disorder report having their first episode during adolescence. Further, manic (type I) episodes have been observed during adolescence, and the earlier the onset, the more likely it is that the patient will have a psychotic form of the illness (Carlson 1990).

A child diagnosed with bipolar I disorder, depressed episode, or MDD may do well treated with an antidepressant. However, if clear symptoms of mania develop, the antidepressant should be quickly stopped and the manic symptoms aggressively treated. Over the years, psychiatrists have worried about precipitating a manic episode by the use of an antidepressant (the phenomenon of switching). In past antidepressant trials with children and adolescents, 6% of patients developed mania; however in recent large trials of SSRIs, no patient developed mania (Emslie 2001). Rates of "activation," the development of apparent hypomanic symptoms precipitated by a medication, were no different in the SSRI and placebo groups in recent large trials (Emslie 2001).

7
Anxiety Disorders in Children and Adolescents

Only a few studies on pediatric clinical psychopharmacology for anxiety disorders have been done (Ambrosini et al. 1993; Klein and Slomkowski 1993; Allen et al. 1995; Campbell and Cueva 1995). Most studies have been open-label and have involved only a small number of patients. Among the anxiety disorders, obsessive-compulsive disorder (OCD) is the condition that has been the best studied in children and adolescents; five double-blind studies have led to the formal labeling of several medications with indications for use in such patients. There have also been four double-blind studies of school phobia/refusal and five of generalized anxiety disorder or mixed diagnostic groups, but none of these studies have produced sufficient data to lead to formal labeling of any medication by the FDA with an indication for the treatment of these conditions in children or adolescents.

7.1
Obsessive-Compulsive Disorder

In the past, patients with OCD were treated with a series of medications, as well as psychotherapy, frequently without substantial improvement. That situation changed with the development of clomipramine and the SSRIs, which appear to have unique efficacy in treating OCD when compared with other types of psychotropics. In addition, increasingly effective behavioral approaches have been developed, which, in combination with medication, can substantially ameliorate this condition.

7.1.1
Clomipramine

The first agent proven to be effective in OCD was clomipramine, a TCA rather than an SSRI. Clomipramine's efficacy in pediatric patients with OCD was demonstrated in two double-blind, placebo-controlled studies (Flament et al. 1985; De-Veaugh-Geiss et al. 1992). These findings were further supported by a double-blind, crossover study with desipramine (Leonard et al. 1989). This latter study provided support for serotonin uptake inhibition being the mechanism of action responsible for the efficacy of clomipramine in OCD. The FDA approved an indication for clomipramine in the treatment of OCD in pediatric patients.

Despite this evidence of efficacy, there are several limitations with clomipramine due to its multiple mechanisms of action, including:

- blockade of α_1-adrenergic receptors, which can cause orthostatic hypotension,
- blockade of histamine receptors, which can produce sedation and possibly weight gain,
- blockade of cholinergic receptors, which can result in a variety of peripheral anticholinergic adverse effects and memory impairment, and
- inhibition of Na^+ fast channels, which can inhibit electrically excitable membranes and can produce intracardiac conduction delays (Preskorn and Fast 1991).

The last action is responsible for the serious central nervous system (CNS) and cardiac toxicity (e.g., delirium, seizures, cardiac arrhythmias, cardiac arrest) that can occur at high plasma drug concentrations of clomipramine just as with any TCA (Preskorn and Fast 1991). These effects are problematic with adults and may be of even greater concern with children and adolescents.

There are other concerns of particular relevance to children and adolescents due to pharmacokinetic differences between these younger patients and adults. While clomipramine is the most potent TCA in terms of inhibiting the neuronal serotonin uptake pump, its major metabolite, desmethylclomipramine, is a potent inhibitor of the neuronal uptake of norepinephrine (Rudorfer and Potter 1987). Depending on the patient's hepatic metabolism profile, either desmethylclomipramine or clomipramine may be the predominant form of the circulating drug. Children tend to be extensive demethylators of TCAs. For this reason, desmethylclomipramine can account for 70% of the circulating drug in children receiving clomipramine. If serotonin uptake inhibition is critical to clomipramine's efficacy in treating OCD, then clomipramine may fail to work as a result of extensive conversion to desmethylclomipramine.

For these reasons, TDM can serve several roles when using clomipramine. First, it can be used to determine whether clomipramine or its demethylated metabolite constitutes the majority of the circulating drug. Second, TDM can be used to guide dose adjustment to ensure equivalent plasma concentrations to those seen in adults whose OCD was successfully treated with clomipramine. This approach also ensures that the patient is not reaching concentrations (i.e., >450 ng/ml) that are above the toxic threshold for TCAs (Preskorn and Fast 1991). Like adults, children and adolescents demonstrate a wide variability in their capacity to metabolize TCAs. As mentioned earlier, children are usually faster metabolizers of these drugs than adults, so that they typically need doses of approximately 2.5–3.5 mg/kg. Once puberty has been reached, the required doses can be reduced by as much as 50%.

7.1.2
Selective Serotonin Reuptake Inhibitors

Both fluvoxamine and sertraline are approved for the treatment of OCD in children and adolescents. Fluvoxamine was proven effective in a 10-week, double-blind, placebo-controlled trial in patients with OCD 8–17 years of age (McConville et al. 1996). Patients in this study were titrated to a total daily dose of fluvoxamine of approximately 100 mg over the first 2 weeks, using a balanced, twice-daily dosing schedule. After that, the dose was adjusted within a range of 50 to 200 mg/day based on clinical assessment of efficacy and tolerability. Fluvoxamine was found to be superior to placebo on the Children's Yale-Brown Obsessive-Compulsive Scale (CY-BOCS) (Riddle et al. 1992) at weeks 1 through 6 and week 10. Response was seen in an average reduction of 10–12 points on the CY-BOCS, but the outcome was not defined as remission. This effect was mainly seen in the 8- to 11-year-old rather than the 12- to 17-year-old age group, although the significance of this age-related difference is not known. A recent study of fluvoxamine in the treatment of OCD by Riddle et al. (2001) demonstrated a 25% reduction in the CY-BOCS but not remission.

The efficacy of sertraline in the treatment of OCD in pediatric patients (6–17 years old) was demonstrated in a 12-week, multicenter, double-blind, placebo controlled study (March et al. 1998). Treatment was initiated at a dose of either 25 mg/day (ages 6–12) or 50 mg/day (ages 13–17) and then titrated over the next 4 weeks to a maximum dose of 200 mg/day as tolerated. The mean dose of completers was 178 mg/day. Dosing was once a day either morning or evening. Patients treated with sertraline showed significantly greater improvement that those on placebo on the CY-BOCS and several other scales. Significant differences in efficacy between sertraline and placebo appeared at week 3 and persisted for the duration of the study.

As expected, children and adolescents were found to metabolize sertraline slightly more efficiently than adults. Relative to adults, the area under the plasma concentration-time curve and the peak plasma concentration of sertraline was on average 22% lower in children and adolescents compared with adults when the dose was adjusted based on body weight. The half-life did not differ between the three groups: 26.2 h for the 6- to 12-year-old age group, 27.8 h for the 13- to 17-year-old age group, and 27.2 h for the 18- to 45-year-old age group (based on results from a separate study involving the same dose).

A small, double-blind, placebo-controlled, crossover trial of fluoxetine in 14 children and adolescents with OCD found a significant decrease in CY-BOCS total scores (Riddle et al. 1992). A recent randomized controlled trial (Geller et al. 2001) using fluoxetine (20–25 mg/day) produced a mean reduction in the CY-BOCS of 9 points compared with a mean reduction of 5 points in the placebo group, and there was a low rate of activation.

A study of sertraline in pediatric OCD demonstrated a further symptom reduction of 20–25% even after 10 weeks of active treatment (McCracken 2001).

7.2
Other Anxiety Disorders

The efficacy of the TCAs, principally imipramine, has also been tested in the treatment of separation anxiety disorder and school phobia. Four placebo-controlled studies involving 140 children have been done (Gittelman-Klein and Klein 1980; Berney et al. 1981; Bernstein et al. 1990; Klein et al. 1992). The first studies were positive but subsequent reports were not. Initially Gittelman-Klein and Klein (1973) demonstrated a significant benefit of imipramine (mean dose=159 mg/day) over placebo with 6 weeks of treatment in 45 children with school phobia. A replication study, however, did not find any significant benefit, due to a high placebo response rate (McCracken 2001). A subsequent study using lower doses of clomipramine (40–75 mg/day) was also negative, but the doses employed make interpretation difficult. Because of its tolerability and safety profile, clomipramine is generally not used as an anxiolytic agent in children or adolescents.

There have been few trials of the potential efficacy of the SSRIs in the treatment of anxiety disorders other than OCD in children and adolescents. Black et al. (1992) found a significant difference between fluoxetine and placebo in 4 out of 6 children with selective mutism (believed to be a variant of social phobia) but only on the global rating of change and parent's rating of mutism change. In another small, open trial, fluoxetine (10–60 mg/day) reduced anxiety and increased speech in 76% of children (ages 5–14 years) with selective mutism (Dummit et al. 1996). The recent Research Unit on Pediatric Psychopharmacology Anxiety Study Group (RUPP) Anxiety Study (Research Unit on Pediatric Psychopharmacology Anxiety Study Group 2001) studied 128 children 6–17 years of age who met criteria for social phobia, separation anxiety disorder, or generalized anxiety disorder. The study included a 3-week open phase of psychosocial intervention to exclude early responders not treated with medication. The participants were then randomly assigned to receive fluvoxamine (maximum dose of 250 mg/day for 6- to 12-year-olds and 300 mg/day for 13- to 18-year-olds) or placebo for 8 weeks. The fluvoxamine group had a mean reduction of 9.7-points in symptoms of anxiety on the Pediatric Anxiety Rating Scale (PARS) (Greenhill et al. 1998; Walkup and Davies 1999) compared with a mean decrease of 3.1 points in the placebo group; 76.2% of the fluvoxamine group had a CGI-I score of less than 4 (at least mild improvement) compared with 29.2% in the placebo group.

8
Attention Deficit Disorder in Children and Adolescents

Attention-deficit/hyperactivity disorder (ADHD) is both one of the best studied and most effectively treated of all disorders in medicine. The evidence of its validity is more compelling than that for many nonpsychiatric medical conditions (Goldman et al. 1998). A quarter century of published treatment studies and clinical experience have documented the short-term effectiveness of pharmacological management using psychostimulants (Greenhill et al. 1996).

Research has also been done on the use of antidepressants in ADHD, and there are a number of studies indicating that some but not all antidepressants are effective in ADHD. Spencer et al. (1996) found 29 studies (involving 1,016 patients) that supported the efficacy of TCAs in the treatment of ADHD. Desipramine is the TCA for which there are the most efficacy data. Based on a meta-analysis of 5 randomized trials involving 170 ADHD patients, desipramine showed efficacy (i.e., effect size) comparable to that of methylphenidate (Biederman et al. 1989; Jadad and Atkins 1998). However, desipramine produced a higher rate of adverse effects compared to psychostimulants. Moreover, there have been several sudden, unexpected deaths reported in children taking desipramine (Biederman et al. 1995; Popper and Ziminitzky 1995). While there are reasons to question whether desipramine had a role in these deaths, these reports have raised considerable concern among child and adolescent psychiatrists and the use of this medication has almost stopped in this population. There is some evidence, although more limited, supporting the efficacy of either imipramine or nortriptyline for ADHD (Spencer et al. 1993; Wilens et al. 1993, 1995; Daly and Wilens 1998).

There are three randomized clinical trials that support the efficacy of bupropion in ADHD. The first used doses up to 6 mg/kg (Casat et al. 1987). The other two used dosages of 100–300 mg/day in equally divided daily doses spaced at least 6 h apart (Barrickman et al. 1995; Conners et al. 1996). The concern with bupropion is the seizure risk associated with it, which requires that the daily dose stay under 450 mg/day in adults (i.e., approximately 6.5 mg/kg). Virtually no work has been done to determine plasma concentrations of bupropion and its three active metabolites in children and adolescents. Hence, it is unknown whether a limit of 6.5 mg/kg is also appropriate for children. There are no data about whether children are more or less sensitive to bupropion in terms of seizure risk at the same drug concentration. Little is also known about pharmacokinetic drug–drug interactions that could reduce the clearance of bupropion. For these reasons, cautious dosing is advised when prescribing bupropion for children on other medications that can reduce oxidative drug metabolism (see chapters entitled "General Principles of Pharmacokinetics" and "Other Antidepressants", this volume, for more details).

The current Texas Children's Medication Algorithm Project calls for the use of an antidepressant (bupropion, imipramine, or nortriptyline) at stages 4 and 5 of the treatment algorithm for uncomplicated ADHD (i.e., after several different stimulant medications have been tried) (Pliszka et al. 2000).

Atomoxetine (a pre-synaptic norepinephrine transporter blocker) has recently been approved by the FDA for the treatment of ADHD in children, adolescents, and adults (www.fda.gov/cder/consumerinfo/druginfo/strattera.htm). Clinical trials to date have included over 1,800 children and 250 adults. Three randomized placebo-controlled trials suggest atomoxetine, a nonstimulant, is efficacious, safe, and well tolerated (Heiligenstein et al. 2001; Leonard 2002).

9
Enuresis

For enuresis, behavioral management is preferable. The use of a modern, portable, battery-operated alarm along with close supervision and contingent reinforcement is the most helpful strategy (Glazener and Evans 2002). Desmopressin acetate (DDAVP) taken at bedtime decreases urine production. It is available as a nasal spray (dose: 10–40 μg) and a tablet (dose: 0.2–0.6 mg). Response rates range from 10%–65% in studies, but the relapse rate upon discontinuation is as high as 80%. A TCA, imipramine, is also used to treat enuresis. Typically, the initial dose is 25 mg/day administered 1 h before bedtime. The dosage may be increased to 50 mg/day in children under 12 years and to 75 mg/day for those over 12 years old, but should not exceed 2.5 mg/kg/day. After a 7-day trial with an adequate dose of imipramine, 40%-60% of the patients will have experienced relief from bed-wetting (Glazener and Evans 2002). However, the relapse rate is as high as 50% upon discontinuation. The mechanism of action of imipramine in enuresis is unknown and is not related to blood level. The adverse effects are the same as those seen in children and adolescents treated with imipramine for MDD.

10
Conclusion

While various childhood disorders have been reported to benefit from drug therapies, including antidepressant medications, systematic data supporting the use of these agents in children and adolescents are usually minimal or lacking. An additional complication is the clinically significant pharmacokinetic (and perhaps pharmacodynamic) differences between adults and younger age groups. Thus, the use of medications, including antidepressants, in any treatment plan must be carefully considered and cautiously monitored to maintain the risk to benefit ratio in favor of the child or adolescent patient.

Acknowledgements. Portions of this chapter are adapted with permission from Janicak PG, Davis JM, Preskorn SH, Ayd FJ, Jr (2001) *Principles and Practice of Psychopharmacotherapy*, 3rd Edition. Lippincott Williams & Wilkins.

References

American Academy of Child and Adolescent Psychiatry (1998). Practice parameters for the assessment and treatment of children and adolescents with depressive disorders. J Am Acad Child Adolesc Psychiatry, 37(10suppl) (Summary and full text available to Academy members at www.aacap.org)
Allen AJ, Leonard H, Swedo SE (1995) Current knowledge of medications for the treatment of childhood anxiety disorders. J Am Acad Child Adolesc Psychiatry 34:976–986

Ambrosini PJ, Bianchi MD, Rabinovich H, Elia J (1993) Antidepressant treatments in children and adolescents: II. Anxiety, physical, and behavioral disorders. J Am Acad Child Adolesc Psychiatry 32:483–493

Ambrosini PJ, Wagner KD, Biederman J, Glick I, Tan C, Elia J, Hebeler JR, Rabinovich H, Lock J, Geller D (1999) Multicenter open label sertraline study in adolescent outpatients with major depression. J Am Acad Child Adolesc Psychiatry 38:566–572

Barrickman LL, Perry PJ, Allen AJ, Kuperman S, Arndt SV, Herrmann KJ, Schumacher E (1995) Bupropion versus methylphenidate in the treatment of attention-deficit hyperactivity disorder. J Am Acad Child Adolesc Psychiatry 34:649–657

Berney T, Kolvin I, Bhate SR (1981) School phobia: a therapeutic trial with clomipramine and short-term outcome. Br J Psychiatry 138:110–118

Bernstein GA, Garfinkel BD, Borchart CM (1990) Comparative studies of pharmacotherapy for school refusal. J Am Acad Child Adolesc Psychiatry 29:773–781

Biederman J (1991) Sudden death in children treated with a tricyclic antidepressant: a commentary. Biological Therapies in Psychiatry 14:1–1

Biederman J, Baldessarini RJ, Wright V, Knee D, Harmatz JS (1989) A double blind placebo controlled study of desipramine in the treatment of ADD I: efficacy. J Am Acad Child Adolesc Psychiatry 28:777–784

Biederman J, Thisted RA, Greenhill LL, Ryan ND (1995) Estimation of the association between desipramine and the risk for sudden death in 5 to 14 year old children. J Clin Psychiatry 56:87–93

Birmaher B, Ryan ND, Williamson DE, Brent DA, Kaufman J (1996a) Childhood and adolescent depression: a review of the past 10 years. Part I. J Am Acad Child Adolesc Psychiatry 35:1427–1439

Birmaher B, Ryan ND, Williamson DE, Brent DA, Kaufman J (1996b) Childhood and adolescent depression: a review of the past 10 years. Part II. J Am Acad Child Adolesc Psychiatry 35:1575–1583

Black B, Uhde TW, Tancer ME (1992) Fluoxetine for the treatment of social phobia. J Am Acad Child Adolesc Psychiatry 29:36–44

Brent DA, Holder D, Kolko D, Birmaher B, Baugher M, Roth C, Iyengar S, Johnson BA (1997) A clinical psychotherapy trial for adolescent depression comparing cognitive, family and supportive treatments. Arch Gen Psychiatry 54:877–885

Briant RH (1978) An introduction to clinical pharmacology. In: Werry JS (ed) Pediatric psychopharmacology: the use of behavior modifying drugs in children. Brunner/Mazel, New York

Campbell M, Cueva J (1995) Psychopharmacology in child and adolescent psychiatry: a review of the past seven years. Part I. J Am Acad Child Adolesc Psychiatry 34:1124–1132

Carlson GA (1990) Bipolar disorders in childrena and adolescents. In: Garfinkel B, Carlson GA, Weller E (eds) Psychiatric disorders in children and adolescents. WB Saunders, Philadelphia, pp 21–36

Casat CD, Pleasants DZ, Van Wyck F (1987) A double-blind trial of bupropion in children with attention deficit disorder. Psycopharmacol Bull 23:120–122

Conners CK, Casat CD, Gualtieri CT, Weller E, Reader M, Reiss A, Weller RA, Khayrallah M, Ascher J (1996) Bupropion hydrochloride in attention deficit disorder with hyperactivity. J Am Acad Child Adolesc Psychiatry 35:1314–1321

Daly JM, Wilens T (1998) The use of tricyclics antidepressants in children and adolescents. Pediatr Clin North Am 45:1123–1135

Derivan A, Aguiar L, Preskorn S, Martin P, D'Amico D, Troy S (1995) A study of venlafaxine in children and adolescents with conduct disorder. Annual Meeting of the American Academy of Child and Adolescent Psychiatry, New Orleans, LA

De-Veaugh-Geiss J, Moroz G, Biederman J (1992) Chlorimipramine hydrochloride in childhood and adolescent obsessive-compulsive disorder: a multicenter trial. J Am Acad Child Adolesc Psychiatry 31:45–49

Dummit ESI, Klein RG, Tancer NK, Asche B, Martin J (1996) Fluoxetine treatment of children with selective mutism: an open trial. J Am Acad Child Adolesc Psychiatry 35:615-621

Poznanski EO, Freeman LN, Mokros HB. Children's Depression Rating Scale-Revised Psychopharmacol Bull 1985; 21:979-989 (revised September 1984).

Emslie GJ, Rush AJ, Weinberg WA, Gullion CM, Rintelmann J, Hughes CW (1997a) Recurrence of major depressive disorder in hospitalized children and adolescents. J Am Acad Child Adolesc Psychiatry 36:785-792

Emslie GJ, Rush AJ, Weinberg WA, Kowatch RA, Hughes CW, Carmody T, Rintelmann J (1997b) A double-blind, randomized, placebo-controlled trial of fluoxetine in children and adolescents with depression. Arch Gen Psychiatry 54:1031-1037

Emslie GJ (2001) Treatment of major depressive disorder. Presented as part of Institute 1: Basic Child Psychopharmacology: Current Best Practices and New Treatment Options. American Academy of Child and Adolescent Psychiatry 48th Annual Meeting, Honolulu, Hawaii, Tuesday, October 23, 2001

Flament MF, Rapoport JL, Berg CJ (1985) Clomipramine treatment of childhood obsessive-compulsive disorder. Arch Gen Psychiatry 42:977-983

Findling RL, Marcus RN, D'Amico F, Reed MD, Preskorn SH, Magnus RD, Marathe P (2000) Nefazodone pharmacokinetics in depressed children and adolescents. J Am Acad Child Adolesc Psychiatry 39:1008-1016

Geller B, Cooper TB, Graham DL, Fetner HH, Marstellar FA, Wells JM (1992) Pharmacokinetically designed double-blind placebo-controlled study of nortriptyline in 6 to 12 year-olds with major depressive disorder. J Am Acad Child Adolesc Psychiatry 31:34-44

Geller DA, Hoog SL, Heiligenstein JH, Ricardi RK, Tamura R, Kluszynski S, Jacobson JG, Fluoxetine Pediatric OCD Study Team (2001) Fluoxetine treatment for obsessive-compulsive disorder in children and adolescents: a placebo-controlled clinical trial. J Am Acad Child Adolesc Psychiatry 40:773-779

Gittelman-Klein R, Klein DF (1973) School phobia: diagnostic consideration in the light of imipramine effects. J Nerv Ment Dis 156:199-215

Gittelman-Klein R, Klein DF (1980) Separation anxiety in school refusal and its treatment with drugs. In: Hersov L, Berg I (eds) Out of school. Wiley, New York, pp 321-341

Glazener CMA, Evans JHC. Alarm interventions for nocturnal enuresis in children (Cochrane Review). In: *The Cochrane Library,* Issue 2, 2002. Oxford: Update Software

Glazener CMA, Evans JHC. Tricyclic and related drugs for nocturnal enuresis in children (Cochrane Review). In: *The Cochrane Library,* Issue 2, 2002. Oxford: Update Software.

Go FS, Malley EE, Birmaher B, Rosenberg DR (1998) Manic behavior associated with fluoxetine in three 12-18 year olds with OCD. J Child Adolesc Psychopharmacol 8:73-80

Goldman LS, Genel M, Bezman RJ, Slanetz PJ (1998) Diagnosis and treatment of attention-deficit/hyperactivity disorder in children and adolescents. JAMA 279:1100-1107

Greenhill LL, Abikoff HB, Arnold LE, Cantwell DP, Conners CK, Elliott G, Hechtman L, Hinshaw SP, Hoza B, Jensen PS, March JS, Newcorn J, Pelham WE, Severe JB, Swanson JM, Vitiello B, Wells K (1996) Medication treatment strategies in the MTA study: relevance to clinicians and researchers. J Am Acad Child Adolesc Psychiatry 35:1304-1313

Greenhill LL, Pine D, March J, Birmaher B, Riddle M (1998) Assessment issues in treatment research of pediatric anxiety disorders: what is working, what is not working, what is missing, and what needs improvement. Psychopharmacol Bull 34:155-164

Guiles JM (1996) Sertraline induced behavior activation during the treatment of an adolescent with MDD. J Child Adolesc Psychopharmacol 6:281-285

Guy W (1976) ECDEU assessment manual for psychopharmacology-revised (DHEW Publ. No ADM 76-338). Rockville, MD, U.S. Department of Health, Education, and

Welfare, Public Health Service, Alcohol, Drug Abuse, and Mental Health Administration, NIMH Psychopharmacology Research Branch, Division of Extramural Research Programs, pp 218-222

Hamilton M (1967) Development of a rating scale for primary depressive illness. Br J Soc Psychol 6:278-296

Hammen C, Rudolph K, Weiss J, Rao UU, Burge D (1999) The context of depression in clinic-referred youth: neglected areas in treatment. J Am Acad Child Adolesc Psychiatry 38:64-71

Hazell P, O'Connell D, Heathcote D, Robertson J, Henry D (1995) Efficacy of tricyclic drugs in treating child and adolescent depression: a meta-analysis. BMJ 310:897-901

Heiligenstein JH, Greenhill LL, Spencer T, Allen AJ, Kratochvil C, Wernicke J (2001) Atomoxetine: a novel non-stimulant treatment for ADHD. Scientific proceedings of the 48th annual meeting of the AACAP, Honolulu, p 105

Hoar W (1998) Prozac rx for children jumps 500%. Mental Health News Alert:13-13

Hughes CW, Preskorn SH, Weller E, Weller R, Hassanein R (1989) A descriptive profile of the depressed child. Psychopharmacol Bull 25:232-237

Hughes C, Preskorn SH, Weller E, Weller R, Hassanein R, Tucker S (1990) The effect of concomitant disorders in childhood depression on predicting treatment response. Psychopharmacol Bull 26:235-238

Jadad A, Atkins D (1998) The treatment of attention deficit/hyperactivity disorder: an evidence report. Hamilton, Ontario

Kashani JH, Shekim WO, Reid JC (1984) Amitriptyline in children with major depressive disorder: a double blind crossover pilot study. J Am Acad Child Psychiatry 23:348-351

Kauffman RE (1998) Drug safety, testing and availability for children. Child Leg Rights J 18:178-185

Kauffman RE, Banner WJr, Berlin CMJr, et al (1995) Guidelines for the ethical conduct of studies to evaluate drugs in pediatric populations. American Academy of Pediatrics 95:169-177

Kaufman J, Birmaher B, Brent D, Rao U, Flynn C, Moreci P, Williamson D, Ryan N (1997) Schedule for affective disorders and schizophrenia for school-age children: present and lifetime version (K-SADS-PL): initial reliability and validity data. J Am Acad Child Adolesc Psychiatry 36:980-989

Keller MB, Ryan ND, Strober M, Klein RG, Kutcher SP, Birmaher B, Hagino OR, Koplewicz H, Carlson GA, Clarke GN, Emslie GJ, Feinberg D, Geller B, Kusumakar V, Papatheodorou G, Sack WH, Sweeney M, Wagner KD, Weller EB, Winters NC, Oakes R, McCafferty JP (2001) Efficacy of paroxetine in the treatment of adolescent major depression: a randomized, controlled trial. J Am Acad Child Adolesc Psychiatry 40:762-772

Klein R, Slomkowski C (1993) Treatment of psychiatric disorders in children and adolescents. Psychopharmacol Bull 29:525-535

Klein RG, Koplewicz HS, Kanner A (1992) Imipramine treatment of children with separation anxiety. J Am Acad Child Adolesc Psychiatry 31:21-28

Korinthenberg R, Haug C, Hannak D (1994) The metabolism of carbamazepine to CBZ-10,11 epoxide in children from the newborn age to adolescence. Neuropediatrics 25:214-224

Kramer A, Feiguine RJ (1981) Clinical effects of amitriptyline in adolescent depression. J Am Acad Child Psychiatry 20:636-644

Kutcher S, Boulos C, Ward B, Marton P, Simeon J, Ferguson HB, Szalai J, Katic M, Roberts N, Dubois C, et al. (1994) Response to desipramine treatment in adolescent depression: a fixed-dose, placebo-controlled trial. J Am Acad Child Adolesc Psychiatry 33(5):686-694

Lake CR, Mikkelsen EJ, Rapoport JL (1979) Effect of imipramine on norepinephrine and blood pressure in enuretic boys. Clin Pharmacol Ther 39:647-647

Leeder JS, Kearns GL (1997) Pharmacogenetics in pediatrics: implications for practice. Pediatr Clin North Am 44:55–77

Leonard, H, ed (2002) New drugs and new indications for children and adolescents. Brown University Child and Adolescent Psychopharmacolgy Update, Vol. 4, July 2002, p.2

Leonard HS, Swedo S, Rapoport JL (1989) Treatment of obsessive-compulsive disorder with clomipramine and desmethylimipramine: a double-blind crossover comparison in children and adolescents. Arch Gen Psychiatry 46:1088–1092

Lubin B, Himelstein P (1976) Reliability of the depression adjective check lists. Percept Mot Skills 43(3 Pt 2):1037–1038

Lubin B, Levitt EE (1979) Norms for the depression adjective check lists: age group and sex. J Consult Clin Psychol 47:192

McCauley E, Myers K, Mitchell J, Calderon R, Schloredt K, Treder R (1993) Depression in young people: initial presentation and clinical course. J Am Acad Child Adolesc Psychiatry 32:714–722

McConville B, Minnery KL, Sorter MT, West SA, Friedman LM, Christian K (1996) An open study of the effects of sertraline on adolescent major depression. J Child Adolesc Psychopharmacol 6:41–51

McCracken JT (2001) Empirically-based pharmacologic treatments for child anxiety. Presented as part of Institute 1: Basic Child Psychopharmacology: Current Best Practices and New Treatment Options. American Academy of Child and Adolescent Psychiatry 48th Annual Meeting, Honolulu, Hawaii, Tuesday, October 23, 2001

March JS, Biederman J, Wolkow R, Safferman A, Mardekian J, Cook EH, Cutler NR, Dominguez R, Ferguson J, Muller B, Riesenberg R, Rosenthal M, Sallee FR, Wagner KD, Steiner H (1998) Sertraline in children and adolescents with obsessive-compulsive disorder: a multicenter randomized controlled trial. JAMA 280:1752–1756

Milavetz G, Vaughan LM, Weinberger E (1986) Evaluation of a scheme for establishing and maintaining dosage of theophylline in ambulatory patients with chronic asthma. J Pediatr 109:351–356

National Institute of Mental Health (1991) Implementation of the national plan for research on child and adolescent mental disorders. PA-91-46. Washington DC, US Dept of Health and Human Services, Public Health Service, Alcohol, Drug Abuse, and Mental Health Administration.

National Institute of Mental Health (1993) Ethical and human subjects issues in mental health research with children and adolescents. Washington DC.

Pliszka SR, Greenhill LL, Crismon ML (July 2000) The Texas children's medication algorithm project: report of the Texas consensus conference panel on medication treatment of childhood attention-deficit/hyperactivity disorder. Part I, J Am Acad Child Adolesc Psychiatry 39(7):908–919

Popper CW, Ziminitzky B (1995) Sudden death putatively related to desipramine treatment in youth: a fifth case and a review of speculative mechanisms. J Child Adolesc Psychopharmacol 5:283–300

Preskorn SH (1998) What happened to Tommy? Journal of Practical Psychiatry and Behavioral Health 4:363–367

Preskorn SH, Fast GA (1991) Therapeutic drug monitoring for antidepressants: efficacy, safety and cost effectiveness. J Clin Psychiatry 52:23–33

Preskorn SH, Weller E, Weller R (1982) Depression in children: relationship between plasma imipramine levels and response. J Clin Psychiatry 43, No. 11:450–453

Preskorn SH, Weller EB, Weller RA, Glotzbach E (1983) Plasma levels of imipramine and adverse effects in children. Am J Psychiatry 140:1332–1335

Preskorn SH, Weller E, Hughes C, Weller R, Bolte K (1987) Depression in prepubertal children: dexamethasone nonsuppression predicts differential response to imipramine vs. placebo. Psychopharmacol Bull 23:128–133

Preskorn SH, Weller E, Jerkovich G, Hughes C, Weller R (1988) Depression in children: concentration-dependent CNS toxicity of tricyclic antidepressants. Psychopharmacol Bull 24:140-142

Preskorn SH, Bupp S, Weller E, Weller R (1989) Plasma levels of imipramine and metabolites in 68 hospitalized children. J Am Acad Child Adolesc Psychiatry 28:373-375

Puig-Antich J, Perel JM, Lupatkin W, Chambers WJ, Shea C, Tabrizi MA, Stiller RL (1979) Plasma levels of imipramine (IMI) and desmethylimipramine (DMI) and clinical response in prepubertal major depressive disorder: a preliminary report. J Am Acad Child Psychiatry 18:616-27

Rao U, Ryan ND, Birmaher B, Dahl RE, Williamson DE, Kaufman J, Rao R, Nelson B (1995) Unipolar depression in adolescents: clinical outcome in adulthood. J Am Acad Child Adolesc Psychiatry 34:566-578

The Research Unit on Pediatric Psychopharmacology Anxiety Study Group (2001) Fluvoxamine for the treatment of anxiety disorders in children and adolescents. N Engl J Med 344:1279-1285

Riddle MA, Nelson JC, Kleinman CS (1991) Sudden death in children receiving norpramin: a review of three reported cases and commentary. J Am Acad Child Adolesc Psychiatry 30:104-104

Riddle MA, Scahill L, King RA (1992) Double blind trial of fluoxetine and placebo in children and adolescents with obsessive compulsive disorder. J Am Acad Child Adolesc Psychiatry 31:1062-1069

Riddle MA, Reeve EA, Yaryura-Tobias JA, Yang HM, Claghorn JL, Gaffney G, Greist JH, Holland D, McConville BJ, Pigott T, Walkup JT (2001) Fluvoxamine for children and adolescents with obsessive-compulsive disorder: a randomized, controlled, multicenter trial. J Am Acad Child Adolesc Psychiatry 40:222-229

Rudorfer MZ, Potter WZ (1987) Pharmacokinetics of antidepressants. In: Metzler JV (ed) Psychopharmacology: the third generation of progress. Raven, New York, pp 1353-1363

Shaffer D, Gould MS, Brasic J, Ambrosini P, Fisher P, Bird H, Aluwahlia S (1983) A children's global assessment scale (CGAS). Arch Gen Psychiatriy 40:1228-1231

Simeon J, Dinicola V, Ferguson HB, Copping W (1990) Adolescent depression: a placebo-controlled fluoxetine treatment study and follow-up. Prog Neuropsychopharmacol Biol Psychiatry 4:791-795

Spencer T, Biederman J, Wilens T, Steingard R, Geist D (1993) Nortriptyline treatment of children with attention deficit hyperactivity disorder and tic disorder or Tourette's syndrome. J Am Acad Child Adolesc Psychiatry 32:205-210

Spencer T, Biederman J, Wilens T, Harding M, O'Donnell D, Griffin S (1996) Pharmacotherapy of attention deficit hyperactivity disorder across the life cycle. J Am Acad Child Adolesc Psychiatry 35:409-432

Tierney E, Joshi PT, Llinas JF, Rosenberg LA, Riddle MA (1995) Sertraline for major depression in children and adolescents: preliminary clinical experience. J Child Adolesc Psychopharmacol 5:13-27

Tosyali MC, Greenhill LL (1998) Child and adolescent psychopharmacology: important developmental issues. Pediatr Clin North Am 45:1021-1035

Walkup J, Davies M (1999) The Pediatric Anxiety Rating Scale (PARS): a reliability study. In: Scientific proceedings of the 46th Annual Meeting of the American Academy of Child and Adolescent Psychiatry, Chicago, October 19-24, 1999; NR78, abstract

Wilens TE, Biederman J, Geist DE, Steingard R, Spencer T (1993) Nortriptyline in the treatment of ADHD: a chart review of 58 cases. J Am Acad Child Adolesc Psychiatry 32:343-349

Wilens TE, Biederman J, Mick E, Spencer TJ (1995) A systematic assessment of tricyclic antidepressants in the treatment of adult attention deficit hyperactivity disorder. J Nerv Ment Dis 183:48-50

Women

K. A. Yonkers[1] · O. Brawman-Mintzer[2]

[1] Department of Psychiatry, Yale School of Medicine, Suite 301, 142 Temple Street, New Haven, CT 06501, USA
e-mail: Kimberly.Yonkers@Yale.edu
[2] University of South Carolina School of Medicine, Charleston, SC, USA

1	Introduction	380
2	Sex Differences in Brain Structure and Function	382
3	Sex Difference in the Pathophysiology of Depression	382
4	Sex Differences in the Metabolism of Antidepressants	384
4.1	Drug Metabolism and the CYP Enzyme System	384
4.2	Plasma Levels and Clearance	385
4.2.1	TCAs	385
4.2.2	SSRIs	386
4.2.3	Nefazodone	386
5	Sex Differences in Antidepressant Treatment Response	386
5.1	Pharmacodynamics	386
5.2	Possible Influence of Exogenous Estrogen and Hypoestrogenic States	387
6	Influence of Gender on the Adverse-Event Profile of Antidepressants	388
7	Summary	389
References		389

Abstract Studies have found (e.g., the Epidemiological Catchment Area study, the National Comorbidity Survey) higher rates of major depressive disorder in women than men. Clinicians need to be knowledgeable about potential sex differences in metabolism of antidepressants and treatment response. Data concerning such differences are limited, since Phase I and early Phase II clinical trials have historically excluded women of child bearing potential. However, the Food and Drug Administration is now encouraging inclusion of women at earlier stages of pharmaceutical research. A number of different factors may lead to sex-related differences in antidepressant response. Differences in brain structure and functioning (e.g., in the responsivity of the serotonin system) may lead to differences in antidepressant response. Sex differences in the pathophysiology of depression may also affect treatment response (e.g., women may be more likely to experience "atypical" depression). Research on mood disorders related to the female reproductive cycle, such as peripartum and perimenopausal depression, also suggests that unique gender-related pathophysiological processes may play a role in some depressive disorders in women and raises the question of the role of estrogen in treating these types of depression. Sex-related differences in the metabolism of antidepressants may also affect treatment response. Although pharmacokinetic data are still limited in this area, it appears that there are some sex-related differences in the activity of various CYP 450 enzymes, especially CYP 1A2, CYP 3A, possibly CYP 2D6, that may affect the metabolism of certain antidepressants and result in different blood levels and hence possible differences in response and side effects. Differences between women and men in the metabolism of the selective serotonin reuptake inhibitors (SSRIs) have been reported (e.g., plasma levels of sertraline have been round to be 27% lower in young men than in women of all ages; older women have been found to have higher plasma levels of nefazodone than younger patients of both sexes). Pharmacodynamic differences between men and women may also lead to differences in antidepressant treatment response. Studies have suggested that women may be more likely to respond to an SSRI than a tricyclic antidepressant (TCA), while men may be more likely to respond to a TCA. Studies have found that such differences in response did not occur in postmenopausal women who were not receiving estrogen replacement therapy, suggesting that estrogen may change blood levels, metabolism, or receptor characteristics in a way that improves the efficacy of the SSRIs in women. The increasing inclusion of women at earlier stages of clinical drug trials will hopefully provide answers to the many questions that remain to be answered regarding sex differences in antidepressant treatment response. Researchers need to consider adequate subgroup sizes and incorporate physiological factors unique to women (reproductive status, phase of menstrual cycle) into the parameters being studied. Such efforts will promote optimal treatment for depression in both women and men.

Keywords Women · Gender · Major depressive disorder · Sex-related differences · Selective serotonin reuptake inhibitors · Tricyclic antidepressants · Monoamine oxidase inhibitors · Estrogen · CYP 450 enzymes · Nefazodone · Peripartum depression · Perimenopausal depression · Pharmacokinetics · Pharmacodynamics

1
Introduction

One of the most replicated statistics in medical epidemiology is the preponderance of depressive disorders among women. In the Epidemiological Catchment Area study, the lifetime prevalence of major depressive disorder (MDD) was 7% in women and 2.6% in men (Weissman et al. 1991). A disproportionate prevalence ratio of approximately 2:1 was found in the National Comorbidity Survey, although this study found that 21% of women and 13% of men will experience MDD at some point in their lives (Kessler et al. 1994). Patient populations in clinical practice settings reflect this preponderance of women with depression, underscoring the need to be knowledgeable about how to treat women as well as men with depression. Part of this knowledge includes an understanding of potential sex differences in the metabolism of antidepressants and in treatment outcome after therapy with these agents.

While the focus of this chapter is a review of sex differences in the metabolism and efficacy of antidepressant therapy, we acknowledge that the data are limited. Historically, the prevailing attitude among efficacy studies of antidepressant agents is that if a medication works for a man, then it is efficacious for a woman. If anything, barriers to testing medications in young women have fostered a lack of sex-specific investigation. The recent history of thalidomide, an agent developed and used to treat women at risk for pregnancy miscarriage that was found to cause severe limb and other deformities in offspring, speaks to the concern of exposing young, potentially fecund women to new medications. After thalidomide, great care was taken not expose women who could become pregnant to new agents that had not yet shown efficacy. Even oral contraceptive agents, medications that are only targeted for use in women, were tested in men during the early stages of development. The approach to limiting phase I and early phase II clinical studies to men and non-fertile women was supported by U.S. Food and Drug Administration (FDA) guidelines (Merkatz et al. 1993). The recognition that this stance resulted in the exclusion of many women from early trials, and hence to a lack of data about drug safety, efficacy, and dosing in women, led the FDA to revise their guidelines in the early 1990s. In revised guidelines, the FDA reversed its position and encouraged the inclusion of women in earlier pharmacokinetic studies (Merkatz et al. 1993).

Encouraging the inclusion of women, as well as individuals of different age, ethnicity, and race was only the first step in understanding more about the relationship between a drug's efficacy and its utility in various subpopulations. In order for potential differences to be identified, data had to be analyzed in a way that response among subgroups, including women, could be identified and contrasted. An important component of the new guidelines was the request that applicants filing new drug applications conduct subset analyses by sex and age, a practice that had previously been uncommon (Merkatz et al. 1993). This initiative strengthened Congressional mandates instructing the National Institutes of Health to include women and minorities and to conduct similar subset analyses in government-sponsored studies (National Institute of Health 1992). While we are only beginning to see results from these initiatives, new data on sex differences in treatment are emerging. The following review summarizes these findings.

If sex differences in treatment response to antidepressants exist, they may be attributable to a number of factors including: (1) unique sex-specific biological processes that underlie the depressive disorder and influence treatment response, such as an abundance of certain target receptor(s) in either sex; (2) the susceptibility to medication treatment of a set of depressive symptoms that are more likely to occur in one sex; (3) sex-specific variations in absorption, distribution, or metabolism that could contribute to differences in the amount of drug that reaches the site of action; (4) pharmacodynamic effects produced by the activity of the drug in a sex-specific environment (e.g., the properties of a neurotransmitter receptor being influenced by endogenous hormones).

2
Sex Differences in Brain Structure and Function

Just as the body of research on sex differences in response to psychotropic treatment is increasing, data regarding sex differences in brain structure are also accumulating at an accelerated pace. Notable findings include differences in the sizes of various hypothalamic nuclei and the corpus callosum (Allen et al. 1991). Results that may relate to antidepressant treatment response include sex differences in the asymmetry of serotonergic receptors in various brain regions, including the orbital cortex (Arato et al. 1991b). This finding coincides with sex differences in the amplitude of P300 evoked potentials (Arato et al. 1991a). There may also be biological differences in the responsivity of the serotonin system (McBride et al. 1990). In all likelihood, such a difference is related to the sexual dimorphism in the endocrine factors that measure serotonergic response, such as prolactin. However, this may also be meaningful in terms of the biology of response to antidepressant treatment.

3
Sex Difference in the Pathophysiology of Depression

With the possible exception of mood disorders related to the reproductive cycle (e.g., peripartum depression), sex differences in the pathophysiology of depressive disorders have not been definitely established. Moreover, in community populations, the symptoms found in women and men do not vary (Kessler et al. 1993). Some note sex differences in various clusters of symptoms. For example, women may be more likely to experience depression characterized by high anxiety and somatic symptoms (Silverstein 1999). Similarly, higher rates of reverse neurovegetative symptoms, sometimes referred to as "atypical depression" have been observed in clinical cohorts (Frank et al. 1988). Such symptoms have been found by some to preferentially respond to monoamine oxidase inhibitors (MAOIs) over tricyclic antidepressants (TCAs) (Quitkin et al. 1993; Thase et al. 2000). As suggested by clinical and community data, if women are more likely to have atypical depression, then they would be more likely to respond to an MAOI than to a TCA. A database re-analysis supports this hypothesis; findings indicate a possible interaction between subtype of depression and gender (Davidson and Pelton 1986).

An early study explored whether age or sex were pertinent variables in patients' response to an MAOI or a TCA (Raskin 1974). Data from several studies in which patients were treated with imipramine, chlorpromazine, phenelzine, diazepam, or placebo were pooled and reanalyzed. Results showed that imipramine was no better than placebo for young women, while imipramine was efficacious for men and older women. On the other hand, the MAOI, phenelzine, was more effective in young women than placebo. The authors did not evaluate the type of depression experienced by the men and women in the studies.

It is also possible that women are more likely to respond to MAOIs whether or not they suffer from the atypical form of depression. In a subsequent database reanalysis of a large clinical trial cohort of patients with depression, including those with the "atypical" subtype ($n=263$), a second group found that young women had a superior response to MAOIs compared to men (Quitkin et al. 2002). The difference in treatment response between men and women was larger than the treatment difference between active treatment and placebo. This effect was not attributable to the subtype of depression, suggesting that the phenomenon is not simply a diagnosis by drug-type interaction. While these results are intriguing, they may be more relevant to our theoretic understanding of the underlying sex differences in the pathophysiology of depression than to clinical practice. The reason is that MAOIs, due to their potentially serious side effects, are now rarely used in patients unless they are treatment-resistant although this may change with new MAO delivery systems under development. Nonetheless, the consistency among these studies in their finding that MAOIs are superior to TCAs for young women, especially in light of the fact that the studies were not necessarily designed nor powered to test this hypothesis, is notable whether it is due to characteristics of the depression experienced or another sex-associated difference.

The possibility that depressive subtype may lend unique attributes to response is suggested by depressive disorders that only occur in women. Preliminary work with mood disorders related to the female reproductive cycle, such as MDD with a peripartum onset and major and minor depression occurring during the perimenopause, suggest that unique gender-related pathophysiological processes may play a role in some depressive disorders experienced by women. Theoretically, these reproductive-related depressive disorders are linked to a withdrawal of gonadal steroids, in particular, β-estradiol, although this has not been definitively established. This hypothesis was tested in a study by Gregoire et al. (1996), in which 64 women who were depressed during the puerperium were treated either with high dose β-estradiol or placebo. As early as the first month of treatment, β-estradiol was found to be superior to placebo. One half of the women in this study were also receiving antidepressant treatment so that it is not known whether the intervention that was superior to placebo involved augmentation or monotherapy with β-estradiol. Similarly, it is not clear how β-estradiol would compare to standard antidepressant therapy.

Mood symptoms occurring during the perimenopause are sex specific because only women go through the menopause. As noted, a decrement in gonadal steroids is thought to be responsible for perimenopausal mood symptoms, a hypothesis bolstered by the fact that mood symptoms occurring during the perimenopause respond to estrogen replacement therapy (Zweifel and O'Brien 1997; Yonkers and Bradshaw 1999). Although estrogen is useful for the treatment of depressive *symptoms*, the benefit of estrogen for the *syndromes* of minor and major depression occurring during the perimenopause remained unclear until of late. Recently, this issue was addressed by two studies in which β-estradiol or placebo was used to treat small cohorts of women with minor or major depression occurring during the perimenopause (Schmidt 2000; Soares 2001). In both investigations, β-estradiol proved to be more effective than placebo. It is not known whether the treatment effect of estrogen would be as enduring as that of antidepressant therapy. Given the unanswered questions about this intervention, it is likely that the value of this work on estrogen may be its illustration of sex-specific processes in the evolution and termination of episodes of MDD.

4
Sex Differences in the Metabolism of Antidepressants

4.1
Drug Metabolism and the CYP Enzyme System

Differences in response to antidepressant treatment may be related to the way in which men and women metabolize medications. However, data on this topic are very limited, as noted by the authors of several recent reviews of sex-related differences in the pharmacokinetics of psychotropic agents (Hamilton et al. 1988; Dawkins and Potter 1991; Yonkers et al. 1992; Hamilton and Grant 1993; Harris et al. 1995; Pollock 1997; Brawman-Mintzer and Book 2002). Fortunately, increasing interest in this area has led to a growing number of publications concerning possible sex-related differences in the pharmacokinetics of antidepressant agents, the findings of which are briefly summarized below.

First, however, a brief review of how medications are metabolized is provided (for a more detailed discussion of these issues, readers are referred to the chapter in this volume by Preskorn and Catterson, "General Principles of Pharmacokinetics"). When a patient ingests an antidepressant, a number of factors influence the degree of drug absorption, including the acid-base properties of the compound and its environment, the amount of food in the gut, and the compound's vulnerability to metabolism at the villous brush border of the gut. Most antidepressants are metabolized in the liver (the mood stabilizer lithium is an exception), and the residua are then distributed to various compartments in the body. Medications undergo oxidative, reductive, and conjugative reactions in the gastrointestinal tract, liver, and other compartments. Oxidative and reductive reactions are mediated largely through the cytochrome P450 (CYP) enzyme system, with enzymes labeled according to their amino acid structure (Arabic numerals are used to identify a family, which has 36% homology; a capital letter is used to identify compounds in the same family, which share 70% homology; and an Arabic numeral is used to identify the gene associated with the enzyme) (Nebert et al. 1991).

Research into potential sex differences in the activities of the various CYP enzymes has been facilitated by the increasing identification of these enzymes (Harris et al. 1995; Pollock 1997; Yonkers and Hamilton 1995; Brawman-Mintzer and Book 2002). Studies have found that women have higher plasma levels of drugs that are metabolized by CYP 1A2 (Ford et al. 1993; Harrter et al. 1998). Other studies (Hunt et al. 1992; Watkins 1992; Harris et al. 1995; Pollock 1997) have found higher levels of CYP 3A activity in women than in men, although not all studies agree (May et al. 1994). There is less support for sex-specific differences in the activity of CYP 2D6; however, since the activity of this enzyme is largely dependent on genetic polymorphisms, any effect related to gender may have been masked. This is supported by a recent investigation that attributed variability in plasma levels of nortriptyline to enzyme genotype *and* gender (Dahl et al. 1996). Although the findings are preliminary, it appears that sex-specific factors may influence several hepatic enzymes that play a role in metabolizing antidepressant medications and result in different blood levels in men and women. These variations in blood levels could then lead to differing response to treatment and differences in side effects.

4.2
Plasma Levels and Clearance

4.2.1
TCAs

Higher plasma levels of the tertiary amines imipramine, amitriptyline, and clomipramine have been found in women in some studies (Moody et al. 1967; Preskorn and Mac 1985; Gex-Fabry et al. 1990), but not in others (Ziegler and Biggs 1977). The finding by Dahl et al. (1996) referred to above that both genetic polymorphisms and gender play a role in the clearance of drugs metabolized by CYP 2D6 may explain some of the inconsistencies reported in earlier studies.

Two studies (Gex-Fabry et al. 1990; Mundo et al., in press) have found evidence for sex-related differences in the metabolism of clomipramine, providing more support for sex differences in the metabolism of this compound compared with the other TCAs. The metabolism of clomipramine is complicated and involves demethylation and hydroxylation, although it appears that hydroxylation is less active in women than in men based on dissection of metabolites. This illustrates how difficult it can be to establish sex differences in the metabolism of a compound if there are multiple metabolic pathways (Nielsen et al. 1994). The fact that both CYP 2C19 and CYP 2D6 are involved in the metabolism of clomipramine must also be taken into account, since the activity of CYP 2C19 may be higher in men than women (for example, men clear methyl phenobarbital, a compound metabolized by CYP 2C19, approximately 1.3 times faster than women) (Hooper and Qing 1990).

Findings concerning CYP 1A2, which is also involved in the metabolism of tertiary amines, may also shed light on sex differences in enzymatic activity. Gonadal steroids inhibit the activity of CYP 1A2 (Abernathy et al. 1982; Lane et al. 1992; Pollock et al. 1999), which has the potential to lower concentrations of certain metabolites in women. Similarly, women who use oral contraceptives may show a lower apparent clearance of medications metabolized by CPY 1A2 compared with men. Pollock et al. 1999 have reported that even small "replacement level" dosages of estrogen can influence plasma levels of compounds in this class.

Studies (Abernathy et al. 1985; Dahl et al. 1996) have found that the clearance of secondary amines desipramine and nortriptyline is lower in women than in men; however, these studies did not correct for body weight, which could neutralize sex differences.

4.2.2
SSRIs

Sexual dimorphism has been reported in the metabolism of the selective serotonin reuptake inhibitor (SSRI) sertraline (Warrington 1991; Ronfeld et al. 1997), with plasma levels of sertraline found to be 27% lower in young men than in women of all ages and in older men. This finding may be clinically meaningful, since there appear to be sex differences in treatment response in a number of trials (see below). However, the failure to find a dose–response relationship between sertraline levels and therapeutic response suggests that a more complex mechanism may be involved in sex differences in treatment response. The metabolism of sertraline is not entirely clear, although one computer model suggested that the predominant metabolic path to the formation of desmethylsertraline is dependent on 2C9 (~23%) with smaller contributions from 3A4 and 2C19 (15% each) (Greenblatt et al. 1999).

4.2.3
Nefazodone

The antidepressant nefazodone, which is putatively metabolized by CYP 3A (Barbhaiya et al. 1996), blocks the reuptake of serotonin and binds to postsynaptic 5-HT$_2$ receptors. Just as reported for sertraline, elderly women have been found to have the highest single-dose and steady-state levels of nefazodone and hydroxynefazodone (Barbhaiya et al. 1996), but there are no significant differences in plasma levels in young men or women. It is possible that an age-related decline in CYP 3A activity may be contributing to these age-related differences in plasma levels of both nefazodone and sertraline.

5
Sex Differences in Antidepressant Treatment Response

5.1
Pharmacodynamics

As discussed earlier in this chapter, revisions to the FDA guidelines recommended that new drug applicants conduct subgroup analyses to identify possible sex-related differences in drug response. Such analyses led to the finding that the 5-HT$_3$ antagonist, alosetron, a compound that had been approved for the treatment of irritable bowel syndrome, is effective in women but not in men (Camilleri et al. 1999). (Note that this medication has since been withdrawn from the market in the United States because of ischemic colitis.) Alosetron produces plasma levels in women that are 30%–50% higher than in men (GlaxoWellcome, product labeling information, September 2000), so that it seemed that pharmacokinetics alone might account for this difference in treatment response; this is not the case, however, since higher doses were found to be no more effective in men than lower doses (Camilleri et al. 1999). Rather, a pharmacodynamic difference may underlie the sex differences in response. One hypothesis suggests that sex differences in endogenous central nervous system levels of serotonin are involved (Nishizawa et al. 1997). Such a hypothesis may be invoked to explain some of the sex-related variability seen with other psychotropic agents. For example, one meta-analysis of all published imipramine trials (35 studies including 342 men and 711 women) (Hamilton et al. 1995) found a relatively weaker response to TCAs among women, with 62% of men but only 51% of women ($p<0.001$) considered imipramine responders.

Other studies suggest that the patient's sex may modify the likelihood of response to a TCA or to an SSRI. A reanalysis of a large pharmaceutical database of trials comparing paroxetine with imipramine found that paroxetine was more effective than imipramine in women with MDD (Steiner et al. 1993). A sex-by-treatment interaction was found in a large study comparing sertraline with imipramine in the treatment of chronic MDD ($n=236$ males and 399 females) (Kornstein et al. 2000); the direction of the effect was the same as reported in other studies, with women in this study 10% more likely to respond to sertraline than to the TCA while the men were 12% more likely to respond to the TCA.

We also found a sex-related differential response in a study comparing sertraline and imipramine in 266 women and 144 men with dysthymia: 64% of the women and 42% of the men ($p=0.02$) treated with the SSRI were responders (K.A. Yonkers, unpublished data). Fluoxetine is the only SSRI for which the findings are not consistent with this picture, since studies have found that it produces a response in men and women at the same rate (Lewis-Hall et al. 1997; Quitkin et al. 2002).

5.2
Possible Influence of Exogenous Estrogen and Hypoestrogenic States

An intriguing component of the study by Kornstein et al. (2000), mentioned earlier, is the finding that sex differences in imipramine and sertraline disappeared in the 74 women who were postmenopausal, again raising questions concerning the unknown role of gonadal steroids in mood disorders (Kornstein et al. 2000). In a recent mega-analysis ("pooled analysis") of placebo-controlled antidepressant trials, the efficacy of the newer antidepressant venlafaxine, which putatively blocks the reuptake of norepinephrine and dopamine in addition to serotonin at higher dosages, was compared with the SSRIs fluoxetine, paroxetine, and fluvoxamine (Entsuah et al. 2001). Outcome was investigated by patient sex and, for women, menopausal status. Women treated with an SSRI who were postmenopausal and were not undergoing estrogen replacement therapy had a lower response rate than women treated with an SSRI who were concurrently taking hormone replacement. The rate of response for postmenopausal women treated with venlafaxine was the same whether or not hormone replacement therapy was used, suggesting a specific interaction between SSRIs and the estrogenic milieu. Post-hoc analyses of other study databases have found similar findings of enhanced response to SSRIs among postmenopausal women undergoing estrogen replacement therapy (Schneider et al. 1997, 2001). It may be that, when present, estrogen changes blood levels, metabolism, or receptor characteristics in a way that improves the efficacy of selective antidepressant agents in these women.

There is some support for the preceding pharmacodynamic explanation and the possibility that the bioactivity at neurotransmitter receptors is altered by the hormonal milieu. One group has found greater cognitive impairment in women administered triazolam in conjunction with progesterone than in women are administered triazolam alone, despite the fact that the addition of progesterone does not change plasma levels of triazolam appreciably (Kroboth et al. 1985; McAuley et al. 1995; Kroboth and McAuley 1997). The sex differences in treatment response to alosetron mentioned above may also fit into such a paradigm: endogenous gonadal steroids may alter the biochemical factors necessary for response.

6
Influence of Gender on the Adverse-Event Profile of Antidepressants

We have suggested the need for further investigation into the relationship between plasma levels of medication and sex differences in response. Sex differences in plasma levels may also lead to variability in the likelihood of side effects in men and women. For example, in one study that compared the efficacy of sertraline and imipramine in the treatment of chronic MDD, women were found to discontinue treatment at a significantly higher rate; unfortunately, data on plasma levels were not obtained in this study, so that correlations with plasma levels could not be investigated (Kornstein et al. 2000). In another study women had higher plasma levels than men of the cholinesterase inhibitor, tacrine, and the incidence of adverse events was strongly correlated with plasma levels (Ford et al. 1993). Other investigations have found that adverse events occur more often in women in association with a wider range of prescribed medications. Female inpatients may be more likely than male inpatients to have adverse drug reactions, especially reactions that are dose related (Domencq et al. 1980; Simpson et al. 1987).

7
Summary

It is clear from the provocative findings presented in this chapter that many questions remain to be answered regarding sex differences in response to antidepressant treatment. To address these questions, researchers need to give increased attention to the effects of sex, reproductive status, and the menstrual cycle in the treatment of depression. It is important that both men and women be included in early pharmacokinetic trials; as part of this effort, larger samples and new study designs will be needed. Researchers will need to consider adequate subgroup sizes and will need to incorporate physiological factors unique to women (e.g., reproductive status, phase of the menstrual cycle) into the parameters being studied. It is also essential that the information obtained from such trials be made widely available, rather than being buried in new drug application files. Such efforts will promote optimal treatment for depression in both women and men.

Acknowledgements. This work is supported in part by funding from grant NIMH-K08-MH01648-01 to Dr. Yonkers.
    Portions of this article are adapted with permission from Yonkers KA, Brawman-Mintzer O (2002) The pharmacologic treatment of depression: is gender a critical factor? J Clin Psychiatry 63:610-615.

References

Abernathy DR, Greenblatt DJ, Shader RI (1985) Imipramine and desipramine disposition in the elderly. J Pharmacol Exp Ther 232:183-188

Abernathy DR, Greenblatt DR, Divoll M, Arendt R, Ochs HR, Shader RI (1982) Impairment of diazepam metabolism by low-dose estrogen-containing oral-contraceptive steroids. N Eng J Med 306:791-792

Allen SS, McBride CM, Pirie PL (1991) The shortened premenstrual assessment form. J Reprod Med 36:769-772

Arato M, Frecska E, Maccrimmon DJ, Guscott R, Saxena B, Tekes K, Tothfalusi L (1991a) Serotonergic interhemispheric asymmetry: Neurochemical and pharmaco-EEG evidence. Prog Neuropsychopharmacol Biol Psychiatry 15:759-764

Arato M, Frecska E, Tekes K, MacCrimmon DJ (1991b) Serotonergic interhemispheric asymmetry: Gender difference in the orbital cortex. Acta Psychiatr Scand 84:110-111

Barbhaiya RH, Buch AB, Greene DS (1996) A study of the effect of age and gender on the pharmacokinetics of nefazodone after single and multiple doses. J Clin Psychopharmacol 16:19-25

Brawman-Mintzer O, Book SW (2002) Sex differences in psychopharmacology. In: Kornstein S, Clayton A (eds) Women's mental health. Guilford, New York, pp 31-48

Camilleri M, Mayer E, Drossman D (1999) Improvement in pain and bowel function in female irritable bowel syndrome patients with alosetron, a 5-HT3 receptor antagonist. Alimentary Pharmacological Therapy 13:1149-1159

Dahl M, Bertilsson L, Nordin C (1996) Steady-state plasma levels of nortriptyline and its 10-hydroxy metabolite: Relationship to the CYP2D6 genotype. Psychopharmacology 123:315-319

Davidson J, Pelton S (1986) Forms of atypical depression and their response to antidepressant drugs. Psychiatry Res 17:87–95

Dawkins K, Potter WZ (1991) Gender differences in pharmacokinetics and pharmacodynamics of psychotropics: Focus on women. Psychopharmacology 27:417–426

Domecq C, Naranjo CA, Ruiz I, Busto U (1980) Sex-related variations in the frequency and characteristics of adverse drug reactions. Int J Clin Pharmacol Ther Toxicol 18:362–366

Entsuah AR, Huang H, Thase M (2001) Response and remission rates in different subpopulations with major depressive disorder administered venlafaxine, selective serotonin reuptake inhibitors, or placebo. J Clin Psychiatry 62:869–877

Ford JM, Truman CA, Wilcock GK, Roberts CJ (1993) Serum concentrations of tacrine hydrochloride predict its adverse effects in Alzheimer's disease. Clin Pharmacol Ther 53:691–695

Frank E, Carpenter LL, Kupfer DJ (1988) Sex differences in recurrent depression: Are there any that are significant? A J Psychiatry 145:41–45

Gex-Fabry M, Balant-Gorgia AE, Balant LP, Garrone G (1990) Clomipramine metabolism: Model-based analysis of variability factors from drug monitoring data. Clin Pharmacokinet 9:241–255

Greenblatt D, von Moltke L, Harmatz J, Shader R (1999) Human cytochromes mediating sertraline biotransformation: seeking attribution. J Clin Psychopharmacol 1999; 19:489–493

Gregoire AJ, Kumar R, Everitt B, Henderson AF, Studd JW (1996) Transdermal oestrogen for treatment of severe postnatal depression. Lancet 347:930–933

Hamilton JA, Grant M (1993) Sex differences in metabolism and pharmacokinetics: Effects on agent choice and dosing. Bethesda, MD, NIMH Conference

Hamilton JA, Grant M, Jensvold MF (1995) Sex and the treatment of depressions: When does it matter? In: Jensvold MF, Halbreich U, Hamilton JA (eds) Psychopharmacology of women: Sex, gender, and hormonal considerations, vol 1. American Psychiatric Press, Washington, DC, pp 241-257

Hamilton JA, Parry BL, Blumenthal SJ (1988) The menstrual cycle in context, I: Affective syndromes associated with reproductive hormonal changes. J Clin Psychiatry 49:474–480

Harris RZ, Benet LZ, Schwartz JB (1995) Gender effects in pharmacokinetics and pharmacodynamics. Drugs 50:222–239

Hartter S, Wetzel H, Hammes E, Torkzadeh M, Hiemke C (1998) Nonlinear pharmacokinetics of fluvoxamine and gender differences. Ther Drug Monit 20:446–449

Hooper WD, Qing MS (1990) The influence of age and gender on the stereoselective metabolism and pharmacokinetics of mephobarbital in humans. Clin Pharmacol Ther 48:633–640

Hunt CM, Westerkam WR, Stave GM (1992) Effect of age and gender on the activity of human hepatic CYP3A. Biochem Pharmacol 44:275–283

Kessler RC, McGonagle KA, Swartz M, Blazer DG, Nelson CB (1993) Sex and depression in the National Comorbidity Survey. I: Lifetime prevalence, chronicity and recurrence. J Affect Disord 29:85–96

Kessler RC, McGonagle KA, Zhao S, Nelson CB, Hughes M, Eshleman S, Wittchen HU, Kendler KS (1994) Lifetime and 12-month prevalence of DSM-III-R psychiatric disorders in the United States. Results from the National Comorbidity Survey. Arch Gen Psychiatry 51:8–19

Kornstein SG, Schatzberg AF, Thase ME, Yonkers KA, McCullough JP, Keitner GI, Gelenberg AJ, Davis SM, Harrison WM, Keller MB (2000) Gender differences in treatment response to sertraline versus imipramine in chronic depression. Am J Psychiatry 157:1445–1452

Kroboth P, McAuley J (1997) Progesterone: Does it affect response to drug? Psychopharmacol Bull 33:297–301

Kroboth PD, Smith RB, Stoehr GP, Juhl RP (1985) Pharmacodynamic evaluation of the benzodiazepine-oral contraceptive interaction. Clin Pharmacol Ther 38:525–532

Lane JD, Steege JF, Rupp SL, Kuhn CM (1992) Menstrual cycle effects on caffeine elimination in the human female. Eur J Clin Pharmacol 43:543–6

Lewis-Hall FC, Wilson MG, Tepner RG, Koke SC (1997) Fluoxetine vs tricyclic antidepressants in women with major depressive disorder. J Women's Health 6:337-343

May DG, Porter J, Wilkinson GR, Branch RA (1994) Frequency distribution of dapsone N-hydroxylase, a putative probe for P4503A4 activity in a white population. Clin Pharmacol Ther 55:492–500

McAuley JW, Reynolds IJ, Kroboth FJ, Smith RB, Kroboth PD (1995) Orally administered progesterone enhances sensitivity to triazolam in postmenopausal women. J Clin Psychopharmacol 15:3–11

McBride PA, Tierney H, DeMeo M, Chen JS, Mann JJ (1990) Effects of age and gender on CNS serotonergic responsivity in normal adults. Biol Psychiatry 27:1143–1155

Merkatz RB, Temple R, Subel S, Feiden K, Kessler DA (1993) Women in clinical trials of new drugs. A change in Food and Drug Administration policy. The Working Group on Women in Clinical Trials. N Engl J Med 329:292–296

Moody JP, Tait AC, Todrick A (1967) Plasma levels of imipramine and desmethylimipramine during therapy. Br J Psychiatry 113:183–193

National Institute of Health (1992) Opportunities for research on women's health. Washington DC, NIH Publication No 92–3457A

Nebert DW, Nelson DR, Coon MJ, Estabrook RW, Feyereisen R, Fujii-Kuriyama Y, Gonzalez FJ, Guengerich FP, Gunsalus IC, Johnson EF (1991) The P450 superfamily: Update on new sequences, gene mapping and recommended nomenclature. DNA Cell Biol 10:1–14

Nielsen KK, Brosen K, Hansen MG, Gram LF (1994) Single-dose kinetics of clomipramine: Relationship to the sparteine and S-mephenytoin oxidation polymorphisms. Clin Pharmacol Ther 55:518–527

Nishizawa S, Benkelfat C, Young SN, Leyton M, Mzengeza S, de Montigny C, Blier P, Diksic M (1997) Differences between males and females in rates of serotonin synthesis in human brain. Proc Natl Acad Sci USA 94:5308–5313

Pollock B (1997) Gender differences in psychotropic drug metabolism. Psychopharmacol Bull 33:235–241

Pollock BG, Wylie M, Stack JA, Sorisio DA, Thompson DS, Kirshner MA, Folan MM, Condifer KA (1999) Inhibition of caffeine metabolism by estrogen replacement therapy in postmenopausal women. J Clin Pharmacol 39:936–40

Preskorn SH, Mac DS (1985) Plasma levels of amitriptyline: Effects of age and sex. J Clin Psychiatry 46:276–277

Quitkin FM, Stewart JW, McGrath PJ, Tricamo E, Rabkin JG, Ocepek-Welikson K, Nunes E, Harrison W, Klein DF (1993) Columbia atypical depression. A subgroup of depressives with better response to MAOI than to tricyclic antidepressants or placebo. Br J Psychiatry 21(Suppl):30–4

Quitkin F, Stewart J, McGrath P, Taylor B, Tisminetzky M, Petkova E, Chen Y, Ma G, Klein D (2002) Are there differences in women's and men's antidepressant response? Am J Psych 159:1848–1854

Raskin A (1974) Age-sex differences in response to antidepressant drugs. J Nerv Ment Dis 159:120–130

Ronfeld RA, Tremaine LM, Wilner KD (1997) Pharmacokinetics of sertraline and its N-demethyl metabolite in elderly and young male and female volunteers. Clin Pharmacokinet 32(Suppl 1):22–30

Schmidt PJ, Nieman L, Danaceau MA, Tobin MB, Roca CA, Murphy JH, Rubinow DR (2000) Estrogen replacement in perimenopause-related depression: A preliminary report. Am J Obstet Gynecol 183:414–420

Schneider LS, Small GW, Hamilton SH, Bystritsky A, Nemeroff CB, Meyers BS (1997) Estrogen replacement and response to fluoxetine in a multicenter geriatric depression trial. Am J Geriatr Psychiatry 5:97–106

Schneider LS, Small GW, Clary CM (2001) Estrogen replacement therapy and antidepressant response to sertraline in older depressed women. Am J Geriatr Psychiatry 9:393–399

Silverstein B (1999) Gender difference in the prevalence of clinical depression: the role played by depression associated with somatic symptoms. Am J Psychiatry 1999;156:480-482

Simpson JM, Bateman DN, Rawlins MD (1987) Using the adverse reactions register to study the effects of age and sex on adverse drug reactions. Stat Med 6:863–867

Soares CN, Almeida OP, Joffe H, Cohen LS (2001) Efficacy of estradiol for the treatment of depressive disorders in perimenopausal women. Arch Gen Psychiatry 58:529–534

Steiner M., Wheadon, D., Kreider, M., et al. Antidepressant response to paroxetine by gender. (1993) Presented at the 146th Annual Meeting of the American Psychiatric Association, San Francisco, CA, May 1993

Thase M, Frank E, Kornstein S, Yonkers, KA (2000) Gender difference in response to treatment of depression. In: Frank E (ed) Gender and its effect on psychopathology. American Psychiatric Press, Washington, DC, pp 103-125

Warrington SJ (1991) Clinical implications of the pharmacology of sertraline. Int Clin Psychopharmacol 6(suppl):11–21

Watkins PB (1992) Drug metabolism by cytochromes P450 in the liver and small bowel. Gastroenterol Clin North Am 21:511–526

Weissman MM, Bruce ML, Leaf PJ, Florio LP, Holzer C (1991) Affective disorders. In: Robins L, Regier D (eds) Psychiatric disorders in America. Free Press, New York, pp 53-80

Yonkers K, Bradshaw K (1999) Hormone replacement and oral contraceptive therapy: Do they induce or treat mood symptoms. In: Leibenluft E (ed) Gender differences in mood and anxiety disorders. American Psychiatric Press, Washington, DC, pp 91-129

Yonkers KA, Hamilton JA (1995) Sex differences in pharmacokinetics of psychotropic medications. Part II: Effects on selected psychotropics. In: Jensvold MJ, Halbreich U, Hamilton JA (eds) Psychopharmacology of women: Sex, gender and hormonal considerations, vol 1. American Psychiatric Association, Washington, DC, pp 11-43

Yonkers KA, Kando JC, Cole JO, Blumenthal S (1992) Gender differences in pharmacokinetics and pharmacodynamics of psychotropic medication. Am J Psychiatry 149:587–595

Ziegler VE, Biggs JT (1977) Tricyclic plasma levels: Effects of age race, sex, and smoking. JAMA 238:2167–2169

Zweifel JE, O'Brien WH (1997) A meta-analysis of the effect of hormone replacement therapy upon depressed mood. Psychoneuroendocrinology 22:189–212

Older Adults

S. Glover[1] · W. F. Boyer[2]

[1] VAMC Atlanta, 1670 Clairmont Rd., Decatur, GA 30033-4004, USA
[2] Behavioral Solutions Inc., 4181 Pleasant Hill Rd Suite 100, Duluth, GA 30096, USA
e-mail: wboyer@emory.edu

1	Introduction	395
2	Diagnostic Considerations	395
3	Management of Geriatric Depression	397
4	Antidepressants	399
4.1	Tricyclic Antidepressants	399
4.1.1	Utilize Secondary Amines (e.g., Desipramine, Nortriptyline) When Possible	400
4.1.2	Monitor Plasma Drug Levels	401
4.1.3	Minimize the Use of Other Medications	401
4.1.4	Educate and Monitor for Side Effects	401
4.1.5	Start Medications for Side Effects If Other Measures Fail	401
4.2	Selective Serotonin Reuptake Inhibitors	401
4.2.1	Fluoxetine	402
4.2.2	Fluvoxamine	403
4.2.3	Paroxetine	403
4.2.4	Sertraline	404
4.2.5	Citalopram	404
4.3	Selective Norepinephrine Uptake Inhibitors	405
4.4	Monoamine Oxidase Inhibitors	405
4.5	Venlafaxine	407
4.6	Bupropion	407
4.7	Trazodone	408
4.8	Nefazodone	408
4.9	Psychostimulants	409
5	Combination and Augmentation Strategies	410
5.1	Antidepressant Augmentation	410
5.1.1	Lithium	410
5.1.2	Thyroid	411
5.1.3	Hormone Replacement Therapy	411
5.2	Antidepressant Combinations	411
5.3	Psychotherapy and Antidepressants	412
6	Other Treatment Considerations	412
6.1	Compliance	412
6.2	Duration of Treatment	413
7	Conclusions	414
	References	414

Abstract Depression in the elderly is a major public health problem associated with increased morbidity, mortality, functional impairment, and a diminished quality of life. Unfortunately, late-life depression often goes unrecognized and untreated. Healthy, ambulatory elderly patients can often be treated in the same way as younger patients, whereas frail elderly patients (usually the "old" old) often need to be approached more conservatively, with special attention to physical status and concomitant illnesses. Clinicians should also consider the patient's environment (e.g., whether living at home or in a nursing home). Elderly patients may underreport psychological symptoms and overreport somatic symptoms (e.g., pain); collateral histories from families, friends, or professional caregivers are invaluable aids in diagnosis. Secondary depressions are also common in the elderly, since numerous medical disorders, medications, and life stresses can lead to depressive syndromes. Before initiating antidepressant treatment, clinicians should screen for and treat any concurrent medical condition(s), provide psychological support for the patient, identify and provide assistance with social or economic difficulties, and involve the patient's family or support network. Social and psychological supportive approaches should preferably precede pharmacologic management in mild or stable cases. Some elderly patients respond well to traditional or time-limited psychotherapies, cognitive-behavioral interventions, or spiritual support in individual or group settings. The ideal antidepressant agent for elderly patients should not cause orthostasis or cardiotoxicity and should cause little sedation or impairment of physical and cognitive abilities. Although data are inconsistent as to whether elderly patients are more likely to develop side effects than younger patients, the aged often do not tolerate side effects as well. Clinicians should take into account the heterogeneity of the elderly population in pharmacokinetic and pharmacodynamic parameters and practice individualized titration of all medications coupled with therapeutic drug monitoring. In general, the rule of "start low, go slow" applies, except when rapid symptom relief is of paramount importance. The SSRIs are important agents for the treatment of depression in the elderly, given their tolerability, wide therapeutic index and efficacy. The tricyclic antidepressants are efficacious in treating depression in the elderly. The secondary amines (e.g., desipramine, nortriptyline) are preferred over the tertiary amines (e.g., amitriptyline and imipramine) because they cause fewer serious side effects. Therapeutic drug monitoring is recommended in using TCAs in older patients, and clinicians should be alert for the risk of drug–drug interactions since many older patients are taking multiple medications. A number of other antidepressants have been shown to be effective and well tolerated in elderly depressed patients, including reboxetine, venlafaxine, and bupropion. In treating more severe depressive illness consideration should be given to ECT or prescribing a TCA or venlafaxine because of possibly increased response rates with these agents. Elderly patients who do not respond adequately to antidepressant monotherapy should be considered for antidepressant combinations or augmentation with lithium, thyroid, or hormone replacement (in perimenopausal depression). The appropriate duration of maintenance antidepressant medication will depend on the patient's history of depressive episodes.

Keywords Depression · Elderly · "Old" old · Antidepressant medications · Side effects · Psychotherapy · Interpersonal psychotherapy · Cognitive-behavioral therapy · Augmentation strategies · Lithium · Thyroid · Hormone replacement · Combination therapy

1
Introduction

The number of aging and aged individuals is rapidly increasing in many parts of the world. In the United States alone, the number of persons 65 years of age and older will likely double in the next 35 years. They will then comprise more than 20% of the U.S. population (Blazer 1989a; Drevets 1994). As the number of elderly grows, issues related to their health care will assume increasing importance. An area of primary concern in this population will be the prompt identification and treatment of psychiatric disorders, which many studies have shown are relatively common in old age (Blazer et al. 1987; Blazer 1989b; Cohen 1990; Skoog 1993).

Depression in the elderly is a major public health problem. It is associated with increased morbidity, mortality, and functional impairment, as well as a diminished quality of life (Evans 1993; Katy et al. 1994). An estimated 1%–2% of community-dwelling elderly experience major depression and 9%–15% have subsyndromal depression (Blazer 1989b; NIH Consensus Development Panel 1992; Reynolds 1994). In contrast, the prevalence of major depression is as high as 20% among those confined to long-term or acute care facilities (Blazer 1989b). Unfortunately, this late-life depression often goes unrecognized and untreated.

The epidemiologic data outlined above point to the diversity of the elderly population, which is an issue that complicates treatment recommendations. Elderly individuals dwelling in the community appear to be substantially different from those who are institutionalized. Furthermore, the aging population displays great heterogeneity in terms of both pharmacokinetic and pharmacodynamic parameters (Ahronheim 1993). Thus, an individualized approach to both diagnosis and treatment is necessary, particularly in the "old" old (age >85) or in those with concurrent physical illness.

2
Diagnostic Considerations

According to the *Diagnostic and Statistical Manual of Mental Disorders*, 4th edition (DSM-IV) (American Psychiatric Association 1994), the diagnostic criteria for major depressive disorder includes depressed mood, anhedonia, weight loss or gain, insomnia or hypersomnia, psychomotor agitation or retardation, fatigue, feelings of worthlessness or guilt, impaired concentration, and recurrent thoughts of death or suicide. Five of these symptoms, one of which must be either depressed mood or anhedonia, are required to make the diagnosis. However, many depressed persons among the elderly exhibit symptoms that are incongruent with current diagnostic categories or coexist with physical illness (Kennedy 1995). Because both patients and clinicians are often more concerned about medical problems, depressive symptoms may be overlooked. As a group, elderly patients may underreport psychological symptoms and focus instead on somatic concerns such as pain. The tendency to underreport psychological symptoms may be further aggravated by age and cognitive decline.

It is important for the clinician to consider other information in addition to patient-reported symptoms when evaluating older persons for depression. Collateral histories from families, friends, or professional caregivers are invaluable. Other people may notice social withdrawal, lack of interest in normal activities, and changes in neurovegetative functioning before they are reported by the patient. Decrements in functional capacity and increasing dependence on others for help with routine activities of daily life should also be investigated, since these problems may be associated with depressive illness. The affect elicited in the interview is also important, particularly when working with patients who exhibit impaired verbal communication.

The diagnosis of depression in the elderly must sometimes be made nearly exclusively on the basis of reports from significant others. This is especially true for patients with concomitant dementia or other conditions that impair communication. Such patients may not respond intelligibly or appropriately to questions concerning their mood. Nevertheless, those familiar with the patient may be able to attest to crying spells, decreased interest in food, difficulty sleeping, and excessive complaints of pain. In severe cases of dementia, the main symptoms may be spells of crying out accompanied by an anguished expression. It can be very helpful in such cases to ask available historians to describe how the patient's behavior may have changed over time.

One may occasionally encounter a situation in which the patient's depression is misinterpreted as an understandable reaction to aging or loss. In such a circumstance, the biggest impediment to treatment may be such a misunderstanding on the part of family, caretakers, or even the consulting physician. In this situation, it can be helpful to first acknowledge the reality of the patient's stresses, but then to draw a parallel with other medical conditions. For example, postoperative pain is "understandable" but this does not imply that it should not be treated; in the same way, even though the clinical syndrome of depression may be an "understandable" complication of loss or stress, it should nevertheless be treated to prevent unnecessary suffering, impairment of functioning, and even suicide.

Secondary depressions are common in the aging population, since numerous medical disorders and medications can lead to depressive syndromes. Common medical etiologies include endocrinopathies (e.g., thyroid disorders, Cushing's disease, diabetes mellitus), cardiovascular disease, collagen-vascular diseases (e.g., rheumatoid arthritis, systemic lupus erythematosus), malignancies, nutritional disorders, hepatic and renal failure, and a plethora of neurological disorders, including Parkinson's disease and cerebrovascular accidents (Koenig 1991; Mendels 1993; Raskind 1993).

Medication causes depressive signs and symptoms more frequently than a full depressive syndrome (Raskind 1993; Dhondt 1995). Antihypertensive drugs (e.g., reserpine, β-blockers, clonidine), steroids, diuretics, and digitalis have all been reported to cause depression (Mendels 1993; Raskind 1993). Among psychotropic drugs, the benzodiazepine sedative-hypnotics may cause lethargy and loss of interest as well as other depressive symptoms. Alcohol and other nonprescription drugs may also be a precipitant for depressive symptoms; unfortunately, the use of multiple prescription and nonprescription drugs is common among elderly individuals in the community and is even higher in acute- and long-term care facilities (Koenig 1991; Turner et al. 1992).

A reasonable first step in the treatment of secondary depressive syndromes is to correct the underlying medical problem(s) or discontinue the medication(s) responsible for the symptoms. However, if the problem is not correctable or if depressive symptoms continue despite these interventions, treatment for the depression should be initiated.

3
Management of Geriatric Depression

It is helpful at this point to distinguish at least two subgroups of the elderly: the "young" old and the "old" old. The latter group is often defined as those over age 85. The important distinction, however, is not so much age as functional capacity and general health. Healthy, ambulatory elderly patients may often be treated in the same fashion as their younger counterparts. The frail, usually "old" old, often need to be approached more conservatively, with special attention to their physical status and concomitant illnesses.

It is also helpful to consider the patient's setting. Patients who live at home with a reliable spouse or other companion are more likely to be able to comply with a complicated medication regimen, if required, and to watch for and report side effects. Nursing homes can vary considerably in the quality of their staff and in ability to monitor patients. The clinician should try to become familiar with local nursing homes in order to tailor the patient's treatment to his or her surroundings. This may mean choosing medication that requires fewer doses per day or that minimizes the risk of side effects, the management of which (e.g., urinary retention) may be especially problematic for the staff. Disruptive patients in a nursing home sometimes face transfer to a more restrictive facility if their symptoms cannot be rapidly treated. With such a patient, the induction of regular sleep patterns may be a high initial priority.

Koenig suggested a five-pronged approach to the management of depressive disorders in the medically ill elderly (Koenig 1991). These steps provide a useful guide for any aging patient.

1. Diagnosis and treatment of any concurrent medical condition(s)
2. Psychological support of the patient
3. Identification and assistance with social or economic difficulties

4. Involvement of the patient's family or support network
5. Consideration of biological therapies, particularly drug therapy, for depression

Ideally, the first four factors will be addressed prior to antidepressant treatment. Failure to do so may directly affect compliance, treatment efficacy, and ultimately outcome.

In addition to treating known medical conditions, one should screen for other conditions that may be present but unrecognized. The typical work-up should include a full history and physical examination, serum chemistries, complete blood count, urinalysis, urine drug screen, thyroid function tests, syphilis serology, serum B-12 and RBC folate levels, chest X-ray, and electrocardiogram. HIV testing should be obtained if there are any risk factors. Neuroimaging, EEG, and/or lumbar puncture may be necessary to rule out some medical etiologies for depression. A complete list of all prescription and nonprescription medications should be obtained from the patient and a collateral historian. Any change in prescription or over-the-counter medicines, especially narcotic analgesics, sedatives, or cardiovascular agents, that coincided with the onset of depression should be suspect.

Often the patient is referred to the psychiatrist by his or her primary physician after a physical evaluation has taken place. Which examinations to request and/or repeat may then pose a problem. In general, if the patient has no personal or family history of affective illness, and if the patient's change in mental status has been relatively acute, the clinician should be especially concerned with ruling out a treatable medical illness. If the patient has a condition that predisposes him or her to certain illnesses, these should be at the top of the list. Examples would be chronic obstructive pulmonary disease complicated by pneumonia and hypertension leading to stroke or renal failure. Occasionally an organic cause for the depression is found, such as a stroke, but antidepressant medication will still be part of the treatment. It is important to realize that antidepressants, in general, may help depressive symptoms regardless of etiology.

Social and psychological supportive approaches should preferably precede pharmacologic management in mild or stable cases. Some elderly patients will respond quite well to traditional or time-limited psychotherapies, cognitive-behavioral interventions, or spiritual support in either individual or group settings. If these interventions are inappropriate or ineffective and an antidepressant trial is indicated, involvement of the family or other caregivers prior to pharmacological treatment is extremely helpful. The patient and family should be educated about depression and about the proposed treatment, including its potential duration, risks versus benefits, and common side effects, as well as economic considerations. In most situations, discussing potential problems before they occur is preferable to taking a "wait and see" attitude.

4
Antidepressants

It is generally accepted that all of the medications approved for the treatment of depression are approximately equally effective in adults (Reynolds 1992; Evans 1993; Mendels 1993). Geriatric patients have been less thoroughly studied, but the principle of equivalence seems to apply. Therefore, the choice of an antidepressant agent is typically based on its pharmacologic profile, including onset of action, ease of administration, safety, and common side effects (Alexopoulos et al. 2001).

Although the data are inconsistent as to whether elderly patients are more likely to develop side effects than their younger counterparts, the aged often do not tolerate side effects as well (Reynolds 1994). The ideal agent should not cause orthostasis or cardiotoxicity and should cause little sedation or impairment of physical and cognitive abilities. There is little information about patient factors that predict response to certain drugs. Previous response to an agent or a biologic relative's response is a consideration. Evidence suggests that response depends on adequate dosage, length of treatment, and blood levels of medication (NIH Consensus Development Panel 1992; Perrel 1994; Reynolds et al. 1994). However, clinicians should be acutely aware of the heterogeneity of the elderly population and practice individual titration of all medications coupled with therapeutic drug monitoring (see Burke and Preskorn, this volume). Standardized dosing guidelines merely that—guidelines. In general, the rule of "start low, go slow" applies. This may be tempered, however, by situations in which rapid symptom relief is of paramount importance.

4.1
Tricyclic Antidepressants

The tricyclic antidepressant agents (TCAs) were the cornerstone of the pharmacologic management of depression for decades beginning in the 1960s. Their efficacy has been well established in treating depressive episodes in the elderly. For example, a literature review of double-blind, controlled TCA trials in patients 55 years of age and older published between 1964 and 1986 showed that, in all but one case, the TCA was superior to placebo (Gerson et al. 1988).

Overall, the bioavailability of all the TCAs is low to moderate. They are highly lipophilic and thus have a large apparent volume of distribution. They are also highly bound to plasma proteins and metabolized via several steps in the liver to facilitate renal elimination. Obviously, the increased body fat, decreased total body water, and decreased glomerular filtration rate which commonly occur with aging may affect the pharmacokinetics of these drugs (Benetello et al. 1990; Furlanut and Benetello 1990; Turner et al. 1992). Although the TCAs themselves have little effect on the pharmacokinetics of other drugs, the action of many other medications may affect the metabolism of TCAs. Drugs that induce certain P450 hepatic isoenzymes (carbamazepine, phenobarbital) may speed the clearance of TCAs, thereby reducing plasma levels (see Preskorn and Catterson, this volume). On the other hand, enzyme-inhibiting medications (e.g., cimetidine, chlorpromazine, fluoxetine, paroxetine) slow TCA metabolism and may increase plasma concentrations, occasionally to seriously toxic levels (Preskorn 1993). This is an important consideration because many elderly individuals are taking numerous medications. Monitoring serum drug levels may be particularly useful in the elderly given the number of variables affecting the pharmacokinetics of these drugs and their narrow therapeutic index. A potential caveat is the paucity of reliable information on optimal effective serum concentrations in the elderly (Maletta et al. 1991). Determination of drug levels may be best applied to rule out unusually low levels (suggesting poor compliance or unusual metabolism) or high levels.

What strategies can clinicians use to minimize the side effects of these drugs and maximize their effectiveness in the older individual?

4.1.1
Utilize Secondary Amines (e.g., Desipramine, Nortriptyline) When Possible

The antidepressant effects of the TCAs are thought to result from their ability to block the reuptake of norepinephrine and/or serotonin by presynaptic terminals (Bressler and Katz 1993). However, they exhibit a number of other pharmacodynamic actions, including inhibition of the α-1 adrenergic receptors, which can cause orthostatic hypotension and reflex tachycardia; inhibition of the muscarinic (M)-1 cholinergic receptor, which can cause dry mouth, constipation, urinary retention, blurred vision, and cognitive dysfunction including delirium (Katz et al. 1988); and inhibition of the H1 histamine receptor, which can cause sedation and weight gain (Nemeroff 1994; Stein et al. 1985). Furthermore, the quinidine-like effects of the TCAs, particularly the hydroxy metabolites, on the heart may result in dangerous or even fatal intraventricular conduction delays. These pharmacodynamic actions are all particularly problematic in the elderly, because their sensitivity to side effects may be exaggerated (Reynolds 1994). In general, the tertiary amines (including amitriptyline, clomipramine, doxepin, and imipramine) cause more of these side effects than the secondary amines (including desipramine, nortriptyline), so that the secondary amines are often a better choice in older patients.

Nortriptyline has proven efficacy in this population and may cause less orthostatic hypotension than the others (Roose and Glassman 1989; Katz et al. 1990). Nortriptyline may also be better tolerated than desipramine (Lazarus et al. 1991). This is beneficial considering that the consequences of a fall due to lowered blood pressure may be devastating (e.g., hip fracture, subdural hematoma). Starting doses of nortriptyline or desipramine should be 10–25 mg/day, given either at night or in divided doses to minimize side effects. Doses should be increased gradually, with routine checking for changes in heart rate and blood pressure as well as other side effects, keeping in mind the need for individual titration

4.1.2
Monitor Plasma Drug Levels

Monitoring plasma levels is especially important in individuals taking medications that may accelerate or slow TCA metabolism. Trough levels should be obtained 10–14 h after the last dose and clinicians should also include an assay for any active metabolites.

4.1.3
Minimize the Use of Other Medications

Clinicians should especially avoid agents with anticholinergic, antihistaminic, or sedative properties. They should also minimize the use of drugs that may themselves induce depression or affect TCA metabolism.

4.1.4
Educate and Monitor for Side Effects

Clinicians should teach the patient and family the means of coping with or minimizing side effects, such as monitoring food intake and increasing exercise to avoid weight gain, sitting up for a few moments before arising from a reclining position to decrease the risk of orthostatic-induced falls, and using sugarless gum or hard candy to relieve dry mouth. It is essential to monitor heart rate and blood pressure during therapy. ECGs are essential in patients with pre-existing cardiac disease or those who become symptomatic while being treated with a TCA.

4.1.5
Start Medications for Side Effects If Other Measures Fail

Medications for side effects may include stool softeners or laxatives for constipation or cholinergic agonists (such as bethanechol 10–25 mg two to four times/day) to decrease anticholinergic side effects.

4.2
Selective Serotonin Reuptake Inhibitors

The selective serotonin reuptake inhibitors (SSRIs) include citalopram, fluoxetine, fluvoxamine, paroxetine, and sertraline. They have achieved considerable importance in the treatment of depression in the elderly in recent years due to their efficacy, tolerability, and wide therapeutic index. All of these agents inhibit the uptake of serotonin into presynaptic terminals, thus enhancing serotonergic neurotransmission. Unlike the TCAs, the SSRIs have highly focused pharmacodynamic effects. They exhibit low or no affinity for adrenergic, histaminergic, and cholinergic receptors and have not been associated with significant alterations in cardiac conduction (Preskorn 1993). All five of these agents have demonstrated efficacy in the elderly (Cohn et al. 1990; Kellett 1991; Dunner et al. 1992; Nyth et al. 1992; Geretsegger et al. 1994). Most studies suggest that the SSRIs have similar or better efficacy and tolerability in the treatment of depression compared with the TCAs (Cohn et al. 1990; Hutchinson et al. 1991; Dunner et al. 1992). However, one study suggests that fluoxetine may be less effective than nortriptyline (Roose et al. 1994) in treating hospitalized elderly patients with unipolar major affective disorder, especially the melancholic subtype.

The side effects typically encountered with the SSRIs are the result of serotonin agonism. They include insomnia, anxiety, restlessness, tremor, and gastrointestinal disturbances, especially nausea, vomiting, and diarrhea. The main late-emergent adverse effects of SSRIs involve sexual dysfunction, including decreased libido, impotence, delayed ejaculation, and anorgasmia (Preskorn 1994). Unfortunately, many practitioners fail to assess for sexual dysfunction, particularly in the elderly. Doing so is important since sexual side effects increase the risk of noncompliance with these generally well-tolerated agents. Management of these side effects is discussed in "Selective Serotonin Reuptake Inhibitors" (Preskorn et al., this volume).

In general, the efficacy and side effect profiles of the SSRIs are similar. All have a wide therapeutic index, generally flat dose–response curves, safety in overdose, and no requirements for routine plasma level monitoring.

4.2.1
Fluoxetine

The extent of gastrointestinal (GI) absorption of fluoxetine has been estimated at 80%. Absorption is slowed with food, but the extent of absorption does not appear to be affected (Finley 1994). Protein binding is quite high (>95% ;van Harten 1993), and concomitant use of other highly protein-bound drugs may cause shifts in plasma concentrations of both medications. Like the other SSRIs, fluoxetine is highly lipophilic. It has a significantly longer elimination half-life (1–4 days) than the other drugs in this class, and its active metabolite norfluoxetine has a half-life of 7–15 days (Preskorn 1994). These features have allowed the development of a once-weekly preparation of fluoxetine. Fluoxetine's long half-life may be of great benefit in noncompliant patients. The maximum plasma concentration of fluoxetine is on average higher in the elderly, supporting reduced dosages (Preskorn and Jerkovich 1990).

All the SSRIs are eliminated via hepatic metabolism and all competitively inhibit the cytochrome P450 2D6 hepatic isoenzyme. Paroxetine, norfluoxetine, and fluoxetine are the most potent inhibitors (Flint 1994). This is relevant because of potential interactions with other drugs metabolized via this system, including the TCAs, phenothiazines, and type IC antiarrhythmics. Fluoxetine may a cause two- to tenfold increase in plasma TCA levels (Preskorn and Burke 1992). Other drugs that induce the P450 enzymes (e.g., phenobarbital, phenytoin) shorten the half-lives of the SSRIs. Drugs that inhibit these enzymes (e.g., cimetidine) increase plasma concentrations of the SSRIs. Once again, drug interactions like these are of major concern in the elderly due to their increased rates of medication usage.

The side-effect profile of fluoxetine is similar to that of the other SSRIs. In addition, it may have a greater propensity to cause anorexia and weight loss. A higher incidence of gastrointestinal side effects, agitation, and insomnia but less somnolence have been reported with fluoxetine and sertraline compared with the other SSRIs (Grimsley and Jann 1992). Initial reports of increased suicidality related to fluoxetine have been discounted (Fava and Rosenbaum 1991).

In older patients, dosing should be at 5–10 mg/day or 20 mg every other day. Although most studies have used doses of 20 mg/day, equal effects on average have been found at daily doses of 5, 20, and 40 mg (Wernicke et al. 1988). It is advisable to treat with the lowest effective dose in order to minimize side effects. Liquid preparations are also available for fluoxetine, paroxetine, and citalopram. Use of a liquid preparation allows even lower starting doses, which can be important in a frail elderly person, or when there is concomitant panic anxiety, which is sometimes aggravated initially if therapy is started with an average dose of an SSRI.

4.2.2
Fluvoxamine

Fluvoxamine is well absorbed from the gastrointestinal tract and has a relatively low percentage of protein binding (77%; van Harten 1993). However, it has been shown to increase plasma levels of warfarin by 65%. Its pharmacokinetic disposition does not appear to be affected by the aging process (Devries et al. 1992; Finley 1994). Fluvoxamine has a relatively short half-life (16 h) and has no active metabolites (Bressler and Katz 1993). It is the only SSRI that causes substantial inhibition of the P450 1A2 isoenzyme (Preskorn 1994). Fluvoxamine commonly causes nausea (Grimsley and Jann 1992) and may cause more sedation than other SSRIs. Triazolam should be used with caution with fluvoxamine since they share the P450 3A4 metabolic isoenzyme.

Fluvoxamine has been found to be as effective as the TCAs desipramine and dothiepin in elderly patients with depression (Mullin et al. 1988; Rahman et al. 1991). Fluvoxamine is as well tolerated as dothiepin and better tolerated than desipramine (Rahman et al. 1991; Tourigny-Rivard 1997). Dosing of fluvoxamine in the elderly should begin at 25–50 mg/day. This is usually given at bedtime to take advantage of the drug's sedating side effects.

4.2.3
Paroxetine

Paroxetine, like fluoxetine and sertraline, is highly protein bound. It is also highly lipophilic. Like fluvoxamine, it has been reported to cause shifts in plasma concentrations of other highly protein-bound drugs, like warfarin. It has a short elimination half-life of around 24 h (Finley 1994) and no active metabolites. It is metabolized by the P450 2D6 isoenzyme and has the highest isoenzyme affinity in this class (Finley 1994). Compared with the other SSRIs, paroxetine's pharmacokinetics are apparently the most dramatically affected by the aging process, with clearance values around 70% lower compared with younger controls (Bayer et al. 1989). Somnolence is a more common side effect with paroxetine than with fluoxetine or sertraline (Grimsley and Jann 1992), but the lower incidence of diarrhea is lower (Finley 1994).

Paroxetine has been tested extensively in the elderly. One study found it superior to fluoxetine beginning in the third week of treatment, indicating possible early effects (Geretsegger et al. 1994). Paroxetine was compared to fluoxetine in another controlled trial of elderly patients (n=106). Both drugs produced equivalent improvement in depression, but there was a significantly higher proportion of responders to paroxetine (Schone and Ludwig 1993). Initial dosing of paroxetine in the elderly should be 10 mg/day. The maximum recommended dose is 40 mg/day, although most patients will respond at a lower dose.

4.2.4
Sertraline

Sertraline is also highly protein bound (>95%) and lipophilic (Finley 1994). It has an elimination half-life of approximately 26 h and has a weakly active metabolite with a half-life of 66 h. Advantages of sertraline in the elderly include its lower inhibition of the P450 2D6 isoenzyme and its linear pharmacokinetics in both young and elderly individuals (Preskorn 1994), reducing the potential for drug interactions.

Arranz and colleagues studied a large sample (n=1,437) of depressed elderly (mean age 68 years) patients treated with sertraline in routine clinical practice. Seventy percent of these patients responded to treatment, which was defined as a Montgomery-Asberg depression scale score less than or equal to 50% of baseline. The presence of concurrent medications and/or concomitant pathologic conditions did not affect this outcome (Arranz and Ros 1997)

Sertraline should be started at a dose of 25–50 mg/day in the elderly patient. The dose may be increased gradually to 200 mg/day if necessary. Increasing the dose too rapidly may increase side effects without hastening response (Preskorn and Lane 1995).

4.2.5
Citalopram

Citalopram has a half-life of about 36 h and undergoes minimal first-pass metabolism in man. It undergoes metabolism through the cytochrome P450 2C family of isoenzymes (Nemeroff et al. 1996). The metabolites of citalopram have the same specificity for serotonin as citalopram, but are significantly less potent, enter the brain less readily, and are present in lower concentrations. Therefore the therapeutic effect of citalopram is essentially due to the parent compound itself. A prolonged half-life and reduced metabolism have been noted in elderly patients receiving citalopram, and a lower dose is therefore recommended in this population (Fredericson-Overo et al. 1985; Leinonen et al. 1996).

A total of 149 patients in seven Scandinavian centers entered a 6-week double-blind trial intended to assess the antidepressant effect and safety of citalopram versus placebo in depressed elderly patients (65 years of age or older) who might also suffer from somatic disorders and/or senile dementia. Citalopram was associated with significantly more improvement than placebo on both depression and dementia ratings (Nyth et al. 1992).

A randomized, double-blind study compared the efficacy and tolerability of citalopram and mianserin in 336 elderly, depressed patients with or without dementia. Patients received either citalopram 20–40 mg/day or mianserin 30–60 mg/day for 12 weeks. Patients in both treatment groups responded equally well and had a relatively low incidence of adverse events (Karlsson et al. 2000).

4.3
Selective Norepinephrine Uptake Inhibitors

Reboxetine is the first non-tricyclic selective norepinephrine reuptake inhibitor (SNRI). Reboxetine has been studied in large acute and long-term studies of elderly depressed patients. It was compared to imipramine in a double-blind 8-week study of 347 depressed patients over the age of 65. The reduction in the Hamilton Rating Scale for Depression (HAM-D) (Hamilton 1960) score was comparable between the treatment groups. There were significantly fewer serious adverse events in the group treated with reboxetine (Katona et al. 1999).

The long-term study was a 52-week open-label investigation of reboxetine in 160 patients age 65 or older with major depression or dysthymic disorder. Of the patients, 139 completed the 6-week run-in period and entered the long-term phase. The mean HAM-D total score showed a reduction from 24.0 at baseline to 10.4 at week 6. The study was completed by 104 patients. Of the original cohort, 25 discontinued treatment due to adverse events. The proportion of patients with clinicians' global impression (CGI)–global improvement ratings of "much" and "very much" improved increased from 15.1% at week 2 to 88.7% at week 6 and to 95.2% at week 52 (Aguglia 2000).

4.4
Monoamine Oxidase Inhibitors

Since monoamine oxidase (MAO) levels increase with age (Robinson et al. 1971), medications that inhibit MAO might theoretically be particularly helpful in treating older, depressed individuals (Georgotas et al. 1981). However, this class of medications has been infrequently used in the elderly because of clinicians' concerns about safety and the troublesome dietary and drug restrictions required with the older, irreversible agents (e.g., isocarboxazid, phenelzine, and tranylcypromine). The advent of reversible MAOIs (e.g., moclobemide, brofaromine, toloxatone, and cimoxatone) may make safe, effective treatment in the elderly less problematic.

The MAOIs presumably exert their antidepressant effects as a result of increasing effective synaptic concentrations of dopamine, norepinephrine, and serotonin by blocking the oxidative enzymes that degrade these monoamines. The irreversible agents mentioned above have all been proven effective in the treatment of depression—particularly "atypical" depression associated with increased appetite, weight gain, hypersomnia, arid anxiety. Several studies have shown that the MAOIs are effective agents in treating depression in the elderly (Robinson 1979; Georgotas et al. 1983; Lazarus et al. 1986). Georgotas and colleagues maintained elderly depressed patients on either phenelzine ($n=13$) or nortriptyline ($n=15$) following successful initial treatment. Patients on phenelzine were significantly less likely to have a recurrence (13.3%) than those on nortriptyline (6.3%; Georgotas and McCue 1989).

These drugs, however, have a number of side effects that may make them unsuitable as first-line treatment in older patients. These include hypotension, sedation, agitation, insomnia, weight gain, peripheral neuropathy, exacerbation of cognitive dysfunction, sexual dysfunction, and hypertensive crisis (Salzman 1992). Hypertensive crises occur when MAOIs interact with stimulant drugs or foods containing pressor amines, primarily tyramine. This necessitates dietary restrictions and limiting the use of certain other drugs, including many over-the-counter cold preparations. Furthermore, use of an irreversible MAOI within 2 weeks of another antidepressant (or 5 weeks for fluoxetine) may precipitate the serotonin syndrome.

Moclobemide and other reversible inhibitors of MAO-A (RIMAs) may provide a safe, effective alternative to the use of irreversible MAOIs in the elderly. Moclobemide has been proven as effective as other antidepressants in older patients and is better tolerated than some TCAs (Nair et al. 1995). There is little tyramine pressor effect with diets containing less than 100 mg of tyramine daily, thus necessitating few dietary restrictions (Norman et al. 1992). Its side-effect profile also compares favorably with the older MAOIs, as it reportedly does not cause orthostatic hypotension, weight gain, or sexual dysfunction. Moclobemide may, however, cause insomnia, headache, nausea, and constipation (Flint 1994; Nair et al. 1995). Usual maintenance doses in the elderly are 300–600 mg/day.

Selegiline is a selective inhibitor of MAO-B at doses less than or equal to 10–20 mg/day (Simpson and De Leon 1989; Sunderland et al. 1994). At these lower doses, it is used to treat Parkinson's disease. At higher doses, it appears to lack MAO-B specificity (i.e., it inhibits both MAO-A and MAO-B) and at 60 mg/day has been found to be effective in treatment-resistant depression in elderly patients (Sunderland et al. 1994). However, as it requires dietary and drug restrictions in this dosing range, it appears to offer few benefits compared to the irreversible, nonselective agents.

4.5
Venlafaxine

Venlafaxine is an antidepressant with a unique chemical structure that inhibits the reuptake of both serotonin and norepinephrine. It has no clinically significant anticholinergic, β-adrenergic, or histaminergic effects. Both venlafaxine and its active metabolite have short half-lives (5 h and 11 h, respectively) and low protein binding (Feighner 1994). Furthermore, venlafaxine has low toxicity compared with the older TCAs and MAOIs as well as a low potential for interaction with other drugs. All these features make venlafaxine a particularly good antidepressant choice for elderly individuals.

Venlafaxine has proven efficacy in treating older patients (Feighner 1994). It has also produced a more rapid resolution of symptoms than some other antidepressants (Guelfi et al. 1992). This may be particularly useful given the longer antidepressant response latency often observed in the elderly. It may also be useful in treatment-resistant depression (Nierenberg et al. 1994; Tsolaki et al. 2000). No dosage adjustment has been recommended based on age alone (Magni 1993). However, when the patient has a concurrent physical illness or when used with other medications, venlafaxine should be started at a lower dose of 12.5–25 mg twice daily. Increasing the dose gradually helps minimize side effects, which may be particularly problematic in older persons. The recently marketed sustained release preparation is generally better tolerated and may be dosed once daily.

4.6
Bupropion

Bupropion is a unique antidepressant that has dopaminergic properties. It has no appreciable effect on serotonin, norepinephrine, or MAO activity. It has been found efficacious in the treatment of elderly, depressed patients (Kirsky and Stern 1984) and has several properties that make it an attractive agent for use in this population. It has no sedative, anticholinergic, or orthostatic hypotensive side effects. It has no known effects on the metabolism of other drugs (Preskorn 1994).

Bupropion has been associated with side effects due to both increased and decreased dopaminergic activity in different individuals. These have included worsening psychosis and parkinsonian symptoms (Koenig 1991; Ames et al. 1992; Strouse et al. 1993). This should be considered before administration to patients with Parkinson's disease, striatal disease, or psychotic symptoms. The concern about bupropion and seizures stems from a study of 55 female patients with bulimia (Horne et al. 1988). Four of these patients (7.3%) experienced a grand mal seizure. This was, as the authors note, "a frequency of seizures much higher than observed in previous studies." The issue of seizures was subsequently re-examined in a 102-center study of 3,341 depressed patients. History of seizures, CNS trauma, and current or past eating disorders were exclusion criteria.

In this study, there was a 0.4% rate of seizures, which was similar to that found in previous studies (Johnston et al. 1991). The sustained-release bupropion preparation, which was introduced later, allows once-daily dosing and may carry a lower risk of seizures because it avoids the sudden absorption of a large bolus of drug. Initial dosing should begin at 50–75 mg twice daily of immediate-release bupropion or 100 mg/day of the sustained-release formulation. The dose may be increased to no more than 450 mg/day administered in divided doses, not to exceed 150 mg per dose (Davidson 1989).

4.7
Trazodone

Trazodone is an antidepressant that blocks the reuptake of serotonin and is an antagonist of the 5-HT_2 and other 5-HT receptor subtypes (Preskorn 1993). In general, its side effects are more benign than those of the TCAs, since it has minimal anticholinergic and cardiac effects. It has also been reported to be better tolerated (Altamura et al. 1989) and to cause less cognitive impairment (Moskowitz and Burns 1986) than some TCAs. It is relatively safe in overdose and its pharmacokinetic profile is not substantially altered by either age or intercurrent illness (Preskorn 1993). Superficially it would appear an excellent antidepressant choice for the older patient.

However, some of the pharmacodynamic properties of trazodone limit its utility in this population. It may cause orthostatic hypotension; it has been reported to induce or exacerbate ventricular arrhythmias (Rudorfer and Potter 1989); and it may cause priapism. Furthermore, it has strong sedative effects that may prevent elderly patients from tolerating an effective antidepressant dose. The efficacy of this medication has also been questioned, with response rates as low as 10%–20% reported in some studies (Shopsin et al. 1981).

When administered to depressed, elderly patients, dosing should be started at 25–50 mg at bedtime and titrated upward as tolerated to at least 150–300 mg/day for antidepressant efficacy. Despite its short half-life (3–9 h), once daily dosing appears comparable in efficacy to multiple daily doses (Fabre 1990). Daytime doses may be useful for decreasing anxiety or symptoms of agitation. Trazodone may be useful in lower doses at bedtime for sedation as well as in controlling agitated behavior associated with dementia.

4.8
Nefazodone

Nefazodone, like trazodone, is a triazolopyridine derivative. It also inhibits reuptake of serotonin and norepinephrine. It has no clinically significant effect on cholinergic, dopaminergic, or benzodiazepine receptors. However, it does antagonize β-adrenergic receptors, which may cause orthostatic hypotension. Nefazodone drug is highly protein bound. It inhibits the P450 3A4 isoenzyme, which may increase plasma concentrations of triazolam and alprazolam if these are coadministered. Nefazodone does not significantly inhibit P450 2D6.

Typical side effects include nausea, dizziness, insomnia, asthenia, and agitation, with side effects more common at higher doses. No unusual adverse events have been reported in elderly patients treated with nefazodone. Nefazodone has not been shown to impair cognition, even at doses above 300 mg/day (Van-Laar et al. 1995). Dosing in the elderly should be begun at 25–50 mg twice daily, approximately half the dose for younger healthy adults. Since few studies have examined the use of nefazodone in the elderly, upward dose titration should proceed slowly to a target dose of 300–600 mg/day.

4.9
Psychostimulants

The psychostimulants, including dextroamphetamine and methylphenidate, are used therapeutically in depressed elderly patients, particularly among those with concurrent physical illness (Satel and Nelson 1989; Warneke 1990; Reynolds 1992; Salzman 1992; Flint 1995). However, there are no prospective controlled trials of these drugs in elderly depressed patients. Subjective side effects are minimal but may include insomnia, nausea, appetite changes, blurred vision, dry mouth, constipation, and dizziness. Objective side effects may include changes in blood pressure, dysrhythmias, and tremor (Satel and Nelson 1989). Of note, in one series, nearly 19% of elderly patents experienced confusion or delusions when treated with dextroamphetamine (Woods et al. 1986). The potential for tolerance and abuse appears to be minimal (Satel and Nelson 1989).

In general, psychostimulants should not be used as a first-line therapy in elderly individuals who clearly have a clinical depressive syndrome, since they have shown no significant advantage over placebo in other age groups in the treatment of primary depression (Satel and Nelson 1989). It appears that response to stimulants is very heterogeneous, with some patients improving, some feeling worse (often due to overactivation), and others experiencing no effect. Therefore, the average response of a group of patients will be minimal.

Psychostimulants may be appropriate for short-term, symptomatic treatment of demoralized, withdrawn, apathetic elderly individuals with concurrent medical illness who do not meet criteria for a true depressive syndrome and for patients who cannot tolerate regular antidepressants or for whom regular antidepressants are contraindicated. One of us (WB) has also found that psychostimulants are often helpful as adjunctive treatment for elderly patients who have significant fatigue or lethargy following otherwise successful SSRI treatment.

Where stimulants are effective, results are generally seen within the first day or two of treatment. The usual starting dose of methylphenidate or dextroamphetamine is 5 mg q A.M. and noon. Dextroamphetamine is generally more potent. If a patient fails to respond to a 1- to 2-day trial of one agent, it is often worthwhile to try the other.

Methylphenidate has a relatively brief duration of action (approximately 3 h), so frequent individualized dosing is often required. Sustained release forms of both methylphenidate and dextroamphetamine are available. However, these preparations sometimes seem to be less potent on a milligram-per-milligram basis and extend the drug's duration of action only minimally. Common maintenance doses of methylphenidate or dextroamphetamine in the elderly are 5–10 mg two to three times per day.

5
Combination and Augmentation Strategies

Some patients clearly require the use of more than one psychotropic drug. For instance, a mood stabilizer is almost always required when starting an antidepressant in an elderly person with a clear history of bipolar disorder. Likewise, depression with psychotic symptoms necessitates an antidepressant/antipsychotic combination for optimal treatment if electroconvulsive therapy is not chosen. Anxiolytic/antidepressant combinations may be needed, at least initially, when treating the depressed, anxious individual. Changing from one class to another (e.g., from an SSRI to a TCA) is often successful (Weintraub 2001) and should be considered prior to a trial of combination or augmentation treatment. However, combination and augmentation strategies may be necessary in the elderly patient who is not responsive to one or more adequate trials of a single antidepressant.

5.1
Antidepressant Augmentation

5.1.1
Lithium

A number of studies have examined lithium's role in augmenting antidepressant response in the elderly. Studies of lithium augmentation in the elderly have reported varying response rates, as high as 60%, but also frequent clinically significant side effects (Lafferman et al. 1988; Finch and Katona 1989; van Marwijk et al. 1990; Zimmer et al. 1991; Flint and Rifat 1994). The side effects of lithium in the elderly may be more pronounced or more poorly tolerated and may occur at lower serum levels than in younger patients.

It is also unclear what serum levels of lithium should be considered therapeutic in the elderly, particularly in augmentation strategies (Flint 1995). In general, serum lithium levels much above 1.0 mEq/l should be avoided and levels as low as 0.3 mEq/l may be helpful. Other mood stabilizers have also been used to augment antidepressant response, including carbamazepine (Cullen et al. 1991) and valproic acid (valproate) (Corrigan 1992). Valproic acid is usually better tolerated than lithium in older patients due to its lower incidence of side effects and wider therapeutic index. The success rates of augmentation with other mood stabilizers seems to be lower than with lithium.

5.1.2
Thyroid

Thyroid hormone has been reported to successfully augment antidepressant response in the elderly (Joffe 1992), but there is little information on the basis of which one can predict who might best respond in the absence of hypothyroidism. Other drugs, including buspirone (Flint 1995) and B vitamins (Bell et al. 1992) may be useful in augmenting antidepressant effects. These are particularly interesting options, since both are well-tolerated with minimal side effects in most older patients.

5.1.3
Hormone Replacement Therapy

There is increasing evidence of the important role of hormone replacement therapy (HRT) in the treatment of depression. Hormone replacement therapy appears to be effective in reducing depressed mood in many menopausal women. It has been shown that estrogen can augment serotonergic activity in postmenopausal women and that HRT may augment SSRI response in elderly women. Therefore, strong consideration should be given to hormone replacement therapy in postmenopausal women if there are no medical contraindications (Palinkas and Barrett 1992; Schneider et al. 1997, 1998).

5.2
Antidepressant Combinations

The use of antidepressant combinations has become increasingly commonplace in younger patients, particularly in patients with treatment-refractory depression. However, there is a lack of systematic studies of antidepressant combinations in the elderly. Using an MAOI with a TCA (starting the two together or adding an MAOI to a TCA) has generally proven to be safe in younger individuals (Flint 1994). However, this strategy is not recommended in the elderly due to the increased potential for orthostatic hypotension and other side effects.

A number of case reports and case series have described the successful use of non-MAOI antidepressants in combination with SSRIs. A review quoted improvement rates of 65%–100% with SSRI/non-MAOl combination therapy (Flint 1994). In addition, most series reported good tolerance rates. Some studies suggest accelerated antidepressant response with TCA/SSRI combinations (McCue and Aronowitz 1994). Caution should be used with TCA/SSRI combination therapy, however, since paroxetine, fluoxetine, norfluoxetine, and to a lesser extent sertraline may elevate plasma TCA levels resulting in potentially life-threatening TCA toxicity. Starting with low doses of TCAs and monitoring plasma drug levels is recommended. However, response may occur independently of therapeutic TCA levels (McCue and Aronowitz 1994).

There have been several reports of younger patients with refractory depression responding to the combination of a psychostimulant and a TCA, MAOI, or SSRI. However, there have been no systematic studies in geriatric patients. Augmentation with a psychostimulant may be more useful in certain subtypes of elderly depression, such as the medically ill and/or female patients (Flint 1995). In general, the combination of irreversible MAOls and psychostimulants should be avoided due to the risk of hypertensive crisis.

5.3
Psychotherapy and Antidepressants

Reviews of antidepressant treatment plus interpersonal psychotherapy (IPT) or cognitive-behavioral therapy (CBT) have quoted success or improvement rates ranging from 50%–90% (Reynolds 1992; Schneider 1993). Elderly patients with less severe depression may be especially good candidates for maintenance IPT after stopping antidepressant therapy (Taylor et al. 1999). Some suggest that these combination therapies are superior to pharmacologic treatment alone; however, more long-term studies are needed. A recent study reported that combination psychotherapy and pharmacotherapy was also 80% effective in treating recurrent episodes of depression (Reynolds et al. 1994). The combination of some type of psychotherapy with pharmacotherapy is considered the standard of care by most clinicians.

6
Other Treatment Considerations

6.1
Compliance

The efficacy and safety of antidepressants are of little consequence if the elderly patient is noncompliant with pharmacotherapy. It has been estimated that 70% of the elderly fail to take 25%–59% of their medication. Furthermore, lack of compliance and the resultant wide fluctuations in plasma drug levels has been a predictor of poor outcome (NIH Consensus Development Panel 1992). Overuse of medications may result from cognitive impairment or the mistaken belief that more drug will speed recovery (Salzman 1995). Underuse may occur if a patient forgets to take medication due to cognitive impairment or if he or she equates pill-taking with drug dependence. Compliance problems may be multiplied in the elderly person taking several drugs. Underuse due to cost and side effects are also common causes of noncompliance (Salzman 1995).

How may compliance be enhanced? One of the most important factors appears to be optimizing good communication between the physician and patient (Reynolds 1994; Salzman 1995). Labeling of medication should be simple and readable, which may include using a larger print style. Timing of medication should be simplified; medication should optimally be administered only 1–2 times daily if possible. Pill storage systems are also available in which a full week's supply of medication may he loaded and then easily dispensed. Education and counseling of the patient and his or her family or caregivers is essential. This should include education on depression itself and medication issues, including side effects and the need for compliance. Enlisting others to facilitate compliance is recommended. This may include not only the family but other healthcare providers (such as a primary care physician), home health services, volunteers, and friends. In summary, compliance is a major issue in the treatment of elderly, depressed persons.

6.2
Duration of Treatment

Latency of antidepressant effects may be prolonged in the elderly (Kennedy 1995), requiring two or more months for full response (Georgotas et al. 1988; McCue 1992) There is currently no unequivocal evidence that one antidepressant acts more rapidly than another. In treating more severe depressive illnesses, consideration should be given to starting a TCA or venlafaxine because of possibly increased response rates with these agents. In educating the patient, family, and/or other caregivers, it is important for the physician to point out the expected latency of response. Sleep disturbances may improve as early as the first week of therapy, followed by improvement of other neurovegetative symptoms and social withdrawal. Others may then begin to notice improvement but it may take up to 2 months for subjective improvement in the patient's mood (Bressler and Katz 1993).

Amelioration of symptoms in the index episode is not the only objective of antidepressant therapy in the elderly. Prevention of recurrence should also be a goal. In younger patients, a number of studies have shown that continuing antidepressant therapy after a full remission of symptoms is beneficial in decreasing relapse rates (Hirschfeld 1994). This appears to be the case in the elderly as well, with lower relapse rates seen in patients on medication continuation therapy (Georgotas et al. 1988).

Some authors have suggested that any patient who has an index episode of depression after the age of 60 years requires long-term maintenance therapy (Old Age Depression Interest Group 1993; Hirschfeld 1994). Since there is still limited knowledge concerning the natural history and long-term treatment risks of depressive episodes in the elderly, it seems reasonable to follow a more cautious approach, as recommended by Reynolds (1992). This entails:

a. At least 6 months of continuation therapy after the first episode of unipolar depression in patients 60 years of age or older.

b. Consideration of longer term maintenance therapy if the index episode is the second of the patients' lifetime, particularly if the previous episode occurred less than 2–3 years before the index episode.
   c. Long-term maintenance therapy should be recommended if the index episode is the third or more lifetime episode.

The dosage should remain equal to that used in treating the acute episode. If/when a decision is made to discontinue medication, withdrawal should be done gradually over several weeks or months. This should be accompanied by education of the patient and family regarding what symptoms might herald a relapse, so that the medication taper may be stopped and more intensive therapy initiated if early depressive symptoms are detected.

7
Conclusions

It is clear that the benefits of rational antidepressant therapy in the elderly outweigh the risks. Unrecognized and untreated depressive disorders result not only in subjective discomfort and suffering for the patient but in increased utilization of health care resources, extended hospitalizations, functional decrements, poor compliance, as well as increased morbidity and mortality due to both suicide and medical illness. It is imperative that clinicians be adept at both recognizing and treating late-life depression. Antidepressants are the cornerstone for the treatment of these syndromes. Safe and effective agents with minimal side effects are available, and their number will increase in the years to come. Further research focusing specifically on different subtypes of the aging population will assist in identifying variables that may predict response to a particular antidepressant or other treatment intervention. Depressive disorders in the elderly present a therapeutic challenge that will intensify in the future with the growth of this segment of the population. Clinicians have the tools to meet that challenge.

References

Aguglia E (2000) Reboxetine in the maintenance therapy of depressive disorder in the elderly: a long-term open study. Int J Geriatr Psychiatry 15:784–793

Ahronheim J (1993) Practical pharmacology for older patients: avoiding adverse drug effects. The Mount Sinai Journal of Medicine 60:497–501

Alexopoulos GS, Katz IR, Reynolds CF, Carpenter D, Docherty JP. The expert consensus guideline series: pharmacotherapy of depressive disorders in older patients. Postgraduate Medicine Special Report 2001;October:1-86

Altamura AC, Mauri MC, Rudas N, Carpiniello B, Montanini R, Perini M, Scapicchio PL, Hadjchristos C, Carucci G, Minervini M, et al (1989) Clinical activity and tolerability of trazodone, mianserin, and amitriptyline in elderly subjects with major depression: a controlled multicenter trial. Clin Neuropharmacology 12(suppl 1):S25–S33

American Psychiatric Association (1994) Diagnostic and statistical manual of mental disorders, fourth edition. American Psychiatric Association, Washington, DC

Ames D, Wirshing WC, Szuba MP (1992) Organic mental disorders associated with bupropion in three patients. J Clin Psychiatry 53:53–55

Arranz FJ; Ros S (1997) Effects of comorbidity and polypharmacy on the clinical usefulness of sertraline in elderly depressed patients: an open multicentre study. J Affect Disord 46:285–291

Bayer AJ, Roberts NA, Allen EA, Horan M, Routledge PA, Swift CG, Byrne MM, Clarkson A, Zussman BD (1989) The pharmacokinetics of paroxetine in the elderly. Acta Psychiatr Scand 80(suppl 350):85–86

Bell IR, Edman JS, Morrow FD, Marby DW, Perrone G, Kayne HL, Greenwald M, Cole JO (1992) Brief communication: Vitamin B1, B2, and B6 augmentation of tricyclic antidepressant treatment in geriatric depression with cognitive dysfunction. J Am Coll Nutrition 11:159–163

Benetello P, Furlanut M, Baraldo ZG (1990) Imipramine pharmacokinetics in depressed geriatric patients. Int J Clin Pharm Res X:191–195

Blazer D (1989a) Depression in the elderly. N Engl J Med 320:164–166

Blazer DG (1989b) The epidemiology of psychiatric disorders in late life. In: Busse EW (ed) Geriatric psychiatry. American Psychiatric Press, Washington, DC, pp 235–260

Blazer D, Hughes DC, George LK (1987) The epidemiology of depression in an elderly community population. The Gerontologist 27:281–287

Bressler R, Katz MD (1993) Drug therapy for geriatric depression. Drugs & Aging 3:195–219

Cohen GD (1990) Prevalence of psychiatric problems in older adults. Psychiatr Ann 20:433–438

Cohn CK, Shrivastava R, Mendels J, Cohn JB, Fabre LF, Claghorn JL, Dessain EC, Itil TM, Lautin A (1990) Double-blind, multicenter comparison of sertraline and amitriptyline in elderly depressed patients. J Clin Psychiatry 51(suppl 12B):28–33

Corrigan FM (1992) Sodium valproate augmentation of fluoxetine or fluvoxamine effects. Biol Psychiatry 31:1178–1179

Cullen M, Mitchell P, Brodaty H, Boyce P, Parker G, Hickie I, Wilhelm K (1991) Carbamazepine for treatment-resistant melancholia. J Clin Psychiatry 52:472–476

Davidson J (1989) Seizures and bupropion: a review. J Clin Psychiatry 50:256–261

DeVries MH, Roghoebar M, Mathlener IS, van Harten J (1992) Single and multiple oral dose fluvoxamine kinetics in young and elderly subjects. The Drug Monit 14:493–498

Dhondt ADF, Hooijer C (1995) Editorial comment: iatrogenic origins of depression in the elderly. Is medication a significant aetiologic factor in geriatric depression? Considerations and a preliminary approach. Int J Geriatr Psychiatry 10:1–8

Drevets WC (1994) Geriatric depression: brain imaging correlates and pharmacologic considerations. J Clin Psychiatry 55(suppl 9A):71–81

Dunner DL, Cohn JB, Walshe T 3rd, Cohn CK, Feighner JP, Fieve RR, Halikas JP, Hartford JT, Hearst ED, Settle EC Jr, et al. (1992) Two combined, multicenter double-blind studies of paroxetine and doxepin in geriatric patients with major depression. J Clin Psychiatry 53(suppl 2):57–60

Evans ME (1993) Depression in elderly physically ill inpatients: a 12-month prospective study. Int J Geriatr Psychiatry 8:587–592

Fabre LF (1990) Trazodone dosing regimen: experience with single daily administration. J Clin Psychiatry 51 (suppl 19):23–26

Fava M, Rosenbaum JF (1991) Suicidality and fluoxetine: is there a relationship? J Clin Psychiatry 52:108–111

Feighner JP (1994) The role of venlafaxine in rational antidepressant therapy. J Clin Psychiatry 55 (suppl 9A):62–68

Finch EJL, Katona CLE (1989) Lithium augmentation in the treatment of refractory depression in old age. Int J Geriatr Psychiatry 4:41–46

Finley PR (1994) Selective serotonin reuptake inhibitors: pharmacologic profiles and potential therapeutic distinctions. Ann Pharmacother 28:1359–1369

Flint AJ (1994) Recent developments in geriatric psychopharmacology. Can J Psychiatry 39(suppl 1):S9–S19

Flint AJ (1995) Augmentation strategies in geriatric depression. Int J Geriatr Psychiatry 10:137–146

Flint AJ, Rifat SL (1994) A prospective study of lithium augmentation in antidepressant-resistant geriatric depression. J Clin Psychopharmacol 14:353–356

Fredericson-Overo K, Toft B, Christophersen L, Gylding-Sabroe JP (1985) Kinetics of citalopram in elderly patients Psychopharmacology (Berlin) 86:253–257

Furlanut M, Benetello P (1990) The pharmacokinetics of tricyclic antidepressant drugs in the elderly. Pharmacological Research 22:15–25

Georgotas A, McCue RE (1989) The additional benefit of extending an antidepressant trial past seven weeks in the depressed elderly. Int J Geriatr Psychiatry 4:191–195

Georgotas A, Mann J, Friedman E (1981) Platelet monoamine oxidase inhibition as a potential indicator of favorable response to MAOIs in geriatric depressions. Biol Psychiatry 16:997–1001

Georgotas A, Friedman E, McCarthy M, Mann J, Krakowski M, Siegel R, Ferris S (1983) Resistant geriatric depressions and therapeutic response to monoamine oxidase inhibitors. Biol Psychiatry 18:195–205

Georgotas A, McCue RE, Cooper TB, Nagachandran N, Chang I (1988) How effective and safe is continuation therapy in elderly depressed patients? Factors affecting relapse rate. Arch Gen Psychiatry 45:929–932

Geretsegger C, Bohmer F, Ludwig M (1994) Paroxetine in the elderly depressed patient: randomized comparison with fluoxetine of efficacy, cognitive and behavioural effects. Int Clin Psychopharmacology 9:25–29

Gerson SC, Plotkin DA, Jarvik LF (1988) Antidepressant drug studies, 1964 to 1986: empirical evidence for aging patients. J Clin Psychopharmacol 8:311–322

Grimsley SR, Jann MW (1992) Paroxetine, sertraline and fluoxetine: new selective serotonin reuptake inhibitors. Clin Pharm 11:930–957

Guelfi JD, White C, Magni G (1992) A randomized double-blind comparison of venlafaxine and placebo in inpatients with major depression and melancholia (abstract). Clin Neuropharmacol 15(suppl 1, pt B):323B

Hamilton M (1960) A rating scale for depression. J Neurol Neurosurg Psychiatry 23:56–62

Hirschfeld RMA (1994) Guidelines for the long-term treatment of depression. J Clin Psychiatry 55(suppl 12):61–69

Horne RL, Ferguson JM, Pope HG, Hudson JI, Lineberry CG, Ascher J, Cato A (1988) Treatment of bulimia with bupropion: a multicenter controlled trial. J Clin Psychiatry 49:262–266

Hutchinson DR, Tong S, Moon CAL, Vince M, Clarke A (1991) A double-blind study in general practice to compare the efficacy and tolerability of paroxetine and amitriptyline in depressed elderly patients. Br J Clin Res 2:43–57

Joffe RT (1992) Triiodothyronine potentiation of fluoxetine in depressed patients. Can J Psychiatry 37:48–50

Johnston JA, Lineberry CG, Ascher JA, Davidson J, Khayrallah MA, Feighner JP, Stark P (1991) A 102-center prospective study of seizure in association with bupropion. J Clin Psychiatry 52:450–456

Karlsson I; Godderis J; Augusto De Mendonca Lima C; Nygaard H; Simanyi M; Taal M; Eglin M 2000 A randomised, double-blind comparison of the efficacy and safety of citalopram compared to mianserin in elderly, depressed patients with or without mild to moderate dementia. Int J Geriatr Psychiatry 15:295–305

Katona C; Bercoff E; Chiu E; Tack P; Versiani M; Woelk H (1999) Reboxetine versus imipramine in the treatment of elderly patients with depressive disorders: a double-blind randomised trial. J Affect Disord 55:203–213
Katz IR, Stoff D, Muhly C, Bari M (1988) Identifying persistent adverse effects of anticholinergic drugs in the elderly. J Geriatr Psychiatry Neurol 1:212–217
Katz IR, Simpson GM, Curlik SM, Parmelee PA, Muhly C (1990) Pharmacologic treatment of major depression for elderly patients in residential care settings. J Clin Psychiatry. 1990;51(Suppl:41–7);discussion 48.
Katz IR, Streim J, Parmalee P (1994) Prevention of depression, recurrences, and complications in late life. Preventive Medicine 23:743–750
Kellett JM (1991) Fluvoxamine: an antidepressant for the elderly? J Psychiatr Neurosci 16(suppl 1):26–29
Kennedy GJ (1995) The geriatric syndrome of late-life depression. Psychiatr Serv 46:43–48
Kirsky DF, Stern WC (1984) Multicenter private practice evaluation of the safety and efficacy of bupropion in depressed geriatric outpatients. Cur Ther Res 35:200–210
Koenig HG (1991) Treatment considerations for the depressed geriatric medical patient. Drugs & Aging 1:266–278
Lafferman J, Solomon K, Ruskin P (1988) Lithium augmentation for treatment-resistant depression in the elderly. J Geriatr Psychiatry Neurol 1:49–52
Lazarus LW, Groves L, Gierl B, Pandey G, Javaid JI, Lesser J, Ha YS, Davis J (1986) Efficacy of phenelzine in geriatric depression. Biol Psychiatry 21:699–701
Lazarus LW, Winemiller D, Blake L, Hartman C, Abbassian M, Kartan U, Langsley P, Ripeckyj A, Markvart V, Fawcett J (1991) Efficacy and side effects of nortriptyline versus desipramine in geriatric inpatients with major depression. New Research Program and Abstracts, American Psychiatric Association, 144th Annual Meeting, 1991
Leinonen E, Lepola U, Koponen H, Kinnunen I (1996) The effect of age and concomitant treatment with other psychoactive drugs on serum concentrations of citalopram measured with a nonenantioselective method. Ther Drug Monit 18:111–117
Magni G (1993) Venlafaxine: tolerance and safety. J Clin Psychiatry 3:124–126
Maletta G, Mattox KM, Dysken M (1991) Guidelines for prescribing psychoactive drugs in the elderly: part 1. Geriatrics 46:40–47
McCue RE (1992) Using tricyclic antidepressants in the elderly. Clinics in Geriatric Medicine 8:323–334
McCue RE, Aronowitz J (1994) Accelerated antidepressant response in geriatric inpatients. Am J Geriatr Psychiatry 2:244–246
Mendels J (1993) Clinical management of the depressed geriatric patient: current therapeutic options. Am J Medicine 94(suppl 5A):13S–18S
Moskowitz H, Burns MM (1986) Cognitive performance in geriatric subjects after acute treatment with antidepressants. Neuropsychobiology 15(suppl 1): 38–43
Mullin JM, Pandita-Gunawardena VR, Whitehead AM (1988) A double-blind comparison of fluvoxamine and dothiepin in the treatment of major affective disorder. Br J Clin Pract 42:51–55
Nair NP, Ahmed SK, Kin NM, West TE (1995) Reversible and selective inhibitors of monoamine oxidase A in the treatment of depressed elderly patients. Acta Psychiatr Scand 91(suppl 386):28–35
Nemeroff CB (1994) Evolutionary trends in the pharmacotherapeutic management of depression. J Clin Psychiatry 55(suppl 12):3–15
Nemeroff CB, DeVane CL, Pollock BG (1996) Newer antidepressants and the cytochrome P450 system Am J Psychiatry 153:311–320
Nierenberg AA, Feighner JP, Rudolph R, Cole JO, Sullivan J ((1994) Venlafaxine for treatment-resistant unipolar depression. J Clin Psychopharmacol 14:419–423
NIH Consensus Development Panel on Depression in Late Life (1992) Diagnosis and treatment of depression in late life. JAMA 268:1018–1024

Norman TR, Judd FK, Burrows GD (1992) New pharmacological approaches to the management of depression from theory to clinical practice. Aust N Z J Psychiatry 26:73–81

Nyth AL, Gottfries CG, Lyby K, Smedegaard-Andersen L, Gylding-Sabroe J, Kristensen M, Refsum HE, Ofsti E, Eriksson S, Syversen S (1992) A controlled multicenter clinical study of citalopram and placebo in elderly depressed patients with and without concomitant dementia. Acta Psychiatr Scand 86:138–145

Old Age Depression Interest Group (1993) How long should the elderly take antidepressants? A double-blind placebo-controlled study of continuation/prophylaxis therapy with dothiepin. Br J Psychiatry 162:175–182

Palinkas LA, Barrett CE (1992) Estrogen use and depressive symptoms in postmenopausal women. Obstet Gynecol 80:30–36

Perrel JM (1994) Geropharmacokinetics of therapeutics, toxic effects, and compliance. In: Schneider LS, Reynolds CF, Lebowitz BD, et al. (eds) Diagnosis and treatment of depression in late life: results of the NIH Consensus Development Conference. American Psychiatric Press, Washington, DC, pp 245-257

Preskorn SH (1993) Recent pharmacologic advances in antidepressant therapy for the elderly. Am J Medicine 94(suppl 5A):2S–12S

Preskorn SH (1994) Antidepressant drug selection: criteria and options. J Clin Psychiatry 55(suppl 9A):6–22

Preskorn S, Burke M (1992) Somatic therapy for major depressive disorder: selection of an antidepressant. J Clin Psychiatry 53(suppl 9):S1–S14

Preskorn SH, Jerkovich GS (1990) Central nervous system toxicity of tricyclic antidepressants: phenomenology, course, risk factors, and role of therapeutic drug monitoring. J Clin Psychopharmacol 10:88-95

Preskorn SH, Lane RM (1995) Sertraline 50 mg daily: the optimal dose in the treatment of depression. Int Clin Psychopharmacol 10:129–141

Raffaitin F (1993) Efficacy and acceptability of tianeptine in the elderly: a review of clinical trials. Eur Psychiatry 8 (suppl 2):117S–124S

Rahman MK, Akhtar MJ, Savla NC, Sharma RR, Kellett JM, Ashford JJ (1991) A double-blind, randomised comparison of fluvoxamine with dothiepin in the treatment of depression in elderly patients. Br J Clin Pract 45:255–258

Raskind MA (1993) Geriatric psychopharmacology: management of late-life depression and the noncognitive behavioral disturbances of Alzheimer's disease. Psychiatr Clin North Am 16:815–827

Reynolds CF (1992) Treatment of depression in special populations. J Clin Psychiatry 53(suppl 9)45–53

Reynolds CF (1994) Treatment of depression in late life. Am J Medicine 97(suppl 6A):39S–46S

Reynolds CF, Frank E, Perel JM, Miller MD, Cornes C, Rifai AH, Pollock BG, Mazumdar S, George CJ, Houck PR, Kupfer DJ (1994) Treatment of consecutive episodes of major depression in the elderly. Am J Psychiatry 151:1687–1690

Robinson DS (1979) Age-related factors affecting antidepressant drug metabolism and clinical response. In: Nandy K (ed) Geriatric psychopharmacology. Elsevier, New York, pp 17–30

Robinson DS, Davis JM, Nies A, Ravaris CL, Sylwester D (1971) Relationship of sex and aging to monoamine oxidase activity of human brain, plasma, and platelets. Arch Gen Psychiatry 24:536–539

Roose SP, Glassman AH (1989) Cardiovascular effects of tricyclic antidepressants in depressed patients with and without heart disease. J Clin Psychiatry Monographs 7:1–18

Roose SP, Glassman AH, Attia E, Woodring S (1994) Comparative efficacy of selective serotonin reuptake inhibitors and tricyclics in the treatment of melancholia. Am J Psychiatry 151:1735–1739

Rudorfer MV, Potter WZ (1989) Antidepressants: a comparative review of the clinical pharmacology and therapeutic use of the "newer" versus the "older" drugs. Drugs 37:713–738

Salzman C (1992) Monoamine oxidase inhibitors and atypical antidepressants. Clinics in Geriatric Medicine 8:335–348

Salzman C (1995) Medication compliance in the elderly. J Clin Psychiatry 56(suppl 1): 18–22

Satel SL, Nelson JC (1989) Stimulants in the treatment of depression: a critical overview. J Clin Psychiatry 50:240–249

Schneider LS (1993) Efficacy of treatment for geropsychiatric patients with severe mental illness. Psychopharmacol Bull 29:501–524

Schneider LS, Small GW, Hamilton SH, Bystritsky A, Nemeroff CB, Meyers BS. Estrogen replacement and response to fluoxetine in a multicenter geriatric depression trial. Fluoxetine Collaborative Study Group (1997) Am J Geriatr Psychiatry 5:97–106

Schneider LS; Small GW; Clary CM (1998) Estrogen replacement therapy status and antidepressant response to sertraline. American Psychiatric Association Annual Meeting, Toronto, May 30-June 4

Schone W, Ludwig M (1993) A double-blind study of paroxetine compared with fluoxetine in geriatric patients with major depression. J Clin Psychopharmacol 13(6 Suppl 2):34S-39S

Shopsin B, Cassano GB, Conti L (1981) An overview of new "second generation" antidepressant compounds: research and treatment implications. In: Enna SJ, et al. (eds) Antidepressants: neurochemical, behavioral and clinical perspectives. Raven, New York, pp 219–245

Simpson GM, De Leon J (1989) Tyramine and new monoamine oxidase inhibitor drugs. Br J Psychiatry 155(suppl 6):32–37

Skoog I, Nilsson L, Landahl S, Steen B (1993) Mental disorders and the use of psychotropic drugs in an 85-year-old urban population. Int Psychogeriatrics 5:33–48

Stein EM, Stein S, Linn MW (1985) Geriatric sweet tooth. J Am Geriatr Soc 33:687–692

Strouse TB, Salehmoghaddam S, Spar JE (1993) Acaute delirium and parkinsonism in a bupropion-treated liver transplant recipient: case report. J Clin Psychiatry 54(12): 489–490

Sunderland T, Cohen RM, Molchan S, Lawlor BA, Mellow AM, Newhouse PA, Tariot PN, Mueller EA, Murphy DL (1994) High-dose selegiline in treatment-resistant older depressive patients. Arch Gen Psychiatry 51:607–615

Taylor MP, Reynolds CF 3rd, Frank E, Cornes C, Miller MD, Stack JA, Begley AE, Mazumdar S, Dew MA, Kupfer DJ (1999) Which elderly depressed patients remain well on maintenance interpersonal psychotherapy alone? report from the Pittsburgh study of maintenance therapies in late-life depression. Depress Anxiety 10:55–60

Tourigny-Rivard M-F (1997) Pharmacotherapy of affective disorders in old age. Can J Psychiatry 42(Suppl 1): 10S-18S

Tsolaki M, Fountoulakis KN, Nakopoulou E, Kazis A (2000) The effect of antidepressant pharmacotherapy with venlafaxine in geriatric depression. Int J Geriatr Psychopharmacol 2:83–85

Turner N, Scarpace PJ, Lowenthal DT (1992) Geriatric pharmacology: basic and clinical considerations. Ann Rev Pharmacol Toxicol 32:271–302

van Harten J (1993) Clinical pharmacokinetics of selective serotonin reuptake inhibitors. Clin Pharmacokinet 24:203–220

van Marwijk HW, Bekker FM, Nolen WA, Jansen PA, van Nieuwkerk JF, Hop WC (1990) Lithium augmentation in geriatric depression. J Affect Disord 20:217–223

Van-Laar MW, Van-Willigenburg APP, Volkerts ER (1995) Acute and subchronic effects of nefazodone and imipramine on highway driving, cognitive functions, and daytime sleepiness in healthy adult and elderly subjects. J Clin Psychopharmacol 15:30–40

Warneke L (1990) Psychostimulants in psychiatry. Can J Psychiatry 35:3–10

Weilburg JB, Rosenbaum JF, Meltzer-Brody S, Shushtari BS (1991) Tricyclic augmentation of fluoxetine. Ann Clin Psychiatry 3:209–213

Weintraub D (2001) Nortriptyline in geriatric depression resistant to serotonin reuptake inhibitors: case series. J Geriatr Psychiatry Neurol 14:28–32

Wernicke JF, Dunlop SR, Domseif BE, Bosomworth JC, Humbert M (1988) Low-dose fluoxetine therapy for depression. Psychopharmacol Bull 24:183–188

Woods SW, Tesar GE, Murray GB, Cassem NH (1986) Psychostimulant treatment of depressive disorders secondary to medical illness. J Clin Psychiatry 47:12–15

Zimmer B, Rosen J, Thornton JE, Perel JM, Reynolds CF (1991) Adjuntive lithium carbonate in nortriptyline-resistant elderly depressed patients. J Clin Psychopharmacol 11:252–256

Bipolar Mood Disorders

R. D. Alarcon[1]

[1] Department of Psychiatry and Psychology, Teaching Unit, Mayo Medical School, Rochester, MN, USA
e-mail: alarcon.renato@mayo.edu

1	Introduction	422
2	Clinical Features of Bipolar Disorders	424
3	Biological Bases of Bipolar Disorders	426
4	Pharmacological Management of Depression in Bipolar Disorder	429
4.1	General Principles	429
4.2	Management of the Acute Depressive Phase	431
4.2.1	SSRIs	431
4.2.2	Lithium	432
4.2.3	Monoamine Oxidase Inhibitors	433
4.2.4	Tricyclic Antidepressants	433
4.2.5	Other Antidepressants	434
4.2.6	Anticonvulsants	434
4.3	Maintenance Treatment	435
4.3.1	Lithium	436
4.3.2	Antidepressants	437
4.3.3	Anticonvulsants	437
4.3.4	Other Agents	438
4.4	Management of Clinical Variants	438
4.4.1	Rapid Cycling	439
4.4.2	Late-Onset/Geriatric Bipolar Disorder	440
5	Conclusion	441
	References	441

Abstract The emotional and socioeconomic impact of bipolar disorders may be comparable to or greater than that of unipolar depression. This chapter examines the clinical features of bipolar depression, reviews the biological bases of the disorder, and discusses its pharmacological management with emphasis on clinical varieties, degrees of severity, phases of treatment, and the use of specific antidepressant agents and mood stabilizers. Bipolar depression is characterized by prominent melancholic features, and carries the risk of quickly evolving into hypomanic or manic symptoms as a result of pharmacological interventions or internal neurophysiological and neurohormonal mechanisms. Its management may also be complicated by rapid cycling or mixed bipolar presentations. The treatment of bipolar depression therefore requires some variations in the conventional use of antidepressants. The choice of initial treatment may be either lithium (whose antidepressant effects appear to be more evident in this patient population) or one of the selective serotonin reuptake inhibitors (SSRIs). At the same time, monoamine oxidase inhibitors (MAOIs) seem to be more effective than the traditional tricyclic antidepressants (TCAs), and some of the second-generation antidepressants such as bupropion are promising. The newer non-SSRI antidepressants have not yet been extensively tested in bipolar depression. The antidepressant effects of anticonvulsants appear to be less marked in bipolar depression than in unipolar or other kinds of depression, but, used in combination with lithium, they reduce the likelihood of relapse or recurrence. Finally, there appears to be some justification for the use of adjuvant agents, such as thyroid replacement, stimulants, or calcium channel blockers. Antipsychotics to treat psychotic symptoms and electroconvulsive therapy (ECT) in the case of severe, refractory depression are also indicated. Comprehensive treatment should generally involve combining pharmacological management with appropriate psychotherapeutic and psychosocial approaches.

Most patients with bipolar disorder require long-term maintenance treatment due to high chances of recurrence. Lithium is the medication of choice for the maintenance phase. There is a great deal of controversy concerning the value of antidepressants as maintenance or prophylactic agents. Close clinical surveillance is needed so that the antidepressant can be quickly discontinued if manic symptoms start to develop. Special concerns arise in the treatment of rapid cycling bipolar disorder, late-onset bipolar disorder, and bipolar disorders in the elderly.

Keywords Bipolar disorder · Bipolar depression · Antidepressants · Mood stabilizers · Lithium · Selective serotonin reuptake inhibitors · Monoamine oxidase inhibitors · Bupropion · Valproate · Lamotrigine

1
Introduction

The impact of depression—including its bipolar variant—on individual lives and on society's functioning makes it a significant public health problem. The Epidemiological Catchment Area (ECA) survey (Robins et al. 1991) found a lifetime prevalence of major depressive episodes of 7.2%–13.1% in both sexes. The most recent National Comorbidity Study (NCS) (Kessler et al. 1994) found a lifetime prevalence of 17.1% and a 1-year prevalence of 10.3% for all kinds of depression among individuals between the ages of 15 and 54. Between 40% and 70% of all suicide victims in the United States suffer from major depression (Brent et al. 1988). Comorbid problems such as alcoholism and drug abuse and socioenvironmental problems such as poverty, unemployment, violence, and ethnic or religious polarizations also contribute to depression. In spite of a greater level of public and professional awareness of depression, only one-third of all cases are detected (Klerman and Weissman 1989; AHCPR 1993). It is also well known that 40%–60% of patients with depressive symptoms are seen initially, and sometimes exclusively, by general practitioners, nonmedical professionals, or providers of alternative treatments (McFarland 1994).

The American economy loses about $44 billion per year due to undiagnosed and untreated depression. Of the 11 million Americans diagnosed with depression, 8 million belong to the country's workforce, so that depression produces an average reduction in income of 26%, with a corresponding loss of productivity (Broadhead et al. 1990; Hall and Wise 1995). Depressed persons make an average of 14.7 visits to professionals per year (Narrow et al. 1993). Of those treated in an inpatient setting, one-third go to general hospital psychiatric units. Depression causes physical and mental disability comparable to or even higher than that caused by severe chronic diseases such as diabetes mellitus, hypertension, rheumatoid arthritis, GI disorders, and orthopedic problems (Stokes 1993). The co-occurrence of depression with anxiety disorders, somatoform disorders, substance abuse, and a significant number of physical illnesses aggravates the overall impact of this disorder.

The prevalence of bipolar mood disorders is considerably lower than that of unipolar depression, but the emotional and socio-economic impact of bipolar disorders is perhaps comparable to if not higher than that of unipolar depression. The ECA survey found a lifetime prevalence of 0.8% for bipolar I disorders and 0.5% for bipolar II disorders, with higher figures (1.3 and 0.7, respectively) among individuals between the ages of 18 and 44 years. Other studies have shown a prevalence range of 0.5% to 4.6% (Tohen and Goodwin 1995). The ECA did not find differences in gender or ethnicity for bipolar disorders. The mean age of onset was 21 years. Although studies have focused more on the prevalence of manic episodes, the consensus is that bipolar depressive episodes are more frequent and pervasive (Weissman and Myers 1978; Regier et al. 1993), and that the depressive component of bipolar mood disorders appears to be clinically and epidemiologically more prominent than its manic counterpart. Bipolar patients with a purely depressive index episode have been found to recover more slowly than those with pure manic index episodes, but faster than patients with mixed or rapid cycling presentations (Keller 1988); their risk of suicide has been reported to be approximately 15%.

The genetic bases of bipolar disorders have been indirectly confirmed by several studies (Blehar et al. 1988), although the mode of transmission remains unclear. There are also pervasive problems concerning the diagnostic validity of this disorder in spite of recent advances (American Psychiatric Association 1994a).

This chapter first examines the clinical features of bipolar depression and then briefly reviews the biological bases of the disorder. This is followed by a discussion of the pharmacological management of bipolar depression, with emphasis on clinical varieties, degrees of severity, phases of treatment, and the use of specific antidepressant agents and mood stabilizers.

2
Clinical Features of Bipolar Disorders

Bipolar disorders take their name from the characteristic clinical course of their symptomatic picture, in which depressive and manic episodes occur in unpredictable sequence. The depressive manifestations include a sense of despair and suffering, anhedonia, and apprehensiveness and tension that are unrelieved by comfort or reassurance. In other cases, depressive complaints include mostly hostility, irritability, self-deprecation, self-blame, severe guilt, anxious self-doubting, concern with unavoidable catastrophes, and pain and other physical symptoms. To make a diagnosis of bipolar disorder, the criteria in the *Diagnostic and Statistical Manual of Mental Disorders* (DSM-IV, American Psychiatric Association 1994a) require that five or more of nine symptoms be present during a 2-week period. Significant weight loss, insomnia, psychomotor agitation, fatigue, feelings of worthlessness, diminished ability to think or concentrate, or recurrent thoughts of death may also be present. Contrary to the traditional view of depression, the biophysiological variations of the depressive picture may also include hyperorexia or increased appetite and hypersomnia.

Bipolar disorders are generally classified as bipolar I or bipolar II. Bipolar I disorder is characterized by full-blown affective episodes (at least one manic or mixed episode, and often one or more major depressive episodes). Bipolar II is characterized by one or more major depressive episodes and at least one hypomanic episode, but no full manic episodes. In order to make the diagnosis of "bipolar I disorder, most recent episode depressed," the patient must have previously had at least one manic episode or a mixed episode (combination of manic and depressive symptoms), and the mood symptoms should not be better accounted for by other Axis I disorders, bereavement, substance abuse, or a general medical condition. A mixed mood episode [also called dysphoric mania by some authors (McElroy et al. 1989)] includes symptoms that meet criteria for a major depressive episode and for a manic episode (except for duration) occurring nearly every day for at least 1 week, and the mood disturbance is sufficiently severe to cause impairment in occupational functioning, social activities, or interpersonal relationships. Mixed episodes should be distinguished from agitated depression in which there is negligible manic symptomatology (Bowden 1993).

Bipolar depression is strongly associated with melancholic features, such as loss of pleasure, lack of reactivity, worsening of symptoms in the mornings, early morning awakening, or excessive or inappropriate guilt, and atypical features, such as excessive mood reactivity, hypersomnia, leaden paralysis, or interpersonal rejection sensitivity. When evaluating a patient with bipolar I disorder, the clinician should also specify a number of other clinical parameters:

1. Level of severity (mild, moderate, or severe), the presence or absence of psychotic features, and the remission status (full or partial).

2. Whether the depressive episode is associated with catatonic features. These include marked psychomotor disturbance, extreme negativism, mutism, movement peculiarities, echolalia, or echopraxia, as well as features such as waxy flexibility, automatic obedience, stupor, stereotypes, mannerisms, mimicry, and risks of malnutrition, exhaustion, or self-inflicted injury.
3. Any temporal relationship between the depressive symptoms and the post-partum period.
4. Course specifiers: longitudinal course (with or without full interepisode recovery), seasonal patterns (e.g., occurring in fall or winter and remitting in the spring), and the presence of rapid cycling (at least four episodes of a mood disturbance in the previous 12 months) (Suppes et al. 2000).

The old notion of "endogeneity" (i.e., originated inside the human body, as opposed to "reactive," a response to external pathogenic factors) applies to the classical description of the depressive phase of manic–depressive illness. The depressions in bipolar affective disorder tend to last longer (median of about 6 months) than the manic episodes. However, they rarely last more than a year, except in elderly patients (World Health Organization 1992). Although the presence of external stressors is not essential for this diagnosis, there are reports that such events may be associated with the occurrence of clinical depression in bipolar disorder. Some authors contend that bipolar disorders are underdiagnosed overall (Ghaemi et al. 1999). It is also well known that remissions tend to be of shorter duration as the individual gets older, so that depressive episodes tend to last longer and occur more frequently after middle age. Furthermore, bipolar patients have significantly shorter cycle lengths when treated with antidepressants alone or in combination with mood stabilizers than with lithium alone (Wehr and Goodwin 1987). Millon (1969) considers active-ambivalent (negativistic), passive-ambivalent (conforming), passive-dependent (submissive), and active-dependent (gregarious) the most frequently seen premorbid personality types in patients with bipolar depression. There are no conclusive findings regarding concomitant DSM-IV Axis II categories. (Clayton 1986).

Leonhard's (1979) views about manic-depressive illness are worth mentioning. This German psychiatrist began to champion the distinction between unipolar and bipolar depression in the mid 1950s. He insisted that the depression in these two "separable" conditions presents with different clinical pictures. Unipolar forms return within a periodic course with the same symptomatology and without transitional stages. On the other hand, in bipolar cases, no clear syndromes can be described since there are many transitions and the picture may even be distorted during one phase. Leonhard consequently called the bipolar form of depression "polymorphic" and the unipolar form "pure." The multiplicity of symptoms coexisting for hours or even days "may not be valued as the expression of separate phases, but it does show the disease potential toward the other pole." In other words, both manic and depressive symptoms only rarely appear in a pure form in bipolar cases. Leonhard also associates some of the manic–depressive symptoms with what he calls "cycloid psychosis," which may include stuporous depression, mutism, "confusion psychosis," and akinesis. It goes without saying that when some depressive symptoms, particularly those of a cognitive nature, are very severe, the clinical picture may reach psychotic dimensions, with an overwhelming sense of guilt, delusions of self-worthlessness and imminent death, self-denial, a sense of doom, and a great variety of physical symptoms.

Leonhard's concept of pure melancholy entails both severity and psychotic-like features. In addition to significant depressed mood and psychomotor inhibition, the concept includes thought inhibition involving slowness of the cognitive process, coupled with indecisiveness, anancastic (obsessive-compulsive) tendencies, and feelings of personal inadequacy.

3
Biological Bases of Bipolar Disorders

The concept that there are biological bases for depression, particularly for bipolar disorder, has existed for centuries. Persistent observations of the familial incidence of the condition, comparative review of data from the pre- and post-pharmacological eras, and analysis of biographic, ethnographic, and even anecdotal accounts have only confirmed this view. The current consensus favors the notion of a significant genetic basis for the predisposition and/or transmission of the disorder, either by inherited genes or potential vulnerabilities expressed into biochemical and neurophysiological abnormalities (Blackwood et al. 1994). Pharmacological, neuroanatomical, and clinical studies underline the notion that, by localizing sites in the central nervous system where biochemical interactions take place and pharmacological agents can act, the diagnosis and treatment of bipolar disorders will change dramatically within the next several decades (Freeman et al. 1993). On the other hand, sleep and chronobiological studies have ascertained that some pathophysiological routes could explain not only the etiological factors but also the pathogenic process that leads to the development of symptoms or syndromes. Among the biochemical studies, animal models of depression are quite promising, although not, of course, completely free from objections. The study of biological rhythms in animals may reflect coupled circadian oscillators related to seasonal mechanisms (Schulz and Lund 1985). Needless to say, the implications for management are extremely relevant.

In their authoritative book on manic–depressive illness, Goodwin and Jamison (1990) subscribe to the notion of a genetic vulnerability "expressed as altered membrane proteins (or lipids) localized in a widely distributed system that subserves the selective integration of cognitive, emotional and motoric functions, particularly in response to stress" (p 593). They also maintain that known neurotransmitter systems (the prime example in current research being serotonin) may present transport abnormalities through membrane uptake mechanisms that also affect second-messenger systems in the postsynaptic region, thus amplifying either inhibitory or stimulatory signals. From a physiological perspective, oscillations of normal compensatory processes may result in "sudden phasic shifts" that increase vulnerability (i.e., a general instability of multiple biological systems that carries both therapeutic and preventive implications).

Clinically, twin studies offer the major support for the genetic transmission of bipolar disorder to date (Gershon 1990). Concordance relates to both transmission and severity of illness. Adoption studies appear to be slightly less conclusive, mostly due to methodological differences. Family studies confirm, however, the familial incidence of bipolar disorders and their excess occurrence in relatives of bipolar patients. Some authors suggest that bipolar II illness is genetically somewhere between bipolar I and unipolar illness. Early onset bipolar disorder appears to have stronger genetic implications than later onset illness, because relatives of patients with early onset bipolar disorder tend to have an increased risk of developing bipolar illness. The presence of a spectrum of other disorders and conditions, which include schizoaffective disorder, substance abuse, anorexia nervosa, and cyclothymia, among those relatives also reinforces this notion.

In spite of some setbacks, the technological progress that has been made in genetic studies holds the promise for solving the methodological difficulties and finally confirming the genetic nature of manic–depressive illness, particularly through linkage research. These linkages could then be applied to clinical risk prediction, genetic counseling, prevention, and management. Research on risk factors includes studies on platelet imipramine binding, lymphocyte β-adrenergic receptors, cholinergic functional measurements, and melatonin suppression by nighttime light exposure. The human genome project is expected to reinforce rather than reject the genetic theories of bipolar depression.

A variety of biochemical hypotheses have been presented, debated, and studied during the last three decades. The first hypothesis (Bunney and Davis 1965; Schildkraut 1965) proposed that depression is based on a functional deficiency of catecholamines (norepinephrine and dopamine). This proposal and its subsequent partial confirmation by the effects of recently discovered pharmacological agents led to the more precise characterization of neurotransmitter and drug receptors in the brain and various peripheral tissues (Jimerson 1984). Neurotransmitters bind themselves selectively to receptors or recognition sites that then allow them to progress through the postsynaptic neurons by becoming their ligands; concomitantly, a number of changes take place that activate a receptor-mediated system through the work of a variety of agonist or antagonist neurochemical compounds (Kilts 1994). Similarly, nonspecific binding has been studied, as have phase alterations in receptor sensitivity; all these processes form the basis for the occurrence of the "switch phenomenon" from mania to depression and vice versa (Bunney et al. 1972).

Another area of research deals with neuroendocrine systems and neuropeptides as their primary components. The study of hormones may offer a "window" into the neurotransmitter system and other systems in the brain. Particular attention has been paid to the thyroid and adrenal glands because of the presence of clinical depression in hypothyroidism and Cushing's disease. The axes of functional and hormonal connections between the hypothalamus and pituitary glands and the target glands (thyroid and adrenal) allow the study of sensitive measures of functions that vary with different mood states. Thus, the most sensitive measure of hypothalamic–pituitary–thyroid (HPT) function, the thyroid-stimulating hormone (TSH) response to intravenous thyrotropin-releasing hormone (TRH), is blunted in patients with major depression. Chronic hypersecretion of TRH from the median eminence of the hypothalamus results in downregulation of anterior pituitary TRH receptors. However, 15% of depressed patients exhibit an exaggerated TSH response to TRH. Patients with this type of response show detectable levels of antimicrosomal thyroid and/or antithyroglobulin antibodies; this symptomless autoimmune thyroiditis occurs at a greater than expected prevalence rate in depressed individuals. Finally, many depressed patients show adrenal axis hyperactivity. Elevated plasma cortisol concentrations in depressed patients are not proportional to increases in ACTH concentration. This can be explained by hypersecretion of corticotropin-releasing factor (CRF) during and/or immediately preceding a depressive episode, with secondary pituitary and adrenal gland hypertrophy. However, this increase is reversed when these individuals recover following electroconvulsive therapy (ECT) or treatment with antidepressants. Thus, elevated CSF CRF concentrations as well as changes in the HPT may very well be associated with changes in putative neurotransmitters; all are state-dependent findings of mood disorders (Nemeroff 1991).

A number of characteristics of bipolar affective illness, including its recurring pattern, the rapid cycling phenomenon, and late onset depressions, are crucial in understanding the concepts of conditioning, sensitization, and kindling as further proof of the biological variations or changes that take place in affective disorders (Post et al. 1984). These mechanisms may underlie biochemical, phenomenological, and clinical changes in the course of the affective illness. Measures of electrical and pharmacological kindling reflect a progressive increase in neuroexcitability that ultimately results in seizures or other clinical phenomena. The proponents of this theory suggest a model of behavioral sensitization to repeated applications of psychomotor stimulants. It is hypothesized that kindling can also be triggered by pharmacological agents and that hormonal changes may form part of a broader learning process in this regard. The hypothesis goes so far as to suggest that depressive ideation could serve as a trigger for the onset of affective episodes, this bidirectional interplay between biological and psychological fields emphasizing the importance of environmental context and conditioning (i.e., psychosocial) variables. Sensitization of specific central nervous system sites to psychosocial or biological events may give way to a gradual presentation of affective symptoms in a relatively short period of time; once these phenomena exhaust themselves as well as their underlying biological factors, the individual may either show clear, specific clinical symptoms or begin moving towards recovery (Goldberg and Harrow 1994).

One of the newest lines of research involve susceptibility studies such as that by Zill et al. (2000), which found polymorphism of the G protein beta 3 subunit

(C825T). Neuroimaging studies also offer evidence of the biological basis of bipolar disorder (Kilts 1994). Physiological stimulation and changes in the "left brain" under conditions such as excessive euphoria or significant depression have been detected through photon emission tomography (PET), single photon emission computed tomography (SPECT), and brain mapping techniques (George et al. 1993). As tools which behavioral scientists and mental health clinicians can use to measure progress and response to treatment, these resources hold great promise.

4
Pharmacological Management of Depression in Bipolar Disorder

4.1
General Principles

The American Psychiatric Association (1994b) has issued a set of practice guidelines for the treatment of patients with bipolar disorder. In addition to performing a careful and thorough psychiatric and general medical evaluation, clinicians should assess for and treat any substance use disorder that may be present, since this is a predictor of noncompliance (Weiss et al. 1998). Comprehensive psychiatric management includes choosing a mood stabilizing medication. When a depressive episode occurs, the use of antidepressant medications is recommended after an assessment of benefits and burdens. The choice of antidepressants "is governed by the prior response of the patient to antidepressants, the side effect profile, the presence of atypical features, the risk of inducing a manic episode, and patient preference" (American Psychiatric Association 1994b, p 28). Antipsychotics to treat psychotic symptoms and ECT in the case of severe, refractory depression are also indicated.

One of the key concepts in the overall management of bipolar affective disorder is what Goodwin and Jamison (1990) call "the pharmacological bridge." This concept is based on the idea that the contributions of the neurosciences to our understanding of the etiopathogenic processes and symptoms of bipolar depression help provide the rationale for the management of the different variations, clinical presentations, and other nuances of this complex condition. It is important to remember that many clinical psychopharmacological findings were either serendipitous or, more important, were the result of clinical research in actual patients *before* they could be formulated as laboratory experiments. For example, the well-known observation that antidepressants take several weeks to produce clinical effects led to the hypothesis that factors other than the availability of neurotransmitters were involved in response. This empirical assumption led to the conceptualization and discovery of receptors, second messengers, and other related mechanisms. Even the formulation of the catecholamine hypothesis of depression was preceded by the clinical observation that hypertensive patients taking reserpine experienced severe bouts of depression without any previous history of depression. Modern clinical research focuses on integrative studies of the action of a given drug on neurotransmitter systems and their attendant circuits. On the other hand, areas such as the distinctions between animal and human studies, phase of drug administration, genetic vulnerability, cycle length, and recurrence patterns still cause conflict and uncertainty.

Treatment of the depressive phase of a bipolar mood disorder entails a careful clinical assessment, personal and family history, history of previous treatments, and assessment of suicide risk. The clinician needs to very carefully delineate the diagnosis and its variations, the degree of severity including course specifiers, syndromal forms (e.g., mixed bipolar disorder, rapid cycling), and the presence of clinical complications such as psychotic symptoms, comorbid substance abuse, or co-occurring physical conditions (Frances et al. 1998; Compton and Nermeroff 2000; Sachs et al. 2000).

The following sections cover (1) management of acute bipolar depression, (2) maintenance treatment and relapse prevention, and (3) management of clinical variants such as rapid cycling and late-onset/geriatric bipolar illness. The following classes of antidepressants are discussed: monoamine oxidase inhibitors (MAOIs), first- and second-generation tricyclic antidepressants (TCAs), and selective serotonin reuptake inhibitors (SSRIs) and other newer antidepressants. The use of lithium and the newer anticonvulsants is also reviewed.

After making a careful diagnostic assessment, clinicians should follow these general management principles in deciding to use a given antidepressant or combination of medications:

- Ascertain the target symptoms within the depressive constellation on the basis of prominence, severity, pervasiveness, and sensitivity to pharmacotherapeutic agents, and then initiate treatment as early as possible (Post 1993).
- Choose pharmacological agent(s) based on the patient's previous history, clinical course, history of side effects, and concurrent physical symptoms or conditions (American Psychiatric Association 1994b) as well as age group and ethnicity (Biederman et al. 2000; Davanzo and McCracken 2000; Soares 2000).
- Educate the patient concerning the kind of medication being administered, with appropriate explanations of its structure, possible mechanisms of action, clinical effects, side effects, time of onset, length of action, and doses (Cole and Bodkin 1990; Danjou et al. 1994).
- Alert both patient and family to the need for continuous monitoring of symptoms and side effects. If possible, educate them in the use of simple symptom tables or clinical assessment instruments to be used on a routine basis for comparative and follow-up purposes. Patient and family should also be educated concerning the key concepts of relapse, recurrence, and recovery (Kupfer 1991). Clinicians should keep in mind the increasing use of antidepressants by primary care practitioners.
- If there is a lack of response or poor response to the initial treatment, consider alternative pharmacological approaches and discuss them fully with the patient and, if necessary, with your treatment team or colleagues

(Solomon and Bauer 1993). Be sure to take into account any factors that may complicate management, such as compliance problems, comorbid substance abuse, metabolic variations, intercurrent psychosocial stressors, cultural considerations, and physical or medical illnesses (Paykel 1995).
- Comprehensive treatment should generally involve combining treatment with antidepressants or other pharmacological management with appropriate psychotherapeutic and psychosocial approaches. The use of antidepressants should not preclude using an interdisciplinary approach to this condition (American Psychiatric Association 1994b).

4.2
Management of the Acute Depressive Phase

Most clinicians concur in the need to initiate pharmacotherapy as soon as possible (Post and Weiss 1996). The agent as well as the dose will depend on the clinician's experience, the level of severity detected, and the current stage of the natural history of the illness. Breakthrough depression can be particularly resistant to treatment in bipolar patients (Hartman 1996). Although some authors suggest starting with lithium in moderate cases of bipolar I or II depression, citing improvement within 7–10 days (Goodwin and Jamison 1990), the general tendency among practitioners is first to use the newer antidepressants, particularly the SSRIs.

4.2.1
SSRIs

This category of drugs has heralded a new era in the management of depression reflected in the use of steady doses without broad variations, a general good response, mildness of side effects, safety in cases with suicidal risk (Feighner and Boyer 1991), and better cost-effectiveness than the traditional antidepressants (McFarland 1994). Currently in the United States, there are four SSRIs approved for the treatment of major depression (fluoxetine, paroxetine, sertraline, and citalopram), plus a fifth (fluvoxamine), approved only for the treatment of obsessive compulsive disorder, but which is primarily an antidepressant and is used extensively as such in Europe. Refer to the dosing table in the Appendix (Table 1) for U.S. package insert dosing recommendations for the management of acute depression with these compounds. It is important to realize that paroxetine and sertraline have elimination half-lives of 24 h and no metabolites. Therefore, they seem to reach steady state serum levels within a week. Fluoxetine has a half-life of 8–12 days and reaches a steady state in 3–4 weeks.

The main side effects of the SSRIs as a group include nausea, diarrhea, tremor, somnolence, dry mouth, and reduction of libido. Additionally, headaches and minor gastrointestinal difficulties have been reported. All SSRIs inhibit the cytochrome isoenzyme B4 52 D6, with paroxetine the most potent and sertraline the least potent. As this enzyme system inhibits other drugs (including neuroleptics, β-blockers, anti-arrhythmics, opiates, antihypertensives, benzodiazepines, and anticonvulsants), the addition of SSRIs may create problems with potential drug toxicity, particularly in sensitive and vulnerable patients. Due to its dopamine-reducing effect, fluoxetine may cause akathisia as a significant side effect; due to an increase in the granular storage of serotonin in platelets, it can increase bleeding times in some patients (Nemeroff 1994).

Generally speaking, SSRI doses are maintained within a relatively narrow range. In most cases, just one dose a day may suffice; nevertheless, after 1–2 weeks, the dose may be adjusted as clinically indicated.

4.2.2
Lithium

It has been suggested that lithium may be more effective in the management of depression in bipolar patients. Most clinicians, however, agree that it could be either a second choice after the SSRIs or a coadjuvant/augmenter in the case of less than optimal response to those medications (Keck and McElroy 1993). Interestingly enough, there are a few studies comparing lithium with more conventional antidepressants in the management of bipolar depression (Janicak and Davis 1992; Soares and Gershon 2000). This agent may be preferred in the management of bipolar II, since the risk of causing pharmacologically induced mania is less evident. Nevertheless, the length of time before onset of action may be similar to or perhaps a little longer for lithium in comparison to conventional antidepressants.

In spite of the more than three decades since it was first introduced, the mechanism of action of lithium still largely remains a mystery. Its effect on circadian rhythms (Wehr et al. 1987) is postulated as the correction of a putative phase advance and/or internal desynchronization in bipolar patients; others (Price et al. 1990) consider that lithium's antidepressant effects are related to its augmenting serotonin function at different levels in the central nervous system. Similarly, lithium may reduce both pre- and postsynaptic dopamine transmission, facilitate the release of norepinephrine through possible effects on presynaptic autoreceptors, change cholinergic-mediated physiological events, and have an impact on γ-aminobutyric acid (GABA)ergic neurotransmission/neuropeptide functions.

The evidence for the antidepressant efficacy of lithium monotherapy is growing, with close to 70% of patients with moderately severe bipolar disorder responding to this compound. In children and adolescents, it may require more frequent dosing to maintain steady plasma levels, due to a more rapid elimination (Fetner and Geller 1992). General side effects include excessive thirst, polyuria, tremor, weight gain, drowsiness, tiredness, diarrhea, and memory problems. Interestingly, patients seem to pay more attention to memory problems and feel less bothered by polyuria or drowsiness (Lenox and Manji 1995).

A factor to be taken into account is the interaction of lithium with a variety of other compounds such as diuretics (some of which increase plasma lithium levels with subsequent risk of lithium toxicity), nonsteroidal antiinflammatory drugs, neuroleptics, and anti-arrhythmics.

4.2.3
Monoamine Oxidase Inhibitors

Himmelhoch et al. (1991) suggest that patients with bipolar depression may be more responsive to MAOIs such as phenelzine and tranylcypromine (the only two currently available on the American market) than to traditional TCAs. Kalin (1996–1997) considers them antidepressants of choice for atypical bipolar depression. In general, however, these agents are used as third-line antidepressants, mostly due to the need for dietary restrictions to prevent hypertensive encephalopathy. It is clear that this risk has been exaggerated in the literature enough to make practitioners overcautious about using MAOIs. Other side effects include dryness of mouth, sexual dysfunction, and physical tiredness. Refer to the dosing table in the Appendix (Table 1) for U.S. package insert dosing recommendations for the management of acute depression with these compounds. Dose increases should be ordered cautiously, and the conventional ceiling for each medication should not be violated, except under special circumstances and by experienced clinicians. Some authors consider that bipolar depression characterized by inertia, hypersomnia, and excessive appetite may be more responsive to MAOIs. In some unresponsive cases, the combination of lithium and an MAOI may be effective. Joffe and Bakish (1994) combined a SSRI with moclobemide, (a reversible MAOI that is not available in the U.S.) in 10 depressed patients, 8 of whom achieved marked to complete remission. Nevertheless, even with this compound, there is the potential for the serotonin syndrome, an occasionally lethal side effect, to occur.

4.2.4
Tricyclic Antidepressants

Conventional TCAs remain useful agents in the treatment of depression. Some "die-hard" practitioners still use them as initial, first-line antidepressants for the management of bipolar patients. Nevertheless, some evidence has accumulated that places them after the SSRIs and even the MAOIs. It is generally recommended that less sedating and more activating TCAs, such as desipramine, nortriptyline, and imipramine, should preferably be used. Refer to the dosing table in the Appendix (Table 1) for U.S. package insert dosing recommendations for the management of acute depression with these compounds. Information on general side effects is presented in the side effects tables in the Appendix (Tables 2 and 3). Sedation, anticholinergic symptoms, hypotension, loss of libido, and weight gain are important undesirable consequences of the use of these compounds.

Onset of action takes the traditional 3–4 weeks; in some severe cases, particularly when the patient needs to be hospitalized, the highest dose may be 3–6 times the initial dose. In patients with treatment-refractory illness, combinations of a TCA and an MAOI or of a TCA, an MAOI, and a mood stabilizer such as lithium have been recommended. Baldessarini (1977) suggests starting the TCA and the MAOI simultaneously or else adding the MAOI after the patient has been started on the TCA, but not adding a TCA to an MAOI. The doses of each medication may be smaller when given in combination.

4.2.5
Other Antidepressants

Dosing recommendations for the management of clinical depression with the so-called "second generation" antidepressants (agents introduced during the 1980s) as well as the most recently introduced antidepressants are presented in the dosing table in the Appendix (Table 1). In general, of the newer agents, bupropion has been the most frequently used in bipolar depression, with maprotiline a distant second. The most recently introduced antidepressants, venlafaxine, nefazodone, and mirtazapine, have not been extensively used in the depressive phase of bipolar mood disorder. Venlafaxine has a three times greater potency for serotonin than for norepinephrine reuptake inhibition. Nefazodone blocks serotonin-2 receptors, and inhibits norepinephrine reuptake. It also inhibits cytochrome P450 (CYP) A34 but not 2D6 (Feighner and Boyer 1991). Wilens et al. (1997) used nefazodone in four adolescents with bipolar depression. Two responded well and two had mild manic activation. Still, reports on the systematic use of these new agents are eagerly awaited. Extrapyramidal symptoms have developed shortly after the addition of nortriptyline to a combination of venlafaxine and valproic acid that had been administered for several months (Conforti et al. 1999).

4.2.6
Anticonvulsants

Until recently, anticonvulsants such as carbamazepine and valproic acid appeared to have only modest antidepressant effects (much less in the acute phase), in spite of their well-accepted role as antimanic agents and mood stabilizers. Their antimanic effects are also quite slow to appear and have found to be closely related to an increase in dose, starting at 200 mg and increasing every 2–4 days by 100 mg as clinically tolerated. Some patients may require doses up to or even exceeding 1,500 mg/day. However, Post et al. (1996) suggest that these drugs are effective alone or in combination with lithium in patients who are less responsive to lithium alone, including those with a greater numbers of prior episodes, rapid cycling, dysphoric mania, comorbid substance abuse or medical problems, and patients without a family history of bipolar illness in first-degree relatives. Swann et al. (1997) reported that depressive symptoms emerging during manic episodes respond better to valproic acid than to lithium. Therapeutic blood levels can be obtained to assist in dosage selection.

Three new anticonvulsants—lamotrigine, gabapentin, and topiramate—have shown early promise as mood-stabilizing agents and seem to be well tolerated by most patients. Clinical experience (Walden et al. 1998; Mitchell 1999) indicates that lamotrigine is more successful at controlling the depressive component of bipolar disorder, and gabapentin is more effective at controlling the manic episodes. Lamotrigine produced considerable improvement in a patient with refractory bipolar disorder after several days of co-administration with valproate (Walden et al. 1996). Daily doses for lamotrigine may range from as little as 25 mg up to 200 or even 300 mg. Doses for gabapentin may vary from 300 mg to as much as 1,800 mg or greater as clinically tolerated.

4.3
Maintenance Treatment

Although the acute treatment of depression in bipolar disorders is a rather dramatic and emergency-like situation, maintenance treatment is crucial for the long-term well-being of patients (American Psychiatric Association 1994b). Its purposes are to maintain a level of clinical stability, optimize the possibilities of personal and social functioning, prevent clinical relapses (including the risk of suicide), regulate doses and control possible side effects, facilitate a continuing educational process of patients and relatives on the nature and characteristics of the illness, substantiate the growing body of knowledge about research issues in the clinical arena, maintain a high level of cooperation with patients and advocacy organizations, and continue the study of the social and economic impact of the disease in the general population.

The basic reason to insist on the need for maintenance treatment is the chronic nature of bipolar disorder. The old axiom that, as the patient gets older, the cyclic episodes become more frequent and severe, has neither been confirmed nor refuted. The familial nature of this disorder makes it imperative to continue the treatment in order to avoid the undermining, deleterious effects of the illness on the family unit. For instance, there is evidence that the discontinuation of lithium can make the condition more severe in terms of frequency of attacks and a poorer response to new cycles of medication (Strober et al. 1990; Post et al. 1992).

There is consensus on the need for maintenance treatment after any episode, be it manic or depressive. It is accepted that, after a first episode, a regular and intensive treatment of about 6–8 months duration may suffice, followed by a period of further observation by the clinician and the patient for the next few years. Unfortunately, the chances of recurrence are high and, therefore, a permanent maintenance treatment becomes almost mandatory for most bipolar patients. The clinician should pay attention to other characteristics of the illness, namely the type of onset (sudden onset necessitates maintenance treatment more than insidious onset), severity (the more severe, the more obvious the need for maintenance), the sex of the patient (males seem to need more prophylactic treatment than females), and psychosocial considerations such as environmental stresses, degree of compliance, family support, intensity of patient's occupation, and personality characteristics.

4.3.1
Lithium

Lithium is the medication of choice for the maintenance phase of treatment. It has been demonstrated that it not only prevents new attacks but also has an antidepressant effect per se. Nevertheless, almost 30% of patients on lithium prophylaxis do not respond adequately (Keck and McElroy 1993).

Once the acute depressive episode has been controlled, lithium should either be initiated or continued at a sufficient dosage to maintain a plasma level between 0.5 and 1.0 mEq/l. The practice of administering lithium in a single daily dose has gained credibility in recent years. Blood levels should be monitored weekly for the first 6 weeks, then twice a month, then once a month, and finally twice a year, once full clinical stability has been reached (i.e., after about 8–12 months of treatment). Monitoring should be more frequent among patients who have a history of frequent relapses or rapid cycling. Monitoring of lithium levels in saliva or sweat has been tried but the results are not convincing enough to do it on a routine basis. Monitoring of lithium plasma levels also has an important psychological component as the patient realizes the seriousness of the maintenance phase and the need to keep track of any physiological variations.

Other laboratory tests that should be done at least once or twice a year while using lithium are thyroid function tests (T4, T3, TSH), urinalysis, and creatinine to monitor kidney function. The role of lithium in causing chronic glomerular toxicity and interstitial nephritis has been found to be less of a problem than was formerly believed (Gitlin 1993; Walker 1993). Nonetheless, even though these problems are quite infrequent, most clinicians feel kidney function should be closely monitored. Additionally, complete blood count (CBC), urine osmolality, creatinine clearance, 24-h urine volume, and electrocardiogram (particularly for people over 50) are recommended (Schou 1989). It should also be kept in mind that some medical illnesses cause significant degrees of dehydration and, like surgical procedures, strict dieting, strenuous exercise, hot weather, old age, and pregnancy, may increase lithium plasma levels.

Although infrequent, perhaps the most serious side effects of lithium treatment are hypothyroidism and diabetes insipidus. The former should be treated with supplemental thyroid, whereas nephrogenic diabetes insipidus requires either a reduction in dosage or the addition of a diuretic such as furosemide. Ultimately, lithium can be replaced by another prophylactic agent (see below). Weight gain, hair loss, and acne are other undesirable effects. Impaired short-term memory, poor muscular coordination, and fatigue (Jamison et al. 1979) are also noteworthy.

Signs of moderate lithium intoxication are drowsiness, disorientation, blurred vision, hand tremor, restlessness, muscle twitches, slurred speech, vomiting, and confusion. Severe intoxication is characterized by intensification of these symptoms evolving into ataxia, severe movement disorders, and generalized muscle spasms. In addition to discontinuation of lithium, management includes renal profile, saline infusion, monitoring of fluid and electrolyte balance, plasma levels every 12 h, and renal or peritoneal dialysis if the patient is comatose, severely dehydrated, the plasma lithium level is higher than 3 mol/l, or if there is a significant clinical deterioration (Johnson 1987).

4.3.2
Antidepressants

There is a great deal of controversy concerning the value of antidepressants as maintenance or prophylactic agents (Frank et al. 1990; Peselow et al. 1991). Initially, the concept of maintenance treatment of bipolar depression with antidepressants was rejected outright, and for decades antidepressants were used only to control acute depressive symptoms, with discontinuation after 6–8 months. In recent years, however, some authors maintain that the use of antidepressants together with lithium, for instance, can enhance the patient's chances to remain asymptomatic for long periods of time. In this sense, the SSRIs have more promise than the TCAs due to their relatively moderate side effects. It has also been suggested that bupropion may be useful in maintenance treatment (Haykal and Akiskal 1990). MAOIs are not highly recommended as maintenance treatment agents (West 1992). Many practitioners use a combination of lithium and an antidepressant in the maintenance phase for pragmatic reasons. It would be fair to say that the issue is not resolved. If a clinician decides to use this combination, close clinical surveillance is needed so that the antidepressant can be quickly discontinued if the patient starts showing manic symptoms. Another criterion could be the frequency and ratio of depressive versus manic attacks in a given patient: if there is a predominance of depressive episodes, the prophylactic use of antidepressants may be more justified. Finally, there appears to be a high risk of switching into mania with TCAs (Calabrese et al. 1999a), while rapid cycling may possibly be aggravated by SSRIs (Laporta et al. 1987; Hon and Preskorn 1989; Howland 1996). Japanese researchers have suggested that serotonin-induced intraplatelet Ca response may be a good predictor of response to antidepressants in bipolar but not in unipolar patients (Kusumi et al. 2000).

4.3.3
Anticonvulsants

Anticonvulsants such as carbamazepine and valproic acid have been introduced in recent years as effective prophylactic antimanic agents, substituting for lithium in cases of lithium toxicity, lack of response, or resistance to lithium effects after a period of time. However, these anticonvulsants are much more controversial and less effective than lithium as antidepressant agents.

Some authors suggest the use of a combination of lithium and carbamazepine, since carbamazepine has shown a somewhat better antidepressant effect than valproate. The initial dose of carbamazepine (100 mg) can be increased by 100 mg every 4–5 days until a blood level of 8–10 µg/ml is reached. Hepatic enzyme induction secondary to carbamazepine may require dose increases. It is important to monitor blood levels as well as CBC, particularly white blood cells and platelets, every 2–3 months, due to the rare but serious risk of aplastic anemia. Hyponatremia has been reported, particularly in elderly patients. A variety of clinically significant pharmacokinetic drug interactions can occur with carbamazepine (Spina et al. 1996). Phenytoin, phenobarbital, and primidone accelerate the elimination of carbamazepine, probably by stimulating CYP 3A4. Inhibition of carbamazepine metabolism and elevation of plasma levels to potentially toxic concentrations can be caused by a variety of analgesics, antibiotics, calcium-channel blockers, and other drugs. Some studies show more frequent relapses after 3–4 years of treatment with carbamazepine (Frankemburg et al. 1988).

Valproic acid appears to have clearer results with manic crises, but it is also said to have antidepressant effects. The initial dose of 300–400 mg/day can be increased until the patient reaches a blood level of 50–100 µg/ml. Side effects are mild to moderate but some of them, such as hair loss, may occur rather quickly during the treatment. Both lamotrigine and gabapentin may also have value as prophylactic agents. In fact, there have been impressive results with lamotrigine (in doses up to 200 mg/day) including findings in large maintenance follow-up studies: the anticonvulsant significantly delayed time to intervention for a depressive episode, suggesting that it has a distinct profile of efficacy, potentially complementary to other mood stabilizers (Dubovsky and Buzan 1997; Calabrese et al. 1999b).

4.3.4
Other Agents

Other agents used to either enhance the effects of antidepressant treatment or contribute to the prophylactic management of bipolar disorder include thyroid hormone (particularly L-triodothyronine, T3), antipsychotics (particularly when depressive symptoms reach psychotic levels or violent behavior occurs; these include the atypical antipsychotics clozapine, risperidone, and olanzapine), benzodiazepines such as clonazepam, and calcium-channel blockers such as verapamil (Dubovsky and Franks 1985; Horst 1990; Prange 1996)

4.4
Management of Clinical Variants

Rapid cycling, late-onset bipolar disorder, and bipolar disorders in the elderly offer unique challenges to the clinician in both diagnosis and management.

4.4.1
Rapid Cycling

Dunner and Fieve (1974) define rapid cyclers as those patients who present with at least four affective episodes (depressive, manic, or hypomanic) per year. DSM-IV (American Psychiatric Association 1994a) has officially sanctioned this clinical subtype whose frequency seems to be increasing. The length of an affective cycle can be as short as 48 h or as long as 12 weeks. The age of onset of rapid cycling generally varies from 30 to 42 years. Rapid cycling occurs more frequently in women, particularly in the postpartum or menopausal periods (Liebenluft 2000). Patients who develop signs and symptoms of lithium-induced hypothyroidism are apparently more prone to develop rapid cycling than other patients. Rapid cyclers may present with mixed affective crises during lithium therapy rather than attacks of mania and depression (Alarcon 1985; Kilzieh and Akiskal 1999). Authors agree on the observation that rapid cycling occurs more frequently in bipolar than in unipolar patients.

The actual clinical symptoms of the rapid cycling phases do not differ from symptoms that occur in regular or "slow" cyclers. There are "spontaneous" rapid cyclers, and those in whom episodes are clearly induced by a pharmacological or nonpharmacological agent. Examples of pharmacological agents that can induce rapid cycling are antidepressants of different kinds, L-dopa, dopamine receptor stimulants such as piribedil, serotonin receptor antagonists such as cyproheptadine, conjugated estrogens, and lithium. Among nonpharmacological inducers of rapid cycling, ECT, pregnancy, physical illnesses such as multiple sclerosis, and withdrawal from TCAs are most frequent. A familial association with rapid cycling has not been demonstrated so far.

No firm conclusions have been reached yet regarding the pathogenesis of rapid cycling. Biological factors such as neurotransmitter abnormalities, circadian desynchronization, cholinergic-monoaminergic interaction, hypothalamic–pituitary–thyroid axis dysfunction ("subclinical hypothyroidism"), neurohormonal connections, membrane phenomena, or neurophysiological vulnerability, as well as features such as age, sex, length of illness, and medications interact to provoke either a clear predisposition toward or an actual appearance of rapid cycling (Bauer and Whybrow 1993). Elevation in phenylethylamine excretion occurs in bipolar patients who show a rapid mood cycling on a daily basis (Semba et al. 1988).

There have been contradictory reports about the efficacy of lithium in the management of rapid cycling (Janicak and O'Connor 1991). Some authors advocate a prolonged lithium treatment in order to ameliorate the intensity of the crises, while others rule it out completely as ineffective or even deleterious to the clinical course of the condition. Dunner (1979) advocates the use of antipsychotic agents or symptomatic antidepressants if deemed necessary; although others suggest avoiding them if possible (American Psychiatric Association 1994b). By far, the most acceptable consensus favors the use of fairly high doses of thyroxine (up to 0.1 mg/day) or, alternatively, anticonvulsants such as carbamazepine (Bauer and Whybrow 1990), valproate (Calabrese et al. 1993), or lamotrigine (Calabrese et al. 1999b; Frye 2002). Other authors have used clorgyline, a selective MAO inhibitor, in the treatment of patients with refractory rapid cycling. Similarly, bupropion, verapamil, magnesium aspartate, ascorbic acid, and low vanadium diets have been tried in rapid cycling with some measure of success. (Kukopulos et al. 1983). Stoll et al. (1996) saw a marked reduction in all mood symptoms in four patients with rapid cycling bipolar disorder with choline augmentation of lithium therapy. Bipolar II patients, who are particularly susceptible to the development of rapid cycling, benefit from bupropion, MAOIs, and low-dose SSRIs, preferably used in conjunction with lithium or mood-stabilizing anticonvulsants (Akiskal 1994).

4.4.2
Late-Onset/Geriatric Bipolar Disorder

The occurrence of late-onset/geriatric bipolar disorder should strongly dictate a work-up for the detection of specific organic problems, such as cerebrovascular accidents, brain tumors, neurological illness, incipient dementia, or systemic conditions. Once such conditions are ruled out, the diagnostician must consider some clinical peculiarities such as greater irritability in the depressive constellation, a slightly more prominent presence of cognitive impairment, more pronounced sleep irregularities, and a higher frequency and longer duration of mood cycles. The risk of suicide is also greater among elderly patients with bipolar disorder and appears to be on the rise (Mitterauer et al. 1988; Roy-Byrne et al. 1988). The lack of consistent social support makes treatment more difficult and prognosis more somber for this type of depression.

As a general rule, the use of antidepressants in this population should be more restricted in doses, length of treatment, and use of combinations; clinicians may not see as great a degree of clinical response, and should also monitor carefully for the occurrence of side effects and the need for interventions to control them. Due to a slower metabolism and age-related deficiencies in cognition, alertness, and habits such as mealtimes and nutrition patterns, the initial and maintenance doses of medications should be about half of those for the younger adult population (see Table 1 in the Appendix). While the onset of action may take a little longer, side effects may appear earlier and lead to a shortening of the overall period of treatment. In some cases, side effects may be so significant that ECT should be considered. The use of antidepressants alone rather than in combination is strongly recommended. The addition of medications such as antihypertensives, diuretics, antiarrhythmics, analgesics, and sedatives should be closely monitored.

SSRIs are still the first choice among antidepressants. Lithium seems to have fewer antidepressant effects among elderly patients with bipolar disorder, but may still be valuable. Bupropion is useful due to its stimulating effects but the risk of seizures is a deterrent, even though the slow-release form of this medication has significantly reduced their incidence. MAOIs and TCAs pose several problems, such as dietary restrictions, oversedation, anticholinergic effects, constipation, tremors, and cognitive deficits. Stimulants such as methylphenidate and amphetamine-like compounds in very low doses have been used but their effect is time-limited. (Feighner et al. 1985; Elkin et al. 1989)

5
Conclusion

Although sharing in most of the main features of a major depressive episode, depression in a bipolar mood disorder has some clinical characteristics that are important to recognize. In addition to the more prominent presence of melancholic features, this depressive syndrome runs the risk of quickly evolving into hypomanic or manic symptoms as a result of pharmacological interventions or internal neurophysiological and neurohormonal mechanisms. Its management may be complicated by rapid cycling or mixed bipolar presentations. Late-onset bipolar disorder and the apparent increase in frequency and severity of affective episodes with old age may also present clinical complications. The clinician, therefore, must be aware that treatment of bipolar depression requires some variations in the conventional use of antidepressants (Reid 1992). For instance, the choice of initial treatment may be either lithium (whose antidepressant effects appear to be more evident in this patient population) or one of the SSRIs. At the same time, MAOIs seem to be more effective than the traditional TCAs, and some of the second-generation antidepressants such as bupropion are promising. The newer non-SSRI antidepressants have not yet been extensively tested in bipolar depression. The antidepressant effects of anticonvulsants appear to be less marked in bipolar depression than in unipolar or other kinds of depression, but, used in combination with lithium, they reduce the likelihood of relapse or recurrence (Solomon et al. 1997). Finally, there appears to be some justification for the use of adjuvant agents, such as thyroid replacement, stimulants, or calcium channel blockers.

The physical manifestations, including somatic symptoms, of depression appear to be more prominent in bipolar affective disorders. Late onset bipolarity requires a full work-up for evidence of macro-organicity. Even as more research provides us with better pharmacological agents for bipolar depression (Horst 1990; Andrews and Nemeroff 1994), the thorough clinician should not forget that these patients also need an adequate, rational, and humane psychotherapeutic treatment.

References

Agency for HealthCare Policy and Research (AHCPR) (1993) Depression in primary care, Vol. 1. Detection and diagnosis. U.S. Department of Health and Human Services, Rockville, MD

Akiskal HS (1994) Dysthymic and cyclothymic depressions: therapeutic considerations. J Clin Psychiatry (Suppl 4): 46-52

Alarcon RD (1985) Rapid cycling affective disorders: a clinical review. Compr Psychiatry 26:522-540
American Psychiatric Association (APA) (1994a) Diagnostic and statistical manual of mental disorders, 4th Edition (DSM-IV). American Psychiatric Association, Washington, DC
American Psychiatric Association (APA) (1994b) Practice guideline for the treatment of patients with bipolar disorder. Am J Psychiatry 151(Suppl):1-36.
Andrews JM, Nemeroff CB (1994) Contemporary management of depression. Am J Medicine 97(Suppl 6A):24S-32S
Baldessarini RJ (1977). Chemotherapy in psychiatry. Harvard University Press, Cambridge, MA
Bauer MS, Whybrow PC (1990) Rapid cycling bipolar affective disorder, II. Treatment of refractory rapid cycling with high dose levothyroxine: a preliminary study. Arch Gen Psychiatry 47:435-440
Bauer MS, Whybrow PC (1993) Validity of rapid cycling as a modifier for bipolar disorder in DSM-IV. Depression 1:11-19
Bierderman J, Mick E, Spencer TJ, Wilens TE, Faraone SV (2000) Therapeutic dilemmas in the pharmacotherapy of bipolar depression in the young. J Child Adolesc Psychopharmacol 10:185-192
Blackwood DHR, Sharp CW, Walker MT, Doody GA, Glaubus MF, Muir WJ (1994) Co-morbidity in a large family with bipolar disorder: implications for genetic studies. In: Langer SZ, Brunello N, Rancagni G, Mendlewicz J (eds) Critical issues in the treatment of affective disorders. Karger, Basel, pp 173-177
Blehar MC, Weissman MM, Gershon ES, Hirschfeld MA (1988) Family and genetic studies of affective disorders. Arch Gen Psychiatry 44:289-292
Bowden CL (1993) The clinical approach to the differential diagnosis of bipolar disorder. Psychiatric Annals 23:57-63
Brent DA, Kupfer BJ, Bromet EJ (1988) The assessment and treatment of patients at risk for suicide. Review of Psychiatry 7:353-385
Broadhead WE, Blazer DG, George LK, Tse CK (1990) Depression, disability days, and days lost from work in a prospective epidemiologic survey. JAMA 264:2524-2528
Bunney WE, Davis JM (1965) Norepinephrine in depressive reactions. Arch Gen Psychiatry 13:483-494
Bunney WE, Goodwin FK, Murphy DL (1972) The "switch process" in manic depressive illness: III: theoretical implications. Arch Gen Psychiatry 27:312-317
Calabrese JR, Woyshville MJ, Kimmel SE, Rapport DJ (1993) Mixed states and bipolar rapid cycling and their treatment with divalproaex sodium. Psychiatric Annals 23:70-78
Calabrese JR, Rapport DJ, Kimmel SE, Shelton MD (1999a). Controlled trials in bipolar I depression: focus on switch rates and efficacy. Euro Neuropsychopharmacol 9:S109-S112
Calabrese JR, Hirschfeld RMA, Wagner KD, Frye MA (1999b) A double-blind placebo-controlled study of lamotrigine monotherapy in outpatients with bipolar I depression. J Clin Psychiatry 60:79–88
Clayton PJ (1986) Bipolar illness. In: Winokur G, Clayton PG (eds) The medical basis of psychiatry. W. B. Saunders Company, Philadelphia, London, Toronto, Mexico City, Rio de Janiero, Sydney, Tokyo Hong Kong, pp 39-59
Cole JO, Bodkin JA (1990) Antidepressant drug side effects. J Clin Psychiatry 51 (Suppl 1):21-26
Compton MT, Nemeroff CB (2000) The treatment of bipolar depression. J Clin Psychiatry 6:57-67
Conforti D, Borgherini G, Fiorellini LA, Magni G (1999) Extrapyramidal symptoms associated with the adjunct of nortriptyline to a venlafaxine-valproic acid combination. Int Clin Psychopharmacol 14:197-198

Danjou P, Weiller E, Richardot P (1994) Onset of action of antidepressants: a literature survey. In, Langer SZ, Brunello N, Racagni G, Mendlewicz J (eds) Critical issues in the treatment of affective disorders. Basel, Karger, pp 136-153

Davanzo PA, McCracken JT (2000) Mood stabilizers in the treatment of juvenile bipolar disorder: advances and controversies. Child Adoles Psychiatr Clin North Am 9:159-182

Dubovsky SL, Franks RD (1985) Intracellular calcium ions in affective disorders: a review and hypothesis. Biol Psychiatry 18:781-797

Dubovsky SL, Buzan RD (1997) Novel alternatives and supplements to lithium and anticonvulsants for bipolar affective disorder. J Clin Psychiatry 58:223-242

Dunner DL (1979) Rapid cycling bipolar manic depressive illness. Psychiatr Clin North Am 2:461-467

Dunner DL, Fieve RR (1974) Clinical factors in lithium prophylaxis failure. Arch Gen Psychiatry 30:229-233

Elkin I, Shea MT, Watkins JT, Imber SD, Sotsky SM, Collins JF, Glass DR, Pilkonis PA, Leber WR, Docherty JP, et al. (1989) National Institute of Mental Health Treatment of Depression Collaborative Research Program. Arch Gen Psychiatry 46:971-982

Feighner JP, Herbstein J, Damlouji N (1985) Combined MAOI, TCA, and direct stimulant therapy of treatment-resistant depression. J Clin Psychiatry 46:206-209

Feighner JP, Boyer WF (eds) (1991) Selective serotonin reuptake inhibitors. John Wiley & Sons, Chichester, New York, Brisbane, Toronto, Singapore

Fetner HH, Geller B (1992) Lithium and tricyclic antidepressants. Psychiatr Clin North Am 15:223-242

Frances AF, Kahn DA, Carpenter D, Docherty JP, Donovan SL (1998). The expert consensus guidelines for treating depression in bipolar disorder. J Clin Psychiatry 59:73-79

Frank E, Kupfer DJ, Perel JM, Cornes C, Jarrett DB, Mallinger AG, Thase ME, McEachran AB, Grochocinski VJ (1990) Three year outcomes for maintenance therapies in recurrent depression. Arch Gen Psychiatry 47:1093-1099

Frankenburg FR, Tohen M, Cohen BM, Lipinski JF Jr (1988) Long-term response to carbamazepine: a retrospective study. J Clin Psychopharmacol 8:130-132

Freeman AM, Stankovic SMI, Bradley RJ, Zhang GZ, Libb JW, Nemeroff CB (1993) Tritated platelet imipramine binding and treatment response in depressed outpatients. Depression 1:20-23

Frye MA (2002) Addressing the underestimated need and treatment challenges of bipolar depression. Presented at the XXII Latin American Congress of Psychiatry. Guatemala City, July 8–12

George MS, Ketter TA, Post RM (1993) SPECT and PET imaging in mood disorders. J Clin Psychiatry 54(Suppl):6-13

Gershon ES (1990) Genetics. In: Goodwin FK, Jamison KR (eds.) Manic depressive illness. Oxford University Press, New York, Oxford, pp 373-401

Ghaemi SN, Sachs GS, Chiou AM, Pandurangi AK, Goodwin FK (1999) Is bipolar disorder still underdiagnosed? Are antidepressants overutilized? J Affect Disord 52:135-144

Gitlin MJ (1993) Lithium-induced renal insufficiency. J Clin Psychopharmacol 13:276-279

Goldberg JF, Harrow M (1994) Kindling in bipolar disorders: a longitudinal follow up study. Biol Psychiatry 35:70-72

Goodwin FK, Jamison KR (1990) Manic-depressive illness. Oxford University Press, New York, Oxford

Hall RCW, Wise MG (1995) The clinical and financial burden of mood disorders: Cost and outcome. Psychosomatics 36:S11-S18

Hartmann PM (1996) Strategies for managing depression complicated by bipolar disorder, suicidal ideation, or psychotic features. J Am Board Fam Pract 9:261-269

Haykal RF, Akiskal HS (1990) Bupropion as a promising approach to rapid cycling bipolar II patients. J Clin Psychiatry 51:450-455

Himmelhoch JM, Thase ME, Mallinger AG, Houck P (1991) Tranylcypromine vs. imipramine in anergic bipolar depression. Am J Psychiatry 148:910-916

Hon DE, Preskorn SH (1989) Mania during fluoxetine treatment for recurrent depression. Am J Psychiatry 146:1638-1639

Horst WD (1990) New horizons in the psychopharmacology of anxiety and affective disorders. Psychiatric Annals 20:634-639

Howland RH (1996) Induction of mania with serotonin reuptake inhibitors. J Clin Psychopharmacol 16:425-427

Jamison KR, Gerner RH, Goodwin FK (1979) Patient and physician attitudes toward lithium: relationship to compliance. Arch Gen Psychiatry 36:866-869

Janicak PG, Davis JM (1992) Advances in the treatment of bipolar disorder. Curr Opinion Psychiatry 5:51-55

Janicak PG, O'Connor E (1991) Prognosis and maintenance treatment in major affective disorders. Curr Opinion Psychiatry 4:60-64

Jimerson VC (1984) Neurotransmitter hypothesis of depression: research update. Psychiatr Clin North Am 7:563-574

Joffe RT, Bakish D (1994) Combined SSRI-moclobemide treatment of psychiatric illness. J Clin Psychiatry 55:24-25

Johnson FN (1987) Depression and mania: modern lithium therapy. IRL Press, Oxford, England

Kalin NH (1996–97) Management of the depressive component of bipolar disorder. Depress Anxiety 4:190-198

Keck PE, McElroy SL (1993) Current perspectives on treatment of bipolar disorder with lithium. Psychiatric Annals 23:64-69

Keller MB (1988) The course of manic-depressive illness. J Clin Psychiatry 49 (Suppl):4-6

Kessler RC, McGonagle KA, Zhao S, Nelson CB, Hughes M, Eshleman S, Wittchen HU, Kendler KS (1994) Lifetime and 12 month prevalence of DSM-IIIR psychiatric disorders in the United States: results from the National Comorbidity Survey. Arch Gen Psychiatry 51:8-19

Kilts CD (1994) Recent pharmacologic advances in antidepressant therapy. Am J Med 97(Suppl 6A): 3S-12S

Kilzieh N, Akiskal HS (1999) Rapid-cycling bipolar disorder: an overview of research and clinical experience. Psychiatr Clin North Am 22:585-607

Klerman GL, Weissman MM (1989) Increasing rates of depression. JAMA 261:2229-2235

Kukopulos A, Caliari B, Tundo A, Minnai G, Floris G, Reginaldi D, Tondo L (1983) Rapid cyclers, temperament, and antidepressants. Compr Psychiatry 24:249-258

Kupfer DJ (1991) Long-term treatment of depression. J Clin Psychiatry 52 (Suppl): 28-34

Kusumi I, Suzuki K, Sasaki Y, Kameda K, Koyama T (2000) Treatment response in depressed patients with enhanced Ca mobilization stimulated by serotonin. Neuropsychopharmacology 23:690-696

Laporta M, Chouinard G, Goldbloom D, Beauclair L (1987) Hypomania induced by sertraline, a new serotonin reuptake inhibitor. Am J Psychiatry 144:1513-1514

Lenox RH, Manji HK (1995). Lithium. In: Schatzberg AF, Nemeroff CB (eds) Textbook of psychopharmacology. American Psychiatric Press, Washington, DC, pp 303-349

Leonhard K (1979) The classification of endogenous psychoses, 5th edition. Irvington Publishers, New York, London, Sydney, Toronto

Liebenluft E (2000) Women and bipolar disorder: an update. Bull Menn Clinic 64:5-17

McElroy SL, Keck PE, Pope HG, Hudson JI (1989) Valproate in the treatment of rapid cycling bipolar disorder. J Clin Psychiatry (Suppl) 23–29

McFarland BH (1994) Cost effectiveness considerations for managed care systems: treating depression in primary care. Am J Med 97(Suppl 6A):47S-58S

Mitchell PB (1999) The place of anticonvulsants and other putative mood stabilizers in the treatment of bipolar disorder. Aust N Z J Psychiatry 33:S99-S107

Mitterauer B, Leibetseder M, Pritz WF, Sorgo G (1988) Comparisons of psychopathological phenomena of 422 manic-depressive patients with suicide-positive and suicide-negative family history. Acta Psychiatr Scand 77:438-442

Narrow WE, Regier DA, Rae DS, Manderscheid RW, Locke BZ (1993) Use of services: findings from the National Institute of Mental Health Epidemiologic Catchment Area Program. Arch Gen Psychiatry 50:95-107

Nemeroff CB (1994) Evolutionary trends in the pharmacotherapeutic management of depression. J Clin Psychiatry 55(Suppl):3-15

Nemeroff CB (1991) Corticotropin-releasing factor. In: Nemeroff CB (ed) Neuropeptides and psychiatric disorders. American Psychiatric Press, Washington, DC, pp 75-92

Paykel ES (1995) Psychotherapy, medication combinations, and compliance. J Clin Psychiatry 56(Suppl 1):24-30

Peselow ED, Dunner DL, Fieve RR, Difiglia C (1991) The prophylactic efficacy of tricyclic antidepressants: a five year followup. Prog Neuropsychopharmacol Biol Psychiatry 15:71-82

Post RM, Rubinow DR, Ballenger JC (1984) Conditioning, sensitization, and kindling: implications for the course of affective illness. In: Post RM, Ballenger JC (eds) Neurobiology of mood disorders. Williams & Wilkins, Baltimore, London, pp 432-466

Post RM, Leverich GS, Altshuler L, Mikalauskas K (1992) Lithium discontinuation-induced refractoriness: preliminary observations. Am J Psychiatry 149:1727-1729

Post RM (1993) Issues in the long-term management of bipolar affective illness. Psychiatric Annals 23:86-93

Post RM, Ketter TA, Denicoff K, Pazzaglia PJ, Leverich GS, Marangell LB, Callahan AM, George MS, Frye MA (1996) The place of anticonvulsant therapy in bipolar illness. Psychopharmacology (Berl) 128:115-129

Post RM, Weiss SR (1996) A speculative model of affective illness cyclicity based on patterns of drug tolerance observed in amygdala-kindled seizures. Mol Neurobiol 13:33-60

Prange AJ (1996) Novel uses of thyroid hormones in patients with affective disorders. Thyroid 6:537-543

Price LH, Charney DS, Delgado PL, Heninger GR (1990) Lithium and serotonin function: implications for the serotonin hypothesis of depression. Psychopharmacology (Berl) 100:3-10

Regier DA, Narrow WE, Rae DS, Manderscheid RW, Locke BZ, Goodwin FK (1993) The de facto US mental and addictive disorders service system. Epidemiologic Catchment Area prospective 1-year prevalence rates of disorders and services. Arch Gen Psychiatry 50:85-94

Reid IC (1992) Treatment strategies in affective disorders. Curr Opinion Psychiatry 5:45-50

Robins LN, Locke BZ, Regier DA (1991) An overview of psychiatric disorders in America. In: Robins LN, Regier DA (eds) Psychiatric disorders in America: The Epidemiological Catchment Area Study. Free Press, New York, pp 328-366

Roy-Byrne PP, Post RM, Hambrick DD, Leverich GS, Rosoff AS (1988) Suicide and course of illness in major affective disorder. J Affect Disord 15:1-8

Sachs GS, Printz DJ, Kahn DA, Carpenter D, Docherty JP. The expert consensus guideline series: medication treatment of bipolar disorder 2000. Postgrad Med Special Report April 2000:1-104

Schildkraut J (1965) The catecholamine hypothesis of affective disorders: a review of supporting evidence. Am J Psychiatry 122:509-522

Schou M (1989) Lithium treatment of manic depressive illness, fourth edition. Karger, Basel, Switzerland

Schulz H, Lund R (1985) On the origin of early REM episodes in the sleep of depressed patients: a comparison of three hypothesis. Psychiatry Res 16:65-77

Semba JI, Nanki M, Maruyama Y, Kaneno S, Watanabe A, Takahashi R (1988) Increase in urinary beta phenylethylamine preceding the switch from mania to depression: a "rapid cycler". J Nerv Ment Dis 176:116-119

Soares JC (2000) Recent advances in the treatment of bipolar mania, depression, mixed states, and rapid cycling. Int Clin Psychopharmacol 15:183-196

Soares JC, Gershon S (2000) The psychopharmacologic specificity of the lithium ion: origins and trajectory. J Clin Psychiatry 61:16-22

Solomon DA, Bauer MS (1993) Continuation and maintenance pharmacotherapy for unipolar and bipolar mood disorders. Psychiatr Clin North Am 16:515-540

Solomon DA, Ryan CE, Keitner GI, Miller IW, Shea MT, Kazim A, Keller MB (1997) A pilot study of lithium carbonate plus divalproex sodium for the continuation and maintenance treatment of patients with bipolar I disorder. J Clin Psychiatry 58:95-99

Spina E, Pisani F, Perucca F (1996) Clinically significant pharmacokinetic drug interactions with carbamazepine: an update. Clin Pharmacokinet 31:198-214

Stoll AL, Sachs GS, Cohen BM, Lafer B, Christensen JD, Renshaw PF (1996) Choline in the treatment of rapid-cycling bipolar disorder: clinical and neurochemical findings in lithium-treated patients. Biol Psychiatry 40:382-388

Stokes PE (1993) A primary care perspective on management of acute and long-term depression. J Clin Psychiatry 54(Suppl.):74-84

Strober M, Morrell W, Lampert C, Burroughs J (1990) Relapse following discontinuation of lithium maintenance therapy in adolescents with bipolar I illness: a naturalistic study. Am J Psychiatry 147:457-461

Suppes T, Dennehy EB, Wells Gibbons E (2000) The longitudinal course of bipolar disorder. J Clin Psychiatry 61:23-30

Swann AC, Bowden CL, Morris D, Calabrese JR, Petty F, Small J, Dilsaver SC, Davis JM (1997) Depression during mania: treatment response to lithium or divalproex. Arch Gen Psychiatry 54:37-42

Tohen M, Goodwin FK (1995) Epidemiology of bipolar disorder. In: Tsuang MT, Tohen M, Zahner GEP (eds.) Textbook in psychiatric epidemiology. Wiley-Liss, New York, pp 301-315

Walden J, Hesslinger B, van Calker D, Berger M (1996) Addition of lamotrigine to valproate may enhance efficacy in the treatment of bipolar affective disorder. Pharmacopsychiatry 29:193-195

Walden J, Normann C, Langosch J, Berger M, Grunze H (1998) Differential treatment of bipolar disorder with old and new antiepileptic drugs. Neuropsychobiology 38:181-184

Walker RG (1993) Lithium nephrotoxicity. Kidney International 44(Suppl 42):S93-S98

Wehr TA, Goodwin FK (1987) Can antidepressants cause mania and worsen the course of affective illness? Am J Psychiatry 144:1403-1411

Wehr TA, Sack VA, Rosenthal NE, Goodwin FK (1987) Sleep and biological rhythms in bipolar illness. In: Hales RE, Francis AJ (eds) Psychiatry update, vol 6. American Psychiatric Association, Washington, DC, pp 61-80

Weiss RD, Greenfield SF, Najavits LM, Soto J, Wyner D, Tohen M, Griffin ML (1998) Medication compliance among patients with bipolar disorder and substance use disorder. J Clin Psychiatry 59:172-174

Weissman MM, Myers JK (1978) Affective disorders in a U.S. urban community: the use of Research Diagnostic Criteria in an epidemiological survey. Arch Gen Psychiatry 35:1304–1311

West R (1992) Depression. Office of Health Economics, London

Wilens TE, Spencer TJ, Biederman J, Schleifer D (1997) Case study: nefazodone for juvenile mood disorders. J Am Acad Child Adolesc Psychiatry 36:481-485

World Health Organization (1992) The International Classification of Diseases (ICD-10): classification of mental and behavioral disorders. Clinical descriptions and diagnostic guidelines. WHO, Geneva, 1992, pp 119-128

Zill P, Baghai TC, Zwanzger P, Schuele C, Minov C, Riedel M, Neumeier K, Rupprecht R, Bondy B (2000) Evidence for an association between a G-protein beta 3-gene variant with depression and response to antidepressant treatment. Neuroreport 11:1893-1897

Treatment-Refractory Depression

M. Trivedi[1] · T. Bettinger[2] · B. Kleiber[3]

[1] University of Texas Southwestern Medical Center at Dallas,
  5959 Harry Hines Blvd., St. Paul POB 1, Suite 600, Dallas, TX 75235-9101, USA
  e-mail: madhukar.trivedi@utsouthwestern.edu
[2] Depression & Anxiety Disorders Clinic,
  University of Texas Southwestern Medical Center at Dallas,
  5959 Harry Hines Blvd, Dallas, TX 75235-9101, USA
[3] University of Texas Southwestern Medical Center at Dallas,
  5959 Harry Hines Blvd., St. Paul POB 1, Suite 520, Dallas, TX 75235-9101, USA

1	Introduction	449
2	Factors Contributing to Treatment Refractory Depression	450
3	Optimizing Treatment	451
3.1	Maximizing Dose	452
3.2	Critical Decision Points in Treatment Planning	452
3.3	Length of Trial	454
3.4	Long-Term Management of Depression	454
3.5	Measuring Improvement: Defining Full Response, Partial Response, and Minimal Response	455
4	Selecting the Next Step	455
4.1	Switching Medication	456
4.1.1	SSRI to SSRI	456
4.1.2	SSRI to or from TCA	458
4.1.3	SSRI to or from Venlafaxine	458
4.1.4	SSRI to or from Nefazodone	459
4.1.5	SSRI to or from Bupropion	459
4.1.6	SSRI to or from Mirtazapine	459
4.1.7	Venlafaxine to or from Nefazodone or Mirtazapine	460
4.1.8	Bupropion to or from Nefazodone, Venlafaxine, or Mirtazapine	460
4.1.9	Switching to or from an MAOI	460
4.2	Augmentation Strategies	461
4.2.1	Lithium	461
4.2.2	Thyroid Hormone	463
4.2.3	Buspirone	464
4.2.4	Pindolol	465
4.2.5	Atypical Antipsychotics	466
4.2.6	Psychostimulants	467
4.2.7	Estrogen	467
4.2.8	Miscellaneous Augmenting Agents	468
4.3	Combination Treatment	469
4.3.1	SSRI/SSRI Combinations	470

4.3.2 SSRI/TCA Combinations 470
4.3.3 Bupropion Combinations 471
4.3.4 Mirtazapine Combinations 471
4.3.5 Nefazodone Combinations 472
4.3.6 Venlafaxine Combinations 472
4.3.7 MAOI Combinations. 473

5   **Alternative Treatments** 473
5.1   Cognitive–Behavioral Therapy 473
5.2   Electroconvulsive Therapy 474
5.3   Vagus Nerve Stimulation 476
5.4   Repetitive Transcranial Magnetic Stimulation. 478

6   **Texas Medication Algorithm Project.** 479

7   **Sequenced Treatment Alternatives to Relieve Depression** 480

8   **Future Directions** 480

      References 481

Abstract Despite the availability of many newer antidepressant medications, a significant percentage of patients are partial responders or have treatment-resistant depression (TRD). In such cases, there are very few data to indicate what step (combination, augmentation, or switching) should be taken next. The first step in optimizing treatment is to provide an antidepressant trial of adequate duration at a dosage that ensures therapeutic plasma levels. Some patients, including previous nonresponders, may benefit from higher doses. Although a medication trial of 4–8 weeks is often adequate to see some treatment response, patients with TRD may require a longer trial (e.g., 12–16 weeks) to achieve a full response. A framework of time-driven prompts called "critical decision points" can be helpful in determining the appropriate "next step" in treatment of TRD. At each decision point, the clinician assesses a patient's overall improvement and side effect burden and may choose to continue the prescribed dose, increase or decrease the dose, augment treatment with another agent, switch or combine medications. Switching medications is advantageous when there is a need to keep treatment simple or patient compliance is an issue. When a patient partially responds to a given treatment, the clinician should consider augmentation before switching. Potential advantages of augmentation include attaining full response without starting a new medication trial, using lower doses of both agents to minimize side effects, treatment of comorbid disorders such as subclinical hypothyroidism, and quicker treatment response. Combination therapy allows the clinician to treat depression with antidepressant medications that have differing mechanisms of action, thus affecting different neurotransmitter systems. The chapter provides a thorough description of medication use in each of these treatment strategies, including appropriate dosing, switching guidelines, augmentation strategies, and possible drug interactions. Clinical evidence in the literature (or the lack thereof) that support these therapies are reviewed. Alternative treatments to medication are discussed, including psychotherapy (e.g., cognitive behavioral therapy), electroconvulsive therapy (ECT), vagus nerve stimulation, and repetitive transcranial magnetic stimulation (rTMS).

Keywords Treatment-refractory depression · Treatment-resistant depression · Antidepressants · Partial response · Augmentation · Combination treatment

1
Introduction

Depression occurs in up to 1 in 8 individuals during their lifetime, making it one of the most prevalent of all medical illnesses. The prevalence rates of major depressive disorder (MDD) are approximately 2%–3% in men and 5%–9% in women, with a lifetime risk for developing an episode of 5%–12% and 10%–25%, respectively (American Psychiatric Association 2000). Despite the availability of many newer antidepressant medications, approximately 50% of patients will not respond to initial treatment, requiring trials of more than one agent to attain remission of symptoms (Fava and Davidson 1996). As many as 20% of patients with MDD are resistant to treatment and another 30% may achieve only a partial response (Keller et al. 1984).

The implications of unremitted, and possibly prolonged, depression are significant. Over 50% of completed suicides involve an episode of depression (Janicak and Martis 1998). The annual cost of depression in the United States is approximately $44 billion (1990 dollars). Of that figure, 72% reflects indirect costs related to functional impairment, such as unemployment, underemployment, and total disability (Greenberg et al. 1993). Continued illness may have a negative impact on social supports. Repeated treatment failures may also influence patients' willingness to continue treatment.

There is no clear consensus among researchers on the definition of treatment-resistant depression. In the literature, the patient populations described as "treatment-resistant" vary from study to study. At times, the term refers to patients who have failed to achieve remission in a single short trial of medication, whereas other studies define treatment-resistant patients as those who have failed to respond to multiple lengthy trials of different classes of antidepressants.

Often, the terms treatment-refractory and treatment-resistant are used interchangeably. In general, patients with treatment-refractory depression (TRD) are viewed as those who have failed to respond to adequate trials of two antidepressants of different classes (Thase et al. 1995; Ananth 1998; Janicak and Martis 1998; Nelson 1998). Patients are considered treatment-resistant if multiple treatments have failed. During an initial assessment, it is difficult to determine treatment status without a thorough medication history that includes information on tolerability or whether medication doses were maximized by previous clinicians. However, it is inappropriate to ascribe treatment resistance until multiple treatment options have been pursued. Often patients are not totally resistant to treatment, but require a more complicated course of treatment with a longer duration. It is therefore important to strive for an adequate sequence of treatment trials prior to diagnosing a patient as having treatment-resistant depression. Although the prevalence of treating depression in primary care clinics has increased over the last decade, studies have not shown clear evidence of improvements in quality or continuity of care, and treatment of depression is often not in accordance with current evidence-based research findings (Simon 2002). In fact, it has been estimated that as many as 40% of patients in treatment do not receive at least a moderate dose of medication for an adequate period of time (Katon et al. 1992). Additionally, these patients often have inadequate follow-up (Ford 2000), particularly during the critical initial stages of treatment (Lin et al. 1995). The result is that desired outcomes (full symptomatic remission and return of premorbid levels of functioning) are often not achieved. Multiple algorithmic approaches are being evaluated to determine the best sequences to utilize for patients with difficult to treat MDD. The recently completed Texas Medication Algorithm Project (TMAP website, accessed 2002) and the currently ongoing Sequenced Treatment Alternatives to Relieve Depression (STAR*D website, accessed 2002) project will provide evidence for the most efficacious algorithms for TRD.

2
Factors Contributing to Treatment Refractory Depression

Despite recent advances in the development of pharmacological treatments for MDD, many patients receiving medication for depression do not experience substantial relief from their symptoms. Approximately 10%–20% of patients cannot tolerate the side effects of antidepressant treatment. Another 25%–35% of patients who complete an adequate trial of an antidepressant do not respond or do not show an acceptable response to treatment, usually defined as a 50% decline in symptom severity as measured by the Hamilton Depression Rating Scale (HAM-D) (Hamilton 1967). In addition, up to 50% of patients who show an "acceptable" response continue to have residual symptoms that interfere with work, family, and social activities (Rush and Trivedi 1995; Thase and Rush 1997; Nierenberg et al. 1999).

Although inadequate treatment (i.e., inadequate doses for inadequate periods of time) of depression is often blamed for poor patient outcomes, patient nonadherence with pharmacotherapy is a substantial contributing factor. Drug trials for the treatment of depression report a 10%–30% rate of nonadherence, and it is thought that nonadherence is much greater in a naturalistic setting (Depression Guideline Panel 1993). Naturalistic studies show that patients often decide independently to discontinue treatment sooner than practice guidelines recommend (Paykel and Priest 1992; American Psychiatric Association 1993, 2001; World Health Organization 1996). A study conducted by Lin et al. (1995) found that 28% of patients discontinued taking antidepressants in the first month of therapy, and that 44% discontinued antidepressants by the third month. Katon et al. (1992) reported that up to 60% of patients in primary care had discontinued their antidepressant medications prior to completion of the recommended 6 months of pharmacotherapy. These studies emphasize the importance of assessing medication adherence prior to classifying a patient as treatment refractory.

In addition to inadequate treatment and patient nonadherence, unsuccessful antidepressant treatment may be a result of misdiagnosis or the presence of an unrecognized comorbid psychiatric disorder or general medical condition (Kornstein and Schneider 2001). Comorbid psychiatric disorders are common in patients with MDD and may increase the likelihood of treatment resistance. If these disorders are missed or inadequately treated, the evaluation and treatment of depression may be complicated. For example, depressed patients with comorbid anxiety disorders tend to be more severely depressed and slower to respond to treatment. They are more likely to have residual symptoms and have increased rates of relapse and recurrence (Fawcett 1997; Kornstein and Schneider 2001). Up to 40% of patients with bipolar disorder are initially diagnosed with major depression (Ghaemi et al. 1999). The acute and chronic effects of substance abuse may worsen symptoms of depression and increase the likelihood of noncompliance (Thase and Rush 1997).

General medical conditions or their treatments may cause or worsen symptoms of depression. Diabetes, coronary artery disease, HIV infection, cancer, and chronic pain may contribute to TRD (Catz et al. 2002; Riedinger et al. 2002; Nichols and Brown 2003). Fibromyalgia, chronic fatigue syndrome, and irritable bowel syndrome also are associated with symptoms of depression (Dwight et al. 1998; Hickie 1999). Endocrine disorders, such as Cushing's disease, Addison's disease, and hypothyroidism in particular also may cause depression (Kornstein and Schneider 2001).

3
Optimizing Treatment

Given the prevalence of depression in our society, there remains a great concern that many patients remain untreated or undertreated. As safer medications have become available, there has been pressure by managed care organizations to treat depression in a primary care setting. However, many patients, both in the primary and specialty care setting, receive less than an adequate treatment trial (Nelson 1997).

Treatment should be addressed on several fronts, including dosage, length of trial, and overall treatment duration. In addition, response should be evaluated objectively at specific intervals. A methodical approach is a critical tool in the effective treatment of TRD. Thase and Howland (1994) suggest that each failed trial reinforces patient demoralization and pessimism, creating a vicious cycle in which pessimism alienates others and increases interpersonal sensitivity, resulting in significant others distancing themselves from the patient. The combination of persistent stressors, eroding social support, and increased demoralization may increase the potential for suicidality (Thase and Howland 1994).

3.1
Maximizing Dose

Undertreatment of MDD patients, particularly in the primary care setting, is a persistent concern of clinicians (Berman et al. 1997; Nelson 1997; Trivedi and Kleiber 2001). It is estimated that fewer than 50% of patients in treatment receive at least a moderate dose of medication (Keller et al. 1986). Patients either may be left at a lower dose due to intolerance of higher doses and/or in an effort to minimize side effects, or inadequately evaluated to optimize outcomes.

In order to ensure an adequate trial, it is essential to attain a therapeutic plasma level for the prescribed medication. Not all patients respond similarly to a medication at a given dose, and the recommended dosage may be inadequate. If possible, plasma levels should be drawn to determine if the patient is at a therapeutic level. The clinician can titrate medication aggressively, in an attempt to reach a therapeutic level more rapidly, as tolerated (Janicak and Martis 1998). Patients may benefit from higher doses, and nonresponders may also improve with an extension of the treatment trial. Among patients who have exhibited some improvement to antidepressant treatment but are still experiencing residual symptoms, titration to the maximum tolerated dose (as opposed to a minimally effective dose) may enable them to achieve a full response, or possibly attain the desired goal of full remission. This initial strategy is especially useful with agents that have a known dose–response relationship. As tolerated, patients should be brought to the maximum therapeutic dose of a medication, and maintained on that dose for an adequate period of time to determine the effectiveness of a given treatment. The Appendix to this volume (Table 1) presents a representative example of appropriate medication doses for various antidepressants.

3.2
Critical Decision Points in Treatment Planning

Due to the increased risk of failed trials in TRD, it is essential to take an organized approach to treatment. A framework of time-driven prompts called "critical decision points" (CDPs) (Trivedi and Kleiber 2001) can be helpful in determining the appropriate "next step" in the treatment of TRD (Fig. 1). This system is designed to assist clinicians in assessing symptom response by establishing a timetable for re-evaluation and suggesting treatment strategies or tactics.

When using such a system, CDPs are set at predetermined intervals of time, for example, at weeks 0, 4, 6, 8, 10, and 12. Week 0, or CDP 1, marks the initiation of a new medication trial. At each decision point, a clinician assesses a patient's overall improvement and side effect burden. Based on that assessment, treatment options are evaluated. The clinician may choose to continue the pre-

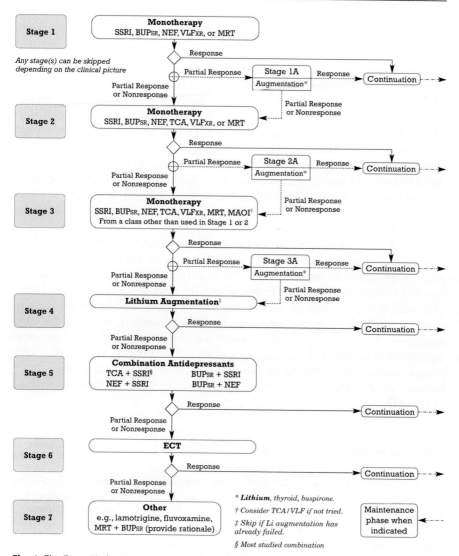

Fig. 1 The Texas Medication Algorithm Project (TMAP) algorithm for nonpsychotic major depressive disorder

scribed dose, increase or decrease the dose, augment treatment with another agent, or switch or combine medications. By evaluating patient outcome and proceeding in a stepwise manner, many of the concerns that are often raised about the management of TRD may be addressed.

3.3
Length of Trial

The average length of a medication trial is 4–8 weeks. For most patients, this is adequate time to see some treatment response. However, there is evidence that additional time may be needed for some patients to show significant improvement of symptoms. A study by Quitkin et al. (1984) challenged the then accepted trial length of 4 weeks. Their results showed that approximately 20% of patients exhibiting minimal response at 4 weeks did have a satisfactory response after 8 weeks of treatment. An analysis of the response profiles of nearly 600 patients treated with mianserin, monoamine oxidase inhibitors (MAOIs), or tricyclic antidepressants (TCAs) revealed that 44% of patients who were showing mild improvement after 5 weeks achieved symptom remission by the end of the 6th week (Quitkin et al. 1996).

A significant study of chronic depression showed that, while about 40% of patients had complete symptom remission following a 12-week trial of either imipramine or sertraline, an additional 47% of partial responders achieved full response to treatment during the extended continuation phase (Keller et al. 1998). This finding suggests that an average trial of 4–6 weeks may not be adequate to produce complete symptom remission in some patients. A retrospective data analysis of patients treated with nefazodone found that 63.6% of those who remitted did so after 6 weeks or more of treatment (Trivedi et al. 2001). Like patients with chronic depression (Koran et al. 2001; Trivedi and Kleiber 2001), patients classified as treatment refractory may need a medication trial of 12–16 weeks to achieve a full response.

3.4
Long-Term Management of Depression

There are three phases of treatment: acute, continuation, and maintenance. The acute phase is generally 12 weeks. Once a patient is progressing satisfactorily, a continuation phase follows during which the therapeutic medication dose is continued for 6–9 months. A major study conducted by Kocsis et al. (1996) found that those patients with chronic depression who were maintained on desipramine throughout the 2-year maintenance phase were significantly less likely to relapse than patients receiving placebo. These results suggest that recurrent or chronic depressions (which may often prove treatment refractory), may require maintenance treatment lasting 2 years or longer to deter relapse or recurrence of significant depressive symptoms. Severe or chronic depressions such as TRD are particularly vulnerable to relapse or recurrences. Premature medication discontinuation may exacerbate symptoms. Conversely, maintenance treatment may enhance remission potential.

3.5
Measuring Improvement: Defining Full Response, Partial Response, and Minimal Response

Another component of thorough treatment is the objective measurement of depressive symptoms. It is important to identify the patient's symptom profile and severity at baseline, and monitor physical and behavioral changes throughout the trial. In this way, the clinician may assess patient progress objectively and avoid pressure from the patient to change treatment prematurely (Janicak and Martis 1998).

After 4 weeks of treatment, patients fall into one of four categories of treatment response: full, partial, minimal, or nonresponse. Full response, or remission, is defined as greater than or equal to 75% reduction in depressive symptoms and a score less than 7 on the 17-item Hamilton Rating Scale for Depression (HRSD-17), partial response as 50%–70% improvement, minimal response as 25%–50% improvement, and nonresponse as less than 25% symptom reduction. Although most clinical trials define adequate treatment response as a greater than 50% reduction in symptoms, the goal of treatment should always be remission of symptoms and restoration of functioning. Many times, patients with MDD face the possibility of relapse (where some symptoms return) or recurrence of significant symptoms after a period of remission (Janicak and Martis 1998). It is suggested that complete symptom remission may help prevent relapse or recurrence for many patients (Rush and Trivedi 1995).

4
Selecting the Next Step

A number of considerations should be taken into account in determining the next step following a failed treatment trial. Switching to another antidepressant is the simplest strategy and may help promote patient adherence (Nelson 1998). This may be the best option if a patient has failed one monotherapy trial. If the patient's depression has partially responded to monotherapy, the clinician may consider augmentation before switching so as to not lose the progress made during the trial (Nelson 1998). The advantages of augmentation include possibly rapid response, no time lost in titration, and the possibility of boosting a partial response. An augmenting agent may have other benefits, such as anxiolytic effects, and can serve as a bridge to another medication (Nelson 1998).

The rationale for both augmentation and combination strategies is to increase the effects of one medication by adding a second medication. Therapeutic pharmacokinetic and pharmacodynamic interactions between the two agents explain, at least in theory, the "boost" in effect. Combination therapy allows the clinician to treat depression with antidepressant medications that have differing mechanisms of action, thus affecting different neurotransmitter systems (i.e., norepinephrine, serotonin, dopamine) thought to be altered in depressed patients. Data to support combination therapy are very limited and most data supporting it come from anecdotal reports, case series, and small open trials. Whether combination treatment is better than switching to a broader spectrum antidepressant, such as venlafaxine or clomipramine, remains to be seen (Thase et al. 1998).

4.1
Switching Medication

When a patient fails to respond to a given medication following a trial of adequate dose and duration, one treatment option is to switch to a different antide-

Table 1 Suggested guidelines for switching between antidepressants

From	To	Plan
SSRI	SSRI	Direct switch from SSRI No. 1 to SSRI No. 2
SSRI	TCA	Discontinue SSRI by taper and then initiate noradrenergic TCA; or direct switch from SSRI to serotonergic/noradrenergic TCA; or taper SSRI and initiate TCA gradually, as tolerated, to therapeutic dose range
SSRI	Nefazodone	Discontinue SSRI by taper and then initiate nefazodone
SSRI	Venlafaxine	Direct switch from SSRI to venlafaxine; or cross-taper venlafaxine with SSRI if patient experiences SSRI withdrawal symptoms
SSRI	Mirtazapine	Cross-taper from SSRI to mirtazapine
SSRI	Bupropion	Cross-taper from SSRI to bupropion
SSRI	MAOI	Discontinue SSRI. After a 5-week washout period for fluoxetine or a 2-week washout period for all others, MAOI therapy can safely be initiated
TCA Venlafaxine Nefazodone Bupropion Mirtazapine	TCA	Discontinue venlafaxine, nefazodone, mirtazapine or bupropion by taper and then initiate TCA; or cross-taper venlafaxine, nefazodone, mirtazapine, or bupropion with TCA
TCA Venlafaxine Nefazodone Bupropion Mirtazapine	SSRI	Discontinue TCA (or venlafaxine, nefazodone, bupropion, mirtazapine) by taper and then initiate SSRI; or taper TCA or bupropion while initiating SSRI at a low dose
Venlafaxine Nefazodone Bupropion Mirtazapine	Venlafaxine Nefazodone Bupropion Mirtazapine	Cross-taper venlafaxine, nefazodone, mirtazapine, or bupropion with nefazodone, venlafaxine, mirtazapine, or bupropion; or discontinue venlafaxine, nefazodone, mirtazapine, or bupropion by taper and initiate nefazodone, venlafaxine, mirtazapine, or bupropion gradually as tolerated to therapeutic dose range
TCA Nefazodone Venlafaxine Bupropion Mirtazapine	MAOI	Discontinue TCA (or nefazodone, venlafaxine, bupropion, mirtazapine). After a 2-week washout, MAOI therapy can be safely initiated
MAOI	MAOI Nefazodone Venlafaxine Bupropion Mirtazapine	Discontinue MAOI No. 1. After a 2-week washout, therapy with MAOI No. 2 (or TCA, venlafaxine, nefazodone, mirtazapine, or bupropion) can be safely initiated

pressant. The clinician may choose to switch within a class or move to another class of antidepressant. The selection may depend on the patient's and family members' response history, health concerns, concomitant medications, and acceptability of known side effect risks.

When the decision is made to switch antidepressants, it is important to consider the mechanism of action of each medication. New generation antidepressants are not as plagued with anticholinergic, antihistaminic, and α-adrenergic receptor blocking effects as the older TCAs, because of the selectivity in their mechanism of action (Ananth 1998). Switching may involve a brief wash-out before starting a new medication, directly switching from one medication to the other, or cross-tapering (gradually decreasing one medication while increasing another).

Some medications appear to be more effective with a specific subtype of depression (Nelson 1997). Venlafaxine or a TCA is indicated for patients with severe depression (Klenk and Rosenbaum 2000). MAOIs have been proven effective in atypical depression (Thase et al. 1995), while venlafaxine and SSRIs are thought to be effective in anxious depression (Davidson et al. 2002). Suggested guidelines for switching between antidepressants are presented in Table 1.

4.1.1
SSRI to SSRI

Some question the advisability of switching from one selective serotonin reuptake inhibitor (SSRI) to another, suggesting that switching to another class is more effective (Nelson 1998). However, there are subtle differences in the neuropharmacology of SSRIs that produce different effects (Janicak and Martis 1998; Nelson 1998). Switching from one SSRI to another has been shown to be efficacious, with response rates of 42%–71% (Brown and Harrison 1995; Joffe et al. 1996; Zarate et al. 1996; Thase and Rush 1997). It should be noted that the patients who were switched had similar side effect profiles with both SSRIs, suggesting that if side effects are problematic on one SSRI, there is little advantage in switching to another.

A direct switch from one SSRI to another will not produce discontinuation effects, with the exception being the switch from paroxetine to fluoxetine (Marangell 2001), because of the short half-life of paroxetine and the long half-life of fluoxetine. While there is no advantage in cross-tapering, a direct switch is better tolerated than washing out the first medication before starting a second. When switching from fluoxetine to any other SSRI, there is a potential for greater serotonin uptake inhibition due to the rapid achievement of steady state of the new SSRI and the slow clearance of fluoxetine. This issue is particularly true in the healthy elderly in whom it can take up to 4 months to clear the active metabolite, norfluoxetine, from the body (Harvey and Preskorn 2000). Both medications will be present for a period of time equal to 5 times the half-life of the first medication, but generally this does not create a problem. Higher levels of either medication will be present if one or both inhibit cytochrome enzymes (such as paroxetine or fluoxetine), and although it is not a safety issue, it may result in a tolerability issue (Marangell 2001).

4.1.2
SSRI to or from TCA

TCAs are clearly efficacious, particularly in severe depression. However, this must be weighed against an increased rate of side effects, possible adverse cardiac effects, and the potential for fatal overdose. It has been suggested that the clinician consider switching to one of the noradrenergic TCAs, such as desipramine or nortriptyline, if a patient does not respond to an SSRI (Nelson 1998). One open-label study conducted by Beasley et al. (1990) found 51%–62% of patients who switched to fluoxetine following a failed trial of a TCA (either intolerance or nonresponse) showed improvement. However, switching from one TCA to another has generally not proven to be advantageous (Nelson 1998).

When switching between an SSRI and a TCA, possible drug interactions should be considered. Abrupt discontinuation of a TCA may cause cholinergic rebound, so that cross-tapering is recommended (Marangell 2001). Fluvoxamine inhibits cytochrome P450 (CYP) 1A2 and CYP 3A3/4, so that concurrent use with imipramine or amitriptyline may lead to decreased clearance of the TCA. SSRIs, particularly paroxetine and fluoxetine, may cause increased TCA levels through CYP 2D6 inhibition, increasing the risk of side effects or toxicity. As a result, it is recommended that the dosage of the TCA be lowered prior to initiating an SSRI.

4.1.3
SSRI to or from Venlafaxine

Venlafaxine acts on both norepinephrine (NE) and serotonin (5-HT) receptors (Ananth 1998). At low doses, it acts similarly to the SSRIs. At high doses, venlafaxine produces both serotonergic and noradrenergic uptake inhibition (a different mechanism of action from that of the SSRIs). However, it is unclear at what exact dose venlafaxine has a dual mechanism of action. There is some evidence that venlafaxine may be more effective in TRD than the SSRIs (Thase et al. 2000). It has been suggested that increasing the dose of venlafaxine may produce a better response than switching to an SSRI (Nelson 1997).

A direct switch between venlafaxine and a SSRI is preferred. There is little benefit in cross-tapering, and washout may cause discomfort for the patient. Caution is warranted when a patient is switched from an SSRI to venlafaxine because CYP 2D6 inhibition by the SSRIs (particularly fluoxetine and paroxetine) can cause decreased metabolism of venlafaxine to O-desmethylvenlafaxine (ODV), increasing the risk for cardiovascular and serotonergic side effects (Marangell 2001).

4.1.4
SSRI to or from Nefazodone

Nefazodone may be a good medication choice because it is less likely than the SSRIs to cause sexual dysfunction and can be beneficial for sleep due to its selective blockage of the 5-HT_2 receptors. However, it may be difficult to switch from a SSRI to nefazodone because of the enzyme-inhibiting effects of fluoxetine and paroxetine on the metachlorophenylpiperazine (mCPP) metabolite of nefazodone, resulting in increased anxiety or restlessness (Nelson 1998).

There have also been reports of nonlethal serotonin syndrome when a SSRI and nefazodone are administered concurrently (Marangell 2001). Because of the differing mechanisms of action, a direct switch will not prevent withdrawal symptoms of the other medication, so it is suggested that the clinician taper the first drug and allow a brief washout period before starting the second medication (Marangell 2001).

4.1.5
SSRI to or from Bupropion

Bupropion blocks the reuptake of both norepinephrine and to a lesser degree, dopamine, but not serotonin. It is generally thought to have fewer adverse effects than the TCAs and theoretically may work particularly well in patients with anergic depression (Nelson 1998). When faced with SSRI-induced sexual dysfunction and partial response, bupropion may be an attractive alternative. However, if anxiety symptoms do not improve with bupropion, an SSRI may be considered (Marangell 2001).

Due to different receptor effects, a direct switch is not advised. Cross-tapering medications is suggested because of possible SSRI discontinuation symptoms (Marangell 2001). SSRI withdrawal symptoms (anxiety, irritability, and flu-like symptoms) are generally transient, but can be quite bothersome to the patient.

4.1.6
SSRI to or from Mirtazapine

Mirtazapine is similar to venlafaxine in that it is a combined-action serotonergic–noradrenergic agent. However, the mechanisms of action differ, with mirtazapine's principal action being that of an α_2-antagonist, increasing the release of serotonin and norepinephrine (Nelson 1998). Its 5-HT_3 antagonistic effects help block nausea, and it does not produce anorexia, sexual dysfunction, or insomnia. Common side effects include sedation and weight gain.

Two open-label trials have shown mirtazapine to be an efficacious option after SSRI treatment failure secondary to nonresponse or sexual side effects (Koutouvidis et al. 1999; Fava et al. 2001). Due to differing mechanisms of action and low potential for drug interactions, cross-tapering is recommended (Fava et al. 2001).

4.1.7
Venlafaxine to or from Nefazodone or Mirtazapine

Venlafaxine is not a significant inhibitor of CYP 450 enzymes. When switching, these medications should be cross-tapered due to differing mechanisms of action (Marangell 2001).

4.1.8
Bupropion to or from Nefazodone, Venlafaxine, or Mirtazapine

Due to differing mechanisms of action, cross-tapering is recommended (Marangell 2001). Bupropion is a substantial inhibitor of CYP 2D6 and has been reported to produce the same magnitude of inhibition as paroxetine and fluoxetine. While that is not an issue when switching to mirtazapine, it may be when switching to nefazodone because the metabolite, mCPP, is dependent on CYP 2D6 for its metabolism (Kennedy et al. 2002). Currently, it is not thought that this influences the safety profile of nefazodone, but caution should still be used, especially in the elderly population with known heart problems (Owen and Nemeroff 1998). Although venlafaxine is metabolized through CYP 2D6 to ODV, the clinical consequences are not known, since venlafaxine and ODV are pharmacologically similar.

4.1.9
Switching to or from an MAOI

MAOIs have been shown to be effective in 50%–75% of patients who have failed to respond to treatment with a TCA (Nelson 1998). However, a note of caution is warranted when switching from an MAOI to another antidepressant. If a patient does not respond to an adequate MAOI trial, the clinician must wait 14 days after discontinuation to start any new medication. The same is true when switching to another MAOI. The clinician can, with caution, add an MAOI while continuing treatment with a TCA (Marangell 2001). After discontinuing fluoxetine, clinicians need to wait 5 weeks before administering an MAOI, unless the patient is over the age of 65; for older patients, waiting at least 2 months is recommended to achieve comparable washout (Nelson 1998). When considering a trial of an MAOI, the special diet required should also be taken into account. A selegiline transdermal patch is under investigation and may soon be approved for use in the United States; this formulation will likely obviate the need to follow an MAOI diet (see Kennedy et al., this volume, for a discussion of this new formulation) (Amsterdam and Bodkin 2000).

4.2
Augmentation Strategies

Switching medications is advantageous when there is a need to keep treatment simple or patient compliance is an issue (Nelson 1998). An alternative treatment option is augmentation. There is evidence that addition of an augmenting agent may enhance the effectiveness of an antidepressant or improve response time. It is recommended that augmentation strategies be considered when the antidepressant alone produces a partial response following an adequate trial. This is especially true if the patient has significant functional impairment due to depression and few if any adverse effects from the current treatment.

There are potential advantages to augmentation, including attaining full response without starting a new medication trial, using lower doses of both agents to minimize side effects, treatment of comorbid disorders such as subclinical hypothyroidism, and quicker response to treatment (Janicak and Martis 1998). There may be increases in side effect burden and medication costs, but whether those issues are of concern depends on the medication selected for augmentation (Nelson 1998). A wide range of medications has been suggested as effective augmentation agents. Some have minimal side effects (e.g., buspirone), while others such as lithium increase the risk of adverse events. Suggested augmentation dosing is shown in Table 2.

Table 2 Suggested augmentation dosing

Augmentation agent	Suggested daily dose/level
Lithium	0.4–1.0 mEq/l
T3	25–50 µg
Buspirone	20–60 mg (10 mg b.i.d.–20 mg t.i.d.)
Pindolol	7.5–15 mg (2.5–5 mg t.i.d.)
Atypical Antipsychotics	
Olanzapine	5–20 mg
Risperidone	0.5–2 mg
Stimulants	
Methylphenidate	30 mg (10 mg t.i.d.)
Dextroamphetamine	15 mg (5 mg t.i.d.)
Modafinil	100–200 mg

4.2.1
Lithium

Over the past two decades, case reports, open studies, and controlled studies have shown that lithium is an effective augmenting agent for TRD. To date, 11 double-blind, placebo-controlled trials investigating lithium as an augmenting agent have been published or presented (de Montigny et al. 1983; Heninger et al. 1983; Cournoyer et al. 1984; Kantor et al. 1986; Zusky et al. 1988; Schopf et al.

1989; Browne et al. 1990; Joffe et al. 1993; Stein and Bernadt 1993; Katona et al. 1995; Baumann et al. 1996). Although some of the earlier studies showed negative results (Kantor et al. 1986; Zusky et al. 1988; Browne et al. 1990), later studies that used more sound methodologic measures resulted in more favorable outcomes (Schopf et al. 1989; Joffe et al. 1993; Stein and Bernadt 1993; Baumann et al. 1996; Katona et al. 1995).

Based on a meta-analysis of nine double-blind, placebo-controlled trials, Bauer and Dopfmer (1999) concluded that lithium augmentation has a statistically significant effect on response rate compared with placebo. With dosing of greater than 800 mg/day of lithium carbonate or a dose sufficient to produce a plasma level of greater than 0.5 meq/l, and a minimum of 2 weeks on lithium therapy, the pooled odds ratio was 3.31 for a response to lithium augmentation compared with a response to placebo. Because there is much more evidence supporting the use of lithium than any other augmenting agent, its use should be considered first.

Several mechanisms have been proposed to explain the action of lithium augmentation. In vitro studies have shown that low levels of lithium (0.1 meq/l) enhance serotonin turnover and induce a short-term effect (Rouillon and Gorwood 1998). De Montigny et al. (1981) hypothesized that lithium's ability to increase presynaptic transmission of serotonin would potentiate TCA-induced post-synaptic serotonin supersensitivity, which would result in enhanced serotonin transmission . There is also the possibility that lithium potentiates or has a synergistic effect that involves other monoamines, such as norepinephrine or acetylcholine.

Some studies have shown a positive correlation between higher lithium plasma levels and response, while others have not been able to find this association. As a general rule, lithium is dosed at a minimum of 600–900 mg/day, with a recommended plasma level above 0.4 meq/l (Nelson 1998; Rouillon and Gorwood 1998). Reports in the literature are also mixed concerning time to treatment response. Some cases have shown a rapid response in less than 48 h (de Montigny et al. 1983; Heninger et al. 1983), while others responded after a period of more than 2 weeks (Stein and Bernadt 1993). Patients should use lithium augmentation for a minimum period of 2 weeks; it is generally unnecessary for trials to extend beyond 6 weeks. However, some patients who are classified as "slow responders" by history may need a longer trial before lithium is determined to be ineffective.

There is even more debate about how long to continue lithium treatment once a response is seen. Although several case reports have addressed this issue, only one double-blind, placebo-controlled study has looked at continuation treatment (Bauer et al. 2000). None of the 14 patients who received lithium augmentation suffered a relapse in the 4 months after remission of their symptoms, while 5 of the 7 patients who were previous responders to lithium and were placed on placebo in the continuation phase of the study experienced a depressive relapse. The authors suggest that lithium augmentation be continued for a minimum of 6 months after remission of depressive symptoms.

Overall, approximately 50% of patients with TRD will respond to lithium augmentation. Currently, no sociodemographic factors have been identified that will predict response. Patients with nonpsychotic melancholic depression or bipolar depression are the most likely to be responders (Rouillon and Gorwood 1998). Thase et al. (1989b) suggested that patients who had a shorter duration of prior treatment would have a more rapid response. In an open label, retrospective study of 71 patients with TCA-resistant depression, it was found that lithium response was associated with more severe depression, a shorter depressive episode, and lower levels of triiodothyronine. Lithium responders were less likely to be diagnosed with a personality disorder or to receive co-administration of antipsychotics and antidepressants (Bschor et al. 2001). Currently, there are no data indicating that any specific antidepressant medication is more likely than another to benefit from lithium augmentation.

Although much of the literature supports the use of lithium as an effective option for TRD, its use has declined. One reason for this may be that it has been best studied with MAOIs and TCAs, which are not considered first-line antidepressant agents by most practitioners. Literature supporting its use with SSRIs is expanding, but information on lithium augmentation of other newer antidepressant agents is limited. Another explanation for its declining use may be the risk of toxicity with lithium and the need to monitor plasma levels as well as thyroid and renal functioning. Lithium is also plagued with many adverse effects (e.g., weight gain, tremor, nausea) and is often not well tolerated by patients. There is also a theoretical risk of developing serotonin syndrome when lithium is combined with other serotonergic agents. Because of these factors, lithium is often not perceived by patients and practitioners as the augmenting agent of choice despite the plethora of data supporting its use.

4.2.2
Thyroid Hormone

Both hyper- and hypothyroidism are associated with psychiatric symptomatology. In particular, hypothyroidism is associated with depression-like symptoms that will commonly resolve after normalization of the thyroid function. This led researchers to investigate the use of thyroid supplementation in depression, although most depressed patients are euthyroid. Triiodothyronine (T_3) was first tested empirically as a treatment for psychiatric illnesses in 1958 (Aronson et al. 1996). This work led to the use of low doses of T_3 to accelerate the antidepressant effects of TCAs. One of the first and most common explanations used to explain the mechanism of action is that thyroid hormone interacts with various neurotransmitters, norepinephrine in particular (Joffe 1998). It also has been suggested that it may directly affect the thyroid axis or reduce thyroxin levels. To date, the exact mechanism has yet to be determined.

Eleven studies (Earle 1970; Ogura et al. 1974; Banki 1975, 1977; Tsutsui et al. 1979; Goodwin et al. 1982; Schwarcz et al. 1984; Gitlin et al. 1987; Thase et al. 1989a; Joffe and Singer 1990; Joffe et al. 1993) have looked at the use of T_3 to augment TCA non-responders; however, only four are double-blind studies (Goodwin et al. 1982; Gitlin et al. 1987; Joffe and Singer 1990; Joffe et al. 1993). Aronson et al. (1996) conducted a meta-analysis of T_3 augmentation in treatment refractory depression. Eight studies totaling 292 patients were included in the analysis, and results showed that patients treated with T_3 were twice as likely to respond as controls. However, among the four double-blind, placebo-controlled trials, pooled effects were not found to be significant. The authors attributed this to the one negative study (Gitlin et al. 1987), which had the most statistical variability. The authors concluded that more controlled studies with larger sample sizes would be necessary before T_3 could be definitively recommended.

T_3 is the preferred agent over T_4 because a controlled study conducted by Joffe and Singer (1990) found T_3 to be more effective. Joffe and Marriott (2000) also found that recurrence of depression was inversely related to T_3 but not T_4 levels. The usual T_3 dose is 25–50 mcg/day or an average of 37.5 mcg/day. Although the duration needed to consider treatment a success or failure has not been definitively determined, a trial of 2–3 weeks is generally perceived as an adequate time period (Olver et al. 2000). As with lithium, it is unclear how long T_3 augmentation should be continued, but it is common practice to continue for the duration of the antidepressant therapy. However, there is no literature available that provides empirical support for this practice. Also, as with lithium, thyroid supplementation is not commonly used in real world clinical practice. This may be due to limited data supporting its use with the newer antidepressants.

4.2.3
Buspirone

Buspirone is an azaspirone that is currently marketed and used as an anxiolytic agent for generalized anxiety disorder (Schweitzer et al. 1986). It is believed to exert its anxiolytic effect through activity as a postsynaptic serotonin-1A ($5-HT_{1A}$) receptor partial agonist. Some early evidence suggested that buspirone also exerts mild antidepressant effects through potentiation of serotonin transmission (Schweitzer et al. 1986; Rickels et al. 1990). In these studies, buspirone monotherapy was superior to placebo in the treatment of depression. This led to several studies looking at buspirone as an augmenting agent.

Four open label studies of buspirone augmentation have reported positive results (Bakish 1991; Jacobson 1991; Joffe and Schuler 1993; Dimitriou and Dimitriou 1998). All four studies involved the addition of buspirone to current SSRI therapy. Sample sizes ranged from 3 to 30 patients and rates of response were usually greater than 50% by week 3, with most patients sustaining that response after 3 months. In a randomized, placebo-controlled study, Landen et al. (1998) were not able to detect a difference between the buspirone plus SSRI group and the placebo plus SSRI group in patients with TRD. Approximately 50% of patients in both groups showed a treatment response at week 4. In the open-label portion of the study, those patients who had initially responded, despite their grouping, continued to show a response. Appelberg et al. (2001) also conducted a randomized, double-blind, placebo-controlled trial of buspirone augmentation of SSRIs in patients with SSRI-resistant depression. After the first week, the SSRI plus buspirone group showed significant improvement compared with the SSRI plus placebo group, but at the end of the 6-week trial, there was no difference between the groups. It is noteworthy that those patients who had an initially high score (>30) on the Montgomery-Asberg Depression Rating Scale (MADRS) (Montgomery and Asberg 1979) showed a significantly greater reduction in symptoms in the buspirone group than the placebo group ($p=0.026$).

When using buspirone as an augmenting agent in TRD, doses may range from 20–50 mg/day, with the most common dose being 10 mg t.i.d. (Nelson 1998). If a response is to occur, it is reasonable to expect it within 3 weeks. Buspirone has a distinct advantage in that it causes very mild side effects and has minimal drug interactions. It also has independent effects on anxiety and may therefore be the treatment of choice in those patients with a comorbid anxiety disorder. Currently, buspirone augmentation has been studied only with serotonergic agents and it is unclear if using it as an augmenting agent with noradrenergic and/or dopaminergic antidepressants would produce similar results. Because of this, and inconsistent findings in studies with SSRIs, additional controlled studies should be conducted before its efficacy and potential clinical utility as an augmenting agent can be fully evaluated.

4.2.4
Pindolol

Pindolol is a β-blocker with 5-HT$_{1A}$ receptor antagonistic properties. Currently, it is thought that the delay in antidepressant effects (specifically with SSRIs) is due to the inhibitory function of the presynaptic 5-HT$_{1A}$ receptors (Martinez et al. 2000). SSRIs cause an initial rise in extracellular serotonin, which feeds back to the 5-HT$_{1A}$ receptors. This action results in an initial decrease of 5-HT firing, synthesis, and release, which may explain the 2–3 week delay in SSRI action (Artigas et al. 1994; Martinez et al. 2000). Over time, there is a downregulation of the 5-HT$_{1A}$ receptors, a decrease in negative feedback autoinhibition, and increased firing and synthesis of 5-HT. This, in conjunction with the inhibition of serotonin reuptake, facilitates 5-HT neurotransmission. The rationale for using pindolol is to block the initial autoinhibitory process and theoretically cause an earlier response to antidepressant therapy and augment antidepressant effects.

Although open-label studies (Blier and Bergeron 1995; Maes et al. 1996; Vinar et al. 1996) suggest that pindolol may enhance or accelerate SSRI response, double-blind, placebo-controlled trials have produced mixed results (Moreno et al. 1997; Perez et al. 1997, 1999; Blier and Bergeron 1998). Investigation of pindolol augmentation of other antidepressants (MAOIs, nefazodone, TCAs) has also shown mixed findings (Blier and Bergeron 1998).

The dose of pindolol that generally has been studied has been 7.5 mg/day (2.5 mg t.i.d.). This dose was chosen in hopes of avoiding cardiovascular effects.

Positron emission tomography (PET) imaging studies conducted by Martinez et al. (2000) showed that 7.5 mg/day was most likely a suboptimal dose when using pindolol to enhance serotonin transmission. They hypothesized that doses of 15–25 mg/day would be necessary to block 50% of the 5-HT$_{1A}$ receptors. Pindolol then becomes less appealing, given the cardiac effects that can occur at these doses. More controlled trials need to be conducted before pindolol can be conclusively ruled in or out as an augmentation agent for TRD.

At doses of 7.5 mg/day, pindolol is generally well tolerated. However, Blier and Bergeron (1995) reported increased irritability, insomnia, and anxiety when using it in TRD patients. Pindolol should be used with caution in patients with severe allergies, asthma, cardiac conduction problems, and diabetes due to its blockade of beta-1 and beta-2 adrenoreceptors. If there is no response after 2 weeks, Blier and Bergeron (1995) recommend that pindolol augmentation should be discontinued, as response is unlikely to occur with continued treatment. Treatment should be continued until remission of symptoms has been achieved, although, with severely resistant cases, continuation of pindolol with the antidepressant should be considered. Because cases of rapid deterioration have been reported when pindolol was suddenly stopped due to adverse effects, pindolol can be tapered off over a period of 2–4 weeks, decreasing by 2.5 mg every 1–2 weeks (Vinar et al. 1996; Blier and Bergeron 1998).

4.2.5
Atypical Antipsychotics

Atypical antipsychotics have been studied primarily for use in psychotic depression, schizoaffective disorder, and depressive symptoms in patients with schizophrenia (Rothschild 1996). Affective symptom improvement in these disorders has prompted some investigators to look at atypical antipsychotics as adjunctive agents in major depression. Risperidone (Osteroff and Nelson 1999; Hirose and Ashby 2002) and olanzapine (Shelton et al. 2001) have both shown positive results in treatment of major depression when augmenting SSRIs in non-responders. The olanzapine study was a double-blind, placebo-controlled trial that showed olanzapine plus fluoxetine was a superior combination to either agent alone. However, it should be noted that while there were significant differences between combination treatment and fluoxetine monotherapy in patient scores on the MADRS scale, no differences were found in the HAM-D scores between the groups. There is one case report of augmentation of venlafaxine with olanzapine 5 mg/day in which the patient with TRD experienced improvement in symptoms within 2–3 days (Pichot and Ansseau 2001). Doses that have been used to augment antidepressants are 0.5–2 mg/day of risperidone and 5–20 mg/day of olanzapine. Results were generally seen within 1 week; therefore, 2–3 weeks should be a sufficient trial period. It appears that these two atypical antipsychotics, risperidone and olanzapine, work effectively to treat insomnia, anxiety, and agitation associated with depression (O'Connor and Silver 1998; Osteroff and Nelson 1999; Kaplan 2000; Pichot and Ansseau 2001; Shelton et al.

2001). Little or no data exist concerning the use of quetiapine or ziprasidone as adjunctive agents in major depression. Drawbacks to the use of atypical antipsychotic medication include increased sedation, potential for weight gain, and higher cost. When treating with atypical antipsychotic medications, particularly risperidone, there is also concern about acute and chronic extrapyramidal symptoms, even though this risk is thought to be less than with the older antipsychotics agents.

4.2.6
Psychostimulants

Psychostimulants have a long history of use in depression. Clinically, dextroamphetamine, methylphenidate, modafinil, and pemoline have all been used as augmenting agents. In a frequently cited study, Wharton et al. (1971) reported successful treatment of seven imipramine-refractory patients with the addition of methylphenidate. Fawcett et al. (1991) looked at the addition of either pemoline or dextroamphetamine to an MAOI and found that 25 of the 32 patients responded and 10 were able to sustain that response for at least 6 months. Although the addition of stimulants to MAOIs, TCAs, SSRIs, and even venlafaxine (Stoll et al. 1996; Bader et al. 1998; Masand et al. 1998; Menza et al. 2000) has been examined, there are no controlled trials evaluating its effectiveness as an augmenting agent.

Modafinil is a stimulant used mostly for narcolepsy. Its exact mechanism of action is unclear. Unlike the amphetamines, it is highly selective for the CNS, has little effect on dopamine activity in the striatum, and appears to have a lower potential for abuse. In a retrospective case series (Menza et al. 2000), seven patients with unipolar or bipolar depression, who were unresponsive or partially responsive to antidepressant monotherapy, were treated with 100–200 mg/day of modafinil. All seven patients achieved full or partial remission within 1–2 weeks.

The usual dosages for methylphenidate and dextroamphetamine are 10 mg three times daily and 5 mg three times daily, respectively (Nelson 1998). Initial doses of the stimulant should be lower when using in conjunction with MAOIs and titrated up to therapeutic response. Common side effects are increased irritability, increased anxiety or agitation, paranoid thinking, and mania (Nelson 1998). Psychostimulants may best be used in anergic patients. They may not be the best augmenting choice in those patients suffering from insomnia or anxiety or in patients with substance abuse problems.

4.2.7
Estrogen

Although it is well established that estrogen affects mood, it is unclear what role it may have in the treatment of MDD. Most research looking at mood fluctuations has dealt with either the normal female life cycle or premenstrual dysphoric disorder versus MDD (Stahl 2001). There is one double-blind, placebo-controlled study comparing the use of 100 mcg transdermal 17b-estradiol versus placebo for treatment of depression in perimenopausal women. Results revealed remission of symptoms in 68% of the estrogen group compared to 20% in the placebo group (Soares et al. 2001). Some critics argue that estrogen is simply alleviating menopausal symptoms, but a study conducted by Schmidt et al. (2000) showed estrogen to be effective even in patients without vasomotor symptoms. Although we are unaware of any studies looking specifically at estrogen as an augmenting agent, it may be a viable augmentation option in this specific patient population.

4.2.8
Miscellaneous Augmenting Agents

Anticonvulsants. It is common in clinical practice to see anticonvulsants (i.e., carbamazepine, oxcarbazepine, valproic acid, lamotrigine, gabapentin, and topiramate) used as augmenting agents in TRD. There are, however, limited data supporting their use in TRD. Several studies have looked at the use of carbamazepine in acute treatment of depression (Shelton 1999), but only one known double-blind trial assessing its use in TRD. Rybakowski et al. (1999) compared carbamazepine to lithium as add-ons to antidepressant therapy and found no difference in therapeutic response between the two groups.

Davis et al. (1996) conducted an open-label study looking at the efficacy of valproate as an antidepressant in MDD. At Week 8, 19 of the 22 completers (86%) showed a significant response, with 66% showing response with an intent-to-treat analysis. These data suggest that valproate may be an effective treatment option for MDD, but double-blind, placebo-controlled trials are necessary to further document its efficacy in the treatment of MDD. It is unclear what its role would be in TRD. There are only two case reports of successful trials of lamotrigine as an augmenting agent for TRD (Maltese 1999); its use in bipolar depression has been much better studied. Because topiramate has been associated with worsening of depressive symptoms, it is most likely a poor choice for augmentation in TRD (Martin et al. 1999; Klufas and Thompson 2001). More studies showing efficacy of the anticonvulsants as augmenting agents are necessary before this group of medications can be recommended.

Inositol. Inositol monotherapy has been shown to be effective in reducing depressive symptoms (Levine et al. 1995). However, in a study conducted by Nemets et al. (1999), inositol was not effective in treating SSRI treatment failures.

Naltrexone. Naltrexone is an opioid antagonist that is mainly used in the treatment of alcohol and opioid dependence. Amiaz et al. (1999) reported a successful case of a woman who responded to naltrexone augmentation (50 mg/day) of paroxetine. Controlled studies are needed in order to establish both efficacy and optimal dosage.

S-Adenosyl-Methionine. A widely available "nutritional supplement" sold in many nutritional/health stores, S-adenosyl-methionine (SAM-e) is produced naturally in the CNS from L-methionine and adenosine triphosphate, and low levels have been correlated with depression (Echols et al. 2000). It has been studied extensively for use in depression, with mixed findings (Echols et al. 2000). In depression trials, the doses generally given are 200–800 mg twice daily. Currently, there are no studies available that have looked at the use of SAMe as an augmenting agent in TRD.

Dehydroepiandrosterone. Its exact physiologic role is currently unknown, but dehydroepiandrosterone (DHEA) is a naturally circulating adrenocorticoid in humans. As well as being a precursor for testosterone and estrogen, it is thought that it may be involved in regulating mood and sense of well-being (Wolkowitz et al. 1999). Studies have shown that DHEA and its sulfated metabolite DHEA-S, decrease over the lifespan and also in times of chronic stress and severe illness (Wolkowitz et al. 1997). Two studies have reported positive results in treating depression in doses up to 90 mg/day (Wolkowitz et al. 1997, 1999). While both these studies showed positive results, long-term effects of DHEA are not known. Theoretically, because it is a precursor for estrogen and testosterone, it could potentially activate hormone-sensitive tumors, such as tumors of the breast, cervix, uterus, or prostate (Wolkowitz et al. 1997). Its use as an augmenting agent in TRD is not recommended at this time.

4.3
Combination Treatment

Current literature often uses the terms combination therapy and augmentation agents interchangeably; there is, however, a clear distinction between the two. Augmentation enhances the effectiveness of an ongoing antidepressant trial, while combined treatment involves the use of separate antidepressant agents to address depressive symptom etiology (Nelson 1998). Another distinction is that, while an augmenting agent is added to the current trial, two antidepressant medications are prescribed at full therapeutic doses/levels in combination treatments. Generally, combination therapy involves using two or more drugs from different groups or classes of antidepressant medications to produce an additive effect (Janicak and Martis 1998).

Disadvantages to combined treatment include the potential for significant drug interactions, increased risk of adverse effects, possible nonadherence, and increased cost (Janicak and Martis 1998). Few double-blind studies of combination therapy have been completed. Most of the combination strategies used by practitioners are supported by very little evidence and are based on theory or case reports. Most reports of combination treatments in the literature involve an

SSRI plus another antidepressant medication. Often the second antidepressant is initiated to help alleviate adverse effects (e.g., sexual dysfunction) and to enhance mood. More research is needed on the use of combination therapy.

Current data suggest that a clinician should not expect improvement in a few days or weeks when starting a combination trial (Fava 2001); improvement may not be seen for 4–6 weeks. It is unclear what would be considered the minimum duration of combined use in treatment responders. In general, maintenance medication should be continued for 6–9 months following symptom remission before gradually discontinuing one of the medications (Fava 2001).

4.3.1
SSRI/SSRI Combinations

SSRI/SSRI combinations are theoretically feasible because of potency variations and differences in SRI specificity. There is, however, little support for this combination. There are only two reports ($n=13$) of SSRI combination showing positive results (Bondolfi et al. 1996; Hunchak 1997) Although recent data suggest that paroxetine and fluoxetine are more potent norepinephrine uptake inhibitors, and sertraline is a more potent dopamine uptake inhibitor than the other SSRIs, the clinical significance of these findings is likely to be of little relevance (Fava 2001). One of the primary disadvantages of combining SSRIs is the risk of patients developing serotonin syndrome or having serotonergic side effects due to the pharmacodynamic drug interaction that would occur when using two of these agents simultaneously. Because there is little clinical support, this type of combination is not recommended.

4.3.2
SSRI/TCA Combinations

There is evidence for improvement in response following combination treatment with an SSRI and TCA. In a study of patients who were unresponsive at the end of an 8-week trial of fluoxetine, some response was seen when patients were treated with a combination of desipramine and fluoxetine (Fava et al. 1994). However, TCA levels will build up as the action of the SSRI interferes with its clearance, and there is risk of cardiac toxicity. In order to manage treatment, it is suggested that clinicians use low doses of the TCA (25–75 mg/day) and monitor TCA plasma levels (Fava 2001). The biggest concern for a pharmacokinetic interaction between TCAs and SSRIs is with CYP 1A2 (fluvoxamine) and CYP 2D6 (paroxetine and fluoxetine) inhibitors and those TCAs that are metabolized through those enzyme systems (imipramine, clomipramine, and amitriptyline). Although sertraline and citalopram are considered less potent inhibitors of the CYP 2D6 enzyme system than paroxetine and fluoxetine, higher doses of these medications increase the potential for a drug–drug interaction. Lower doses of an antidepressant such as trazodone (50–100 mg) may be combined with the usual dose of the primary antidepressant for sleep disturbance (Janicak and

Martis 1998). Again, it is important to evaluate for possible drug interactions as well as a possible increase in side effects when combining medications. Trazodone is cleared by CYP 3A, producing mCPP, which is an active metabolite with a pharmacologic profile virtually the opposite of trazodone. That metabolite is cleared by CYP 2D6. Coadministration with fluvoxamine can slow the clearance of trazodone leading to daytime drowsiness, whereas coadministration with fluoxetine or paroxetine can lead to a build-up in the levels of mCPP, which can lead to anxiety and agitation. (For a more detailed discussion of pharmacokinetic issues and therapeutic drug monitoring, see chapters "General Principles of Pharmacokinetics" and "Therapeutic Drug Monitoring of Antidepressants.")

4.3.3
Bupropion Combinations

Although bupropion is frequently combined with SSRIs in clinical practice because of the different pharmacologic effects of the two agents, there are only a few open-label trials examining this combination (Marshall and Liebowitz 1996; Bodkin et al. 1997; Kennedy 2002). Sustained-release bupropion 100–150 mg/day is often used to help treat SSRI-induced sexual dysfunction (Fava 2001). Clinical lore suggests bupropion may be helpful as an add-on in patients whose anergic symptoms do not respond adequately to treatment with an SSRI. A disadvantage in using bupropion is its potential for inducing seizures. The risk for seizures increases as the bupropion dose (level) increases. Currently, the metabolism of bupropion through the CYP enzymes is not clearly understood. Because of uncertainty about its metabolism and the seizure potential, it is probably best to avoid using bupropion in combination with antidepressants known to be potent CYP enzyme inhibitors. If the agents are used together, it is recommended that one use bupropion doses at the lower end of the therapeutic range. These issues are discussed further in chapters "General Principles of Pharmacokinetics" and "Therapeutic Drug Monitoring of Antidepressants."

4.3.4
Mirtazapine Combinations

Mirtazapine is a dual action antidepressant that increases both serotonergic and noradrenergic activity by blocking α_2-adrenergic autoreceptors and heteroreceptors and serotonergic 5-HT$_2$ and 5-HT$_3$ receptors. Combining mirtazapine (15–30 mg/day) with a SSRI has been reported to help nonresponders (Fava 2001). There is also evidence of a significantly higher response rate with combination therapy than monotherapy with either mirtazapine or a SSRI (Fava 2001). In an open-label study conducted by Carpenter et al. (1999), there was a 55% response rate at week 4 when mirtazapine (15–30 mg/day) was combined with another antidepressant following a failed trial of monotherapy. Unfortunately, at these doses in particular, mirtazapine has the potential to cause weight gain and sedation. Mirtazapine does have a lower potential to cause sexual dysfunction than the SSRIs, so practitioners will often switch to mirtazapine if patients experience this adverse effect from a SSRI. There is, however, little to no evidence to support the combined use of these two agents to relieve SSRI-induced sexual dysfunction. Few data exist concerning the combination of mirtazapine and other antidepressant medications at this time.

4.3.5
Nefazodone Combinations

There is no strong evidence in the literature for combining nefazodone and an SSRI, but nefazodone does have a lower propensity to cause sexual dysfunction. However, as for mirtazapine, there is no evidence indicating that this combination would alleviate SSRI-induced sexual dysfunction. With this combination (specifically nefazodone plus paroxetine or fluoxetine), patients may experience increased irritability and anxiety because of an accumulation of mCPP (active metabolite of nefazodone). This situation is created because paroxetine and fluoxetine inhibit the metabolism of mCPP through the CYP 2D6 enzyme (Fava 2001).

4.3.6
Venlafaxine Combinations

There are anecdotal reports of using 75–300 mg/day venlafaxine with SSRI nonresponders, but no clear evidence that combination treatment with venlafaxine and an SSRI is efficacious (Fava 2001). There are concerns regarding combined use of these medications since venlafaxine is a substrate of the CYP 3A (minor pathway) and is broken down into O-desmethylvenlafaxine (ODV), which is a substrate of CYP 2D6 (major pathway) (Ereshesky and Dugan 2000). When venlafaxine is combined with an agent that inhibits the CYP2D6, venlafaxine will accumulate. Initially this was not thought to be of great concern because ODV and venlafaxine are considered to be equal in terms of their mechanisms of action; however, some evidence exists that venlafaxine may be a more potent inhibitor of the ion channels than ODV, causing a theoretical increase in the risk of *torsades de pointes*. Although the clinical significance of this discovery has not yet been determined, the combination of venlafaxine and 2D6 inhibitors should be used cautiously. For more detailed discussion of these issues, see chapters "General Principles of Pharmacokinetics" and "Therapeutic Drug Monitoring of Antidepressants." Combined use of venlafaxine and other antidepressant agents may also lead to serotonin syndrome (Bhatara et al. 1998) or marked elevation of blood pressure and severe anticholinergic effects (Benazzi 1999).

4.3.7
MAOI Combinations

Clinicians should avoid combining an SSRI and an MAOI, because this combination may produce potentially fatal serotonin syndrome (Janicak and Martis 1998). TCAs and MAOIs are no longer combined due to the risk of a lethal hypertensive crisis, particularly since newer, safer antidepressants are available (Fava 2001). Combining venlafaxine with an MAOI should also be avoided because of the risk of both serotonin syndrome (due to serotonin uptake inhibition) and hypertensive crisis (due to norepinephrine uptake inhibition).

5
Alternative Treatments

Advances in neuroscience and improved psychotropic medications provide strong support for pharmacotherapeutic treatment of depression. However, a number of factors, such as poor social support, marital difficulties, or a distorted worldview or attitude may have a negative impact on treatment (Thase et al. 2001). Patients with TRD often present with problem behaviors such as reduced social activity and increased time spent alone, diminished ability to experience enjoyment or pleasure, and negative thinking (Thase et al. 2001). Consequently, a number of studies have investigated the efficacy of psychotherapy in TRD patients. The development of a strong therapeutic alliance may also promote treatment adherence, thereby improving chances for remission.

5.1
Cognitive–Behavioral Therapy

In one long-term study by Fava et al. (1997), 20 patients with TRD were assigned to a minimum of ten 40-min cognitive–behavioral therapy (CBT) sessions conducted every other week. In the course of treatment, patients were tapered off medication until symptom exacerbation occurred. The patients were tracked quarterly for up to 24 months, and the rate of full response was 63%. Half of those who responded had no residual symptoms. Eight of the 12 responders discontinued medication and, at the 24-month follow-up, only 1 patient had relapsed. However, when medication was discontinued, 4 of the 12 responders had increased symptoms, suggesting a need to maintain pharmacotherapy (Fava et al. 1997).

A recurrent depression study found CBT to be helpful in reducing residual symptoms and frequency of relapse (Fava et al. 1998). Study patients had been successfully treated with antidepressants prior to enrollment, but had residual symptoms such as anxiety and irritability. Comparing pharmacotherapy and CBT to pharmacotherapy alone, there was significant improvement in residual symptoms in the group receiving CBT. After 2 years, 80% of the pharmacotherapy group had relapsed compared to 25% of those receiving combination treatment (Fava et al. 1998).

Recently, a large, multi-site study was conducted comparing the use of nefazodone, psychotherapy, and combination treatment in chronic depression (Keller et al. 2000). The form of cognitive therapy used was Cognitive Behavioral Analysis System for Psychotherapy (CBASP). This therapy was designed for use with chronic depression and takes a structured, directive approach to interpersonal interactions. Results revealed that for the first 4 weeks, response to combined treatment was not significantly different from pharmacotherapy. In fact, nefazodone produced a quicker response than CBASP. However, the overall response rate for combined treatment was significantly higher than that of either nefazodone or CBASP alone (Keller et al. 2000). After 12 weeks, the response rates for CBASP and nefazodone were 52% and 55%, respectively, while the response rate for combined treatment was 85%. These results indicate a clear treatment advantage for combined pharmacotherapy and psychotherapy. The authors argue that the similar efficacy of combined treatment and psychotherapy alone in the first 4 weeks suggests independent action of those treatments rather than synergic interaction (Keller et al. 2000).

There is little evidence to support the use of one form of psychotherapy over another (Thase et al. 2001). However, cognitive therapies seem to have a more lasting effect (Thase et al. 2001), because they often address behaviors that are problematic for the patient and interfere with social functioning. As the treatment goals for depression are full remission of symptoms and a return to premorbid levels of functioning, psychotherapy provides a useful tool to address patient needs. Evidence is coming to light that each treatment has its individual place, but that, in moderate to severe depression, a combination of pharmacotherapy and psychotherapy may be the best treatment option (Keller et al. 2000).

5.2
Electroconvulsive Therapy

Long considered the preferred treatment for severe, treatment-resistant depression, electroconvulsive therapy (ECT) involves inducing modified seizures (passing a direct brief-pulse current through the scalp via electrodes while under general anesthesia) (Nelson 1997; Janicak and Martis 1998). Effective and relatively safe, ECT should be considered when there is high suicide risk, a rapid physical decline, medication nonresponse or intolerability, or a history of response to ECT (Janicak and Martis 1998). ECT has been shown effective in treating severe depression, psychosis, catatonia, neuroleptic malignant syndrome, and parkinsonism (Fink 2001). There are no restrictions in the use of ECT due to age or systemic conditions. The major limitations to the use of ECT include a high relapse rate and possible effects on memory (Sylvester et al. 2000; Fink 2001).

ECT was introduced in the mid-1930s and used extensively through the 1950s (Sylvester et al. 2000; Fink 2001). At that time, ECT was the primary treatment available to hospitalized patients. As pharmacotherapies were developed, the use of ECT declined. While it continued to be used to treat refractory depression in the 1960s, there was a social and political movement to discourage ECT treatment due to perceived misuse or abuse. The use of ECT declined, so that, by 1980, only 0.3% of patients hospitalized for mood disorders received ECT (Sylvester et al. 2000). Today, ECT is primarily used in private facilities, although it has been suggested that state facilities reconsider its use in treating patients with TRD (Sylvester et al. 2000).

ECT may impair neurocognitive functioning. Neuropsychological testing has shown that declarative memory may be impaired, while immediate memory is generally preserved (Rami-Gonzalez et al. 2001). Due to disruption of specific brain regions, patients may experience impairment of selective memory as well. Another concern in the use of ECT is comorbidity. TRD frequently involves more than one psychiatric diagnosis. For example, patients with TRD who are also diagnosed with borderline personality disorder may not respond well to ECT, since the available data suggest that ECT is effective for only a portion of depressed patients with borderline personality disorder (de Battista and Mueller 2001).

A number of recent studies have evaluated the effectiveness of ECT. Case studies were completed on three patients with lengthy, refractory major depressive episodes (Fox 2001) in which all three patients were treated with ECT, and there were no medical complications. Memory difficulty was a common complaint, but Mini-Mental Status Examinations done 6 weeks post-treatment revealed no impairment in any of the patients. The patients did have relapses over the study period, but the frequency, severity, and duration of relapses diminished with continued treatment. In addition, the frequency of relapse diminished over the course of individual episodes and were typically reversed by 3–4 closely spaced ECT sessions (Fox 2001).

Another study evaluated the use of ECT in a Pennsylvania state hospital over a 10-year period (Sylvester et al. 2000). Researchers reviewed records of 17 patients who were treated with ECT during that time. ECT was used only in 0.4% of cases at this facility. Findings indicated that ECT was safe overall, with no significant or lasting complications in any of the study patients. All but one patient was switched from unilateral to bilateral treatments due to TRD status. ECT was shown to be an effective treatment in this population, with more than half of the study patients discharged within 10 days of completing treatment. Within 6 months, two-thirds of them had been discharged. Even if the patients remained hospitalized, they showed significant improvement based on a reduction in the frequency of seclusion and restraint episodes. Those patients with shorter hospitalizations prior to treatment showed the most improvement. The authors suggest ECT should be considered earlier in the treatment of depression, avoiding a lengthy series of failed medication trials (Sylvester et al. 2000).

An examination of changes in regional cerebral blood flow (rCBF) in depressed patients prior to treatment showed hypoperfusion of the frontal region and multiple areas of altered perfusion throughout the brain compared to normal controls (Milo et al. 2001). Another scan was done following ECT, in which changes toward normal rCBF were seen in those patients who had a good response, while those patients who had a minimal to moderate clinical response to ECT showed no significant changes in rCBF (Milo et al. 2001).

While the presence of antidepressant medication may not affect the duration of seizures (Dursun et al. 2001), the response rate appears to be influenced by the quality of prior medication trials (Nelson 1997). An 86% response rate to ECT has been seen in patients who have not received an adequate drug trial. However, if a patient has received adequate prior antidepressant drug treatment, response to ECT is about 50%. This response rate is similar to the rate that is expected when a patient switches from one class of drug to another. Consequently, ECT is most effective when used as first-line treatment (Nelson 1997).

A high relapse rate has been noted with ECT. This rate is particularly high in the first month after treatment. In one continuation study, the use of pharmacotherapy following ECT produced marked reduction in relapse in patients who received nortriptyline plus lithium compared to those who received nortriptyline alone or placebo (Sackeim et al. 2001a). Without continuation treatment, almost all remitted patients relapsed within 6 months. Over a 24-week trial, the relapse rate for placebo was 84%, for nortriptyline was 60%, and for combination treatment with nortriptyline and lithium was 39%.

5.3
Vagus Nerve Stimulation

Vagus nerve stimulation (VNS) refers to stimulation of the left cervical vagus nerve using a device called the NCP (NeuroCybernetic Prosthesis) (George et al. 2000; Rush et al. 2000). VNS has been available in Europe since 1994 and in the United States since 1997 for use with treatment-resistant partial-onset epileptic seizures (Rush et al. 2000).

The procedures surrounding the use of VNS make it similar to a cardiac pacemaker. Both use a subcutaneous generator that transmits an electrical signal to the affected organ via an implanted electrode (George et al. 2000). VNS is delivered by way of a multi-programmable bipolar pulse generator that is implanted in the left chest wall. The generator sends electrical signals to the left vagus nerve through a bipolar lead. An electrode is wrapped around the vagus nerve near the carotid artery by means of a separate incision and then connected to the generator subcutaneously. Once done only by neurosurgeons, now VNS implantation is also being done by vascular and ENT surgeons in an outpatient setting using local anesthesia (George et al. 2000).

As a supplement to medication, VNS has been shown to be effective in treatment-resistant epilepsy (George et al. 2000). Some adverse events reported in epilepsy trials include voice alteration/hoarseness, cough, throat pain, nonspecific pain, dyspnea, paresthesia, dyspepsia, vomiting, and infection. Generally, these side effects diminish over time. There have been no reports of significant cardiac changes as a result of stimulation. VNS has been shown to be well tolerated over time, and some patients have had active implants for up to 10 years (George et al. 2000).

In recent years, VNS has been considered as an alternative treatment for depression. Epilepsy patients have displayed improved cognition and mood following VNS, even if there has not been significant seizure reduction (Rush et al. 2000). In addition, anticonvulsants such as carbamazepine and gabapentin are commonly used to treat mood disorders. PET scans reveal limbic system effects as a result of VNS, and it is thought VNS alters monoamine concentrations in the central nervous system (George et al. 2000).

The basic mechanism of action for VNS is unknown. It is suspected that the stimulation produces changes in serotonin, norepinephrine, γ-aminobutyric acid (GABA), and glutamate (George et al. 2000). Due to its inconvenience and the surgery required, VNS is most often used with patients who have not responded to other therapies.

There has been one multisite study that examined the efficacy and safety of VNS in treatment-resistant, chronic, or recurrent nonpsychotic MDD and type I or II bipolar disorder (Rush et al. 2000). This study utilized the same techniques as prior epilepsy VNS studies. The NCP system was used, which involves an implantable and multiprogrammable pulse generator delivering electrical signals to the left vagus nerve via a bipolar lead. After implantation, a programming wand attached to a computer was used to program the generator and to set or adjust stimulation parameters.

Patients enrolled in the study were required to have had more than four major depressive episodes in their lifetime, with the current episode persisting longer than 2 years. They also had to have failed multiple pharmacotherapy trials. Hamilton Depression Rating Scale (HDRS)-28 (Hamilton 1967) scores greater than 20 were required at both baseline visits. Patients continued on medication for 4 weeks prior to implantation. Following a 2-week recovery period, patients were re-assessed and those with HDRS scores exceeding 18 continued in the study. At that point, the NCP system was activated and the output current (mA setting) was progressively increased over a 2-week period to the maximum level that was comfortably tolerated. At 4 weeks, stimulation parameters were set and remained constant for 8 weeks. The total duration of treatment was 10 weeks. Patients were allowed to continue VNS following the acute phase and were monitored for 9–12 months post-implantation.

Thirty-eight patients were enrolled in the study, with 30 patients receiving implants. Results showed a 40%–50% response rate (based on HDRS or MADRS scores), with 17% of patients achieving full remission. According to final Clinical Global Impression (CGI) scores, 30% of patients were rated as minimally improved, 20% were as much improved, and 20% as very much improved. There was some early response, but the majority of symptom reduction was seen during the first 6 weeks of the 8-week trial.

Mild side effects were seen. All 10 responders sustained their response throughout the 4- to 9-month follow-up period. The results of this study should be interpreted with caution, because there was no control group included in the study. However, the response rate was significantly higher than would be expected in patients with TRD (Rush et al. 2000), suggesting a promising treatment option for this population.

Another study evaluated cognitive performance in VNS patients (Sackeim et al. 2001b). Neuropsychological batteries were completed pre- and post-implantation, and no evidence of cognitive deterioration was seen. When compared to baseline performance, improvement was found in motor speed, psychomotor function, language, and executive function. The researchers suggest that the use of VNS in treatment-resistant depression may enhance neurocognitive function, particularly when patients show clinical improvement (Sackeim et al. 2001b).

5.4
Repetitive Transcranial Magnetic Stimulation

Similar to ECT in neurophysiological, neurochemical, and behavioral effects, repetitive transcranial magnetic stimulation (rTMS) is another treatment option for treatment-resistant depression (Janicak and Martis 1998). First introduced in 1985, rTMS is a noninvasive procedure for stimulation of the central nervous system. A small, figure-eight-shaped insulated coil is placed on the scalp, and a rapidly alternating electrical current is transmitted through the wire (Grunhaus et al. 2000). The resultant magnetic pulse depolarizes neurons in a localized area of the brain, inducing ionic flow (Janicak and Martis 1998; Grunhaus et al. 2000). Multiple trains of magnetic pulses at various frequencies (1–20 Hz) are applied for several seconds (Janicak and Martis 1998). The use of rTMS has several advantages compared with ECT. The patient may receive treatment as an outpatient and continue normal activities, since there is no need to induce seizure, no use of anesthesia, and no significant cognitive disruption (Janicak and Martis 1998).

The clinical applicability of rTMS has been greatly enhanced by the development of stimulators capable of delivering frequencies up to 60 Hz (Grunhaus et al. 2000). Several studies have shown significant antidepressant effects for rTMS. One study found a 42% response rate to rTMS in patients with TRD (Figiel et al. 1998).

In a randomized trial, 40 patients referred for ECT were assigned to either ECT or rTMS treatment (Grunhaus et al. 2000). rTMS was performed at 90% motor threshold (MT), and 20 trains were administered at 10 Hz for either 2 or 6 s for 20 treatment days. Study results indicated that patients with psychotic MDD responded significantly better to ECT, while patients with MDD without psychosis responded equally well to both treatments. Comparing across groups, ECT was determined to be a better treatment for depression. Also, research findings indicate that continuation of rTMS for 4 weeks increased its efficacy, but if the patient had minimal response at 2 weeks, an additional 2 weeks generally produced little improvement. The primary advantage to rTMS over ECT is its less invasive nature.

In another randomized clinical trial, the efficacy of rTMS was evaluated in 20 unmedicated patients with treatment-resistant MDD (Berman et al. 2000). Patients were randomly assigned to active or sham rTMS treatment, in which they received 20 2-s, 20-Hz trains delivered at 80% MT with 58-s inter-train intervals for 10 consecutive days. Results showed a significant reduction in depressive symptoms following active treatment, compared to the sham rTMS group. When the response was modest, the effect was short-lived. However, one patient had a robust, enduring treatment response. Assessments of side effects and EEG tracings showed rTMS to be safe, with no persistent adverse events. The overall study findings are consistent with other studies, revealing a small, but statistically significant symptom improvement with rTMS. The authors suggest more work is needed to determine optimal treatment parameters for a more effective and lasting response (Berman et al. 2000).

Additional work has been done to assess the neurocognitive effects of rTMS. In a recent study, 46 normal volunteers were given one session of either right or left prefrontal active rTMS or sham rTMS (Koren et al. 2001). Neuropsychological testing results indicated that all three groups showed improvement over time in processing speed (reaction time) and efficiency. The authors concluded that rTMS, unlike ECT, does not interfere with neuropsychological function, since no adverse cognitive effects were seen. However, it should be noted that this finding is based on a single rTMS session, not the usual treatment course of 10 or more daily sessions.

6
Texas Medication Algorithm Project

The Texas Medication Algorithm Project (TMAP website, accessed 2002) was a multi-site research project that was completed in April 2000. The algorithms that were used were developed by an expert consensus panel, which was made up of national experts, Texas public mental health sector practitioners who were to implement the algorithms, patients, and family members. TMAP was designed to determine the clinical and economic value of the treatment algorithms for schizophrenia, bipolar disorder, and MDD as compared with treatment as usual. An example of the depression algorithm may be seen in Fig. 1. In addition to the medication algorithms, an extensive patient/family education piece and additional clinic support staff were also utilized as interventions.

Over 1,400 patients in 17 public mental health sectors were enrolled in the third phase of the study and followed for a year. Analysis of the results of the project is ongoing. The patient/family education materials, physician manuals, and publications related to this project can be found on the website: http://www.mhmr.state.tx.us/centraloffice/medicaldirector/tmaptoc.html.

7
Sequenced Treatment Alternatives to Relieve Depression

Sequenced Treatment Alternatives to Relieve Depression (Star*D website, accessed 2002) is an ongoing project using an algorithmic approach for the treatment of non-psychotic MDD. Star*D differs from the TMAP algorithm in that its aim is to examine the next "next-step" treatments. All patients enter into treatment at level 1 with the SSRI citalopram. Based on patient response and acceptability, those patients whose symptoms do not remit may progress through three additional levels. Patients are randomly assigned to a "switch therapy" or augmentation therapy. When a patient has a satisfactory response at any given level, they are then entered into the 12-month naturalistic follow-up portion of the study. Included in the treatment protocol is extensive patient and family education.

The researchers in the Star*D project hope to enroll over 4,000 patients in 14 different regional centers over a 5-year period. Independent assessors, blinded to treatment, will administer clinician ratings of patients' depressive symptoms at the entrance and exit from each treatment level. Outcome measures to be evaluated include symptom severity, level of functioning, side effect burden, patient satisfaction/quality of life, as well as health care utilization and cost. The timing of relapse will be assessed during the naturalistic portion of the study. More information on the study can be accessed through the Star*D website: http://www.edc.gsph.pitt.edu/stard.

8
Future Directions

With the availability of newer antidepressant medications, clinicians are better able to treat depression than ever before. Even with these new medications, however, a significant portion of patients remain symptomatic (partial responders) or are classified as having TRD. There are very few data available concerning what step (combination, augmentation, or switching) should be taken next. Research continues to attempt to answer the question of what is "best practice." For example, the Star*D project is attempting to answer questions concerning best "next step" treatment. Data from TMAP and Star*D are likely to form the groundwork for future research examining the use of medications and/or cognitive therapy to treat TRD. A computerized version of the TMAP algorithms, with a decision support portion to the program, will soon be available for clinicians to use (Trivedi et al. 2000). This program is driven by patient response, side effect burden, and length of treatment and will provide prompts to the clinician to promote faster and fuller treatment response. The practice of using an algorithmic approach in treating depression is likely to continue as more economic and efficacy data supporting the use of such algorithms become available in the literature.

References

American Psychiatric Association (1993) Practice guideline for major depressive disorder in adults. Am J Psychiatry 150(suppl 4):1-26

American Psychiatric Association (2000) Practice guideline for major depressive disorder in adults. Am J Psychiatry 157(suppl 4):1-45

American Psychiatric Association (20002) Diagnostic and statistical manual of mental disorders, 4th edition, text revision. American Psychiatric Association, Washington, DC

Amiaz R, Stein O, Dannon PN, Grunhaus L, Schreiber S (1999) Resolution of treatment-refractory depression with naltrexone augmentation of paroxetine: a case report. Psychopharmacology (Berl) 143:433-434

Amsterdam and Bodkin (2000) Transdermal selegiline in the treatment of patients with major depression: a double-blind, placebo-controlled trial.(Abstract). Presented at: the 40th annual meeting of the New Clinical Drug Evaluation Unit (NCDEU), June 1, 2000, Boca Raton, FL. Abstract N 15

Ananth J (1998) Treatment-resistant depression. Psychother Psychosom; 67:61-70

Appelberg BG, Syvalahti EK, Koskinen TE, Mehtonen OP, Muhonen TT, Naukkarinen HH (2001) Patients with severe depression may benefit from buspirone augmentation of selective serotonin reuptake inhibitors: results from a placebo-controlled, randomized, double-blind, placebo wash-in study. J Clin Psychiatry 62:448-52

Aronson R, Offman HJ, Joffe RT (1996) Triiodothyronine augmentation in the treatment of refractory depression. Arch Gen Psychiatry 53:842-848

Artigas F, Perez V, Alvarez E (1994) Pindolol induces a rapid improvement of depressed patients treated with serotonin reuptake inhibitors. Arch Gen Psychiatry 51:248-251

Bader G, Hawley JM, Short DD (1998) Venlafaxine augmentation with methylphenidate for treatment-refractory depression: a case report. J Clin Psychopharmacol 18:255-256

Bakish D (1991) Fluoxetine potentiation by buspirone: three case histories. Can J Psychiatry 36:749-50

Banki CM (1975) Triiodothyronine in the treatment of depression. Orv Hetil 116:2543-2547

Banki CM (1977) Cerebrospinal fluid amine metabolites after combined amitriptyline triiodothyronine treatment of depressed women. Eur J Clin Pharmacol 11:311-315

Bauer M, Dopfmer S (1999) Lithium augmentation in treatment resistant depression: a meta-analysis of placebo controlled studies. J Clin Psychopharmacol 19:427-434

Bauer M, Bschor T, Kunz D, Berghofer A, Strohle A, Muller-Oerlinghausen B (2000) Double-blind, placebo-controlled trial of the use of lithium to augment antidepressant medication in continuation treatment of unipolar major depression. Am J Psychiatry 157:1429-1435

Baumann P, Nil R, Souche A, Montaldi S, Baettig D, Lambert S, Uehlinger C, Kasas A, Amey M, Jonzier-Perey M (1996) A double-blind, placebo controlled study of citalopram with and without lithium in the treatment of therapy-resistant depressive patients: a clinical, pharmacokinetic, and pharmacogenetic investigation. J Clin Psychopharmacol 16:307-314

Beasley CM Jr., Sayler ME Cunningham GE, Weiss AM, Masica DN (1990) Fluoxetine in tricyclic refractory major depressive disorder. J Affect Disord 20:193-200

Benazzi F (1999) Venlafaxine-fluoxetine interaction. J Clin Psychopharmacol 19:96-98

Berman RM, Narasimhan M, Charney DS (1997) Treatment-refractory depression: definitions and characteristics. Depress Anxiety 5:154-164

Berman RM, Narasimhan M, Sanacora G, Miano AP, Hoffman RE, Hu XS, Charney DS, Butros NN (2000) A randomized clinical trial of repetitive transcranial magnetic stimulation in the treatment of major depression. Biol Psychiatry 47:332-337

Bhatara VS, Magnus RD, Paul KL, Preskorn SH (1998) Serotonin syndrome induced by venlafaxine and fluoxetine: a case study in polypharmacy and potential pharmacodynamic and pharmacokinetic mechanisms. Ann Pharmacother 32:432–6

Blier P, Bergeron R (1995) Effectiveness of pindolol with selected antidepressant drugs in the treatment of major depression. J Clin Psychopharmacol 15:217-222

Blier P, Bergeron R (1998) The use of pindolol to potentiate antidepressant medication. J Clin Psychiatry 59(suppl 5):16-23

Bodkin JA, Lasser RA, Winmes JD Jr, Gardner DM, Baldessarini RJ (1997) Combining serotonin reuptake inhibitors and bupropion in partial responders to antidepressant monotherapy. J Clin Psychiatry 58:137–145

Bondolfi G, Chautems C, Rochat B, Bertschy G, Baumann P (1996) Non-response to citalopram in depressive patients: pharmacokinetic and clinical consequences of a fluvoxamine augmentation Psychopharmacology (Berl) 128:421-425

Brown WA, Harrison W (1995) Are patients who are intolerant to one serotonin selective reuptake inhibitor intolerant to another? J Clin Psychiatry; 56:30–34

Browne M, Lapierre YD, Hrdina PD, Horn E (1990) Lithium as an adjunct in the treatment of major depression. Int Clin Psychopharmacol 5:103-110

Bschor T, Canata B, Muller-Oerlinghaussen B, Bauer M (2001) Predictors of response to lithium augmentation in tricyclic antidepressant-resistant depression. J Affect Disord 64:261-5

Carpenter LL, Jocie Z, Hall JM, Rasmussen SA, Price LH. (1999) Mirtazapine augmentation in the treatment of refractory depression. J Clin Psychiatry; 60:45–49

Catterson ML, Preskorn SH (1996) Double-blind crossover study of mirtazapine in depressed patients with major depression (abstract). Presented at: the 149th annual meeting of the American Psychiatric Association. May 6, 1996; New York, NY. Abstract NR 157:110

Catz SL, Gore-Felton C, McClure JB (2002) Psychological distress among minority and low-income women living with HIV. Behav Med 28:53-60

Cournoyer G, deMontigny D, Oullette J, Leblare G, Langlois R, Elie R (1984) Lithium addition in tricyclic-resistant unipolar depression: a placebo-controlled study. Presented at the Congress Internationale Neuro-Psychopharmacologicum Congress June 19–23, Florence, Italy, 14:F-177

Davidson JR, Meoni P, Haudiquet V, Cantillon M, Hackett D (2002) Achieving remission with venlafaxine and fluoxetine in major depression: its relationship to anxiety symptoms. Depress Anxiety 16:4-13

Davis LL, Kabel D, Patel D, Choate AD, Foslien-Nash C, Gurguis GN, Kramer GL, Petty F (1996) Valproate as an antidepressant in major depressive disorder. Psychopharmacol Bull 32:647-652

de Battista C, Mueller K (2001) Is electroconvulsive therapy effective for the depressed patient with comorbid borderline personality disorder? J ECT 17:91-98

de Montigny CF, Grunberg AF, Deschenes JP (1981) Lithium induces rapid relief of depression in tricyclic antidepressant non-responders. Br J Psychiatry 138:252-256

de Montigny C, Cournoyer G, Morissette R, Langlois R, Caille G (1983) Lithium carbonate addition in tricyclic antidepressant-resistant unipolar depression: correlations with the neurobiologic actions of tricyclic antidepressant drugs and lithium ion on the serotonin system. Arch Gen Psychiatry 40:1327-1334

Depression Guideline Panel (1993) Clinical practice guideline. Depression in primary care: Volume 2: Treatment of major depression. U.S. Department of Health and Human Services-Agency for Health Care Policy Research, Publication No. 93–0551

Dimitriou EC, Dimitriou CE (1998) Buspirone augmentation of antidepressant therapy. J Clin Psychopharmacol 18:465-469

Dursun SM, Patel JK, Drybala T, Shinkwin R, Drybala G, Reveley MA (2001) Effects of antidepressant treatments on first-ECT seizure duration in depression. Prog Neuropsychopharmacol Biol Psychiatry 25:437-443

Dwight MM, Arnold LM, O'Brien H, Metzger R, Morris-Park E, Keck PE (1998) An open clinical trial of venlafaxine treatment of fibromyalgia. Psychosomatics 39:14-17

Earle BV (1970) Thyroid hormone or tricyclic antidepressants in resistant depression. Am J Psychiatry 126:1667-1669

Echols JC, Naidoo U, Salzman C (2000) SAMe (S-adenosylmethionine). Harvard Rev Psychiatry 8:84-90

Ereshefsky L, Dugan D. (2000) Review of the pharmacokinetics, pharmaogenetics, and drug interaction potential of antidepressants: focus on venlafaxine. Depression and Anxiety; 12 (suppl 1): 30–44

Fava M (2001) Augmentation and combination strategies in treatment-resistant depression. J Clin Psychiatry 62(suppl 18):4-11

Fava M, Davidson KG (1996) Definition and epidemiology of treatment resistant depression. Psychiatr Clin North Am 19:179–200

Fava M, Rosenbaum JF, McGrath PJ, Stewart JW, Amsterdam JD, Quitkin FM (1994) Lithium and tricyclic augmentation of fluoxetine treatment for resistant major depression: a double-blind, controlled study. Am J Psychiatry 151:1372-1374

Fava GA, Savron G, Grandi S, Rafanelli C (1997) Cognitive-behavioral management of drug-resistant major depressive disorder. J Clin Psychiatry 58:278-282

Fava GA, Rafanelli C, Grandi S, Conti S, Belluardo P (1998) Prevention of recurrent depression with cognitive behavioral therapy: preliminary findings. Arch Gen Psychiatry 55:816-820

Fava M, Dunner DL, Greist JH, Preskorn SH, Trivedi MH, Zajecka J, Cohen M (2001) Efficacy and safety of mirtazapine in major depressive disorder patients after SSRI treatment failure: an open-label trial. J Clin Psychiatry 62:413–420

Fawcett J (1997) The detection and consequences of anxiety in clinical depression. J Clin Psychiatry 58(Suppl 8):35-40

Fawcett J, Kravitz HM, Zajecka JM, Schaff MR (1991) CNS stimulant potentiation of monoamine oxidase inhibitors in treatment-refractory depression. J Clin Psychopharmacol 11:127-132

Figiel GS, Epstein C, McDonald WM, Amazon-Leece J, Figiel L, Saldivia A, Glover S (1998) The use of rapid-rate transcranial magnetic stimulation (rTMS) in refractory depressed patients. J Neuropsychiatry Clin Neurosci 10:20-25

Fink M (2001) Convulsive therapy: a review of the first 55 years. J Affect Disord; 63:1-15

Ford DE (2000) Managing patients with depression: is primary care up to the challenge? J Gen Intern Med 15(5):344–345

Fox HA (2001) Extended continuation and maintenance ECT for long-lasting episodes of major depression. J ECT 17:60-64

George MS, Sackeim HA, Rush AJ, Marangell LB, Nahas Z, Husain MM, Lisanby S, Burt T, Goldman J, Ballenger JC (2000) Vagus nerve stimulation: a new tool for brain research and therapy. Biol Psychiatry 47:287-295

Ghaemi SN, Sachs GS, Chiou AM, Pandurangi AK, Goodwin K (1999) Is bipolar disorder still underdiagnosed? Are antidepressants overutilized? J Affect Disord 52:135-44

Gitlin MJ, Weiner H, Fairbanks L, Hershman JM, Friedfeld N (1987) Failure of T3 to potentiate tricyclic antidepressant response. J Affect Disord 13:267-272

Goodwin FK, Prange AJ Jr, Post RM, Muscettola G, Lipton MA (1982) Potentiation of antidepressant effects by L-triiodothyronine in tricyclic non-responders. Am J Psychiatry 139:34-38

Greenberg PE, Stiglin LE, Finkelstein SN, Berndt ER (1993) The economic burden of depression in 1990. J Clin Psychiatry 54:405–418

Grunhaus L, Dannon PN, Schreiber S, Dolberg OH, Amiaz R, Ziv R, Lefkifker E (2000) Repetitive transcranial magnetic stimulation is as effective as electroconvulsive therapy in the treatment of nondelusional major depressive disorder: an open study. Biol Psychiatry 47:314-324

Hamilton M (1967) Development of a rating scale for primary depressive illness. Br J Soc Clin Psychol 6:278-296
Harvey AT, Preskorn SH (2001) Fluoxetine pharmacokinetics and efect on CYP2C19 in young and elderly volunteers. J Clin Psychopharmacol 2001;21:161-166
Heninger GR, Charney DS, Sternberg DE (1983) Lithium carbonate augmentation of antidepressant action: an effective prescription for treatment refractory depression. Arch Gen Psychiatry 40:1335-1342
Hickie I (1999) Nefazodone for patients with chronic fatigue syndrome. Aust N Z J Psychiatry 33:278-280
Hirose S, Ashby CR (2002) An open pilot study combining risperidone and a selective serotonin reuptake inhibitor as initial antidepressant therapy. J Clin Psychiatry 63:733–736
Hirschfeld KM (1999) Efficacy of SSRIs and newer antidepressants in severe depression: comparison with TCAs. J Clin Psychiatry; 60:326-35
Hunchak J (1997) SSRI combination treatment for depression. Can J Psychiatry 42:531-532 (letter)
Jacobson FM (1991) Possible augmentation of antidepressant response by buspirone. J Clin Psychiatry 52:217-220
Janicak PG, Martis B (1998) Strategies for treatment-resistant depression. Clinical Cornerstone, Depression 1:58-67Joffe RT (1998) The use of thyroid supplements to augment antidepressant medication. J Clin Psychiatry 59(suppl 5):26-29
Joffe RT, Marriott M (2000) Thyroid hormone levels and recurrence of major depression. Am J Psychiatry 157:1689-1691
Joffe RT, Schuler DR (1993) An open label study of buspirone augmentation of serotonin reuptake inhibitors in refractory depression. J Clin Psychiatry 54:269-271
Joffe RT, Singer W (1990) A comparison of triiodothyronine and thyroxine in the potentiation of tricyclic antidepressants. Psychol Res 32:241-251
Joffe RT, Singer W, Levitt AJ, MacDonald C (1993) A placebo-controlled comparison of lithium and triiodothyronine augmentation of tricyclic antidepressants in unipolar refractory depression. Arch Gen Psychiatry 50:387-393
Joffe RT, Levitt AJ, Sokdov ST, Young LT (1996) Response to an open trial of a second SSRI in major depression. J Clin Psychiatry; 57:114–5
Kantor D, McNevin S, Leichner P, Harper D, Krenn M (1986) The benefit of lithium carbonate adjunct in refractory depression: fact or fiction? Can J Psychiatry 31:416-418
Katon W, Von Korff M, Lin E, Bush T, Ormel J (1992) Adequacy and duration of antidepressant treatment in primary care. Med Care 30:67-76
Kaplan M (2000) Atypical antipsychotics for treatment of mixed depression and anxiety. J Clin Psychiatry 61:388-389
Katona CL, Abou-Saleh MT, Harrison DA, Nairac BA, Edwards DR, Lock T, Burns RA, Robertson MM (1995) Placebo controlled trial of lithium augmentation of fluoxetine and lofepramine. Br J Psychiatry 166:80-86
Keller MB, Klerman GL, Lavori PW, Coryell W, Endicott J, Taylor J (1984) Long term outcome of episodes of major depression: clinical and public health significance. JAMA 252:788–793
Keller MB, Lavori PW, Klerman GL, Andreasen NC, Endicott J, Coryell W, Fawcett J, Rice JP, Hirschfeld RM (1986) Low levels and lack of predictors of somatotherapy and psychotherapy received by depressed patients. Arch Gen Psychiatry 43:458-466
Keller MB, Gelenberg AJ, Hirschfeld RM, Rush AJ, Thase ME, Kocsis JH, Markowitz JC, Fawcett JA, Koran LM, Klein DN, Russell JM, Kornstein SG, McCullough JP, Davis SM, Harrison WM (1998) The treatment of chronic depression, part 2: A double-blind, randomized trial of sertraline and imipramine. J Clin Psychiatry 59:598-607
Keller MB, McCullough JP, Klein DN, Arnow B, Dunner DL, Gelenberg AJ, Markowitz JC, Nemeroff CB, Russell JM, Thase ME, Trivedi MH, Zajecka J (2000) A comparison of

nefazodone, the cognitive behavioral analysis system of psychotherapy, and their combination for the treatment of chronic depression. New Engl J Med; 342:1462-1470

Kennedy SH, McCann SM, Masellis M, McIntyre RS, Raskin J, McKay G, Baker GB (2002) Combining bupropion SR with venlafaxine, paroxetine, or fluoxetine: a preliminary report on pharmacokinetic, therapeutic, and sexual dysfunction effects. J Clin Psychiatry 63:181-186

Klenk AS, Rosenbaum JF (2000) Efficacy of venlafaxine in the treatment of severe depression. Depression and Anxiety; 12 (suppl 1): 50–4

Klufas A, Thompson D (2001) Topiramate-induced depression. Am J Psychiatry 158:1736

Kocsis JH, Friedman RA, Markowitz JC, Leon AC, Miller NL, Gniwesch L, Parides M (1996) Maintenance therapy for chronic depression: a controlled clinical trial of desipramine. Arch Gen Psychiatry 53:769-774

Koran LM, Gelenberg AJ, Kornstein AJ, Howland RH, Friedman RA, DeBattista C, Klein D, Kocsis JH, Schatzberg AF, Thase ME, Rush AJ, Hirshfeld RM, LaVange LM, Keller MD (2001) Sertraline versus imipramine to prevent relapse in chronic depression. J Affect Disord 65:27-36

Koren D, Shefer O, Chistyakov A, Kaplan B, Feinsod M, Klein E (2001) Neuropsychological effects of prefrontal slow rTMS in normal volunteers: a double-blind sham-controlled study. J Clin Exp Neuropsychol 23:424-430

Kornstein SG, Schneider RK (2001) Clinical features of treatment-resistant depression. J Clin Psychiatry 62(Suppl 16):18-25

Koutouvidis N, Pratikakis M, Fotiadou A (1999) The use of mirtazapine ina group o f11 patients following poor compliance to selective serotonin reuptake inhibitor treatment due to sexual dysfunction. Int Clin Psychopharmacol 14:253–255

Landen M, Bjorling G, Agren H, Fahlen T (1998) A randomized, double-blind, placebo-controlled trial of buspirone in combination with an SSRI in patients with treatment-refractory depression. J Clin Psychiatry 59:664-668

Levine J, Barak Y, Gonzalves M, Szor H, Elizur A, Kofman O, Belmaker RH (1995) A double-blind controlled study of inositol treatment of depression. Am J Psychiatry 152:792-794

Lin EH, Von Korff M, Katon W, Bush T, Simon GE, Walker E, Robinson P (1995) The role of the primary care physician in patients' adherence to antidepressant therapy. Med Care 33:67–74

Maes M, Vandoolaeghe E, Desnyder R (1996) Efficacy of treatment with trazadone in combination with pindolol or fluoxetine in major depression. J Affect Disord 41:201-210

Maltese TM (1999) Adjunctive lamotrigine treatment for major depression. Am J Psychiatry 156;1833

Marangell LB (2001) Switching antidepressants for treatment-resistant major depression. J Clin Psychiatry 62(suppl 18):12-17

Marshall JD, Liebowitz MR (1996) Paroxetine/bupropion combination treatment for refractory depression. J Clin Psychopharmacol 16:80–1

Martin R, Kuzniecky R, Ho S, Hetherington H, Pan J, Sinclair K, Gilliam F, Faught E. (1999) Cognitive effects of topiramate, gabapentin and lamotrigine in healthy young adults. Neurology 52:321-327

Martinez D, Broft A, Laruelle M (2000) Pindolol augmentation of antidepressant treatment: recent contributions from brain imaging studies. Biol Psychiatry 48:844-853

Masand PS, Anand VS, Tanquary JF (1998) Psychostimulant augmentation of second generation antidepressants: a case series. Depress Anxiety 7:89-91

Menza MA, Kaufman KR, Castellanos AM (2000) Modafinil augmentation of antidepressant treatment in depression. J Clin Psychiatry 61:378-381

Milo TJ, Kaufman GE, Barnes WE, Konopka LM, Crayton JW, Ringelstein JG, Shirazi PH (2001) Changes in regional cerebral blood flow after electroconvulsive therapy for depression. J ECT 17:15-21

Montgomery SA, Asberg M (1979) A new depression scale designed to be sensitive to change. Br J Psychiatry 134:382-389

Moreno FA, Gelenberg AJ, Bachar K, Delgado PL (1997) Pindolol augmentation of treatment-resistant depressed patients. J Clin Psychiatry 58:437-439

Nelson JC (1997) Treatment of refractory depression. Depress Anxiety; 5:165-174

Nelson JC (1998) Overcoming treatment resistance in depression. J Clin Psychiatry 59(suppl 16):13-19

Nemets B, Mishory A, Levine J, Belmaker RH (1999) Inositol addition does not improve depression in SSRI treatment failures. J Neural Transm 106:795-798

Nichols GA, Brown JB (2003) Unadjusted and adjusted prevalence of diagnosed depression in type 2 diabetes. Diabetes Care 26:744-749

Nierenberg AA, Keefe BR, Leslie VC, Alpert JE, Pava JA, Worthington JJ 3rd, Rosenbaum JF, Fava M (1999) Residual symptoms in depressed patients who respond acutely to fluoxetine. J Clin Psychiatry 60:221-225

O'Connor M, Silver H (1998) Adding risperidone to selective serotonin reuptake inhibitors improves chronic depression. J Clin Psychopharmacol 18:89-91

Ogura C, Okuma T, Uchida Y, Imai S, Yogi H, Sunami Y (1974) Combined thyroid (triiodothyronine)-tricyclic antidepressant treatment in depressive states. Folia Psychiatr Neurol Jpn 28:179-186

Olver JS, Cryan JF, Burrows GD, Norman TR (2000) Pindolol augmentation of antidepressants: a review and rationale. Aust. N Z J Psychiatry; 34:71–79

Osteroff RB, Nelson JC (1999) Risperidone augmentation of selective serotonin reuptake inhibitors in major depression. J Clin Psychiatry 60:256-259

Owen JR, Nemeroff CB (1998) New antidpressant and the cytochrome P450 system: focus on venlafaxine, nefazadone and mirtazapine. Depression and Anxiety 7 (suppl 1):24–32

Paykel ES, Priest RG (1992) Recognition and management of depression in general practice: consensus statement. BMJ 305:1198-1202

Paykel ES, Scott J, Teasdale J, Johnson AL, Garland A, Moore R, Jenaway A, Cornwall PL, Hayhurst H, Abbott R, Pope M (1999) Prevention of relapse in residual depression by cognitive therapy: a controlled trial. Arch Gen Psychiatry 56:829–835

Perez V, Gilaberte I, Faries D, Alvarez E, Artigas F (1997) Randomized, double-blind, placebo-controlled trial of pindolol in combination with fluoxetine antidepressant treatment. Lancet 349:1594-1597

Perez V, Soler J, Puigdemont D, Alvarez E, Artigas F (1999) A double-blind, placebo-controlled trial of pindolol augmentation in depressive patients resistant to serotonin reuptake inhibitors. Arch Gen Psychiatry 56:375-379

Pichot W, Ansseau M (2001) Addition of olanzapine for treatment-resistant depression. Am J Psychiatry 158;1737-1738

Quitkin FM, Rabkin JG, Ross D, McGrath PJ (1984) Duration of antidepressant drug treatment: What is an adequate trial? Arch Gen Psychiatry 41:238-245

Quitkin FM, McGrath PJ, Stewart JW, Ocepek-Welikson K, Taylor BP, Nunes E, Deliyannides D, Agosti V, Donovan SJ, Petkova E, Klein DF (1996) Chronological milestones to guide drug change. When should clinicians switch antidepressants? Arch Gen Psychiatry 53:785-792

Rami-Gonzalez L, Bernardo M, Boget T, Salamero M, Gil-Verona JA, Junque C (2001) Subtypes of memory dysfunction associated with ECT: characteristics and neurobiological bases. J ECT 17:129-135

Rickels K, Amsterdam J, Clary C, Hassman J, London J, Puzzuoli G, Schweizer E (1990) Buspirone in depressed outpatients: a controlled study. Psychopharmacol Bull 26:163-167

Riedinger MS, Dracup KA, Brecht ML (2002) Quality of life in women with heart failure, normative groups, and patients with other chronic conditions. Am J Crit Care 11:211-219

Rothschild AJ (1996) Management of psychotic treatment-resistant depression. Psychiatr Clin North Am 19:237-251

Rouillon F, Gorwood P (1998) The use of lithium to augment antidepressant medication. J Clin Psychiatry 59(suppl 5):32-39

Rush AJ, Trivedi MH (1995) Treating depression to remission. Psychiatr Ann 25:704-705, 709

Rush AJ, George MS, Sackeim HA, Marangell LB, Husain MM, Giller C, Nahas Z, Haines S, Simpson RK Jr, Goodman R (2000) Vagus nerve stimulation (VNS) for treatment-resistant depression: a multicenter study. Biol Psychiatry 47:276-286

Rybakowski JK, Suwalska A, Chlopocka-Wozniak M (1999) Potentiation of antidepressants with lithium or carbamazepine in treatment-resistant depression. Neuropsychobiology 40:134-139

Sackeim HA, Haskett RF, Mulsant BH, Thase ME, Mann JJ, Pettinati HM, Greenberg RM, Crowe RR, Cooper TB, Prudic J (2001a) Continuation pharmacotherapy in the prevention of relapse following electroconvulsive therapy: a randomized controlled trial. JAMA 285:1299-1307

Sackeim HA, Keilp JG, Rush AJ, George MS, Marangell LB, Dormer JS, Burt T, Lisanby SH, Husain M, Cullum CM, Oliver N, Zboyan H (2001b) The effects of vagus nerve stimulation on cognitive performance in patients with treatment-resistant depression. Neuropsychiatry, Neuropsychology and Behavioral Neurology 14:53-62

Schmidt PJ, Nieman L, Danaceau MA, Tobin MB, Roca CA, Murphy JH, Rubinow DR (2000) Estrogen replacement in perimenopausal-related depression: a preliminary report. Am J Obstet Gynecol 183:414-420

Schopf J, Baumann P, Lemarchand T, Rey M (1989) Treatment of endogenous depressions resistant to tricyclic antidepressants or related drugs by lithium addition. Pharmacopsychiatry 22:183-187

Schwarcz G, Halaris A, Baxter L, Escobar J, Thompson M, Young M (1984) Normal thyroid function in desipramine non-responders converted to responders by the addition of L-triiodothyronine. Am J Psychiatry 141:1614-1616

Schweizer EE, Amsterdam J, Rickels K, Kaplan M, Droba M (1986) Open trial of buspirone in the treatment of major depressive disorder. Psychopharmacol Bull 22:183-185

Scott J, Teasdale JD, Paykel ES, Johnson AL, Abbott R, Hayhurst H, Moore R, Garland A (2000) Effects of cognitive therapy on psychological symptoms and social functioning in residual depression. Br J Psychiatry 177:440-446

Sequenced Treatment Alternatives to Relieve Depression (STAR*D) Available at: http://www.edc.gsph.pitt.edu/stard/. (Accessed October 17, 2002)

Shelton RC (1999) Mood-stabilizing drugs in depression. J Clin Psychiatry 60(suppl 5):37-40

Shelton RC, Tollefson GD, Tohen M (2001) A novel augmentation strategy for treating resistant major dpression. Am J Psychiatry 158:131-134

Simon GE (2002) Evidence review: efficacy and effectiveness of antidepressant treatment in primary care. Gen Hosp Psychiatry 24:213–24

Soares CN, Almeida OP, Joffe H, Cohen L (2001) Efficacy of estrodial for the treatment of depressive disorders in perimenopausal woman. Arch Gen Psychiatry 58:529-534

Stahl SM (2001) Sex and psychopharmacology: is natural estrogen a psychotropic drug in women? Arch Gen Psychiatry 58:537-538

Stein G, Bernadt M (1993) Lithium augmentation therapy in tricyclic-resistant depression: a controlled trial using lithium in low and normal doses. Br J Psychiatry 162:634-640

Stoll AL, Pillay SS, Diamond L, Workum SB, Cole JO (1996) Methylphenidate augmentation of serotonin selective reuptake inhibitors: a case series. J Clin Psychiatry 57:72-76

Sylvester AP, Mulsant BH, Chengappa KNR, Sandman AR, Haskett RF (2000) Use of electroconvulsive therapy in a state hospital: a 10-year review. J Clin Psychiatry 61:534-539

Texas Medication Algorithm Project (TMAP) Available at http://www.mhmr.state.tx.us/centraloffice/medicaldirector/tmaptoc.html (accessed October 17, 2002)

Thase ME, Howland RH (1994) Refractory depression: relevance of psychosocial factors and therapies. Psychiatric Annals 24:232-240

Thase ME, Rush AJ (1997) When at first you don't succeed: sequential strategies for antidepressant nonresponders. J Clin Psychiatry. 1997; 58(suppl 13): 23–29

Thase Me, Kupfer DJ, Jarrett DB (1989a) Treatment of imipramine-resistant recurrent depression, I: an open clinical trial of adjunctive L-triiodothyronine. J Clin Psychiatry;50:385–388

Thase ME, Kupfer DJ, Frank E, Jarrett DB (1989b) Treatment of imipramine-resistant recurrent depression, II: an open clinical trial of lithium augmentation. J Clin Psychiatry 50:413-417

Thase ME, Trivedi MH, Rush AJ (1995) MAOIs in the contemporary treatment of depression. Neuropsychopharmacology; 12:185–219

Thase ME, Howland RH, Friedman ES (1998) Treating antidepressant nonresponders with augmentation strategies: an overview. J Clin Psychiatry 59 (suppl 5): 5–12

Thase ME, Friedman ES, Howland RH (2000) Venlafaxine and treatment-resistant depression. Depress Anxiety 12 (Suppl 1):55–62

Thase ME, Friedman ES, Howland RH (2001) Management of treatment-resistant depression: psychotherapeutic perspectives. J Clin Psychiatry 62(suppl 18):18-24

Trivedi MH, Kleiber BA (2001) Algorithm for the treatment of chronic depression. J Clin Psychiatry 62(suppl 6):22-29

Trivedi MH, Kern JK, Baker SM, Altshuler KZ (2000) Computerized medication algorithms and decision support systems in major psychiatric disorders. J Psychiatr Pract 6:237-246

Tsutsui S, Yamazaki Y, Namba T, Tsushima M (1979) Combined therapy of T3, and antidepressants in depression. J Int Med Res 7:138-146

Vinar O, Vinarova E, Horacek J (1996) Pindolol accelerates the therapeutic action of selective serotonin reuptake inhibitors (SSRI) in depression. Homeostasis 37:93-95

Wharton RN, Perel JM, Dayton PG, Malitz S (1971) A potential clinical use for methylphenidate with tricyclic antidpressants. Am J Psychiatry 127:1619-1625

Wolkowitz OM, Reus VI, Roberts E, Manfredi F, Chan T, Raum WJ, Ormiston S, Johnson R, Canick J, Brizendine L, Weingartner H (1997) Dehydroepiandrosterone (DHEA) treatment of depression. Biol Psychiatry 41:311-318

Wolkowitz OM, Reus VI, Keebler A, Nelson N, Friedland M, Brizendine L, Roberts E (1999) Double-blind treatment of major depression with dehydroepiandosterone. Am J Psychiatry;156:646-649

World Health Organization (1996) Diagnostic and management guidelines for mental disorders in primary care ICD-10 chapter V primary care version. Hogrefe and Huber, Bern

Zarate CA, Kando JC, Tohen M, Weiss MK, Cole JO (1996) Does intolerance or lack of response with fluoxetine predict the same will happen with sertraline? J Clin Psychiatry; 57:67-71

Zusky PM, Biederman J, Rosenbaum JF, Manschreck TC, Gross CC, Weilberg JB, Gastfriend DR (1988) Adjunct low dose lithium carbonate in treratment-resistant depression: a placebo controlled double bind study. J Clin Psychopharmacol 8:120-124

Personality Disorders

M. L. Wainberg[1] · A. J. Kolodny[2] · L. J. Siever[3]

[1] Columbia College of Physicians and Surgeons, Unit 112, 1051 Riverside Drive, New York, NY 10032, USA
e-mail: mlw35@columbia.edu

[2] The Mount Sinai Medical Center, One Gustave L Levy Place, New York, NY 10029, USA

[3] Mount Sinai School of Medicine, One Gustave L Levy Place, New York, NY 10029, USA

1	Introduction	490
2	**Cognitive Disorganization**	492
2.1	Deficit-Like Symptoms: Psychobiological Studies	492
2.2	Deficit-Like Symptoms: Psychopharmacological Studies	494
3	**Impulsivity and Affective Instability**	494
3.1	Impulsivity/Aggressive Behavior: Psychobiological Studies	495
3.2	Impulsivity/Aggressive Behavior: Psychopharmacological Studies	497
3.3	Affective Instability: Psychobiological Studies	500
3.4	Affective Instability: Psychopharmacological Studies	502
4	**Anxiety Threshold**	504
4.1	Anxiety Threshold: Psychobiological Studies	504
4.2	Anxiety Threshold: Psychopharmacological Studies	505
5	**Practical Issues in the Treatment of Personality Disorders**	506
6	**Conclusions**	507
	References	507

Abstract Recent years have witnessed an important growth in the understanding and treatment of persons with Axis II disorders. Research into the biological bases of Axis II disorders increasingly suggests that these disorders ought to be regarded less as a discrete category and more as existing on a spectrum with their Axis I cohorts. This chapter summarizes these studies as well as the studies into what pharmacological agents can contribute to the amelioration of symptoms. Each section of the chapter takes up a different "dimension," or grouping, of behavioral traits exhibited by persons with Axis II disorders. Each section first summarizes the studies to date into the psychobiology underlying these traits and then discusses the various psychopharmacological approaches to target these traits. The chapter's primary focus, however, is on the use of antidepressants to treat many of the common symptoms underlying personality disorders. Studies evaluating the use of selective serotonin reuptake inhibitors (SSRIs), tricyclic antidepressants (TCAs), and monoamine oxidase inhibitors (MAOIs) in treating impulsivity, affective instability, and anxiety are described. Although the emphasis is on antidepressants, studies that have looked into the treatment of these traits with other medications, such as anticonvulsants, beta-blockers, and benzodiazepines, are also discussed. Many agents have shown promise in curbing the traits that often accompany personality disorders, although the results have often been mixed. The findings reported on in this chapter suggest potential interventions and point the way toward future avenues of research into the clinical psychopharmacological management of patients with personality disorders.

Keywords Personality disorders · Axis II · Antidepressants · Psychopharmacology · Psychobiology · Cognitive disorganization · Impulsivity · Affective instability · Anxiety

1
Introduction

Pharmacotherapy for the treatment of Axis I disorders is considered an essential part of standard treatment. Personality disorders have been explained mostly in terms of psychodynamic and developmental models, implying a treatment that does not include psychopharmacological agents. However, an important body of research now suggests that biological factors are implicated in the pathogenesis of personality disorders, and that pharmacological agents can contribute to the amelioration of symptoms, constituting a suitable therapeutic intervention.

The pharmacological treatment of personality disorders is a difficult task, due in part to the heterogeneity of patients meeting DSM-IV criteria (American Psychiatric Association 1994) for a particular personality disorder. Successful pharmacological approaches have been obtained in treating symptom clusters, targeting, for example, patients' depression or psychotic-like symptoms that resemble those of their Axis I cohorts. Research into the neuropsychopharmacology of personality disorders has suggested that the enduring traits found in these disorders actually reflect an underlying biology—similar to that of the Axis I disorders—and that they are treatable with psychopharmacological interventions. Siever and Davis (1991) suggested a dimensional model, conceptualizing Axis II personality disorders on a continuum with Axis I disorders. The personality clusters, such as the "odd," the "dramatic," and the "anxious" clusters, can be translated into dimensions of "cognitive disorganization," "impulsivity and affective instability," and "anxiety" respectively. These dimensions can then provide a framework for investigating the psychopharmacological treatment of Axis II disorders, taking into consideration psychological traits and neurobiology (Table 1).

The dimension of cognitive disorganization—referring primarily to the schizophrenia spectrum disorders—is characterized by impairment in the ca-

Table 1 Dimensional model of Axis II personality disorders

	Mood regulation	Impulse control	Cognitive organization	Anxiety threshold
Schizotypal			*	
Schizoid			*	
Paranoid			*	
Borderline	*	*		*
Histrionic	*	*		
Antisocial		*		
Narcissistic		*		
Avoidant	*			*
Dependent	*			*
Obsessive–Compulsive				*

pacity to attend and select relevant information from the environment. Serious distortion of this capacity is observed in patients with chronic schizophrenia, yet is also manifested in the psychotic-like symptoms of patients with schizotypal personality disorder. Impulsivity, a low threshold for psychomotor action or aggressive behavior in reaction to environmental stimuli, is present in borderline, histrionic, and antisocial personality disorders. Similarly, borderline, narcissistic, and histrionic personality disorders are characterized by affective instability, with dramatic shifts in affect over short periods of time and hypersensitivity to changes in the environment. In anxiety spectrum disorders, such as generalized anxiety disorder, obsessive–compulsive disorder, and phobias, anticipation of the feared consequences of actions and thoughts results in anxiety with concomitant inhibition of the behaviors and/or associated autonomic arousal, which is also prominent in avoidant, dependent, and obsessive–compulsive personality disorder. Biological research on these phenomenological clinical dimensions may offer a basis for pharmacological interventions targeting the specific symptoms defined in each dimension.

This review concentrates on evidence regarding the biology of personality disorders and symptom-targeted psychopharmacotherapy for these dimensions, and focuses specifically on presentations for which the appropriate pharmacological intervention may be the use of an antidepressant. Few placebo-controlled pharmacological trials in personality disorders have been done. This chapter addresses the possible use of antidepressants in patients with personality disorders; it describes trends in the data, offers recommendations for interventions, and discusses possible future directions for research on the use of antidepressants in the clinical psychopharmacological management of patients with personality disorders. While this chapter addresses the use of antidepressants in the treatment of personality disorders, an extensive range of psychotherapies, which are beyond the scope of this chapter, are suitable for use in combined psychotherapy and psychopharmacology treatment.

2
Cognitive Disorganization

The DSM-IV personality disorders within the schizophrenia spectrum include schizotypal, paranoid, and schizoid personality disorders. Schizotypal personality disorder is the best studied, since it is the most severe and most closely resembles schizophrenia in biology, phenomenology, and genetics (Siever et al. 1992). A core finding in schizotypal personality disorder is a dysfunction in perceptual and/or cognitive organization with consequent impairment in attentional processes. Both psychotic-like symptoms and deficit-related symptoms/social detachment are clinical characteristics of schizotypal personality disorder (Kendler 1985). Initial studies, which focused on defining the similarities between schizophrenia and schizotypal personality disorder, have been superseded by studies which center on correlates of psychotic-like and deficit-like symptoms of the schizophrenia-related disorders (Siever et al. 1993a).

Plasma homovanillic acid (HVA), a major dopamine metabolite, has been found to be significantly higher and to be associated with psychotic-like symptoms in patients with schizotypal personality disorder compared with patients with other personality disorders and normal controls (Siever et al. 1991, 1993a).

Since the psychopharmacological trials investigating the treatment of prominent psychotic-like symptoms in moderately to severely impaired patients with schizotypal personality disorder have mainly used low-dose antipsychotics targeting the dopamine system (Goldberg et al. 1986; Soloff et al. 1986b, 1989; Coccaro 1993), they will not be discussed in this review. It is worth mentioning that psychostimulants such as amphetamines can worsen pre-existing psychotic-like symptoms in patients with schizotypal personality disorder (Schulz et al. 1988) and clinical experience suggests that antidepressants may, in some instances, also have this deleterious effect.

2.1
Deficit-Like Symptoms: Psychobiological Studies

The interpersonal deficits that are found in the schizophrenia-related personality disorders may be due to impaired cortical processing of complex stimuli required for the synchronized reciprocal interaction and selection of information from others, as well as deficits in attachment behavior. Familial/genetic studies suggest an increased prevalence of schizophrenia-related disorders in the relatives of those with schizophrenia compared with control groups (Kety et al. 1975; Kendler et al. 1981, 1991; Gottesman and Shields 1982; Gunderson et al. 1983; Kendler and Gruenberg 1984; Baron et al. 1985; Kety 1988; Onstad et al. 1991), as well as increased morbid risk for schizophrenia-related disorders (Siever et al. 1990) and schizophrenia (Schulz et al. 1986; Battaglia et al. 1991) in relatives of patients with schizotypal personality disorder, compared with patients with non-schizophrenia-related personality disorders. The interactional patterns might be inappropriately set in infancy in a neurologically immature child impairing the development of appropriate patterns of attachment and interaction (Fish 1987; Siever and Davis 1991).

Multiple studies have reported an association between the deficit-like symptoms in schizophrenia and the schizophrenia spectrum personality disorders and deficiencies in information processing and performance on tests of prefrontal and frontal functioning (Wainberg et al. 1993). Patients with schizotypal personality disorder perform worse on the Wisconsin Card Sort Test (WCST) (Berg 1948; Siegel et al. 1996) and the California Verbal Learning Test (Bergman et al. 1998) than normal subjects and patients with other non-odd cluster personality disorders (Biederman et al. 1991). An alteration in brain structure as found in imaging studies (Siever et al. 1993a; Kiranne and Siever 2000), and a decrease in the dopaminergic activity required for a functioning working memory (Sawaguchi and Goldman-Rakic 1991) have been proposed as possible hypotheses.

The increased ventricular/brain ratio (frontal horn and lateral ventricle) found in patients with schizotypal personality disorder with deficit-like schizotypal traits (Cazzulo et al. 1991; Rotter et al. 1991; Raine et al. 1992; Silverman et al. 1992) has been found to be associated with poor performance on the WCST (Siever et al. 1993b), a test of prefrontal function, similar to what is seen in patients with schizophrenia (Lyons et al. 1991; Raine et al. 1992; Siever et al. 1993a). In contrast, non-specific frontal tasks such as verbal fluency, Wechsler Adult Intelligence Scale (WAIS)-R Vocabulary, and Block Design, (Wechsler 1984) do not distinguish patients with schizotypal personality disorder from normal controls (Trestman et al. 1995; Bergman et al. 1998). A positive correlation between deficit-like symptoms and impairment on tests of prefrontal and frontal function has been found in patients with schizotypal personality disorder (Wainberg et al. 1993). Impaired performance on several tasks that depend on attentional capacities and working memory, such as impaired eye-movement tracking of a smoothly moving target (Siever et al. 1993b) and poor performance on the Continuous Performance Task (CPT) (Nuechterlein 1991; Cornblatt et al. 1992; Siever et al. 1993b) and on the Backward Masking Task (BMT) (Braff 1986; Merritt and Baloh 1989), have been consistent findings in schizophrenia spectrum personality disorders and have been found to be specifically associated with deficit-like symptoms.

In parallel with theories of schizophrenia, it has been hypothesized that deficit-like symptoms in patients with schizotypal personality disorder may be related to impairment in frontal cortical cognitive processing associated with hypodopaminergia, as measured by concentrations of cerebrospinal fluid and plasma homovanillic acid (pHVA) (Siever et al. 1991, 1993a). Both increased ventricular size (Siever et al. 1993a,b) and poor performance on the WCST (Siever et al. 1993a; Wainberg et al. 1993) have been associated with reduced concentrations of pHVA.

Therefore, patients with schizotypal personality disorder with deficit-like symptoms can be characterized as having increased ventricular size, impairment on several pre-frontal and frontal cortical processing tasks, and hypodopaminergia.

2.2
Deficit-Like Symptoms: Psychopharmacological Studies

Patients with schizophrenia with predominant deficit-like symptoms have commonly been treated with combinations of antipsychotics and antidepressants. However, no formal research studies have looked at the use of antidepressants in the schizophrenia spectrum personality disorders. Some studies (Wainberg et al. 1993; Siegel et al. 1996; Kirrane et al. 2000) suggest that amphetamine may improve cognitive performance on tests sensitive to prefrontal function, such as WCST, and visuospatial working memory performance in patients with schizotypal personality disorder (Table 2), yet further research is necessary to establish this. The use of catecholaminergic agents and atypical antipsychotics, such as risperidone (Koenigsberg et al. 2003), to improve deficit-like symptoms and cognitive performance in schizotypal personality disorder also warrants further investigation.

Table 2 Deficit-like symptoms: treatment data studies

Study	Disorder	Pharmacological Agent	Effect
Wainberg et al. 1993	Deficit-like symptoms/schizotypal personality disorder	d-Amphetamine	May improve cognitive performance (preliminary results)
Siegel et al. 1996	Deficit-like symptoms/schizotypal personality disorder	d-Amphetamine	Improved cognitive performance
Kirrane et al. 2000	Deficit-like symptoms/schizophrenia spectrum personality disorder	d-Amphetamine	Improved visuospatial working memory
Koenigsberg et al. 2003	Deficit-like symptoms/schizotypal personality disorder	Risperidone	Improved deficit-like symptoms

3
Impulsivity and Affective Instability

Impulsivity, a prominent feature both in borderline and antisocial personality disorders, which is also present in histrionic and narcissistic personality disorders, is defined as action without reflection, particularly as an expression of anger. Consequently, individuals with these disorders are more prone to assaultive behavior, substance abuse, self-damaging acts, and promiscuity. Affective instability is a common feature of several of the personality disorders in the dramatic and anxious cluster. It is one of the DSM-IV criteria for borderline personality disorder, while exaggerated displays of emotion and rapidly shifting emotions characterize histrionic personality disorder, and depressive symptoms associated with rejection sensitivity are present in dependent and avoidant personality disorders. Impulsivity and affective instability may or may not be coupled. The former is the case in the borderline patient whose impulsive behaviors (e.g., self-injurious behavior) are often preceded by strong negative emotions, while the latter is the case in the antisocial patient, whose impulsive acts are not triggered or accompanied by an affective shift.

3.1
Impulsivity/Aggressive Behavior: Psychobiological Studies

Some studies have suggested that previous findings regarding the familial relationship between major mood disorder and some personality disorders, specifically borderline personality disorder, seem to be secondary to the presence of mood disorder in probands with personality traits and personality disorder (Zanarini et al. 1988; Silverman et al. 1991). Family members of patients with borderline personality disorder without a history of depression had the same morbid risk for depression as did family members of individuals with other personality disorders without a history of depression. However, family members of patients with borderline personality disorder had an increased morbidity risk of impulsive personality disorder traits and affective personality disorder traits (Silverman et al. 1991) than family members of individuals with other personality disorders. Individuals with impulsive personality disorder traits were characterized as having at least three of the following chronic symptoms: physical fighting with others (not associated with alcohol), non-premeditated stealing (e.g., shoplifting), problems with drinking or drugs, binge eating, problems with gambling, sexual promiscuity, self-damaging acts (e.g., wrist slashing), and irrational angry outbursts or overreaction to minor events (not associated with alcohol). Individuals with affective personality disorder traits were characterized as having chronic dysphoria (e.g., depression, anxiety) or fluctuations in mood not associated with very severe mood disturbances, psychomotor agitation or retardation, psychotic features, or extreme guilt, with at least one of the following chronic symptoms: easily disappointed or self-pitying attitude, low self-esteem, a pessimistic outlook, and the absence of satisfactory intimate relationships (Silverman et al. 1991). The presence of these characteristics in family members of individuals with borderline personality disorder suggests a familial relationship between borderline personality disorder and personality disorder traits consistent with the core features of borderline personality disorder rather than with major mood disorder.

Research with twins who were reared apart suggests that impulsivity may be an independent heritable trait in healthy, non-psychiatric populations (Tellegen et al. 1988; Coccaro and Bergeman 1993), as well as in twins in whom one twin had borderline personality disorder (Torgerson 1984). Even though there appears to be a familial aggregation of borderline personality disorder (Zanarini et al. 1988), impulsivity and affective instability seem to aggregate independently in relatives of those with borderline personality disorder (Silverman et al. 1991), thus raising the possibility of specific biological correlates. Evidence for different biological models may determine a more specific pharmacological approach for these two core features.

Impairment in the serotonergic system, which is known for its behavioral inhibition properties, as well as attentional disorders, epileptiform disorders, and elevated circulating levels of testosterone and/or endorphins have been associated with impulsivity/aggression in personality disorders. Concentrations of cerebrospinal fluid (CSF) 5-hydroxyindoleacetic acid (5-HIAA) concentrations, a serotonin (5-HT) metabolite, seem to be heritable and associated with aggressive, dominant behavior in primates (Higley et al. 1992a,b). Lesions in serotonergic neurons of rodents disinhibit aggression (Coccaro et al. 1989), which is reversed by serotonergic agents (Stark et al. 1985). In humans, decreased serotonin, its metabolites, and its receptors have been associated with aggression against others (Brown et al. 1982; Linnoila et al. 1983; Coccaro et al. 1990b), or towards the self, as occurs in completed suicide, suicide attempts, or parasuicidal behavior (Arango et al. 1990; Coccaro et al. 1990a; Mann and Arango 1992). Other associated findings have been decreases in serotonin at the pre-synaptic transport site, as measured by platelet imipramine binding capacity (Meltzer and Arora 1986; Marazziti et al. 1989) and 5-HT$_2$ post-synaptic receptors (Pandey et al. 1990).

Neuroendocrine studies also imply the association of impulsivity in patients with personality disorders with a decreased central serotonergic system. Serotonergic functioning may be measured by the prolactin (PRL) response to fenfluramine, a 5-HT-releasing/uptake inhibiting agent assessing net serotonergic activity. A decreased PRL response to fenfluramine suggests a diminished serotonergic function, which has been associated with suicide, irritability, and aggression in personality disorders (Coccaro 1989). Metabolic activity in frontal brain regions believed to play a role in the inhibition of impulsive behavior can now be measured with positron emission tomography (PET) and it appears that, in response to fenfluramine, there is decreased activation of the orbital and ventromedial frontal cortices in impulsive–aggressive patients with personality disorders (Siever et al. 1999). This observation suggests that impulsive–aggressive patients demonstrate diminished serotonergic activity in relevant brain regions.

The noradrenergic system, with nerve cell bodies in the locus coeruleus in charge of regulating arousal and responsiveness to the environment (Aston-Jones and Bloom 1981; Levine et al. 1990), appears to play a role in impulsivity and aggression. Increased activity may be associated with increased irritability and aggression (Lamprecht et al. 1972; Stolk et al. 1974), while decreased activity has been associated with withdrawal from the environment (McKinney et al. 1984). Noradrenergic activity can be documented with the neuroendocrine challenge with clonidine, a central alpha$_2$ agonist which, through its action in the hypothalamus, causes release of plasma growth hormone in normal subjects. Irritability, impulsivity, and verbal hostility have been positively correlated with the augmented release of growth hormone in neuroendocrine challenges with clonidine in patients with personality disorders (Coccaro et al. 1991; Trestman et al. 1992). A hyperadrenergic state may be associated with impulsive behaviors; it may be hypothesized that agents that stabilize or reduce noradrenergic activity might be expected to improve the irritability and reactivity of these patients.

Although one study questions the specificity of findings of electroencephalographic (EEG) abnormalities in patients with borderline personality disorder (Cornelius et al. 1986), preliminary data suggest epileptiform activity in the limbic system of patients with borderline personality disorder (Snyder and Pitts 1984; Cowdry et al. 1985–86) and improvement in their dyscontrol behavior with the use of anticonvulsants (Cowdry and Gardner 1988).

3.2
Impulsivity/Aggressive Behavior: Psychopharmacological Studies

The hypothesis of diminished central serotonin activity led to treatment trials with selective serotonin reuptake inhibitors (SSRIs), which may enhance serotonin availability in the central nervous system. Open-label trials of fluoxetine (Prozac) in patients with borderline personality disorder with dosages ranging from 5–80 mg/day (Norden 1989; Coccaro et al. 1990a; Cornelius et al. 1990, 1991; Markowitz et al. 1991) suggest that SSRIs may be helpful in diminishing impulsivity and aggression in patients with borderline personality disorders. Treatment response was observed as early as 1 week (Norden 1989) to 4 weeks (Coccaro et al. 1990a; Cornelius et al. 1990, 1991; Markowitz et al. 1991). A double-blind, placebo-controlled study with fluoxetine demonstrated a significant decrease in impulsivity and aggression in patients with personality disorders who were treated for up to 12 weeks (Coccaro and Kavoussi 1995). From a practical point of view, SSRIs have clear advantages when used to treat impulsivity: they are well tolerated, have low lethality potential in overdose, and do not require management of blood levels.

Lithium has been reported to clinically reduce impulsivity in antisocial personality disorder. In a double-blind, placebo-controlled trial, lithium significantly reduced impulsive aggressive behavior in a prison population at doses producing levels of 0.6–1.0 mEq/l, with a return to the baseline frequency of these behaviors with the blind cross-over to placebo after 3 months (Sheard et al. 1976). Only impulsive aggressive behavior was affected by lithium treatment; no changes were observed in such areas as "conning" or "disregard for the rights of others." Lithium's effect on impulsivity has been hypothesized to be due either to an increase in serotonergic activity (Schiff et al. 1982; Price et al. 1989) or to a dampening of catecholaminergic activity (Bunney and Bunney-Garland 1987). Lithium's disadvantages are that it is less well tolerated than the SSRIs and has a greater risk of toxicity in overdose, so that periodic blood level assessment is required.

Carbamazepine was initially reported to decrease the frequency and severity of episodes of behavioral dyscontrol in patients with DSM-III borderline personality disorder (Cowdry and Gardner 1988). However, this finding was not replicated in a later study, which found that carbamazepine might actually worsen impulsive and aggressive behavior in some patients with borderline personality disorder (de la Fuente and Lotstra 1994). Other anticonvulsants, such as valproic acid and lamotrigine, have shown modest benefits in reducing impulsivity in borderline patients in small open-label trials (Stein et al. 1995; Pinto and Akiskal 1998). Diphenylhydantoin has been reported to decrease "anger, irritability, impatience and anxiety" in patients with DSM-II neurosis with prominent histories of hostility (Stephens and Schaffer 1970). It has been suggested that carbamazepine may enhance serotonin activity in humans, as measured by prolactin response to tryptophan during carbamazepine challenge (Elphick et al. 1990).

Beta-blockers, especially propranolol (Volavka 1988), have been reported to diminish aggressive behavior. Yet many of these studies have been done in patients with schizophrenia who were taking antipsychotics, with the possibility of the beta-blocker treating the akathisia more than the aggressive behavior. Nadolol, which minimally penetrates the blood–brain barrier, has also been reported to diminish the frequency of aggressive behaviors in a chronic psychiatric population (Yudofsky et al. 1987; Ratey et al. 1992). It is unclear if the mechanism of action is central or peripheral, if it is through noradrenergic blockade or serotonergic agonism at higher doses, or what the effect would be in subjects with personality disorders.

Exacerbation of impulsive aggressive behavior can be an unwanted side effect of tricyclic antidepressants (TCAs). Patients with borderline personality disorder with significant pre-treatment ratings for hostility who were treated with amitriptyline demonstrated increased ratings for impulsivity not correlated with their depression or psychoticism (Soloff et al. 1986a). Increases in impulsivity/aggression have also been reported in healthy volunteers (Gottschalk et al. 1965) and depressive patients (Rampling 1978). TCAs may enhance serotonergic activity, yet they can also enhance noradrenergic activity.

Benzodiazepines have been found to disinhibit some individuals (Rickles and Downing 1974; Hall and Zisook 1981). More specifically, alprazolam has been reported to significantly worsen behavioral dyscontrol by increasing impulsive–aggressive behaviors in patients with borderline personality disorder compared with controls (Gardner and Cowdry 1986; Cowdry and Gardner 1988). Taking into consideration the risk of drug dependency and abuse, and the reported possibility of behavioral dyscontrol, benzodiazepines should not be administered to patients with impulsive–aggressive behavior.

Studies on the treatment of impulsivity/aggression are summarized in Tables 3 and 4.

Table 3 Impulsivity/aggression: treatment data studies

Study	Disorder	Pharmacological agent	Effect
Coccaro and Kavoussi 1995; Salzman et al. 1992	Borderline PD	Fluoxetine (double-blind, placebo-controlled)	Decreased impulsivity
Coccaro et al. 1990; Cornelius et al. 1990; Cornelius et al. 1991; Markowitz et al. 1991	Borderline PD	Fluoxetine (open-label)	Decreased impulsivity
Sheard et al. 1976	Antisocial PD	Lithium (double-blind, controlled)	Decreased impulsivity
Stein et al. 1995	Borderline PD	Valproic Acid (open-label)	Decreased impulsivity
Pinto and Akiskal 1998	Borderline PD	Lamotrigine (open-label)	Decreased impulsivity
Cowdry and Gardner 1988	Borderline PD	Carbamazepine >tranylcypromine= trifluoperazine>placebo	Decreased behavioral dyscontrol
Soloff et al. 1986	Borderline PD	Amitriptyline	Increased impulsivity
Cowdry and Gardner 1988; Gardner and Cowdry 1986	Borderline PD	Alprazolam	Increased impulsivity

Table 4 Impulsivity/aggression studies: treatment choices

Impulsivity/aggression (studied in borderline and antisocial personality disorders)			
Decreased		Increased	
Fluoxetine and SSRIs	Coccaro and Kavoussi 1995; Salzman et al. 1992: Coccaro et al. 1990a; Cornelius et al. 1990; Cornelius et al. 1991; Markowitz et al. 1991	Alprazolam and other benzodiazepines	Cowdry and Gardner 1988; Gardner and Cowdry 1986
Lithium (but lithium levels needed and risk of overdose)	Sheard et al. 1976	Amitriptyline and noradrenergic agents	Soloff et al. 1986a
Valproic acid	Stein et al. 1995		
Lamotrigine	Pinto and Akiskal 1998		
Carbamazepine (but carbamazepine levels needed and risk of overdose)	Cowdry and Gardner 1988		
Tranylcypromine (but risk of overdose and tyramine crisis)	Cowdry and Gardner 1988		
Trifluoperazine	Cowdry and Gardner 1988		

3.3
Affective Instability: Psychobiological Studies

Depressive symptoms can present in patients with personality disorders either as a part of their usual personality profile or as a more sustained and discrete depressive syndrome indistinguishable from major depression. Even though depressive symptoms can occur in any personality disorder, affective-related traits are particularly characteristic of personality disorders of the dramatic and anxious dimensions. The affective instability characteristic of borderline personality disorder as well as the exaggerated displays of emotions and rapidly shifting emotions that characterize histrionic personality disorder and the rejection sensitivity often associated with depressive symptoms in dependent and avoidant personality disorders are some of these affective-related traits. In some patients, particularly in patients with histrionic and borderline personality disorders, these episodes of abrupt dysphoria predispose them to self-damaging behavior.

The discordant findings from dexamethasone non-suppression in patients with borderline personality disorder, which range from 9% to 73% (Soloff et al. 1982; Sternbach et al. 1983; Baxter et al. 1984; Beeber et al. 1984; Krishnan et al. 1984; Siever et al. 1985; Lahmeyer et al. 1988; Korzekwa et al. 1991), seem to be related to the comorbidity of major depression in most of these studies (Soloff et al. 1982; Sternbach et al. 1983; Baxter et al. 1984; Beeber et al. 1984; Krishnan et al. 1984; Korzekwa et al. 1991). Low comorbidity of major depression corresponded to low rates of dexamethasone non-suppression (Siever et al. 1985; Lahmeyer et al. 1988). Similarly, studies suggesting abnormalities in the thyrotropin-stimulating hormone (TSH) with thyrotropin-releasing hormone (TRH) stimulation testing in patients with borderline and other personality disorders were done in cohorts with 80%–100% comorbidity of major depression or alcoholism (Loosen and Prange 1982; Garbutt et al. 1983; Sternbach et al. 1983). These findings suggest that abnormalities in dexamethasone non-suppression and in TSH responses to TRH stimulation in patients with personality disorders are attributable to the comorbidity of major depression or alcoholism rather than to the personality disorder itself.

Whereas heightened noradrenergic activity may be associated with increased irritable aggression (Lamprecht et al. 1972; Stolk et al. 1974), decreased noradrenergic activity seems to be associated with relative withdrawal, such as is observed in separation of a primate infant from its mother (McKinney et al. 1984). This finding is consistent with the noradrenergic abnormalities of major depression, particularly endogenous depression, where reductions of the responsiveness of the noradrenergic system are suggested by responses to noradrenergic challenges such as clonidine (Siever 1987). Yet the growth hormone response to clonidine challenge, an index of alpha$_2$-noradrenergic receptor sensitivity, tends to be similar or increased in patients with personality disorders compared with normal controls and to be correlated with irritability rather than with depressive symptoms (Coccaro et al. 1991). Similarly, assessment of the serotonergic system, for example, by measuring the prolactin response to D,L-fenfluramine challenge, does not seem to differentiate between patients with personality disorders with or without current depression or a history of depression, whereas correlations were found between this response and assaultiveness and irritability (Coccaro 1989; Coccaro et al. 1989). This may explain the often unsuccessful response observed in patients with personality disorders when treated for mood dysregulation with TCAs, which tend to increase the availability of both norepinephrine and serotonin in the synaptic cleft. However, improvement of global function, including mood, has been reported with amphetamine challenges in a subgroup of patients with borderline personality disorder (Schulz et al. 1988), which may correlate with trait indices of affective lability (Kavoussi and Coccaro 1993). It seems possible that mood instability or dysregulation may be related to more than one neurotransmitter system, including both norepinephrine and serotonin (Siever and Davis 1991). The response to amphetamine, which triggers the release of different monoaminergic substances, may be modulated both through noradrenergic and serotonergic activity, reflecting a more complex biological substrate associated with affective lability and dysregulation. However, noradrenergic enhancement may not be advisable when impulsivity is prominent.

Increased cholinergic receptor responsiveness seems to be associated with major depressive disorders (Janowsky and Risch 1987). Increasing the postsynaptic cholinergic receptor stimulation with physostigmine, an acetylcholinesterase inhibitor, and arecoline, a muscarinic agonist, induces a behavioral syndrome resembling depression in animals (Janowsky et al. 1972; Janowsky and Risch 1987) and depressive symptoms in acute and remitted depressed patients and normal controls (Janowsky et al. 1974; Risch et al. 1981; Janowsky and Risch 1987) and appears to have antimanic properties (Davis et al. 1978). In normal males, dysphoric responses to physostigmine correlate with traits of irritability and emotional lability (Fritze et al. 1990). Physostigmine has also been shown to produce greater depressive responses in patients with borderline personality disorder than in controls (Steinberg et al. 1995). Similarly, the cholinergic activity in rapid eye movement (REM) latency studies suggests that patients with borderline personality disorder share disturbances in REM regulation with patients with major depressive disorder, showing both decreased and more variable REM latency than normal controls and possibly exaggerated reduction in REM latency in response to a cholinomimetic agent (Lahmeyer et al. 1988). Preliminary studies suggest that patients with personality disorders with prominent affective instability, particularly patients with borderline personality disorder, have a significantly greater dysphoric response to physostigmine than patients with other personality disorders (Steinberg et al. 1994). Thus, enhanced cholinergic receptor sensitivity may be associated with dysphoric symptoms and the susceptibility to affective shifts observed in the dramatic cluster (e.g., borderline or histrionic personality disorder).

3.4
Affective Instability: Psychopharmacological Studies

Clinical trials investigating the use of psychotropic medication targeting depressive symptoms have mostly been done in patients with borderline and/or schizotypal personality disorder, with minimal to modest improvements found with antipsychotics [chlorpromazine (Leone 1982), thiothixene (Goldberg et al. 1986), haloperidol (Soloff et al. 1986b, 1989, 1993)], and TCAs [amitriptyline (Soloff et al. 1986a), imipramine (Links et al. 1990)] and more significant improvements found in some studies utilizing monoamine oxidase inhibitors (MAOIs) [phenelzine (Hedberg et al. 1971; Liebowitz and Klein 1981; Parsons et al. 1989), tranylcypromine (Cowdry and Gardner 1988)].

Few placebo-controlled trials have been designed to target affective lability. The first study in 1968 compared chlorpromazine with imipramine in a group of patients with "emotionally unstable character disorders," described as rapid autonomous affective lability, with chronic personality traits such as difficulty with authority, job instability, and problems in interpersonal relationships (Klein 1968). Chlorpromazine was significantly superior to imipramine and both chlorpromazine and imipramine were significantly superior to placebo, with a subset of patients who exhibited increased anger with the use of imipramine. The modest efficacy shown by the TCAs, with the increased agitation and irritability that some patients with borderline personality disorder may experience (Soloff et al. 1986b), together with the disadvantage of significant anticholinergic side effects and very high lethality potential in overdose, does not make these agents a first-line option in the treatment of depressive symptoms in a population prone to impulsivity, such as those with personality disorders in the dramatic cluster.

MAOIs, especially phenelzine, have been more extensively studied in patients with personality disorders in the dramatic and/or anxious cluster. An early open-label study of patients with "hysteroid dysphoria," a syndrome involving rejection sensitivity and affective lability similar to some of the traits of borderline personality disorder, found phenelzine efficacious in treating this syndrome (Liebowitz and Klein 1981). A retrospective study suggested that phenelzine was three times more effective than imipramine in treating atypical depression that was comorbid with borderline personality disorder, whereas phenelzine and imipramine were equally effective in treating atypical depression when borderline personality disorder was not present (Parsons et al. 1989). In a placebo-controlled study, phenelzine was reported to be ineffective in treating both the typical and atypical signs of depression in hospitalized subjects with borderline and/or schizotypal personality disorder (Soloff et al. 1993). The lack of efficacy can be attributed to lower phenelzine dosages secondary to adverse side effects (average dose of 60 mg/day in the inpatient study versus 90 mg/day in the previous study) and the shorter duration of the trial. Yet the lack of efficacy for phenelzine in a well-defined hospitalized sample suggests that phenelzine may not be efficacious in the treatment of depressive symptoms in moderately to severely impaired patients with personality disorders. On the other hand, tranylcypromine has been associated with significant antidepressant/mood enhancing effect in a small group of females with treatment-resistant borderline personality disorder (Cowdry and Gardner 1988). It is possible that the efficacy of tranylcypromine can be attributed to its amphetamine-like structure, which is consistent with preliminary data showing mood enhancement with amphetamine challenge in some patients with borderline personality disorder (Schulz et al. 1988). MAOIs have a number of side effects and risks, such as agitation, orthostatic hypotension, and hypertensive crisis after non-compliance with low tyramine diet, making their use difficult in the impulsive population. Deprenyl, a MAO-B inhibitor with milder side effects and a lower risk of hypertensive crisis, has been reported to be an effective antidepressant (Faltus and Janeckova 1985; Mann et al. 1989), yet no studies have been done in patients with personality disorders.

5-HT uptake inhibitors may treat depressive symptoms in patients with personality disorders. Although no placebo-controlled studies have been reported on the efficacy of SSRIs in treating depression in the dramatic cluster, open-label trials of fluoxetine in borderline personality disorder suggest that it may be effective (Norden 1989; Cornelius et al. 1990; Markowitz et al. 1991). Trials with paroxetine and sertraline have not been reported, but these drugs may be as effective as fluoxetine. Advantages of the SSRIs include fewer side effects and low lethality in overdose.

Mood stabilizers used for bipolar disorder, such as lithium and anticonvulsants, may be effective in treating mood lability in other disorders (Van der

Table 5 Affective instability: treatment data studies

Study	Disorder	Pharmacological agent	Effect
Klein 1968	Emotionally unstable character disorder	Chlorpromazine imipramine	Decreased lability Chlorpromazine>Imipramine= placebo (imipramine subset with increased anger)
Soloff et al. 1986a	Borderline PD	Amitriptyline	Slightly decreased depression, increased anger
Liebowitz and Klein 1981	Hysteroid dysphoria	Phenelzine	Decreased depression
Parsons et al. 1989	Borderline PD+Atypical depression	Phenelzine; imipramine (retrospective outpatients)	Decreased depression Phenelzine>imipramine
Soloff et al. 1993	Borderline PD+Depression (atypical and typical)	Phenelzine (placebo-controlled, inpatient)	Ineffective
Cowdry and Gardner 1988	Borderline PD	Tranylcypromine	Decreased depression
Norden 1989; Cornelius et al. 1990, 1991; Markowitz et al. 1991	Borderline PD	Fluoxetine open-label trial	Decreased depression
Rifkin et al. 1972	Emotionally unstable PD	Lithium open-label	Decreased depression

Kolk 1986). Lithium has been shown to stabilize mood in "emotionally unstable personality disorder" (Rifkin et al. 1972). A placebo-controlled trial of lithium in patients with borderline personality disorder found a trend for lithium to be superior to desipramine in terms of decreasing anger, but did not examine its potential use for affective lability. A small uncontrolled study with patients with emotionally unstable character disorders showed no efficacy with the use of diphenylhydantoin (Klein and Greenberg 1967).

Since enhanced cholinergic receptor sensitivity has been associated with dysphoric symptoms and the susceptibility to affective shifts in patients with dramatic cluster personality disorders, such as borderline or histrionic personality disorder, future trials for the treatment of affective instability in patients with personality disorders might include the testing of anticholinergic agents, although these have not generally been successful for patients with affective disorders.

Studies concerning the treatment of affective instability are summarized in Table 5.

4
Anxiety Threshold

The DSM-IV anxiety-related personality disorders include avoidant, dependent, and obsessive–compulsive. Individuals with these personality disorders are hypersensitive to anxiety and eschew potential anxiety-provoking situations. Thus, individuals with avoidant personality disorder directly avoid situations due to excessive anticipatory anxiety regarding the prospect of future rejection. Individuals with dependent personality disorder maintain themselves attached to their protective caretakers, avoiding conflict with them, and acquiescing to their wishes for fear of rejection. Patients with obsessive–compulsive personality disorder occupy themselves with a rigid structure of thoughts or behaviors, thus blinding themselves from anxious thoughts or situations.

Although no systematic research has focused on these personality disorders, several studies suggest a 90% comorbidity of avoidant personality disorder and social phobia (Schneier et al. 1991; Widiger 1992). Research studies on social phobia may be useful in understanding the biology and treatment of avoidant personality disorder.

4.1
Anxiety Threshold: Psychobiological Studies

Familial transmission of "anxious personality traits and disorders" has been implied by family studies (Reich 1989, 1991), while the heritability of social anxiety has been reported in twin studies (Torgerson and Kringlen 1978). First-degree blood relatives of patients with social phobia have a significantly greater prevalence of social phobia (Schneier et al. 1991). Longitudinal studies suggest stable measurements of fearfulness and inhibition in children (Kagan et al. 1988) with an increased prevalence of childhood anxiety phenomena at age 8 in toddlers who exhibited behavioral inhibition as early as 21 months old, whose parents also had increased prevalence of social phobia and anxiety disorders (Rosenbaum et al. 1991), suggesting a familial association for anxiety disorders (Biederman et al. 1991).

Besides these data for the probable genetic basis for social anxiety and its longitudinal stability, there are limited biological studies concerning the anxious dimension of personality disorders, except for those studies in social phobia with comorbid avoidant personality disorder. Hypothalamo-pituitary function studies have not found a difference between patients with social phobia and normal controls either in the hypothalamo-thyroid (Tancer et al. 1990) or the hypothalamo-adrenal axis (Uhde et al. 1991). Unlike patients with panic disorder in whom sodium lactate infusion, caffeine, and norepinephrine have been used as probes, inconsistent responses have been observed in patients with social phobia (Liebowitz et al. 1985; Tancer 1993). However, like patients with panic disorder, patients with social phobia have shown a blunted growth hormone response to challenges with clonidine (Amies et al. 1983; Steinberg et al. 1994; Uhde 1994). Social phobia patients also exhibit an increased cortisol response to fenfluramine when compared with controls (Uhde 1994), suggesting that the serotonergic system may play a role in anxiety and the avoidance response. The response in patients with social phobia when challenged with clonidine and fenfluramine suggests that patients with avoidant personality disorder, in contrast to impulsive patients with borderline personality disorder, have decreased noradrenergic activity and increased serotonergic activity.

4.2
Anxiety Threshold: Psychopharmacological Studies

Few pharmacological studies have targeted anxiety in personality disorders. A report of four cases suggests similar efficacy using phenelzine, tranylcypromine, or fluoxetine in avoidant personality disorder (Deltito and Stam 1989). Placebo-controlled, double-blind studies have reported the efficacy of MAOIs in the treatment of social phobia (Liebowitz 1989, 1992; Liebowitz et al. 1990a,b, 1991, 1992). More specifically, phenelzine was associated with a significant decrease in avoidant personality features in patients with social phobia, whereas atenolol and placebo were not (Liebowitz et al. 1992). In a double-blind, controlled study of social phobia, moclobemide (200–600 mg/day), a reversible MAOI with a lower incidence of side effects, was found to be statistically comparable to phenelzine (30–90 mg/day) after 16 weeks of treatment, with both agents demonstrating efficacy by week 8 (Versiani et al. 1992). Relapse after discontinuation of either agent was also noted (Versiani et al. 1992). Other trials with moclobemide in patients with social phobia suggest weak efficacy (Marshall 1994).

Double-blind placebo-controlled studies have demonstrated that SSRIs are effective in the treatment of social phobia. Paroxetine was compared with placebo in a 12-week multi-site study of 93 patients with social phobia (Stein et al. 1999)

Table 6 Anxiety threshold: treatment data studies

Study	Disorder	Pharmacological agent	Effect
Stein et al. 1999	Social phobia	Paroxetine (placebo-controlled, double-blind)	Decreased anxiety
Liebowitz 1989, 1992; Liebowitz et al. 1990a,b, 1991	Social phobia	Phenelzine, tranylcypromine (placebo-controlled, double-blind)	Decreased anxiety
Versiani et al. 1992	Social phobia	Moclobemide=phenelzine	Decreased anxiety
Marshall 1994	Social phobia	Moclobemide	Decreased anxiety (modest efficacy)
Liebowitz et al. 1992	Social phobia	Phenelzine; atenolol	Phenelzine>atenolol=placebo (decreased avoidant traits)
Tancer and Golden 1993	Social phobia/avoidant personality disorder	SSRIs>TCAs (open-label, preliminary result)	Decreased anxiety

and found to produce a significant decrease in anxiety symptoms as measured by the Liebowitz Social Anxiety Scale (LSAS) (Liebowitz 1987) and the Clinical Global Impression Scale (Guy 1976). Patients with social phobia treated with paroxetine in an 11-week open-label trial also demonstrated significant improvements as measured by LSAS scores and the Duke Social Phobia Scale (Davidson et al. 1991; Stein et al. 1996); 16 responders were then followed up in a placebo-controlled, double-blind discontinuation study in which only one relapse occurred among 8 patients continued on paroxetine compared with five relapses in the group of 8 patients randomized to receive placebo.

Treatment studies concerning anxiety symptoms are summarized in Table 6.

5
Practical Issues in the Treatment of Personality Disorders

The pharmacotherapy of the different personality disorders can be complicated by the various medication regimens and the interpersonal (transferential) issues which develop within the treatment. Therapeutic techniques selected based on specific personality dimensions will be required to develop a therapeutic alliance and ensure successful treatment, especially when more than one provider is present (split treatment).

A complicating element is the presence of substance abuse in many patients with personality disorders, which could result in sporadic treatment compliance and may confuse the clinical presentation as well as psychobiological assessment. Proper assessment should be part of the initial evaluation and follow-up of patients, and prompt intervention is required if substance abuse is suspected.

There are specific therapeutic techniques for the treatment of the different personality disorders, which are beyond the scope of this chapter. Yet the therapeutic alliance in all of these cases is enhanced if the treatment goals and limitations are explicit from the beginning, including the clarification of contact between therapist and psychopharmacologist. It is helpful for the patient to assume responsibility for the treatment, giving the patient an active role in selecting among treatment options. The expected side-effect profile should be explained prior to starting treatment, as well as the expected time course of action, target symptoms, and limitations of the treatment.

It is particularly important to be aware of safety issues in medicating an impulsive individual, especially if the patient has a history of self-damaging acts. Low toxicity in overdose is an essential precaution to be taken in such cases. As mentioned above, SSRIs, with their low toxicity in overdose, together with the fact that they do not require periodic blood levels, make them safer for use when impulsivity and a history of self-damaging acts are present.

6
Conclusions

Research into the biological and neuropsychopharmacological correlates of the dimensions in personality traits has provided important knowledge which can aid in the pharmacological treatment of patients with personality disorders. The understanding of the heritability and biological characteristics of the personality disorders may provide a rational basis for future treatment interventions for patients with personality disorders.

Despite the important advances in this area, specific definitive recommendations for the pharmacological treatment of personality disorders are still premature. Placebo-controlled, double-blind clinical trials in well-defined populations within the personality disorder clusters are required before such recommendations can be made.

References

American Psychiatric Association (1994) Diagnostic and statistical manual of mental disorders, 4th ed. American Psychiatric Association, Washington, DC
Amies PL, Gelder MG, Shaw PM (1983) Social phobia: A comparative clinical study. Br J Psychiatry 142:174–179
Arango V, Ernsberger P, Marzuk PM, Chen JS, Tierney H, Stanley M, Reis DJ, Mann JJ (1990) Autoradiographic demonstration of increased serotonin 5-HT$_2$ and beta-adrenergic receptors binding sites in the brain of suicide victims. Arch Gen Psychiatry 47:1038–1047
Aston-Jones G, Bloom FE (1981) Norepinephrine-containing locus coeruleus neurons in behaving rats exhibiting pronounced responses to non-noxious environmental stimuli. J Neurosci 1:887–890
Baron M, Gruen R, Rainer JD, Kane J, Asnis L, Lord S (1985) A family study of schizophrenic and normal control probands: Implications for the spectrum concept of schizophrenia. Am J Psychiatry 142:447–455

Battaglia M, Gasperini M, Sciuto G, Scherillo P, Diaferia G, Bellodi L (1991) Psychiatric disorders in the families of schizotypal subjects. Schizphr Bull 17:659–668

Baxter L, Edell W, Gerner R, Fairbanks L, Gwirtsman H (1984) Dexamethasone suppression test and Axis I diagnoses of inpatients with DSM-II borderline disorder. J Clin Psychiatry 45:150–153

Beeber AR, Kline MD, Pies RW, Manring JM Jr (1984) Dexamethasone suppression test in hospitalized depressed patients with borderline personality disorder. J Nerv Ment Dis 172:301–303

Berg E (1948) A simple objective technique for measuring flexibility in thinking. J Gen Psychol 39:15–22

Bergman AJ, Harvey PD, Roitman SL, Mohs RC, Marder D, Silverman JM, Siever LJ (1998) Verbal learning and memory in schizotypal personality disorder. Schizophr Bull 24:635–641

Biederman J, Newcorn J, Sprich S (1991) Comorbidity of attention deficit hyperactivity disorder (ADHD) with conduct, depressive, anxiety and other disorders. Am J Psychiatry 148:564–577

Braff DL (1986) Impaired speed of information processing in non-medicated schizotypal patients. Schizophr Bull 7:499–508

Brown GL, Ebert MH, Goyer PF, Jimerson DC, Klein WJ, Bunney WE, Goodwin FK (1982) Aggression, suicide, and serotonin relationships to CSF metabolites. Am J Psychiatry 139:741–745

Bunney WE Jr, Bunney-Garland BL (1987) Mechanism of action of lithium in affective illness: Basic and clinical implications. In: Meltzer HY (ed) Psychopharmacology: Third generation of progress. Raven Press, New York, pp 553–563

Cazzulo C, Vita A, Giobbio G, Dieci M, Saccheti E (1991) Cerebral structural abnormalities in schizophreniform disorder and in schizophrenia spectrum personality disorders. In Schizophrenia Research: Advances in neuropsychiatry and psychopharmacology, vol 1. Edited by Tamminga C, Schultz S. New York, Raven Press, pp 209–217

Coccaro EF (1989) Central serotonin in impulsive aggression. Br J Psychiatry 155(suppl 8):52–62

Coccaro EF (1993) Psychopharmacologic studies in patients with personality disorder: Review and perspective. Journal of Personality Disorders (Spring suppl):181–192

Coccaro EF, Bergeman CS (1993) Heritability of irritable impulsiveness: A study of twins reared together and apart. Psychiatry Res 48:229–242

Coccaro EF, Kavoussi RJ (1995) Fluoxetine in aggression in personality disorders [New Research Abstracts]. Presented at the American Psychiatric Association 148th annual meeting, Miami, Florida, May 20–25

Coccaro EF, Siever LJ, Klar HM, Maurer G, Cochrane K, Cooper TB, Mohs RC, Davis KL (1989) Serotonergic studies in patients with affective and personality disorders: Correlates with suicidal and impulsive aggressive behavior. Arch Gen Psychiatry 46:587–599

Coccaro EF, Astill JL, Herbert JL, Schut AG (1990a) Fluoxetine treatment of impulsive aggression in DSM-III-R personality disorder patients. J Clin Pharmacol 10:373–375

Coccaro EF, Gabriel S, Siever LJ (1990b) Buspirone challenge: Preliminary evidence for a role for 5-HT$_{1a}$ receptors in impulsive aggressive behavior in humans. Psychopharmacol Bull 26:393–405

Coccaro EF, Lawrence T, Trestman R, Gabriel S, Klar HM, Siever LJ (1991) Growth hormone responses to intravenous clonidine challenge correlates with behavioral irritability in psychiatric patients and in healthy volunteers. Psychiatry Res 39:129–139

Cornblatt BA, Lenzenweber MF, Dworkin RH, Erlenmeyer-Kimling L (1992) Childhood attention dysfunctions predict social deficits in unaffected adults at risk for schizophrenia. Br J Psychiatry 161(suppl 18):59–64

Cornelius JR, Brenner RP, Soloff PH, Schulz SC, Tumuluru RV (1986) EEG abnormalities in borderline personality disorder patients: Specific or non-specific. Biol Psychiatry 21:977-980

Cornelius JR, Soloff PH, Perel JM, Ulrich RF (1990) Fluoxetine trial in borderline personality disorder. Psychopharmacol Bull 26:151-154

Cornelius JR, Soloff PH, Perel JM, Ulrich RF (1991)A preliminary trial of fluoxetine in refractory borderline patients. J Clin Pharmacol 11:116-120

Cowdry RW, Gardner DL (1988) Pharmacotherapy of borderline personality disorder: Alprazolam, carbamazepine, trifluoperazine, and tranylcypromine. Arch Gen Psychiatry 45:111-119

Cowdry RW, Pickar D, Davies R (1985-86) Symptoms and EEG findings in the borderline syndrome. Int J Psychiatry Med 15:201-211

Davidson JRT, Potts NLS, Richichi EA, Ford SM, Krishnan KR, Smith RD, Wilson W (1991) The brief social phobia scale. J Clin Psychiatry 52(suppl 11):48-51

Davis KL, Berger PA, Hollister LE, Defraites E (1978) Physostigmine in mania. Arch Gen Psychiatry 35:119-122

De la Fuente JM, Lotstra F (1994). A trial of carbamazepine in borderline personality disorder. Euro Neuropsychopharmacol 4:479-486

Deltito J, Stam M (1989) Psychopharmacological treatment of avoidant personality disorder. Compr Psychiatry 30:498-504

Elphick M, Yang JD, Cowen PJ (1990) Effects of carbamazepine on dopamine- and serotonin-mediated neuroendocrine responses. Arch Gen Psychiatry 47:135-140

Faltus F, Janeckova E (1985) The antidepressant effect of deprenyl. Jesenik, Czechoslovakia, 27th Annual Psychopharmacology Meeting

Fish B (1987) Infants predictors of the longitudinal course of schizophrenic development. Schizophren Bull 13:395-409

Fritze J, Sofic E, Muller T, Pfuller H, Lanczik M, Riederer P (1990) Cholinergic-adrenergic balance: Part 2, relationship between drug sensitivity and personality. Psychiatry Res 34:271-279

Garbutt JC, Loosen PT, Tipermas A, Prange AJ Jr (1983) The TRH test in patients with borderline personality disorder. Psychiatry Res 9:107-113

Gardner DL, Cowdry RW (1986) Alprazolam induced dyscontrol in borderline personality disorder. Am J Psychiatry 142:98-100

Goldberg SC, Schulz SC, Schulz PM, Resnick RJ, Hamer RM, Friedel RO (1986) Borderline and schizotypal personality disorders treated with low-dose thiothixine versus placebo. Arch Gen Psychiatry 43:680-686

Gottesman II, Shields J (1982) Schizophrenia: The epigenetic puzzle. Cambridge University Press, New York

Gottschalk LA, Gleser GC, Wylie HW Jr, Kaplan SM (1965) Effects of imipramine on anxiety and hostility levels. Psychopharmacology (Berlin) 7:303-310

Gunderson JG, Siever LJ, Spaulding E (1983) The search for the schizotypy: Crossing the border again. Arch Gen Psychiatry 40:15-22

Guy W (1976) ECDEU assessment manual for psychopharmacology-revised (DHEW Publ No ADM 76-338). Rockville, MD, U.S. Department of Health, Education, and Welfare, Public Health Service, Alcohol Drug Abuse, and Mental Health Administration, NIMH Psychopharmacology Research Branch, Division of Extramural Research Programs, pp 218-222

Hall RCW, Zisook S (1981) Paradoxical reactions to benzodiazepines. Br J Clin Pharmacol 11:99S-194S

Hedberg DC, Hauch JH, Glueck BC (1971) Tranylcypromine-trifluoperazine combination in the treatment of schizophrenia. Am J Psychiatry 127:1141-1146

Higley JD, Mehlman PT, Taub DM, Higley SB, Suomi SJ, Vickers JH, Linnoila M (1992b) Cerebrospinal fluid monoamine and adrenal correlates of aggression in free-ranging rhesus monkeys. Arch Gen Psychiatry 49:436-441

Janowsky DS, Risch CS (1987) Role of acetylcholine mechanisms in the affective disorders. In: Meltzer HY (ed) Psychopharmacology: The third generation of progress. Raven Press, New York

Janowsky DS, El-Yousef MK, Davis JM, Sekerke HJ (1972) Cholinergic antagonism of methylphenidate-induced stereotyped behavior. Psychopharmacology 27:297–314

Janowsky DS, El-Yousef MK, Davis JM (1974) Acetylcoline and depression. Psychosom Med 36:248–257

Kagan J, Reznick JS, Snidman N, Gibbons J, Johnson MO (1988) Childhood derivatives inhibition and lack of inhibition to the unfamiliar. Child Dev 59:1580–1589

Kavoussi RJ, Coccaro EF (1993) The amphetamine challenge test correlates with affective lability in healthy volunteers. Psychiatry Res 48:219–228

Kendler K (1985) Diagnostic approaches to schizotypal personality disorder: A historical perspective. Schizophren Bull 11:538–553

Kendler KS, Gruenberg AM (1984) An independent analysis of the Copenhagen sample of the Danish study of schizophrenia. VI. The relationship between psychiatric disorders as defined by DSM III in the relatives and adoptees. Arch Gen Psychiatry 41:555–564

Kendler KS, Gruenberg AM, Strauss JS (1981) An independent analysis of the Copenhagen sample of the Danish adoption study of schizophrenia. Arch Gen Psychiatry 38:982–984

Kendler KS, Ochs AL, Gorman AM, Hewitt JK, Ross DE, Mirsky AF (1991)The structure of schizotypy: A pilot multitrait twin study. Psychiatry Res 36:19–36

Kety SS (1988) Schizophrenic illness in the families of schizophrenic adoptees: Findings from the Danish national sample. Schizophr Bull 14:217–222

Kety SS, Rosenthal D, Wender PH (1975) Mental illness in the biological and adoptive families of adopted individuals who have become schizophrenics: a preliminary report based on psychiatric interviews. E. Fieve, D. Rosenthal and H. Brill (eds) Genetic research in psychiatry. John Hopkins University Press, Baltimore, pp 147–165

Kirrane RM, Siever LJ (2000) New perspectives on schizotypal personality disorder. Curr Psychiatry Rep 2:62–66

Kirrane RM, Mitropoulou V, Nunn M, New AS, Harvey PD, Schopick F, Silverman J, Siever LJ (2000) Effects of amphetamine on visuospatial working memory performance in schizophrenia spectrum personality disorder. Neuropsychopharmacology 22(1):14–18

Klein DF (1968) Psychiatric diagnosis and a typology of clinical drug effects. Psychopharmacology 13:359–386

Klein DF, Greenberg IM (1967) Behavioral effects of diphenylhydantoin in severe psychiatric disorders. Am J Psychiatry 124:847–849

Koenigsberg HW, Reynolds D, Goodman M, New A, Mitropoulou V, Trestman R, Silverman J, Siever LJ (2003) Risperidone in the treatment of schizotypal personality disorder. J Clin Psychiatry 64:628–634

Korzekwa M, Steiner M, Links P, Eppel A (1991) The dexamethasone suppression test in borderlines: Is it useful? Can J Psychiatry 36:26–28

Krishnan KR, Davidson JR, Rayasam K, Shope F (1984) The dexamethasone suppression test in borderline personality disorder. Biol Psychiatry 19:1149–1153

Lahmeyer HW, Val E, Gaviria FM, Prasad RB, Pandey GN, Rodgers P, Weiler MA, Altman EG (1988) EEG sleep, lithium transport, dexamethasone suppression, and monoamineoxidase activity in borderline personality disorder. Psychiatry Res 25:19–30

Lamprecht F, Eichelman B, Thoa NB, Williams RB, Kopin IJ (1972) Rat fighting behavior: Serum dopamine-B-hydroxylase and hypothalamic tyrosine hydroxylase. Science 177:1214–1215

Leone NF (1982) Response of borderline patients to loxapine and chlorpromazine. J Clin Psychiatry 43:148–150

Levine ES, Litto WJ, Jacobs BL (1990) Activity of cat locus coeruleus noradrenergic neurons during the defense reaction. Brain Res 531:189-195

Liebowitz, MR (1987) Social phobia. Modern Problems in Pharmacopsychiatry 22:141-173

Liebowitz MR (1989) Phenelzine versus atenolol in social phobia: A placebo controlled study. J Clin Psychiatry 49:498-504

Liebowitz MR (1992) Reversible MAO inhibitors in social phobia, bulimia and other disorders. Clin Neuropharmacol (suppl 1):434A-435A

Liebowitz MR, Klein DF (1981) Interrelationship of hysteroid dysphoria and borderline personality disorder. Psychiatr Clin North Am 4:67-87

Liebowitz MR, Fyer AJ, Gorman JM, Dillon D, Davies S, Stein JM, Cohen BS, Klein DF (1985) Specificity of lactate infusions in social phobia versus panic disorders. Am J Psychiatry 142:947-950

Liebowitz MR, Hollander E, Schneier F, Campeas R, Welkowitz L, Hatterer J, Fallon B (1990a) Reversible and irreversible monoamine oxidase inhibitors in other psychiatric disorders. Acta Psychiatr Scand Suppl 360:29-34

Liebowitz MR, Schneier F, Campeas R, Gorman J, Fyer A, Hollander E, Hatterer J, Papp L (1990b) Phenelzine and atenolol in social phobia. Psychopharmacol Bull 26:123-125

Liebowitz MR, Schneier FR, Hollander E, Welkowitz LA, Saoud JB, Feerick J, Campeas R, Fallon BA, Street L, Gitow A (1991) Treatment for social phobia with drugs other than benzodiazepines. J Clin Psychiatry 52(suppl 11):10-15

Liebowitz MR, Schneier F, Campeas R, Hollander E, Hatterer J, Fyer A, Gorman J, Papp L, Davies S, Gully R, et al. (1992) Phenelzine vs atenolol in social phobia: A placebo-controlled comparison. Arch Gen Psychiatry 49:290-300

Links PS, Steiner M, Boiago I, Irwin D (1990) Lithium therapy for borderline patients: Preliminary findings. Journal of Personality Disorders 4:173-181

Linnoila M, Virkkunen M, Sheinin M Nuutila A, Rimon R, Goodwin FK (1983) Low cerebrospinal fluid 5-hydroxyindolacetic acid concentration differentiates impulsive from nonimpulsive violent behavior. Life Sci 33:2609-2614

Loosen PT, Prange AJ (1982) Serum thyrotropin response to thyrotropin-release hormone in psychiatric patients. Am J Psychiatry 139:405-415

Lyons MJ, Merla ME, Young L, Kremen WS (1991) Impaired neuropsychological functioning in symptomatic volunteers with schizotypy: Preliminary findings. Biol Psychiatry 30:424-426

Mann JJ, Arango V (1992) Integration of neurobiology and psychopathology in a unified model of suicidal behavior. J Clin Psychopharmacol 12(suppl 2):2S-7S

Mann JJ, Aarons SF, Wilner PJ, Keilp JG, Sweeney JA, Pearlstein T, Frances AJ, Kocsis JH, Brown RP (1989) A controlled study of the antidepressant efficacy and side-effects of L-deprenyl: A selective monoamine oxidase inhibitor. Arch Gen Psychiatry 46:45-50

Marazziti D, De Leo D, Conti L (1989) Further evidence supporting the role of serotonin system in suicidal behavior: A preliminary study of suicide attempters. Acta Psychiatr Scand 80:322-324

Markovitz PJ, Calabrese JR, Schulz SC, Meltzer HY (1991) Fluoxetine treatment of borderline and schizotypal personality disorder. Am J Psychiatry 148:1064-1067

Marshall JR (1994) Practical approaches to the treatment of social phobia. J Clin Psychiatry 55:367-374

McKinney WT, Moran EC, Kraemer GW (1984) Separation in nonhuman primates as a model for human depression: neurobiological implications. In: Post RM, Ballenger JC (eds) Neurobiology of mood disorders. Williams and Wilkins, Baltimore, pp 393-406

Meltzer HY, Arora RC (1986) Platelet markers of suicidality. Ann N Y Acad Sci 487:271-280

Merritt RD, Baloh DW (1989) Backward masking spatial frequency effects among hypothetically schizotypal individuals. Schizophr Bull 15:573-583

Norden MJ (1989) Fluoxetine in borderline personality disorder. Prog Neuropsychopharmacol Biol Psychiatry 13:885-893

Nuechterlein KH (1991) Vigilance in schizophrenia and related disorders. In: Handbook of schizophrenia, Vol 5: Neuropsychology, psychophysiology and information processing, Steinhauer SR, Gruzelier JH, Zubin J, eds. New York, Elsevier, 397-443

Onstad S, Skre I, Edvardsen J, Torgersen S, Kringlen E (1991) Mental disorders in first degree relatives of schizophrenics. Acta Psychiatr Scand 83:463-467

Pandey GN, Pandey SC, Janicak PG, Marks RC, Davis JM (1990) Platelet serotonin-2 receptor binding sites in depression and suicide. Biol Psychiatry 28:215-222

Parsons B, Quitkin FM, McGrath PJ, Stewart JW, Tricamo E, Ocepek-Welikson K, Harrison W, Rabkin JG, Wager SG, Nunes E (1989) Phenelzine, imipramine, and placebo in borderline patients meeting criteria for atypical depression. Psychopharmacol Bull 25:524-534

Pinto OC, Akiskal HS (1998). Lamotrigine as a promising approach to borderline personality: an open case series without concurrent DSM-IV major mood disorder. J Affect Disord 51:333-343.

Price LH, Charney DS, Delgado PL, Heninger GR (1989) Lithium treatment and serotonergic function. Br J Psychiatry 46:13-19

Raine A, Sheard C, Reynolds GP, Lemcz T (1992) Pre-frontal structural and functional deficits associated with individual differences in schizotypal personality. Schizophr Res 7:237-247

Rampling D (1978) Aggression: a paradoxical response to tricyclic antidepressants. Am J Psychiatry 135:117-118

Ratey JJ, Sorgi P, O'Driscoll GA, Sands S, Daehler ML, Fletcher JR, Kadish W, Spruiell G, Polakoff S, Lindem KJ, et al. (1992) Nadolol to treat aggression and psychiatric symptomatology in chronic psychiatric inpatients: A double-blind, placebo-controlled study. J Clin Psychopharmacol 53:41-46

Reich JH (1989) Familiarity of DSM-III dramatic and anxious personality clusters. J Nerv Ment Dis 177:96-100

Reich JH (1991) Avoidant and dependent personality traits in relatives of patients with panic disorder, patients with dependent personality disorder, and normal controls. Psychiatr Res 39:89-98

Rickles K, Downing RW. Chlordiazepoxide and hostility in anxious outpatients. Am J Psychiatry 1974; 131:442-444

Rifkin A, Quitkin F, Carrillo C, Blumberg AG, Klein DF (1972) Lithium carbonate in emotionally unstable character. disorder. Arch Gen Psychiatry 27:519-523

Risch SC, Cohen RM, Janowsky DS, Kalin NH, Sitaram N, Gillin JC, Murphy DL (1981) Physostigmine induction of depressive symptomatology in normal human subjects. Psychiatr Res 4:89-94

Rosenbaum JF, Biederman J, Hirshfeld DR, Bolduc EA, Chaloff J (1991) Behavioral inhibition in children: A possible precursor to panic disorder or social phobia. J Clin Psychiatry 52(suppl 1):5-9

Rotter M, Kalus O, Losonczy M, Guo L, Trestman RL, Coccaro E Davidson M, Davis KL, Siever LJ (1991) Lateral ventricle enlargement in schizotypal personality disorder. Biol Psychiatry 29:43a-185a

Salzman C, Wolfson AN, Miyawaki, E: Fluoxetine treatment of anger in borderline personality disorder. Proceedings of the American College of Neuropharmacol 1992, p. 24

Sawaguchi T, Goldman-Rakic PS (1991) D_1 dopamine receptors in prefrontal cortex: Involvement in working memory. Science 251:947-950

Schiff HB, Sabin TD, Geller A, Alexander L, Mark V (1982) Lithium in aggressive behavior. Am J Psychiatry 139:1346-1348

Schneier FR, Spitzer RL, Gibbon M, Fyer AJ, Liebowitz MR (1991) The relationship of social phobia subtypes and avoidant personality disorders. Compr Psychiatry 32:496–502

Schulz PM, Schulz SC, Goldberg SC, Ettigi P, Resnick RJ, Friedel RO (1986) Diagnoses of the relatives of schizotypal outpatients. J Nerv Ment Dis 174:457–463

Schulz SC, Cornelius J, Schulz PM, Soloff PH (1988) The amphetamine challenge test in patients with borderline personality disorder. Am J Psychiatry 145:809–814

Sheard M, Marini J, Bridges C, Wapner A (1976) The effect of lithium on impulsive aggressive behavior in man. Am J Psychiatry 133:1409–1413

Siegel BV, Trestman RL, O'Flaithbheartaigh SO, Mitropoulou V, Amin F, Kirrane R, Silverman JM, Schmeidler J, Keefe RSE, Siever LJ (1996) D-Amphetamine challenge effects on Wisconsin Card Sort Test performance in schizotypal personality disorder. Schizophr Res 20:29–32

Siever LJ (1987) The role of noradrenergic mechanisms in the etiology of the affective disorders. In: Meltzer HY (ed) Psychopharmacology: The third generation of progress. Raven Press, New York, pp 493–504

Siever LJ, Davis KL (1991) A psychobiological perspective of the personality disorders. Am J Psychiatry 148:1647–1658

Siever LJ, Klar H, Coccaro EF (1985) Psychobiological substrates of personality. In: Klar H, Siever LJ (eds) Biological response styles: Clinical implications. American Psychiatric Press, Washington DC, pp 38–66

Siever LJ, Silverman JM, Horvath TB, Klar H, Coccaro E, Keefe RS, Pinkham L, Rinaldi P, Mohs RC, Davis KL (1990) Increased morbid risk for schizophrenia-related disorders in relatives of schizotypal personality disorder patients. Arch Gen Psychiatry 47:634–640

Siever LJ, Amin F, Coccaro EF, Bernstein D, Kavoussi RJ, Kalus O, Horvath TB, Warne P, Davidson M, Davis KL (1991) Plasma homovanilic acid in schizotypal personality disorder patients and controls. Am J Psychiatry 148:1246–1248.

Siever LJ, Trestman RL, Coccaro EF, Bernstein D, Gabriel SM, Owen K, Moran M, Lawrence T, Rosenthal J, Horvath TB (1992) The growth hormone response to clonidine in acute and remitted depressed male patients. Neuropsychopharmacology 6:165–177

Siever LJ, Kalus O, Keefe RSE (1993a). The boundaries of schizophrenia. Psychiatr Clin North Am 16:217–244

Siever LJ, Rotter M, Trestman RL, Coccaro EF, Losconzy MF, Davis KL (1993b) Increased ventricular brain ratio in schizotypal personality disorder. 146th Annual Meeting of the American Psychiatric Association, NR335

Siever LJ, Buchsbaum MS, New AS, Spiegel-Cohen J, Wei T, Hazlett EA, Sevin E, Nunn M Mitropoulou V (1999) d,l-fenfluramine response in impulsive personality disorder assessed with [18F]flourodeoxyglucosepositron emission tomography. Neuropsychopharmacology 20:413–423

Silverman JM, Pinkham L, Horvath TB, Coccaro EF, Klar H, Schear S, Apter S, Davidson M, Mohs RC, Siever LJ (1991) Affective and impulsive personality disorder traits in the relatives of borderline personality disorder. Am J Psychiatry 148:1378–1385

Silverman JM, Keefe RSE, Losonczy MF, LI G, O'Brian V. Mohs RC, Siever LJ (1992) Schizotypal and neuro-imaging factors in relatives of schizophrenic probands. Society of Biological Psychiatry Annual Meeting 31:70a, April 1992

Snyder S, Pitts WM (1984) Electroencephalography of DSM-III borderline personality disorder. Acta Psychiatr Scand 69:129–134

Soloff PH, George A, Nathan RS (1982) The dexamethasone suppression test in patients with borderline personality disorder. Am J Psychiatry 139:1621–1623

Soloff PH, George A, Nathan RS, Schulz PM, Perel JM (1986a) Paradoxical effects of amitriptyline in borderline patients. Am J Psychiatry 143:1603–1605

Soloff PH, George A, Nathan RS, Schulz PM, Ulrich RF, Perel JM (1986b) Progress in the psychopharmacotherapy of borderline disorders: A double-blind study of amitriptyline, haloperidol, and placebo. Arch Gen Psychiatry 43:691–697

Soloff PH, George A, Nathan S, Schulz PM, Cornelius JR, Herring J, Perel JM (1989) Amitriptyline versus haloperidol in borderlines: Final outcomes and predictors to response. J Clin Psychopharmacol 9:238–246

Soloff PH, Cornelius J, George A, Nathan S, Perel JM, Ulrich RF (1993) Efficacy of phenelzine and haloperidol in borderline personality disorder. Arch Gen Psychiatry 50:377–385

Stark P, Fuller RW, Wong DT (1985) The pharmacologic profile of fluoxetine. J Clin Psychiatry 46:1647–1658

Stein DJ, Simeon D, Frenkel M, Islam M, Hollander E (1995) An open trial of valproate in borderline personality disorder. J Clin Psychiatry 56:506–510

Stein MB, Chartier MJ, Hazen AL, Kroft CD, Chale RA, Cote D, Walker JR (1996). Paroxetine in the treatment of generalized social phobia: Open-label treatment and double-blind placebo-controlled discontinuation. J Clin Psychopharmacol 16:218–222

Stein DJ, Berk M, Els C, Emsley RA, Gittelson L, Wilson D, Oakes R, Hunter B (1999) A double-blind placebo controlled trial of paroxetine in the management of social phobia (social anxiety disorder) in South Africa. S Afr Med J 89:402–6.

Steinberg BJ, Trestman RL, Siever LJ (1994) The cholinergic and noradrenergic neurotransmitter systems in affective instability in borderline personality disorder. In: Silk KR (ed) Biological and neurobehavioral studies in borderline personality disorder. American Psychiatric Press, Washington DC, pp 57–63

Steinberg BJ, Trestman R, Mitropolous V, Serby M, Coccaro E, Weston S, DeVegvar M, Siever LJ (1997) Depressive response to physostigmine challenge in borderline personality disorder patients. Neuropsychopharmacology 17:264–273

Stephens JH, Schaffer JW (1970) A controlled study of the effects of diphenylhydantoin on anxiety, irritability, and anger in neurotic outpatients. Psychopharmacologia (Berlin) 17:169–181

Sternbach HA, Fleming J, Extein I, Pottash AL, Gold MS (1983) The dexamethasone suppression and thyrotropin-releasing hormone tests in depressed borderline patients. Psychoneuroendocrinology 8:459–462

Stolk JM, Conner RL, Levine S, Barchas JD (1974) Brain norepinephrine metabolism and shock-induced fighting behavior in rats: Differential effects of shock and fighting on the neurochemical response to a common footshock stimulus. J Pharmacol Exp Ther 190:193–209

Tancer ME (1993) Neurobiology of social phobia. J Clin Psychiatry 54(suppl 12):26–30

Tancer ME, Golden RN (1993) A neuropharmacologic test of the tridimensional personality questionnaire in social phobia. Biol Psychiatry 33:48A

Tancer ME, Stein MB, Gelernter CS, Uhde TW (1990) The hypothalamic-pituitary-thyroid axis in social phobia. Am J Psychiatry 147:929–933

Tellegen A, Lykken DT, Bouchard TJ Jr, Wilcox KJ, Segal NL, Rich S (1988) Personality similarity in twins reared apart and together. J Pers Soc Psychol 54:1031–1039

Torgerson S (1984) Genetic and nosological aspects of schizotypal and borderline personality disorders. Arch Gen Psychiatry 41:546–554

Torgerson AM, Kringlen E (1978) Genetic aspects of temperamental differences in infants: A study of same-sexed twins. J Am Acad Child Psychiatry 17:438–444

Trestman RL, Coccaro EF, Weston S, Mitropoulou V, Ramella F, Gabriel S, Siever LJ (1992) Impulsivity, suicidal behavior, and major depression in personality disorder: Differential correlates with noradrenergic and serotonergic function. Biol Psychiatry 31:68A

Trestman RL, Keefe RS, Mitropoulou V, Harvey PD, deVegvar ML, Lees-Roitman S, Davidson M, Aronson A, Silverman J, Siever LJ (1995) Cognitive function and biolog-

ical correlates of cognitive performance in schizotypal personality disorder. Psychiatry Res 59:127–136
Uhde TW (1994) A review of biological studies in social phobia. J Clin Psychiatry 55(suppl 6):17—27
Uhde TW, Tancer ME, Black B, Brown TM (1991) Phenomenology and neurobiology of social phobia: comparison with panic disorder. J Clin Psychiatry 52(suppl 11):31–40
Van der Kolk BA (1986) Uses of lithium in patients without major affective illness. Hosp Community Psychiatry 37:675
Versiani M, Nardi AE, Mundim FD, Alves AB, Liebowitz MR, Amrein R (1992) Pharmacotherapy of social phobia: A controlled study with moclobemide and phenelzine. Br J Psychiatry 161:353–360
Volavka J (1988) Can aggressive behavior in humans be modified by beta-blockers? Postgrad Med Feb 29:163–168
Wainberg ML, Trestman RL, Keefe RS, Cornblatt B, deVegvar M, Siever LJ (1993) CPT in schizotypal personality disorder. American Psychiatric Association Annual Meeting 1993, San Francisco, CA
Wechsler D (1981) Wechsler adult intelligence scale-revised. New York: Psychological Corporation.
Widiger TA (1992) Generalized social phobia versus avoidant personality disorder: A commentary on three studies. J Abnorm Psychol101:340–343
Yudofsky SC, Silver JM, Schneider SE (1987) Pharmacological treatment of aggression. Psychiatric Annals 17:397–406
Zanarini MC, Gunderson JG, Marino MF, Schwartz EO, Frankenburg FR (1988) DSM-III disorders in the families of borderline outpatients. Journal of Personality Disorders 2:292–302

Part 5
Future Directions in the Treatment of Major Depressive Disorder

Part 5
Future Directions in the Treatment
of Major Depressive Disorder

New Hypotheses to Guide Future Antidepressant Drug Development

I. Nalepa[1] · F. Sulser[2]

[1] Institute of Pharmacology, Polish Academy of Sciences, Krakow, Poland
[2] Departments of Psychiatry and Pharmacology, Vanderbilt University Medical Center, Nashville, TN 37203, USA
e-mail: sulserf@comcast.net

1	The Evolution of Hypotheses on the Mode of Action of Antidepressants . . .	521
1.1	Monoamine Hypotheses of Depression	521
1.2	The β-Adrenoceptor Down-Regulation Hypothesis and Other Adaptive Changes in Aminergic Receptor Systems	521
1.3	The "5-HT/NE/Glucocorticoid Link" Hypothesis of Affective Disorders and the Action of Antidepressants	523
1.4	Some Puzzles Resolved and a Reinterpretation of the β-Adrenoceptor Desensitization Hypothesis	525
2	New Vistas on the Mode of Action of Antidepressants.	526
2.1	A Role for NMDA Receptors in Antidepressant Action	526
2.2	CRF Antagonists as Antidepressants	528
2.3	Substance P Receptor Antagonists as Putative Antidepressants	530
2.4	Neurotransmitter-Induced Intracellular Processes: The Importance of the Crosstalk at the Level of Protein Kinases	530
2.5	Protein Kinase C-Related Processes as a Target for Antidepressants	532
2.6	G Proteins as a Target for Antidepressants.	534
2.7	The Convergence of Neurotransmitter Signals Beyond the Receptors at the Level of Protein Kinase-Mediated Phosphorylation	536
2.8	Transcription Factors as Targets for Antidepressants.	540
3	Towards the Discovery of the Next Generation of Antidepressants.	544
3.1	Models of Depression with Increased Disease Validity	545
3.2	Antidepressants and the Program of Gene Expression	547
References		548

Abstract The emergence of molecular neurobiology is rapidly changing the traditional focus of antidepressant drug research with emphasis on effector sites *beyond* the receptors. This switch in emphasis is leading to new conceptual and methodological approaches to understanding the mode of action of antidepressants. Events beyond the receptors—intracellular signal transduction pathways and regulation of programs of gene expression—are promising new and exciting targets for antidepressants. A short historical review of the evolution of hypotheses (e.g., monoamine, β-adrenoceptor down-regulation, and the 5-HT/NE/ glucocorticoid link) concerning the mode of action of antidepressants is first presented. The chapter then presents an overview of new hypotheses concerning the mode of action of antidepressants and discusses the role of N-methyl-D-aspartate (NMDA) receptors in antidepressant action and the possible role of corticotropin releasing factor (CRF) antagonists and substance P receptor antagonists as antidepressants. Neurotransmitter-induced intracellular processes and the importance of crosstalk at the level of protein kinases are described. The chapter than discusses protein kinase C-related processes, G proteins, and transcription factors as potential targets for antidepressants and considers the convergence of neurotransmitter signals beyond the receptors at the level of protein kinase-mediated phosphorylation. To discover the next generation of antidepressants, two avenues of interrelated investigations seem promising: (1) a more rigorous elucidation of the molecular psychopathology of affective disorders and the development of animal models of depression with greater disease validity and (2) the development of new methodology to explore mechanisms beyond the receptors and second messengers in animal models of depression and in patients with affective disorders. Possible models of depression that may have increased disease validity are described. The chapter concludes with a discussion of how programs of gene expression are likely to affect antidepressant drug development. Differentially expressed genes and their protein products can be used as novel drug targets for the development of the next generation of antidepressants, which hopefully will meet the yet unmet criteria of greater efficacy, shorter onset of therapeutic action, and efficacy in therapy-resistant depression.

Keywords Antidepressants · Drug development · Mode of action · N-methyl-D-aspartate (NMDA) · Corticotropin releasing factor (CRF) antagonists · Substance P receptor antagonists · G proteins · Transcription factors · Animal models of depression · Gene expression

Since affective disorders are predominantly human disorders, it is not surprising that the prototypes of clinically effective antidepressant drugs—monoamine oxidase inhibitors (MAOIs) and tricyclic antidepressants (TCAs)—were discovered by astute clinical observation. Only after their clinical efficacy had been established did research on the possible mechanism(s) of action of these drugs begin. Comprehensive and authoritative monographs on the discovery of antidepressants and on their behavioral and biochemical pharmacology have appeared (Costa and Racagni 1982a,b; Lehmann and Kline 1983; Sulser and Mishra 1983; Zeller 1983; Porter et al. 1986; Briley and Fillion 1988; Leonard and Spencer 1990; Wong et al. 1995; Blakely et al. 1997; Briley and Montgomery 1998).

The emergence of molecular neurobiology is now rapidly changing the traditional focus of antidepressant drug research with emphasis on effector sites *beyond* the receptors. This switch in emphasis leading to new conceptual and methodological approaches to understanding the mode of action of antidepressants is the main focus of this chapter.

1
The Evolution of Hypotheses on the Mode of Action of Antidepressants

1.1
Monoamine Hypotheses of Depression

The future development of antidepressants and new approaches to their possible mode of action are based to a large extent on previously generated knowledge and hypotheses. A short historical review of the evolution of hypotheses on the mode of action of antidepressants is thus justified as a prelude to the main topic of this chapter.

Two observations provided the scientific basis for the development of biological hypotheses of depression: MAO inhibitors, which increased the levels of monoamines in the brain, alleviated the symptoms of depression, while reserpine, a drug used for the treatment of hypertension, could precipitate a depressive syndrome (Quetsch et al. 1959) and decreased the levels of norepinephrine (NE) and serotonin (5-HT) in the brain (Pletscher et al. 1955; Holzbauer and Vogt 1956). These observations led to the monoamine hypothesis of depression, which stated that deficits of monoaminergic transmission in brain play an important role in the etiology of endogenous depression (Bunney and Davis 1965; Schildkraut 1965; Coppen 1967; Schildkraut and Kety 1967). This view was expanded into a cholinergic-noradrenergic hypothesis of mania and depression, which suggested that depression was the consequence of reduced noradrenergic transmission coupled with cholinergic predominance, while mania was due to the reverse events (Janowsky et al. 1972).

These observations were followed by the discovery of imipramine as an effective antidepressant (Kuhn 1958), even though neither imipramine nor other TCAs increased the levels of NE and/or 5-HT in the brain. This led to the discovery of the role of uptake blockade of NE and/or 5-HT, leading to an enhanced synaptic availability of the amines, in the pharmacological action of the TCAs (Axelrod et al. 1961; Carlsson et al. 1968). However, these acute effects of antidepressants did not explain the delay in the therapeutic action of antidepressant treatments (Oswald et al. 1973) and prompted the search for delayed adaptive changes caused by antidepressants.

1.2
The β-Adrenoceptor Down-Regulation Hypothesis
and Other Adaptive Changes in Aminergic Receptor Systems

In the mid-1970s, it was demonstrated that chronic but not acute administration of different types of antidepressants and also electroconvulsive therapy (ECT) reduced the sensitivity of the β-adrenoceptor-coupled adenylate cyclase system to NE (Vetulani and Sulser 1975; Vetulani et al. 1976b). This reduced sensitivity was generally accompanied by a decrease in the density of β-adrenoceptors (Banerjee et al. 1977). Thus, the "β-adrenoceptor down-regulation" hypothesis was born. Subsequent papers stressed the adaptive character of these changes (Vetulani et al. 1976a,b). These findings concerning β-adrenoceptor subsensitivity (desensitization of the β-adrenoceptor-coupled adenylate cyclase) and down-regulation (reduction in the density of β-adrenoceptors) were quickly confirmed (Schultz 1976; Wolfe et al. 1978; Bergstrom and Kellar 1979a,b; Pandey et al. 1979; Kellar et al. 1981a,b; Kellar and Bergstrom 1983). In due course, new biochemical hypotheses on the pathophysiology of affective disorders have been proposed. These hypotheses are by and large an extension of existing hypotheses at the level of second messenger function. For example, Wachtel (1989) suggested that affective disorders arise from the imbalance of the two major intraneuronal signal amplification systems, the adenylate cyclase and the phospholipase C systems, with depression resulting from underfunctioning of cyclic AMP effector cell responses associated with an absolute or relative dominance of inositol triphosphate/diacylglycerol-mediated responses and mania resulting from the converse effects.

It has also been proposed that the dopaminergic system is involved in depression and in the mode of action of antidepressants (Randrup et al. 1975; Willner 1983; Maj 1984). Studies demonstrated that chronic treatment with ECT (Wielosz 1981) and with antidepressant drugs increased the behavioral responses to dopaminergic stimulation (e.g., Spyraki and Fibiger 1981; Maj 1984; Maj et al. 1984, 1987; Willner 1983). Consequently, it was postulated that D_2 dopamine receptors were involved in antidepressant action (Maj et al. 1989a,b). However, no correlation between the enhancement of dopaminergic responses and dopamine receptor binding could be found when dopaminergic antagonists were used as ligands (Willner 1983; Klimek et al. 1985). Five subtypes of dopamine receptors have since been cloned and characterized (for a review, see Kapur and Mann 1992). In addition to the initially known and pharmacologically well-defined D_1 and D_2 dopamine receptors, D_3, D_4, and D_5 subtypes were also characterized. In turn, new and more specific compounds for dopamine receptors have been developed. Recent studies, employing agonists as ligands, suggest that antidepressant treatments may increase the density of D_2 and D_3 receptors (Maj et al. 1996, 1998; Ainsworth et al. 1998) as well as the steady-state level of D_2 receptor mRNA (Dziedzicka-Wasylewska et al. 1997; Ainsworth et al. 1998). In addition, D_2/D_3 receptor agonists were shown to evoke antidepressant-like effects in some animal models of depression and to exert potential antidepressant activity in man (for reviews, see Willner 1997a; Willner and Papp 1997). Thus, the development of more selective compounds directed towards subtypes of dopamine receptors will help define the role of dopamine in depression and the action of antidepressant drugs.

Avissar and Schreiber (1992a,b) have pointed out the role of G proteins in the etiology of affective disorders and in the mechanism of antidepressant drug action. The role of G proteins and G protein function as possible targets of antidepressants is discussed later in the chapter.

1.3
The "5-HT/NE/Glucocorticoid Link" Hypothesis of Affective Disorders and the Action of Antidepressants

All currently available and clinically effective antidepressant drugs affect central noradrenergic and/or serotonergic neuronal systems at various levels of the aminergic signal transduction cascades (Fig. 1) or are converted in vivo to metabolites which, in concert with the parent drug, affect the synaptic availability of NE and/or 5-HT. Since stressful life events and the vulnerability to stress are believed to be predisposing factors in the precipitation of affective disorders and chronic mild stress is often used to construct animal models of depression with disease validity (Willner et al. 1987), it is of interest that alterations in circulating glucocorticoids alter noradrenergic receptor sensitivity in brain (Mobley et al. 1983; Roberts et al. 1984; Harrelson et al. 1987). Moreover, glucocorticoid receptors have been identified in the nuclei of NE- and 5-HT-containing cell bodies in the brain (Harfstrand et al. 1986) and in limbic structures such as the amygdala (Honkaniemi et al. 1992). An important finding is that glucocorticoid receptor immunoreactivity (Kitayama et al. 1988), density (Przegalinski and Budziszewska 1993; Budziszewska et al. 1994), and expression (Pepin et al. 1989; Peiffer et al. 1991; Seckl and Fink 1992; Rossby et al. 1995) have been shown to be increased after chronic treatment with some antidepressants. The regulation of glucocorticoid receptors in the brain by antidepressants is of considerable neurobiological interest as these cytoplasmic steroid receptors are hormone-activated transcription factors which can act as positive or negative regulators of gene expression (Burnstein and Cidlowsky 1989). It thus became imperative to integrate the glucocorticoid receptor system into any modern amine hypothesis of affective disorders (Pryor and Sulser 1991). It is tempting to speculate that glucocorticoids may affect the diffusely projecting stress-responsive monoamine systems in the brain via changes in the transcription of pivotal proteins.

In view of the effects of the pituitary–adrenal axis on central nervous system function and mood, the finding that chronic foot shock stress can desensitize the NE sensitive adenylate cyclase system in rat cortex (Stone 1979a) is of particular interest. Stone has proposed that this decrease in sensitivity may play a role in the adaptation to emotional and physiological stress and that various antidepressant treatments mimic the desensitizing action of stress on central NE receptor systems (Stone 1979b). The findings that psychotropic drugs that precipitate depressive reactions increase the sensitivity of the system, while drugs that alleviate depressive mood decrease its sensitivity raise new questions about the psychobiology of depressive illness. One is tempted to speculate that people who are prone to affective disorders may suffer from an inability to down-regulate the central noradrenergic receptor cascade in response to enhanced neuronal input (lack of proper adaptation) and, therefore, successful treatment with antidepressant drugs and/or ECT would depend on the successful induction of subsensitivity of this system (Sulser 1978).

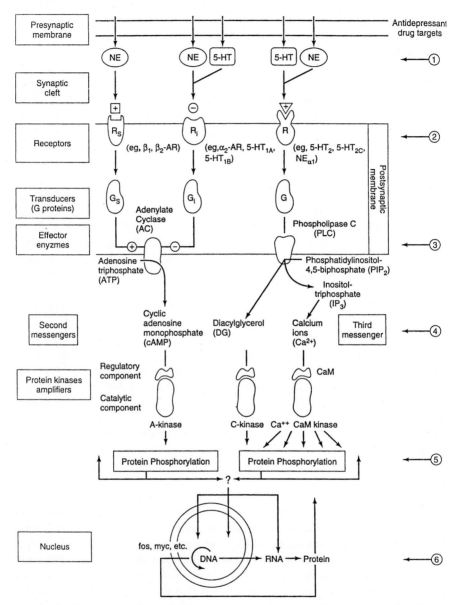

Fig. 1 Neurotransmitter signal transduction cascades as a target for antidepressant drugs. Antidepressant drugs can influence the information flow at various levels of the signal transduction cascade: *1.* Change in the synaptic availability of the primary signals NE and/or 5HT (MAO inhibitors; blockade of reuptake; autoreceptor subsensitivity). *2.* Change in receptor number or sensitivity. *3.* Change in the function of G proteins. *4.* Change in the formation of second messengers. *5.* Change in the activity of protein kinases (amplifier function). *6.* Modification of nuclear events (e.g., gene transcription). From Rossby and Sulser (1993)

It has been suggested that the desipramine (DMI)-induced increase in glucocorticoid receptor mRNA and glucocorticoid receptor density could be responsible—via feedback inhibition—for the normalization of the hypothalamic–pituitary–adrenocortical (HPA) axis hyperactivity that is often seen in depression (Pepin et al. 1992; Barden et al. 1995). While such an antidepressant-induced increase in feedback inhibition might normalize the endocrine abnormality often seen in depression, this apparent correction of the endocrine abnormality is, however, not a prerequisite for the therapeutic efficacy of antidepressants, since non-tricyclic antidepressants, particularly selective serotonin reuptake inhibitors (SSRIs) such as citalopram and fluoxetine do not increase either the density or the expression of glucocorticoid receptors in brain in vivo (Brady et al. 1992; Seckl and Fink 1992; Budziszwska et al. 1994; Rossby et al. 1995). The suggestion that antidepressants stabilize mood through actions on the HPA system—although attractive at first glance—is thus too general and is not supported by the available experimental evidence. The findings that some currently available and clinically effective antidepressants can influence the expression of genes are, however, of considerable heuristic value (see the discussion of antidepressants and the program of gene expression at the end of this chapter).

1.4
Some Puzzles Resolved and a Reinterpretation of the β-Adrenoceptor Desensitization Hypothesis

For some time, the antidepressant bupropion appeared to be an exception to the rule described above because this drug did not alter either serotonergic or noradrenergic transduction cascades. However, recent studies have revealed a striking difference in the metabolic disposition of bupropion between rat and man. Unlike in the rat, in man, bupropion is metabolized mainly to hydroxybupropion, which is a potent uptake inhibitor of NE (Ferris and Cooper 1993), thus placing bupropion in the category of noradrenergic antidepressants. Nefazodone has been demonstrated to inhibit both the NE and 5-HT transporters with moderate potency (Owens et al. 1995), thus providing a rationale for its antidepressant action. Among the newly developed drugs is reboxetine, a selective NE reuptake inhibitor. Unlike the non-selective TCAs, it acts specifically at noradrenergic sites (Leonard 1997; Montgomery 1997) and exerts antidepressant activity in man, with low affinity for β-adrenergic and muscarinic receptors and low toxicity in animals (Riva et al. 1989; Szabadi et al. 1998; Wong et al. 2000).

Mirtazapine is an antagonist of somatodendritic as well as terminal α_2- adrenergic auto- and heteroreceptors (Haddjeri et al. 1996). The antagonism of α_2-autoreceptors enhances the release of NE. Also, mirtazapine enhances the release of NE in the raphe nuclei so that α_1-adrenoceptors on serotonergic soma or dendrites are activated to increase the firing rate of serotonergic neurons and the release of 5-HT. The mirtazapine-induced blockade of α_2-adrenergic receptors located on serotonergic terminals can also contribute to the facilitation of 5-HT release (Frazer 1997).

Studies with the dual uptake inhibitor venlafaxine have provided strong evidence for a link in the brain between the two aminergic signal transduction pathways beyond the β-adrenoceptors. Venlafaxine does not desensitize the β-adrenoceptor-coupled adenylate cyclase system in the brain after chronic administration, except in the absence of 5-HT (Nalepa et al. 1998). It has been suggested that a crosstalk between 5-HT and NE receptor-mediated activation of protein kinases and their counter regulation at the level of agonist-induced desensitization of β-adrenoceptors may explain the unexpected pharmacology of this drug. It is hypothesized that this crosstalk is the consequence of stimulation by venlafaxine of the inositol pathway by increased synaptic 5-HT leading to increased diacylglycerol (DAG) production and thus, increased protein kinase C activation. It has been shown that activation of protein kinase C in a fibroblast cell line down-regulates the activity of protein kinase A (Döbbeling and Berchtold 1996). However, extrapolation to other cell lines and to action in the brain in vivo is difficult. It is of interest that milnacipran, a 5-HT and NE dual reuptake inhibitor, has also been reported not to down-regulate β-adrenoceptors or to desensitize the β-adrenoceptor-coupled adenylate cyclase system (Assie et al. 1988; Neliat et al. 1996). The results with venlafaxine and milnacipran indicate that the desensitization of the β-adrenoceptor-coupled adenylate cyclase system is not a prerequisite for the therapeutic action of antidepressant drugs and mandate a reinterpretation of the β-adrenoceptor desensitization hypothesis. β-Adrenoceptor desensitization which occurs after chronic administration of noradrenergic antidepressants is now viewed as being the consequence of receptor phosphorylation by protein kinase A and perhaps the cyclic AMP independent β-adrenoceptor kinase 1 (BARK), caused by persistent signaling via the NE β-adrenoceptor interaction. The new data with venlafaxine have mandated a shift in the emphasis on the mode of action of antidepressants from changes in β-adrenoceptor sensitivity to the convergence of aminergic signals *beyond* the receptors and have lent support to the "5-HT/NE link" hypothesis of affective disorders and mode of action of antidepressants. Theoretically, the dual signaling by drugs such as venlafaxine and milnacipran should lead to a potentiation of signal transduction downstream from the receptors. It remains a challenge to demonstrate experimentally how dual signaling (NE and 5-HT) as opposed to single signaling (NE or 5-HT) is translated into differential expression of programs of genes believed to be ultimately responsible for the therapeutic action in man.

2
New Vistas on the Mode of Action of Antidepressants

2.1
A Role for NMDA Receptors in Antidepressant Action

N-methyl-D-aspartate (NMDA) antagonists have repeatedly been demonstrated to be active in many preclinical tests thought to be predictive of antidepressant action (Maj et al. 1992a,b; Panconi et al. 1993; Papp and Moryl 1994). Both competitive and noncompetitive antagonists at NMDA receptors have also been shown to significantly down-regulate β-adrenoceptors (Paul et al. 1992; Klimek and Papp 1994; Layer et al. 1995; Wedzony et al. 1995) and to be effective in the Willner model of depression (Papp and Moryl 1994), suggesting that the behavioral results may not be "false" positives.

Interestingly, eliprodil, an NMDA antagonist acting at polyamine sites, has also been reported to exert antidepressant activity, both behaviorally and neurochemically (Layer et al. 1995). These findings prompted Skolnick and colleagues to examine the effects on NMDA receptors of chronic treatment with a large variety of antidepressants. They found that chronic but not acute antidepressant treatments of mice and rats altered the ligand-binding profile of NMDA receptors, producing a reduction in the potency of glycine to inhibit [^{3}H]-CGP 39653 binding to strychnine-insensitive glycine receptors and a reduction in the proportion of high affinity glycine sites inhibiting [^{3}H]-CGP 39653 binding to NMDA receptors (Nowak et al. 1993; Paul et al. 1993). In addition to glycine and glutamatergic stimulation, the ion channel of the NMDA receptor complex is subject to a voltage-dependent regulation by $Mg2^+$ cations. Under physiological conditions, this channel is supposed to be blocked by a high concentration of magnesium in extracellular fluids. Recently, it has been demonstrated that oral administration of magnesium to normal animals can antagonize NMDA-mediated responses and lead to antidepressant-like effects that are comparable to those of MK-801 (Decollogne et al. 1997). The ability of a structurally diverse group of antidepressants as well as ECT to produce adaptive changes in NMDA receptors (Paul et al. 1994) suggests that this family of ligand-gated ion channels may be an attractive novel target for antidepressant action—i.e., substances capable of reducing neurotransmission at NMDA receptors with their individually encoded subunits (Monyer et al. 1992) may represent a new class of antidepressants. To date, studies from several laboratories have indeed demonstrated that NMDA antagonists are potential antidepressants (for reviews, see Skolnick et al. 1996; Huang et al. 1997).

Recently, Skolnick (1999) proposed a hypothesis that links, at the molecular level, the action of most conventional antidepressants to reductions in NMDA receptor function. According to Skolnick's hypothesis, two different treatment strategies—conventional antidepressants and compounds that reduce transmission at NMDA receptors—converge at an identical endpoint target to produce a region-specific dampening of NMDA receptor function. Moreover, he suggests a role for brain-derived neurotrophic factor (BDNF) as one of the putative molecular links between conventional antidepressants and NMDA receptors. The hypothesis is based on evidence showing that (1) BDNF reduces mRNA and protein levels of NMDAR-2A and -2C (Brandoli et al. 1998), and (2) repeated administration of antidepressants to mice alters the regional expression of mRNA that encode multiple NMDA receptor subunits (Boyer et al. 1998), and (3) data indicate that antidepressants increase the expression of BDNF (Nibuya et al. 1995, 1996; reviewed in Duman et al. 1997). The hypothesis that glutamatergic dysfunction may be involved in the pathophysiology of major depression in man has been supported by studies using autopsy material from suicide victims (Nowak et al. 1995) and by data from in vivo proton magnetic resonance spectroscopy (3H-MRS) (Auer et al. 2000). The latter study provided evidence of a reduced glutamate concentration within the anterior cingulate of depressed patients. More recently, Berman et al. (2000) have reported that the infusion of a low-dose of ketamine (a potent NMDA receptor antagonist) was associated with decreases in depressive symptoms. These results are compatible with the hypothesis of NMDA receptor dysfunction in depression. Attractive as this hypothesis is, it remains to be seen whether or not NMDA antagonists will indeed be viable, clinically effective antidepressants and whether or not possible psychotomimetic effects, which have been reported with some NMDA antagonists (e.g., MK-801) after long-term administration, will make treatment of depressed patients with such drugs feasible.

2.2
CRF Antagonists as Antidepressants

Increased concentrations of the corticotropin releasing factor (CRF) in cerebrospinal fluid (CSF) have been reported in patients with major depression (Nemeroff et al. 1984) and in those with posttraumatic stress disorders (Bremner et al. 1997). In experimental animals, centrally administered CRF produces a wide spectrum of behavioral changes reminiscent of depression and/or anxiety disorders (Dunn and Berridge 1990a,b). The observation of a decrease in CRF-binding sites in the frontal cortex of suicide victims compared to controls is consistent with the hypothesis that CRF is hypersecreted in major depression (Nemeroff et al. 1988). These results have led to the suggestion that CRF hypersecretion may underlie not only certain endocrine abnormalities observed in some patients with major depression but might also perhaps be involved in the pathophysiology of mood disorders (Holsboer et al. 1992), although an unequivocal link between CRF hyperdrive and human depression has not yet been established.

It has been reported that the central administration of a competitive CRF antagonist, α-helical CRF_{9-41} (Rivier et al. 1984) reverses or attenuates the suppression of ingestive, exploratory, and operant behaviors produced by administration of CRF or by exposure to various stressors (see Dunn and Berridge 1990a,b). A major biological question is the specificity of the role of CRF in relationship to the type or severity of the stressor (Heinrichs et al. 1994). CRF plays a major role in stress physiology and in the pathophysiology of disorders of the HPA axis. It stimulates the release of adrenocorticotropin (ACTH) and proopiomelanocortin-derived peptides from the corticotrophic anterior pituitary cells (Rivier and Plotsky 1986). The major co-releaser of ACTH, which potentiates or synergizes the effect of CRF, is vasopressin (sometimes oxytocin).

The arginine–vasopressin receptor antagonists have been found to prevent arginine–vasopressin-stimulated CRF secretion (Bernardini et al. 1994). It has been suggested that there is a separate mechanism or cell type for vasopressin-stimulated ACTH release distinct from that responsible for CRF-induced ACTH release. It should be pointed out that, in this way, two different second messenger systems are involved in the control of CRF release. CRF receptors are positively coupled via G_s to adenylate cyclase, cyclic AMP production, and protein kinase A (PKA) stimulation whereas vasopressin through V_1 receptors activates phospholipase C and via IP_3/diacylglycerol (DG) stimulates protein kinase C (PKC) (Birnbaumer et al. 1990). Among adrenal steroids, corticosterone is the major negative feedback signal in the regulation of ACTH, affecting hippocampus, hypothalamus, and pituitary at a genomic and nongenomic level. Hypersecretion of cortisol and its resistance to dexamethasone suppression are biological markers of major depression.

CRF release is controlled by neurotransmitters and neuropeptides. γ-Aminobutyric acid (GABA) (Calogero et al. 1988a) and β-endorphin are inhibitory, while acetylcholine, 5-HT (in low doses), and NE stimulate CRF release (Antoni et al. 1983; Calogero et al. 1988b). However, NE has a biphasic effect; at low doses it acts through α_1-adrenergic receptors, stimulating CRF release, while at higher doses, it causes inhibition mediated by α_2- and β-adrenergic receptors (Plotsky 1987). Considerable evidence suggests a functional interaction between CRF and NE in the locus coeruleus (LC).

Taken together, it is not surprising that drugs which affect the synaptic availability of NE and/or 5-HT, GABA, or steroids can influence the HPA axis. Thus, long-term antidepressant drug treatment decreases the activity of the HPA axis (Brady et al. 1991, 1992). Moreover, antidepressants may also block the activation of the hypothalamic CRF system that is induced by chronic stress (Brady 1994). It has been found, using electrophysiological and morphological methods, that stress-induced depression (long-term forced walking stress) causes degenerative changes in LC neurons, and this effect is reversed by repeated treatment with imipramine (Kitayama et al. 1994). It should be pointed out that antidepressant drugs seem to interfere with CRF neurotransmission in the LC only under abnormal conditions. None of the antidepressant drugs, given chronically, altered the LC activation by intracerebroventricularly administered CRF. However, chronic administration of desmethylimipramine (DMI) and mianserin decreased CRF release in the LC, thereby inhibiting LC activation by a hypotensive stress that requires endogenous CRF release (Curtis and Valentino 1994). Moreover, chronic treatment with sertraline (an SSRI) and phenelzine (a nonselective MAOI) antagonized CRF in the LC, changing LC responses to repeated sciatic nerve stimulation in a manner opposite to CRF effects (Curtis and Valentino 1994). Tianeptine, a novel tricyclic drug that has been reported to increase 5-HT uptake, has been found to reduce stress-evoked stimulation of the HPA axis (Delbende et al. 1991). Recently, Skutella et al. (1994) reported that socially defeated rats treated with CRF antisense oligodeoxynucleotides (injected into the lateral ventricle) displayed markedly reduced anxiety-related behavior. They suggest that pharmacological interventions directed against activation of CRF neurons might be valuable strategies for treating disorders associated with CRF hyperactivity. Assuming that we are dealing with a causal relationship between CRF and mood disorders rather than just an endocrine abnormality in the pleiotropic spectrum of depression, the development of selective CRF-1 and CRF-2 receptor antagonists and their careful clinical testing in major depression will provide some answers to many yet unresolved questions.

Results obtained by Mansbach et al. (1997) implicate stress systems in the pathophysiology of depression and suggest the potential efficacy of CRF receptor antagonists in the treatment of affective disorders. Thus, CP-154,526, a selective corticotropin releasing factor (CRF)-1 receptor antagonist, exerted an antidepressant-like effect in the learned helplessness procedure, a putative model of depression.

Moreover, Okuyama et al. (1999) found that CRA1000 and CRA1001, novel and selective CRF-1 receptor antagonists, showed anxiolytic- and antidepressant-like properties in various experimental animal models. Thus, both CRA1000 and CRA1001, when administered orally, reversed the effects of CRF infusion on time spent in the open arms in the elevated plus-maze in rats and inhibited the hyperemotionality induced by lesioning of olfactory bulbs. Recently, Arborelius et al. (1999) reviewed findings concerning the hyperactivity of CRF neuronal systems in depression, which indicated that CRF receptor antagonists may represent a novel class of antidepressants.

2.3
Substance P Receptor Antagonists as Putative Antidepressants

The demonstration that some cell bodies in the brain that contain norepinephrine and serotonin express substance P (Hökfelt et al. 1987) and that the administration of some antidepressants reduces substance P biosynthesis (Shirayama et al. 1996) triggered the development of nonpeptide substance P receptor antagonists as putative antidepressants. MK-869 [bis (trifluoromethyl) morpholine], a long-acting substance P antagonist, and a number of structurally related agents have shown a pharmacological profile in preclinical behavioral tests that resembles that of clinically effective antidepressants, although, biochemically, these substance P antagonists did not share the action of established antidepressants on either noradrenergic or serotonergic neuronal systems. Although early clinical trials reported clinical efficacy of MK-869 in moderate to severe major depression (Kramer et al. 1998), this therapeutic action of the substance P receptor antagonist turned out to be rather disappointing in follow-up studies.

2.4
Neurotransmitter-Induced Intracellular Processes: The Importance of the Crosstalk at the Level of Protein Kinases

Extracellular signals—neurotransmitters as first messengers—are selectively recognized at the receptor level and translated into intracellular second messengers. The signals become less specific at the cytoplasmic level, only to retrieve their specificity in the nucleus (nuclear receptors at the promoter level). All presently known neurotransmitters acting through their specific metabotropic

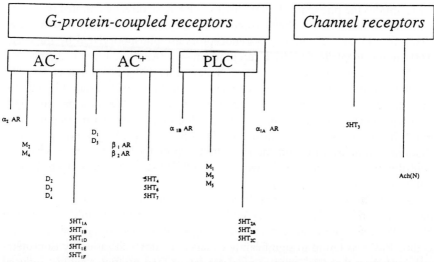

Fig. 2 The mode of coupling among neurotransmitter receptors and their intracellular effector systems. Neurotransmitters can operate through two classes of receptors: metabotropic and ionotropic (channel) receptors. The same neurotransmitter can stimulate different second messenger systems depending on the type of receptor it occupies. Receptors for norepinephrine, acetylcholine, dopamine, and serotonin are indicated as examples. Metabotropic receptors are coupled with second messenger systems through G proteins activating phospholipase C (PLC) and activating (AC$^+$) or inhibiting (AC$^-$) adenylate cyclase. Stimulation of α_2-adrenergic (α_2 AR), acetylcholine muscarinic (M_2 and M_4), dopaminergic (D_2, D_3, D_4), and serotonergic (5HT$_{1A, 1B, 1D, 1E, 1F}$) receptors inhibit AC and cyclic AMP generation, while dopaminergic (D_1, D_5), β_1 and β_2 adrenergic (β_1 AR, β_2 AR), and serotonergic (5HT$_{4,6,7}$) receptors activate AC and cyclic AMP generation. Two additional serotonergic receptors 5HT$_{5A}$, 5HT$_{5B}$ (not indicated in the figure) have been cloned. Although, their physiological roles are not clear, at least one of the two, 5HT$_{5A}$, is negatively coupled to the cyclic AMP pathway. PLC is positively linked with α_{1B} adrenergic (α_{1B} AR), muscarinic ($M_{1,3,5}$) and serotonergic (5HT$_{2A, 2B, 2C}$) receptors, stimulating the generation of the second messengers diacylglycerol and inositol trisphosphate (IP$_3$). The α_{1A}- adrenergic receptor (α_{1A} AR) and the α_{1D}-adrenoceptor (not shown) operating through G proteins may modulate a voltage and dihydropyridine-dependent calcium channel. Recent data indicate that they are able to stimulate PLC-coupled second messenger systems, as well. Acetylcholine and serotonin can act through channel receptors as well

receptors (coupled with G proteins) utilize at least two intracellular messenger systems: the adenylate cyclase/cyclic AMP/PKA system and the phospholipase C/inositol trisphosphate (IP$_3$)/diacylglycerol (DG)/PKC system (Fig. 2). In addition, they may affect the cytoplasmic calcium level operating through G protein-linked ion channels. The intracellular increase of calcium ions activates the calcium/calmodulin-dependent protein kinase. The strength of the neurotransmitter transduction signal reaching the nucleus is modulated by PKA- and PKC-induced changes in the sensitivity of the respective receptors (covalent modification by phosphorylation).

Catecholamines activate both PKA (via β-adrenergic receptors) and PKC (via α_1-adrenergic receptors), and α- and β-adrenoceptors can be directly phosphorylated in vitro by both PKA and PKC. However, agonist occupancy of the receptor facilitates the phosphorylation only by the kinase directly coupled to its own signaling transduction pathway (Leeb-Lundberg et al. 1985, 1987; Bouvier et al. 1987). In addition, β_2- and β_1-adrenoceptors are phosphorylated by the cyclic AMP-independent β-adrenergic receptor kinase (BARK) in a totally agonist-dependent fashion (Benovic et al. 1986; Sibley et al. 1986; Freedman et al. 1995; Premont et al. 1995). Thus, β-adrenoceptors can be phosphorylated by at least three kinases: PKC in an agonist-independent fashion, PKA in a partially agonist-dependent fashion, and BARK in a predominantly agonist-dependent fashion. It has been suggested that the agonist-independent phosphorylation of the β-adrenoceptor by PKC and the α_1- adrenoceptor by PKA occurs in vivo and contributes to the crosstalk between the different adrenoceptor transduction pathways (Bouvier et al. 1987).

Recently, the PKC contribution in β-adrenergic agonist-induced desensitization has been demonstrated in A-431 and DDT_1MF-2 cell lines (Shih and Malbon 1994). The knockout of PKC by antisense oligodeoxynucleotides to the mRNA encoding PKC was found to significantly enhance the desensitization by isoproterenol, indicating that PKC counter-regulates rather than promotes agonist induced β-adrenoceptor desensitization via PKA-mediated receptor phosphorylation.

2.5
Protein Kinase C-Related Processes as a Target for Antidepressants

Contrary to the down-regulation of the density of β-adrenoceptors and the desensitization of the cyclic AMP responses to NE, chronic treatment with ECT and imipramine has been reported to increase the density of α_1-adrenoceptors (Vetulani et al. 1983, 1984a,b). Moreover, after chronic administration, antidepressants reduce the density of 5-HT_2 receptors (Bergstrom and Kellar 1979a; Blackshear and Sanders-Bush 1982; Kendall and Nahorski 1985), while ECT increases it (Kellar et al. 1981b; Vetulani et al. 1981; Kellar and Stockmeier 1986; Pandey et al. 1992). Both the α_1-adrenergic and 5-HT_2 serotonergic receptor families, although activated by different neurotransmitters, operate through IP_3/DAG-PKC signaling pathways. However, changes in receptor density induced by antidepressants are not consistently correlated with changes in the inositol phosphate (IP) response. Thus, chronic ECT or imipramine treatment has been found either not to affect the IP response to NE (Li et al. 1985; Nalepa and Vetulani 1993a; Nalepa et al. 1993) or to increase it (Newman and Lerer 1989). Mianserin, an atypical antidepressant, and citalopram, a selective 5-HT reuptake inhibitor, augmented the IP response after chronic treatment (Nalepa and Vetulani 1993b, 1994). In addition, chronic ECT, imipramine, and mianserin treatment attenuate the negative feedback between PKC and the α_1-adrenergic receptor (Nalepa et al. 1993, 1996). In summary, chronic treatment with antidepressants enhances α_1-adrenoceptor-related transmission affecting at least three levels of signal transduction cascades. In addition to the increase in receptor density and/or IP generation, antidepressants may counteract the PKC-dependent reduction in the α_1-adrenoceptor response.

Szmigielski and Gorska (1997) found that chronic treatment with imipramine augmented the α_1-adrenoceptor-mediated redistribution of PKC from the cytosolic to the membrane fraction in the rat frontal cortex and hippocampus. In animals treated with imipramine, much lower doses of α_1-adrenergic agonists were effective for PKC redistribution. Moreover, prolonged treatment with imipramine markedly increased the basal activity of PKC in the membrane fraction of both cortex and hippocampus.

Recent research has focused on the regulatory role of α-adrenoceptors in the responsiveness of the cyclic AMP generating system to stimulation of β-adrenoceptors. Such a regulatory role was initially suggested by the discovery of an α-adrenergic potentiation of the β-adrenergic response (Daly et al. 1980). Recent studies suggest that PKC is responsible for the crosstalk between α_1- and β-adrenoceptors (Nalepa and Vetulani 1991a,b, 1993a,b, 1994; Nalepa et al. 1993). Also, the shift of the balance between α_1- and β-adrenoceptors towards α_1-adrenoceptors seems to be a characteristic effect of chronic antidepressant treatment (Vetulani et al. 1984a) and it has been suggested that PKC-related processes play the main role in this phenomenon (Nalepa 1994).

It seems that the action of antidepressants is to a large extent related to the adrenergic system, even when their action on this system is not direct and is accomplished by influencing another neurotransmitter system (e.g., the serotonergic system).

An important finding was that ACTH, like restraint stress, was found to reduce only the α-induced potentiation of the β-response (Stone et al. 1986). Moreover, in chronic, mild, stress-induced depression, only the cyclic AMP response to NE was changed (Papp et al. 1994a), while the response to the β-agonist isoproterenol was unaffected (I. Nalepa, unpublished data). Thus, it is suggested that PKC and/or PKC-related processes might be targets for both antidepressant drugs and stress.

PKC can affect neurotransmitter release. Two major substrates for PKC are GAP-43 and MARCKS, proteins that are associated with neurotransmitter release and cellular plasticity (Manji and Lenox 1994). However, it is unlikely that changes in GAP-43 phosphorylation are involved in antidepressant-induced modulation of 5-HT release, as the extent of phosphorylation of GAP-43 by native PKC in synaptosomes of rats treated with either fluoxetine or DMI was not significantly different from that observed in control animals (Li and Hrdina 1997). A significant relationship exists between the release of NE and 5-HT. These two neurotransmitters can affect each other's levels in a reciprocal way. There is a central noradrenergic facilitatory influence, mediated by α_1-adrenoceptors, on serotonergic neurons projecting to the hippocampus (Rouquier et al. 1994). Evidence also exists that 5-HT$_2$ receptor activation decreases NE release in the rat hippocampus in vivo (Done and Sharp 1992). One can hypothesize that, since adrenergic neurons are under the inhibitory control of 5-HT (Newman and Lerer 1989; Plaznik et al. 1989), an increase in the level of 5-HT should lead to an attenuation of the NE interaction with the α_1-adrenergic receptor and consequently a decrease of 5-HT release. Recently, the atypical antipsychotic drug clozapine has been shown to inhibit serotonergic transmission by its action at α_1-adrenoceptors (Lejeune et al. 1994). On the other hand, the blockade of cortical α_1-adrenergic receptors appears to facilitate cortical D_1 receptor-mediated transmission (Tassin et al. 1986). The blockade of cortical α_1-adrenergic receptors by prazosin prevented the appearance of the EEDQ (1-ethoxycarbonyl-2-ethoxy-1,2-dihydroquinoline)-induced supersensitivity of dopamine-sensitive adenylate cyclase (Trovero et al. 1992). This suggests that the stimulation of α_1-prazosin-sensitive receptors by NE plays an inhibitory role on signals linked to cortical dopamine D_1 transmission.

Taken together, PKC and/or the activation pathway for PKC could be one of the targets for future antidepressant drugs. It is noteworthy that lithium is used as the drug of choice in the therapy of mania. Lithium is a potent inhibitor of inositol monophosphatase that results in an accumulation of IP as well as a reduction in free inositol (Hallcher and Sherman 1980). It might be expected that PKC activation is reduced as a consequence of selective suppression of phosphoinositide hydrolysis and DAG availability. For instance, chronic administration of lithium reduces the expression of PKC α and PKC ε, as well as a major PKC substrate, MARCKS, which has been implicated in long-term neuroplastic events in the developing and adult brain (Manji and Lenox 1999).

However, recent evidence suggests that agonist-induced activation of phospholipases D (PLD) and A_2 (PLA$_2$) may contribute significantly to the potentiation of endogenous DG stimulation of PKC (Nishizuka 1992). The activation has been shown to enhance phosphatidylcholine (PC) hydrolysis mediated by PLA$_2$ and PLD. Several reports indicate that PKC plays a role in the modulation of neurotransmitter transport from the synapse into presynaptic terminals. Huff et al. (1997) demonstrated that the PKC-dependent phosphorylation of the rat dopamine transporter (stably expressed in LLC-PK1 cells) induced a decrease in transporter functioning. Moreover, they suggested that PKC-induced phosphorylation of the dopamine transporter could be involved in rapid neuroadaptive processes in dopaminergic neurons. Qian et al. (1997) showed that the stimulation of PKC caused a time-dependent reduction in 5-HT uptake. However, they indicated that altered surface abundance, rather than reduced catalytic transport efficiency, mediates acute PKC-dependent modulation of 5-HT uptake. Also, the activity of the rat GABA transporter 1 and mouse glycine transporter 1 (expressed in human embryonic kidney 293 cells) was down-regulated by PKC activation (Sato et al. 1995a,b). An initial transient increase in the level of DG arising from phosphatidylinositol-4,5-biphosphate (PIP2) breakdown may act as a trigger for hydrolysis of PC through activation of PKC. Consequently, the DG generated from PC hydrolysis results in a more prolonged activation of PKC and PKC-mediated events (Manji and Lenox 1994).

2.6
G Proteins as a Target for Antidepressants

Receptors for neurotransmitters such as NE and 5-HT are coupled via heterotrimeric G proteins consisting of α, β, and γ sub-units, in a stimulatory (G$_s$,

G_q) or inhibitory (G_i) fashion with effector systems such as adenylate cyclase, phospholipase C, phospholipase A_2, or ion channels (Gilman 1987; Birnbaumer 1990), generating second messengers (e.g., cyclic AMP, diacylglycerol, IP_3, and arachidonic acid), which in turn activate various protein kinases. Protein phosphorylation by protein kinases and dephosphorylation by protein phosphatases represent one of the major mechanisms of signal integration of eukaryotic cells. Protein kinases activated by second messengers are "pleiotropic" enzymes that regulate via phosphorylation a large number of neuronal proteins [e.g., enzymes involved in neurotransmitter synthesis and degradation, ion channels, G proteins, receptors and the regulation of their number and sensitivity, transcription factors such as the cyclic AMP response element-binding protein (CREB), the cyclic AMP response element-modulating protein (CREM), steroid receptors, and immediate early genes] (Nestler and Greengard 1984). For these reasons, the development of novel antidepressant drugs by targeting G proteins and their function remains an exciting possibility (Target No. 3 in Fig. 1). Antidepressant drugs could change (1) the levels and expression of G proteins and their subunits, (2) the coupling between receptors and G proteins, (3) the coupling between G proteins and their effectors, and (4) the intrinsic properties of G proteins [e.g., affinity of guanosine 5'-triphosphate (GTP) to α-subunits, rate of GTPase activity of the α-subunit that determines the length of effector activation].

Alterations in the levels of G proteins and their mRNAs and in G protein function after treatment with currently available antidepressants have been reported, but the results so far have been rather equivocal. For example, chronic administration of imipramine has been reported to decrease the levels of $G\alpha_s$ immunolabelling, choleratoxin ADP ribosylation, and $G\alpha_s$ mRNA in rat brain (Duman et al. 1989; Lesch and Manji 1992). Lesch et al. (1991) also reported a decrease in G_s and G_i proteins but an overall increase in G_o proteins in cerebral cortex following chronic treatment with TCAs. However, other investigators did not find changes in $G\alpha_s$ or $G\alpha_i$ mRNAs or the levels of G proteins following chronic treatment with antidepressants (Li et al. 1994; Rasenick 1994; Dwivedi et al. 1995). Emanghoreishi et al. (1996) have concluded that, after chronic treatment with TCAs and MAOIs, the adaptive changes of the β-adrenoceptor–adenylate cyclase system, often seen after chronic antidepressant treatments, are not accompanied by changes in the abundance and gene expression of $G\alpha_s$, $G\alpha_i$, or $G\beta$ proteins. When G protein function was assayed, Rasenick's group reported that chronic treatment with antidepressants increased the stimulation of adenylate cyclase by a non-hydrolyzable GTP analog without changing the content of G proteins (Menkes et al. 1983; Rasenick 1994; Chen and Rasenick 1995). Since direct activation of adenylate cyclase by Mn^{++} was not altered after chronic treatment with antidepressants, a modification of G protein function, rather than of adenylate cyclase, seems to be responsible for the enhanced activation of adenylate cyclase (Menkes et al. 1983). The mechanism of this enhanced $G\alpha_s$-adenylate cyclase interaction is not known with certainty, although it has been suggested that changes in the nature of the cytoskeleton-membrane interface are responsible. Using agonist–receptor-induced increased guanine nucleotide binding by G proteins, β-adrenergic coupled G_s protein function has been reported to be attenuated by chronic treatment with antidepressants (Avissar and Schreiber 1992a). Avissar et al. (1997b) postulate that the measurement of G protein function and quantity in non-neuronal tissue of patients with mood disorders may serve as biochemical markers for the affective state of these patients. They found a significant reduction in the levels of alpha subunits of $G\alpha_s$ and $G\alpha_i$ proteins in mononuclear leukocytes of depressed patients, and an elevation in $G\alpha_s$ and $G\alpha_i$ in manic patients (Avissar et al. 1997a). The low levels of G protein function and immunoreactivity in depressed patients were normalized by ECT (Avissar et al. 1998) and light therapy (Avissar et al. 1999).

The discrepancy in data on G protein function and cyclic AMP accumulation after chronic administration of antidepressants may be explained by the two fundamentally different experimental conditions: β-adrenoceptor-mediated G protein function and cyclic AMP formation versus GTP-mediated adenylate cyclase activity. Under the former conditions, the β-adrenoceptor-coupled adenylate cyclase system is deamplified following chronic administration of antidepressants due to β-adrenoceptor desensitization and uncoupling of the β-adrenoceptors from the G protein (reduced cyclic AMP formation). In the later case, G protein-adenylate cyclase coupling seems to be enhanced (increased cyclic AMP formation). Since the stimulation of adenylate cyclase by $G\alpha_s$ in vivo depends on agonist–receptor-mediated exchange of GTP for guanosine 5′-diphosphate (GDP), the antidepressant-induced desensitization of the β-adrenoceptor coupled adenylate cyclase system (down-regulation of beta adrenoceptors and uncoupling of receptors from G proteins) seems to be the pharmacologically more relevant action. Recent crystallographic studies of G proteins and their α, β, and γ subunits should lead to a better understanding of how G proteins actually function (Coleman and Sprang 1996). Such understanding may then catalyze the development of pharmaceutical agents tailored selectively to G protein subunits and their function, which in turn can modify overall transmembrane signaling mediated by heterotrimeric GTP binding proteins.

2.7
The Convergence of Neurotransmitter Signals Beyond the Receptors at the Level of Protein Kinase-Mediated Phosphorylation

Considering the activation of the aminergic signal transduction cascades by antidepressant drugs (Fig. 1) and the evidence that transcriptional activities of DNA-binding proteins are regulated by the convergent activities of various protein kinases (Hoeffler et al. 1989), the demonstration that antidepressant drug mechanisms involve changes in gene expression (Brady et al. 1991; Peiffer et al. 1991; Hosoda and Duman 1993; Toth and Schenk 1994; Rossby et al. 1995, 1996; Schwaninger et al. 1995; Nibuya et al. 1996) is not surprising. Table 1 illustrates some of the reported antidepressant-induced changes in mRNA steady-state levels. The recent findings by Nibuya et al. (1995) on the effect of various prototypes of antidepressants—after their chronic but not acute administration—on

Table 1 Changes in steady-state levels of mRNAs by chronic administration of some antidepressants (modified from Rossby and Sulser 1997)

Tyrosine Hydroxylase	Rat brain	↓↓	Nestler et al. 1990
	Rat brain	↓↓	Brady et al. 1991
CRH	Rat hippocampus	↓↓	Brady et al. 1991
β_1-Adrenoceptor	Rat cortex	↑↑↓↓	Hosoda and Duman 1993
Preproenkephalin	Rat amygdala	↑↑	Rossby et al. 1996
Glucocorticoid Receptors	Primary neuronal cultures	↑↑	Pepin et al. 1989
	Rat hippocampus	↑↑	Peiffer et al. 1991
	Rat hippocampus	↑↑	Seckl and Fink 1992
	Rat hippocampus	↑↑	Rossby et al. 1995
Mineralocorticoid Receptor	Rat hippocampus	↑↑	Brady et al. 1991
c-fos induction by restraint stress	Rat cortex	↓↓	Morinobu et al. 1995
NGFI-A induction by restraint stress	Rat cortex	↓↓	Morinobu et al. 1995
CREB	rat hippocampus	↑↑0	Nibuya et al. 1996; Rossby et al. 1999
BDNF and trk B	Rat hippocampus and frontal cortex	↑↑0	Nibuya et al 1995; Russo-Neustadt et al. 1999

0, no significant change.
The data have been compiled from the recent literature (years 1989 to 1999).

the expression of brain-derived neurotrophic factor (BDNF) and its receptor, trk B, in brain are provocative and suggest that this neurotrophin system (and perhaps other growth factors) could be a novel target of antidepressant treatments. Based on these and other studies, a heuristic neurotrophic hypothesis of depression has been proposed (Duman 2001).

However, the up-regulation of CREB mRNA is not a common finding after chronic treatment with antidepressants. For example, studies from our laboratory have shown that venlafaxine, one of the more efficacious antidepressants, does not up-regulate CREB mRNA, and that nuclear CREB-P is actually down-regulated, at least in the frontal cortex (Rossby et al. 1999). Since it is the phosphorylation state of CREB rather than its total amount or its mRNA steady-state level that regulates CREB-CRE-directed transcription (Meyer and Habener 1993), the down-regulation by antidepressants of nuclear CREB-P (Manier et al. 2002) may be the pharmacologically and perhaps therapeutically more relevant action after chronic administration of antidepressants. The findings that some CRE-containing genes are down-regulated—and not up-regulated—following chronic treatment with antidepressants (e.g., β-adrenoceptor, tyrosine hydroxylase, CRF) would be more consistent with a down-regulation of nuclear CREB-P (Nestler et al. 1990; Brady et al. 1991, 1992; Hosoda and Duman 1993). Moreover, utilizing the β-adrenoceptor cyclic AMP-mediated formation of melatonin in the pineal, Heydorn et al. (1982) have shown that chronic but not acute treatment with TCAs and MAOIs results in a reduced dark-induced norepinephrine-mediated formation of melatonin, thus providing evidence of a *net* deamplification of the norepinephrine signal. These studies and those of Friedman et al. (1984), using the pineal as a model system, clearly indicate that the down-regulation of the norepinephrine receptor-coupled adenylate cyclase system by antidepressants does not merely reflect a compensatory mechanism to offset the increased availability of synaptic norepinephrine with the overall rate of signal transduction being unchanged. Thus, the neurotrophic hypothesis of depression (Duman 2001) linked to an increase in CREB mRNA and also BDNF mRNA (Russo-Neustadt et al. 1999) after chronic administration of antidepressants remains equivocal. Moreover, GABA B receptor antagonists cause a very robust increase in the steady-state level of BDNF mRNA and protein in neocortex and hippocampus after a single administration (Heese et al. 2000), indicating that this effect is pharmacologically not selective for antidepressants.

Since some of the changes in gene expression appear to be orchestrated—at least in part—by 5-HT, NE, and glucocorticoids (Yoshikawa and Sabol 1986; Eiring et al. 1992; Rossby et al. 1996), it is tempting to suggest that the antidepressant-sensitive, 5-HT-linked, and glucocorticoid-responsive β-adrenoceptor signal transduction system in the brain is involved in a much more general way as an amplification–adaptation system of stimulus-transcription coupling and the regulation of brain-specific gene expression. The demonstration that DMI can increase promoter activity of a GRII reporter construct transfected into various cell lines (Pepin et al. 1992) and that DMI increases GRII mRNA in brain in vivo by a NE-independent mechanism (Rossby et al. 1995; Eiring and Sulser 1997; Fig. 3) raises important questions about the possibility of developing novel antidepressants by targeting cytoplasmic (other than G proteins) and/or nuclear components. A drug such as DMI (and possibly other lipid-soluble pharmaceuticals) is taken up into cells in a dose- and time-dependent manner

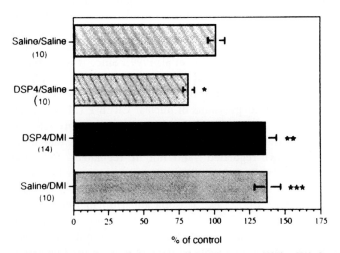

Fig. 3 Steady-state levels of glucocorticoid type II receptor (GRII) mRNA. Rats were treated with DSP4/saline ($n=10$), DSP4/DMI ($n=14$), and saline/DMI ($n=10$). The hippocampal GRII mRNA levels are expressed as a percentage of controls treated with saline/saline. Significance was determined by ordinary ANOVA followed by a Newman-Keuls test: *$p<0.05$ (DSP4/saline vs. saline/saline); **$p<0.001$ (DSP4/DMI vs. DSP4/saline); ***$p<0.01$ (saline/DMI vs. saline/saline). From Rossby et al. (1995)

Table 2 Putative targets for antidepressants beyond the receptors

G proteins: G_s, G_i, G_o
Second Messengers:
Cyclic AMP, cyclic GMP, diacylglycerol, inositol trisphosphate (IP_3) arachidonic acid
Second messenger-dependent protein kinases:
Cyclic AMP-dependent protein kinase (PKA)
Cyclic GMP-dependent protein kinase (PKG)
Ca^{++}/calmodulin-dependent protein kinase
Protein kinase C (PKC)
Second messenger independent protein kinase:
Family of G protein receptor-coupled kinases (GRKs) e.g., β-adrenergic receptor kinase (BARK)
Isoforms of protein kinase inhibitor proteins
Protein phosphatases (2A, 2B, 2C)
Transcription factors (see Table 3)

Table 3 Transcription factors in brain as putative targets of antidepressants (adapted from Hyman and Nestler 1993)

Leucine zipper proteins
CREB/ATF family
Fos/Jun family
Fos-related antigens (FRAs) Jun B, Jun D
Zinc finger proteins
E.g., Zif 268
Steroid hormone receptors
E.g., glucocorticoid receptors, $1,25(OH)_2 D_3$ receptors
Thyroid hormone receptors
Retinoic acid receptor family

(Honegger et al. 1983), which, after chronic administration, leads to intracellular drug concentrations high enough to exert biological actions beyond the receptors. Hypothetical targets beyond the receptors for future antidepressants are listed in Tables 2 and 3.

Putative actions of antidepressants on G proteins were discussed earlier in the chapter. Steps in the transduction cascade beyond G proteins include receptors for second messengers. For example, the regulatory subunits of PKA function as cyclic AMP receptors which, upon occupancy by cyclic AMP, promote the dissociation of the tetrameric PKA complex and the release of the active catalytic subunits (Taylor 1989). Chronic treatment with various prototypes of antidepressant drugs has been reported to enhance the covalent binding of [^{32}P]-cyclic AMP to the regulatory subunits of PKA in the soluble fraction of the rat cerebral cortex (Perez et al. 1989, 1994). Although the mechanism of this increased cyclic AMP binding to regulatory subunits of PKA remains to be elucidated, the results suggest that PKA could be an intracellular target for the action of antidepressants. Indeed, such a covalent binding of cyclic AMP to the regulatory subunit of PKA could promote the release of the catalytic subunits and hence an increase in the phosphorylation of specific phosphoproteins and, in doing so, it could conceivably compensate for the blunted β-adrenoceptor linked PKA activity observed in fibroblasts from patients with major depression (Manier et al. 1996; Shelton et al. 1996, 1999). The reported reduced [^{3}H]-cyclic AMP binding in postmortem brain from subjects with bipolar affective disorder (Rahman et al. 1997) is of interest in this regard, particularly since the blunted β-adrenoceptor-linked PKA activity in fibroblasts from patients with major depression is also associated with a significant reduction in the Bmax value of [^{3}H]-cyclic AMP binding to the regulatory subunit of PKA (Manier et al. 2000). Recently, Dwivedi et al. (2000) reported decreased PKA activity associated with a reduction in the Bmax value of [^{3}H]-cyclic AMP binding in the brains of suicide victims with a history of major depression. Since protein kinase inhibitor (PKI) isoforms are widely distributed in brain, it has been suggested that PKI may serve an important modulatory role for the cyclic AMP second messenger system in the nervous system (Seasholtz et al. 1995). Although speculative at this time, isoforms of endogenous PKI could be another target for novel antidepressant drugs to modulate PKA-mediated signal transduction. Repeated administration of DMI and fluoxetine has been reported to significantly decrease the basal activity of PKC in rat brain cortex and hippocampus (Mann et al. 1995), although the reason for this decrease is not immediately apparent. Since the state of phosphorylation of a substrate protein is determined by the relative activities of a protein kinase (phosphorylation) and a protein phosphatase (dephosphorylation), an altered regulation of the activity and expression of protein phosphatases is a provocative target for future antidepressants. Antidepressants could also enhance the biological response of phosphorylated effector proteins (e.g., transcription factors) indirectly by decreasing the dephosphorylation of the substrate protein through protein kinase-mediated phosphorylation and activation of a phosphatase inhibitor. The complexity and paramount importance of protein phosphorylation as a final regulatory mechanism of cell function in the nervous system has been authoritatively reviewed by Nestler and Greengard (1984).

2.8
Transcription Factors as Targets for Antidepressants

Because gene-specific transcription factors are composed of functionally distinct domains, predisposing them to the effects of pharmaceutical agents (Petersen and Tupy 1994), it is conceivable that drugs could alter gene expression by modifying the activity of gene-specific transcription factors by (1) changing the formation of homo- and/or heterodimers between CREB, CREM, fos, jun, junB, and other leucine zipper proteins, (2) altering their nuclear import from the cytoplasm, or (3) changing their affinity to specific response promoter elements. A modulation of the function of gene-specific transcription factors, as opposed to that of general transcription factors, would be advantageous because of their specificity for regulatory elements in the promoter region of a

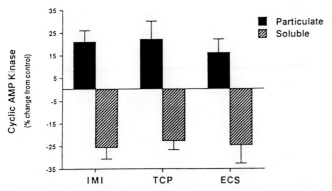

Fig. 4 Regulation by antidepressant treatments of cyclic AMP-dependent protein kinase activity in rat frontal cortex. Rats were treated chronically with imipramine (IMI; 18 days), tranylcypromine (TCP; 18 days), or ECT (10 days). Cyclic AMP-dependent protein kinase activity was assayed in soluble and particulate fractions of frontal cortex isolated from control and treated animals. Data are shown as mean±SEM (*bars*) percentage changes from control and represent the results from 6 to 9 animals in each treatment group. Cyclic AMP-dependent protein kinase activity (in pmol/min/mg of protein; mean±SEM) in particulate and soluble fractions of control brain samples was 199±6 and 227±11, respectively. All of the changes shown were statistically significant ($p<0.05$) by χ^2 test. From Nestler et al. (1989)

subset of genes. Some eukaryotic transcription factors that are putative targets for antidepressants are listed in Table 3.

During the next decade, drugs that suppress or enhance the action of transcription factors associated with specific gene expression can be developed (i.e., transcription factors can be targeted for the development of pharmaceuticals including antidepressants). In this regard, the reported translocation of PKA from the cytoplasmic to the particulate and particularly to the nuclear fraction (Fig. 4) in frontal cortex after chronic treatment with various prototypes of antidepressants (Nestler et al. 1989) is of considerable interest. CREB is a member of the ATF1 family of transcription factors (Lee and Masson 1993) and is located in the nucleus prior to its phosphorylation by PKA (Gonzales and Montminy 1989). Translocation of PKA to the nucleus by antidepressants could lead to the activation of a vast array of promoters via phosphorylation of CREB and, because many of the primary target genes are transcription factors (e.g. immediate early gene products Fos, Jun), the potential for CREB to amplify the effects of cyclic AMP is further increased. Moreover, since CREB is targeted by other signaling pathways (protein kinase C; calcium/calmodulin dependent kinase), it may serve to integrate distinct cellular pathways in the nucleus (Sheng et al. 1991; Lee and Masson 1993). Evidence has been presented that distinct cellular signal transduction pathways involving PKA and PKC modulate gene transcription through common as well as distinct *cis*-acting elements and DNA-binding proteins (Hoeffler et al. 1989). A hypothetical model for the convergence of PKA and PKC pathways at the level of transcriptional activation is depicted in Fig. 5.

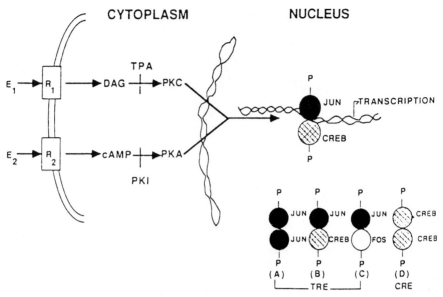

Fig. 5 Hypothetical model for the convergence of protein kinase A and C pathways at the level of transcriptional activation. The two distinct protein kinase signaling pathways converge at the site of transcriptional activation by the phosphorylation of hypothetical individual (*A* and *D*) or separate (*B* and *C*) *trans*-acting DNA-binding proteins. From Hoefler et al. (1989)

Using the nuclear transcription factor CREB in human fibroblasts as a target, both PKA and PKC cause phosphorylation of CREB in an additive manner (Manier et al. 2001). These results have considerable pharmacotherapeutic implications, providing a rationale for the apparently superior efficacy of dual-uptake inhibitor antidepressants. When each individual neurotransmitter signal (norepinephrine or serotonin) is relatively weak, convergence of the signals beyond the receptors at the level of protein kinase-mediated phosphorylation maybe a critical mechanism in gene regulation as the ultimate mode of action of antidepressant drugs.

Since most, if not all, transcription factors are phosphoproteins, phosphorylation and dephosphorylation processes play a major role in altering transcription factor protein conformation into an active/inactive DNA-binding structure (Hunter and Karin 1992). Thus, antidepressants could directly or indirectly modify transcription factor and DNA-binding activities by phosphorylation via PKA, PKC, or calcium/calmodulin-dependent protein kinase (target No. 5 in Fig. 1). A direct correlation between the DMI-induced modification of CREB phosphorylation and the activation of CREB/CRE-directed gene transcription has recently been demonstrated in HIT and PC-12 cells (Schwaninger et al. 1995). Rossby et al. (1996) have suggested that the fluoxetine-induced serotonergic regulation of the preproenkephalin gene expression in the rat amygdala could be mediated via phosphorylation of CREB by the calcium/calmodulin-dependent protein kinase, stimulated via 5-HT_{2A} and/or 5-HT_{2C} receptor cascades. In fact, some evidence exists that the activity of calcium/calmodulin-dependent protein kinase II (CaM-II kinase) is regulated by antidepressants. Thus, Popoli et al. (1995) have found that long-term treatment with paroxetine, fluvoxamine, and venlafaxine induced a large increase of CaM-II kinase autophosphorylation in synaptic vesicles and synaptic cytosol in the hippocampus, but not in synaptosomal membranes. Imipramine and ECT have been reported to induce a significant increase in CaM-II activity in the particulate fraction of the hippocampus (Pilc et al. 1999). These findings suggest that various classes of currently available antidepressants may affect pre- and postsynaptic CaM-II kinase activity differently. Recently, Nibuya et al. (1996) have demonstrated that chronic administration of several classes of antidepressants, including 5-HT- and NE-reuptake inhibitors, increase the expression of CREB mRNA in the rat hippocampus, suggesting that CREB could be an intracellular target of antidepressants. The replication of these data has been rather difficult, however, due to the small changes in the CREB messages. Transcriptional effects of CREB are modulated by CREM (Foulkes et al. 1991). The CREB:CREM dimer can act either as a positive or negative regulator of transcription through the β-adrenoceptor–cyclic AMP–PKA system. It has become increasingly evident that antidepressant-induced modifications of gene expression may ultimately be mediated by indirectly or directly altering transcription factor activity. While the pharmacology of these gene-modulating proteins has long been recognized in the fields of immunology and chemotherapy of neoplastic diseases, neuropsychopharmacology is just beginning to appreciate their role. Although the target genes regulated by transcription factors in the CNS remain largely unknown, it is conceivable that the antidepressant-induced changes in transcription factors and their activity (Tables 2 and 3) play a role in the regulation of other target proteins involved in long-term adaptive responses to antidepressant treatment (Morinobu et al. 1995). Recently, it has been suggested that the transcription factor CREB is a mediator of long-term memory from molluscs to mammals by controlling the activation of genes that encode proteins critical for the generation of memory (Frank and Greenberg 1994).

Recently, Yuan et al. (1998) reported that lithium modulates activator protein (AP)-1 DNA binding activity in human neuroblastoma SH-SY5Y cells and the expression of genes regulated by the AP-1 transcription factor pathway. The lithium-induced increases in AP-1 DNA binding were accompanied by increases in p-c-Jun and c-Jun levels and in the expression of cJun-mediated reporter genes (Yuan et al. 1999). The latter effect was more pronounced in the absence of *myo*-inositol and was blocked by PKC inhibitors. In addition, chronic lithium administration in vivo increased AP-1 DNA binding activity in frontal cortex and hippocampus, and also increased the levels of the phosphorylated, active forms of c-Jun NH2-terminal kinases (JNKs) in both brain regions. [JNKs belong to the subgroup of the mitogen-activated protein (MAP) kinases and phosphorylate specific sites on the amino-terminal transactivation domain of the transcription factor c-Jun. JNKs are known to be major regulators of AP-1 DNA binding.]

Recent studies by Frechilla et al. (1998) suggest that antidepressant treatments differentially modify the transcription factor binding activity of DNA response elements and that these effects are region-specific. Thus, chronic treatment with fluoxetine and DMI produced a similar effect on DNA response elements in the frontal cortex [i.e., increasing the binding activity of nuclear proteins to the cyclic AMP response element (CRE) and decreasing SP1 consensus binding]. (SP1 belongs to the SP family of transcription factors, which recognize the GC-rich box present in a wide variety of promoters.) In the hippocampus, only fluoxetine enhanced CRE and SP1 consensus binding. In contrast, only DMI caused an increase in hippocampal binding activity of the glucocorticoid receptor response element (GRE). It is known that chronic treatment with DMI but not fluoxetine increases the expression of GR mRNA in the hippocampus (Rossby et al. 1995). Thus, Frechilla et al. (1998) suggest that the increased GR level followed by interactions between transcription factors (i.e., GR-CREB and GR-SP1) may cause the lack of changes in binding activity of CREB and SP1 after treatment with DMI.

3
Towards the Discovery of the Next Generation of Antidepressants

Since the serendipitous discovery of the MAOIs and the TCAs 40 years ago, three goals are still unmet and remain a challenge for the future:

1. The discovery of antidepressants that are therapeutically more efficacious and effective in a larger percentage of the patient population
2. The discovery of antidepressants with a shorter latency in the onset of their therapeutic action
3. The discovery of antidepressants that are efficacious in therapy-resistant depression

The current antidepressant discovery process is still based chiefly on modifications of steps 1 to 4 of the neurotransmitter transduction cascades (Fig. 1). It seems unlikely that screening procedures based solely on the modification of any of these steps will lead to the discovery of novel types of antidepressants but rather to the proliferation of more "me-too" drugs, with none of them displaying more efficacy than the original prototypes. To discover the next generation of antidepressants, which will hopefully meet the challenges listed above, two avenues of interrelated investigations seem promising:

- A more rigorous elucidation of the molecular psychopathology of affective disorders and the development of animal models of depression with greater disease validity

- The development of new methodology to explore steps 5 and 6 of the signal transduction cascade (Fig. 1) (i.e., the exploration of mechanisms beyond the receptors and second messengers in animal models of depression with increased disease validity and, importantly, in patients with affective disorders)

3.1
Models of Depression with Increased Disease Validity

Depression is a syndrome—a collection of symptoms that occur together with sufficient frequency to constitute a recognizable clinical condition (Deakin 1991). The diagnostic system in the *Diagnostic and Statistical Manual of Mental Disorders* (DSM-IV) (American Psychiatric Association 1994) defines two core symptoms for the diagnosis of major depression: loss of interest or inability to experience pleasure (anhedonia) and depressed mood. Among animal models of depression (Porsolt et al. 1991; Willner 1991; see Porsolt and Lenergre 1992), chronic mild stress (CMS)-induced anhedonia seems to be a promising model of depression with disease validity. According to this model, chronic sequential administration of a variety of mild stressors causes a decrease in responsiveness to rewards (sucrose consumption; intracranial self-stimulation) in rats, which is reversed by chronic administration of antidepressant drugs (Willner et al. 1992; Moreau et al. 1996). Antidepressants apparently have no effect on sucrose consumption in nonstressed animals (antidepressants have no mood-elevating effects in normal human subjects), but following the reduction of sucrose intake by stress, normal behavior is restored by chronic treatment with TCAs, fluoxetine, maprotiline, and mianserin, while chlordiazepoxide has been shown to be ineffective (Willner et al. 1987; Muscat et al. 1992; Sampson et al. 1991). Stress-induced anhedonia has also been reversed by the MAOIs moclobemide (Moreau et al. 1993) and brofaromine (Papp et al. 1996), ECT (Moreau 1997), lithium (Sluzewska and Nowakowska 1994), and buspirone (Przegalinski et al. 1995). Recently, Papp and Moryl (1994) reported that chronic administration of non-competitive and competitive antagonists of NMDA receptors reverse stress-induced anhedonia. In addition, they have found that chronic treatment with L-amino cyclopropanecarboxylic acid (ACPC) also caused a gradual reversal of CMS-induced anhedonia. (ACPC is a partial agonist at strychnine-insensitive glycine sites of the NMDA receptor complex and acts as a functional NMDA antagonist.) Moreover, the animals treated with ACPC recovered from the stress-induced anhedonia much faster than those treated with imipramine and NMDA antagonists [i.e., MK-801 (non-competitive); CGP 37849 and CGP 40116 (competitive)] (Papp 1996; Papp and Moryl 1996; see also the earlier section of this chapter, "A Role for NMDA Receptors in Antidepressant Action").

The CMS procedures produce not only behavioral but biochemical changes as well. Thus, CMS-induced anhedonia increases dopamine release in the nucleus accumbens (Stamford et al. 1991; Willner et al. 1991), accompanied by a decrease in the sensitivity of post-synaptic D_2 receptors (Papp et al. 1994b). Recently, an increase in the density of β-adrenoceptors with no changes in affinity and an increase in the cyclic AMP response to NE in cortical slices have been found in rats treated with CMS (Papp et al. 1994c). Imipramine, which in nonstressed rats reduced β-adrenoceptor density and the cyclic AMP response but did not affect sucrose intake, reversed the effect of stress (Papp et al. 1994c). CMS increased not only the B_{max} values of β-adrenoceptors, but also those of 5-HT_2 and 5-HT_{1A} receptors. Imipramine reversed the effect of CMS on the density of β-adrenoceptors and 5-HT_2 receptors while having no effect on 5-HT_{1A} receptors (Papp et al. 1994a). Recently, Cheeta et al. (1997) provided further support for the validity of the CMS paradigm as an animal model for the study of mechanisms underlying endogenous depression. Their studies of the effects of CMS on sleep architecture indicated that the stressed animals demonstrated decreases in active waking and deep sleep, as well as disruptions of REM sleep. The changes in REM sleep included increases in the duration of and transitions into REM sleep over the sleep part of the sleep–wake cycle and a reduced latency to the onset of the first REM period. These sleep abnormalities, and in particular the decrease in REM latency, are consistent with those reported in endogenous depression (Benca et al. 1992; Berger and Rieman 1993). The availability of an animal model of depression with predictive validity (correspondence between drug actions in the model and in the clinic), face validity (phenomenological similarities between the model and the disorder in man), and construct validity (sound theoretical rationale) should greatly "enhance drug discovery and increase insight into the nature of the disorder modeled" (Willner 1997b).

Interestingly, a similar increase in β-adrenoceptor density and the cyclic AMP response to NE have been found in the hippocampus in a learned helplessness animal model of depression (Henn 1989; Martin et al. 1990). The development of learned helplessness is influenced by a strong genetic component. It has been reported that a selective breeding strategy generated a strain of rats that is highly vulnerable and another that is resistant to the development of learned helplessness (Edwards et al. 1990, 1991a,b). The highly vulnerable rats showed an up-regulation of 5-HT_{1B} receptors in the cortex, hippocampus, and septum (Edwards et al. 1991b), an increase in the B_{max} value of β-adrenoceptors not responsive to antidepressants (F. Henn, personal communication), and an increase in hippocampal opioid receptors (Edwards et al. 1990). The increased vulnerability was correlated with an alteration in glucocorticoid-mediated gene expression in limbic structures (Lachman et al. 1993). Moreover, Lachman et al. (1992) reported a decrease in basal hippocampal neuropeptide Y mRNA in rats resistant to the development of learned helplessness compared with control and highly vulnerable rats. They suggest that the regulation of neuropeptide Y gene expression may be involved in the reduced vulnerability of those rats which develop learned helplessness. The neuropeptide Y gene expression is influenced by glucocorticoids and by dopamine (Salin et al. 1990; White et al. 1990) and modulated in vitro by phorbol esters and cyclic AMP inducers (Angel et al. 1987; Higuchi et al. 1988). The brain neuropeptide Y is involved in the regulation of a number of behaviors known to be altered in clinical depression (Widerlov et al. 1988; Wahlestedt et al. 1989). Recently, the antidepressant-like effect of neuropeptide Y has been demonstrated in the forced swim test in rats (Stogner and Holmes 2000).

3.2
Antidepressants and the Program of Gene Expression

Nerve cells integrate signals and regulate them to maintain a state of homeostasis. Thus, the state of homeostasis depends on the efficacy of intracellular modulation processes that involve positive or negative feedback, which ensure correct functioning of the organism. Nalepa (1994) has made the following suggestions:

- If the amplitude of received stimuli is too high and surpasses the proper modulation capability, the organism, struggling to maintain homeostasis, will undergo adaptive changes that increase certain intracellular biochemical processes at the expense of others
- This process ensures survival but causes a number of pathologies.
- The development of these pathologies may lead to affective disorders, whose symptoms develop slowly and whose treatment is a long-term process.

Based on the information provided in previous sections of this chapter, it is evident that events beyond the receptors—intracellular signal transduction pathways and regulation of programs of gene expression—are promising new and exciting targets for antidepressants. Rossby and Sulser (1997) have suggested (1) that the loss of plasticity at the level of gene expression accounts for the persistence of major depression (melancholia), and (2) that the ultimate mechanisms of antidepressant drug effects involve restoration of this lost plasticity. The initial actions of currently available antidepressants on reuptake transporter systems (i.e., the availability of 5-HT and/or NE at various receptors and/or subtypes of receptors) occur rapidly within minutes or hours and are obviously not directly responsible for the delayed therapeutic effects of these agents. Rather, the clinically relevant therapeutic actions of antidepressants can be viewed as being the consequence of slowly developing adaptive changes in neuronal plasticity involving "changes in programs of gene expression that determine the intensities of incoming signals, the sensitivities of neuronal systems to those signals and the nature, amplitude and duration of CNS responses" (Rossby and Sulser 1997). The successful development of the next generation of antidepressants—more efficacious, truly faster acting, effective for therapy-resistant depression—cannot ignore this new conceptual framework, which demands that novel sophisticated molecular biological technology be applied within a functional context.

The differential display and cDNA microarray technologies represent major advanced techniques for studying the role of altered programs of gene expression in affective disorders (Rossby et al. 2001; Piétu et al. 2001). Differentially expressed genes and their protein products can then be used as novel drug targets for the development of the next generation of antidepressants, which will hopefully meet the yet unmet criteria of greater efficacy, shorter onset of therapeutic action, and efficacy in therapy-resistant depression.

Acknowledgements. Original research referred to in this chapter has been supported by statutory funds of the Institute of Pharmacology, Polish Academy of Sciences, by a M. Sklodowska-Curie grant PAN/NIH-97-312, by USPHS grants MH-29228 and MH52339, and by a grant-in-aid from Wyeth-Ayerst Research.

References

Ainsworth K, Smith SE, Zetterström TS, Pei Q, Franklin M, Sharp T (1998) Effect of antidepressant drugs on dopamine D1 and D2 receptor expression and dopamine release in the nucleus accumbens of the rat. Psychopharmacology (Berl) 140:470–477

American Psychiatric Association (1994) Diagnostic and statistical Manual of Mental Disorders: 4th Ed., American Psychiatric Association, Washington, D.C.

Angel P, Imagawa M, Chiu R, Stein B, Imbra RJ, Rahmsdorf HJ, Jonat C, Herrich P, Karin M (1987) Phorbol ester-inducible genes contain a common cis element recognized by a TPA-modulated trans-acting factor. Cell 49:729–739

Antoni FA, Palkovits M, Makara GB, Linton EA, Lowry PJ, Kiss JZ (1983) Immunoreactive corticotropin-releasing hormone in the hypothalamoinfundibular tract. Neuroendocrinology 36:415–423

Arborelius L, Owens MJ, Plotsky PM, Nemeroff CB (1999) The role of corticotropin-releasing factor in depression and anxiety disorders. J Endocrinol 160:1–12

Assie MB, Broadhurst A, Briley M (1988) Is down-regulation of β adrenoceptors necessary for antidepressant activity? In: Briley M, Fillion G (eds) New concepts in depression. MacMillan Press, London, pp 161–166

Auer DP, Putz B, Kraft E, Lipinski B, Schill J, Holsboer F (2000) Reduced glutamate in the anterior cingulate cortex in depression: an in vivo proton magnetic resonance spectroscopy study. Biol Psychiatry 47:305–313

Avissar S, Schreiber G (1992a) Interaction of antibipolar and antidepressant treatments with receptor-coupled G proteins. Pharmacopsychiatry 25:44–50

Avissar S, Schreiber G (1992b) The involvement of guanine nucleotide binding proteins in the pathogenesis and treatment of affective disorders. Biol Psychiatry 31:435–459

Avissar S, Nechamkin Y, Barki-Harrington L, Roitman G, Schreiber G (1997a) Differential G protein measures in mononuclear leukocytes of patients with bipolar mood disorder are state dependent. J Affect Disord 43:85–93

Avissar S, Nechamkin Y, Roitman G, Schreiber G (1997b) Reduced G protein functions and immunoreactive levels in mononuclear leukocytes of patients with depression. Am J Psychiatry 154:211–217

Avissar S, Nechamkin Y, Roitman G, Schreiber G (1998) Dynamics of ECT normalization of low G protein function and immunoreactivity in mononuclear leukocytes of patients with major depression. Am J Psychiatry 155:666–671

Avissar S, Schreiber G, Nechamkin Y, Neuhaus I, Lam GK, Schwartz P, Turner E, Matthews J, Naim S, Rosenthal NE (1999) The effects of seasons and light therapy on

G protein levels in mononuclear leukocytes of patients with seasonal affective disorder. Arch Gen Psychiatry 56:178–183

Axelrod J, Whitby LG, Hertting G (1961) Effect of psychotropic drugs on the uptake of ^{3}H-norepinephrine by tissues. Science 133:383–384

Banerjee SP, Kung LS, Riggi SJ, Chanda SK (1977) Development of β-adrenergic receptor subsensitivity by antidepressants. Nature 268:455–456

Barden N, Reul JMHM, Holsboer F (1995) Do antidepressants stabilize mood through actions on the hypothalamic-pituitary-adrenocortical system? TINS 18:6–11

Benca RM, Obermeyer WH, Thisted RA, Gillin C (1992) Sleep and psychiatric disorders: A meta-analysis. Arch Gen Psychiatry 49:651–668

Benovic JL, Strasser RH, Caron MG, Lefkowitz RJ (1986) Beta-adrenergic receptor kinase: Identification of a novel protein kinase that phosphorylates the agonist-occupied form of the receptor. Proc Natl Acad Sci USA 83:2797–2801

Berger M, Rieman D (1993) REM sleep in depression: An overview. J Sleep Res 2:211–223

Bergstrom DA, Kellar KJ (1979a) Adrenergic serotonergic receptor binding in rat brain after chronic desmethylimipramine treatment. J Pharmacol Exp Ther 209:256–261

Bergstrom DA, Kellar KJ (1979b) Effect of electroconvulsive shock on monoaminergic receptor binding sites in rat brain. Nature 278:464–466

Berman RM, Cappiello A, Anand A, Oren DA, Heninger GR, Charney DS, Krystal JH (2000) Antidepressant effects of ketamine in depressed patients. Biol Psychiatry 47:351–354

Bernardini R, Chiarenza A, Kamilaris TC, Renaud N, Lempereur L, Demitrack M, Gold PW, Chrousos GP (1994) In vivo and in vitro effects of arginine-vasopressin receptor antagonists on the hypothalamic-pituitary-adrenal axis in the rat. Neuroendocrinology 60:503–508

Birnbaumer L (1990) G protein in signal transduction. Ann Rev Pharmacol Toxicol 30:675–705

Birnbaumer L, Abramowitz J, Brown AM (1990) Receptor-effector coupling by G proteins. Biochim Biophys Acta 1031:163–224

Blackshear MA, Sanders-Bush E (1982) Serotonin receptor sensitivity after acute and chronic treatment with mianserin. J Pharmacol Exp Ther 221:303–308

Blakely RD, Ramamoorthy S, Qian Y, Schroeter SR, Bradley ChC (1997). Regulation of antidepressant-sensitive serotonin transporters. In: Reith MEA (ed) Neurotransmitter transporters: Structure, function and regulation. Humana Press, Totowas, NJ, pp 29–72

Bouvier M, Leeb-Lundberg LMF, Benovic JL, Caron MG, Lefkowitz RJ (1987) Regulation of adrenergic receptor function by phosphorylation. II. Effects of agonist occupancy on phosphorylation of α_1- and β_2-adrenergic receptors by protein kinase C and the cyclic AMP-dependent protein kinase. J Biol Chem 62:3106–3113

Boyer PA, Skolnick P, Fossom LH (1998) Chronic administration of imipramine and citalopram alters the expression of NMDA receptor subunit mRNAs in mouse brain. A quantitative in situ hybridization study. J Mol Neurosci 10:219–233

Brady LS (1994) Stress, antidepressant drugs, and the locus coeruleus. Brain Res Bull 35:545–556

Brady LS, Whitfield HJ Jr, Fox RJ, Gold PW, Herkenham M (1991) Long-term antidepressant administration alters corticotropin releasing hormone, tyrosine hydroxylase and mineralocorticoid receptor gene expression in rat brain. J Clin Invest 87:831–837

Brady LS, Gold PW, Herkenham M, Lynn AB, Whitfield HJ (1992) The antidepressants fluoxetine, idazoxan and phenelzine alter corticotropin-releasing hormone and tyrosine hydroxylase mRNA levels in rat brain: Therapeutic implications. Brain Res 572:117–125

Briley M, Fillion G (eds) (1988) New concepts in depression. MacMillan Press, London

Briley M, Montgomery ST (eds) (1998) Antidepressant therapy at the dawn of the third millennium. Martin, Dunitz, Ltd., London

Brandoli C, Sanna A, De Bernardi MA, Follesa P, Brooker G, Mocchetti I (1998) Brain-derived neurotrophic factor and basic fibroblast growth factor downregulate NMDA receptor function in cerebellar granule cells. J Neurosci 18:7953–7961

Bremner JD, Licinio J, Darnell A, Krystal JH, Owens MJ, Southwick SM, Nemeroff CB, Charney DS (1997) Elevated CSF corticotropin–releasing factor concentrations in posttraumatic stress disorder. Am J Psychiatry 154:624–629

Budziszewska B, Siwanowicz J, Przegaliński E (1994) The effect of chronic treatment with antidepressant drugs on the corticosteroid receptor levels in the rat hippocampus. Pol J Pharmacol 46:147–152

Bunney WE, Davis JM (1965) Norepinephrine in depressive reactions. Arch Gen Psychiatry 13:483–494

Burnstein KL, Cidlowsky JA (1989) Regulation of gene expression by glucocorticoids. Ann Rev Physiol 51:683–699

Calogero AE, Galluci WT, Tomai P, Loriaux DL, Chrousos GP, Gold P (1988a) Inhibition of corticotropin releasing hormone secretion by $GABA_A$ and $GABA_B$ receptor action in vitro: Clinical implications. In: D'Agata R, Chrousos GP (eds) Recent advances in adrenal regulation and function. Raven, New York, pp 279–284

Calogero AE, Galluci WT, Chrousos GP, Golg PW (1988b) Catecholamine effects upon rat hypothalamic corticotropin-releasing hormone secretion in vitro. J Clin Invest 82:839–846

Carlsson A, Fuxe K, Ungerstedt U (1968) The effect of imipramine on central 5-hydroxytryptamine neurons. J Pharmacy Pharmacol 20:150–151

Cheeta S, Ruigt G, van Proosdij J, Willner P (1997) Changes in sleep architecture following chronic mild stress. Biol Psychiatry 41:419–27

Chen J, Rasenick MM (1995) Chronic antidepressant treatment facilitates G protein activation of adenylyl cyclase without altering G protein content. J Pharmacol Exp Ther 275:509–517

Coleman DE, Sprang SR. How G proteins work: a continuing story.Trends Biochem Sci. 1996;21:41-44

Coppen A (1967) The biochemistry of affective disorders. Br J Psychiatry 113:1237–1264

Costa E, Racagni G (eds) (1982a) Typical and atypical antidepressants: Molecular mechanisms. Adv Biochem Psychopharmacol vol 31

Costa E, Racagni G (eds) (1982b) Typical and atypical antidepressants: Clinical practice. Adv Biochem Psychopharmacol vol 32

Curtis AL, Valentino RJ (1994) Corticotropin-releasing factor neurotransmission in locus coeruleus: A possible site of antidepressant action. Brain Res Bull 35:581–587

Daly JW, Padgett W, Creveling CR, Cantacuzene D, Kirk KL (1980) Fluoroepinephrines: Specific agonists for the activation of alpha and beta adrenergic-sensitive cyclic AMP-generating system in rat slices. J Pharmacol Exp Ther 212:382–389

Deakin JFW (1991) The clinical relevance of animal models of depression. In: Willner P (ed) Behavioural models in psychopharmacology: theoretical, industrial and clinical perspectives. Cambridge University Press, Cambridge, pp 157–174

Decollogne S, Tomas A, Lecerf C, Adamowicz E, Seman M (1997) NMDA receptor complex blockade by oral administration of magnesium: Comparison with MK-801. Pharmacol Biochem Behav 58:261–268

Delbende C, Contesse V, Mocaer E, Kamoun A, Vaudry H (1991) The novel antidepressant, tianeptine, reduces stress-evoked stimulation of the hypothalamo-pituitary-adrenal axis. Eur J Pharmacol 202:391–396

Döbbeling U, Berchtold MW (1996) Down-regulation of the protein kinase A pathway by activators of protein kinase C and intracellular Ca^{2+} in fibroblast cells. FEBS Letters 391:131–133

Done CJG, Sharp T (1992) Evidence that $5HT_2$ receptor activation decreases noradrenaline release in rat hippocampus in vivo. Br J Pharmacol 107:240-245

Duman R (2001). Regulation of neural plasticity by stress and antidepressant treatment. In: Briley M, Sulser F (eds) Molecular genetics of mental disorders. Martin Dunitz, London, pp 171-198

Duman RS, Terwilliger RZ, Nestler EJ (1989) Chronic antidepressant regulation of GS μ and cyclic AMP-dependent protein kinase. Pharmacologist 31:182

Duman RS, Nibuya M, Vaidya VA (1997) A role for CREB in antidepressant action. In: Skolnick P (ed) Antidepressants: New pharmacological strategies. Humana Press, Totowa, NJ, pp 173-194

Dunn AJ, Berridge CW (1990a) Is corticotropin-releasing factor a mediator of stress responses? Ann NY Acad Sci 579:183-191

Dunn AJ, Berridge CW (1990b) Physiological and behavioral responses to corticotropin-releasing factor administration: Is CRF a mediator of anxiety or stress responses? Brain Res Rev 15:71-100

Dwivedi Y, Pandey SC, Pandey GN (1995) Effect of chronic administration of antidepressants on the levels of various subtypes of G-proteins in rat brain. Soc Neurosci 21:731.8 (Abstract)

Dwivedi Y, Roberts R, Conley RC, Tamminga C, Pandey GN (2000) Protein kinase A in the post mortem brain of suicide victims. Biol. Psychiatry 47:75S (Abstract)

Dziedzicka-Wasylewska M, Rogoz R, Klimek V, Maj J (1997) Repeated administration of antidepressant drugs affects the levels of mRNA coding for D1 and D2 dopamine receptors in the rat brain. J Neural Transm 104:515-524

Edwards E, Muneyyrci J, Van Houten P, Michel C, Henn FA (1990) The effect of learned helplessness breeding on opioid mechanisms. Am Coll Neuropsychopharmacol 29:222 (Abstract)

Edwards E, Harkins K, Henn FA (1991a) Learned helplessness modulation of ^{3}H-paroxetine binding in the rat brain. J Neurochem 56:1581-1586

Edwards E, Harkins K, Wright G, Henn FA (1991b) $5HT_{1b}$ Receptors in an animal model of depression. Neuropharmacology 30:101-105

Eiring A, Sulser F (1997) An increased synaptic availability of norepinephrine is *not* essential for antidepressant induced increases in hippocampal GR mRNA. J Neural Transm 104:1255-1258

Eiring A, Manier DH, Bieck PR, Howells RD, Sulser F (1992) The 'Serotonin/ Norepinephrine/ Glucocorticoid Link' beyond the beta adrenoceptors. Molec Brain Res 16:211-214

Emanghoreishi M, Warsh JJ, Sibony D, Li PP (1996) Lack of effect of chronic antidepressant treatment on Gs and Gi α-subunit protein and mRNA levels in the rat cerebral cortex. Neuropsychopharmacol 15:281-287

Ferris RM, Cooper BR (1993) Mechanism of antidepressant activity of bupropion. J Clin Psychiatry 11:1-14

Foulkes NS, Laoide BM, Schlotter F, Sassone-Corsi P (1991) Transcriptional antagonist cAMP-responsive element modulator down-regulates c-fos cAMP-induced expression. Proc Natl Acad Sci USA 88:5448-5452

Frank DA, Greenberg ME (1994) CREB: A mediator of long-term memory from molluscs to mammals. Cell 79:5-8

Frazer A (1997) Pharmacology of antidepressants. J Clin Psychopharmacol 17:2S-18S

Frechilla D, Otano A, Del Rio J (1998) Effect of chronic antidepressant treatment on transcription factor binding activity in rat hippocampus and frontal cortex. Prog Neuropsychopharmacol Biol Psychiatry 22:787-802

Freedman NJ, Liggett SB, Drachman DE, Caron MG, Lefkowitz RJ (1995) Phosphorylation and desensitization of the human $β_1$-adrenergic receptor. J Biol Chem 270:17953-17961

Friedman E, Yocca FD, Cooper TD (1984) Antidepressant drugs with varying pharmacological profiles alter rat pineal beta adrenergic mediated function. J Pharmacol Exp Ther 228:545–550

Gilman AG (1987) G proteins in signal transduction. Ann Rev Biochem 56:615–649

Gonzalez GA, Montminy MR (1989) Cyclic AMP stimulates somatostatin gene transcription by phosphorylation of CREB at serine 133. Cell 59:675–680

Haddjeri N, Blier P, de Montigny C (1996) Effect of the alpha-2 adrenoceptor antagonist mirtazapine on the 5-hydroxytryptamine system in the rat brain. J Pharmacol Exp Ther 277:861–871

Hallcher LM, Sherman WR (1980) The effects of lithium ion and other agents on the activity of myoinositol-1-phosphatase from bovine brain. J Biol Chem 255:10896–10901

Harfstrand A, Fuxe K, Cintra A, Agnati LF, Zim I, Wirkstrom AC, Okret S, Yu ZL, Goldstein M, Steinbusch H, Verhofstadt A, Gustafsson JA (1986) Demonstration of glucocorticoid receptor immunoreactivity in monoamine neurons of the rat brain. Proc Nat Acad Sci USA 83:9779–9783

Harrelson AL, Rosterre W, McEwen BS (1987) Adrenocortical steroids modify neurotransmitter stimulated cyclic AMP accumulation in the hippocampus and limbic brain of the rat. J Neurochem 48:1648–1655

Heese K, Otten U, Mathivet P, Raiteri M, Marescaux C, Bernasconi R (2000) Gaba$_B$ receptor antagonists elevate both mRNA and protein levels of the neurotrophins nerve growth factor (NCG) and brain-derived neurotophic factor (BDNF) but not neurotrophin-3 (NT-3) in brain and spinal cord of rats. Neuropharmacology 39:449–462

Heinrichs SC, Menzaghi FM, Pich EM, Baldwin HA, Rassnick S, Britton KT, Koob GF (1994) Anti-stress action of a corticotropin-releasing factor antagonist on behavioral reactivity to stressors of varying type and intensity. Neuropsychopharmacology 11:179–185

Henn FA (1989) Animal models. In: Mann JJ (ed) Models of depressive disorders. Plenum, New York, pp 93–107

Heydorn WE, Brunswick FT, Frazer A (1982) Effect of treatment of rats with antidepressants on melatonin concentrations in the pineal gland and serum. J Pharmacol Exp Ther 222:534–543

Higuchi H, Yang H-YT, Sobol S (1988) Rat neuropeptide Y precursor gene expression. J Biol Chem 262:6288–6295

Hoeffler JP, Deutsch PJ, Lin J, Habener JF (1989) Distinct adenosine 3',5'-monophosphate and phorbolester-responsive signal transduction pathways converge at the level of transcriptional activation by the interaction of DNA-binding proteins. Molec Endocrinol 3:868–880

Hokfelt T, Johansson O, Holets V, Meister B, Melander T (1987) Distribution of neuropeptides with special reference to their coexistence with classical transmitters. In: Meltzer HY (ed) Psychopharmacology: The third generation of progress. Raven Press, New York, pp 401–416

Holsboer F, Spengler D, Heuser I (1992) The role of corticotropin-releasing hormone in the pathogenesis of Cushing's disease, anorexia nervosa, alcoholism, affective disorders and dementia. Prog Brain Res 93:385–417

Holzbauer M, Vogt M (1956) Depression by reserpine of the noradrenaline concentration in the hypothalamus of the cat. J Neurochem 1:8–11

Honegger UE, Roscher AA, Wiesmann UN (1983) Evidence for lysosomotropic action of desipramine in cultured human fibroblasts. J Pharmacol Exp Ther 225:436–441

Honkaniemi J, Pelto-Huikko M, Rechardt L, Isola J, Lammi A, Fuxe K, Gustaffson J, Wikström AC, Hokfelt T (1992) Colocalization of peptide and glucocorticoid receptor immunoreactivity in rat central amygdaloid nucleus. Neuroendocrinol 55:451–459

Hosoda K, Duman RS (1993) Regulation of β_1-adrenergic receptor mRNA and ligand binding by antidepressant treatments and norepinephrine depletion in rat frontal cortex. J Neurochem 69:1335-1343

Huang N-Y, Layer RT, Skolnick P (1997) Is an adaptation of NMDA receptors an obligatory step in antidepressant action? In: Skolnick P (ed) Antidepressants: New pharmacological strategies. Humana Press, Totowa, NJ, pp 125-143

Huff RA, Vaughan RA, Kuhar MJ, Uhl GR (1997) Phorbol esters increase dopamine transporter phosphorylation and decrease transport Vmax. J Neurochem 68:225-232

Hunter T, Karin M (1992)The regulation of transcription by phosphorylation. Cell 70:375-387

Hyman SE, Nestler EJ (1993) The molecular foundations of psychiatry. American Psychiatric Press, Washington DC

Janowsky DS, El-Yousef MK, Davis JM, Sekerke HJ (1972) A cholinergic-adrenergic hypothesis of mania and depression. Lancet 2:632-635

Kapur S, Mann JJ (1992) Role of the dopaminergic system in depression. Biol Psychiatry 32:1-17

Kellar KJ, Bergstrom DA (1983) Electroconvulsive shock: Effects on bichemical correlates of neurotransmitter receptors in the brain. Neuropharmacology 22:401-406

Kellar KJ, Stockmeier CA (1986) Effects of electroconvulsive shock and serotonin axon lesions on beta-adrenergic and serotonin-2 receptors in rat brain. Ann NY Acad Sci 462:76-90

Kellar KJ, Cascio CS, Bergstrom DA, Butler JA, Iaradola P (1981a) Electroconvulsive shock and reserpine: Effects on β-adrenergic receptors in rat brain. J Neurochem 37:830-836

Kellar KJ, Cascio CS, Butler JA, Kurtzke RN (1981b) Differential effects of electroconvulsive shock and antidepressant drugs on serotonin-2 receptors in rat brain. Eur J Pharmacol 69:515-518

Kendall DA, Nahorski SR (1985) 5-hydroxytryptamine-stimulated inositol phospholipid hydrolysis in rat cerebral cortical slices: Pharmacological characterization and effects of antidepressants. J Pharmacol Exp Ther 233:473-479

Kitayama I, Janson AM, Cintra A, Fuxe K, Agnati LF, Ögren SO, Harfstrand A, Eneroth P, Gustafsson JA (1988) Effects of chronic imipramine treatment on glucocorticoid receptor immunoreactivity in various regions of the rat brain. J Neural Transm 73:191-203

Kitayama I, Nakamura S, Yaga T, Murase S, Nomura J, Kayahara T, Nakano K (1994) Degeneration of locus coeruleus axons in stress-induced depression model. Brain Res Bull 35:573-580

Klimek V, Papp M (1994) The effects of MK-801 and imipramine on beta adrenergic and 5HT2 receptors in the chronic mild stress model of depression in rats. Pol J Pharmacol Pharm 46:67-69

Klimek V, Nielsen M, Maj J (1985) Repeated treatment with imipramine decreased the number of [3H]piflutixol binding sites in the rat striatum. Eur J Pharmacol 109:131-132

Kramer MS, Cutler N, Feighner J, Shrivastava R (1998). Distinct mechanisms for antidepressant activity by blockade of central substance P receptors. Science 281:1640-1645

Kuhn R (1958) The treatment of depressive states with G22355 (imipramine hydrochloride). Am J Psychiatry 115:459-464

Lachman HM, Papolos DF, Weiner ED, Ramazankhana R, Hartnick C, Edwards E, Henn FA (1992) Hippocampal neuropeptide Y mRNA is reduced in a strain of learned helpless resistant rats. Mol Brain Res 14:94-100

Lachman HM, Papolos DF, Boyle A, Sheftel G, Juthani M, Edwards E, Henn FA (1993) Alterations in glucocorticoid inducible RNAs in the limbic system of learned helpless rats. Brain Res 609:110-116

Layer RT, Popik P, Olds T, Skolnick P (1995) Antidepressant-like actions of the polyamine site NMDA antagonist, eliprodil (SL-82.0715). Pharmcol Biochem Behav 52:621–627

Lee KAW, Masson N (1993) Transcriptional regulation by CREB and its relatives. Biochem Biophys Acta 1174:221–233

Leeb-Lundberg LMF, Cotecchia S, Lomasney JW, DeBernardis JF, Lefkowitz RJ, Caron MG (1985) Phorbol esters promote α_1-adrenergic receptor phosphorylation and receptor uncoupling from inositol phospholipid metabolism. Proc Natl Acad Sci USA 82:5651–5655

Leeb-Lundberg LMF, Cotecchia S, DeBlasi A, Caron MG, Lefkowitz RJ (1987) Regulation of adrenergic receptor function by phosphorylation. I. Agonist-promoted desensitization and phosphorylation of α-adrenergic receptors coupled to inositol phospholipid metabolism in DDT_1 MF-2 smooth muscle cells. J Biol Chem 262:3098–3105

Lehmann HE, Kline NS (1983) Clinical discoveries with antidepressant drugs. In: Parnham MJ, Bruinvels J (eds) Discoveries in pharmacology, vol 1. Elsevier, Amersterdam, pp 209–221

Lejeune F, Audinot V, Gobert A, River JM, Spedding M, Millan MJ (1994) Clozapine inhibits serotoninergic transmission by an action at a_1-adrenoceptors not at $5HT_{1A}$ receptors. Eur J Pharmacol 260:79–83

Leonard BE (1997) Noradrenaline in basic models of depression. Eur Neuropsychopharmacol 7 (Suppl 1):S11–S16

Leonard BE, Spencer P (eds) (1990) Antidepressants: Thirty years on. CNS Publishers, London

Lesch KP, Manji HK (1992) Signal-transducing G proteins and antidepressant drugs: evidence of modulation of µ subunit gene expression in rat brain. Biol Psychiatry 32:549–579

Lesch KP, Anlakh CS, Tollives TG, Hill JL, Murphy DL (1991) Regulation of G proteins by chronic antidepessant drugs in rat brain: tricyclics but not clorgyline increase Go µ subunits. Eur J Pharmacol 207:361–364

Li PP, Warsh JJ, Sibony D, Chiu A (1985) Assessment of rat brain alpha 1-adrenoceptor binding and activation of inositol phospholipid turnover following chronic imipramine treatment. Neurochem Res 13:1111–1118

Li PP, Young LT, Warsh JJ (1994) Effects of antibipolar and antidepressant drugs on the levels of signal transducing G proteins and their messenger ribonucleic acid transcripts. Neuropsychopharmacology 10:380S

Li Q, Hrdina PD (1997) GAP-43 phosphorylation by PKC in rat cerebrocortical synaptosomes: effect of antidepressants. Res Commun Mol Pathol Pharmacol 96:3–13

Maj J (1984) Central effects following repeated treatment with antidepressant drugs. Pol J Pharmacol Pharm 36:87–99

Maj J, Rogóz Z, Skuza G, Sowińska H (1984) Repeated treatment with antidepressant drugs potentiates the locomotor response to (+)-amphetamine. J Pharm Pharmacol 36:127–130

Maj J, Wedzony K, Klimek V (1987) Desipramine given repeatedly enhances behavioural effects of dopamine and d-amphetamine injected into the nucleus accumbens. Eur J Pharmacol 140:179–185

Maj J, Papp M, Skuza G, Bigajska K, Zazula M (1989a) The influence of repeated treatment with imipramine, (+)- and (-)-oxaprotiline on behavioural effects of dopamine D-1 and D-2 agonists. J Neural Transm 76:29–38

Maj J, Rogóz Z, Skuza G, Sowińska H (1989b) Antidepressants given repeatedly increase the behavioural effect of dopamine D-2 agonist. J Neural Transm Gen Sect 78:1–8

Maj J, Rogóz Z, Skuza G, Sowińska H (1992a) The effect of CGP 37849 and CGP 39551, competitive NMDA receptor antagonists, in the forced swimming test. Pol J Pharm 44:337–346

Maj J, Rogóz Z, Skuza G, Sowińska H (1992b) Effects of MK-801 and antidepressant drugs in the forced swimming test in rats. Eur Neuropsychopharmacology 2:37–41

Maj J, Dziedzicka-Wasylewska M, Rogoz R, Rogóz Z, Skuza G (1996) Antidepressant drugs given repeatedly change the binding of the dopamine D2 receptor agonist [^{3}H]N-0437 to dopamine D2 receptors in the rat brain. Eur J Pharmacol 304:49–54

Maj J, Dziedzicka-Wasylewska M, Rogoz R, Rogóz Z (1998) Effect of antidepressant drugs administered repeatedly on the dopamine D3 receptors in the rat brain. Eur J Pharmacol 351:31–37

Manier DH, Eiring A, Shelton RC, Sulser F (1996) Beta adrenoceptor-linked protein kinase A (PKA) activity in human fibroblasts from normal subjects and patients with major depression. Neuropsychopharmacology 15:555–561

Manier DH, Shelton RC, Ellis T, Peterson CS, Eiring A, Sulser F (2000) Human fibroblasts as a relevant model to study signal transduction in affective disorders. J Affect Disord 2000:61:51-58

Manier DH, Shelton RC, Sulser F (2001) Cross-talk between PKA and PKC in human fibroblasts: What are the pharmacotherapeutic implications? J Affect Disord 65:275-279

Manier DH, Shelton RC, Sulser F (2002) Noradrenergic antidepressants: does chronic treatment increase or decrease nuclear CREB-P? J Neural Transm 109:91-99

Manji HK, Lenox RH (1994) Long-term action of lithium: A role for transcriptional and postranscriptional factors regulated by protein kinase C. Synapse 16:11–28

Manji HK, Lenox RH (1999) Protein kinase C signaling in the brain: Molecular transduction of mood stabilization in the treatment of manic-depressive illness. Biol Psychiatry 46:1328–1351

Mann CD, Vu TB, Hrdina PD (1995) Protein kinase C in rat brain cortex and hippocampus: Effect of repeated administration of fluoxetine and desipramine. Br J Pharmacol 115:595–600

Mansbach RS, Brooks EN, Chen YL (1997) Antidepressant-like effects of CP-154,526, a selective CRF1 receptor antagonist. Eur J Pharmacol 323:21–26

Martin JV, Edwards E, Johnson JO, Henn FA (1990) Monoamine receptors in an animal model of affective disorders. J Neurochem 55:1142–1148

Menkes DB, Rasenick MM, Wheeler MA, Bitensky MW (1983) Guanosine triphosphate activation of brain adenylate cyclase: Enhancement by long-term antidepressant treatment. Science 219:65–67

Meyer TE, Habener JF (1993) Cyclic adenosine 3',5'-monophosphate response element binding protein (CREB) and related transcriptional-activating deoxyribonucleic acid-binding proteins. Endocrine Reviews 14:269–290

Mobley PL, Manier DH, Sulser F (1983) Adrenal corticoids regulate the norepinephrine sensitive adenylate cyclase system in brain. J Pharmacol Exp Ther 226:71–77

Montgomery SA (1997) Is there a role for a pure noradrenergic drug in the treatment of depression? Eur Neuropsychopharmacol 7 (Suppl 1):S3–S9

Monyer H, Sprengel R, Schoepfer R, Herb A, Higuchi M, Lomeli H, Burcracher N, Sakmann B, Seeberg PH (1992) Heteromeric NMDA receptors: Molecular and functional distinction of subtypes. Science 256:1217–1221

Moreau JL (1997) Validation of an animal model of anhedonia, a major symptom of depression. Encephale 23:280–289

Moreau JL, Jenck F, Martin JR, Mortas P, Haefely W (1993) Effects of moclobemide, a new generation reversible MAO-A inhibitor, in a novel animal model of depression. Pharmacopsychiatry 26:30–33

Moreau JL, Bös M, Jenck F, Martin JR, Mortes P, Wichmann J (1996). 5HT2C receptor agonists exhibit antidepressant-like properties in the anhedonia model of depression in rats. Eur Neuropsychopharmacol 6:169–175

Morinobu S, Nibuya M, Duman RS (1995) Chronic antidepressant treatment down-regulates the induction of c-fos mRNA in response to acute stress to rat frontal cortex. Neuropsychopharmacology 12:221–228

Muscat R, Papp M, Willner P (1992) Reversal of stress-induced anhedonia by the atypical antidepressants, fluoxetine and maprotiline. Psychopharmacology (Berl) 109:433-438

Nalepa I (1994) The effect of psychotropic drugs on the interaction of protein kinase C with second messenger systems in the rat cerebral cortex. Pol J Pharmacol Pharm 46:1-14

Nalepa I, Vetulani J (1991a) Different mechanisms of β-adrenoceptor downregulation by chronic imipramine and electroconvulsive treatment: possible role for protein kinase C. J Neurochem 57:904-910

Nalepa I, Vetulani J (1991b) Involvement of protein kinase C in the mechanisms of in vitro effects of imipramine on generation of second messengers by noradrenaline in the cerebral cortical slices of the rat. Neuroscience 44:585-590

Nalepa I, Vetulani J (1993a) The effect of calcium channel blockade on the action of chronic ECT and imipramine on responses of α_1- and β-adrenoceptors in the rat cerebral cortex. Pol J Pharmacol Pharm 45:201-205

Nalepa I, Vetulani J (1993b) Enhancement of the responsiveness of cortical adrenergic receptors by chronic administration of the 5-hydroxytryptamine uptake inhibitor citalopram. J Neurochem 60:2029-2035

Nalepa I, Vetulani J (1994) The responsiveness of cerebral cortical adrenergic receptors after chronic administration of atypical antidepressant mianserin. J Psychiatry Neurosci 19:120-128

Nalepa I, Chalecka-Franaszek E, Vetulani J (1993) The antagonistic effect of separate and consecutive chronic treatment with imipramine and ECT on the regulation of α_1-adrenoceptor activity by protein kinase C. Pol J Pharmacol Pharm 45:521-532

Nalepa I, Chalecka-Franaszek E, Vetulani J (1996) Modulation by mianserin pretreatment of the chronic electroconvulsive shock effects on the adrenergic system in the cerebral cortex of the rat. Human Psychopharmacology 11:273-282

Nalepa I, Manier DH, Gillespie DG, Rossby SP, Schmidt DE, Sulser F (1998) Lack of beta adrenoceptor desensitization in brain following the dual noradrenaline and serotonin reuptake inhibitor venlafaxine. Eur Neuropsychopharmacol 8:227-232

Neliat G, Bodinier MC, Panconi E, Briley M (1996) Lack of effect of milnacipran, a double noradrenaline and serotonin reuptake inhibitor, on the β-adrenoceptor-linked adenylate cyclase system in the rat cerebral cortex. Neuropharmacology 35:589-593

Nemeroff CB, Widerlöv E, Bissette G, Karlsson I, Eklund K, Kilts CD, Loosen P, Vale W (1984) Elevated concentrations of CSF corticotropin-releasing factor-like immunoreactivity in depressed patients. Science 226:1342-1344

Nemeroff CB, Owens MJ, Bissette G, Andorn AC, Stanley M (1988) Reduced corticotropin releasing factor binding sites in the frontal cortex of suicide victims. Arch Gen Psychiatry 45:577-579

Nestler EJ, Greengard P (1984) Protein phosphorylation in the nervous system. Wiley, New York

Nestler EJ, Terwilliger RZ, Duman RS (1989) Chronic antidepressant administration alters the subcellular distribution of cyclic AMP-dependent protein kinase in rat frontal cortex. J Neurochem 53:1644-1647

Nestler EJ, McMahon A, Sabban EL, Tallman JT, Duman RS (1990) Chronic antidepressant administration decreases the expression of tyrosine hydroxylase in the rat locus coeruleus. Proc Natl Acad Sci USA 87:7522-7526

Newman ME, Lerer B (1989) Modulation of second messenger function in rat brain by in vivo alteration of receptor sensitivity: relevance to the mechanism of action of electroconvulsive therapy and antidepressants. Prog Neuropsychopharmacol Biol Psychiatry 13:1-30

Nibuya M, Morinobu S, Duman RS (1995) Regulation of BDNF and trkB mRNA in rat brain by chronic electroconvlusive seizure and antidepressant drug treatment. J Neurosci 15:7539-7547

Nibuya M, Nestler EJ, Duman RS (1996) Chronic antidepressant administration increases the expression of CREB in rat hippocampus. J Neurosci 16:2365–2372

Nishizuka Y (1992) Intracellular signaling by hydrolysis of phospholipids and activation of protein kinase C. Science 258:607–614

Nowak G, Trullas R, Layer RT, Skolnick P, Paul IA (1993) Adaptive changes in the N-methyl D-aspartate receptor complex after chronic treatment with imipramine and 1-aminocyclopropane-carboxylic acid. J Pharmacol Exp Ther 265:1380–1386

Nowak G, Ordway GA, Paul IA (1995) Alterations in the N-methyl-D-aspartate (NMDA) receptor complex in the frontal cortex of suicide victims. Brain Res 675:157–164

Okuyama S, Chaki S, Kawashima N, Suzuki Y, Ogawa S, Nakazato A, Kumagai T, Okubo T, Tomisawa K (1999) Receptor binding, behavioral, and electrophysiological profiles of nonpeptide corticotropin-releasing factor subtype 1 receptor antagonists CRA1000 and CRA1001. J Pharmacol Exp Ther 289:926–935

Oswald I, Brezinova V, Dunleavy DLF (1973) On the slowness of action of tricyclic antidepressant drugs. Br J Psychiatry 120:673–677

Owens MJ, Morgan WN, Plott SE, Jeni J, Nemeroff CB (1995) In vitro inhibition of the rat and human serotonin and norepinephrine transporters by the antidepressant nefazodone and its metabolites. Soc Neurosci 21:309.1 (abstract)

Panconi E, Roux J, Altenbaumer M, Hampe S, Porsolt RD (1993) MK-801 and enantiomers: Potential antidepressants or false positives in classical screening models? Pharmacol Biochem Behav 46:15–20

Pandey GN, Heinze WJ, Brown BD, Davis JM (1979) Electroconvulsive shock treatment decreases β-adrenergic receptor sensitivity in rat brain. Nature 280:234–235

Pandey GN, Pandey SC, Isaac L, Davis M (1992) Effect of electroconvulsive shock on $5HT_2$ and α_1-adrenoceptors and phosphoinositide signaling system in rat brain. Eur J Pharmacol 226:303–310

Papp M (1996) Comparison of antidepressant and psychomimetic effects of agents acting at various sites of the NMDA receptor complex. Behav Pharmacol 7 (suppl 1):82–83

Papp M, Moryl E (1994) Antidepressant activity of noncompetitive and competitive NMDA receptor antagonists in a chronic mild stress model of depression. Eur J Pharmacol 263:1–7

Papp M, Moryl E (1996) Antidepressant-like effects of 1-aminocyclopropanecarboxylic acid and D-cycloserine in an animal model of depression. Eur J Pharmacol 316:145–151

Papp M, Klimek V, Willner P (1994a) Effects of imipramine on serotonergic and beta-adrenergic receptor binding in a realistic animal model of depression. Psychopharmacology (Berlin) 114:309–14

Papp M, Klimek V, Willner P (1994b) Parallel changes in dopamine D2 receptor binding in limbic forebrain associated with chronic mild stress-induced anhedonia and its reversal by imipramine. Psychopharmacology-Berl 115:441–446

Papp M, Nalepa I, Vetulani J (1994c) Reversal by imipramine of beta adrenoceptor upregulation induced in a chronic mild stress model of depression. Eur J Pharm 261:141–147

Papp M, Moryl E, Willner P (1996) Pharmacological validation of the chronic mild stress model of depression. Eur J Pharmacol 296:129–136

Pariante CM, Pearce BD, Pisell TL, Owens MJ, Miller AH (1997) Steroid-independent translocation of the glucocorticoid receptor by the antidepressant desipramine. Mol Pharmacol 52:571–581

Paul IA, Trullas R, Skolnick P, Nowak G (1992) Down-regulation of cortical β-adrenoceptors by chronic treatment with functional NMDA antagonists. Psychopharmacology 106:285–287

Paul IA, Layer RT, Skolnick P, Nowak G (1993) Adaptation of the NMDA receptor in rat cortex following chronic electroconvulsive shock or imipramine. Eur J Pharmacol 247:305–312

Paul IA, Nowak G, Layer RT, Popik P, Skolnick P (1994) Adaptation of the N-methyl-D-aspartate receptor complex following chronic antidepressant treatments. J Pharmacol Exp Ther 269:95–102

Peiffer A, Veilleaux S, Barden N (1991) Antidepressant and other centrally acting drugs regulate glucocorticoid receptor messenger RNA levels in rat brain. Psychoneuroendocrinology 16:505–515

Pepin MC, Beaulieu S, Barden N (1989) Antidepressants regulate glucocorticoid receptor messenger RNA concentrations in primary neuronal cultures. Mol Brain Res 6:77–83

Pepin MC, Pothier F, Barden N (1992) Antidepressant drug action in a transgenic mouse model of the endocrine changes seen in depression. Mol Pharmacol 42:991–995

Perez J, Tinelli D, Brunello N, Racagni G (1989) CAMP-dependent phosphorylation of soluble and crude microtubule fractions of rat cerebral cortex after prolonged desmethylimipramine treatment. Europ J Pharmacol 172:305–316

Perez J, Moris, Caivano M, Fumagalli I, Pezzetta B, Tascedda F, Brunello N, Racagni G (1994) cAMP protein kinase as a intracellular target for the action of antidepressant drugs. Neuropyschopharmacology 10:171S (Abstract)

Peterson MG, Tupy JL (1994) Transcriptional factors: A new frontier in pharmaceutical development. Biochem Pharmacol 47:127–128

Piétu G, Decraene C, Fayerin N, Manage-Samson R, Eveno E, Matingan C, Devigues M, Auffray C (2001) Characterization of expression profiles of genes involved in brain functions by quantitative hybridization of high-density cDNA arrays. In: Briley M, Sulser F (eds) Molecular genetics of mental disorders. Martin Dunitz Ltd, London pp 1-19

Pilc A, Branski P, Palucha A, Aronowski J (1999) The effect of prolonged imipramine and electroconvulsive shock treatment on calcium/calmodulin-dependent protein kinase II in the hippocampus of rat brain. Neuropharmacology 38:597–603

Plaznik A, Kostowski W, Archer T (1989) Serotonin and depression: Old problems and new data. Prog Neuropsychopcharmacol Biol Psychiatry 13:623–633

Pletscher A, Shore PA, Brodie BB (1955) Serotonin release as a possible mechanism of reserpine action. Science 122:374–375

Plotsky PM (1987) Facilitation of immunoreactive corticotropin-releasing factor secretion into the hypophyseal-portal circulation after activation of catecholaminergic pathways of central norepinephrine injection. Endocrinology 121:924–930

Popoli M, Vocaturo C, Perez J, Smeraldi E, Racagni G (1995) Presynaptic Ca2+/calmodulin-dependent protein kinase II: Autophosphorylation and activity increase in the hippocampus after long-term blockade of serotonin reuptake. Mol Pharmacol 48:623–629

Porsolt RD, Lanegre A (1992) Behavioral models of depression. In: Elliot JM, Heal DJ, Marsden CA (eds) Experimantal approaches to anxiety and depression. John Wiley: Chichester, pp 73–86

Porsolt RD, Lenegre A, McArthur RA (1991) Pharmacological models of depression. In: Oliver B, Mos J, Slangen JL (eds) Animal models in psychopharmacology. Birkhauser, Basel, pp 137–159

Porter R, Bock G, Clark S (eds) (1986) Antidepressants and receptor function. Ciba Foundation Symposium 123, Wiley, Chichester, UK

Premont RT, Inglese J, Lefkowitz RJ (1995) Protein kinases that phosphorylate activated G protein-coupled receptors. FASEB J 9:175–182

Pryor JC, Sulser F (1991) Evolution of monoamine hypotheses of depression. In: Horton RW, Katona C (eds) Biological aspects of affective disorders. Academic Press, London, pp 77–94

Przegaliñski E, Budziszewska B (1993) The effect of long-term treatment with antidepressant drugs on the hippocampal mineralocorticoid and glucocorticoid receptors in rats. Neurosci Lett 161:215–218

Przegaliński E, Moryl E, Papp M (1995) The effect of 5-HT1A receptor ligands in a chronic mild stress model of depression. Neuropharmacology 34:1305–1310

Quetsch RM, Achor RWP, Litin EM, Faucett RL (1959) Depressive reactions in hypertensive patients. A comparison of those treated with rauwolfia and those receiving no specific antihypertensive treatment. Circulation 19:366–375

Qian Y, Galli A, Ramamoorthy S, Risso S, DeFelice LJ, Blakely RD (1997) Protein kinase C activation regulates human serotonin transporters in HEK-293 cells via altered cell surface expression. J Neurosci 17:45–57

Rahman S, Li PP, Young LT, Kofman O, Kish SJ, Warsh JJ (1997) Reduced [3H]cyclic AMP binding in postmortem brain from subjects with bipolar affective disorder. J Neurochem 68:297–304

Randrup A, Munkvad I, Fog R, Gerlach J, Molander R, Kjellenberg B, Scheel-Krueger J (1975) Mania, depression and brain dopamine. Curr Develop Psychopharmacol 2:207–229

Rasenick MM (1994) G proteins as the molecular target of antidepressant action: Chronic treatment increases coupling between Gs and adenylate cyclase. Neuropsychopharmacology 10:580S (Abstract)

Riva M, Brunello N, Rovescalli AC, Galimberti R, Carfagna N, Carminati P, Pozzi O, Ricciardi S, Roncucci R, Rossi A, Racagni G (1989) Effect of reboxetine, a new antidepressant drug, on the central noradrenergic system: behavioural and biochemical studies. J Drug Development 1:243–253

Rivier CL, Plotsky PM (1986) Mediation by corticotropin-releasing factor (CRF) of adenohypophyseal hormone secretion. Ann Rev Physiol 48:475–494

Rivier J, Rivier C, Vale W (1984) Synthetic competitive antagonists of corticotropin-releasing factor: Effect on ACTH secretion in the rat. Science 224:889–891

Roberts VJ, Singhal RL, Roberts DCS (1984) Corticosterone prevents the increase in noradrenaline stimulated adenyl cyclase activity in rat hippocampus following adrenalectomy or metopirone. Eur J Pharmacol 103:235–240

Rossby SP, Sulser F (1993) Die Wirkmechanismen von Antidepressiva: Ein historischer Rückblick und neue neurobiologische Aspekte. ZNS Journal, Forum für Psychiatrie und Neurologie 1:10–19

Rossby SP, Sulser F (1997) Antidepressants: Events beyond the synapse. In: Skolnick P (ed) Antidepressants: New pharmacological strategies. Humana Press, Totowa, NJ, pp 195–212

Rossby SP, Nalepa I, Huang M, Burt A, Perrin C, Schmidt DE, Sulser F (1995) Norepinephrine-independent regulation of GRII mRNA in vivo by a tricyclic antidepressant. Brain Res 687:79–82

Rossby SP, Perrin C, Burt A, Nalepa I, Schmidt DE, Sulser F (1996) Fluoxetine increases steady-state levels of preproenkephalin mRNA in rat amygdala by a serotonin dependent mechanism. J Serotonin Res 3:69–74

Rossby SP, Manier DH, Liang S, Nalepa I, Sulser F (1999) Venlafaxine: Pharmacological actions beyond aminergic receptors. Int J Neuropsychopharmacol 2:1–8

Rossby SP, Liang S, Manier DH, Chakrabarti A, Shelton RC, Sulser F (2001) Molecular psychopharmacology as a prelude to a molecular psychopathology of affective disorders: The significance of differential display methodology to study programs of gene expression. In: Briley M, Sulser F (eds) Molecular genetics of mental disorders. Martin Dunitz, Ltd., London, pp 31–46

Rouquier L, Claustre Y, Benavides J (1994) α_1-Adrenoceptor antagonists differentially control serotonin release in the rat hippocampus and striatum: A microdialysis study. Eur J Pharm 261:59–64

Russo-Neustadt A, Beard RC, Cotman CW (1999) Exercise, antidepressant medications, and enhanced brain derived neurotrophic factor expression. Neuropsychopharmacology 21:679–682

Salin P, Kerkerian L, Nieoullon A (1990) Expression of neuropeptide Y immunoreactivity in the rat nucleus accumbens is under the influence of the of the dopaminergic mesencephalic pathway. Exp Brain Res 81:363–371

Sampson D, Muscat R, Willner P (1991) Reversal of antidepressant action by dopamine antagonists in an animal model of depression. Psychopharmacology 104:491–495

Sato K, Adams R, Betz H, Schloss P (1995a) Modulation of a recombinant glycine transporter (GLYT1b) by activation of protein kinase C. J Neurochem 65:1967–1973

Sato K, Betz H, Schloss P (1995b) The recombinant GABA transporter GAT1 is downregulated upon activation of protein kinase C. FEBS Lett 375:99–102

Schildkraut JJ (1965) The catecholamine hypothesis of affective disorders: a review of supporting evidence. Am J Psychiatry 122:509–522

Schildkraut JJ, Kety SS (1967) Biogenic amines and emotion. Science 156:21–30

Schultz J (1976) Psychoactive drug effects on a system which generates cyclic AMP in brain. Nature 261:417–418

Schwaninger M, Schöfl C, Blume R, Rössig L, Knepel W (1995) Inhibition by antidepressant drugs of cyclic AMP response element-binding protein/cyclic AMP response element-directed gene transcription. Mol Pharmacol 47:1112–1118

Seasholtz AF, Gamm DM, Ballestero RP, Scarpetta MA, Uhler MD (1995) Differential expression of mRNAs for protein kinase inhibitor isoforms in mouse brain. Proc Natl Acad Sci USA 92:1734–1738

Seckl JR, Fink G (1992) Antidepressants increase glucocorticoid and mineralocorticoid receptor mRNA expression in rat hippocampus in vivo. Neuroendocrinology 55:621–626

Shelton R, Manier DH, Sulser F (1996) Cyclic cAMP-dependent protein kinase activity in major depression. Am J Psychiatry 153:1037–1042

Shelton RC, Manier DH, Ellis T, Peterson CS, Sulser F (1999) Cyclic AMP dependent protein kinase in subtypes of major depression and normal volunteers. Int J Neuropsychopharmacol 3:187–192

Sheng M, Thompson MA, Greenberg ME (1991). CREB: A Ca^{++} regulated transcriptional factor phosphorylated by calmodulin-dependent kinases. Science 252:1427–1430

Shih M, Malbon CC (1994) Oligodeoxygnucleotides antisense to mRNA encoding protein kinase A, protein kinase C and β-adrenergic receptor kinase reveal distinctive cell-type specific roles in agonist-induced desensitization. Proc Natl Acad Sci USA 91:12193–12197

Shirayama Y, Mitsushio H, Takashima M, Ichikawa H (1996) Reduction of substance P after chronic antidepressants treatment in the striatum, substantia nigra and amygdala of the rat. Brain Res 739:70–78

Sibley DR, Strasser RH, Benovic JL, Daniel K, Lefkowitz RJ (1986) Phosphorylation/dephosphorylation of the b-adrenergic receptor regulates its functional coupling to adenylate cyclase and subcellular distribution. Proc Natl Acad Sci USA 83:9408–9412

Skolnick P (1999) Antidepressants for the new millennium. Eur J Pharmacol 375:31–40

Skolnick P, Layer RT, Popik P, Nowak G, Paul IA, Trullas R (1996) Adaptation of N-methyl-D-aspartate (NMDA) receptors following antidepressant treatment: implications for the pharmacotherapy of depression. Pharmacopsychiatry 29:23–26

Skutella T, Montkowski A, Stoehr T, Probst JC, Landgraf R, Holsboer F, Jirikowski GF (1994) Corticotropin-releasing hormone (CRF) antisense oligodeoxynucleotide treatment attenuates social defeat-induced anxiety in rats. Cell Mol Neurobiol 14:579–588

Sluzewska A, Nowakowska E (1994) The effects of carbamazepine, lithium and ketoconazole in chronic mild stress model of depression in rats. Behav Pharmacol 5 (Suppl 1):86 (Abstract)

Spyraki C, Fibiger HC (1981) Behavioural evidence for supersensitivity of postsynaptic dopamine receptors in the mesolimbic system after chronic administration of desipramine. Eur J Pharmacol 74:195–206

Stamford JA, Muscat R, O'Connor JJ, Patel JJ, Wieczorek WJ, Kruk ZL, Willner P (1991) Voltammetric evidence that subsensivity to reward following chronic mild stress is associated with increased release of mesolimbic dopamine. Psychopharmacology 105:275–282

Stogner KA, Holmes PV (2000) Neuropeptide-Y exerts antidepressant-like effects in the forced swim test in rats. Eur J Pharmacol 387:R9–R10

Stone EA (1979a) Reduction by stress of norepinephrine-stimulated accumulation of cyclic AMP in rat cerebral cortex. J Neurochem 32:1335–1337

Stone EA (1979b) Subsensitivity to norepinephrine as a link between adaptation to stress and antidepressant therapy: A hypothesis. Res Commun Psychol Psychiat Behav 4:241–255

Stone EA, Platt JE, Herrera AS, Kirk KL (1986) Effect of repeated restraint stress, desmethylimipramine or adrenocorticotropin on the alpha and beta adrenergic components of the cyclic AMP response to norepinephrine in rat brain slices. J Pharm Exp Ther 237:702–707

Sulser F (1978) Functional aspects of the norepinephrine receptor coupled adenylate cyclase system in the limbic forebrain and its modification by drugs which precipitate or alleviate depression: Molecular approaches to an understanding of affective disorders. Pharmakopsychiatry 11:43–52

Sulser F, Mishra R (1983) The discovery of tricyclic antidepressants and their mode of action. In: Parnham MJ, Bruinvels J, (eds) Discoveries in pharmacology, vol 1. Elsevier, Amersterdam, pp 233–247

Szabadi E, Bradshaw CM, Boston PF, Langley RW (1998). The human pharmacology of reboxetine. Human Psychopharmacology 13;S3–S12

Szmigielski A, Gorska D (1997) The effect of prolonged imipramine treatment on the alpha 1-adrenoceptor-induced translocation of protein kinase C in the central nervous system in rats. Pharmacol Res 35:569–576

Tassin JP, Studler JM, Herve D, Blanc G, Glowinski J (1986) Contribution of noradrenergic neurons to the regulation of dopaminergic (D1) receptor denervation supersensitivity in rat prefrontal cortex. J Neurochem 46:243–248

Taylor SS (1989) cAMP-dependent protein kinase. J Biol Chem 264:8443–8446

Toth M, Shenk T (1994) Antagonist-mediated down-regulation of 5-hydroxytryptamine type 2 receptor gene expression: Modulation of transcription. Mol Pharmacol 45:1095–1100

Trovero F, Herve D, Blanc G, Glowinski J, Tassin JP (1992) In vivo partial inactivation of dopamine D1 receptors induces hypersensitivity of cortical dopamine-sensitive adenylate cyclase: permissive role of alpha 1-adrenergic receptors. J Neurochem 59:331–337

Vetulani J, Sulser F (1975) Action of various antidepressant treatment reduces reactivity of noradrenergic cyclic AMP generating system in limbic forebrain. Nature 257:495–496

Vetulani J, Stawarz RJ, Sulser F (1976a) Adaptive mechanisms of the noradrenergic cyclic AMP generating system in the limbic forebrain of the rat: Adaptation to persistent changes in the availability of norepinephrine (NE). J Neurochem 27:661–666

Vetulani J, Stawarz RJ, Dingell JV, Sulser F (1976b) A possible common mechanism of action of antidepressant treatments. Reduction in the sensitivity of the noradrenergic cyclic AMP generating system in the rat limbic forebrain. Naunyn-Schmiedeberg's Arch Pharmacol 293:109–114

Vetulani J, Lebrecht U, Pilc A (1981) Enhancement of responsiveness of the central serotonergic system and serotonin-2 receptor density in the rat frontal cortex by electroconvulsive treatment. Eur J Pharmacol 76:81–85

Vetulani J, Antkiewicz-Michaluk L, Rokosz-Pelc A, Pilc A (1983) Chronic electroconvulsive treatment enhances the density of [^{3}H]prazosin binding sites in the central nervous system of the rat. Brain Res 275:392–395

Vetulani J, Antkiewicz-Michaluk L, Rokosz-Pelc A, Pilc A (1984a) Alpha up-beta down adrenergic regulation: A possible mechanism of action of antidepressant treatments. Pol J Pharmacol Pharm 36:231–248

Vetulani J, Antkiewicz-Michaluk L, Rokosz-Pelc A (1984b) Chronic administration of antidepressant drugs increases the density of cortical ^{3}H-prazosin binding sites in the rat. Brain Res 310:360–362

Wachtel H (1989) Dysbalance of neuronal second messenger function in the aetiolgy of affective disorders: A pathophysiological concept hypothesising defects beyond first messenger receptors. J Neural Transm 75:21–29

Wahlestedt C, Ekman R, Widerlov E (1989) Neuropeptide Y and the central nervous system: Distribution and possible relationship to neurological and psychiatric disorders. Prog Neuropsychopharmacol Biol Psychol 13:31–54

Wedzony K, Klimek V, Nowak G (1995) Rapid down-regulation of beta-adrenergic receptors evoked by combined forced swimming test and CGP 37849—a competitive antagonist of NMDA receptors. Pol J Pharmacol Pharm 47:537–540

White BD, Dean R, Martin RJ (1990) Adrenalectomy decreases neuropeptide Y mRNA levels in the arcuate nucleus. Brain Res Bull 25:711–715

Widerlov E, Lindstom LM, Wahlestedt C, Ekman R (1988) Neuropeptide Y and peptide YY as possible cerebrospinal markers for major depression and schizophrenia, respectively. J Psychiatric Res 22:69–79

Wielosz M (1981) Increased sensitivity to dopaminergic agonists after repeated electroconvulsive shock (ECT) in rats. Neuropharmacology 10:941–945

Willner P (1983) Dopamine and depression: A review of recent evidence. III. The effects of antidepressant treatments. Brain Res 287:237–246

Willner P (1991) Animals models as simulation of depression. Trends Pharmacol Sci 12:131–136

Willner P (1997a) The mesolimbic dopamine system as a target for rapid antidepressant action. Int Clin Psychopharmacol 12 (Suppl 3):S7–S14

Willner P (1997b) Validity, reliability and utility of the chronic mild stress model of depression: A 10 year review and evaluation. Psychopharmacology 134:319–329

Willner P, Papp M (1997) Animal models to detect antidepressants: Are new strategies necessary to detect new agents? In: Skolnick P (ed) Antidepressants: New pharmacological strategies. Humana Press, Totowa, NJ, pp 213–234

Willner P, Towell A, Sampson D, Sophokleus S, Muscat R (1987) Reduction of sucrose preference by chronic mild stress and its restoration by a tricyclic antidepressant. Psychopharmacology 93:358–364

Willner P, Klimek V, Golembiowska K, Muscat R (1991) Changes in mesolimbic dopamine may explain stress-induced anhedonia. Psychobiology 19:79–84

Willner P, Muscat R, Papp M (1992) Chronic mild stress-induced anhedonia: a realistic animal model of depression. Neurosci Biobehav Rev 16:525–534

Wolfe BB, Harden TK, Sporn JR, Molinoff PB (1978) Presynaptic modulation of beta-adrenergic receptors in rat cerebral cortex after treatment with antidepressants. J Pharmacol Exp Ther 207:446–457

Wong DT, Bymaster FP, Engleman EA (1995) Prozac (fluoxetine, Lilly 110140), the first selective serotonin uptake inhibitor and an antidepressant drug: Twenty years since its first publication. Life Sci 57:411–441

Wong EHF, Sonders MS, Amara SG, Tinholt PM, Piercey MF, Hoffmann WP, Hyslop DK, Franklin S, Porsolt RD, Bonsignori A, Carfagna N, McArthur RA (2000) Reboxetine: A pharmacologically potent, selective and specific norepinephrine reuptake inhibitor (NRI). Biol Psychiatry 47:818–829.

Yoshikawa K, Sabol SL (1986) Expression of the enkephalin precursor gene in C6 glioma cells: Regulation by β-adrenergic agonists and glucocorticoids. Molec Brain Res 1:75–83

Yuan PX, Chen G, Huang LD, Manji HK (1998) Lithium stimulates gene expression through the AP-1 transcription factor pathway. Mol Brain Res 58:225–230

Yuan P, Chen G, Manji HK (1999) Lithium activates the c-Jun NH2-terminal kinases in vitro and in the CNS in vivo. J Neurochem 73:2299–2309

Zeller EA (1983) Monoamine oxidase and its inhibitors in relation to antidepressive activity. In: Parnham MJ, Bruinvels J (eds) Discoveries in pharmacology, vol 1. Elsevier, Amersterdam, pp 223–232

Promising New Directions in Antidepressant Development

V. Garlapati[1] · W. F. Boyer[2] · J. P. Feighner[3]

[1] Department of Psychiatry and Behavioral Sciences,
University of Kansas School of Medicine, 1100 North St. Francis, Suite 200,
Wichita, KS 67214, USA
e-mail: gpati@hotmail.com
[2] Behavioral Solutions Inc., 4181 Pleasant Hill Rd., Suite 100, Duluth, GA 30096, USA
[3] President and Director of Research, Feighner Research Institute,
5375 Mira Sorrento Place, Suite 210, San Diego, CA 92121, USA

1	Introduction	566
2	Neuropeptide Neurotransmitters of Potential Relevance to the Pathophysiology of Clinical Depression for Which Data on Investigational Agents Are Available	568
2.1	Corticotropin-Releasing Hormone	568
2.1.1	Role in the Pathophysiology of Depression	568
2.1.2	Investigational Antidepressants	569
2.2	Substance P (NK1)	570
2.2.1	Role in the Pathophysiology of Depression	570
2.2.2	Investigational Antidepressants	570
3	Neuropeptide Neurotransmitters of Potential Relevance to the Pathophysiology of Clinical Depression for Which Data on Investigational Agents Are Not Available	570
3.1	Neuropeptide Y: Role in the Pathophysiology of Depression	571
3.2	Galanin: Role in the Pathophysiology of Depression	571
3.3	Vasopressin: Role in the Pathophysiology of Depression	572
3.4	Oxytocin: Role in the Pathophysiology of Depression	572
4	Novel Peptides	573
4.1	Historical Review	573
4.2	Nemifitide	574
5	Conclusion	576
	References	577

Abstract This chapter extends the discussion of novel mechanisms of action that may mediate antidepressant efficacy presented in the previous chapter and reviews findings concerning investigational antidepressants that are currently either in phase II or III clinical testing and for which data are available in the public domain. Six neuropeptides that are currently of interest in antidepressant drug development are discussed: corticotropin releasing hormone (CRH), substance P [also known as neurokinin 1 (NK1)], neuropeptide Y (NPY), galanin, vasopressin (VPN), and oxytocin. These six were chosen because there are animal and/or human data that support a potential role for each of them in the pathophysiology of clinical depression. The role of each of the six neuropeptides in the pathophysiology of depression is reviewed. Results in the public domain concerning investigational antidepressants affecting one of these mechanisms of action are available only for the first two neuropeptides—CRH and substance P. Data are described for R121919, a CRH-1 receptor antagonist, and for the substance P antagonist, MK-869. Note that there are other investigational antidepressants in phase I or early phase II testing for which efficacy data are either not available or not yet in the public domain which are not discussed in this chapter. The naturally occurring neuropeptides described in the first part of the chapter have multiple functions throughout the body and thus drugs acting on these peptides could produce multiple unwanted effects. For that reason, it might be desirable to design novel peptides that have therapeutic potential with minimal side effects. Findings concerning one such novel peptide are presented: INN 00835 (nemifitide) is a synthetic pentapeptide that has shown promise in preclinical and clinical trials as a future antidepressant.

Keywords Antidepressants · Drug development · Neuropeptides · Corticotropin-releasing hormone (CRH) · Substance P · Neuropeptide Y (NPY) · Galanin · Vasopressin (VPN) · Oxytocin · Nemifitide

1
Introduction

Earlier chapters (Part 3) reviewed the history of modern antidepressant pharmacotherapy from the chance discovery antidepressants [i.e., tricyclic antidepressants (TCAs) and monoamine oxidase inhibitors (MAOIs)] to the rationally developed selective serotonin reuptake inhibitors (SSRIs) and other newer antidepressants. This chapter will extend the previous chapter's discussion of novel mechanisms of action that may mediate antidepressant efficacy and review findings concerning investigational antidepressants that are currently in clinical testing and for which there are published data. Although the authors are aware of other investigational antidepressants, no efficacy data have been presented in the public domain on these agents to allow for scientific review at this time.

As explained in Part 3, the chance discovery of the TCAs and the MAOIs not only produced effective treatments for patients with clinical depression but also provided tools that enabled researchers to begin to understand mechanisms that might underlie antidepressant efficacy. The study of the pharmacology of these antidepressants formed the basis for the biogenic amine theories of depression, which in turn were the basis for the development of the newer antidepressants described in the chapters entitled "Selective Serotonin Reuptake Inhibitors" and "Other Antidepressants".

While SSRIs and other newer antidepressants were rationally developed, they represent a refinement of the pharmacology of the TCAs and MAOIs rather than a new direction in terms of mechanism of action. They were a refinement principally in terms of designing out the unnecessary effects of the TCAs and MAOIs. Nevertheless, the mechanisms of action believed to mediate their antidepressant efficacy remained the same as those of the earlier antidepressants (i.e., promotion of the activity of one or more than one of the biogenic amine neurotransmitter systems). For this reason, the newer agents have better safety and tolerability profiles compared with the TCAs and MAOIs but not necessarily better efficacy. The drug-specific response to the newer antidepressants that are available, as determined by formal clinical response (i.e., overall drug response minus the parallel placebo response), is only 15%–30%.

In the chapter on treatment-refractory depression, Trivedi et al. reviewed the problems posed by the limited efficacy of existing antidepressants. As he points out in that chapter, a sizable percentage of patients with clinical depression do not respond to any available antidepressant. In fact, approximately one third of patients treated in published clinical trials of newer antidepressants do not achieve a meaningful, clinical response in terms of relief of their depressive syndrome. Therefore, there is considerable need for improved antidepressants with greater overall efficacy in terms of both the percentage of the population who responds and the completeness of the response (i.e., full remission), more rapid onset of antidepressant action (days versus weeks), and better tolerability, particularly with regard to reduced adverse effects on sexual function, which is a problem with serotonin reuptake inhibitors such as the SSRIs and venlafaxine. One way to achieve one or more than one of these goals is to develop antidepressants with new mechanism(s) of antidepressant action that go beyond effects on biogenic amine neurotransmission.

Trivedi et al. reviewed a number of different strategies that are currently used when faced with patients who do not respond to the first antidepressant tried. Despite all of these efforts, too many patients are still not helped. In part, that is likely to reflect the fact that many of these options are still limited to direct effects on biogenic amine neurotransmission. For this reason, there is a need for antidepressants that have novel mechanisms of action directed at different neurotransmitter systems and that are perhaps more fundamentally involved in the pathophysiology of specific biochemically defined forms of clinical depression.

The previous chapter by Nalepa and Sulser reviewed the amazing recent expansion in our knowledge of new mechanisms and systems in the brain of potential relevance to the syndrome of clinical depression. A number of hypotheses involving novel mechanisms have been advanced to explain the pathophysiology of depression (for more detailed discussions of these hypotheses, see Horst, Connor and Leonard, and Nalepa and Sulser, this volume). Many of these hypotheses are based on the well-established abnormalities in hypothalamic–pituitary axis (HPA) function found in a sizable percentage of patients with clinical depression. A number of pharmaceutical companies have ongoing programs aimed at discovering antidepressants capable of correcting these abnormalities through neuropeptide rather than biogenic amine mechanisms of action.

This chapter will review how antidepressant drug discovery is likely to change the options available to clinicians and their patients over the next 5–10 years. Six neuropeptides that are currently of interest in antidepressant drug development are discussed in this chapter:

- Corticotropin releasing hormone (CRH)
- Substance P [also known as Neurokinin 1 (NK1)]
- Neuropeptide Y (NPY)
- Galanin
- Vasopressin (VPN)
- Oxytocin

These six were chosen because there are animal and/or human data that support a potential role for each of them in the pathophysiology of clinical depression; these data are presented in the sections that follow. Results in the public domain concerning investigational antidepressants affecting one of these specific mechanisms of action are available only for the first two neuropeptides—CRH and Substance P—and these findings are reviewed in the following section after the data on their role in the pathophysiology of depression are presented. Note that the discussion in this chapter is limited to those agents that are either in phase II or III and for which data are available in the public domain. However, there are other investigational antidepressants in phase I or early phase II for which efficacy data are either not available or not yet in the public domain.

2
Neuropeptide Neurotransmitters of Potential Relevance to the Pathophysiology of Clinical Depression for Which Data on Investigational Agents Are Available

2.1
Corticotropin-Releasing Hormone

2.1.1
Role in the Pathophysiology of Depression

As noted above, the dysfunction of the HPA axis has been widely studied with regard to its possible relevance to the pathophysiology of clinical depression. Numerous studies have documented abnormalities in one or more components of the HPA axis in patients with clinical depression. This work began with the clinical observation that Cushing's syndrome and clinical depression shared many symptoms and that an adverse effect of exogenous corticosteroids was the development of clinical depression. Subsequent research found a high incidence of abnormal dexamethasone suppression in patients with clinical depression.

Corticotropin releasing hormone (CRH) is an important component of the HPA axis. CRH is synthesized in the hypothalamus and a portal system carries it to the anterior pituitary. There, CRH facilitates secretion of adrenocorticotropic hormone (ACTH) into the systemic circulation. ACTH, in turn, stimulates adrenal secretion of cortisol. The plasma concentration of cortisol through a feedback loop regulates CRH synthesis/secretion from the hypothalamus.

When CRH is administered centrally to animals, it produces changes in the rate of firing of locus coeruleus (LC) neurons similar to those seen in both depression and anxiety (Mitchell 1998; Weiss et al. 1994). Based on animal studies, a number of currently marketed antidepressants enhance the sensitivity of corticosteroid receptors (Holsboer 1999). Animal and human cerebrospinal fluid (CSF) studies have documented decreased CRH secretion following electroconvulsive therapy (ECT) or antidepressant treatment (Nemeroff et al. 1991; De Bellis et al. 1993; Kling et al. 1994; Heuser et al. 1998). Stressful early life experiences in nonhuman primates have been found to produce persistently elevated CRH in CSF compared with that in peers raised under predictable circumstances (Coplan et al. 1996). CRH may therefore be an important link between "nature" and "nurture" in this area.

In terms of clinical studies, suppression of cortisol production is associated with amelioration of depression in a significant portion of depressed individuals (Murphy et al. 1991; Thakore and Dinan 1995; Iizuka et al. 1996). In postmortem brain samples, the total number of CRH-expressing neurons in depressed patients was found to be four times higher than that of the controls (Hoogendijk et al. 2000). This finding raises the question of why CRH levels are increased in individuals with affective disorders. One possibility is an impairment in the negative feedback loop that normally prevents hypercortisolemia.

Consistent with this hypothesis, there is evidence that corticosteroid receptor function is impaired in many patients with depression and in many healthy persons who are at genetic risk for a depressive disorder (Pariante et al. 1995). This decreased glucocorticoid receptor sensitivity may be the primary pathology underlying some forms of clinical depression or may be a secondary phenomenon resulting from prolonged production of steroid hormones in the face of prolonged stress. CRH may also influence the activity of tyrosine hydroxylase, the rate-limiting step in the synthesis of catecholamines, which in turn has been hypothesized to be associated with the neurobiology of depression (Redmond and Leonard 1997).

2.1.2
Investigational Antidepressants

Various tissue samples, including human samples, have shown subtypes of CRH receptors as identified by radio-ligand binding and cloning techniques (Grigoriadis et al. 1996). The CRH-1 receptor appears to mediate the anxiogenic effects of CRH. Patients with clinical conditions that are causally related to HPA hyperactivity may therefore benefit from treatment with a CRH-1 receptor antagonist.

R121919, an CRH-1 receptor antagonist, was administered to 24 patients with a major depressive episode. The patients were enrolled in two dose-escalation panels: in one group ($n=10$), the dose was increased from 5–40 mg, while in the other group ($n=10$), the dose was increased from 40–80 mg within 30 days. Both groups showed significant reductions in depression and anxiety scores on patient and clinician ratings, with greater reductions seen in the second group. Depressive symptoms worsened in both groups after drug discontinuation (Zobel et al. 2000).

2.2
Substance P (NK1)

2.2.1
Role in the Pathophysiology of Depression

Substance P, also called neurokinin 1 (NK1), is found throughout the body. In the central nervous system (CNS), it is usually co-localized with classical neurotransmitters such as serotonin and norepinephrine (Sergeyev et al. 1999). Given this fact and its widespread distribution in the brain, substance P may have a role in a wide range of psychiatric conditions, including schizophrenia, depression, and anxiety (Argyropoulos and Nutt 2000).

Substance P stimulates the HPA axis and produces anxiety-like responses in the rat (Gavioli et al. 1999). Substance P antagonists suppress isolation-induced vocalizations in guinea pigs, an animal model of depression (Kramer et al. 1998). The blood concentration of substance P in the peripheral blood correlates with anxiety scores in humans (Fehder et al. 1997; Hasenohrl et al. 1998; Gavioli et al. 1999).

2.2.2
Investigational Antidepressants

The substance P antagonist, MK-869, has been studied in clinical trials that compared its potential antidepressant activity with that of paroxetine and placebo in outpatients with moderate to severe major depression with anxiety. In this study, MK-869 was well tolerated and as effective as paroxetine (Kramer et al.1998). Other NK-1 antagonists are under development and this technology appears promising.

3
Neuropeptide Neurotransmitters of Potential Relevance to the Pathophysiology of Clinical Depression for Which Data on Investigational Agents Are Not Available

Findings concerning the following endogenous neurotransmitters in most animal and some human studies suggest that they are relevant either to the causality or treatment of depression. There are no data available in the public domain concerning investigational agents in phase II or III testing related to any of the neurotransmitters discussed in this section.

3.1
Neuropeptide Y: Role in the Pathophysiology of Depression

Neuropeptide Y (NPY) is relatively abundant in the mammalian brain. It affects a wide variety of functions, including emotions, eating, memory, response to stress, arterial blood pressure, cardiac contractility, and intestinal secretions (Munglani et al. 1996). Several types of NPY receptors have been identified (Sanglard and Tavares 1998).

NPY promotes sleep and inhibits the HPA axis in humans (Antonijevic et al. 2000). These effects may involve NPY-mediated inhibition of CRH (Antonijevic et al. 2000). NPY immunoreactivity is decreased in the CSF of patients with clinical depression compared with controls (Widerlov et al. 1988).

A variety of animal studies have suggested that NPY may have significant antidepressant and anxiolytic activity (Ehlers et al. 1997). First, fluoxetine treatment normalizes NPY-related gene expression in a rat model of depression (Caberlotto et al. 1998; Mathe et al. 1998; Mathe 1999). Second, chronic treatment with lithium, ECT, or citalopram increases NPY neurotransmission in the hippocampus of rat brains (Husum et al. 2000). Third, NPY produces an antidepressant-like effect in the forced swim test (Stogner and Holmes 2000).

There are also significant links between NPY and central serotonin activity. NPY also interacts with other neurotransmitters, including noradrenaline, somatostatin, nitric oxide, and glutamate. In the samples of rat brains, long-term treatment with 5-hydroxytryptamine, a serotonin precursor, reduces NPY activity in the hypothalamus. Conversely, treatment with serotonin-specific neurotoxins or inhibitors elevates NPY activity in the hypothalamus (Kakigi and Maeda 1992). Thus, this neuropeptide system appears to be interacting with serotonin and possibly the noradrenaline biogenic amine system, although the precise nature of the relationship has not been fully elucidated and additional work is needed.

Work with ECT treatment in both animals and humans also suggests that enhanced NPY activity may play a role as a potential mechanism mediating antidepressant efficacy. In rats, ECT treatment reproducibly elevated concentrations of NPY-like immunoreactivity (NPY-LI) in hippocampus, frontal, and occipital cortex. In comparable studies in humans, ECT treatment increased CSF concentrations of NPY-like immunoreactivity in parallel with recovery from clinical depression.

3.2
Galanin: Role in the Pathophysiology of Depression

Galanin is produced in the tuberomammillary nucleus of the hypothalamus (Swaab et al. 1993). Galanin can have anxiolytic-like action in animal models

(Bing et al. 1993). Galanin inhibits CRH release (Cimini 1996) and may inhibit norepinephrine and serotonin activity (Kask et al. 1997; Fuxe et al. 1998). Weiss and colleagues have postulated that galanin may also inhibit the activity of dopamine cell bodies in the ventral tegmentum. These neurons send axons to the forebrain and their down-regulation may therefore cause at least two of the principal symptoms seen in depression: decreased motor activity and decreased appreciation of pleasurable stimuli (anhedonia). These data suggest that galanin antagonists could be of therapeutic benefit in the treatment of depression (Weiss et al. 1998).

3.3
Vasopressin: Role in the Pathophysiology of Depression

Vasopressin (VPN) is released from two sites in the hypothalamus: from the paraventricular nucleus, where CRH is formed, and from the supraoptic nucleus (Scott and Dinan 1998). As shown in human volunteers, VPN stimulates CRH release and acts synergistically with CRH (Gaillard et al. 1988; Hauger and Aguilera 1993; Perraudin et al. 1993; Scott and Dinan 1998). In animal studies, VPN secretion has been found to be inhibited by central serotonin activity (Ferris and Delville 1994) and fluoxetine treatment has been found to be associated with a decrease in VPN levels in plasma and hypophysial portal (Gibbs and Vale 1983; Nemeroff et al. 1991; Altemus et al. 1992).

VPN levels and the number of VPN-expressing neurons are elevated in postmortem brain samples of patients with major depression (Purba et al. 1996). In patients with clinical depression, increased plasma concentrations of arginine vasopressin (AVP) were noted and the elevation was correlated with greater degrees of psychomotor retardation (Van Londen et al. 1997, 1998).

3.4
Oxytocin: Role in the Pathophysiology of Depression

Oxytocin production, like VPN, is influenced by central serotonin activity (Saydoff et al. 1993). Oxytocin administration lowers CRH and cortisol levels and has anxiolytic and antidepressant action in animal models (Arletti and Bertolini 1987; Uvnas-Moburg et al. 1994; Windle et al. 1997; Mitchell 1998; Uvnas-Moburg 1998). Oxytocin inhibits stress-induced elevations of ACTH in laboratory animals as well as inhibiting the increases in ACTH which normally follow angiotensin II challenge (intravenous administration of endothelin I and/or exercise in humans) (Coiro et al. 1988; Bianconi et al. 1990; Volpi et al. 1995; Windle et al. 1997). Oxytocin levels in humans are decreased in clinical depression and in individuals experiencing "normal" sad emotion (Frasch et al. 1996; Turner et al. 1999).

4
Novel Peptides

The discussion so far has shown that a number of naturally occurring peptides or their antagonists may have a role in the treatment of clinical depression. However, these peptides have multiple functions throughout the body and thus drugs acting on these peptides could produce multiple unwanted effects. For that reason, it might be desirable to design novel peptides that have therapeutic potential with minimal side effects. One such novel peptide, R121919, was discussed above ("Investigational Antidepressants"). INN 00835 (Nemifitide) is another novel peptide that has promise as a future antidepressant compound.

4.1
Historical Review

Melanocyte-stimulating hormone inhibiting factor (MIF)-1 is a naturally occurring hypothalamic tripeptide that inhibits release of melanocyte stimulating hormone (MSH) in certain assays (Nair et al. 1971).

Earlier studies with MIF-1 in hypophysectomized animals have demonstrated its ability to exert a direct effect on the CNS (Plotnikoff et al. 1974). MIF-1 was found to be more effective than the TCAs in the dopa-response-potentiation test (a standard depression screening test in animals) and was found to antagonize the tremor induced by oxotremorine (animal model of parkinsonism) (Plotnikoff et al. 1971, 1972). MIF-1 has also been found to reverse the sedative effects of deserpidine in animal studies (Plotnikoff et al. 1973).

With the background of these animal studies, MIF-1 was tested in patients with Parkinson's disease, where it showed some efficacy (Kastin et al. 1972). However, administration of MIF-1 in patients with Parkinson's disease was also correlated with improvement in mood (Fischer et al. 1974; Barbeau 1975). Another hypothalamic tripeptide, protirelin[thyrotropin-releasing hormone (TRH)], was tried in clinically depressed patients due to the limited availability of MIF-1, with somewhat encouraging results (Kastin and Barbeau 1972; Prange and Wilson 1972; Prange et al. 1972).

Antidepressant drugs are known to increase water-wheel turning in animals and MIF-1 and another structurally related compound, Tyr-MIF-1, were effective in producing similar results (Kastin et al. 1984). Environmental stress, particularly chronic low-grade stressors, appear to be more significant than single occurrences in depressive illness (Leff et al. 1970; Thomson and Hendric 1972; Lloyd 1980); consequently, the chronic stress model of depression used in animal studies was thought to be highly valid. MIF-1 was found to produce improvement comparable to the TCAs in a chronic stress model of depression in animals (Pignatiello et al. 1989). The response of animals with nonclinical stress-induced depression to an experimental compound to some extent predicts depressed individuals' clinical response to that compound (Willner 1984, 1990).

Based on the encouraging findings in patients with Parkinson's disease and in experimental models in animals described above, a double-blind clinical trail of MIF-1 was done in depressed patients. In that study, low doses of MIF-1 showed efficacy compared with high doses of MIF-1 and placebo. It is interesting to note the curvilinear dose–response curve seen with MIF-1. Effects were observed within the first 48 h after administration of MIF-1 (Ehrensing et al. 1974). The difference between response to the low dose compared with response to higher doses (with lower doses more efficacious) was found to be significant in a second study (Ehrensing et al. 1978). In another double-blind study, which compared the antidepressant efficacy of MIF-1 and imipramine, MIF-1 was found to be comparable to imipramine, with a rapid onset of action shown by significant differences in rating scales being evident as early as day 8 (Van der Velde 1983).

MIF-1 offers a great advantage, since it is rapid, safe, and effective. However, certain factors limit its use for clinical purposes. Since it is a natural compound, it cannot be patented. Although it crosses the blood–brain barrier, MIF-1 has a brief half-life and there is reason to suspect it has low bioavailability orally (Redding et al. 1973, 1976; Kastin et al 1975). Because of these limitations, research has been undertaken to develop similar synthetic antidepressant compounds that would offer rapid onset of action, higher potency, and greater bioavailability. This quest has produced a series of newer synthetic peptides. Nemifitide is one of the first of these that has shown promise in several clinical trials.

4.2
Nemifitide

Nemifitide (INN 00835) is a synthetic pentapeptide and the lead compound in a series of related, novel, highly potent small-chain peptides synthesized at Innapharma (Hlavka et al. 1997). These peptides appear to have antidepressant activity based on both animal and human studies and may have a rapid onset of antidepressant action. Nemifitide has been shown to be effective in a variety of tests and animal models of clinical depression and indicated an inverted U-shape dose response curve in rats (Hlavka et al. 2002).

Nemifitide was metabolized in vitro by liver and intestinal enzymes. Three major metabolites (hydrolysis and oxidation product of the parent drug) were identified by liquid chromatography/mass spectrometry/mass spectrometry (LC/MS/MS) and were also found in animal and human plasma, following its in vivo administration (Nicolau et al. 2001). One of the metabolites (5-Tryp OH nemifitide) was pharmacologically active in the rat. Nemifitide did not show significant in vitro induction or inhibition for the following cytochrome P450 (CYP) isoforms investigated: 1A2, 2A6, 2C9, 2C19, 2D6, 2E1, and 3E4. The effect of these CYP isoforms on the metabolism of nemifitide indicated that CYP 1A2, 2C19, and 2D6 were mainly responsible for its metabolism. The metabolism of nemifitide by multiple pathways and the lack of significant inhibition or induction of CYP isoforms indicate that it is not likely to have significant in vivo interactions with known CYP inhibitory drugs (Nicolau et al. 2002).

Several studies were conducted in an effort to elucidate the mechanism of action of nemifitide. Following subcutaneous administration nemifitide rapidly crosses the blood–brain barrier in rats, and was measured at nanomolar concentrations in the hippocampus, amygdala, striatum, and frontal cortex, sites that express high concentrations of receptors involved with the pathophysiology of depression (Feighner 2003a). In vitro binding assays indicate that nemifitide binds at micromolar concentrations to several receptors, including 5-HT$_{2A}$, NPY1, bombesin, and melanocortin MC4 and 5. In vivo studies in rats evaluating the interactions of nemifitide with several psychoprobes indicated the participation of the serotonergic pathway in the mode of action of nemifitide in a different way than seen with SSRI antidepressants. Significant interaction was found between nemifitide and the serotonin-releaser D-fenfluramine (Kelly et al. 2002). Nemifitide showed 5-HT$_{2A}$ antagonist properties by blocking the hyperthermic effect of the 5-HT$_{2A}$ agonist DOI (Overstreet 2003; University of NC; data on file).

In both preclinical and clinical studies, nemifitide showed linear pharmacokinetics and an excellent safety profile over a wide range of doses administered subcutaneously (SC). Single and multiple doses of nemifitide ranging from 9 to 320 mg were administered SC to healthy volunteers in five phase I studies and to depressed patients in four phase II studies. Pharmacokinetic parameters were calculated from plasma concentrations of unchanged nemifitide measured by LC/MS. Nemifitide was rapidly absorbed (C_{max} at 10–15 min) and eliminated from plasma ($t_{1/2}$ 15–30 min) in most subjects. PK parameters were close to dose proportional in the dosage range investigated. There was no systemic accumulation of drug following five daily doses (Feighner et al. 2002a).

The efficacy of nemifitide has been investigated in two pilot studies in which over 100 unipolar depressed patients diagnosed with major depression received either nemifitide or placebo (Feighner et al. 2000a, 2001a,b). In one study, the drug was administered subcutaneously at a dose of 0.2 mg/kg for five consecutive days. In the second study, the dose was fixed at 18 mg/patient/day. The subjects were evaluated during treatment and weekly for four consecutive weeks after completion of treatment. Efficacy was evaluated using the Hamilton Rating Scale for Depression (HAM-D) (Hamilton 1960), Montgomery-Asberg Depression Rating Scale (MADRS) (Montgomery and Asberg 1979), Carroll Self-Rating Scale (CSRS) (Carroll et al. 1981), Clinical Global Impression (Guy 1976), and a total Visual Analog Scale (VAS). The effect of treatment with nemifitide was also evaluated using a biochemical marker, changes in blood platelet serotonin (5-HT) uptake rates in patients treated with nemifitide compared with those in the placebo group (Kelly et al. 1999).

In addition to these measures, plasma concentrations of nemifitide were measured by LC/MS and were found to be statistically correlated with reduction in the severity of depression as measured by the psychiatric rating scales. Based on this analysis, the minimum effective plasma concentration (MEC) of nemifitide appeared to be around 46 ng/ml (C_{max}) about 15 min after dosing. That conclusion was based on a post hoc analysis that found statistically significant differences in antidepressant response in patients treated with nemifitide who had plasma concentrations above this value compared with patients treated with nemifitide who had plasma concentrations below this value and patients treated with placebo. In addition to traditional statistical analysis, the separation effect was confirmed with multivariate cluster (Feighner et al. 2000b) and discriminant (Feighner and Sverdlov 2002b) analysis. The difference in response was seen as early as day 3 of treatment. The peak clinical antidepressant effect of nemifitide was observed approximately 1 week after treatment and the response to treatment persisted during the 4-week follow-up period.

The long-term open-label extension study (Feighner et al. 2003b) enrolled both responders and non-responders from the initial short-term study and followed them for up to 36 months. Among 27 enrolled patients, 13 were from the placebo group in the initial study and 14 from the drug group in the initial study. Mean duration between re-treatment of responders in the extension study was 3.3 months, which was similar to the length of response in the initial study. Of the 27 patients, 18 (66.7%) responded in the extension study and 9 patients (33.3%) were terminated for lack of efficacy. Of the 14 patients from the drug-treated group in the initial study, 12 patients (85.7%) experienced the same response or non-response to treatment in the extension study as they had in the initial study with nemifitide.

The results from another double-blind study without placebo group showed greater effect at 160 mg and 40 mg versus 80 mg, suggesting a non-statistically significant trend toward a potential U-shape curve of dose response (Feighner 2002; Innapharma, data on file).

In addition to these conventional studies, preliminary results from an open-label pilot study in 22 patients with severely refractory depression are encouraging (45.5% responders after 10–20 doses of nemifitide) and warrant further study (Feighner 2003; Innapharma, data on file).

Based on these preclinical and clinical studies, nemifitide appears to be a promising drug for the treatment of major depression.

5
Conclusion

There have been a significant number of animal and human studies that identified the involvement of the neuropeptide system in the pathogenesis of depression. This finding will certainly influence the development of treatment for depression during the next decade. Phase II studies have provided encouraging evidence for possible new avenues of psychopharmacotherapy, and we anticipate more exciting data coming out in this area. Completion of the human genome project will open up a whole new era of research and will hopefully reveal new potential avenues of treatment. The next chapter discusses the promise of genetics, pharmacogenomics, and the human genome project for the future of antidepressant drug development.

References

Altemus M, Cizza G, Gold PW (1992) Chronic fluoxetine treatment reduces hypothalamic vasopressin secretion in vitro. Brain Res 593: 311–313
Antonijevic IA, Murck H, Bohlhalter S, Frieboes RM, Holsboer F, Steiger A (2000) Neuropeptide Y promotes sleep and inhibits ACTH and cortisol release in young men. Neuropharmacology 39:1474–1481
Argyropoulos SV, Nutt DJ (2000) Substance P antagonists: novel agents in the treatment of depression. Expert Opin Investig Drugs 9:1871–1875
Arletti R, Bertolini A (1987) Oxytocin acts as an antidepressant in two animal models of depression. Life Sci 41:1725–1730
Barbeau A. (1975) Potentiation of levodopa effect by intravenous l-prolyl-l-leucyl-glycine amide in man. Lancet 2:683-684
Bianconi L, Chiodera P, Capretti, L, Volpi R, Marcato A, Cavazzini U, Camellini L, Rossi G, Caiazza A, Coiro V (1990) Oxytocin reduces angiotensin II-induced ACTH release in man. Neuroendocrinology Letters 12:391–400
Bing O, Moller C, Engel JA, Soderpalm B, Heilig M (1993) Anxiolytic-like action of centrally administered galanin. Neurosci Lett 164:17–20
Caberlotto L, Fuxe K, Overstreet DH, Gerrard P, Hurd YL (1998) Alterations in neuropeptide Y and Y1 receptor mRNA expression in brains from an animal model of depression: region specific adaptation after fluoxetine treatment. Brain Res Mol Brain Res 59:58–65
Carroll BJ, Feinberg M, Smouse PE, Rawson SG, Greden JF(1981) The Carroll rating scale for depression I: development, reliability and validation. Br J Psychiatry 138:194–200
Cimini V (1996) Galanin inhibits ACTH release in vitro and can be demonstrated immunocytochemically in dispersed corticotrophs. Exp Cell Res 228:212–215
Coiro V, Passeri M, Davoli C, Bacchi-Modena A, Bianconi AL, Volpi R, Chiodera P (1988) Oxytocin reduces exercise-induced ACTH and cortisol rise in man. Acta Endocrinol 119:405–412
Coplan JD, Andrews MW, Rosenblum LA, Owens MJ, Friedman S, Gorman JM, Nemeroff CB (1996) Persistent elevations of cerebrospinal fluid concentrations of corticotropin-releasing factor in adult nonhuman primates exposed to early-life stressors: implications for the pathophysiology of mood and anxiety disorders. Proc Natl Acad Sci U S A 93:1619–1623
De Bellis MD, Gold PW, Geracioti Jr, TD, Listwak SJ, Kling MA (1993) Association of fluoxetine treatment with reductions in CSF concentrations of corticotropin-releasing hormone and arginine vasopressin in patients with major depression. Am J Psychiatry 150:656–657
Ehlers CL, Somes C, Seifritz E, Rivier JE (1997) CRF/NPY interactions: a potential role in sleep dysregulation in depression and anxiety. Depress Anxiety 6:1–9
Ehrensing RH, Kastin AJ (1974) Melanocyte-stimulating hormone-release inhibiting hormone as an antidepressant: a pilot study. Arch Gen Psychiatry 30:63-65
Ehrensing RH, Kastin AJ (1978) Dose-related biphasic effect of prolyl-leucyl-glycinamide (MIF-1) in depression. Am J Psychiatry 135:562-566
Fehder WP, Sachs J, Uvaydova M, Douglas SD (1997) Substance P as an immune modulator of anxiety. Neuroimmunomodulation 4:42–48
Feighner JP, Ehrensing RH, Kastin AJ, Leonard BE, Sverdlov L, Nicolau G, Patel SA, Hlavka J, Abajian H, Noble JF (2000a) A double-blind, placebo-controlled efficacy, safety, and

pharmacokinetic study of INN 00835, a novel antidepressant peptide, in the treatment of major depression. J Affect Disord 61:119-126

Feighner JP, Sverdlov L (2000b) Cluster analysis of clinical data to identify subtypes within a study population following treatment with a new pentapeptide antidepressant. International Journal of Neuropsychopharmacology. 3:237-242.

Feighner JP (2001a) Nemifitide(INN 00835), a novel pentapeptide, in severe treatment-resistant depression. 2nd international forum on mood and anxiety disorders, Monte Carlo, Nov 28-Dec 01, 2001

Feighner JP, Ehriensing RH, Kastin AJ, Patel A, Sverdlov L, Hlavka J, Abajian HB, Noble JF, Nicolau G (2001b) Double-blind, placebo-controlled study of INN 00835 (nemifitide) in the treatment of outpatients with major depression. Int Clin Psychopharmacol 16:345-352

Feighner JP, Nicolau G, Abajian H, Marricco NC, Morrison J, Sverdlov L, Hlavka J, Tonelli G Jr., Di Spirito C, Faria G (2002a) Clinical pharmacokinetic studies with INN 00835 (nemifitide), a novel pentapeptide antidepressant. Biopharm Drug Dispos. 23:33-39.

Feighner JP, Sverdlov L (2002b) The use of discriminant analysis to separate a study population by treatment subgroups in a clinical trial with a new pentapeptide antidepressant. The Journal of Applied Research. Vol. 2, 1:50-57.

Feighner JP (2003a) Clinical and preclinical overview of nemifitide (INN 00835), a novel pentapeptide antidepressant. EuroΩConference, Depression: Emerging Research and Treatment Approaches, Paris, January 16-17. Extensive Abstract, Lecture #9.

Feighner JP, Sverdlov L, Nicolau G, Abajian H, Hlavka J, Freed J, Tonelli G Jr. (2003b) Clinical effectiveness of nemifitide, a novel antidepressant, in depressed outpatients: comparison of follow-up re-treatment with initial treatment. International Journal of Neuropsychopharmacology. (In press.)

Ferris CF, Delville Y (1994) Vasopressin and serotonin interactions in the control of agonistic behavior. Psychoneuroendocrinology 19:593-601

Fischer PA, Schneider E, Jacobi P, Maxion H (1974) Effect of melanocyte-stimulating hormone release inhibiting factor(MIF) in Parkinson's syndrome. Eur Neurol 12:360-368

Frasch A, Zetzsche T, Steiger A, Jirikowski GF (1996) Reduction of plasma oxytocin levels in patients suffering from major depression. Adv Exp Med Biol 395:257-258

Fuxe K, Jansson A, Diaz-Cabiale Z, Andersson A, Tinner B, Finnman UB, Misane I, Razani H, Wang FH, Agnati LF, Ogren SO (1998) Galanin modulates 5-hydroxytryptamine functions: focus on galanin and galanin fragment/5-hydroxytryptamine 1A receptor interactions in the brain. Ann NY Acad Sci 863:274-290

Gaillard RC, Riondel AM, Ling N, Muller AF (1988) Corticotropin releasing factor activity of CRF 41 in normal man is potentiated by angiotensin II and vasopressin but not by desmopressin. Life Sci 43:1935-1944

Gavioli EC, Canteras NS, De Lima TCM (1999) Anxiogenic-like effect induced by substance P injected into the lateral septal nucleus. Neuroreport 10:3399-3403

Gibbs DM, Vale W (1983) Effect of the serotonin reuptake inhibitor fluoxetine on corticotropin-releasing factor and vasopressin secretion into hypophysial portal blood. Brain Res 280:176-179

Grigoriadis DE, Lovenbert TW, Chalmers DT, Liaw C, De Souza EB (1996) Characterization of corticotropin releasing factor receptor subtypes. Ann. NY Acad Sci 780:60-80

Guy W (1976) ECDEU Assessment Manual for Psychopharmacology-Revised (DHEW Publ No ADM 76-338). U.S. Department of Health, Education, and Welfare, Public Health Service, Alcohol, Drug Abuse, and Mental Health Administration, NIMH Psychopharmacology Research Branch, Division of Extramural Research Programs, Rockville, MD, pp 218-222

Hamilton M (1960) A rating scale for depression. J Neurol Neurosurg Psychiatry 23:56-62

Hasenohrl RU, Jentjens O, De Souza Silva MA, Tomaz C, Huston JP (1998) Anxiolytic-like action of neurokinin substance P administered systemically or into the nucleus basalis magnocellularis region. Eur J Pharmacol 354:123-133

Hauger RL, Aguilera G (1993) Regulation of pituitary corticotropin releasing hormone (CRH) receptors by CRH: interaction with vasopressin. Endocrinology 133:1708-1714

Heuser I, Bissette O, Dettling M, Schweiger U, Gotthardt U, Schmider J, Lammers C-H, Nemeroff CB, Holsboer F (1998) Cerebrospinal fluid concentrations of corticotropin-releasing hormone, vasopressin, and somatostatin in depressed patients and healthy controls: response to amitriptyline treatment. Depress Anxiety 8:71-79

Hlavka JJ, Nicolau G, Noble JF, Abajian H (1997) INN 00835: antidepressant. Drugs of the Future 22:1314-1318

Hlavka JJ, Abajian H. Morrison J. Overstreet D, Kelly J, Nicolau G, Feighner JP (2002) Chemistry and pharmacology of nemifitide, a novel antidepressant peptide. The International Journal of Neuropsychopharmacology, Vol. 5, Suppl. 1, S97. Abstracts from XXIII CINP Congress, Montreal, June 23-27

Holsboer F (1999) The rationale for corticotropin-releasing hormone receptor (CRH-R) antagonists to treat depression and anxiety. J Psychiatr Res 33:181-214

Hoogendijk WJG, Meynen G, Eikelenboom P, Swaab DF (2000) Brain alterations in depression. Acta Neuropsychiatrica 12:54-58

Husum H, Mikkelsen JD, Hogg S, Mathe AA, Mork A (2000) Involvement of hippocampal neuropeptide Y in mediating the chronic actions of lithium, electroconvulsive stimulation and citalopram. Neuropharmacology 39:1463-1473

Iizuka H, Kishimoto A, Nakamura J, Mizukawa R (1996) Clinical effects of cortisol synthesis inhibition on treatment-resistant depression. Japanese Journal of Psychopharmacology 16: 33-36

Kakigi T, Maeda K (1992) Effect of serotonergic agents on regional concentrations of somatostatin- and neuropeptide Y-like immunoreactivities in rat brain. Brain Res 599:45-50

Kask K, Berthold M, Bartfai T (1997) Galanin receptors: involvement in feeding, pain, depression and Alzheimer's disease. Life Sci 60:1523-1533

Kastin AJ, Barbeau A (1972) Preliminary clinical studies with l-prolyl-l-leucyl-glycine amide in Parkinson's disease. Canad Med Assoc J 107:1079-1081

Kastin AJ, Ehrensing RH, Schalch DS, Anderson MS (1972) Improvement in mental depression with decreased thyrotropin response after administration of thyrotropin-releasing hormone. Lancet 2:740-742

Kastin AJ, Plotnikoff NP, Sandman CA, et al. (1975) The effects of MSH and MIF on the brain. In: Stumpt WE. Grant LD (eds) Anatomical neuroendocrinology. S Karger, Basel, pp 290-297

Kastin AJ, Abel DA, Ehrensing RH, Coy DH, Graf MV (1984) Tyr-MIF-1 and MIF-1 are active in the water wheel test for antidepressant drugs. Pharmacology, Biochemistry and Behavior 21:767-771

Kelly JP, Nicolau G, Redmond A, Leonard BE, Noble J, Sverdlov L, Molinar R, Kastin AJ, Ehrensing RH, Feighner JP (1999) The effect of treatment with a new antidepressant, INN 00835, on platelet serotonin uptake in depressed patients. J Affect Disord 55:231-235

Kelly J, Harkin A, Nicolau G, Feighner JP, Leonard BE (2002) A preclinlical investigation of the mechanism of action of nemifitide: a possible role for serotonin? The International Journal of Neuropsychopharmacology, Vol. 5, Suppl. 1, S97. Abstracts from XXIII CINP Congress, Montreal, June 23-27

Kling MA, Geracioti TD, Licinio J, Michelson D, Oldfield EH, Gold PW (1994) Effects of electroconvulsive therapy on the CRH-ACTH-cortisol system in melancholic depression: preliminary findings. Psychopharmacol Bull 30:489-494

Kramer MS, Cutler N, Feighner J, Shrivastava R, Carman J, Sramek JJ, Reines SA, Liu G, Snavely D, Wyatt-Knowles E, Hale JJ, Mills SG, MacCoss M, Swain CJ, Harrison T, Hill RG, Hefti F, Scolnick EM, Cascieri MA, Chicchi GG, Sadowski S, Williams AR, Hewson L, Smith D, Carlson EJ, Hargreaves RJ. Rupniak NMJ (1998) Distinct mecha-

nism for antidepressant activity by blockade of central substance P receptors. Science 281:1640-1645
Leff MJ, Roatch JF, Bunney WE (1970) Environmental factors preceding the onset of severe depression. Psychiatry 33:293-311
Lloyd C (1980) Life events and depressive disorders reviewed. II. Events as predisposing factors. Arch Gen. Psychiatry 37:529-535
Mathe AA (1999) Neuropeptides and electroconvulsive treatment. J ECT 15:60-75
Mathe AA, Jimenez PA, Theodorsson E, Stenfors C (1998) Neuropeptide Y, neurokinin A and neurotensin in brain regions of Fawn Hooded 'depressed', Wistar, and Sprague Dawley rats: effects of electroconvulsive stimuli. Prog Neuropsychopharmacol Biol Psychiatry 22:529-546
Mitchell AJ(1998) The role of corticotropin releasing factor in depressive illness: a critical review. Source Neuroscience & Biobehavioral Reviews 22:635-651
Montgomery SA, Asberg M (1979) A new depression scale designed to be sensitive to change. Br J Psychiatry 134:382-389
Munglani R, Hudspith MJ, Hunt SP (1996) The therapeutic potential of neuropeptide Y: analgesic, anxiolytic and antihypertensive. Drugs 52:371-389
Murphy REP, Dhar V,. Ghadirian AM, Chouinard G, Keller, R (1991) Response to steroid suppression in major depression resistant to antidepressant therapy. J Clin Psychopharmacol 11:121-126
Nair RM, Kastin AJ, Schally AV (1971) Isolation and structure of hypothalamic MSH-release inhibiting hormone. Biochem Biophys Res Commun 43:1376-1381
Nemeroff CB, Bissette G, Akil H, Fink M (1991) Neuropeptide concentrations in the cerebrospinal fluid of depressed patients treated with electroconvulsive therapy: corticotrophin-releasing factor, beta-endorphin and somatostatin. Br J Psychiatry 158:59-63
Nicolau G, Feighner JP, Reimer M, Li A, Hu Z, Hlavka J (2002) Metabolism of the antidepressant nemifitide and its potential for drug interactions. The International Journal of Neuropsychopharmacology, Vol. 5, Suppl. 1, S204. Abstracts from XXIII CINP Congress, Montreal, June 23-27
Nicolau G, Brown N, Reimer M, Tsikos T, Guilbaud R, Hlavka J, Abajian H, Morrison J, Feighner JP (2001) Metabolism of an antidepressant pentapeptide: INN 00835 in liver, in-testinal and brain preparations. Drug Metabolism Reviews. Vol.33, p.99. Abstracts from the 6th International ISSX Meeting, October 7-11, 2001, Munich, Germany
Pariante CM, Nemeroff CB, Miller AH (1995) Glucocorticoid receptors in depression. Isr J Med Sci 31:705-712
Perraudin V, Delarue C, Lefebvre H, Contesse V, Kuhn J-M, Vaudry H (1993) Vasopressin stimulates cortisol secretion from human adrenocortical tissue through activation of V1 receptors. J Clin Endocrinol Metab 76:1522-1528
Pignatiello MF, Olson GA, Kastin AJ, Eherensing RH, McLean JH, Olson RD (1989) MIF-1 is active in a chronic stress animal model of depression. Pharmacol Biochem Behav 32:737-742
Plotnikoff NP, Kastin AJ, Anderson MS, Schally AV (1971) DOPA potentiation by a hypothalamic factor MSH release inhibiting hormone (MIF). Life Sci 10:1279-1283
Plotnikoff NP, Kastin AJ, Anderson MS, Schally AV (1972) Oxotremorine antagonism by a hypothalamic hormone: melanocyte stimulating hormone release-inhibiting factor (MIF). Proc Soc Exp Biol Med 140:811-814
Plotnikoff NP, Kastin AJ, Anderson MS, Schally AV (1973) Deserpidine antagonism by a tripeptide, l-prolyl-l -leucyl glycinamide. Neuroendocrinology 11:67-71
Plotnikoff NP, Minard FN, Kastin AJ (1974) DOPA potentiation in ablated animals and brain levels of biogenic amines in intact animals after prolye-leucyl-glycinamide. Neuroendocrinology 14:271-279
Prange AJ, Wilson IC (1972) Thyrotropin releasing hormone(TRH) for immediate relief of depression: a preliminary report. Psychopharmacologia 25(suppl):82

Prange AJ Jr, Lara PP, Wilson IC, Alltop LB, Breese GR (1972) Effects of thyrotropin-releasing hormone in depression. Lancet 2:990-1002

Purba JS, Hoogendijk WJG, Hofman MA, Swaab DF (1996) Increased number of vasopressin- and oxytocin-expressing neurons in the paraventricular nucleus of the hypothalamus in depression. Arch Gen Psychiatry 53:137-143

Redding TW, Kastin AJ, Nair RM, Schally AV. Distribution, half-life, and excretion of 14 C- and 3 H-labeled l-prolyl-l-leucyl-glycinamide in the rat. Neuroendocrinology. 1973;11:92-100

Redding TW, Kastin AJ, Gonzalez-Barcena D, et al. (1976) The disappearance, excretion, and metabolism of tritiated prolyl-L-Leucyl-glycinamide in man. Neuroendocrinol 16:119-126

Redmond AM, Leonard BE (1997) An evaluation of the role of the noradrenergic system in the neurobiology of depression: a review. Human Psychopharmacology 12:407-430

Sanglard M, Tavares A (1998) Neuropeptide Y and psychiatry. J Bras Psiquiatr 47:441-444

Saydoff JA, Carnes M, Brownfield MS (1993) The role of serotonergic neurons in intravenous hypertonic saline-induced secretion of vasopressin, oxytocin, and ACTH. Brain Res Bull 32:567-572

Scott LV, Dinan TG (1998) Vasopressin and the regulation of hypothalamic-pituitary-adrenal axis function: Implications for the pathophysiology of depression. Life Sci 62:1985-1998

Sergeyev V, Hokfelt T, Hurd Y (1999) Serotonin and substance P co-exist in dorsal raphe neurons of the human brain. Neuroreport 10:3967-3970

Stogner KA, Holmes PV (2000) Neuropeptide-Y exerts antidepressant-like effects in the forced swim test in rats. Eur J Pharmacol 387:R9-R10

Swaab DF, Hofman MA, Lucassen PJ, Purba JS, Raadsheer FC, Van de Nes JA (1993) Functional neuroanatomy and neuropathology of the human hypothalamus. Anat Embryol Berl 187:317-330

Thakore JH, Dinan TG (1995) Cortisol synthesis inhibition: a new treatment strategy for the clinical and endocrine manifestations of depression. Biol Psychiatry 37:364-368

Thomson K, Hendric H.(1972) Environmental stress in primary depressive illness. Arch Gen Psychiatry 26:130-132

Turner RA, Altemus M, Enos T, Cooper B, McGuinness T (1999) Preliminary research on plasma oxytocin in normal cycling women: investigating emotion and interpersonal distress. Psychiatry 62:97-113

Uvnas-Moberg K (1998) Oxytocin may mediate the benefits of positive social interaction and emotions. Psychoneuroendocrinology 23:819-835

Uvnas-Moberg K, Ahlenius S, Hillegaart V, Alster P (1994) High doses of oxytocin cause sedation and low doses cause an anxiolytic-like effect in male rats. Pharmacol Biochem Behav 49:101-106

Van der Velde CD (1983) Rapid clinical effectiveness of MIF-1 in the treatment of major depressive illness. Peptides 4:297-300

Van Londen L, Goekoop JG, Van Kempen GMJ, Frankhuijzen-Sierevogel AC, Wiegant VM, Van der Velde EA, De Wied D (1997) Plasma levels of arginine vasopressin elevated in patients with major depression. Neuropsychopharmacology 17:284-292

Van Londen L, Kerkhof GA, Van den Berg F, Goekoop JG, Zwinderman KH, Frankhuijzen-Sierevogel AC, Wiegant VM, De Wied D (1998) Plasma arginine vasopressin and motor activity in major depression. Biol Psychiatry 43:196-204

Volpi R, Chiodera P, Caiazza A, Magotti MG, Caffarri G, Coiro V (1995) Inhibition by oxytocin of endothelin-1-induced ACTH secretion in normal men. Neuroendocrinology Letters 17:289-294

Weiss JM, Stout JC, Aaron MF, Quan N, Owens MJ, Butler PD, Nemeroff CB (1994) Depression and anxiety: role of the locus coeruleus and corticotropin- releasing factor. Brain Res Bull 35:561-572

Weiss JM, Bonsall RW, Demetrikopoulos MK, Emery MS, West CHK (1998) Galanin: a significant role in depression? Ann NY Acad Sci 863:364–382

Widerlov E, Lindstrom LH, Wahlestedt C, Ekman R (1988) Neuropeptide Y and peptide YY as possible cerebrospinal fluid markers for major depression and schizophrenia, respectively. J Psychiatr Res 22:69–79

Willner P (1990) Animal models of depression: an overview. Pharmacol Ther 45:425-455

Willner P (1984) The validity of animal models of depression. Psychopharmacol 83:1-16

Windle RJ, Shanks N, Lightman SL, Ingram CD (1997) Central oxytocin administration reduces stress-induced corticosterone release and anxiety behavior in rats. Endocrinology 138:2829–2834

Zobel AW, Nickel T, Kunzel HE, Ackl N, Sonntag A, Ising M, Holsboer F (2000) Effects of the high-affinity corticotropin-releasing hormone receptor 1 antagonist R121919 in major depression: the first 20 patients treated. J Psychiatr Res 34:171–181

Role of Pharmacogenetics/Pharmacogenomics in the Development of New Antidepressants

S. H. Preskorn[1]

[1] Department of Psychiatry and Behavioral Sciences,
University of Kansas School of Medicine and Psychiatric Research Institute,
1010 N. Kansas, Wichita, KS 67214, USA
e-mail: spreskor@kumc.edu

1	Introduction	584
2	The Human Genome Project	585
3	The Stages of Drug Development	586
4	The SSRIs as an Example of the Changing Landscape of Drug Discovery	586
5	A New Era of Drug Discovery	587
6	Drug Discovery in Psychiatry and Neurotransmitters	588
6.1	Neuropeptides	590
6.2	Genetic Approaches to the Identification of Transmitters and Receptors	592
6.3	Orphan Receptors and Reverse Pharmacology	592
6.4	Selecting the Best Targets	593
6.5	Challenges Associated with These New Developments	594
References		594

Abstract The three chapters in this last section of the book are devoted to where the future of antidepressants may lie. Nalepa and Sulser discussed emerging new theories about the mechanisms that may underlie antidepressant efficacy. Such theories are guiding current drug development. Garlapati and colleagues discussed drugs currently in development for which information is in the public domain. This chapter is focused the furthest in the future and describes processes that will define new targets and mechanisms that might mediate antidepressant efficacy. These processes are the result of advances being made in molecular biology and neuroscience, which are leading to an improved understanding of the biological processes that subserve the regulation of higher brain function, including cognition, emotion, and sensory processing. Disturbances in these functions constitute psychiatric signs and symptoms. Given its focus on process, this chapter is as relevant to the development of drugs for psychotic, anxiety, and dementing illnesses as for affective illnesses, even though the focus of this book is antidepressants. This chapter describes how new findings from the human genome project will change the field of psychiatric drug discovery. The author first reviews the status of the human genome project and considers how findings from that project will (1) increase the number of potential targets for drug discovery, (2) help researchers understand the mechanisms that determine the drug concentration that is achieved on a given dose of a drug, and (3) help researchers determine the causes of variability in patients' responses to a given dose of a drug. New findings concerning neuropeptides and other neurotransmitters that appear to be promising future targets for drug discovery are reviewed. The chapter describes the use of molecular biological approaches (e.g., transfecting sequences into cell lines or single cell organisms that do not normally contain them and studying their expression) to identify new receptors and neurotransmitters. The chapter also describes how the human genome is being searched for orphan receptors that can be used as "bait" to troll for and identify natural ligands (neurotransmitters), enabling researchers to discover biological processes that were either previously unknown or poorly understood and to map the distribution of neurotransmitters and receptors in the brain and identify their functions. Such knowledge will help pinpoint potential new targets (i.e., regulatory proteins) for drug discovery. The chapter concludes with a consideration of the challenges these new developments pose for researchers and clinicians. With the ever-expanding pool of potential targets for drug discovery, researchers need to conserve resources by carefully identifying targets that appear likely to have the greatest potential clinical utility. With the development of increasingly focused and targeted drugs that are likely to affect only the brain rather than peripheral systems, prescribers of antidepressants and other psychotropic drugs will need to be increasingly aware of the behavioral effects of the medications they prescribe as well as of the potential interactions that may occur with the increasing use of multiple, narrowly targeted medications.

Keywords Human genome project · Rational drug development · Drug discovery · Orphan receptors · High-throughput screening · Structure–activity relationships · Molecular targeting · Stages of drug development · Reverse pharmacology · Bridging studies

1
Introduction

As discussed in earlier chapters on the selective serotonin reuptake inhibitors and other newer antidepressants (see the chapters by Preskorn et al. and by Preskorn and Ross), the field of psychotropic drug development has moved from chance discovery to drug development based on targeted rational exploration (Preskorn 2001b). As far back as antiquity, the health-enhancing properties of certain herbs and other natural products had been observed. Such observations included the mind-altering properties of substances such as alcohol, opium, and other plant products (Preskorn 2001c). The first organized period of drug development dates to the last half of the nineteenth century and was built on a series of chance observations. The process of drug development accelerated significantly with Fleming's chance discovery of penicillin. During the first half of the twentieth century, scientists focused on altering the structure of drugs that had been discovered by chance. The next phase of drug development, which began in the late 1950s, continues today. It involved the use of receptor binding studies and other in vitro techniques to refine structure–activity relationships in order to synthesize compounds that had a specific desired neural mechanism(s) while avoiding other, undesirable mechanisms of action. The third phase of drug development began in the early 1990s and has involved the use of techniques derived from molecular biology to discover completely new sites of action (targets) for drug development in psychiatry. This process is receiving an enormous impetus from ongoing work made possible by the human genome project. This final chapter of this handbook reviews how exciting new developments in our understanding of human genetics are likely to revolutionize the development of medications to treat depression and other mental disorders.

2
The Human Genome Project

Although the number has not yet been determined with certainty, there are an estimated 40,000–50,000 genes in the human genome (Preskorn 2001c). Each gene codes for a protein, and these proteins may be structural (e.g., collagen) or functional (i.e., regulatory), such as enzymes and receptors. Although the human genome project has now sequenced the entire human genome in several individuals, only perhaps 10% of the genes that code for human proteins have so far been identified. One goal of the next phase of the human genome project is to identify the approximately 40,000 remaining genes and the structural or regulatory proteins they produce (Preskorn 2000a,b). Another goal will be to identify mutations in these genes that are biologically meaningful—i.e., that may represent disease mechanisms or may pharmacodynamically or pharmacokinetically influence drug action (Preskorn 2000b).

After a new human gene has been identified, researchers can then deduce the amino acid sequence of the protein from the nucleotide sequence. The three-dimensional configuration of the protein can then be deduced, which in turn will define the structure–activity relationship needed to affect that target.

The targets of drug development are primarily regulatory proteins, in particular receptors. Three different types of drugs can potentially be developed for a regulatory protein such as a receptor. The drug can be (1) an agonist, (2) an antagonist, or (3) an inverse agonists. Agonists have the same effect on the receptor as the endogenous neurotransmitter (i.e., they turn the receptor on); antagonists occupy but do not activate the receptor and thus block the action of the endogenous neurotransmitter (i.e., put the receptor in neutral); and inverse agonists have an effect on the receptor opposite to that of the endogenous neurotransmitter.

3
The Stages of Drug Development

A brief review of the stages of drug development may help readers follow the subsequent discussion of the implications of the human genome project for the future of drug development. There are six major phases of drug development:

1. Drug discovery: identifying compounds that appear promising for preclinical testing
2. Preclinical testing: in vitro testing, testing in isolated organs, animal studies
3. Phase I studies: testing in normal volunteers and mildly symptomatic volunteers (often referred to as "bridging studies")
4. Phase II studies: establishing the efficacy of the agent for the condition being studied and determining the optimal dose to be used in later studies (may be open-label or double-blind, uncontrolled or placebo-controlled, generally involving fairly small samples, short-term)
5. Phase III studies: accumulating sufficient evidence for FDA approval, obtaining supportive data on efficacy in long-term maintenance treatment, and increasing the database on human exposure
6. Phase IV studies: postmarketing studies of varied designs

4
The SSRIs as an Example of the Changing Landscape of Drug Discovery

When the selective serotonin reuptake inhibitors (SSRIs) were developed, the genes coding for the cytochrome P450 (CYP) enzymes had not yet been identified and the drugs could therefore not be screened against such targets. For this reason, three of the SSRIs (fluoxetine, fluvoxamine, and paroxetine) that cause substantial inhibitory effects on CYP enzymes would probably not have been developed today. If these drugs were in development today, the medicinal chemists in the drug discovery departments would be advised to modify the structure of these drugs to affect the desired target (the serotonin receptor) without affecting these CYP enzymes.

That these three highly successful agents would likely not make it into clinical testing today—much less be marketed—is illustrative of the continued refinement in the ability to test and develop structure–activity relationships in an iterative fashion, which can produce truly selective, new chemical entities. The term "structure" refers to the molecular configuration needed to affect a specific regulatory protein in a specific way (e.g., as an agonist, antagonist, or inverse agonist). With the completion of the human genome project and the resultant increased knowledge of brain-specific enzymes, this process will doubtless accelerate still further. In other words, the human genome project will yield both more targets to hit *and* more targets to avoid (Preskorn 2000a).

5
A New Era of Drug Discovery

The development of the SSRIs and other newer antidepressants was the result of a process of refining existing treatments (i.e., of finding antidepressant agents that would share the desired effects of the tricyclic antidepressants and monoamine oxidase inhibitors without their undesirable side effects). In other words, the newly developed agents to date share the same mechanisms of action as the older treatments, although in a refined form. With the elucidation of the remaining 90% of genes that code for biologically relevant proteins in humans, the process of drug discovery will undergo radical changes.

As discussed in earlier chapters of this volume, the effect of a drug is determined by the following equation:

Effect = potency at site of action × drug concentration × biological variance
        (Variable 1)   (Variable 2)   (Variable 3)

The findings from the human genome project are relevant to all three variables in this equation.

First, the findings from the human genome project have the potential to enable researchers to define truly novel mechanisms of action (Kuhlmann 1999; McCarthy and Hilfiker 2000) based on newly discovered regulatory proteins (e.g., enzymes, receptors) coded for by human genes. Each regulatory protein could represent a potentially useful target for drug action and a novel mechanism of action (Landro et al. 2000). Thus, the project will yield a multitude of new targets for drug development (Variable 1). Based on findings to date, perhaps as many as 5,000 human genes code for brain-specific proteins. If half of those code for regulatory proteins, that represents 2,500 targets for which up to three different types of drugs could be developed: agonist, antagonist, and inverse agonist. That translates into thousands of potential brain-specific, mechanism-based classes of drugs (Preskorn 2000a).

Second, the findings from the human genome project will help researchers understand the mechanisms that determine the drug concentration that is achieved on a given dose of a drug (Variable 2).

Third, the findings from the project will help researchers determine why some patients respond to a given dose of a drug while others do not and why still others have toxic side effects at that same dose (Preskorn 1998) (Variable 3). Up to now, the third variable in the equation has primarily been the focus of the later phases of drug development. For example, phase III studies may examine the pharmacodynamics and pharmacokinetics of the agent in special populations, such as patients with liver or renal impairment, or young or old patients. However, the knowledge being gained from the human genome project will lead to earlier, more extensive, and more specific investigation of how human biological variance is likely to modify the expression of a drug's effect (Preskorn 2000b).

The goal of drug development is to reduce uncertainty about both the good and bad effects of a drug. The findings that are becoming available from the human genome project will improve researchers' ability to detect potential liability problems early in the drug development process, which will allow companies to discontinue development on a problematic agent and design a structural analog with a safer profile. That is important because it costs approximately $500 million to successfully bring a new drug to market in the United States. This ability to identify potential problems early in the drug development process should lead to significant reductions in costs (Bierrum 2000; Reichert 2000).

Once the gene that codes for a regulatory protein has been identified, it can be isolated and transfected into a single-cell organism or a cell line. These transfected cells will express the regulatory protein of interest and can thus be used to refine the structure–activity relationship to produce an agent that selectively affects that target. Today, an automated assay process known as high-throughput screening enables pharmaceutical developers to screen very large libraries of molecules against a large number of clinically important human proteins in a relatively short time (e.g., 500,000–1,000,000 compounds in approximately 2 months) (Bu et al. 2000; Preskorn 2000a). The results of this screening enable medicinal chemists to refine structure–activity relationships in order to synthesize compounds for preclinical testing that have a high affinity for the desired target and low affinity for undesired targets (Kuhlmann 1999; Johnson and Wolfgang 2000; Panchagnula and Thomas 2000; Preskorn 2000a). Such drug design can be done concurrently with work to elucidate the specific function of a newly discovered regulatory protein (e.g., a receptor).

6
Drug Discovery in Psychiatry and Neurotransmitters

Three types of regulatory proteins are the most common sites of action for psychiatric drugs:

- Enzymes involved in the synthesis or degradation of specific neurotransmitters
- Receptors that are the targets of specific neurotransmitters
- Uptake pumps that conserve specific neurotransmitters

Since these mechanisms of neurotransmission are crucial to the organization and function of the brain, drugs that alter these sites of action can influence specific areas of brain functioning. As noted above, the human genome project will enable researchers to identify new neurotransmitters and the regulatory proteins associated with them. Although slightly more than 70 neurotransmitters have been identified in mammalian brain to date (Table 1), most research on clinical psychopharmacology today focuses on only six of these neurotransmitters.

Table 1 Transmitters in the mammalian brain (copyright Preskorn 2003)

Amines
  Acetylcholine
  Dopamine
  Epinephrine
  Histamine
  Norepinephrine
  Serotonin
Amino Acids
  γ-Aminobutyric acid
  Glycine
  l-glutamate
Neuropeptides
  Adrenocorticotrophic hormone (ACTH)
  Adrenomedullin
  Amylin
  Angiotensin II
  Apelin
  Bradykinin
  Calcitonin
  Calcitonin gene-related peptide (CGRP)
  Cholecystokinin (CCK)
  Corticotropin releasing factor (CRF) (urocortin)
  Dynorphins, neoendorphins
  Endorphins, [lipotropic hormones (LPHs)]
  Endothelins
  Enkephalins
  Follicle stimulating hormone (FSH)
  Galanin
  Gastric inhibitory peptide (GIP)
  Gastrin
  Gastrin releasing peptide
  Glucagon-like peptides (GLPs)
  Gonadotropin releasing hormone (GnRH)
  Growth hormone-releasing factor (GHRF)
  Lipotropin hormone (LPH)
  Luteinizing hormone (LH)
  Melanin concentrating hormone (MCH)
  Melanin stimulating hormone (MSH)
  Motilin
  Neurokinins
  Neuromedins
  Neurotensin (NT)
  Neuropeptide FF (NPFF)
  Neuropeptide Y (NPY)
  Orexins/hypocretins
  Orphanin
  Oxytocin
  FQ/nociceptin
  Pituitary adenylate cyclase activating polypeptide (PACAP)
  Pancreatic polypeptide (PP)
  Peptide histidine isoleucine (PHI)

Table 1 (continued)

Parathyroid hormone (PTH)
Peptide YY (PYY)
Prolactin releasing peptide (PrRP)
Secretin/PHI
Somatostatin (SS) (cortistatin)
Tachykinins
Thyroid stimulating hormone (TSH)
Thyroid releasing hormone (TRH)
Urotensin II
Vasopressin
Vasoactive intestinal peptide (VIP)
Others
Adenosine
Adenosine triphosphate
Anandamide (arachidonylethanolamide)
Arachidonic acid
Nitric oxide

6.1
Neuropeptides

Among the neurotransmitters, there is great interest in the neuropeptides as potential targets for drug discovery. These short sequences of amino acids exert trophic actions and appear to serve developmental and adaptive functions (Fontaine et al. 1987; Hokfelt 1991; Strand et al. 1991; Gressens et al. 1993). Based on these findings, neuropeptides may play an important role in the pathophysiology of a number of psychiatric disorders, including bipolar disorder and schizophrenia.

Increased knowledge of the trophic function of neuropeptides suggests the potential for somatic therapies that can influence biological processes that are far more complex than simply synaptic transmission, including the plasticity of the brain itself. The brain, after all, is designed to adapt to its environment. (i.e., "learn"). New targets defined by the human genome project are likely to include mechanisms that mediate such plasticity, which represents the ability to reshape the brain in response to environmental input (i.e., the basic mechanisms underlying learning).

The neurobiology underlying some specific psychiatric disorders may turn out to be due more to fundamental deficits in plasticity than to deficits in classic transmitter–receptor interactions, even though the latter has been the dominant concept/model guiding psychiatric drug development during the last half of the twentieth century. While that approach has been fruitful and has yielded many effective medications, it may not truly be addressing the fundamental nature of illnesses such as major depression, but instead simply ameliorating the signs and symptoms of the illness. If deficiency in plasticity is the primary problem, then the identification of these mechanisms has the potential to change the approach to psychiatric treatment as radically as did the chance discovery of the monoamine oxidase inhibitors and tricyclic antidepressants (see the chapters by Lader and Kennedy et al., this volume).

If this happens, then it is probable that such treatments will make psychotherapeutic approaches both more effective and more necessary (Preskorn 2001a). While psychopharmacology and psychotherapy have frequently been viewed as an either/or phenomenon in the last half of the twentieth century, that may well change as drugs are developed which enhance the ability of the brain to re-shape itself in response to environmental input such as various specific forms of psychotherapy. In this model, the psychiatric medications will address hardware problems underlying specific psychiatric disorders. However, these approaches will be of limited value without concomitant software programming (psychotherapy) to take advantage of the enhanced brain plasticity made possible by the medication (hardware) intervention.

Neuropeptides are just one group of neuronal trophic and differentiation factors that are involved in the maturation and reshaping of the brain. Other proteins (i.e., gene products) that have been identified in the human brain are listed in Table 2. These proteins also exert potent and specific effects on neurotransmitter systems in the brain and thus have the potential to modify behavior, so that they are also potential targets for drug discovery.

Table 2 Neuronal differentiation and growth factors in mammalian brain (copyright Preskorn 2003)

Neurotrophins
Brain-derived neurotrophic factor (BDNF)
Nerve growth factor (NGF)
Neurotrophins 3 and 4/5 (NT-3, NT-4/5)
Neuropoietic factors
Cholinergic differentiation factor/leukemia inhibitory factor (CDF/LIF)
Ciliary neurotrophic factor (CNTF)
Growth-promoting activity (GPA)
Interleukins 6 and 11 (IL-6, IL-11)
Oncostatin M (OSM)
Sweat gland factor (SGF)
Transforming Growth Factors (TGF) family
Activin A
Epidermal growth factor (EGF)
Glial-cell-line-derived neurotrophic factor (GDNF)
Transforming growth factors a and b (TGF-α, TGF-β)
Fibroblast Growth Factors (FGF) family
Acidic fibroblast growth factor (aFGF)
Basic fibroblast growth factor (bFGF)
Fibroblast growth factor-5 (FGF-5)
Insulin-like growth factors
Insulin
Insulin-like growth factor (IGF)
Others
Platelet-derived growth factors (PDGF)

6.2
Genetic Approaches to the Identification of Transmitters and Receptors

Genetic studies have the potential to accelerate the pace of neuropeptide discovery and clarify the structure of these compounds. Based on studies by Masu et al. (1987) and subsequent researchers in the late 1980s and early 1990s (Yokota et al. 1989; Hershey and Krause 1990; Tanaka et al. 1990; Meyerhof et al. 1993), it was found that many neuropeptide receptors share a common structure (being coupled to a G protein and having seven transmembrane spanning segments), suggesting that such receptors for different neuropeptides may belong to a supergene family of receptors (i.e., may be a genetic variation on a theme analogous to the case of the CYP enzymes). This superfamily of G-coupled receptors could thus be subdivided into smaller families based on the degree of shared sequence homology and function in the same way as are the CYP enzymes. These findings have raised the possibility of scanning the genome for DNA sequences that have a high probability of coding for previously unidentified neuropeptide receptors; such sequences could then be tested using molecular biological approaches (e.g., transfecting these sequences into cell lines or single cell organisms that do not normally contain them and studying their expression) to determine whether the expressed product is incorporated into the cell membrane in a fashion consistent with being a receptor and whether it is linked to a G protein. The investigation of the family of G-protein coupled receptors is producing a fruitful array of targets for drug discovery (Ballesteros et al. 2001; Brauner-Osborne et al. 2001; Dahl et al. 2001, 2002; Foord 2002; Gough 2001; Howard et al. 2001; Kenakin 2002; Kroeze et al. 2002; Lee et al. 2001; Sadee et al. 2001).

6.3
Orphan Receptors and Reverse Pharmacology

Proteins that are identified by these methods are referred to as "orphan" receptors, since they have no identified natural ligand (i.e., neurotransmitter) or function; however, such orphan receptors can be used to fill gaps in our knowledge using a technique that is sometime referred to as "reverse pharmacology" (Civelli et al. 1999; Preskorn 2001a). Traditionally, pharmacology has gone from first observing the function of an agent to identifying the neurotransmitter and receptor involved. However, this process can now be reversed with structure used to determine physiology (Civelli et al. 1998; Lembo et al. 1999; Sautel and Milligan 2000).

Orphan receptors can be used as "bait" to troll for and identify natural ligands (neurotransmitters) (Preskorn 2001a). This process may enable researchers to discover biological processes that were either previously unknown or poorly understood and to map the distribution of neurotransmitters and receptors in the brain and identify their functions. Such knowledge will help pinpoint potential new targets (i.e., regulatory proteins) for drug discovery. This is why researchers are currently searching the human genome for additional "orphan" G protein-coupled receptors (see preceding section).

The human genome project is making great progress in identifying genes that code for previously unknown receptors, so that the number of identified receptors (see Table 1) may double over the next 10 years. Hopefully, these developments will enable researchers to use the process of reverse pharmacology described above to gain a much greater understanding of normal physiological mechanisms in the brain as well as of the abnormal mechanisms that may underlie psychiatric disease.

6.4
Selecting the Best Targets

As discussed above, several superfamilies of brain receptors have been identified (e.g., the G protein-coupled receptors). These superfamilies of receptors demonstrate considerable homology in the sequence of their gene-encoded nucleotides and it is this homology that determines how they are classified and grouped within each family (Nelson et al. 1993; Hoyer et al. 1994). The more homologous the sequences of any two regulatory proteins, the more closely related they are likely to be, both genetically and evolutionarily (Wittenberger et al. 2001). That is because the genes that code for these proteins differ in the nucleotides that code for the different amino acids. These differences, which are the result of mutations in the common ancestral gene that evolved into the current genes in the genome for a species, affect the function of the regulatory protein, including what neurotransmitter will affect it. Since different receptor subtypes for a single neurotransmitter share considerable sequence homology and structure, a drug may affect several different members of the same receptor family even though the desired effect of the drug is related to (mediated by) its action on only one of these receptors.

In terms of the drug discovery process, knowing the specific amino acid sequence of a regulatory protein (e.g., a specific serotonin receptor) is analogous to knowing the tumbler system of a lock one wants to pick. The sequence determines the conformation of the receptor and where and how the drug can alter that conformation. Therefore, the increasing knowledge of the structures of receptor proteins that will be gained through the human genome project will enable researchers to refine structure–activity relationships to rationally develop new chemical entities that are selective for *only one* receptor subtype within a family (Preskorn 2001a,b).

Researchers also now have the ability to exchange a given amino acid in the sequence of a regulatory protein or a nucleotide in a gene and see how that alters the function or conformation of the regulatory protein—the resulting information can then be used to test and further refine structure–activity relationships for drug discovery.

6.5
Challenges Associated with These New Developments

As noted above, the advances brought by the human genome project are greatly expanding the pool of potential targets for drug discovery. Given limited resources, scientists in drug discovery must carefully select the targets that are most likely to be clinically useful, based on the presumed function the target subsumes in the brain.

In addition, as the pharmacology of psychiatric medications becomes more focused on and limited to neural mechanisms of action, the adverse effects associated with psychotropic medications are likely to primarily affect the brain (e.g., to present as behavioral problems such as confusion rather than as peripheral symptoms such as dry mouth or cardiac arrhythmias). That means that those who prescribe psychiatric medications will need to be accomplished behavioral pharmacologists who are sensitive to all the possible effects of a drug and able to take into account the effects and interactions of the multiple function-specific drugs that may need to be used to treat a single patient.

Finally, because the adverse effects of newly developed psychotropic agents are more likely to affect the human brain than other somatic systems, it will be important that researchers place more emphasis on testing the behavioral pharmacology of new drugs in volunteers and mildly symptomatic human subjects (that is, make increased use of "bridging studies") to identify potential effects in humans.

Acknowledgements. The author gratefully acknowledges permission from the Lippincott, Williams and Wilkins to adapt material that appeared in a series of columns in the *Journal of Psychiatric Practice* in 2000 and 2001.

References

Ballesteros JA, Shi L, Javitch JA (2001) Structural mimicry in G protein-coupled receptors: implications of the high-resolution structure of rhodopsin for structure-function analysis of rhodopsin-like receptors. Mol Pharmacol 60:1–19

Bierrum OJ (2000) New safe medicines faster: A proposal for a key action within the European Union's 6th Framework Programme. Pharmacol Toxicol 86(suppl 1):23-26

Brauner-Osborne H, Jensen AA, Sheppard PO, Brodin B, Krogsgaard-Larsen P, O'Hara P (2001) Cloning and characterization of a human orphan family C G-protein coupled receptor GPRC5D. Biochim Biophys Acta 1518:237–48

Bu HZ, Knuth K, Magis L, Teitelbaum P (2000) High-throughput cytochrome P450 inhibition screening via cassette probe-dosing strategy. IV. Validation of a direct injection on-line guard cartridge extraction/tandem mass spectrometry method for simultaneous CYP3A4, 2D6 and 2E1 inhibition assessment. Rapid Commun Mass Spectrom 14:1943–8

Civelli O, Nothacker HP, Reinscheid R. (1998) Reverse physiology: discovery of the novel neuropeptide, orphanin FQ/nocieceptin. Critical Reviews in Neurobiology 12:163–176

Civelli O, Reinscheid RK, Nothacker HP (1999) Orphan receptors, novel neuropeptides and reverse pharmaceutical research. Brain Res 848:63–65

Dahl SG, Edvardsen O, Kristiansen K, et al. (2001) Bioinformatics and receptor mechanisms of psychotropic drugs. Biotechnol Annu Rev 7:165–177

Dahl SG, Kristiansen K, Sylte I (2002) Bioinformatics: from genome to drug targets. Ann Med. 34:306–312

Fontaine B, Klarsfeld A, Changeux J-P (1987) Calcitonin gene-related pep-tide and muscle activity regulate acetycholine receptor alpha-subunit mRNA levels by distinct intracellular pathways. J Cell Biol 105:1337–1342

Foord SM (2002) Receptor classification: post genome. Curr Opin Pharmacol 2:561–566

Gough NR (2001) Signal transduction pathways as targets for therapeutics. Sci STKE 76:PE1

Gressens P, Hill JM, Gosez I, et al. (1993) Growth factor function of vasoac-tive intestinal peptide in whole cultured mouse embryos. Nature 362:155–158

Hershey AD, Krause JE (1990) Molecular characterization of a functional cDNA encoding the rat substance P receptor. Science 247:958–962

Hokfelt T (1991) Neuropeptides in perspective: the last ten years. Neuron 7:867–9

Howard AD, McAllister G, Feighner SD, et al. (2001) Orphan G-protein-coupled receptors and natural ligand discovery. Trends Pharmacol Sci 22:132–140

Hoyer D, Clarks D, Fozard J, et al. (1994) VII. International union of pharmacology classification of receptors for 5-hydroxytroptamine (serotonin). Pharmacol Rev 46:157–203

Johnson DE, Wolfgang GH (2000) Predicting human safety: screening and computational approaches. Drug Discov Today 5:445–54

Kenakin T (2002) Drug efficacy at G protein-coupled receptors. Annu Rev Pharmacol Toxicol 42:349–379

Kroeze WK, Kristiansen K, Roth BL (2002) Molecular biology of serotonin receptors structure and function at the molecular level. Curr Top Med Chem 2:507–528

Kuhlmann J (1999) Alternative strategies in drug development: Clinical pharmacological aspects. Int J Clin Pharmacol Ther 37:575–583

Landro JA, Taylor IC, Stirtan WG, et al. (2000) HTS in the new millennium: The role of pharmacology and flexibility. J Pharmacol Toxicol Methods 44:273–289

Lee DK, George SR, Cheng R, et al. (2001) Identification of four novel human G protein-coupled receptors expressed in the brain. Brain Res Mol Brain Res 86:13–22

Lembo PM, Grazzini E, Cao J, et al. (1999) The receptor for the orexigenic peptide melanin-concentrating hormone is a G-protein-coupled receptor. Nat Cell Biol 1:267–271

McCarthy JJ, Hilfiker R (2000) The use of single-nucleotide polymorphism maps in pharmacogenomics. Nat Biotechnol 18:505–508

Masu Y, Nakayama K, Tamaki H, et al. (1987) cDNA cloning of bovine substance-K receptor through oocyte expression system. Nature 329:836–838

Meyerhof W, Darlison MG, Richter D (1993) The elucidation of neuropeptide receptors and their subtypes through the application of molecular biology. In: Hucho F, ed. Neurotransmitter receptors. Elsevier, Amsterdam, pp. 335–353

Nelson DR, Kamataki T. Waxman DJ, et al. (1993) The P450 superfamily: Update on new sequences, gene mapping, accession numbers, early trivial names of enzymes, and nomenclature. DNA and Cell Biology 1993;12:1–51

Panchagnula R, Thomas NS (2000) Biopharmaceutics and pharmacokinetics in drug research. Int J Pharm 201:131–50

*Preskorn SH (1998) Why did Terry fall off the dose-response curve? J Pract Psychiatry Behav Health 4:363-367

*Preskorn SH (2000a) The human genome project and modern drug development in psychiatry. J Pract Psychiatry Behav Health 6:272-276

*Preskorn SH (2000b) The stages of drug development and the human genome project: drug discovery. J Pract Psychiatry Behav Health 6:341-344

*Preskorn SH (2001a) The human genome project and drug discovery in psychiatry: identifying novel targets. Journal of Psychiatric Practice 7:133–137

*Preskorn SH (2001b) Drug discovery in psychiatry: drilling down on the target of interest. Journal of Psychiatry Practice 7:267–272

*Preskorn SH (2001c) Drug development in psychiatry and the human genome project: the explosion in knowledge and potential targets. Journal of Psychiatry Practice 7:336–340

Reichert JM (2000) New biopharmaceuticals in the USA: Trends in development and marketing approvals 1995–1999. Trends Biotechnol 18:364–9

Sadee W, Hoeg E, Lucas J, et al (2001) Genetic variations in human G protein-coupled receptors: implications for drug therapy. AAPS PharmSci 3:E22

Sautel M, Milligan G (2000) Molecular manipulation of G-protein-coupled receptors: a new avenue into drug discovery. Curr Med Chem 1 889–896

Strand FL, Rose KJ, Zuccarelli LA, et al (1991) Neuropeptide hormones as neurotrophic factors. Physiol Rev 71:1017–1037

Tanaka K, Masu M, Nakanishi S (1990) Structure and functional expression of the cloned rat neurotensin receptor. Neuron 4:847–854

Wittenberger T, Schaller HC, Hellebrand S (2001) An expressed sequence tag (EST) data mining strategy succeeds the discovery of new G-protein coupled receptors. J Mol Biol 307:799–813

Yokota Y, Sasai Y, Tanaka K, et al. (1989) Molecular characterization of a functional cDNA for rat substance P receptor. J Biol Chem 264:17649–7652

*All reference citations marked with an asterisk are available at the author's website: www.preskorn.com

Appendix

S. H. Preskorn

Department of Psychiatry and Behavioral Sciences,
University of Kansas School of Medicine and Psychiatric Research Institute,
1010 N. Kansas, Wichita, KS 67214, USA
e-mail: spreskor@kumc.edu

Selected publications by preskorn can be accessed at his website
http://www.preskorn.com

Table 1 Summary of package insert dosing guidelines

Generic/*trade* drug name	Recommended dose start/max (mg/day)	Dosage guidelines for specific patients				
		Children	Adolescents	Elderly	Hepatic*	Renal*
Mixed reuptake inhibitors and neuroreceptor blockers[a,b]						
Amitriptyline/*Elavil*	75/300[c]	↓	NA	↓	↓	↓
Amoxapine/*Ascendin*	100/600[c,d]	NR	NA	↓	↓	↓
Clomipramine/*Anafranil*[e]	25/250[c,f]	NA	NA	↓	↓	↓
Doxepin/*Sinequan*	75/300[c]	NA	NA	NA	↓	↓
Imipramine/*Tofranil*	75/300[c]	NA	↓	↓	↓	↓
Norepinephrine selective reuptake inhibitors[a,b]						
Desipramine/*Norpramin*	100/300[c]	↓	NA	↓	↓	↓
Maprotiline/*Ludiomil*	75/225[c,f]	NA	NA	↓	↓	↓
Nortriptyline/*Pamelor*	50/150[c]	NA	NA	↓	↓	↓
Serotonin selective reuptake inhibitors						
Citalopram/*Celexa*	20/60	NA	NA	↓	↓	↓
Fluoxetine/*Prozac*	20/80	NA	NA	↓	↓	↓
Fluvoxamine/*Luvox*[e]	50/300[g]	NR	NA	↓	↓	↓
Paroxetine/*Paxil*	20/50	NA	NA	↓	↓	↓
Sertraline/*Zoloft*	50/200	↓	Same	Same	↓	↓
Serotonin and norepinephrine reuptake inhibitors						
Venlafaxine-IR/*Effexor IR*	75[g,h]/375[g]	NA	NA	NA	↓	↓
Venlafaxine-XR/*Effexor XR*	75[h]/375	NA	NA	NA	↓	↓
Serotonin (5-HT$_{2A}$) receptor blockers and weak serotonin uptake inhibitors						
Nefazodone/*Serzone*	200[g]/600[g]	NA	NA	↓	↓	↓
Trazodone/*Dyserel*	150[g]/600[g]	NA	NA	↓	↓	↓
Serotonin (5-HT$_{2A}$ and 5-HT$_{2C}$) and norepinephrine receptor blockers						
Mirtazapine/*Remeron*	15/45	NA	NA	↓	↓	↓
Dopamine and norepinephrine reuptake inhibitors						
Bupropion-IR/*Wellbutrin IR*	200[g]/450[c,f,i]	NA	NA	↓	↓	↓
Bupropion-SR/*Wellbutrin SR*	150/400[c,f,i]	NA	NA	↓	↓	↓
Monoamine oxidase inhibitors						
Moclobemide (Manerix)[j]	300/600	NA	NA	↓	↓	↓
Phenelzine/*Nardil*	45[g]/90[g]	NA	NA	↓	↓	↓
Selegiline or l-deprenyl (Eldepryl)	10/30	NA	NA	↓	↓	↓
Tranylcypromine/*Parnate*	30[g]/60[g]	NA	NA	↓	↓	↓

Table 1 (continued)

This table was adapted with permission from Preskorn SH (1999) *Outpatient Management of Depression.* Professional Communications, Caddo, p 130.

IR, immediate release; NA, not available; NR, not recommended; SR, sustained release; XR, extended release.

* Impairment.

[a] Starting dose may be given either as a once-a-day dose or on a divided schedule. Once an effective and tolerated dose has been established, it may be given on a once-a-day basis, but a divided dose may still be more prudent with a higher total dose and in patients who are elderly or debilitated. The maximum once-a-day dose of doxepin is 150 mg.

[b] Usual dose may be given either as a once-a-day dose or on a divided schedule.

[c] Therapeutic drug monitoring has been either demonstrated to increase the safe and efficacious use of this drug or theoretically should; demonstrated for amitriptyline, clomipramine, desipramine, imipramine, and nortriptyline. Theoretical for the rest, but has not been adequately studied.

[d] Doses should exceed 400 mg/day only in hospitalized patients who do not have a history of seizures and who have not benefited from an adequate trial of 400 mg/day.

[e] Not formally labeled by the FDA for the treatment of clinical depression but rather for obsessive-compulsive disorder; labeled for use as an antidepressant in other countries.

[f] Maximum daily dose should not be exceeded due to an increased risk of seizures.

[g] Dose should be given on a divided schedule (b.i.d. or t.i.d.).

[h] For some patients, it may be desirable to start at half the dose for 4–7 days to improve tolerance, particularly in terms of nausea.

[i] It is particularly important to administer in a manner most likely to minimize the risk of seizures. Dose increases should not exceed 100 mg/day in a 3-day period. Cautious dose titration can also minimize agitation, motor restlessness, and insomnia. Time between doses should be at least 4 h for 100 mg IR doses, 6 h for 150 mg IR doses, and 8 h for SR doses. Increases above 300 mg/day should only be done in patients with no clinical effects after several weeks of treatment at 300 mg/day. Bupropion should be discontinued in patients who do not experience an adequate response after an adequate period on maximum recommended daily dose. Dosing in the elderly, the debilitated, and patients with hepatic and/or renal impairment has not been adequately studied so increased caution may be prudent.

[j] Not available by prescription in the United States.

Additional comments on dose titration: The package inserts for the following drugs indicate that they can be started at a dose which is usually effective to treat clinical depression: fluoxetine, mirtazapine, paroxetine, tranylcypromine, sertraline, venlafaxine. The following comments apply regarding the use of higher doses with these antidepressants. For *fluoxetine, paroxetine,* and *sertraline*: although fixed-dose studies in patients with clinical depression found no advantage on average to higher doses, an increase may be considered after several weeks on the starting dose if no clinical improvement has been observed. For *mirtazapine,* dose escalation should not be made at intervals of less than 1–2 weeks to adequately evaluate therapeutic response to a given dose. For *tranylcypromine*, improvement can be seen between 48 h and 2 weeks of starting therapy; if not, dose increases in 10 mg/day increments may be made at intervals of 1–3 weeks.

The package inserts for the following drugs recommend starting at a lower than usually effective dose and titrating up to a dose which is usually effective to treat clinical depression in order to minimize tolerability or safety problems: amitriptyline, amoxapine, bupropion, citalopram, clomipramine, fluvoxamine, doxepin, imipramine, nefazodone, phenelzine, trazodone, and trimipramine. The following are additional comments about dose titration with these antidepressants: For the *tricyclic antidepressants*, the dose should be gradually increased during the first 2 weeks based on therapeutic drug monitoring and clinical assessment of efficacy and tolerability. For *fluvoxamine*, a lower than usually effective starting dose is recommended to improve tolerability. The dose should be increased every 4–7 days as tolerated until maximum therapeutic benefit is achieved. For *citalopram*, the starting dose is 20 mg/day

Table 1 (continued)

with the recommendation to generally increase to 40 mg/day. While doses above 40 mg/day are not ordinarily recommended, some patients may require a dose of 60 mg/day. For *nefazodone*, a lower than usually effective starting dose is recommended to improve tolerability. Dose titration should occur in increments of 100–200 mg/day as determined by tolerability and the need for further clinical improvement. These incremental advances should be done using divided doses and at intervals of at least 1 week. It may be advisable to titrate up more slowly in elderly and debilitated patients. For *phenelzine*, a lower than usually effective starting dose is recommended to improve tolerability. Its dose should be increased to at least 60 mg/day at a fairly rapid pace consistent with good tolerability. For *trazodone*, the same comments apply as for nefazodone when trazodone is used as an antidepressant; however, it is now mainly used as a nonhabit-forming sedative given as a single bedtime dose of 50–200 mg as needed for sleep.

Table 2 Comparison of the placebo-subtracted incidence rate (%) of frequent adverse effects for citalopram, fluoxetine, fluvoxamine, paroxetine, and sertraline*†

Adverse effect	Citalopram (n=1063, n=446)[a]	Fluoxetine (n=1730, n=799)[a]	Fluvoxamine (n=222, n=192)[a]	Paroxetine (n=421, n=421)[a]	Sertraline (n=861, n=853)[a]
Anorexia	2	7.2	8.6	4.5	1.2
Confusion[b]	NA	1.5	NA	1	0.8
Constipation	<Placebo	1.2	11.2	5.2	2.1
Diarrhea[c]	3	5.3	−0.4	4	8.4
Dizziness[d]	<Placebo	4	1.3	7.8	5
Drowsiness[e]	8	5.9	17.2	14.3	7.5
Dry mouth	6	3.5	1.8	6	7
Dyspepsia	1	2.1	3.2	0.9	3.2
Fatigue[f]	2	5.6	6.2	10.3	2.5
Flatulence	NA	0.5	NA	2.3	0.8
Frequent micturition	NA	1.6	0.6	2.4	0.8
Headache	<Placebo	4.8	2.9	0.3	1.3
Increased appetite	NA	NA	NA	NA	NA
Insomnia	1	6.7	4	7.1	7.6
Nausea[g]	8	11	25.6	16.4	14.3
Nervousness[h]	3	10.3	7.6	4.9	4.4
Palpitations[i]	<Placebo	−0.1	NA	1.5	1.9
Paresthesia[j]	NA	−0.3	NA	2.1	1.3
Rash[k]	NA	0.9	NA	1	0.6
Respiratory[l]	8	5.8	−1.3	0.8	0.8
Sweating	2	4.6	−1.3	8.8	5.5
Tremors	2	5.5	6.1	6.4	8
Urinary retention[m]	<Placebo	NA	NA	2.7	0.9
Vision disturbances	<Placebo	1	0	2.2	2.1
Weight gain	NA	NA	NA	NA	NA

This table was adapted with permission from Preskorn SH (1999) *Outpatient Management of Depression*. Professional Communications, Caddo, p 68.
NA, not available.
* Data for fluoxetine, paroxetine, and sertraline are from Preskorn SH (1995) *J Clin Psychiatry* 56 [Suppl 6]:12–21; data for fluvoxamine are from *Compendium of Pharmaceuticals and Specialties*. 33rd ed (1998) pp. 922–924; data for citalopram are from Forest Pharmaceuticals prescribing information, 1998. Incidence of each respective adverse effect for patients taking each drug minus the incidence for each drug's parallel placebo control in double-blind, placebo-controlled studies.
† The above adverse effect data come from product labeling as opposed to head-to-head trials. Such data may not necessarily reflect the actual rate of these adverse effects in clinical practice or the actual differences between these various drugs.

Table 2 (continued)

[a] The first value is the number of patients on that medication, while the second represents those treated in the parallel, placebo group.
[b] Includes decreased concentration, memory impairment, abandoned thinking concentration.
[c] Includes gastroenteritis.
[d] Includes lightheadedness, postural hypotension, and hypotension.
[e] Includes somnolence, sedation, and drugged feeling.
[f] Includes asthenia, myasthenia, and psychomotor retardation.
[g] Includes vomiting.
[h] Includes anxiety, agitation, hostility, akathisia, and central nervous system stimulation.
[i] Includes tachycardia and arrhythmias.
[j] Includes sensation disturbances and hypesthesia.
[k] Includes pruritus.
[l] Includes respiratory disorder, upper respiratory infection, flu, dyspnea, pharyngitis, sinus congestion, oropharynx disorder, fever, and chill.
[m] Includes micturition disorder, difficulty with micturition, and urinary hesitancy.

Table 3 Comparison of the placebo-subtracted incidence rate (%) of frequent adverse effects for bupropion, imipramine, mirtazapine, nefazodone, and venlafaxine*†

Adverse Effect	Bupropion (n=323, n=185)[a]	Imipramine (n=367, n=672)[a]	Mirtazapine (n=453, n=361)[a]	Nefazodone (n=393, n=394)[a]	Venla-faxine-IR (n=1033, n=609)[a]	Venla-faxine-XR (n=357, n=285)[a]
Anorexia	−0.1	NA	NA	NA	9	4
Confusion[b]	2.8	NA	2	9	1	2
Constipation	8.7	17.4	6	6	8	3
Diarrhea[c]	−1.8	−2.7	<Placebo	1	1	<Placebo
Dizziness[d]	6.8	22.7	4	23	12	11
Drowsiness[e]	0.3	12	36	11	14	9
Dry mouth	9.2	47.1	10	12	11	6
Dyspepsia	0.9	NA	<Placebo	2	1	<Placebo
Fatigue[f]	−3.6	7.6	3	7	6	1
Flatulence	NA	NA	<Placebo	NA	1	1
Frequent micturition	0.3	NA	1	1	1	NA
Headache	3.5	−8.7	<Placebo	3	1	<Placebo
Increased appetite	NA	NA	15	NA	1	2
Insomnia	5.3	0.4	<Placebo	2	8	6
Nausea[g]	4	1.3	<Placebo	11	26	21
Nervousness[h]	13.9	3.6	<Placebo	NA	12	7
Palpitations[i]	4.7	NA	<Placebo	NA	2	<Placebo
Paresthesia[j]	0.8	NA	NA	2	1	NA
Rash[k]	3.7	NA	NA	2	1	NA
Respiratory[l]	−2.5	−2.3	3	9	NA	1
Sweating	7.7	11.2	<Placebo	NA	9	11
Tremors	13.5	10	1	1	4	3
Urinary retention[m]	−0.3	4	NA	1	2	NA
Vision disturbances	4.3	5.4	<Placebo	12	4	4
Weight gain	NA	NA	10	NA	NA	NA

This table was adapted with permission from Preskorn SH (1999) *Outpatient Management of Depression.* Professional Communications, Caddo, p. 72.
IR, immediate release; NA, not available; XR, extended release.
* Data from Preskorn SH (1995) *J Clin Psychiatry* 56 [Suppl 6]:12–21; Remeron (mirtazapine). *Physicians' Desk Reference* (1999) pp 2147–2149; and Effexor (venlafaxine hydrochloride). *Physicians' Desk Reference* (1999) pp 3298–3302.
† The above adverse effect data come from product labeling as opposed to head-to-head trials. Such data may not necessarily reflect the actual rate of these adverse effects in clinical practice or the actual differences between these various drugs.

Table 3 (continued)

[a] The first value is the number of patients on that medication, while the second represents those treated in the parallel, placebo group.
[b] Includes decreased concentration, memory impairment, abandoned thinking concentration.
[c] Includes gastroenteritis.
[d] Includes lightheadedness, postural hypotension, and hypotension.
[e] Includes somnolence, sedation, and drugged feeling.
[f] Includes asthenia, myasthenia, and psychomotor retardation.
[g] Includes vomiting.
[h] Includes anxiety, agitation, hostility, akathisia, and central nervous system stimulation.
[i] Includes tachycardia and arrhythmias.
[j] Includes sensation disturbances and hypesthesia.
[k] Includes pruritus.
[l] Includes respiratory disorder, upper respiratory infection, flu, dyspnea, pharyngitis, sinus congestion, oropharynx disorder, fever, and chill.
[m] Includes micturition disorder, difficulty with micturition, and urinary hesitancy.

Table 4 Pharmacokinetic parameters for selected antidepressants in adults*

Drug	Oral bio-availability $(F)^a$	T_{max}^b (h)	Volume of distribution^b (V_d) (l/kg)	Protein binding^b (%)	Clearance^b (ml/min)	Half-life^b ($t_{1/2}$) (h)	Comments
Tricyclic antidepressants							
Amitriptyline	0.4–0.6	2–4	15.5	96	700–1,000	9–25	V_d and $t_{1/2}$ increased with age>65; nortriptyline is an active metabolite
Clomipramine	0.2–0.8	1.5–4	7–20	97	–	19–37	Clearance may be dose dependent; desmethylclomipramine is an active metabolite with $t_{1/2}$ of 54–77 h
Desipramine	0.5	4–6	15–37	90	1,600–2000	14–25	Clearance decreased with age>65
Doxepin	0.2–0.4	0.5–1	20	80	75–110	6–8	Has an active metabolite
Imipramine	0.8–1.0	1–2	21	89	750–1,300	8–6	Clearance decreased with age>65; desipramine is an active metabolite
Nortriptyline	0.6	7–8.5	21–27	93	375–625	18–35	Clearance decreased with age>65
Selective serotonin reuptake inhibitors							
Citalopram	>0.9	2–4	12–16	50	80–400	33	Citalopram metabolites with no clinically meaningful serotonin uptake inhibition
Fluoxetine	<0.9	4–8	14–100	94	94–704	48–96	Food slows rate but not extent of absorption. Active metabolite, norfluoxetine, has $t_{1/2}$=7–15 days and is equipotent to parent drug in terms of serotonin uptake inhibition and inhibition of CYP2D6
Fluvoxamine	>0.9	4–8	25	77	65	15	Multiple metabolites without clinically relevant serotonin reuptake inhibition
Paroxetine	>0.9	3–8	13	95	15–92	21	No metabolite with clinically relevant activity in terms of serotonin uptake inhibition. M2 metabolite is 1/3 as potent as paroxetine at 2D6 inhibition
Sertraline	>0.9	5–8	20	98	96	26	No metabolite with clinically relevant activity in terms of serotonin uptake inhibition; desmethylsertraline is equipotent to sertraline at 2D6 inhibition

Table 4 (continued)

Drug	Oral bio-availability (F)[a]	T_{max}[b] (h)	Volume of distribution[b] (V_d) (l/kg)	Protein binding[b] (%)	Clearance[b] (ml/min)	Half-life[b] ($t_{1/2}$) (h)	Comments
Other antidepressants							
Bupropion	≈0.8	1–3	19–21	82–88	–	10–14	Three active metabolites. High levels associated with reduced efficacy and greater toxicity
Maprotiline	>0.9	8–24	22–52	88	1,060	27–58	Has an active metabolite
Mirtazapine	0.5	2	–	85	–	20–40	Low likelihood of CYP enzyme inhibition based on in vitro modeling
Nefazodone	0.2	1–3	N/A	>95	N/A	2–4	Hydroxynefazodone and triazole-dione metabolites may contribute to clinical efficacy
Trazodone	0.6–0.8	1–2	–	89–95	–	5–9	Food delays, but increases the extent (up to 20%) of absorption
Venlafaxine	0.92	2	5–19	27	50–144	5	Major metabolite O-desmethylvenlafaxine with similar pharmacologic profile to venlafaxine

* Adapted and updated from Scott G, Nierenberg D (1992) Pharmacokinetic data for commonly used drugs. In: Melmon KL, Morrelli HF, Hoffman BB, et al. (eds) Melmon and Morrelli's *Clinical Pharmacology: Basic Principles in Therapeutics*, 3rd edition. New York, McGraw Hill, 1029–1072; from Preskorn SH (1993) Pharmacokinetics of antidepressants: why and how they are relevant to treatment. *J Clin Psychiatry* 54 [Suppl 9]:14–34.
A dash [–] means that reliable data were not available.
[a] Oral bioavailability (F) is the fraction of the dose administered orally that is absorbed and reaches the systemic circulation as active drug. F therefore has a value between 0 and 1.0.
[b] For a detailed discussion of these terms, readers are referred to the chapter "General Principles of Pharmacokinetics" by Preskorn and Catterson (this volume).

Table 5 Optimum plasma drug levels for commonly prescribed antidepressants[a]

Drug or metabolite	Optimum therapeutic range	Minimum necessary to produce a clinical effect
Tricyclics[b]		
Amitriptyline[c]	75–175 ng/ml	
Desipramine[c]	100–160 ng/ml	
Imipramine[d]	265–300 ng/ml	
Nortriptyline[c]	50–150 ng/ml	
Clomipramine and desmethylclomipramine	175–400 ng/ml	
SSRIs[e]		
Citalopram		85 ng/ml (40 mg/day)
Fluoxetine and norfluoxetine		120-300 ng/ml (20 mg/day)
Fluvoxamine		100 ng/ml
Paroxetine		70–120 ng/ml (20 mg/day)
Sertraline		10–50 ng/ml (50 mg/day)
Other antidepressants		
Bupropion[f]	10–50 ng/ml (parent drug only)	
Venlafaxine	195–400 ng/ml	
Nefazodone	Not established	
Trazodone	Not established	
Maprotiline	Not established	
Mirtazapine	Not established	
Monoamine oxidase inhibitors (MAOIs)	TDM measuring plasma drug concentration not applicable to MAOIs. Optimal efficacy associated with approximately 80% inhibition of platelet MAO[g]	

[a] The data given in this table are based on:
Literature cited in the chapter "Therapeutic Drug Monitoring of Antidepressants" by Burke and Preskorn in this volume.
Preskorn SH (1991) Should bupropion dosage be adjusted based on therapeutic drug monitoring? *Psychopharmacol Bull* 27:637–643.
Goodnick PJ (1992) Blood levels and acute response to bupropion. *Am J Psychiatry* 149:399–400.
Preskorn SH (1993) Pharmacokinetics of antidepressants: why and how they are relevant to treatment. *J Clin Psychiatry* 54 [Suppl 9]:14–34.
Preskorn SH (1999) Two in one: the vanlafaxine story. J Psychiatr Pract 5:346–350
Preskorn SH (2000) Bupropion: what mechanism of action? J Psychiatr Pract 6:39–44
Table 3.7 in Sheldon SH (1996) *Clinical Pharmacology of Selective Serotonin Reuptake Inhibitors*. Professional Communications, Caddo, p 50.
Preskorn S, Othmer S (1984) Evaluation of bupropion hydrochloride: the first of a new class of atypical antidepressants. *Pharmacotherapy* 4:20–34.
Preskorn SH, Mac DS (1985) Plasma levels of amitriptyline: effect of age and sex. *J Clin Psychiatry* 46:276–277.
Preskorn SH, Katz SE (1989) Bupropion plasma levels: intraindividual and interindividual variability. *Ann Clin Psychiatry* 1:59–61.
Preskorn SH, Jerkovich GS (1990) Central nervous system toxicity of tricyclic antidepressants: phenomenology, course, risk factors, and role of therapeutic drug monitoring. *J Clin Psychopharmacol* 10:88–95.
Davidson J, McLeod MN, White HL (1978) Inhibition of platelet monoamine oxidase in depressed subjects treated with phenelzine. *Am J Psychiatry* 135:470–472.
Preskorn SH, Silkey B, Beber JH, Dorey C (1991) Antidepressant response and plasma concentration of fluoxetine. *Ann Clin Psychiatry* 3:147–151.

Table 5 (continued)

[b] For these TCA plasma levels, daily dose should fall between 0.5 ng/ml/mg and 1.5 ng/ml/mg. Values less than 0.5 indicate noncompliance or rapid metabolizers. Values greater than 1.5 indicate poor metabolizers due to lack of 2D6 capacity from genetic deficiency or coprescribed drugs that inhibit 2D6 such as fluoxetine or paroxetine (Preskorn and Mac 1985; Preskorn and Jerkovich 1990; Preskorn 1993).

[c] Data for amitriptyline, desipramine, and nortriptyline consistently demonstrate curvilinear concentration-antidepressant response relationships, with reduced efficacy occurring below the concentration at which toxicity begins to be an issue.

[d] Imipramine demonstrates a linear rather than curvilinear relationship between concentration and antidepressant response. Because there is an increased risk of TCA-induced delirium at imipramine levels above 300 ng/ml and an increased risk of first-degree atrioventricular block at levels above 350 ng/ml, this drug has a very narrow therapeutic range and a relatively poor response rate compared with the other TCAs.

[e] The concentrations listed for the SSRIs given are an estimate of the minimum concentration necessary to produce an antidepressant effect (e.g., 70–120 ng/ml of paroxetine on 20 mg/day). TDM can be used to check compliance for all except fluoxetine. TDM can be used to determine whether fluoxetine and paroxetine have cleared sufficiently so that they will not interfere with the metabolism of other drugs dependent upon 2D6 (Preskorn et al 1991).

[f] Note that the average levels of bupropion, hydroxybupropion, erythrohydrobupropion, and threohydrobupropion achieved on 450 mg/day are 33, 1452, 138, and 671, respectively. Thus, the total for parent drug plus metabolites is over 2,000 ng/ml (2 ug/ml) (Preskorn and Katz 1989; Preskorn 2000). Recommended dosing guidelines must be followed pending more TDM data regarding safety and efficacy as it correlates with metabolites (Preskorn and Othmer 1984; Preskorn and Katz 1989; Preskorn 1991; Goodnick 1992).

[g] The platelet assay for MAOI activity is cumbersome (involving two samples), expensive, and not always readily available. The main applications are for research, to check compliance, or support need for unusually high doses (Davidson et al 1978).

Subject Index

absorption 37, 38, 47, 49, 52, 62, 65
– rapid absorption 39
additive effect 469
adolescents 355–378
– psychopharmacology 357
β-adrenergic receptor kinase 532
α_1-adrenoceptor 532
α_2-adrenoceptor 121
β-adrenoceptor 120
– kinase 1 (BARK) 526
– down-regulation 521
affective instability 490, 494, 500, 502
agmatine 228
alcohol abuse (see also substance abuse) 125, 133, 161, 186, 195, 214, 305, 308, 362, 422, 468, 500
algorithm 479
allergy, severe 466
γ-aminobutyric acid (GABA) 214
amitryptiline (AT) 99, 191, 365
amoxapine 193
amphetamines (see also substance abuse) 46, 467, 492
anhedonia 123
α_2-antagonist 459
anticholinergic effect 188, 199, 472
antidepressant
– classification 38, 173
– development 177
– efficacy 52, 58
– mechanisms of action 174
antihistaminic agent 187
antiparkinsonian agent 187
antipsychotic 186

anxiety 337, 338, 490, 491, 495, 498, 504
anxiolytic agent 464
anxious depression 457
arginine vasopressin (AVP) 126
asthma 466
atomoxetine 372
atypical depression 406, 457
augmentation strategies 394, 410
axis II disorders 489–491

Beck Depression Inventory (BDI) 156
befloxatone 212
benzylamine oxidase enzyme 226
biogenic amine theory 187
biological markers of depression 117–148
– lymphocyte 127, 131
– neuroendocrine marker 133
– trace amine 214
bipolar depression 126, 224, 421, 423, 424, 441
– genetic theories 427
– maintenance treatment 437
– management 432
bipolar disorders 421–446
– biological bases 426
– clinical features 424
– pharmacological management of depression 429
BNDF mRNA 538
borderline personality disorder 475
brain-derived neurotrophic factor (BDNF) 527, 537
bridging study 586, 594
brofaromine 215, 219

bulimia nervosa 187, 224
bupropion 61, 103, 225, 265, 267, 269, 310, 372, 407, 422, 437, 440, 441, 525

calcium/calmodulin-dependent protein kinase II (CaM-II kinase) 543
cardiac arrhythmia 198
cardiac toxicity 470
cardiovascular effect 465
catecholamine 120
cheese reaction 210
children 355–378
- psychopharmacology 357
cholinergic rebound 458
chronic depression 474
chronic mild stress (CMS) 545
citalopram 102, 243, 246–248, 250, 252, 254, 256–258, 404
clinical judgment 138
clomipramine 135, 192, 368, 371
clonidine 195
- [^{3}H]-clonidine 128
CNS 130
cocaine (see also substance abuse) 188
Cognitive Behavioral Analysis System for Psychotherapy (CBASP) 474
cognitive
- disorganization 490, 492
- performance 478
combination therapy 225, 411
comorbid
- anxiety disorder 465
- psychiatric disorder 451
compliance 96, 159
concordance 159
conduction defect 198
coping skills 159
corticotropin releasing factor (CRF) 125, 528
- receptor 529
- receptor antagonist 530
corticotropin releasing hormone 568
CREB 542
- mRNA 537, 538
- P 537
CYP
- 2D6 194, 460, 470
- 450 enzymes 380
- isoenzymes 219
cytokine 130

decision support 480
delay in the onset 190
delirium 197
dementia 225
depression 118, 202, 203, 332, 379, 380
- atypical depression 223, 406
- bipolar depression 224, 421, 423, 424, 463
- clinical syndromes 396
- endogenous depression 189
- in adolescents 355–378
- in children 355–378
- in the elderly 393–420
- in the community 204
- in women 379–392
- major depression (also major depressive disorder, MDD) 121, 129, 136, 151, 202, 203, 223, 253–257, 266, 272, 294, 307, 331–337, 343, 362–365, 395, 405, 431, 451, 453, 466, 500, 527–530, 540, 547, 572, 575, 590
- melancholic depression, nonpsychotic 463
- models 545
- monoamine hypothesis 521
- neuroendocrine marker 133
- nonpsychotic melancholic depression 463
- perimenopausal depression 379, 394
- peripartum depression 379, 382
- psychotic MDD 478
- reactive depression 189
- severe depression 457
- subsyndromal depression 395
- treatment-refractory depression 219, 223, 447–488
- unipolar depression 126, 413
desipramine (DMI) 99, 188, 192, 365, 372, 525
desmopressin acetate (DDAVP) 373
dexamethasone suppression 133
diabetes 466
disability 150
discontinuation syndrome 222, 457
distribution 37–39, 49, 52, 62, 65
dopamine 123
- receptor D_1–D_3 522
dothiepin 193
downregulation 465
doxepin 193

drowsiness 196
drug development 568, 576, 584, 588, 590
– rational 179, 242, 244
– stages 586
drug discovery 586, 587, 590, 594
– in psychiatry 588
drug–drug interaction 38, 51, 221, 458
– pharmacokinetically-mediated 60, 63
DSM III/IV 151
duloxetine 64, 265, 267, 274, 291, 296, 297

educational background 191
elderly 393–420
– bipolar disorder 440
– compliance 412
– depression (late-life depression) 395, 397, 414
– – treatment 401
– pharmacodynamic actions 400
– subgroups 397
elimination 37, 38, 43, 44, 46, 47, 50, 62
– aging and disease 46
– anticonvulsants 438
– bupropion 61
– fluoxetine 402
– gender 47
– mirtazapine 73, 274
– monoamine oxidase inhibitors 71
– of psychotropic medications 46
– paroxetine 58, 404
– phenelzine 217
– phenylpiperazine agents 68
– selective serotonin reuptake inhibitors 53, 431
– serotonin norepinephrine specific reuptake inhibitors 66
– sertraline 58, 404
– tricyclic antidepressants 50, 399
endogenous depression 189
epilepsy 188, 476
escitalopram 243, 246–248, 250, 252, 256
estrogen 380, 383, 384, 386, 388
excitatory amino acid 124
extrapyramidal symptoms 467

Feighner criteria 151
D-fenfluramine 135
flavoprotein 212

fluoxetine 70, 101, 243, 246–250, 252–258, 360, 363, 364, 370, 371, 403
fluvoxamine 102, 243, 246, 248–252, 254–258, 370, 371, 403
functional impairment 449

G protein 4, 5, 11, 17–19, 21, 23, 27, 534
– signal coupling 22
gabapentin 435
GABA-transaminase (GABA-T) 214
galanin 568, 571
gastric emptying 201
GDNF (glial cell line-derived neurotrophic factor) 3, 8
gender, see women
gene expression 536, 547
General Health Questionnaire (GHQ) 160, 162
general
– medical condition 451
– practitioner 204
generic quality of life scale 159
gene-specific transcription factor 540
genitourinary 198
geriatric patient, see elderly
ginkgo biloba 326, 327, 344, 346, 347
glial cell 3, 4, 6, 17
glial cell line-derived neurotrophic factor (GDNF) 3, 8
glucocorticoid 538
– receptor 523
GRII mRNA 538
growth hormone 135

half-life 38, 43, 51, 58, 63, 72
HAM-D17 154
Hamilton Depression Rating Scale (HAM-D) 152, 405
harmane 228
health-related quality of life 159
herbal preparation 325–351
5-HIAA (5-hydroxyindole acetic acid) 121, 122
homovanillic acid (HVA) 124
Hopkins Symptom Checklist (HSCL) 157, 158
hormone replacement 394, 411
5-HT$_1$ 217
human genome project 585, 586, 590, 593, 594

HVA (homovanillic acid)   124
hydrazine   212
5-hydroxyindole acetic acid (5-HIAA)   121
5-hydroxytryptophan   135
hypercorticosolemia   135
hypothalamic-pituitary-adrenal (HPA)   131
hypertensive
– crisis   221, 473
– patient   195
hypomania   197
hypomanic swing   191
hypothalamic-pituitary-adrenal (HPA)   131
hypothyroidism   463

ICD-10   151
imidazoline   128, 228
imipramine (IMI)   99, 187, 365, 371–373, 521
– [^{3}H]-imipramine   130
immune system   326, 342, 347
impulsivity   490, 491, 494, 507
International Quality of Life Assessment (IQOLA)   160
inter-rater reliability   152
Inventory of Depressive Symptomatology (IDS)   155
– C   155
– Self-Rating Version (IDS-SR)   155, 157
investigational antidepressant   179, 566, 569, 570, 573
[^{125}I]-p-iodoclonidine   129
iproniazid   210
isocarboxazid (ISO)   211

kava kava   326, 327, 339, 341, 347
Kielholz classification system   154

lamotrigine   435, 438, 440
late-life depression, see elderly
learned helplessness   546
limbic system   477
lithium   361, 394, 410, 422, 425, 430–432, 436, 439, 440
– augmentation   220
lofepramine   193
loxapine   193

maintenance therapy   202, 203, 225, 413, 414

Major Depression Inventory (MDI)   157, 161
major depressive disorder (MDD), see depression
managed care   204
Mania Scale (MAS)   154
maprotiline   193
mCPP   471
mechanism of action   16, 17, 27, 48, 50, 61, 72, 101, 171, 174, 175, 178–180, 187, 242, 250, 254, 255, 266, 267, 270, 273, 276, 290, 302, 310, 368, 373, 432, 457, 458, 463, 467, 477, 498, 567, 606
medication
– adherence   451
– trial   454
Melancholia Scale (MES)   154
melatonin   136
memory difficulty   475
metabolic-endocrine   199
metabolism   7, 37–39, 41–47, 50, 53, 62, 66, 68, 71, 73, 76, 88, 92, 103, 108, 191, 193, 215, 217, 221, 226, 247, 256, 269, 273, 277, 280, 286, 291, 296, 301, 335, 361, 384, 385, 399, 438, 458, 471, 472, 574, 607
– 3A4   70
– bupropion   64
– first-pass metabolism   39, 49, 65
– oxidative drug metabolism   60
– phase I/II   41
– TCA   43, 51
3-methoxy-4-hydroxy phenyl glycol (MHPG)   120
MHPG (3-methoxy-4-hydroxy phenyl glycol)   120
mianserin   265, 273, 307
milnacipran   265, 267, 291, 296, 301, 307, 526
mirtazapine   72, 106, 265, 269, 273, 309, 310, 364, 525
moclobemide   212, 215, 219, 257
monoamine hypothesis of depression   521
monoamine oxidase inhibitor (MAOI)   71, 72, 106, 172, 186, 209–239, 382, 383, 405, 422, 430, 433, 437, 440, 441, 521
– A   213
– B   213, 214

Subject Index

– diet 460
– side-effects 222
Montgomery-Åsberg Depression Scale 404
mood stabilizer 421, 423, 425, 434, 438
MRI 136
multiple sclerosis 225

NCP system 477
nefazodone 67, 105, 265, 266, 269–271, 277, 366, 380, 386, 408, 525
nemifitide 573, 574
neurocognitive/neuropsychological functioning 475, 479
neuron 3, 4, 16, 23
– electrical nature 8
neuron-to-neuron communication 10
neuropeptide 568, 576
– neurotransmitters 570
– Y 546, 568, 571
neurotransmitter 3, 4, 7, 9, 10, 120
– inactivation 13
– receptors 17
– storage and release 12
– synthesis 11
neutrophil phagocytosis 131
nifedipine 222
nitric oxide (NO) 216
NMDA receptor 125
N-methyl-D-aspartate (NMDA) 526
nocturnal enuresis 187
noradrenaline 120
norfluoxetine 252, 257, 457
Norrie disease 213
nortryptiline (NT) 99, 191, 365, 372

obsessive-compulsive disorder (OCD) 187
O-desmethylvenlafaxine 458
older adults, see elderly
orphan receptor 592
overdose 200
oxytocin 568, 572

panic disorder 224
paroxetine 101, 243, 246, 248–250, 252, 254–258, 364, 403
– [^{3}H]-paroxetine 130
patient/family education 479
perimenopausal depression 379, 394

personality disorders 489–515
PET 136
pharmacodynamics 381, 386, 388
pharmacogenetics 583–596
pharmacogenomics 583–596
pharmacokinetics 47, 88, 89, 104–107, 174, 190, 191, 194, 242, 246, 252, 269, 282, 293, 305, 330, 356, 360, 362, 366, 380, 384, 387, 389, 399, 404, 587, 605, 606
– clinical relevance 48
– general principles 35–86
– interactions 49
– linear 38, 44, 51, 58, 59, 63, 298
– monoamine oxidase inhibitors 71
– multiple-dose study 53
– nonlinear 38, 44, 45, 51, 58, 59, 63
– of psychotropic agents 384
– of SSRI 61
– volume of distribution 47
phenelzine (PLZ) 211, 217
phentolamine 222
phenylacetic acid (PAA) 218
phenylethylamine (PEA) 213, 215
– 2-phenylethylamine 215, 218
phenylpiperazine agent 67
PKA 531, 539
– translocation 541
PKC 531, 533
placebo response 96, 102, 362, 363, 371, 567
plasma amine oxidase (PAO) 226
platelet 127
– 5-HT uptake 129
postmortem brain tissue 120
postsynaptic neuron 9, 11
postural hypotension 222
premenstrual dysphoric disorder 467
presynaptic neuron 9, 10, 13, 17
primary care 204, 450
prolactin 135
protein
– kinase 532, 535
– – inhibitor (PKI) 540
– phosphatase 535
protriptyline 193
psychobiology 489
Psychological General Well-Being Scale (PGWB) 160, 161
psychometric triangle 151

psychostimulants 409
psychotherapy 412
public mental health 479

quality of life 149–167
Quality of Life in Depression Scale
  (QLDS) 160, 162

rating scales for depression 149–167
reboxetine 265–267, 280, 299, 307
regional cerebral blood flow (rCBF)
  476
relapse 152, 202, 455
– predictors 202
REM sleep 188, 222
residual symptoms 452, 473
reverse pharmacology 592, 593
reversibility 212
risperidone 225

scales
– Alzheimer's Disease Assessment
  Scale 344
– Beck Depression Inventory 156, 162
– Carroll Self-Rating Scale 575
– Clark Personal and Social Adjustment
  Scale 159
– Children's Depression Rating Scale
  Revised 363
– Children's Global Assessment
  Scale 363
– Children's Yale-Brown Obsessive-
  Compulsive Scale 370
– Clinical Global Impressions Scale 332,
  506, 575
– Duke Social Phobia Scale 506
– General Health Questionnaire 160,
  162
– Hamilton Anxiety Rating Scale 340
– Hamilton Depression Rating Scale 99,
  132, 152, 289, 330, 364, 405, 450, 455,
  477, 575
– Hopkins Symptom Checklist 157
– International Quality of Life
  Assessment 160
– Liebowitz Social Anxiety Scale 506
– Locus of Control Scale 159
– Major Depression Scale 154, 155, 161
– Mania Scale 154
– Melancholia Scale 154

– Montgomery-Åsberg Scale 152, 290,
  332, 404, 465, 575
– Pediatric Anxiety Rating Scale 371
– Psychological General Well-Being
  Scale 160, 162
– Quality of Life in Depression Scale 160,
  162
– Short Form (SF-36) 157, 159
– Visual Analog Scale 575
– Wechsler Adult Intelligence Scale 493
– Zung Self-Report Depression
  Scale 157, 162
seasonal affective disorder 326, 336
second messenger 4, 7, 11, 17, 19, 21, 23,
  27, 535
sedation 187, 195, 471
seizure 471
– frequency 196
selective serotonin reuptake inhibitor
  (SSRI) 241–262, 380, 386, 387, 422,
  430, 431, 433, 437, 440, 575
– withdrawal symptoms 459
selectivity 212
selegiline (SEL) 211, 216, 218
semicarbazide-sensitive amine oxidase
  (SSAO) 226, 227
sensitivity 152
serotonin 121
– syndrome 219, 459, 470, 473
sertraline 101, 243, 246, 248, 250, 252,
  254–256, 258, 364, 370, 404
sex-related differences (see also women)
  379, 380, 384–386
sexual dysfunction 459, 471
sexual effect 199
SF-36 Health Survey 157
side effect 244, 246, 250, 394, 397, 398,
  463
– burden 452
site of action (SOA) 174, 176
sleep 326, 329, 347
– disturbance 331, 336, 342
– impairment 346
– in depression 343
– rapid eye movement (REM) 343
social phobia 224
SPECT 136
St. John's wort 327, 329, 334
standardization 160
Star*D 480

state and trait marker   119
stimulant effect   195, 196
stress   125, 326, 337–339, 345
– insomnia   343
structure–activity relationship   585, 586, 588, 593
substance abuse   102, 358, 363, 423, 427, 434, 451, 466, 467, 494, 506,
substance P   530, 568, 570
– receptor antagonist   530
suicide   42, 98, 120–126, 226, 284, 307, 336, 395, 414, 422, 430, 440, 452, 474, 496, 528
– suicidal ideation   200, 364
superoxide dismutase (SOD)   215
synapse   5, 9, 13, 16, 17, 27
synergistic effect   462

teratogenicity   201
test–retest reliability   152
therapeutic alliance   473
therapeutic
– dose   452
– drug monitoring (TDM)   87–114
thyroid   394, 411
– disorders   396
– function test   398
tianeptine   265, 267, 299, 301
TMAP   480
tolerability   242, 249, 250, 254
topiramate   435
torsades de pointes   472
transcription factor   540
translation procedure   160
tranylcypromine (TCP)   210, 217, 218
traumatic brain injury   225

trazodone   67, 105, 265, 266, 277, 278, 408
tremor   196
tricyclic antidepressant (TCA)   48, 98, 172, 185–208, 380, 382, 385, 387, 422, 433, 498
– combinations   202
– dosages   204
– withdrawal   197
triiodothyronine   463
tryptamine   228
tryptophan   122
– L-tryptophan   135
tuberculosis   210
tyramine (TA)   188, 213
tyrosine hydroxylase (TH)   121

valerian   326, 327, 342, 347
validity   152
valproate   438, 440
vasopressin   468, 572
venlafaxine   64, 105, 265–267, 274, 277–279, 286, 291, 296, 310, 366, 497, 526

weight gain   199, 471
well-being
– negative   160
– positive   159
worthlessness   190
women   379–392

$[^3H]$-yohimbine   128

Zung Self-Report Depression Scale (SRDS)   157

Printing: Saladruck, Berlin
Binding: Stein+Lehmann, Berlin